PHYSICS OF SOLAR AND STELLAR CORONAE:

G.S. VAIANA MEMORIAL SYMPOSIUM

PROCEEDINGS OF A CONFERENCE OF THE INTERNATIONAL ASTRONOMICAL UNION, HELD IN PALERMO, ITALY, 22-26 JUNE, 1992

Edited by

JEFFREY F. LINSKY

Joint Institute for Laboratory Astrophysics,
National Institute of Standards and Technology
and University of Colorado,
Boulder, Colorado, U.S.A.

and

SALVATORE SERIO

Instituto ed Osservatorio Astronomico di Palermo,
Palermo, Italy

SPRINGER SCIENCE+BUSINESS MEDIA, B.V.

Library of Congress Cataloging-in-Publication Data

G.S. Vaiana Memorial Symposium (1992 : Palermo, Italy)
Physics of solar and stellar coronae : G.S. Vaiana Memorial Symposuim : proceedings of a conference held in Palermo, Italy, 22-26 June 1992 / edited by Jeffrey L. Linsky, Salvatore Serio.
p. cm. -- (Astrophysics and space science library ; v. 183)
"In celebration of the bicentennial of the Palermo Astronomical Observatory"--CIP t.p.
Includes index.
ISBN 978-0-7923-2346-4 ISBN 978-94-011-1964-1 (eBook)
DOI 10.1007/978-94-011-1964-1
1. Sun--Corona--Congresses. 2. Stars--Corona--Congresses. 3. Astronautics in astronomy--Congresses. 4. Yohkoh (Solar-A) Mission (Project)--Congresses. 5. Vaiana, G. I. Linsky, J. L. (Jeffrey L.), 1941- . II. Serio, Salvatore. III. Title. IV. Series.
QB529.G18 1992
523.7'5--dc20 93-17752

ISBN 978-0-7923-2346-4

Printed on acid-free paper

PHYSICS OF SOLAR AND STELLAR CORONAE: G.S. VAIANA MEMORIAL SYMPOSIUM

ASTROPHYSICS AND SPACE SCIENCE LIBRARY

A SERIES OF BOOKS ON THE RECENT DEVELOPMENTS OF SPACE SCIENCE AND OF GENERAL GEOPHYSICS AND ASTROPHYSICS PUBLISHED IN CONNECTION WITH THE JOURNAL SPACE SCIENCE REVIEWS

VOLUME 184

PROCEEDINGS

Giuseppe Salvatore Vaiana

(January 1, 1935 - August 25, 1991)

Rapian gli amici una favilla al Sole
a illuminar la sotterranea notte

(Ugo Foscolo, *A Ippolito Pindemonte*, 119-120)

"His friends would steal a spark from the Sun
to lighten the night underground."

1 G. Foderà Serio, **2** B. Vittorelli Palma, **3** S. Vaiana, **4** M. Giacconi, **5** S. Verga, **6** ??, **7** U. Palma, **8** F.R. Harnden Jr., **9** R. Giacconi, **10** S. Vaiana, **11** A. Vaiana, **12** V. Vajana, **13** F. Pinello, **14** E. Schreier, **15** G. Oertl, **16** O. Gingerich, **17** R. Ruffini, **18** ??, **19** G. D'Angelo, **20** S. Saguto, **21** J. Linsky, **22** F. Verga, **23** G. D'Angelo, **24** G. D'Angelo, **25** P. D'Angelo, **26** G. Bromage, **27** J. Truemper, **28** T. Oda, **29** J. Lemen, **30** M. Oda, **31** J. Pye, **32** R. Oda, **33** L. Vallone, **34** N. Sangiorgio, **35** D. Priori, **36** S. Serio, **37** D. Battista Spinelli, **38** M. Capaccioli, **39** L. Golub, **40** R.N. Smartt, **41** G. Halberstadt, **42** M. Corrao, **43** S. Pinello, **44** D. Duncan, **45** D. Barrado, **46** D. Montes, **47** ??, **48** M. Elvis, **49** G. Guerrero, **50** J.P. Goedbloed, **51** G. Tagliaferri, **52** G. Fabbiano, **53** G. Bignami, **54** F. Walter, **55** R. Mewe, **56** B. Sylwester, **57** J. Sylwester, **58** B. Somov, **59** A. Maggio, **60** A. Collura, **61** F. Favata, **62** L. Matthews, **63** V. D'Angelo, **64** R. Traina, **65** V. D'Angelo, **66** ??, **67** A. Ciaravella, **68** A. Lanzafame, **69** G. Costa, **70** M. Tschinke, **71** W. Gan, **72** C. Cheng, **73** J. Butler, **74** P. Ulmschneider, **75** R. Freire Ferrero, **76** M. Guainazzi, **77** T. Kosugi, **78** B. Montesinos, **79** L. La Fata, **80** ??, **81** R. Lorente, **82** R. Thomas, **83** F. Damiani, **84** A. Peters, **85** M. Hoskin, **86** L. Prestinenza, **87** D. Luzio, **88** B. Byrne, **89** J.P. Caillault, **90** R. Mutel, **91** G. Gerardi, **92** S. Hodgkin, **93** R. Martin, **94** T. Fleming, **95** B. Foing, **96** C. Boynton, **97** G. Noci, **98** L. Paternó, **99** L. Pastori, **100** S. Randich, **101** S. Catalano, **102** M. Tuscano, **103** A. Bocchino, **104** G. Bocchino, **105** S. Tsuneta, **106** F. Drago, **107** C. Barbieri, **108** R. Pallavicini, **109** G. Cutispoto, **110** ??, **111** ??, **112** V. Iuliano, **113** R. Dempsey, **114** S.A. Drake, **115** N. Weiss, **116** R.B. Dahlburg, **117** Y.M. Wang, **118** L. Di Trapani, **119** L. Cigna, **120** G. Fiorentino, **121** K. Tsinganos, **122** R. Lojacono, **123** C. Tardanico, **124** G.J. Doyle, **125** M. Guedel, **126** F. Grillo, **127** L. Militello, **128** I. Pagano, **129** M. Peres, **130** G. Peres, **131** Y. Uchida, **132** I. Vajana, **133** E. Bocchino, **134** M. Katsova

TABLE OF CONTENTS

V: Other Observations

VI: Theoretical Aspects of Coronal Physics

LIST OF PARTICIPANTS

ANTONUCCI Ester
Istituto di Fisica, Università di Torino, Via Giuria 1, 10125 Torino, Italy
BALLESTER Jose Luis
Departament de Fisica, Universidad de les Illes Balears, 07071 Palma de Mallorca, Spain
BARBERA Marco
Istituto ed Osservatorio Astronomico di Palermo, Piazza del Parlamento 1, 90134 Palermo, Italy
BARBIERI Cesare
Osservatorio Astronomico di Padova, Vicolo dell'Osservatorio 5, 35122 Padova, Italy
BARRADO David
Dpto. Astrofisica, Facultad de Fisicas, Universidad Complutense, E-28040 Madrid, Spain
BELVEDERE Gaetano
Istituto di Astronomia, Città Universitaria, Viale A. Doria 6, 95125 Catania, Italy
BIGNAMI Giovanni
Istituto di Fisica Cosmica del CNR, Via Bassini 15a, 20133 Milano, Italy
BOCCHINO Fabrizio
Istituto ed Osservatorio Astronomico di Palermo, Piazza del Parlamento 1, 90134 Palermo, Italy
BOYNTON G. Chris
Sterrekundig Instituut, Princetonplein 5, TA 3508 Utrecht, The Netherlands
BROMAGE Gordon
Rutherford Appleton Laboratory, Didcot, Oxfordshire OX11 0QX, United Kingdom
BUTLER C. John
Armagh Observatory, College Hill, Armagh, BT61 9DG, Northern Ireland
BYRNE P. Brendan
Armagh Observatory, College Hill, Armagh, BT61 9DG, Northern Ireland
CAILLAULT Jean-Pierre
Department of Physics and Astronomy, University of Georgia, Athens, GA 30602, USA
CAPACCIOLI Massimo
Dipartimento di Astronomia, Università di Padova, Vicolo dell'Osservatorio 5, 35122 Padova
CASANOVA Sophie
Service d'Astrophysique, Centre d'Etude de Saclay, 911991 Gif-sur-Yvette, France
CATALANO Santo
Osservatorio Astrofisico di Catania, Città Universitaria, Via Andrea Doria 6, 95100 Catania, Italy

CHENG Chung-Chieh
E.O. Hulbert Center for Space Research, Naval Research Laboratory, Washington, DC 20375, USA
CHINNICI Ileana
Istituto ed Osservatorio Astronomico di Palermo, Piazza del Parlamento 1, 90134 Palermo, Italy
CHIUDERI Claudio
Dipartimento di Astronomia e Scienza dello Spazio, Università di Firenze, Largo E. Fermi 5, 50125 Firenze, Italy
CIARAVELLA Angela
Istituto ed Osservatorio Astronomico di Palermo, Piazza del Parlamento 1, 90134 Palermo, Italy
COLLURA Alfonso
Istituto per le Applicazioni Interdisciplinari della Fisica del CNR, Via Archirafi 36, 90100 Palermo, Italy
COSTA Giuseppina
Osservatorio Astrofisico di Catania, Città Universitaria, Via A. Doria 6, 95100 Catania, Italy
CUTISPOTO Giuseppe
Osservatorio Astrofisico di Catania, Città Universitaria, Via A. Doria 6, 95100 Catania, Italy
DAHLBURG Russell B.
Laboratory for Computational Physics, Naval Research Laboratory, Washington, DC 20375, USA
DAMIANI Francesco
Istituto ed Osservatorio Astronomico di Palermo, Piazza del Parlamento 1, 90134 Palermo, Italy
DEMPSEY Robert C.
Joint Institute for Laboratory Astrophysics, University of Colorado, Boulder, CO 80309-0440, USA
DOYLE J.Gerry
Armagh Observatory, College Hill, Armagh, BT61 9DG, Northern Ireland
DRAGO CHIUDERI Franca
Dipartimento di Astronomia e Scienza dello Spazio, Università di Firenze, Largo E. Fermi 5, 50125 Firenze, Italy
DRAKE Stephen A.
HEASARC, Code 668, NASA/Goddard Space Flight Center, Greenbelt, MD 20771, USA
DUNCAN Douglas
Space Telescope Science Institute, 3700 San Martin Drive, Baltimore, MD 21218, USA

ELVIS Martin
Harvard-Smithsonian Center for Astrophysics, 60 Garden Street, Cambridge, MA 02138, USA
FABBIANO Giuseppina
Harvard-Smithsonian Center for Astrophysics, 60 Garden Street, Cambridge, MA 02138, USA
FAVATA Fabio
European Space Agency, Astrophysics Division, P.O. Box 299, 2200 AG Noordwijk, The Netherlands
FIELD George
Harvard-Smithsonian Center for Astrophysics, 60 Garden Street, Cambridge, MA 02138, USA
FLEMING Thomas A.
Max-Planck Institut für Extraterrestrische Physik, D-8046 Garching bei München, Germany
FODERÀ SERIO Giorgia
Istituto ed Osservatorio Astronomico di Palermo, Piazza del Parlamento 1, 90134 Palermo, Italy
FOING Bernard
Institut d'Astrophysique Spatiale (IAS), CNRS/Université Paris Sud, Bat. 121, F-91405 ORSAY Cedex, France
FREIRE FERRERO Rubens
Strasbourg Observatory, 11 Rue de l'Université, F-67000, France
GAN Wei-qun
Max-Planck Institut für Extraterrestrische Physik, D-8046 Garching bei München, Germany
GIACCONI Riccardo
Space Telescope Science Institute, 3700 San Martin Drive, Baltimore, MD 21218, USA
GINGERICH Owen
Harvard-Smithsonian Center for Astrophysics, 60 Garden Street, Cambridge, MA 02138, USA
GOEDBLOED J.P.
FOM-Instituut voor Plasmafysica, P.O. Box 1207, 3430 BE Nieuwegein, The Netherlands
GOLUB Leon
Harvard-Smithsonian Center for Astrophysics, 60 Garden Street, Cambridge, MA 02138, USA
GRILLO Flora
Istituto ed Osservatorio Astronomico di Palermo, Piazza del Parlamento 1, 90134 Palermo, Italy
GUAINAZZI Matteo
Istituto di Fisica, Università di Palermo, Via Archirafi 36, 90100 Palermo, Italy

GÜDEL Manuel
Joint Institute for Laboratory Astrophysics, University of Colorado, Boulder, CO 80309-0440, USA
GUERRERO Giannantonio
Osservatorio Astronomico di Brera, Via E. Bianchi 46, 22055 Merate, Como, Italy
HALBERSTADT G.
FOM-Instituut voor Plasmafysica, P.O. Box 1207, 3430 BE Nieuwegein, The Netherlands
HAMMER Reiner
Kiepenheuer-Institut für Sonnenphysik, Schoneckstr. 6, D-7800 Freiburg, Germany
HARNDEN F. Rick Jr.
Harvard-Smithsonian Center for Astrophysics, 60 Garden Street, Cambridge, MA 02138, USA
HAYRAPETYAN Vladimir
Byurakan Observatory, Ashrakskii raion, Armenian Academy of Sciences, 378433, Armenia
HEMPELMANN A.
Astrophysikalisches Institut Potsdam, Rosa Luxembourg Str. 17a, D-O-1590 Potsdam-Babelsberg, Germany
HODGKIN Simon T.
Department of Physics and Astronomy, Leicester University, Leicester, LE1 7RH, United Kingdom
HOSKIN Michael
Churchill College, Cambridge, CB3 0DS, United Kingdom
KATSOVA Maria
Sternberg State Astronomical Institute, Moscow State University, 119899 Moscow V-234, Russia
KOSUGI Takeo
National Astronomical Observatory, Mitaka, Tokyo 181, Japan
KUDRITZKI Rolf-Peter
Institute of Astronomy and Astrophysics, Scheinerstrasse 1, W-8000 Munich 80, Germany
KÜRSTER Martin
Max-Planck Institut für Extraterrestrische Physik, D-8046 Garching bei München, Germany
LANZAFAME Alessandro
Armagh Observatory, College Hill, Armagh, BT61 9DG, Northern Ireland
LEMEN James R.
Lockheed Research Laboratory, 3251 Hannover St., Palo Alto, CA 94306, USA
LINSKY Jeffrey
Joint Institute for Laboratory Astrophysics, University of Colorado, Boulder, CO 80309-0440, USA

LIVSHITS Misha
Sternberg State Astronomical Institute, Moscow State University, 119899 Moscow V-234, Russia
LORENTE Rosario
Dpto. Astrofisica, Facultad de Fisicas, Universidad Complutense, E-28040 Madrid, Spain
MAGGIO Antonio
Istituto ed Osservatorio Astronomico di Palermo, Piazza del Parlamento 1, 90134 Palermo, Italy
MARTIN Renato
Istituto di Fisica, Università di Torino, Via Giuria 1, 10125 Torino, Italy
MATTHEWS Lee
Astrophysics Group, Imperial College, Prince Consort Road, London SW7 2BZ, United Kingdom
MEWE Rolf
SRON-Laboratory for Space Research, Sorbonnelaan 2, 3584 CA Utrecht, The Netherlands
MICELA Giuseppina
Istituto ed Osservatorio Astronomico di Palermo, Piazza del Parlamento 1, 90134 Palermo, Italy
MONTES David
Dpto. Astrofisica, Facultad de Fisicas, Universidad Complutense, E-28040 Madrid, Spain
MONTESINOS Benjamin
Department of Physics (Theoretical Physics), University of Oxford, 1 Keble Road, Oxford OX1 3NP, United Kingdom
MUTEL Robert
Department of Physics and Astronomy, University of Iowa, Iowa City, IA 52242, USA
NOCI Giancarlo
Osservatorio Astrofisico di Arcetri, Largo E. Fermi 50, 50125 Firenze, Italy
ODA Minoru
The Institute of Physical and Chemical Research, Wako, Saitama, 351-01 Japan
OERTEL Goetz
AURA Inc., 1625 Mass. Avenue, NW 701, Washington, DC 20036, USA
OLIVER Ramon
Departament de Fisica, Universidad de les Illes Balears, 07071 Palma de Mallorca, Spain
ORLANDO Salvatore
Istituto ed Osservatorio Astronomico di Palermo, Piazza del Parlamento 1, 90134 Palermo, Italy

OTTMANN, Renate
Max-Planck Institut für Extraterrestrische Physik, D-8046 Garching bei München, Germany
PAGANO Isabella
Istituto di Astronomia, Città Universitaria, Via A. Doria 6, 95100 Catania, Italy
PALLAVICINI Roberto
Osservatorio Astrofisico di Arcetri, Largo E. Fermi 50, 50125 Firenze, Italy
PASTORI Livio
Osservatorio Astronomico di Brera, Via E. Bianchi 46, 22055 Merate, Como, Italy
PATERNÒ Lucio
Istituto di Astronomia, Città Universitaria, Via A. Doria 6, 95100 Catania, Italy
PERES Giovanni
Osservatorio Astrofisico di Catania, Città Universitaria, Via A. Doria 6, 95100 Catania, Italy
PITERS Ankie J.M.
Astronomical Institute "Anton Pannekoek", Kruislaan 403, 1098 SJ Amsterdam, The Netherlands
PRIEST Eric R.
Mathematical and Computational Sciences Dept., The University, St. Andrews KY16 9SS, Scotland, United Kingdom
PYE John P.
Department of Physics and Astronomy, Leicester University, Leicester, LE1 7RH, United Kingdom
RANDICH Sofia
Dipartimento di Astronomia, Università di Firenze, Largo E. Fermi 5, 50125 Firenze, Italy
REALE Fabio
Istituto ed Osservatorio Astronomico di Palermo, Piazza del Parlamento 1, 90134 Palermo, Italy
REIMERS Dieters
Hamburger Steinwarte, Gojenbergsweg 112, D-2050 Hamburg 80, Germany
ROBBA Renato
Istituto di Fisica, Università di Palermo, Via Archirafi 36, 90100 Palermo, Italy
RODONÒ Marcello
Istituto di Astronomia e Osservatorio Astrofisico di Catania, Città Universitaria, Via A. Doria 6, 95100 Catania, Italy
ROSNER Robert
Department of Astronomy and Astrophysics, University of Chicago, Chicago, IL 60637, USA
RUFFINI Remo
Dipartimento di Fisica, Università "La Sapienza", Piazzale A. Moro 2, 00100 Roma, Italy

SALAMONE Vincenzo
Istituto di Fisica, Università di Palermo, Via Archirafi 36, 90100 Palermo, Italy
SCHMITT Jurgen
Max-Planck Institut für Extraterrestrische Physik, D-8046 Garching bei München, Germany
SCHREIER Ethan J.
Space Telescope Science Institute, 3700 San Martin Drive, Baltimore, MD 21218, USA
SCIORTINO Salvatore
Istituto ed Osservatorio Astronomico di Palermo, Piazza del Parlamento 1, 90134 Palermo, Italy
SERIO Salvatore
Istituto ed Osservatorio Astronomico di Palermo, Piazza del Parlamento 1, 90134 Palermo, Italy
SMARTT Raymond N.
National Solar Observatory, Sacramento Peak, Sunspot, NM 88349, USA
SOMOV Boris V.
Sternberg State Astronomical Institute, Moscow State University, 119899 Moscow V-234, Russia
SURLANTZIS George
Department of Physics, University of Crete, GR-71409, Heraklion, Crete, Greece
SYLWESTER Barbara
Space Research Centre, Polish Academy of Sciences, ul. Kopernika 11, 51-622 Wroclaw, Poland
SYLWESTER Janusz
Space Research Centre, Polish Academy of Sciences, ul. Kopernika 11, 51-622 Wroclaw, Poland
TAGLIAFERRI Giampiero
ISO-Observatory, Astrophysics Division, ESA-SSD, ESTEC, Keplerlaan 1, 2200 AG Noordwijk, The Netherlands
THOMAS, Roger J.
Laboratory for Astronomy and Solar Physics, Code 680, NASA/Goddard Space Flight Center, Greenbelt, MD 20771, USA
TRÜMPER, Joachim
Max-Planck Institut für Extraterrestrische Physik, D-8046 Garching bei München, Germany
TSINGANOS Kanaris
Department of Physics, University of Crete, GR-71409, Heraklion, Crete, Greece
TSUNETA Saku
Institute of Astronomy, The University of Tokyo, Mitaka, Tokyo 181, Japan
UCHIDA Yutaka
Institute of Astronomy, The University of Tokyo, Mitaka, Tokyo 181, Japan

ULMSCHNEIDER Peter
Institut für Theoretische Astrophysik, Universität Heidelberg, Im Neuenheimer Feld 561, D-6900 Heidelberg, Germany
VAHIA M.N.
Tata Institute of Fundamental Research, Space Physics Group, Homi Bhabha Road, Bombay 400 005, India
VENTURA Rita
Osservatorio Astrofisico di Catania, Città Universitaria, Via A. Doria 6, 95100 Catania, Italy
WALTER Frederick M.
Department of Earth and Space Sciences, State University of New York, Stony Brook, NY 11794-2100, USA
WANG Yi-Ming
E.O. Hulbert Center for Space Research, Naval Research Laboratory, Washington, DC 20375, USA
WEISS Nigel
Department of Applied Mathematics and Theoretical Physics, University of Cambridge, Cambridge CB3 9EW, United Kingdom

FOREWORD

The original plans for a meeting to celebrate the second centenary of the Astronomical Observatory of Palermo were for a celebration with a double character. The gathering was to have both a historical character, appropriate for a bicentennial, and a technical character, to note and chronicle the new phase of the history of the Observatory, which has prospered in parallel with the development of this fairly recent topic in astronomical research, the physics of stellar and solar coronae.

After the untimely death of the Observatory's Director, Giuseppe S. Vaiana (Pippo to his many friends), a number of colleagues and friends insisted that the celebration should nevertheless be held and should be dedicated to this farsighted scientist who stimulated the development of coronal physics from the early x-ray observations of the solar corona to the recognition of coronae as an observable feature of nearly all stars.

This memorial dedication did not change the character of the meeting, which was held in Palermo from 22 to 26 June 1992; as his contributions are very alive in the papers presented at the meeting and collected here, Pippo Vaiana has certainly achieved his place in the history of Astronomy.

Without doubt, the presence of so many qualified scientists and the level of their contributions served as a personal homage to Pippo by his many friends, pupils, associates and colleagues. They brought to the symposium fresh data from the ROSAT and *Yohkoh* satellites, the latest theoretical developments, and their wisdom and insights gleaned from nearly twenty years of progress in coronal physics.

The meeting was made possible by the generous contributions of Presidenza della Regione Sicilia, Assessorato Regionale ai Beni Culturali, Ambientali e Pubblica Istruzione, Assemblea Regionale Siciliana, Cassa Centrale di Risparmio per le Provincie Siciliane, Consiglio Nazionale delle Ricerche, Osservatorio Astronomico di Palermo, Università di Palermo, Edwards Alto Vuoto, Digital Equipment, Ordine degli Ingegneri di Palermo. We thank them all heartily, and also Società Italiana per l'Esercizio delle Comunicazioni, which provided support for the data link.

We greatly appreciate the efforts of the many individuals who provided critical support before, during, and after the meeting. In particular, we thank Lorraine Volsky and Juanita Crane for their help in editing the book.

Scientific Organizing Committee: L. Acton, R. Bonnet, O. Gingerich, L. Golub, M. Hoskin, C. Jordan, J. Linsky (co-chairman), R. Mewe, R. Pallavicini, J. Pye, M. Rodonò, R. Rosner, J. Schmitt, S. Serio (co-chairman), Y. Uchida.

Local Organizing Committee: M. Barbera, F. Bocchino, I. Chinnici, A. Ciaravella, A. Collura, F. Damiani, G. Foderà Serio, F. Grillo, A. Maggio, G. Micela, S. Orlando, G. Peres, D. Randazzo, F. Reale, F. Salemi, S. Sciortino (chairman).

HONOUR COMMITTEE:

Domenica Battista Spinelli	Director, ufficio XIV Ministero dell'Università
Alberto Bombace	Director, Assessorato Regionale Beni Culturali Ambientali e Pubblica Istruzione
Alexander Boyarchuck	President, International Astronomical Union
Renato Cannarozzo	President, Ordine degli Ingegneri di Palermo
Massimo Capaccioli	President, Società Astronomica Italiana
Gaetano Di Fresco	Secretary General, Presidenza Regionale Siciliana
George Field	Professor of Astronomy, Harvard University
Riccardo Giacconi	Director, Hubble Space Telescope Science Institute
Angelo Guarino	V. President, Comitato CNR Beni Culturali
Silvestre Liotta	Secretary General, Assemblea Regionale Siciliana
Francesco Maggio	Dean, Facoltà di Scienze, Università di Palermo
Ignazio Melisenda Giambertoni	President, Università di Palermo
Minoru Oda	President, RIKEN Institute of Physical and Chemical Research
Goetz Oertel	President, Association of Universities for Research in Astronomy
Massimo Ugo Palma	Director, Istituto per le Applicazioni Interdisciplinari della Fisica, CNR
Francesco S. Persico	Director, Istituto di Fisica Università di Palermo
Ken Pounds	Director, Physics Department, University of Leicester
Bruno Rossi	Professor Emeritus, Massachusetts Institute of Technology
Remo Ruffini	President, Comitato Scientifico Agenzia Spaziale Italiana
Giorgio Salvini	President, Accademia Nazionale dei Lincei
Giancarlo Setti	V. President, Consiglio delle Ricerche Astronomiche
Karl Teeter	Professor of Linguistics, Harvard University
Joachim Trümper	Director, Max Planck Institut für Extraterrestrische Physik
Maria Beatrice Vittorelli	President, Corso di Laurea in Fisica, Università di Palermo

PART I

G.S. VAIANA AND THE PALERMO ASTRONOMICAL OBSERVATORY

G. S. VAIANA MEMORIAL LECTURE

RICCARDO GIACCONI
Space Telescope Science Institute
3700 San Martin Drive
Baltimore, MD 21218 USA

1. Introduction

The bicentennial of the Palermo Astronomical Observatory gives us the opportunity to honor the memory of Professor Giuseppe Vaiana who certainly will rank among its most distinguished Directors. It is an especially poignant occasion for me because Pippo (as he was known to his friends) was not only a close scientific collaborator in the early days of X-ray astronomy and for almost 30 years thereafter, but also a close personal friend.

I first became acquainted with Pippo in 1964 when he joined the X-ray astronomy group of American Science and Engineering a small research firm in Cambridge, Massachusetts. We worked closely together until 1981, first at AS&E, and then at the Harvard-Smithsonian Center for Astrophysics and we continued to exchange information and ideas about our work until just last year.

A number of colleagues from all over the world are participating in this Symposium, and will certainly discuss in more detail than I can Vaiana's contribution to many fields of solar and stellar astronomy and the importance of his legacy to current research endeavors. I will only attempt to trace in broad outlines the evolution of his work and highlight some of his contributions which appear to me particularly important and with which I am familiar. They include:

- The early development of X-ray telescopes and the first high-resolution solar X-ray pictures (1968).
- The Skylab-ATM Study of formation and evolution of solar X-ray emitting features over several solar rotations (1973).
- The study of the X-ray coronae of all main-sequence stars with the Einstein Observatory (1979).
- Vaiana's plans for a spectroscopic stellar X-ray mission and his scientific methodology (1981).

2. Early X-Ray Telescope Development (1964–1973) and the First High-Resolution Pictures of the Sun

Soon after Vaiana's arrival in 1964 the AS&E group, including Reidy, Zehnpfennig, and myself, had completed the first rather primitive rocket payload shown in Figure 1 in a collaborative effort with Lindsay and Muney of Goddard Space Flight Center. We flew that payload on March 17, 1965 and obtained the picture of the SUN shown in Figure 2. (Giacconi *et al.* 1965). Primitive as this payload was, it already represented the result of a development effort of several years (since

J.F. Linsky and S. Serio (eds.), Physics of Solar and Stellar Coronae, 3–19.

Figure 1. Solar pointing control for the Aerobee rocket flown on March 17, 1965. The apertures of three grazing incidence X-ray telescopes can be seen.

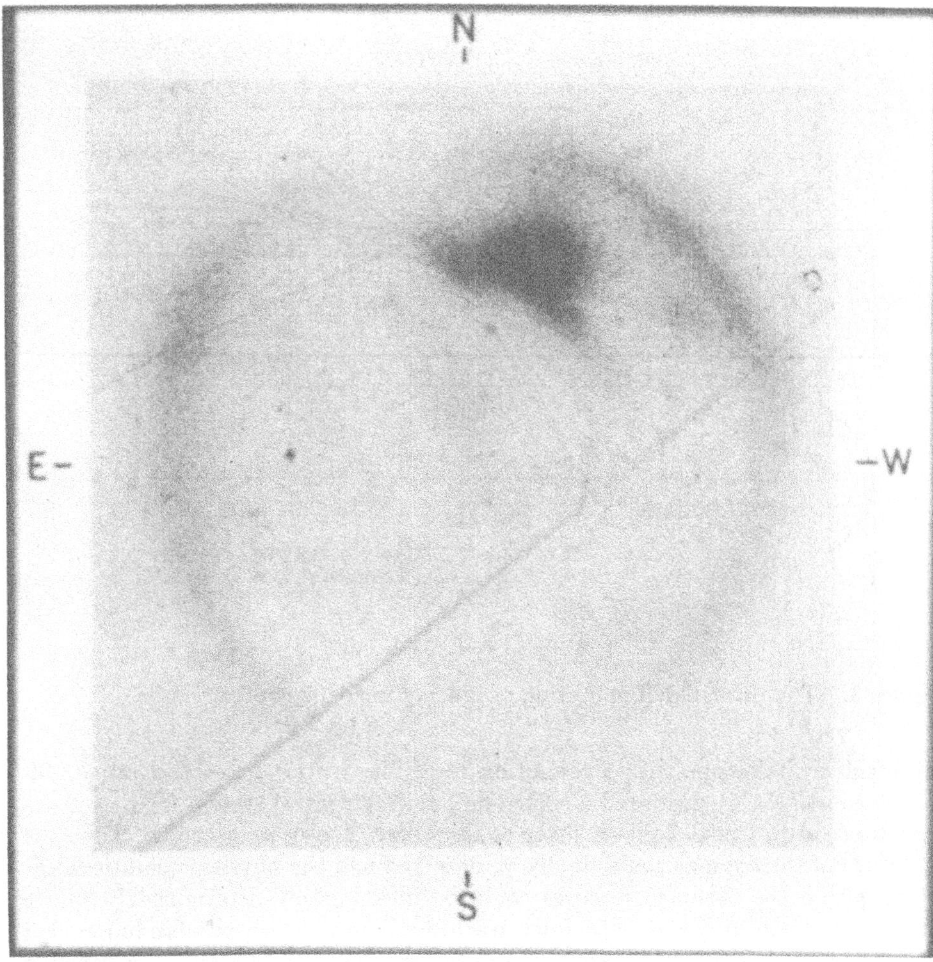

Figure 2. First X-ray picture of the Sun obtained on March 17, 1965 with grazing incidence optics.

1960) to obtain grazing incidence mirrors of increasing sophistication. Figure 3 illustrates the principle of operation of X-ray optics and Figure 4 shows one of the first mirrors which had been obtained by electro-deposition of nickel on a polished aluminum mandrel and was used to obtain the picture in Figure 2 showing an angular resolution of one arc minute.

From 1965 on Pippo assumed increasingly the scientific leadership of the solar X-ray astronomy group at AS&E and quickly achieved important results. I quote his words from a 1973 paper (Vaiana *et al.* 1973)

"In 1968 a breakthrough occurred with the development, at AS&E, of a graz-

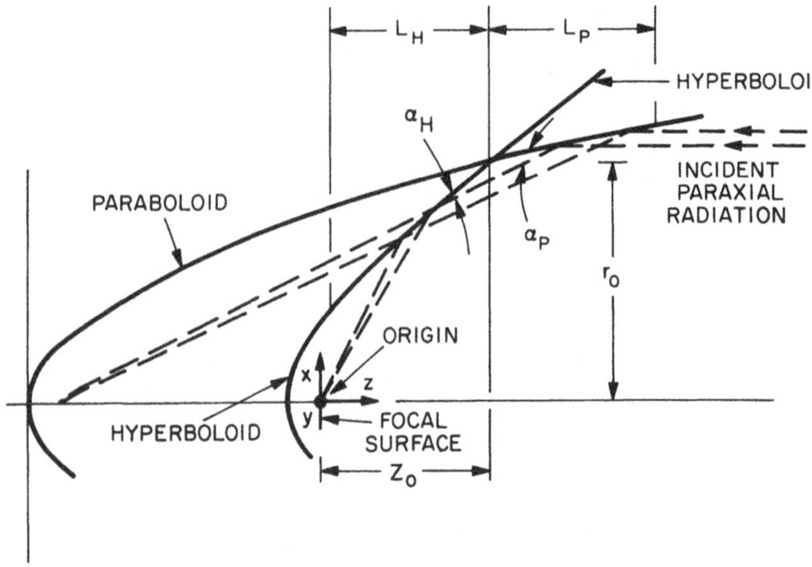

Figure 3. The principle of operation of grazing incidence optics.

ing incidence telescope with a resolution capability in the arc second range. Simultaneously NASA produced a solar fine pointing control system for rocket observations with a peak-to-peak jitter of the order of two arc seconds. Thus the structure of the coronae could finally be observed and the physical conditions existing within the observed features could be meaningfully determined." "Since that time a further six successful, high resolution, X-ray telescope sounding rocket payloads have been flown by our group each involving additional improvements in technique. Individual experiments were designed with specific observations in mind, such as the study of active region structure, flare dynamics, or the determination of the morphology of the quiet coronal features." Yet in each successive flight new, previously unsuspected features were discovered: The existence of *large scale coronal features* was detected in the June 8, 1968 flare observation. *Bright points* were first detected in April 8, 1969. Filament cavities were observed in the two 1970 rocket flights which first detected *coronal holes*. *Coronal loops* surrounding the developing active regions were discovered in the March 1973 flight. Figure 5 shows the results of this stunningly successful series of rocket flights.

2.1. The ATM Experiment on "Skylab"

This extensive work paved the way for the much more ambitious experiment by Vaiana and the American Science and Engineering Group on "Skylab." For me the interest in the study of the Sun was only as a proving ground for technology to be

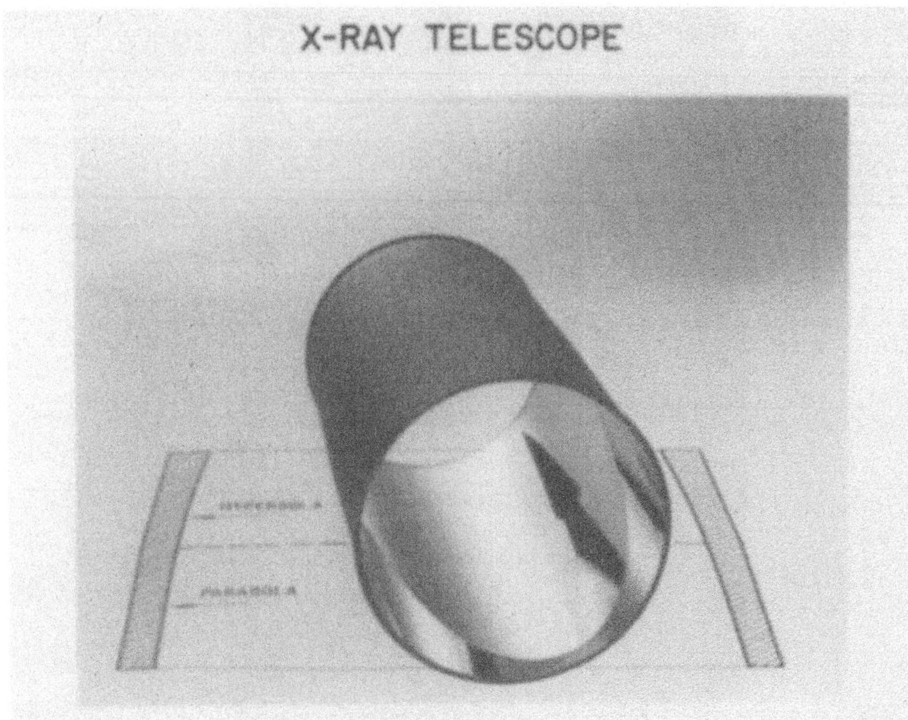

Figure 4. The electroformed nickel X-ray mirror used in the March 17, 1965 solar observations.

used later on the study of X-ray stars. Pippo took scientific as well as managerial leadership of the solar X-ray astronomy group and succeeded brilliantly in developing the first high resolution nested X-ray mirror and in obtaining a veritable motion picture of the formation, evolution, and dynamics of plasma features on the Sun over several solar rotations. Figure 6 and shows the Skylab mirror in comparison with some earlier ones. Figure 7 shows a view of Skylab during the second rendezvous. An example of changing coronal structures observed by Skylab is shown in Figure 8. The profound changes in our understanding of the solar corona that this work brought about are summarized by Vaiana and Rosner in the 1978 volume of Annual Reviews of Astronomy and Astrophysics, (Vaiana 1978).

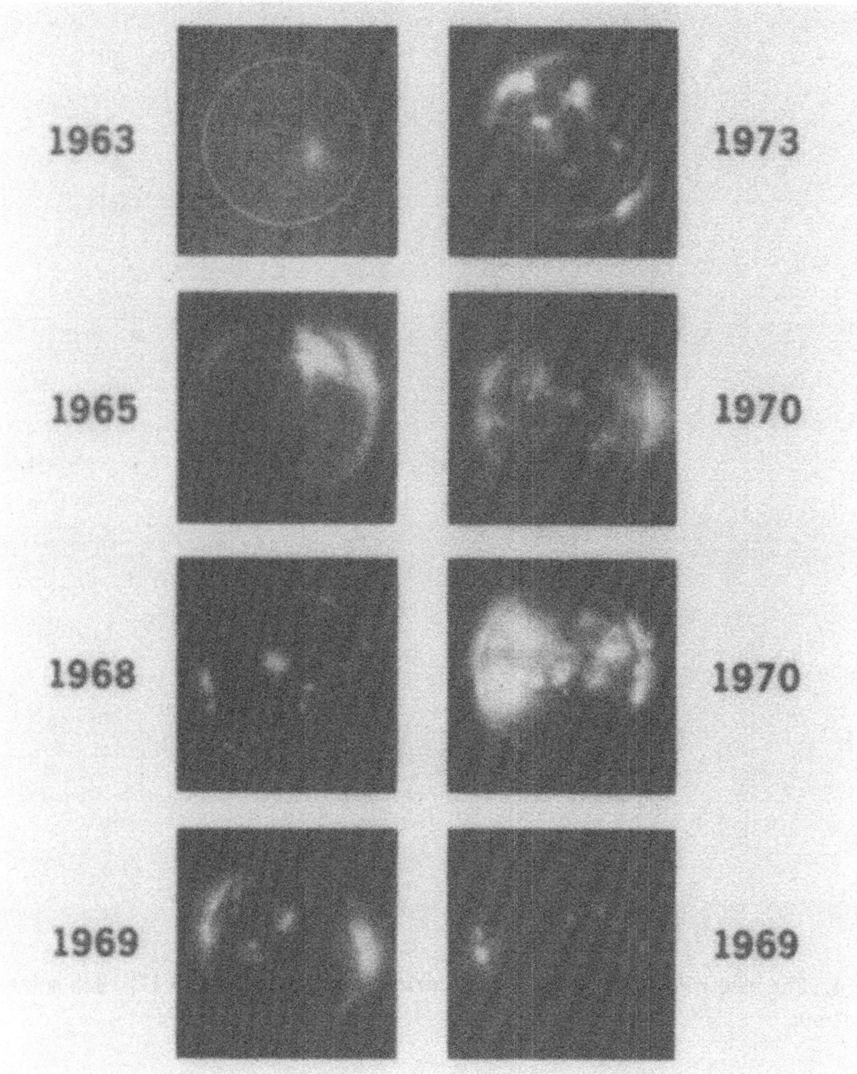

Figure 5. Progress in high resolution X-ray studies of the solar corona accomplished by Vaiana and his collaborators in the rocket flights of 1968 through 1973. Angular resolution was improved by more than a factor of 10 from 1965 to 1968.

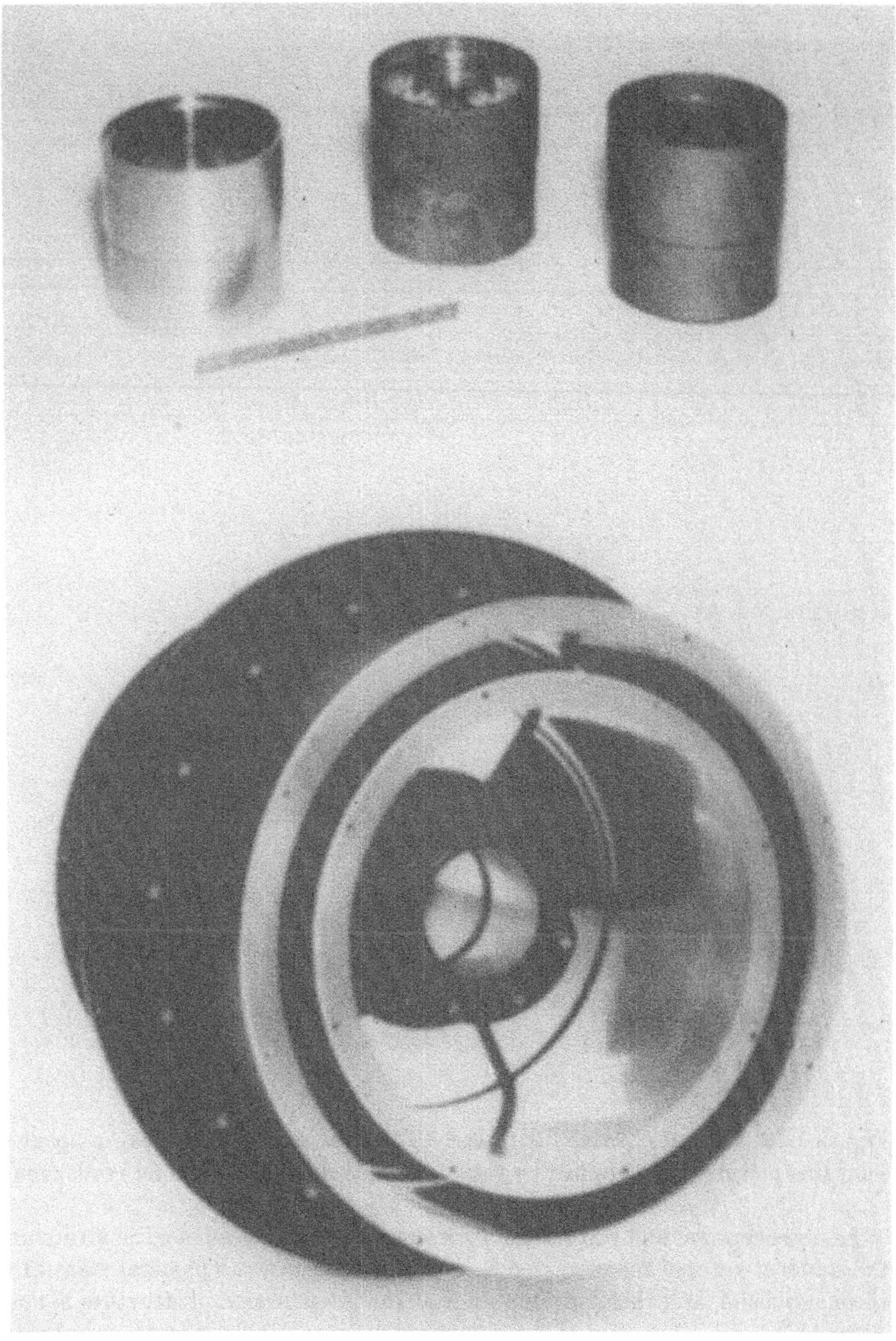

Figure 6. The 30 cm diameter X-ray mirror for ATM. Note the nesting of two surfaces to obtain higher sensitivity: the increase in collecting area was a factor of about 30 with respect to earlier X-ray mirrors.

Figure 7. A view of "Skylab" with the ATM solar experiment looking up at the Sun; this picture was obtained by the astronauts during the second rendezvous.

"The observational and theoretical work suggests that the topological structure of the ambient coronal magnetic fields largely determines the physical state of the coronal plasma, and therefore implies that the introduction of structure is not to be regarded as a refinement of theories based upon spatial homogeneity, but rather as a fundamental change in our understanding of the physics of the corona."

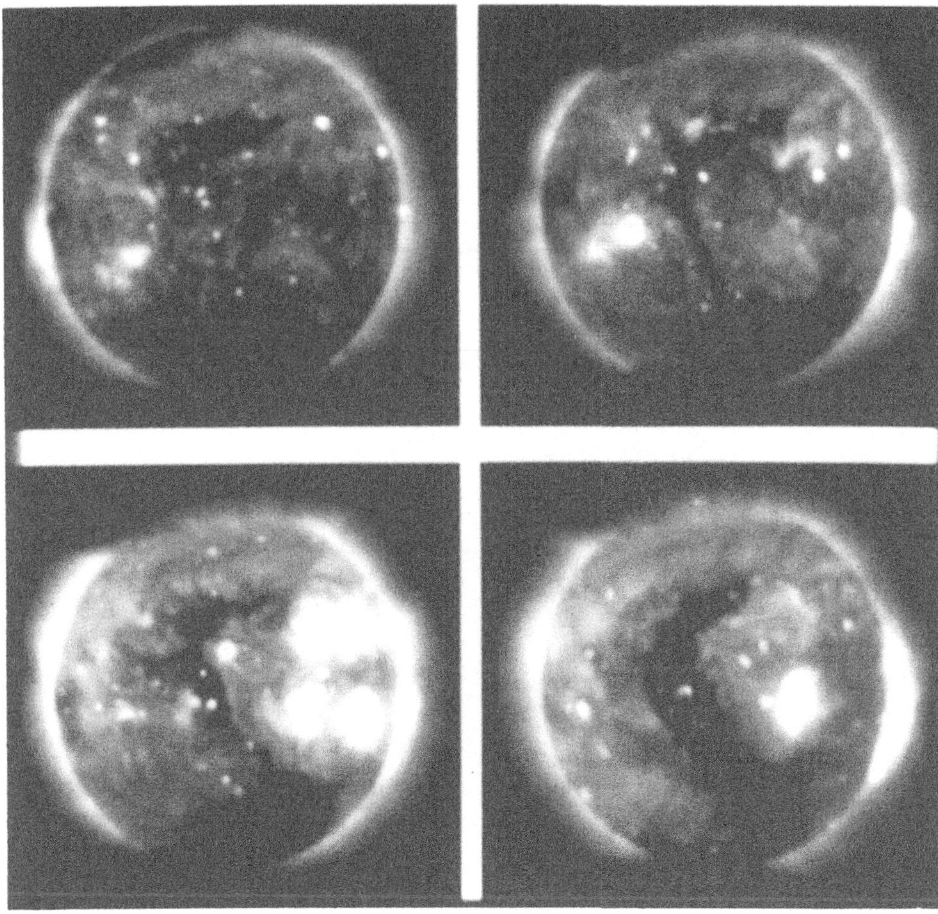

Figure 8. Examples of high resolution X-ray pictures of the Solar corona obtained during several solar rotations. Note the coronal holes, the bright points, the coronal loops, and flare regions.

2.2. Stellar X-Ray Astronomy

The next major step in Pippo's scientific interest came about with the flight of the first X-ray telescope mission dedicated to the study of extra solar objects: the Einstein Observatory. Figures 9, 10, 11 show the Einstein Observatory mirror, the test chamber at Marshall Space Flight Center and the entire spacecraft and experiment just prior to assembly at TRW. Pippo's work on X-ray optics contributed greatly to the development of this mission. Leon van Speybroeck, the telescope

Figure 9. The 60 cm diameter X-ray mirror for "Einstein." Note the four nested surfaces.

scientist for Skylab, had the responsibility for "Einstein," and many of the expert engineers and technicians who had worked on Skylab with Pippo brought invaluable know how to the design and development of Einstein.

But the most significant contribution of Pippo to the mission came from his deep interest in the study of the X-ray emission from the coronas of normal stars. While most of us were interested in the study of extra galactic objects such as quasars, BL Lacs, clusters of galaxies, active galactic nuclei or exotic galactic objects such as binary X-ray stars, X-ray sources in globular clusters, supernova and supernova remnants, Pippo fully understood the great potential for discovery which existed in the study of normal stars. He and his collaborators planned and carried out the research program which resulted in some of the most unexpected and exciting discoveries of the "Einstein" Observatory.

Figure 10. "Einstein" observatory being tested at Marshall Space Flight Center. A 300 meter long vacuum pipe provided a quasi parallel X-ray beam at the test chamber.

Figure 11. "Einstein" Observatory ready for assembly at TRW Corporation.

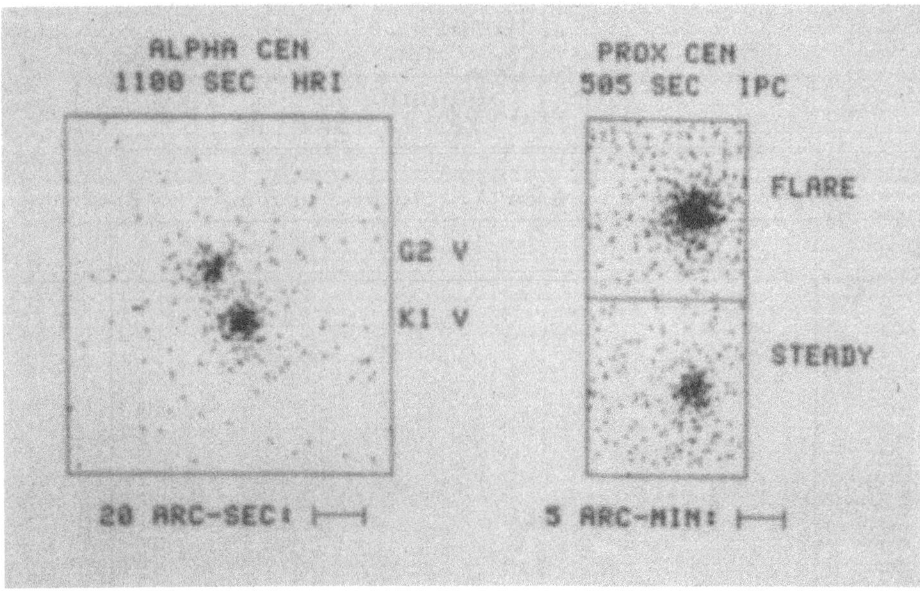

Figure 12. Detection of Alpha Cen and Prox Cen from "Einstein." Each dot represents a detected photon.

Figure 12 gives some examples of the detection of nearby stars. Figure 13 shows that stars of all spectral types are being detected and Figure 14 shows the comparison with the predictions of coronal heating through acoustic waves.

It was clear, even on the basis of the preliminary survey data, that a new theory was necessary to explain the observed effects. Vaiana summarized the results in a 1979 paper which he delivered at a joint discussion of the IAU General Assembly in Montreal, Canada (Vaiana 1979).

"The primary result of this survey is that all categories of stars have been detected as X-ray emitters, with the sole exception of very late-type supergiants. Except within a very narrow range of spectral types, both the levels of X-ray emission and the general behavior of the median luminosities disagree with theoretical predictions. The disagreement appears not only in the behavior of median X-ray luminosity as a function of spectral type, but also in the very wide range of observed X-ray luminosities at each spectral type. This finding is quite general: within each spectral type and luminosity class there is observed a large range of X-ray luminosities (two to three orders of magnitude).

"This range is also large when compared to the variation of bolometric luminosity within each spectral type and to the change in median X-ray luminosity. The question therefore is raised: what stellar characteristic determines the total level of X-ray emission? It clearly is not spectral type, luminosity class, or mass,

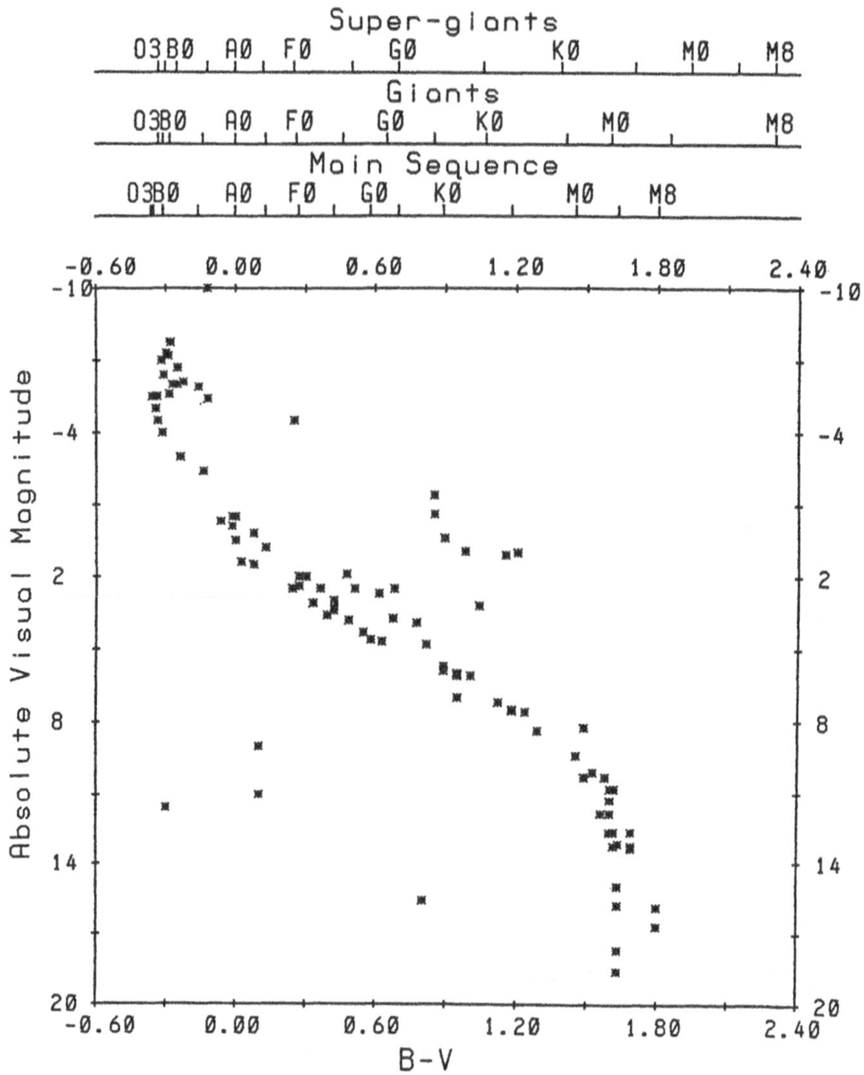

Figure 13. Detected X-ray sources in the visual magnitude versus color index diagram. Stars of all spectral types are observed.

those very parameters that locate main-sequence stars on the H-R diagram. What then is it that determines the level of X-ray emission?"

These questions formulated only a few months after the launch of Einstein set the direction for the study of stellar X-ray emission for years to come. Many of the papers that will be presented at this Symposium deal with one aspect or

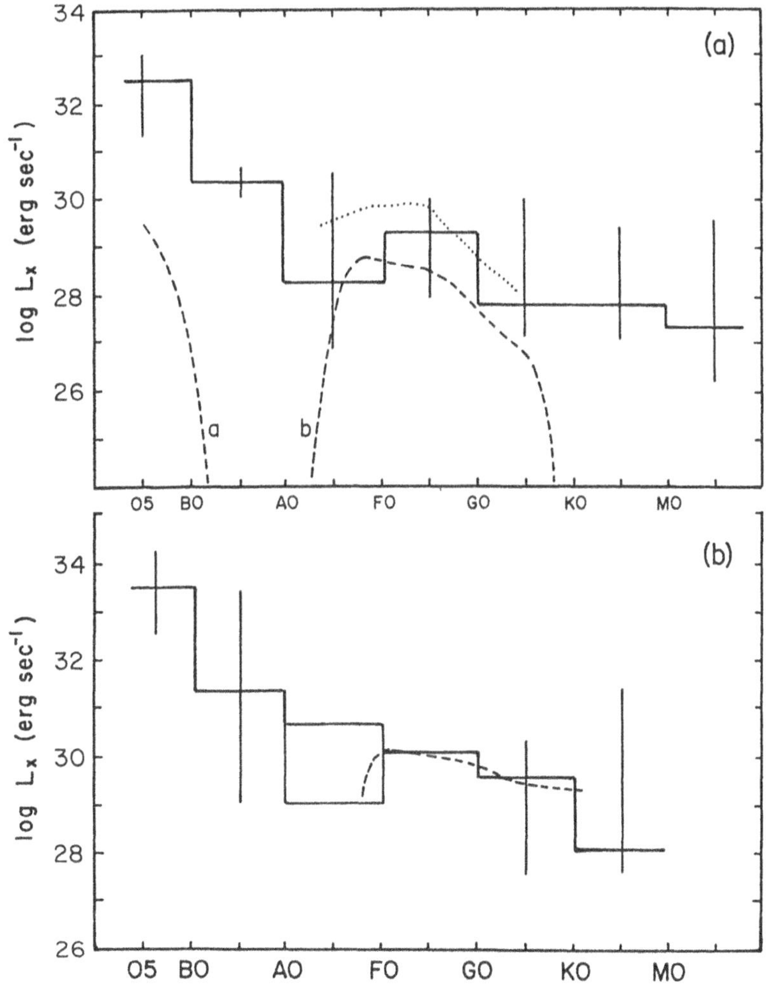

Figure 14. The discrepancy between observed X-ray emission and theoretical predictions for main sequence (a) and giant and supergiant (b) stars.

the other of these queries. Pippo himself continued working energetically on this subject up until his untimely death; first with Einstein archival data and more recently with ROSAT results. His pioneering work in solar and stellar coronas represents a landmark in astronomy for which he will always be remembered.

3. Conclusion

I would like to conclude these remarks by discussing one aspect of Pippo's activity as scientist, teacher, and mentor to young people. In particular the unity between observational and theoretical efforts which distinguished Vaiana's work and his careful understanding of the importance of process.

Separation between observationalists and theorists is growing in astronomy, and to almost a ludicrous degree in space astronomy. Instrument builders, data analysts, and theoreticians are rarely the same person. Pippo and I spent a great deal of time in the evenings over cognacs discussing how science should be done. Possibly because neither of us had been trained as an astronomer, we found it natural to take inspiration from the giants of the Italian Physics school such as Fermi, Rossi, Segre, etc., and even earlier from the immortal Galileo. "Provare e riprovare" does not mean to try twice. It means to observe nature, formulate theories, predict effect in new conditions and test them by experiment or observation. Thus all the technology development and the engineering effort and the experimental work that Pippo did was not separate from, but the expression of, his love of nature and his desire to understand it. Each of the early solar rocket experiment was designed to study a specific aspect of solar activities. Skylab was carefully conceived ab initio to obtain the resolution and dynamic range required to elucidate the formation, evolution and dynamics of coronal structure. The study of the data, the analysis system and the theoretical efforts to elucidate the physics of the phenomenon were part and parcel of Pippo's activities.

Pippo was prescient and ready when the first stellar X-ray data returned from the Einstein Observatory. He was almost alone in our group to fully understand and vigorously pursue this new discipline in X-ray astronomy: the study of stellar coronae. He was certainly one of the pioneers in the field.

His last experiment (a proposal for a small X-ray satellite which combined imaging and dispersive spectroscopy) to further study the stars would have been a wonderful logical extension of the line of inquiry he had initiated. He fought with courage and perseverance to bring this effort to fruition. In the U.S. the opportunities for small satellite investigations were just then (1980) closing rather then increasing; in that decline of the U.S. Space Program from which we have not yet recovered. In Italy (1981) politics, industrial and institutional interests rather then scientific worth led to different choices. Here again it will take years to recover from these mistakes.

Pippo never wavered, never accepted to do science except at the highest level, never compromised. In our evening discussions on the process of doing science we had agreed that often the process is more important then the results. A piece of hardware, no matter how advanced or refined is just a lifeless thing. It soon becomes obsolete even if successful or it might fail altogether. What is important is the personal growth and deepening of our understanding and appreciation of nature. This we can achieve in pursuing our studies with mutual trust and ruthless honesty and with joy in truth. Pippo leaves us a moral as well as a scientific legacy. His untimely death impoverishes us all.

4. References

Giacconi R., Reidy, W. P., Zehnpfenning, T., Lindsay, J. C., Muney, W. S.: 1965, *Ap. J.*, **142**, 1274.

Vaiana, G. S., Krieger, A. S., Timothy, A. F.: 1973, *Solar Physics* **32**, 81–116.

Vaiana, G. S., and Rosner, R.: 1978, "Recent Advances in Coronae Physics," *Ann. Rev. Astron. Astrophysics*, **16**, 393–428.

Vaiana, G. S.: 1976, "Stellar Coronae," Joint Discussion on Very Hot Plasmas in Circumstellar, Interstellar, and Intergalactic Space, IAU General Assembly, August 22, 1976, Montreal, Canada.

ON THE HISTORY OF THE PALERMO ASTRONOMICAL OBSERVATORY

GIORGIA FODERÀ SERIO
Istituto ed Osservatorio Astronomico, Palermo

Fig. 1. The Palermo Astronomical Observatory as it appeared in 1804

The history of the Palermo Astronomical Observatory is the history of an institution that had its beginning two hundred years ago, actually two hundred and two years ago, since the Observatory was officially established in 1790.

For two hundred years, therefore, it has been possible in actual fact, or at least in principle, to practice Astronomy continuously, at the institutional level, in Palermo. Such a long continuity is a unique feature for any specific branch of Science in Sicily. Perhaps it was the main goal, and certainly constitutes the major achievement of those who fostered the birth of the Observatory.

True enough, institutions may, and usually do, function over long time scales at varying levels of efficiency. Yet, it is their very existence and persistence that allows bridging the gaps of dark periods, whenever inspired and capable scientists are allowed - or invited - to take over.

J.F. Linsky and S. Serio (eds.), Physics of Solar and Stellar Coronae, 21–33.

In the recent past, for about forty years, from the early thirties up to the mid seventies, the life of the Observatory was in serious danger. During those years the Observatory struggled for mere existence with scarce personnel and funds, and with limited scientific guidance and objectives. Yet, the Institution was, at least formally, extant, and this challenged the great inspiration and stamina of Pippo Vaiana. Quietly and vigorously, with great scientific depth and endless energy, he revitalized the Observatory and guided it towards new objectives in the world scene of research. It took a short time for the Observatory to find renewed recognition.

Likely, without Pippo Vaiana there would have been no Bicentennial Celebration for the Palermo Astronomical Observatory, certainly not a Celebration like this one, with such a distinguished participation. We are here today to pay homage to the scientist who has given important contributions to Astronomy. The rebuilding of the Palermo Astronomical Observatory is one of these.

In 1790, the Palermo Observatory was established as the National Observatory of the Kingdom of Two Sicilies, a kingdom that included essentially all of Italy South of Rome. It was the National Observatory not because it was the first and only Observatory to have ever existed in this part of Italy,[1] but because it was the King that gave his consent, that is, secured the necessary funds for the erection of the building, the provision of instruments and staff, and the prosecution of long term research programmes.

It may seem strange that such an Observatory was built in Palermo rather than in Naples. Naples was in fact the Capital of the kingdom, and certainly the center of whatever scientific life was active in this part of Europe. It was the King's and Government's residence, with obvious consequences in terms of expenditures.[2]

Palermo on the other hand, and more generally Sicily, had reached, in the second half of the eighteenth century, their lowest level, as far as science is concerned. This was particularly true for Astronomy. It is impossible to find a single work on Astronomy, let aside gnomonics, by a Sicilian author published during the XVIII century before 1790. To speculate on this point would take too long. I can only suggest to keep in mind that, among other factors, the *Inquisition*, active in Sicily from 1601 to 1782, undeniably played a major and tragic role.[3]

Further, Sicily was run through a Viceroy, whose discretional powers, especially concerning expenditures, were extremely limited. This notwithstanding, the Palermo Observatory was established and, since its very inception, and for a period of about thirty years around the turn of the century, rivaled, for equipment and scientific results, the most venerable institutions of its kind. In those specific sociological, political and financial conditions, this could hardly be due to anything but shear luck. I mean that a number of circumstances, each of which was not by itself determinant, clustered together, at just the right time and right place, to produce such a result.

The first relevant circumstance was the banning of the Jesuits from the kingdom in 1767. I must clarify this point. I am by no means giving or implying a negative or positive judgement on the role played by the Jesuits in the dissemination and progress of science, and of Astronomy in particular. I am just stating a fact: the

Jesuits in the Kingdom of the two Sicilies retained, until 1767, the monopoly of public education up to University level. For this reason, the ban against their order caused the immediate collapse of the entire educational system i.e., on the very same day the ban was promulgated, all schools in the kingdom were closed.

This forced the Government to assume responsibility for the organization of public education. The Government was not at all ready to face the emergency it had caused. Nobody had been charged to prepare a reform plan. The only action taken, at first, was to reopen the old schools staffing them with make-shift personnel, and to use the income of the confiscated Jesuit estates to run them. For quite a few years public education was thus a complete disaster. Nevertheless, things slowly got on the move.

As for higher education, in 1779 the "Accademia dei Regj Studi" was established, which was to become, in 1805, the University of Palermo. This was the first success obtained by a small but influential reformist party that found strong and active support in Domenico Caracciolo and Francesco D'Aquino Prince of Caramanico, the most enlightened Viceroys Sicily has ever had. The aim of the reformist party was to create in Palermo a University up to the standards of France and England, in which the teaching of "things," that is practical and scientific disciplines, had to be privileged over the teaching of "words."

Experienced teachers and scientists for this new University were eagerly sought, and they had to come from "abroad," to bring into the scientifically deprived Sicily their experience, their knowledge, their new ideas. In line with this programme, they planned for structures like a Botanical Garden, a Theatre of Anatomy, and an Astronomical Observatory.

Astronomy was in fact thought to be central in this project, if only because of its practical fallouts, from the improvement of navigation to geodesy and geography. Suffice it to mention here that, up to 1791, when Piazzi redetermined it, the latitude of Palermo was known with an error of 4 arcminutes, an incredibly bad value for the time, not to mention geographical maps.[4]

The large scale programme set up by the Deputies of the Accademia was a failure: countless approaches were made to try to hire reputed scientists like Spallanzani, for the chair of Physics, or Lagrange for that of Mathematics. All of them politely refused. The main reason for such refusals is efficaciously expressed in this comment by D'Alembert:

> *Je n'ai pas oui dire que Mr de la Grange songe à quitter Berlin, où il me paroit qu'il se trouve bien, mais comme je n'ai pas très souvent de ses nouvelles, j'ignore ses dispositions. Je doute que le Marquis de Caracciolo trouve à Palermo de quoi composer une acadèmie. S'il y appelloit Mr de La Grange, ce seroit un Président sans conseillers.*[5]

Sicily was considered, and rightly so, a scientific periphery.

As for Astronomy, Barnaba Oriani, Astronomer in Milan, was invited and, like the others, declined the offer. Oriani's refusal provided a unique opportunity for Giuseppe Piazzi (1746-1826). When he was appointed to the chair of Astronomy early in 1787, Piazzi, already in his forties, had only made a mediocre career as a teacher of Theology and Mathematics, migrating from one city to another. He had

Fig. 2. Giuseppe Piazzi pointing at Ceres

taught in Ravenna, Genoa, Rimini, Rome, and Malta, before landing in Palermo in 1781 as lecturer of Mathematics. Extensive searches have indicated that he had not published anything, neither does it seem that he had any interest in research. More than that, he was not an astronomer, not even an amateur. Yet, in a matter of a few years he became one of the most respected astronomers of his time.

I have, for some time, tried to find a rationale for his sudden success, but I have found none. I can only say that, as soon as he was charged with the project for the Observatory, he showed a strong drive, and started on the right footing, with a keen insight for the most appropriate project to be carried out at the time.

He left for a three year stay in Paris and London. From Paris, where he arrived at the beginning of April 1787, he wrote to his friend Prince of Torremuzza a long letter in which he neatly stated his programme:

> *[...] esaminerò da principio quelli Osservatorj tutti di Parigi, che non son pochi, osserverò io stesso, prenderò cognizione degli Antichi, ed infine assisterò all'assemblee dell'Accademia delle Scienze, ove siffatte materie vengono trattate. Dopo di ciò unitamente a M.r Lalande M.r Condorcet, e Cassini si tratterà di quanto sarà necessario per istabilire codest'Osservatorio che certamente potrà essere uno de' migliori.*[6]

After seven months in Paris, during which he gained the esteem and friendship of the most famous astronomers of the time, including Lalande, Messier, and Cassini, he moved to London where in Piazzi's words, he was

> *graziosamente accolto dal Dr. Maskelyne che mi usa alla giornata molte cortesie e mi procurerà la conoscenza de' migliori artefici e di Ramsden principalmente.*[7]

Piazzi had, in fact, already made up his mind: for the new Observatory he was going to have only the best, and best at that time meant Ramsden. Now, to obtain an instrument from Ramsden was not an easy task. His disregard of time and commitments were in fact legendary. Nevertheless, Piazzi obtained from him, in less than two years, a complete set of astronomical instruments, including an instrument which was considered at the time *le plus bel instrument d'Astronomie qu'on ait fait jusqu'ici.*[8]

The Palermo circle is relevant to the history of Astronomy because it is the first example of a new generation of astronomical instruments: namely instruments with circular scales. In a matter of a few decades around the turn of the century, circular scale instruments replaced the traditional quadrants and transits, allowing a substantial increase in the accuracy of astronomical measurements.

Back from London, in August 1789, Piazzi immediately started to search for a suitable place to build the Observatory. This he found on the norman tower of Santa Ninfa of the Royal Palace. Viceroy Prince of Caramanico was instrumental in obtaining from the King both the permission to build the Observatory over the Royal Palace and, most important, the necessary funds.

Illogical as it may seem, King Ferdinand at this point was not at all eager to have the Observatory built. He maintained that Piazzi could be content with the beautiful and expensive instruments he had purchased. What else did he want? Why an Observatory? As late as February 4th 1790 the King ordered that, if an Observatory had to be built, the funds had to come from Piazzi's salary![9] It took all of the Viceroy's consummate diplomacy to persuade the King to secure the funds for the building, and on July 1st 1790 finally the King's consent came. At this point Piazzi got immediately on the move, perhaps in fear that the King might change his mind and, within eight months, in February 1791, the Observatory was completed.

The plan of the Observatory reflects both the tightness of the budget obtained from the King, and Piazzi's eagerness to have the Observatory built anyway. It included a room for Ramsden's circle and a meridian room joined by a gallery where portable instruments were stored. Living quarters for the astronomer and his assistants were obtained at a lower level by adapting pre-existent rooms.

Fig. 3. The Palermo Circle, made by J. Ramsden

Encouraged by the possession of the 5 foot circle, whose accuracy was believed to be much superior to that of any other instrument in existence,[10] Piazzi centered his scientific programme on the accurate measurement of stellar positions. His observational technique implied that each star had to observed for at least four nights, before its position could be established. This painstaking work resulted in the publication, in 1803, of his first star catalogue. For this work, received by Lalande as "the most important astronomical work published since a long time," he was awarded the prize of the French Acadèmie des Sciences, and was elected to the Royal Society.

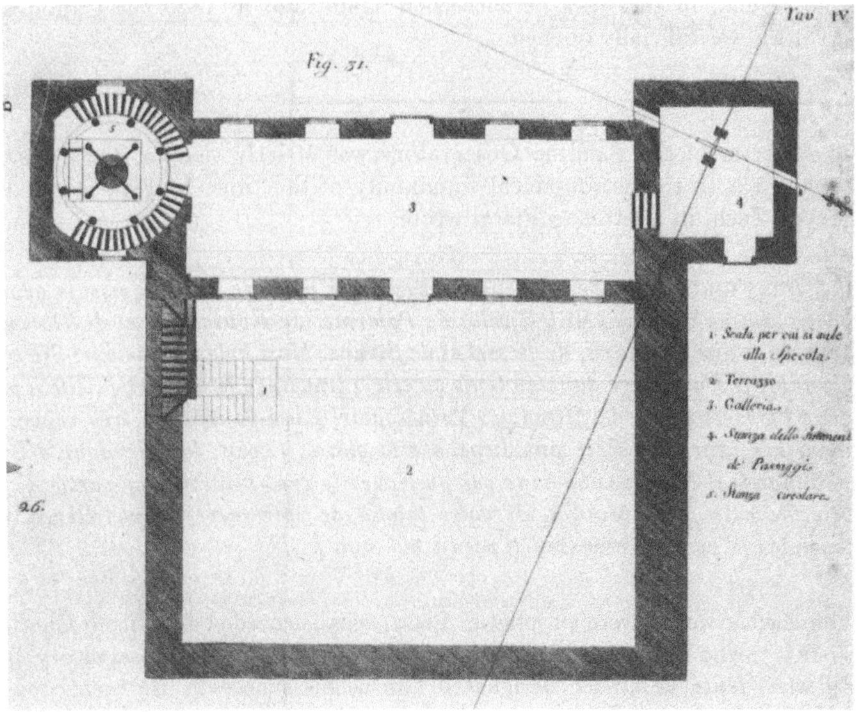

Fig. 4. Plan of the Observatory, from G. Piazzi, Della Specola Astronomica ..., Palermo, 1792

The discovery, January first 1801, of Ceres, the "planet" long sought for between the orbits of Mars and Jupiter, came as an unexpected reward of his observational technique. In line with his programme, Piazzi investigated at length proper motions and parallaxes. His commitment to the measurement of proper motions resulted, over the years, in the publication of several valuable catalogues of the proper motions of hundreds of stars.

It is perhaps little known that in 1806, well before the appearance of Bessel's famous paper "Uber den Doppel-Stern N.ro 61 Cygni", Piazzi had already correctly determined and published the flying star's large proper motion, pointing it out to the astronomical community as a good candidate for parallax measurement.[11]

His scientific career culminated in 1814 with the publication of his second star catalogue, which is not an augmented edition of his first one, as it has been often stated. Indeed, as Piazzi himself pointed out, it is a completely new catalogue. For this work he was again awarded the prize of the Acadèmie des Sciences.

Piazzi died in Naples in 1826. To the capital he had been summoned in 1817 by King Ferdinand with the task to bring to completiòn the Naples Observatory, a project that, in the preceding 20 years, had consumed enormous amounts of money with no results. In this task he succeeded again, and in 1820 the Capodimonte Observatory was officially opened.

The success of the Palermo Observatory was strictly tied to Piazzi's person. This was clear to the astronomical community of the time. In 1824 Baron Franz Xaver Von Zach, in a letter to Piazzi wrote:

> *[...] Dans toute l'Europe (Greenwich compté à part) je ne connaissais que trois observatoires vraiment utiles, celui de Palerme, de Königsberg et de Dorpat, ou pour miex dire de Piazzi, de Bessel et de Struve. Mais helas Monsieur! Sic transit gloria! [...] J'ai esperé dans un tems qu'elle a traversée la mer, et qu'elle a passée dans l'Observatoire de Miradois. Vain espoir! Avec vous, mon trés révèrend et mon trés venerable Père ont disparues la gloire, l'éclat, la splendeur d'Urania Ferdinandea! N'avez vous donc pas pu léguer la cent-millionieme partie de votre zéle, de votre persévérance, de votre talent, de votre esprit à vos élèves, à vos disciples, à vos successeurs? Il parait que non [...].*[12]

Von Zach's words were prophetic. Piazzi was succeeded by Nicolò Cacciatore (1780-1841), who had served for many years as assistant in the Observatory. It was Piazzi who, *faute de mieux*, designated him as his successor. He even consented that his latinized name be introduced in the 1814 star catalogue,[13] probably as a recognition of his contribution.

Cacciatore did not have the prestige necessary to guarantee the maintenance of the Palermo Observatory at the level reached by Piazzi. On the other hand, the French Revolution had put an end to the Bourbonic reforms, and the Kingdom of Sicily emerged from the Vienna's Congress as one of the provinces of the Kingdom of two Sicilies. All circumstances that had played in favor of the establishment of the Observatory were now reversed and acting against its consolidation. The scientific decline of the Palermo Observatory that followed Piazzi's death was thus due both to the incapability of its head and to the unfavorable political situation.

This lasted roughly until 1850, when Domenico Ragona (1820-1890) became Director of the Observatory. To the extent to which History of Science is the history of winners, Ragona does not have a place in it, yet he played an important role in the history of the Palermo Observatory. He was able to profit from the final convulsions of the moribund regime, and, with the support of the Lieutenant General for Sicily, Prince of Satriano, managed to completely renew the observatory's instrumentation, which had remained the same since Piazzi's time. He acquired in Berlin a meridian circle from Pistor & Martins, and a 25 cm equatorial telescope from Merz in Munich.

Ragona's direction was short lived, so short that he could not see the new instruments in place. His work was stopped in 1860 by the overthrowing of the Bourbons. Ragona, considered close to the old regime, was forced to leave his post, and the directorship was assumed by Gaetano Cacciatore (1814-1889), the son of Nicolò. Gaetano Cacciatore was more a politician of science than a scientist himself. Under his long directorship, almost thirty years until his death in 1889, the Observatory reached its maximum expansion in terms of personnel. It did in fact employ, besides the director, three astronomers, three assistants, a research assistant, and six technicians. This rendered the Palermo Observatory, by far, the largest in Italy. Accordingly, work at the Observatory was organized in three different sections: Positional Astronomy, Astrophysics or, as it was called at the time, Physical Astronomy, and Meteorology.

Meteorological observations had been, since Piazzi's time, part of the routine work of the Observatory, as in almost any other Observatory. Yet, since Italy up to 1860 had been divided into so many sovereign states, it had been impossible to organize a national network of "comparable" observations, as had been done since the beginning of the XIX century in many other European countries. With Italian Unity this became possible, and Cacciatore was among the most influential promoters of what was to become, in the late 70's, the *Ufficio Centrale di Meteorologia* in Rome. Meteorology thus acquired in Italy the status of a branch of Science in its own right, rather than being considered subsidiary to Astronomy, and in Palermo Cacciatore established in 1880 a Meteorological Observatory, which was a section of the Astronomical Observatory, but had its own separate buildings, personnel and, of course, instruments. Positional Astronomy and Astrophysics remained in the old Observatory building.

In 1863 Cacciatore managed to have Pietro Tacchini 1838-1905[14] assigned to the Observatory as Primo Astronomo Aggiunto.

Tacchini was a young engineer, in 1863 he was only 25, who had practiced Astronomy in Padua and in 1859 had been appointed director of the small Observatory of Modena, his native town. Judging from the fact that at the same time that Tacchini came to Palermo, Ragona took his post in Modena, one can suspect a sort of political *changez les dames* in his coming to Palermo. Whatever the reasons, young Tacchini quickly became the scientific leader of the Palermo Observatory. His first task was to put into operation the 25 cm equatorial telescope by Merz. This took almost 2 years, until early 1865. As soon as the instrument was set up, he concentrated on observations of the Sun. At first he could only observe sunspots. Of these, he produced innumerable detailed drawings. As is well known, in fact, it was only in 1868 that Janssen and Lockyer independently discovered that spectroscopic observations of solar prominences were possible outside eclipses.

In 1871, as soon as Tacchini obtained his first spectroscope from Leipzig, he started a daily monitoring of the solar limb. Tacchini firmly believed that the accumulation of a large mass of hard data was one of the keys to the understanding of the unsuspected features that just in those years were being carefully observed for the first time. Accordingly, he organized with Secchi in Rome and Lorenzoni in Padua a network of coordinated observations: each astronomer had to observe the solar limb every day at the same time, so that the results would be comparable. Eager to get observations during the times of the year in which it was impossible to observe in Italy, he went so far as to organize an Astrophysical Observatory in Calcutta, where he had gone to observe the 1874 transit of Venus over the Sun; this was Father Lafont Observatory, for which Tacchini himself made the drawings.

MEMORIE

SOCIETÀ DEGLI SPETTROSCOPISTI ITALIANI

PER CURA DEL PROF. P. TACCHINI

PALERMO

1872.

Fig. 5. Cover page of the first volume of the *Memorie della Società degli Spettroscopisti Italiani*

He also conceived with Secchi the idea of a specialized Journal where most, if not all of the work that was being done on the Sun could be brought together and, in 1872, he founded in Palermo the *Memorie della Società degli Spettroscopisti*, the first specialized Journal in Physical Astronomy. He edited this journal until 1905.

Tacchini has been frequently referred to as a pupil of Secchi, the leading Italian solar physicist of the time. Not only is this not true but, on the contrary, they had frequent scientific quarrels. For instance, since they were using identical instruments, Secchi accused Tacchini of making visionary drawings of solar prominences. The fact is that Tacchini was able to use his instrument much better than Secchi, not to mention his superior ability in drawing.

He left Palermo in 1879, when he was called to Rome to succeed Secchi at the head of the Osservatorio del Collegio Romano, and soon after, in 1881, at the head of the *Ufficio Centrale di Meteorologia*. Before leaving Sicily he succeeded in persuading the Government to establish an Astronomical Observatory on Mount Etna, convinced that with a high altitude telescope, spectroscopic observations of the Sun would improve and in the hope that perhaps some of the solar corona could be observed outside of eclipses.

The Bellini Observatory on Mount Etna was completed in 1880, while in 1885, again upon his recommendation, the Catania Observatory was established. This was the first Astrophysical Observatory in Italy. After Tacchini's departure, solar

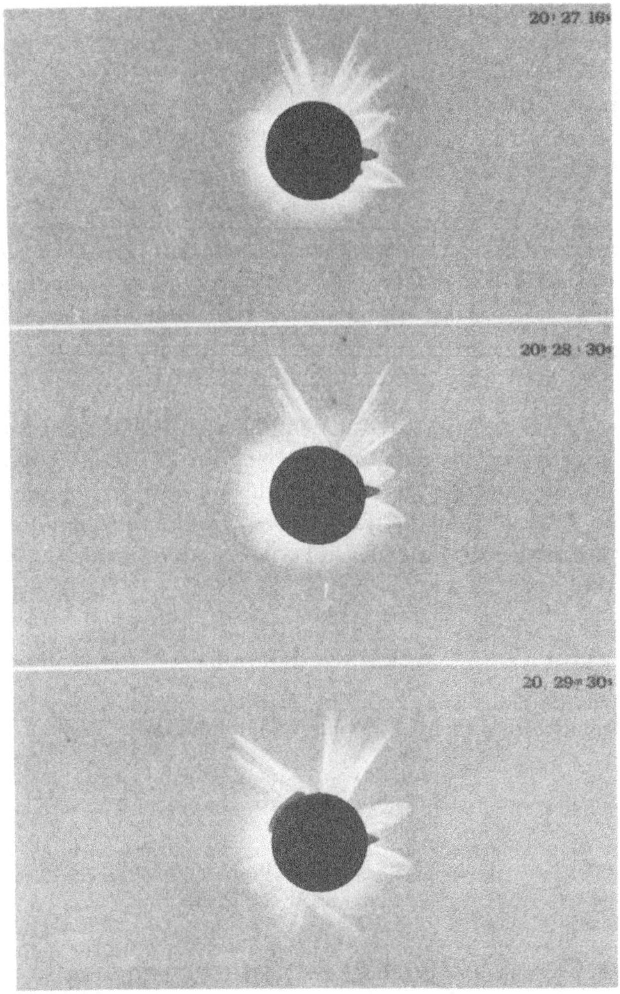

Fig. 6. Drawings of the 1887 eclipse published by P. Tacchini in his report of the Italian expedition to Russia.

observations in Palermo where continued by Annibale Riccò, who later became the first director of the Catania Observatory.

From the beginning of the XX century, the relevance of the Palermo Observatory started declining once again. The institution had to wait some 70 years. But there it was, and it waited. Then, the social, economic, technological evolution of the western world called for a new season of the institution.

At that time Pippo Vaiana came to Palermo. He was the right man in the right place at the right time. He brought here his unlimited energy, his record of achievements, his direct involvement in the birth of imaging X-ray Astrophysics. Being back home, returning to his very roots, had prompted him to fresh and powerful inspiration. The new season started.

How shall we view the micro-history of the Observatory? Let me quote from Hesse's *Magister Ludi*:

> *[...] Of course, one should bring order into history. [...]. Every science is, among other things, a method of ordering, simplifying, making the indigestible digestible for the mind [...]. To study history means recognizing and accepting chaos, and nevertheless retaining faith in order and meaning [...].*

The fluctuation (that is chaos) that caused the unlikely clustering of circumstances which gave Piazzi his beautiful and yet singular opportunities was bound to thermalize and die out. On the other hand, I venture the opinion (not only out of my deep hope, I trust) that the wave started by Pippo Vaiana at the cost of his immense energy, has already gone past its point of no return (that is retaining faith in order and meaning).

Acknowledgements

This research was partially supported by *Ministero dell'Università e della Ricerca Scientifica e Tecnologica* and by *Consiglio Nazionale delle Ricerche.*

Footnotes and References

1 Before the establishment of the Palermo Observatory I have been able to trace in Sicily only two other observatories, in the sense of places specifically devoted to astronomical observations. The first was in the rooms that Marquess Giovanni II of Ventimiglia had adapted to that purpose in his family's castle in Pollina, for his friend and mathematician Francesco Maurolico (1497-1575). The second was the rooms used by Giovan Battista Hodierna (1597-1660) in the castle of his patron Carlo Tomasi et Caro, in Palma di Montechiaro. Both these "observatories" were short lived and disappeared at their users' deaths.

2 As a matter of fact, in 1786 King Ferdinand had consented that an astronomical observatory be built in Naples and, for this purpose, had allotted substantial amounts of money. The building of the Naples Observatory took more than thirty years, and was finally brought to completion in 1819. It was Piazzi who succeeded in the task of getting it completed in two years, between 1817 and 1819. On the history of the Naples Observatory and on Piazzi's role, see the chapter by G. Foderà Serio in *L'Osservatorio Astronomico di Capodimonte*, M. Rigutti ed., F. Fiorentino, Napoli 1992.

3 It is certainly a rough oversimplification to attribute to the Inquisition alone the fact that Sicily through almost the entire eighteen century suffered a severe scientific deprivation, especially for what concerns Astronomy, but also Physics and Mathematics. On this subject see for instance C. Dollo, *Filosofia e Scienze in Sicilia*, A. Milani, Padova 1979.

4 It was only after Italian Unity (1860) that a "modern" geographical survey of Sicily was accomplished. The need for such a survey was deeply felt at the turn of the eighteen century, by the reformist party. Reliable cadastral data were in fact needed,

if only for taxation purposes. On Piazzi's efforts and failure to produce such a map see: G. Foderà Serio and L. Indorato, *The Matthew Berge's instruments at the Palermo Astronomical Observatory*, Annali dell'Istituto e Museo di Storia della Scienza di Firenze, VI, 1981, pp. 217-24; G. Foderà Serio and P. Nastasi, *Giuseppe Piazzi's survey of Sicily: the chronicle of a dream*, Vistas in Astronomy, vol. 28, 1985, pp. 269-276.

5 D'Alembert clearly mistakes the *Accademia de' Regj Studi* for a scientific society. This however does not diminish the weight of his words. J. D'Alembert to P. Frisi, Feb. 24 1782 - For this reference I am grateful to John Pappas.

6 G. Piazzi to Prince of Torremuzza, Paris, April 21 1787, in *Carteggio Letterario del sig. Principe di Torremuzza Gabriele Lancellotto Castelli*, mss. Biblioteca Comunale di Palermo, Qq.E.136, ff. 222-224. The Prince was one of the most active deputies of the Accademia.

7 G. Piazzi to Prince of Torremuzza, London 8 October 1787, cit., ff. 224-225.

8 Journal des Scavants, August 1789.

9 See, for instance, *Documenti relativi alla nomina di G. Piazzi alla cattedra di Astronomia ed all'erezione della Specola di Palermo*, G. Foderá Serio ed., Elementi Astronomici dell'Osservatorio Astronomico di Palermo, Stass, Palermo 1990.

10 It must be pointed out that the Palermo circle did not perform any better than, say, Bird's quadrant at Greenwich. Piazzi, after working with it, estimated between 1 and 3 arcseconds the possible systematic error. However this does not lessen the relevance of the instrument and it must be credited to Piazzi to have clearly foreseen that the circular instruments were the instruments of the future: "After what I have been arguing none can possibly doubt the superior advantages of an entire circle over the most perfect quadrant of equable radius. The hinderance lays essentially in the tooling, but if astronomers will set their minds to use such an instrument, and will gode the tool artists, in a short time the construction of such instruments shall became facile and plain... A circle of six foot radius in the hands of an effective observer will yield the position of the main stars with such an accuracy that only an uncertainty of a fraction of an arcsecond would be left. This epoch is perhaps not far in the future and when it will arrive it will beautifully lighten the 18th century for the ages to come" [G. Piazzi, *Della Specola Astronomica de' Regj Studj di Palermo*, Palermo 1792, p. 46].

11 G. Foderà Serio, *Giuseppe Piazzi and the discovery of the proper motion of 61 Cygni*, JHA, xxi (1990) pp. 275-182.

12 Von Zach to Piazzi. Genova 28 September 1824, Archivio Museo Copernicano, Osservatorio Astronomico, Monte Porzio Catone, Roma.

13 The names Sualocin and Rotanev for the stars α and β Delphini appear on page 147. They spell Nicolaus Venator when read backward. Obviously this is a sort of designation by G. Piazzi of Cacciatore as his *Douphin*, i.e. his successor.

14 Tacchini is one of the leading Italian scientists of the second half of the XIX century. Yet he is very little known. The only work on his life and scientific activity is: Ileana Chinnici - Pietro Tacchini ingegnere, astrofisico, meteorologo. Una prima ricostruzione biografica. 1992, Tesi di Laurea in Fisica, Università di Palermo.

BODE'S LAW AND THE DISCOVERY OF CERES

MICHAEL HOSKIN
Churchill College, Cambridge

1. Copernicus

When Copernicus's *De revolutionibus* appeared in 1543, it was valued by the professionals for its innovative planetary models rather than for anything it might have to say about which body is at the centre of the universe. In a volume dominated by complex geometry, and introduced by a misleading preface inserted without the author's authority to the effect that what followed was guided by the search for accuracy and convenience rather than the quest for truth, the cosmological Book I was largely overlooked. In Book I Copernicus shows in broad, qualitative terms, how so many of the hitherto-puzzling features of the observed motions of these 'wandering stars' – such as their retrogressions – are readily explained if one begins from the assumption that the Earth is an ordinary planet orbiting the Sun.

Another consequence of the heliocentric hypothesis outlined in Book I, and one especially satisfying to its author, was that the planets at last formed a single, integrated system. In the accepted Ptolemaic astronomy, even the very order of the planets was uncertain. It was supposed that the planets whose movements differed least from the daily spinning of the fixed stars – Saturn, and then Jupiter and Mars – were physically closest to the stars and so furthest from the central Earth. But since Mercury and Venus appear to accompany the Sun around the sky, all three seemed to have the same period of one year, and so their order of distances was a matter of guesswork. But the heliocentric hypothesis revealed that the supposed equal periods were no more than an illusion resulting from the status of Mercury and Venus as inner planets. It also became clear that the circular movements with a period of one year that occurred in the various traditional models of the planetary orbits were no more than the reflection of the terrestrial motion; this being so, the radius of each of these circles should now be equated to the astronomical unit, giving a common scale to the models and permitting them to be seen as components in an integrated planetary system. From this it transpired that the further a planet from the central Sun, the longer it took to complete a circuit of sky – an harmonious relation that strongly appealed to Copernicus's Platonic intuition.

The planetary system is represented by Copernicus in simplified form in his famous diagram in Book I. But his figure is not to scale. A scale representation would have made it obvious that there is an astonishing gap between the orbit of the fourth planet, Mars, and that of the fifth, Jupiter.

2. Kepler

Towards the end of the century, the young Johannes Kepler, in one of the first publications that were irrevocably heliocentric, his ***Mysterium cosmographicum*** (1596),

J.F. Linsky and S. Serio (eds.), Physics of Solar and Stellar Coronae, 35–46.

sought to make sense of the dimensions of the planetary system. Why, he asked himself, had God been motivated mathematically to select the planetary orbits in just the way he had. "There were", he tells the reader in the Preface, "three things in particular about which I persistently sought the reasons why they were such and not otherwise: the number, the size, and the motions of the circles." The gap between Jupiter and Mars was especially awkward to explain. After various attempts, he tried a novel and bold approach.

> Between Jupiter and Mars I placed a new planet, and also another between Venus and Mercury, which were to be invisible on account of their tiny size, and I assigned periodic times to them. For I thought that in this way I should produce some agreement between the ratios, as the ratios between the pairs would be respectively reduced in the direction of the Sun and increased in the direction of the fixed stars.... Yet the interposition of a single planet was not sufficient for the huge gap between Jupiter and Mars; for the ratio of Jupiter to the new planet remained greater than is the ratio of Saturn to Jupiter. [1]

3. Dynamical Explanations of the Mars/Jupiter Gap

Eventually Kepler found suitable motivation for the divine geometer in a totally different approach, the nesting of spheres and regular solids; but it was one that commended itself to few in the generations that followed. Some found an acceptable explanation of the gap in the sheer size of the outer planets. Isaac Newton for example regarded the gap as part of the divine plan for the stable and clockwork universe: the massive planets, Jupiter and Saturn, had been located by Providence at the outside of the planetary system, well clear of the smaller planets whose orbits their gravitational force would otherwise disrupt. [2] In the middle of the eighteenth century Immanuel Kant also sees a dynamical justification for the gap in the great mass of Jupiter: "The width between the orbit of Jupiter and Mars is so great that the space enclosed there exceeds the regions of all lower planetary orbits taken together ... that space is worthy of the greatest among all planets, namely, of that which has more mass than all the others together." [3] Johann Heinrich Lambert in 1761 likewise remarks on the gap. Lambert in general is as committed to an eternal, unchanging clockwork universe as was Newton, but at the level of the solar system he is prepared to accept that change has been brought about by the attractive power of Jupiter: "And who knows whether already planets are missing which have departed from the vast space between Mars and Jupiter? Does it then hold of celestial bodies as well as of the Earth, that the stronger chafe the weaker, and are Jupiter and Saturn destined to plunder forever?" [4]

4. The Possibility of Undiscovered Planets

These speculative dynamical explanations of the 'gap' took place in the context of a surprising willingness on the part of professionals and informed amateurs alike to accept that there may exist planets as yet undiscovered, perhaps inside Mercury, but more plausibly beyond Saturn: surprising, because no primary planet had been discovered since history began. Significantly, just as the most interesting late seventeenth and early eighteenth century speculations on cosmology came from writers whose interests had a theological dimension, so the same is true of speculations about additional planets. So William Wall, cited in the Postscript to the second edition (1727) to Tobias Swinden's *An Enquiry into the Nature and Place of Hell*, [5] wrote:

> I think it very probable, that there are, belonging to the Sun, a great many more planets, than what we see, some perhaps *within* the Orb of *Mercury*, never seen nor to be seen by us; but a great number *without*, or beyond the Orb of *Saturn*, which we can never see ... partly by reason of the distance from us, and partly because they, being very remote from the Sun, do receive but a weak Light from him, and do much more weakly reflect it.

William Whiston hints at the same in his *Astronomical Principles of Religion, Natural and Reveal'd* (London, 1717), when he says carefully that "Mercury is the nearest to the Sun of all the known Planets", and that Saturn is "the highest and most remote of all the known Planets". [6] As so often, Whiston's views in this book are reflected in the writings of that well-known maverick in both astronomy and theology, Thomas Wright of Durham. In his *Clavis Coelestis* (1742) he speaks of Mercury as "the first Planet *we know of* in the System" [7][italics supplied], and Venus as "the second Planet known in the System", while "Saturn is the last and highest known Planet in the System". [8] In his more famous *An Original Theory or New Hypothesis of the Universe* (1750) he again refers [9] to "the known Planets". And in his often bizarre *Second Thoughts*, which remained in manuscript until our own time, he is explicit: "...I am far from supposing our present knowledge of ye solar system perfect and fully known". [10]

> Is it not more reasonable to imagine a coelum in cognito beyond ye known planets than to suppose a terra in cognito at present upon Earth. It is therefor my opinion, that there are or may be, many more bodies belonging to the system of ye Sun, whose more feeble light has not been able yet to reach us at ye Earth, besides others perhaps within ye orbit of Mercury, though lost or lying hid to us in a too radiant state of light. [11]

Another of the mid-eighteenth century speculators to anticipate an undiscovered planet was Immanuel Kant. In his *Universal Natural History* ...he writes:

> ...we see even in our solar world the members of a system which stand immensely apart from one another and between which one has not yet discovered the intervening parts. Should there be between Saturn, the outermost of planets which we know, and the least excentric comet which descends to us from perhaps a distance 10 and more times greater, no more planet whose motion would come closer to the cometary motion than that [of Saturn]...? [12]

These suggestions concerning planets as yet undiscovered relate mostly to the regions outside Saturn. Only a few were concerned with the gap between Mars and Jupiter. One who did 'surmise' the presence of one or more planets in the gap was apparently the Scottish mathematician Colin Maclaurin. [13] Another to focus on the gap was Thomas Wright. In one of those unexpectedly insightful speculations that make him so fascinating a figure, Wright actually suggests in his unpublished manuscript that the gap between Mars and Jupiter results from a planet having broken up following collision with a comet:

> That comets are capable of distroying such worlds as may chance to fall in their way, is, from their vast magnitude, velocity, firey substance, not at all to be doubted, and it is more than probable from the great and unoccupied distance betwixt ye planet Mars and Jupiter some world may have met with such a final dissolution. [14]

Yet the gap was readily apparent to anyone who glanced at the data for the planetary orbits. Near the beginning of the eighteenth century, for example, William Whiston, Newton's successor in Cambridge, gives the actual distances of the planets in millions of miles as 32, 59, 81, 123, 424, 777. [15] We note that there are four planets within 123 million miles of the Sun, but the gap before the next planet, Jupiter, is over 300 million miles.

5. The First Statement of the Law

Whiston's contemporary, David Gregory, in his widely-read *The Elements of Astronomy* [16] puts the planetary distances into proportional numbers: "...supposing the distance of the Earth from the Sun to be divided into ten equal Parts, of these the distance of Mercury will be about four, of Venus seven, of Mars fifteen, of Jupiter fifty two, and that of Saturn ninety five." Gregory's work was published in Latin in 1702 and again in 1726, and an English translation appeared in 1715 with a second edition in 1726. The words quoted appear at the very beginning of the work, in Proposition 1 of Section 1 of Book I, and are therefore in a very prominent position. But they have been overlooked by historians, who have found exactly the same numbers – indeed, a paraphrase of the same sentence – in a work published in 1724 by Christian Wolff: *Vernünfftige Gedanken von den Absichten der natürlichen Dinge*, which was to go through several editions. [17] In 1764, the French natural philosopher Charles Bonnet published his *Contemplation de la Nature*, a successful work that was quickly translated into other European languages. The German translation was undertaken by Johann Daniel Titius of Wittenberg. It had long been common for translators to supplement the text they were translating, usually to bring it up to date, for in those days when book publishing was even slower than it is today, many years often elapsed between first publication and translation. Translators, that is, took a greater initiative than is now thought proper; indeed, it was not unknown for a translator to conduct a running battle with his author through the medium of footnotes. Titius, probably because he was by nature self-effacing, not only left his additions unsigned but actually incorporated them in the text itself, with no hint that they were not the original work of the author. He chose to make such an addition to the paragraph where Bonnet remarks that "We know seventeen planets that enter into the composition of our solar system [that is, major planets and their satellites]; but we are not sure that there are no more", going on to anticipate more discoveries as telescopes improve. Titius then inserts what we now know as Bode's Law:

> Take notice of the distances of the planets from one another, and recognize that almost all are separated from one another in a proportion which matches

> their bodily magnitudes. Divide the distance from the Sun to Saturn into 100 parts; then Mercury is separated by four such parts from the Sun, Venus by 4+3=7 such parts, the Earth by 4+6=10, Mars by 4+12=16. But notice that from Mars to Jupiter there comes a deviation from this so exact progression. From Mars there follows a space of 4+24=28 such parts, but so far no planet was sighted there. But should the Lord Architect have left that space empty? Not at all. Let us therefore assume that this space without doubt belongs to the still undiscovered satellites of Mars, let us also add that perhaps Jupiter still has around itself some smaller ones which have not been sighted yet by any telescope. Next to this for us still unexplored space there rises Jupiter's sphere of influence at 4+48=52 parts; and that of Saturn at 4+96=100 parts. What a wonderful relation! [18]

It is interesting to note that these numbers are not exactly the ones listed by Gregory and Wolff; nor do they follow from the actual distances published by Whiston, which would give 96 for Saturn in place of the 95 of Gregory and the 100 of Titius. But it seems that Wolff was indeed the immediate source for Titius, for in the fourth edition of his translation, by which time he was clearly identifying his own contributions as such, he adds the comment: "This relationship and the related considerations which Herr Bonnet thought had first been observed by Herr Lambert had already been recited by Freyherr von Wolf in his German Physics more than forty years earlier." [19] How Titius could declare that Bonnet had drawn his ideas of unknown planets from Lambert is not clear, though perhaps Titius and Bonnet may have corresponded over the translation; but this reference of Titius to Wolff suggests that Wolff had indeed been Titius's original source.

As it happened, Titius published a second edition of his translation – with the law now properly located in a footnote – just as the promising young astronomer Johann Elert Bode was putting the finishing touches to the second edition of his introduction to astronomy, *Anleitung zur Kenntniss des gestirnten Himmels*, which he had published in 1768 when he was only nineteen. Bode came across the relationship proposed by Titius, was convinced by it, and inserted it as a footnote in his text:

> This latter point seems in particular to follow from the astonishing relation which the known six planets observe in their distances from the Sun. Let the distance from the Sun to Saturn be taken as 100, then Mercury is separated by 4 such parts from the Sun. Venus is 4+3=7. The Earth 4+6=10. Mars 4+12=16. Now comes a gap in this so orderly progression. After Mars there follows a space of 4+24=28 parts, in which no planet has yet been seen. Can one believe that the Founder of the universe had left this space empty? Certainly not. From here we come to the distance of Jupiter by 4+48=52 parts, and finally to that of Saturn by 4+96=100 parts. [20]

It is clear from the wording that Bode is following Titius, although he of course realized that the suggestion that the missing planet was a moon of Mars was preposterous, a fact he emphasized in the third edition of his book. But he makes no acknowledgement to Titius; indeed, it is only in later editions that Bode identifies his source (possibly because Titius had pressed him to do so). In the hands of Bode the relationship assumed a new importance, for Bode was a professional astronomer soon to take on international stature, and he was well-placed to act as apostle of the new law.

6. The Discovery of Uranus

Given the willingness on the part of many astronomers to believe that there were planets as yet undiscovered, and especially so in orbit beyond Saturn, it is a little surprising that it never crossed the mind of William Herschel in March 1781 that the "curious either nebulous star or perhaps a comet" he had noticed in his telescope might indeed be a major planet. [21] It is often said that this failure of imagination was because of the total novelty of his discovery – that no primary planet had ever been discovered in historic times, which of course is true. But in view of the numerous references we have seen to the "known" planets, including that of Bode just cited, it seems more likely that Herschel – an isolated and self-taught amateur – was simply unaware of the professional astronomers' openness to new discoveries among the planets. As early as 4 April the Astronomer Royal, Nevil Maskelyne, wrote to their mutual friend William Watson of Herschel's "comet or new planet", and on the 23rd he wrote to Herschel:

> I am to acknowledge my obligations to you for the communication of your discovery of the present Comet, or planet, I don't know which to call it. It is as likely to be a regular planet moving in an orbit nearly circular round the sun as a Comet moving in a very excentric ellipsis.

How to calculate the orbit, however, was a difficult problem, for the body had been observed for only a very tiny fraction of a complete orbit. If it was a comet, then it would be simplest to assume a parabolic orbit. P.-F.-A. Méchain, a French mathematician who had discovered several comets, being misled by the earliest observations which made it likely that the object was indeed a comet, sent Herschel a letter in which he gave the perihelion distance of the supposed comet as 0.46 AU and perihelion date as 23 May 1781; Anders Johan Lexell, a Finnish-born professor of mathematics at St Petersburg who was visiting England at the time, soon after proposed a perihelion distance of 16 AU with perihelion not to be reached until 1789.

On the other hand, if it was a planet, then it was simplest to assume a circular orbit, and Lexell was one of a number of astronomers who, finding that parabolic orbits were incompatible with the observations, investigated circular orbits. Lexell derived for the radius of the orbit the excellent value of 18.93 AU – that is, with the radius of Saturn's orbit put at 100, a distance that compared well with the prediction of 196 from the Titius-Bode relation. More sophisticated calculations followed, some of them taking into account observations made years earlier when the planet had been mistaken for a star; and soon it was clear that the object was indeed a planet and, moreover, one that fitted well the Titius-Bode relation.

7. The Search for the Planet between Mars and Jupiter

This remarkable confirmation of the relation naturally reinforced Bode's belief, and it likewise persuaded Baron Francis Xaver von Zach, the court astronomer at Gotha. Both men were convinced there was an undiscovered planet between Mars and Jupiter, and in 1787 Zach began to search for it. Not unreasonably, he limited his investigation to the Zodiac, and believing that only a methodical search offered hope of success, he produced for himself a catalogue of zodiacal stars arranged by right ascension; but without success. The autumn of 1799 found him visiting astronomers in Celle, Bremen and Lilienthal, and there the idea of a cooperative attack on the problem emerged:

> It was the opinion of these men of discernment, that to get onto the trail of this so-long-hidden planet, it cannot be a matter for one or two astronomers to scrutinise the entire Zodiac down to the telescopic stars. [22]

It was on 21 September the following year that the cooperative attack – probably without precedent in the history of science – became a reality. On that day six astronomers met in Lilienthal: von Zach himself; J.H. Schröter, the chief magistrate of Lilienthal, whose world-famous collection of instruments included a Herschel reflector of 27ft focal length; H.W.M. Olbers, physician from nearby Bremen and longtime collaborator with Schröter; C.L. Harding, who was employed by Schröter and who was himself to discover the third asteroid in 1804; F.A. Freiherr von Ende; and Johann Gildemeister. They decided that even six observers were too few for the task ahead, and nominated instead a group of twenty-four practising astronomers chosen from throughout Europe. Schröter was to be president and Zach secretary. The entire Zodiac was divided up into twenty-four zones each of 15 degrees in longitude, and extending some 7 or 8 degrees north and south of the ecliptic in latitude. The zones were allocated to the members by lot. Each member was to draw up a star chart for his zone, extending to the smallest telescopic stars,

> and through repeated examination of the sky was to confirm the unchanging state of his district, or the presence of each wandering foreign guest. Through such a strictly organized policing of the heavens, divided into twenty-four sections, we hoped eventually to track down this planet, which had so long escaped our scrutiny. [23]

8. Piazzi and the Discovery of Ceres

Zach accordingly sent out the invitations to join this society of celestial cops. One of those chosen was, naturally, Giuseppe Piazzi of Palermo, the southernmost of the European observatories. Piazzi had been born in 1746 in Valtellina, in what was then part of Switzerland but is now northern Italy. [24] As a young man, Piazzi joined the Theatine Order, and afterwards taught mathematics in a number of Italian cities. In 1780 he was invited to take the chair of higher mathematics at the Academy of Palermo. Arriving at Palermo, Piazzi, although inexperienced in astronomy, expressed a wish to found an astronomical observatory: Palermo was further south than any existing European observatory and so offered access to regions of the sky inaccessible elsewhere. His royal patron was in favour, and prepared to forego Piazzi's services while he equipped himself and his observatory for the task ahead. Piazzi accordingly set off for England where he might obtain good advice, from such disparate figures as William Herschel and the Astronomer Royal, Nevil Maskelyne, and – equally important – good instruments. Jesse Ramsden was an instrument-maker without peer, but he was notorious for failing to produce on time. Piazzi persuaded him to attempt a 5ft vertical circle of unique design. The circle, which has been described as "a masterpiece of eighteenth-century technology", was twice abandoned by Ramsden, and he completed it eventually in August 1789 only because Piazzi himself was present in London – indeed, in Ramsden's workshop, and quite literally breathing down Ramdsen's neck as the work proceeded. The circle gave readings in azimuth by micrometer microscope, and readings in altitude by two diametrically opposed microscopes. The divisions on the circles were illuminated by an inclined silver mirror fixed to each microscope, and the wires in the telescope eyepiece by transmitting light from a small lamp through the hollow tube-axis. [25]

Once the Ramsden circle was installed in Palermo, Piazzi found himself in a privileged situation. He had an instrument of unique quality, a good climate, and the southernmost latitude of any European observatory. He very rightly set to work to exploit these advantages in the compilation of a star catalogue better than any that had gone before. A feature of his painstaking work was the repeated measurement of stellar positions on different nights, so that the final coordinates were accurate to a few seconds of arc. The first of Piazzi's two great catalogues, with the coordinates of some 6,748 stars, was to appear in 1803. [26] The accuracy of his work gave astronomers once more the confidence to tackle the question of stellar parallax, which had been largely in abeyance since it became clear around 1730 that parallax could not be much more than a second of arc.

The beginning of 1801 found Piazzi patiently at work on the star catalogue. As he wrote a few months later,

> ...on the evening of the 1st of January of the current year, together with several other stars, I sought for the 87th of the Catalogue of the Zodiacal stars of Mr la Caille. I then found it was preceded by another, which, according to my custom, I observed likewise, as it did not impede the principal observation. The light was a little faint, and of the colour of Jupiter, but similar to many others which generally are reckoned of the eighth magnitude. Therefore I had no doubt of its being any other than a fixed star. In the evening of the 2d I repeated my observations, and having found that it did not correspond either in time or in distance from the zenith with the former observation, I began to entertain some doubts of its accuracy. I conceived afterwards a great suspicion that it might be a new star. The evening of the third, my suspicion was converted into certainty, being assured it was not a fixed star. Nevertheless before I made it known, I waited 'till the evening of the 4th, when I had the satisfaction to see it had moved at the same rate as on the preceding days. From the fourth to the tenth the sky was cloudy. In the evening of the 10th it appeared to me in the

> Telescope, accompanied by four others, nearly of the same magnitude. In the uncertainty which was the new one, I observed them all, as exactly as possible, and having compared these observations with the others which I made in the evening of the 11th, by its motion I easily distinguished my star from the others. Mean while however I greatly wished to see it out of the meridian, to examine and to contemplate it more at leisure. But with all my labour, and that of my assistant D. Niccola Cacciatore and [of] D. Niccola Carioti belonging to this Royal Chapel both enjoying a sharp sight, and very expert in the knowledge of the heavens, neither with the night Telescope, nor with another achromatic one of 4 inches aperture, was it possible to distinguish it from many others among which it was moving. I was therefore obliged to content myself with seeing it on the meridian, and for the short time of two minutes, that is to say the time it employed in traversing the field of the Telescope; other observations, which were making at the same time, not permitting the instrument to be moved from its position.
>
> In the mean time, in order to render the observations more certain, while I was observing with the Circle, D. Niccola Carioti observed with the transit instrument. The sky was so hazy, and often cloudy, that the observations were interrupted 'till the 11th of February; when the star having approached so near the Sun, it was not possible to see it any longer at its passage over the meridian. I intended to search for it, out of it [the meridian], by means of the Azimuth; but having fallen ill on the thirteenth of February, I was not able to make any further observations. These, however, which have been made, though they are not at the necessary distance from one another in order to assure us of the true course which the star describes in the heavens, are, notwithstanding, sufficient in my opinion, to make us know the nature of the same, as one may collect from the results, which I have deduced from them. [27]

Piazzi had in fact measured the position of the object on a total of 24 nights between 1 January and 11 February, though some positions were marked as 'doubtful' or even 'very uncertain'. On 24 January, Piazzi had announced his discovery in letters to fellow astronomers, among them his fellow-countryman, Barnaba Oriani of Milan. In it, Piazzi confided to him that

> I have announced this star as a comet, but since it is not accompanied by any nebulosity and, further, since its movement is so slow and rather uniform, it has occurred to me several times that it might be something better than a comet. But I have been careful not to advance this supposition to the public. [28]

To the others, he claimed nothing more than the discovery of a comet, though making it clear that the 'comet' had no nebulosity or tail.

When after 11 February he could no longer see the object, Piazzi set to work to investigate its orbit, though such mathematical investigations were not his strength. He began with the assumption that it was indeed a comet, and fitted a parabola to three of the observations to see if the orbit would account for the others. It did not. A second attempt with a different group of three observations likewise failed:

> From the parabolic hypothesis I passed then to the circular; and having made a few suppositions, I found two radii, 2.7067 and 2.6862; with each of which all the observations were represented a great deal better than any parabola. The planets describing ellipses more or less eccentric, and not circles, it is to be believed that ours will not deviate from this rule. In an ellipsis I should then have continued my calculations; but as the arch observed is very small, the results would be very uncertain, and the labour long and painful. I have therefore preferred the circle....
>
> The agreement of the observed longitudes with the calculated ones in the circular hypothesis, its motion in the Zodiac, from which it only departs a little way in the greatest latitudes, and its position between Mars and Jupiter, leave no doubt that this new star is a true planet.... [29]

But how to recover the now-lost planet at some future date? In Piazzi's opinion, the best hope lay in identifying some past occasion when the planet had been observed in the belief that it was a star. He thinks it may well have been the object observed by Bode in 1772, and that it had probably been listed at some time or other by la Caille or by Tobias Mayer:

> In the catalogues of the Zodiacal stars of these two Astronomers, there are some observed only once, which I could never find, though I have sought them several times, and on different occasions. If the original observations of Mayer are preserved at Göttingen, and those of la Caille at Paris, it is possible that some light may be thrown by them on this matter. At the end of my work on the position of the fixed stars, ... I shall give a catalogue of lost stars, which will much facilitate this research. [30]

We can well believe Piazzi when he says that not only Oriani, but more especially Bode of the celestial cops, "were instantly of the opinion that it was a new planet; and settled nearly the same elements of its orbit, as I have done". One can imagine the German's delight that the hoped-for planet had been found, even if the discovery owed nothing to the celestial cops themselves.

But now Piazzi himself was beginning to have doubts. He had estimated the size of the object from the fact that it was almost, but not quite, covered by one of the wires of his telescope, and his conclusion was that it was larger than the Earth. However, it would seem that in the hazy nights that followed, the true (and much smaller) size of the object became more evident to the Palermo astronomer, who began to think that the object was diminishing in size and therefore moving rapidly away, so that it must be a comet after all:

> As after the 23rd [of January] the star began sensibly to diminish in size and brightness, uncertain whether it was to be attributed to its rapid receding from the Earth, or rather to the state of the atmosphere, which became after that still more dark and hazy, I began to doubt of its nature, so as even to believe it was a comet and not a planet. [31]

Eventually, in April, when illness had prevented him from making further progress in the investigation of the object's orbit, Piazzi sent his complete observations to Oriani, Bode, and Lalande in Paris, together with his suspicions that it might be a comet after all. And with that, Piazzi's own role in the story comes to an end, save for his naming the body – should it ever be recovered – Ceres Ferdinandea, Ceres for the patron goddess of Sicily, and Ferdinandea for Piazzi's royal patron.

It is one of the problems of writing history, that no story ever has a tidy ending. One would wish to go on to discuss the mathematical analysis of Piazzi's observations by Gauss that enabled Zach to recover it at the end of the year; the discovery of Pallas by Olbers in March 1802; the announcement by William Herschel [31] in May that these bodies were tiny compared to the planets – he estimated Ceres had a diameter of only 162 English miles (though this is perhaps a quarter of what we consider the true value), and proposed that these bodies should be termed asteroids rather than planets, much to Piazzi's annoyance; the discovery of Juno by Harding in 1804, and of Vesta by Olbers in 1807; and indeed the role of Bode's Law (or better, as we have seen, the Titius-Bode Law) in the discovery of Neptune in 1846. But in this celebration of the bicentenary of Palermo Observatory, we are perhaps justified in ending this story with the most famous discovery ever made here – but a discovery made possible by the fine instrument Piazzi had managed to acquire, and by Piazzi's dedication in using it towards the compilation of his two great star catalogues – catalogues that raised European standards of precision astronomy in the opening years of the new century.

References

1 Johannes Kepler, *Mysterium cosmographicum* (Tübingen, 1596), pp. 7-8; transl. by A.M. Duncan, *The Secret of the Universe* (New York, 1981), pp. 63-64.

2 Isaac Newton, letter to Richard Bentley, 10 Dec. 1692 (*Four Letters from Isaac Newton to Doctor Bentley* (London, 1756), p. 9.

3 Immanuel Kant, *Allgemeine Naturgeschichte und Theorie des Himmels* (Königsberg and Leipzig, 1755), p. 163; transl. by S.L. Jaki, *Universal Natural History and Theory of the Heavens* (Edinburgh, 1981), p. 177.

4 J.H. Lambert, *Cosmologische Briefe* (Augsberg, 1761), p. 7; transl. by S.L. Jaki, *Cosmological Letters* (Edinburgh, 1976), p. 57.

5 Tobias Swinden, *An Enquiry into the Nature and Place of Hell*, 2nd edn (London, 1727), pp. 354-5.

6 William Whiston,*Astronomical Principles of Religion, Natural and Reveal'd* (London, 1717), pp. 15, 19.

7 Thomas Wright of Durham, *Clavis Coelestis* (London, 1742), p. 16.

8 *Ibid.*, pp. 17, 33.

9 Thomas Wright of Durham, *An Original Theory or New Hypothesis of the Universe* (London, 1750), p. 31.

10 Thomas Wright of Durham, *Second or Singular Thoughts upon the Theory of the Universe*, ed. by M.A. Hoskin (London, 1968), p. 45.

11 *Ibid.*, p. 50.

12 Kant, *op. cit.* (ref. 3), p. 17; transl. by Jaki, pp. 108-9.

13 James Ferguson, *Astronomy Explained upon Sir Isaac Newton's Principles*, 12th edn (London, 1809), p. 37: "By comparing the great interval between the Orbits of Mars and Jupiter, it was surmised upwards of seventy years ago, by Mr. Maclaurin and others, and lately by C. Loft, Esq that there must, at least, be one planet, whose orbit is exterior to that of Mars, and interior to the Orbit of Jupiter." I have not located the work in which Maclaurin makes this suggestion.

14 Wright, *Second or Singular Thoughts*, p. 24.

15 William Whiston, *Praelectiones astronomicae* (London, 1707), Lectio VII.

16 David Gregory, *Astronomiae elementa* (Oxford, 1702), Book I, Section I, Prop. I; transl. from the English edn (London, 1715), p. 2.

17 The best source for the post-Gregory story of this section is M.M. Nieto, *The Titius-Bode Law of Planetary Distances* (Oxford, 1972). As the distances quoted by Whiston would convert (to the nearest integer) to give Saturn the figure 96 rather than 95, it is likely that Wolff took the numbers directly from Gregory rather than deriving them himself. Roger Long, *Astronomy* (2 vols, Cambridge, 1742, 1754), vol. 1, p. 339, gives 32, 59, 82, 125, 426 and 780 millions of miles. Ferguson, *Astronomy Explained...*, 1st edn (London, 1756), gives Whiston's values.

18 Johann Daniel Titius, *Betrachtung über die Natur, vom Herrn Karl Bonnet* (Leipzig, 1766), pp. 7-8; transl. by Stanley Jaki in "The early history of the Titius-Bode Law", *American Journal of Physics*, vol. 40 (1972), pp. 1014-23.

19 Titius, *Betrachtung...*, 4th edn (Leipzig, 1783), p. 13; transl. by Nieto, *The Titius-Bode Law*, p. 11.

20 Johann Elert Bode, *Anleitung zur Kenntnis des gestirten Himmels*, 2nd edn (Hamburg, 1772), p. 462; transl. by Jaki, *op. cit.* (ref. 18).

21 For a facsimile of the original entry in Herschel's observing book, see *The Scientific Papers of Sir William Herschel*, ed. by J.L.E. Dreyer (2 vols, London, 1912), vol. 1, p. xxviii. In the following pages Dreyer gives further details of the discovery and its aftermath. For futher information, see the articles by R. Porter, J.A. Bennett, M. Hoskin, E.G. Forbes and R.W. Smith in the section on "History of the Discovery of Uranus" in *Uranus and the Outer Planets*, ed. by Garry Hunt (Cambridge, 1982). For early attempts to determine the orbit of Uranus see A.F. O'D. Alexander, *The Planet Uranus: A History of Observation, Theory and Discovery* (London, 1865), chap. 2.

22 The story is told by F.X. von Zach, "Über einen zwischen Mars und Jupiter längst vermutheten nun wahrscheinlich entdeckten neuen Hauptplaneten unseres Sonnen-Systems", *Monatliche Correspondenz*, June 1801, 592-623, quotation from p. 602.

23 *Ibid.*, pp. 602-3.

24 On Piazzi see the article by Giorgio Abetti in *Dictionary of Scientific Biography*, and the bibliography therein.

25 W. Pearson, *Introduction to Practical Astronomy*, vol. 2 (London, 1829), pp. 413-17.

26 G. Piazzi, *Praecipuarum stellarum inerrantium positiones mediae ineunte saeculo decimonono ex observationibus habitis in specula panoramitana ab anno 1792 ad annum 1802* (Palermo, 1803).

27 G. Piazzi, *Risultati delle Osservazioni della Nuova Stella* (Palermo, 1801), pp. 3-6. I have used (without amendment) the English translation by one Antonio Parachinatti, "teacher of the Italian language", prepared for Nevil Maskelyne (Cambridge University Library, RGO ms 4/221). Quoted by courtesy of the Syndics of Cambridge University Library and of the Director of the Royal Greenwich Observatory.

28 Cited by Abetti, *op. cit.* (ref. 24).

29 Piazzi, *op. cit.* (ref 27), pp. 7-8, 13.

30 *Ibid.*, pp. 13-14.

31 *Ibid.*, p. 16.

32 W. Herschel, "Observations on the Two Lately Discovered Celestial Bodies", *Philosophical Transactions*, vol. 92 (1802), 187-98. As early as 18 February 1802, and before the discovery of Pallas, Herschel had told the Royal Society that Ceres was much smaller than the Moon: "Observations of the New Planet", first published in *The Scientific Papers of Sir William Herschel* (ref. 21), vol. 1, pp. cix-cxi.

THE NINETEENTH-CENTURY BIRTH OF ASTROPHYSICS

OWEN GINGERICH
Harvard-Smithsonian Center for Astrophysics
60 Garden St., MS9, Cambridge, MA 02138

"What astronomy is expected to accomplish," wrote the Astronomer Royal George Biddell Airy around 1840,

> is at all times the same. It may lay down rules by which the movements of the celestial bodies, as they appear to us upon earth, can be computed. All else which we may learn respecting these bodies, as, for example, their appearance, and the character of their surfaces, is, indeed, not undeserving of attention, but possesses no proper astronomical interest. [1]

Airy's rather circumscribed view of astronomy echoed the definition given by August Comte in his *Cours Philosophie Positive* (1835). Comte defined astronomy as "the science by which we discover the laws of the geometrical and mechanical phenomena presented by the heavenly bodies," and in a well-worn quotation went on to say that, while the motions, sizes, and shapes of distant stars could be analyzed,

> we would never know how to study by any means their chemical composition, or their mineralogical structure.... In a word, our positive knowledge with respect to stars is necessarily limited solely to geometric and mechanical phenomena, without being able to encompass at all those other lines of physical or chemical research....[2]

1. The Pioneering Discoveries by Kirchhoff and Bunsen

Within a few decades the work of Gustav Kirchhoff in collaboration with the chemist Robert Bunsen would send Comte's opinion to the scrap-heap. Their dramatic breakthrough was described by Bunsen in a letter sent from Heidelberg to his English colleague, Henry Roscoe, on 15 November 1859:

> At this moment Kirchhoff and I are engaged in a common work that doesn't let us sleep. Kirchhoff has made a wonderful, entirely unexpected discovery in finding the origin of the dark lines in the solar spectrum, increasing them artificially in the solar spectrum, and producing them in the continuous spectrum of a flame, and indeed in the identical positions as the Fraunhofer lines. Hence a way has been found to determine the composition of the sun and fixed stars with the same certainty as we determine sulfuric acid, chlorine, etc., with our [chemical] reagents. Terrestrial substances can be clearly differentiated by this method just as easily as those on the sun, so that, for example, we can detect lithium in twenty grams of sea water. [3]

The background for their discovery was Bunsen's chemical research, which eventually led to his isolation of the elements caesium and rubidium. Around 1853 Heidelberg had been piped for a central gas distribution, and very quickly Bunsen developed the burner known nowadays to every student of chemistry. By controlling the input of oxygen, Bunsen achieved a transparent flame that could be used to

J.F. Linsky and S. Serio (eds.), Physics of Solar and Stellar Coronae, 47–58.

examine the colors when various compounds were introduced into the flame. At first Bunsen used filters to help distinguish the colors, but his younger colleague Kirchhoff suggested that a direct examination of the spectrum would be more accurate. As a result, they developed the original spectroscope. It is true that Fraunhofer before them had united a prism with an eyepiece in a single instrument, but at Heidelberg the entire combination of slit, collimator, prism, and eyepiece was realized for the first time.

In those days Kirchhoff and Bunsen liked to take off time from the laboratory to enjoy the so-called philosopher's walk in the hills north of the Neckar River, which flows past Heidelberg on its way to the Rhine. On one such walk, Bunsen reflected on an event a few days earlier when they had looked out from their laboratory window across the Rhine plain toward Mannheim and had noticed a major fire raging. With the aid of their spectroscope, they had detected barium and strontium in the flames. "If we could determine the nature of the substances burning in Mannheim," mused Bunsen, "why should we not do the same with regard to the sun?" [4]

The problem was, of course, that the solar spectrum had dark lines, in distinction to the bright lines found in the laboratory. But would the dark D line in the Fraunhofer spectrum match the bright sodium line of the laboratory? A simple test would be to reflect sunlight into the lab, and to pass the beam through a flame doped with sodium; if the match was exact, one might expect the dark solar line to filled in by the bright line from the flame. We can well imagine Kirchhoff's and Bunsen's surprise when they discovered that the dark line was actually enhanced rather than being balanced out. Very quickly Kirchhoff realized that he had the key to the sun's dark-line spectrum.

In retrospect, it is fascinating to notice how close several other scientists had come to making this discovery. Foucault in 1849 had actually produced an absorption lines in his laboratory, and in England G.G. Stokes and William Thomson (who later became Lord Kelvin) both knew of this experiment. But none of them made the connection to the possibility of spectral analysis, perhaps because of the confusing contamination of sodium in so many chemical compounds prior to the 1850s. It takes very little contamination by ordinary salt to add the sodium D line to a spectrum, and until salt could be eliminated, it was not obvious that each element gave its own characteristic spectrum. This understanding was achieved in the Heidelberg laboratory, where Bunsen had achieved excellent control on the purity of his chemical samples.

Kirchhoff's discovery was reported in a memorable and widely reprinted paper [5] given before the Berlin Academy of Sciences on 27 October 1859; a second, more complete paper was presented in December. These dates mark the birth of astrophysics, though the name itself did not emerge until 1890 (in the *Saturday Review*, and two years later in the title of the newly-named journal, *Astronomy and Astro-physics*, the immediate forerunner to the *Astrophysical Journal*).

2. Eclipses and Early Solar Physics

Once Kirchhoff's hypotheses were announced, an obvious test suggested itself. If the Fraunhofer spectrum really did arise from an atmosphere of glowing gas that surrounded the sun's glowing globe, then at the time of eclipse, just a moment

Fig. 1. Angelo Secchi's drawing of the eclipse of 18 July 1860, which helped establish that the prominences were physically part of the sun. *Memorie dell' Osservatorio del Collegio Romano*, NS 2 (1863), Plate 3.

before totality, the emission line spectrum might be glimpsed. This would provide an empirical verification of the idea of a reversing layer where the dark lines formed.

The man who would soon become Italy's leading astronomical spectroscopist would no doubt have tried to find the reversing layer at the Spanish eclipse of 1860, had he known about Kirchhoff's ideas soon enough. As it was, Angelo Secchi discovered something else, also quite important, from drawings made at his eclipse site on the shore of the Mediterranean and the photoheliographs made by Warren De la Rue at his station some 400 kilometers to the northwest. By comparing the prominences observed at the two locations, Secchi was able to establish that these were genuine atmospheric phenomena on the sun, and not merely the distorting effects of the earth's own atmosphere. As Agnes Clerke described it, the evidence strongly confirmed the inference "that an uninterrupted stratum of prominence-matter encompasses the sun on all sides, forming a reservoir from which gigantic jets issue, and into which they subside." [6]

At the time, young Secchi was just beginning to make his mark on Italian astronomy. A Jesuit, he had briefly gone to Georgetown University in America when his order was temporarily suppressed. Upon his return to Rome in 1849 he was made director of the Gregorian Observatory of the Collegio Romano. Probably it was his 1860 eclipse observations that attracted the notice of Jules Janssen, the French solar astronomer, who visited Rome in 1863. Janssen drew Secchi's attention to the possibilities of spectroscopic stellar work, and very soon Secchi launched on a program of observing the spectra of stars. What he found was that most stellar spectra could be binned into comparatively few categories. Some spectra were like the sun's, another large group resembled that of Sirius, and yet another group showed the shaded spectra we now associate with the molecular bands of the cooler stars. Relatively few fell into his fourth group, which he called "very bizarre," and which were represented by relatively faint carbon stars such as Y Canes Venatici. [7]

Concerning his program, Secchi wrote, "In substance I wanted to see whether, just as the stars are countless, their composition is also proportionately varied. This has been my query, and having been fortunate enough to perfect the observation instrument, the harvest was abundant, even more than I had hoped." [8] Altogether he examined 4,000 stars, looking to see if any were different; the number he actually classified was considerably smaller.

Meanwhile, for several years there were no well-situated solar eclipses, but at the end of the decade three came in swift succession. The first of these, in August, 1868, crossed the Indian and Malay peninsulas. During the eclipse Janssen realized that prominences could be seen spectroscopically outside of eclipse simply by opening the spectroscope slit and carefully setting it adjacent to the solar limb, which he demonstrated the following morning. It seemed unlikely that he would be anticipated in his discovery, so not until a month later did he communicate the news back to Europe. Yet just moments before his report reached the secretary of the French Academy of Sciences, a very similar communication had come from Norman Lockyer, who had independently conceived of the technique. The French mint issued a medal with both profiles, and today both men share equally in credit for the discovery.

The eclipse of August, 1869, swept across North America, and the most important discovery was made by Dartmouth's Charles Young, who observed it from Burlington, Iowa. He found the green coronal line, which by 1889 was attributed to "coronium," and is today recognized as Fe XIV following the brilliant deductive work of Bengt Edlén in 1941. Young's description of the brevity of totality, made in his report of that eclipse, is wonderfully graphic: "I cannot describe the sensation of surprise and mortification, of personal imbecility and wasted opportunity, that overwhelmed me when the sunlight flashed out." [9]

What might have been the best observed of the three eclipses (because its path fell across Spain, North Africa, and Sicily), that of 22 December 1870, was hampered by clouds along most of its path. With her splendid Victorian prose, Agnes Clerke has given a vivid description of the event:

> A further trophy was carried off by American skill sixteen months after the determination due to it of the distinctive spectrum of the corona. The eclipse ..., though lasting only two minutes and ten seconds, drew observers from the New, as well as from the Old World to the shores of the Mediterranean. Janssen issued from beleaguered Paris in a balloon, carrying with him the *vital parts* of

> a reflector specially constructed to collect evidence about the corona. But he reached Oran only to find himself shut behind a cloud-curtain more impervious than the Prussian lines. [German troops had just surrounded Paris in the opening stages of the Franco-Prussian War.] Everywhere the sky was more or less overcast. Lockyer's journey from England to Sicily, and shipwreck in the *Psyche*, were recompensed with a glimpse of the solar aureola during *one second and a half!* Three parties stationed at various heights on Mount Etna saw absolutely nothing. Nevertheless important information was snatched in despite of the elements.
>
> The prominent event was Young's discovery of the "reversing layer." As the surviving solar crescent narrowed before the encroaching moon, "the dark lines of the spectrum," he tells us, "and the spectrum itself, gradually faded away, until all at once, as suddenly as a bursting rocket shoots out its stars, the whole field of view was filled with bright lines more numerous than one could count. The phenomenon was so sudden, so unexpected, and so wonderfully beautiful, as to force an involuntary exclamation." Its duration was about two seconds, and the impression produced was that of a complete reversal of the Fraunhofer spectrum–that is, the substitution of a bright for every dark line. [10]

Agnes Clerke's breathless account, while correctly focusing on Young's achievement, makes it sound as if the Sicilian observers were clouded out, but quite the contrary. The Italian astronomers in Augusta (just north of Syracuse on the eastern side of Sicily) and in Terranova (on the western shore, today Gela) produced a well-illustrated report of over 200 pages. [11] Secchi published a photograph of the inner corona, and images of prominences and coronal streamers drawn by his team in Augusta. Palermo's Pietro Tacchini, who headed the station 80 km to the west, obtained a magnificently detailed record of the prominences.

3. Early Astrophysics in Palermo

Perhaps stimulated by the three eclipses, as well as the ongoing observations of prominences and of stellar spectra, Secchi and Tacchini got together in 1871 to found the Società degli Spettroscopisti Italiani. Tacchini had come to Sicily in 1863 after having served briefly as director of the observatory in his home town of Modena; altogether he would spend sixteen productive years at the Royal Observatory in Palermo. An active solar observer, he took a major part in organizing the observational campaign in Sicily for the 1870 elipse, the first scientific expedition organized by the Italian government. While in Palermo Tacchini organized an expedition to Bengal to observe the transit of Venus in 1874, and another to the Nicobar Islands in 1875 to observe the solar eclipse. In 1871 Tacchini published in volume 7 of the *Bullettino Meteorologico* of the Royal Observatory in Palermo his most important work on the sun, a series of articles on solar physics that gave one of the first classifications of prominences. He supported the idea that prominences were jets of chromospheric hydrogen stirred up by the action of atmospheric currents, and later he won both the Rumford Medal of the Royal Society of London and the Prix Janssen of the French Academie des Sciences for his solar researches.

In Palermo Tacchini edited the *Memorie* of the Society of Italian Spectroscopists;

this became the first journal of astronomical spectroscopy, antedating the Astrophysical Journal by just over two decades. Though its pages were dominated by articles on sunspots and prominences observed from Palermo and Rome, its international set of authors would eventually include such astrophysicists as Henry Draper, Charles Young, William Huggins, and Jules Janssen. Today the paper of its early issues has become dangerously brittle, but the wonderful illustrations, often printed in black and red, are on rag paper and still retain their brilliance and grandeur. And the journal occasionally includes a new technique that had begun to make its mark with respect to astronomical spectroscopy: actual photographs. These are used, for example, to illustrate the memorable paper by Giovanni Schiaparelli on the "canali" of Mars (not photographs of Mars, of course, but of Schiaparelli's exquisite drawings). [12]

4. Photographic Spectroscopy and Spectral Classifications

Inevitably photography and spectroscopy were joined. The two great pioneers, and competitors, of spectroscopic photography were William Huggins in Great Britain and Henry Draper in America. Huggins had attempted to photograph the spectrum of Sirius with wet collodion plates as early as 1863, but the result was unsatisfactory because the emulsion dried and lost its sensitivity before a sufficiently long exposure could be obtained. In Hastings-on-Hudson, north of New York City, Henry Draper completed his own 28-inch silvered-glass reflector in 1872, and with the greater light-gathering power finally succeeded in photographing the spectrum of Vega. Huggins promptly seized upon the advances in photographic sensitivity to record his own stellar spectra. In the spring of 1879 Draper visited Huggins in England, and learned that further important improvement was now possible with the new dry plates, which retained their sensitivity for exposures of more than an hour. The following year Draper succeeded in capturing the Orion nebula photographically with a dry plate. He was about to embark on a major program of photographic spectroscopy when he fell ill and died prematurely, at age 45.

Henry Draper's widow, Anna Palmer Draper, hoped to carry on his researches, and when this proved too formidable for her personally, she endowed a program at Harvard College Observatory so that its director, Edward C. Pickering, could undertake a vast spectrographic survey. Pickering organized a comprehensive census of the heavens, using both Draper's telescopes and others. By outfitting the telescopes with thin glass prisms placed over the objective end of the instruments, the spectra of an entire field of stars could be recorded simultaneously, a technique allowed the mass production of photographic spectra.

Pickering assigned a series of capable women astronomers to study prototype spectra and to arrange them into an orderly system. In the process of examining the stars for the Draper project, the women astronomers at Harvard discovered dozens of novae and other variable stars on the photographic plates, many recognized from their peculiar spectral characteristics. The classification scheme that emerged was pioneered by Williamina Fleming and further developed by Annie Jump Cannon, who eventually classified over 250,000 spectra on the Harvard system. The classification succeeded so well that astronomers worldwide eventually adopted it. Finding

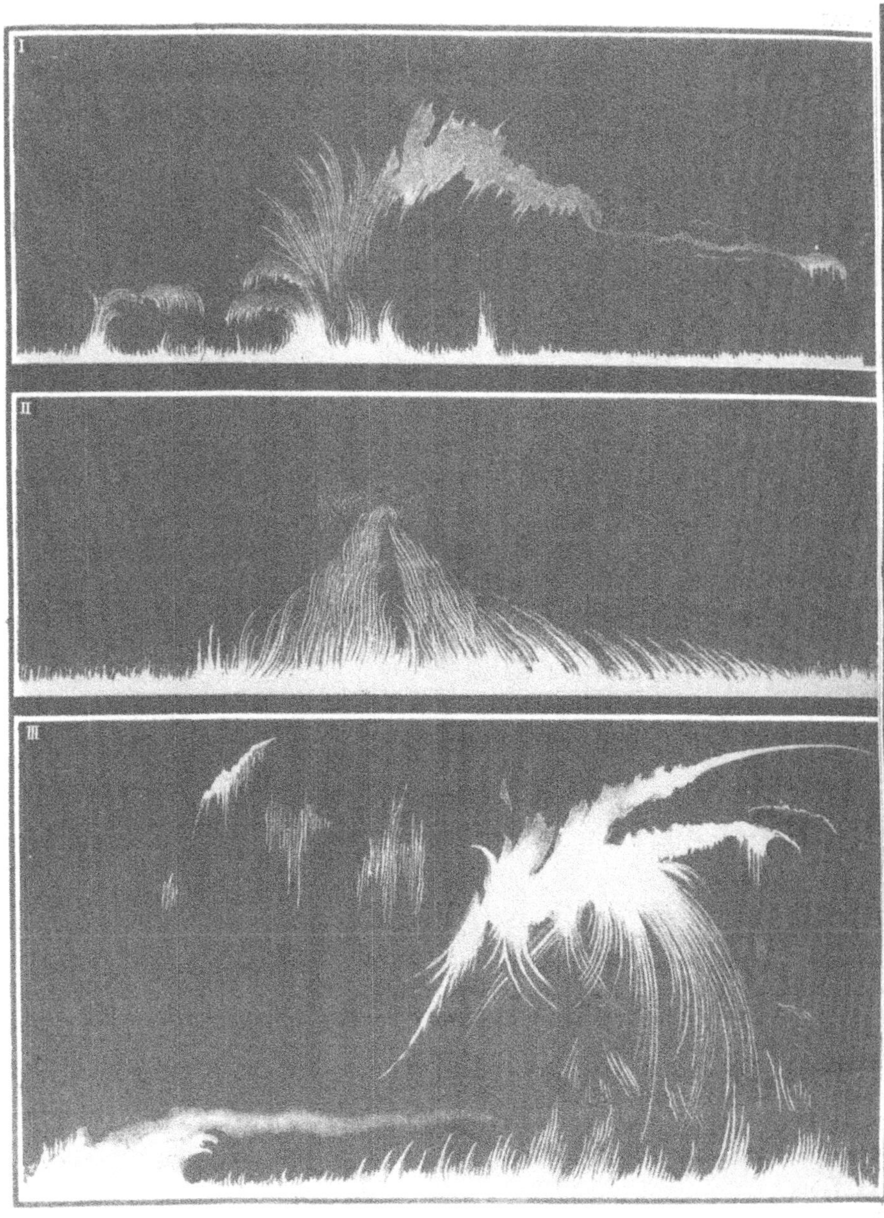

Fig. 2. Pietro Tacchini's drawings of the solar prominences, observed at Palermo, 22 December 1871, from the first volume of the *Memorie della Società degli Spettroscopisti Italiani*, (1872), Plate 10.

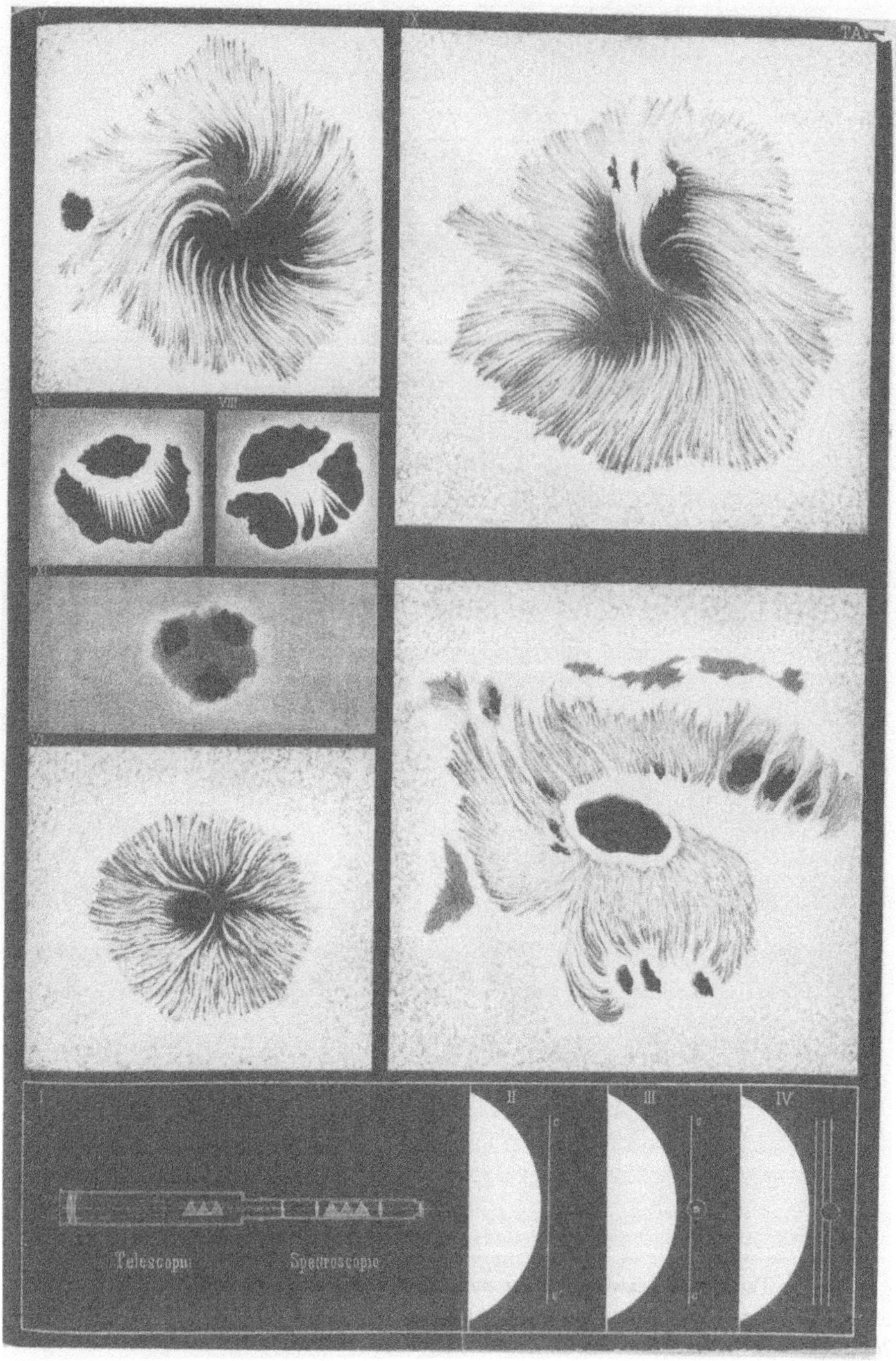

Fig. 3. Pietro Tacchini's drawings of sunspots, observed on 7 and 25 July 1872, with details of the nuclear region, from the first volume of the *Memorie della Società degli Spettroscopisti Italiani*, (1872), Plate 17 (detail).

Fig. 4. Pietro Tacchini's drawings of the eclipse of May 1882 and the eclipse comet, observed from Egypt. *Memorie della Società degli Spettroscopisti Italiani*, 11 (1882), Plate 146.

a universally-acceptable classification was essential in Pickering's game plan, and that was the primary reason why he insisted on a scheme simple enough for mass production (i.e., this is why he failed to promote Antonia Maury's far more sophisticated but overly intricate system). His search for a strictly empirical scheme helps explain why he discouraged any local attempts to link the Harvard spectral sequence with any sort of evolutionary picture. Miss Cannon fitted into Pickering's regime particularly well because, in the words of Cecilia Payne Gaposchkin, she "never published a controversial word or speculative thought." [13]

Elsewhere, however, astrophysicists had been eager to exploit the information about stellar types and to place them into a broader picture of cosmic evolution. Norman Lockyer, the founding editor of Nature and the co-discoverer of the method of observing prominences outside of an eclipse, envisioned an evolutionary sequence in which stars began as swarms of metoritic material that condenses, heated up to brilliant iridescence, and then declined in an extended period of senescent cooling. These ideas he summarized in the 1880s in a graphic "temperature arch." By 1914 he added the spectral types according to the Harvard sequence, starting with M stars, going up through K, and peculiar F, A and B stars, to the summit of his arch; on the downward cooling branch he placed normal B, A and F types, ending with other K and N stars. In one form or another Lockyer's scheme continued to be influential until Eddington's work in the 1920s showed that it was untenable.

5. Spectroscopy and the Nebulae

Meanwhile George Ellery Hale, who as an undergraduate at MIT had been inspired by Pickering and the spectrographic work at Harvard, grew keen to track the evolutionary history not only of stars, but of nebulae as well. He was to become the ablest astronomical entrepreneur of the twentieth century after a good start already in the 1890s. In 1892 he found philanthropic support, from the Chicago streetcar magnate Charles Yerkes, to build the world's largest refractor. The Yerkes Observatory, northwest of Chicago, was dedicated in 1897. Its 40-inch refractor could collect enough light for detailed, high- dispersion spectra, which Hale hoped would unlock the evolutionary secrets. A fundamental auxiliary to his observatory was the physical laboratory, where his associates obtained the basic spectroscopic data on many elements. Daring as a fund raiser, Hale was cautious as a theorist and never in the same speculative league with Lockyer, Percival Lowell, or Harlow Shapley. Yet his vision of applied astrophysics–the telescope combined with the laboratory–paved the way for much of today's astonishing astronomical edifice.

Out of the great proliferation of astrophysics by the end of the nineteenth century, let me choose just one more thread, one that leads to Hale's later observatory on Mount Wilson. Sidereal structure, including the role and nature of the nebulae, had blossomed in that century as a new addition to the astronomical repertoire. The practitioners of "the new astronomy"– the spectroscopists–contributed as much as anyone else toward the understanding of the nebulae and their place in modern cosmology.

Only a few years after Kirchhoff's pioneering papers, in 1864, Huggins showed that a planetary nebula in Draco (now known as NGC 6543) exhibited bright lines characteristic of a glowing gas, [14] and in a survey two years later he found that ten more planetary or diffuse nebulae had very similar spectra. [15] On the other hand, he recorded even more (including a number of spirals) that showed only a faint continuous spectra. Not until 1898 did Julius Scheiner at Potsdam announce the detection of the dark Fraunhofer lines in spectrum of the Andromeda spiral M31. At this point Huggins admitted that he had had similar spectra in hand for a decade, but that he had been "very uncertain whether the dark lines were really due to the Nebula itself, or were not rather produced by some faint traces of solar light; though the nights on which they were taken were free from moonlight at the times of taking the photographs." [16] Despite his caveat that "our results are lacking in the definiteness and in the certainty that we could wish," the obvious inference was that the spirals were stellar systems. Yet nearly a decade later as eminent an astronomer as Hale could still question whether the spiral nebulae really showed Fraunhofer lines in their spectra; subsequent work by E. A. Fath with the 36-inch Crossley reflector at Lick Observatory and with Hale's new 60-inch reflector on Mount Wilson finally settled the matter.

The appearance in 1885 of a bright nova near the nucleus of the Andromeda Nebula provided one reason for the reluctance of astronomers to admit the stellar nature of the spirals. One of the contemporary records of this remarkable event was made in Palermo, by A. Riccò, and published in the *Memorie* in 1888. [17] Not until the 1920s did astronomers sort out the difference between novae and supernovae, but before that the 1885 outburst continued to baffle astronomers and helped to persuade some (such as Harlow Shapley) that the Andromeda Nebula could not be a distant island universe.

Confusion about the nature of the spirals continued so long that James Jeans, in his *Problems of Cosmogony and Stellar Dynamics* of 1919, could illustrate M64 and NGC 7217 as examples of stellar and planetary systems in the process of formation, the nebular hypothesis in action: "In this nebula [NGC 7217] we appear to have a close approach to the manner of evolution imagined by Laplace." [18] Percival Lowell, ever interested in the formation of planetary systems, followed the same reasoning when he assigned to his assistant at Lowell Observatory, V. M. Slipher, the task of photographing the spectra of spirals, and especially of determining their radial velocities. Lowell had hoped to learn more about the evolution of the solar system. The stunning consequences of Slipher's investigations are well known: instead of finding the modest speeds of solar systems in the making, by 1914 he had found velocities of hundreds of kilometers per second.

It was these extraordinary velocities of the spirals, in the hands of Edwin Hubble at Mount Wilson Observatory, that led to the discovery of the expansion of the universe. By then astrophysics had indeed come of age, and no one doubted that spectroscopy not only unlocked the chemical composition of the universe (as the commemorative plaque on chemistry laboratory in Heidelberg states), but much, much more regarding the physical nature of the celestial realms.

6. References

1. E. W. Maunder, *The Royal Observatory Greenwich* (The Religious Tract Society: London, 1900), pp. 266-67.

2. Auguste Comte, *Cours de Philosophie Positive*, II, 19th lesson (1835), quoted by J. B. Hearnshaw, *The Analysis of Starlight* (Cambridge University Press: Cambridge, 1986), p. 1.

3. Bunsen to Roscoe, translated by Roscoe in H. E. Roscoe, *The Life and Experiences of Sir Henry Enfield Roscoe, D.C.L., LL.D., F.R.S.* (Macmillan: London and New York, 1906), p. 81; translated rather differently by him in "Bunsen Memorial Lecture," *Journal of the Chemical Society*, **77**, pt. 1 (1900), 531; also translated by Mary Elvira Weeks, *Discovery of the Elements* (Journal of Chemical Education: Easton, 1945), p. 367; my translation is based on the facsimile of the original letter reproduced in the Bunsen Memorial Lecture.

4. "Some Scientific Centres: The Heidelberg Physical Laboratory," *Nature*, **65** (1902), 587-90.

5. G. R. Kirchhoff, "Über die Fraunhofer'schen Linien," *Monatsberichte der Königlichen Preussischen Akademie der Wissenschaft zu Berlin*, (1859) 662-65; translation in *Philosophical Magazine*, series 4, **19** (1860), 193-97 and in W. F. Magie (ed.), *A Source Book in Physics* (Harvard University Press: Cambridge, 1963), pp. 354-60.

6. Agnes Clerke, *A Popular History of Astronomy during the Nineteenth Century* (Adam and Charles Black: London, 1902), p. 167.

7. Angelo Secchi, *Les étoiles* (Librairie Germer Baillière et Cie: Paris, 1879), t. 1, p. 97.

8. Angelo Secchi, report to the Pontificia Accademia Tiberina, 27 January 1868, cited in *Dictionary of Scientific Biography*, **12** (1975), p. 269.

9. "Report of Prof. C. A. Young," p. 48 in J. H. C. Coffin, *Reports of Observations of the Total Eclipse of the Sun, August 7, 1869* (Washington: Authority of the Secretary of the Navy [1870]).

10. Agnes Clerke, *A Popular History of Astronomy during the Nineteenth Century*, (Adam and Charles Black: London, 1902), p. 171.

11. Giovanni Santini, *Rapporti sulle Osservazioni dell'eccliss Totale di Sole del 22 Dicembre 1870* (Stabilimento Tipgrafico Lao: Palermo, 1872).

12. G. V. Schiaparelli, "Osservazioni sulla Topografia del Planeta Marte," *Memorie della Società degli Spettroscopisti Italiani*, **11** (1882), Tav. CXLII.

13. Owen Gingerich, "Cannon, Annie Jump," *Dictionary of Scientific Biography*, **3** (1971), 50.

14. William Huggins, "On the Spectra of Some Nebulae," *Philosophical Transactions*, **154** (1864), 437-44.

15. William Huggins, "Further Observations on the Spectra of Some of the Nebulae," *Philosophical Transactions*, **156** (1866), 381-.

16. Sir William Huggins and Lady Huggins, *An Atlas of Representative Stellar Spectra*, (Wesley and Son: London, 1899), p. 120.

17. A. Riccò, "*Nova* nella Nebulosa di Andromeda," *Memorie della Società degli Spettroscopisti Italiani*, **17** (1888) = *Pubblicazioni del Real Osservatorio di Palermo*, **4** (1889), Fig. 1.

18. J. H. Jeans, *Problems of Cosmogony and Stellar Dynamics* (Cambridge University Press: Cambridge, 1919), p. 215.

HALF A CENTURY OF SOLAR PHYSICS
- SOME EPISODES -

M. ODA

Riken Institute of Physical and Chemical Research, Wako, Saitama, Japan

1. Prologue

I am honored to have been invited as a speaker to this memorable symposium for the bicentennial of the Astronomical Observatory of Palermo, particularly since the symposium is dedicated to our beloved friend Pippo Vaiana. Indeed, I clearly remember that a couple of years ago Pippo told me about this bicentennial symposium and we discussed on my possible attendance. But I would have never thought of this symposium as an occasion for the commemoration of Prof. Vaiana himself.

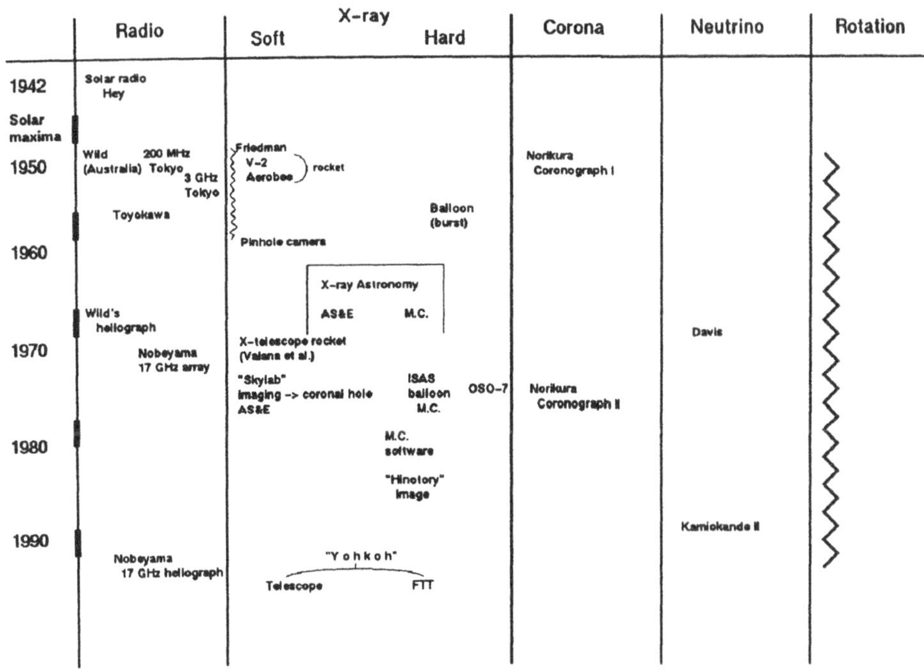

Fig. 1. Matrix of subjects of observation as a function of time.

As the title of my talk implies, I will review several episodes in the history of Solar Physics in the latter half of the XXth century, which represent examples of the rapid development of Physics. As I am not an expert in this field and am not

J.F. Linsky and S. Serio (eds.), Physics of Solar and Stellar Coronae, 59–68.

qualified to describe the subject in deep details, I will limit myself to a number of selected topics, tailored to the progress in Japan according to my own recollection.

Figure 1 shows a matrix of time versus subjects of observations. Following this matrix I will describe the history of solar radio, X-ray and other observations by showing some historical photographs.

Fig. 2. Informal gathering of solar radio astronomers around the 200 MHz TAO antenna.

Fig. 3. The 3.3 GHz manually oriented antenna in Osaka.

2. Radio observations

After the end of World War II, many radar engineers in Australia lost their job. This, I believe, is part of the reason why radio astronomy developed remarkably in Australia, as well as in Britain and in the Netherlands. In early 1950s, Wild discovered in Australia the solar radio burst and found that it propagated into the corona.

Fig. 4. Observation of a partial solar eclipse on May 9, 1949 with an adapted 1930 acoustic reflector (K. Shimoda, Tokyo).

Fig. 5. Toyokawa Interferometer built by the late Haruo Tanaka et al. (Nagoya University).

In Japan, recovering from ashes of the war, the late Prof. Hatanaka of the Tokyo Astronomical Observatory (TAO) pioneered japanese radio astronomy. Fig. 2 shows the first informal gathering of solar radio astronomers around the 200 MHz antenna of TAO with Prof. Hatanaka, on the occasion of my visit to TAO from Osaka University in 1949. Fig. 3 shows Japan's first 3.3 GHz manually-handed observing station in Osaka, which was made with the war-surplus radar parts and started observations in 1949. In 1950 it detected an extraordinarily large solar radio outburst. In the next few years, a primitive concept of a radio heliograph with 25 horns was proposed, but not realized.

Fig. 6. The Nobeyama radio heliograph

Fig. 7. A picture of the corona synthesized by the Nobeyama radio heliograph

Shimoda, of the University of Tokyo, made a radio parabola from a military acoustic reflector produced in 1930 and observed a partial solar eclipse in May 9, 1948 as indicated in Fig. 4. This was likely one of the few meaningful observations during this difficult period for the (almost starving) Japanese scientists after the war.

Fig. 5 shows the Toyokawa Radio Interferometer at 7.5 cm, made by the late Haruo Tanaka and others of Nagoya University in 1957. This was later expanded to a much larger scale.

In the late 1960s Prof. Wild and his group in Eastern Australia, including Dr. Morimoto visiting from TAO, built an advanced heliograph at GHz frequencies. Nobeyama Observatory of TAO constructed a 17 GHz array in the early 1970s, which was followed by the plan developed in the 1980s by the late Kai and his group to construct an advanced radio heliograph in time for the forthcoming solar maximum around 1990. This plan took a while to be funded, until 1988, but finally the heliograph, which is shown in Fig. 6, started to work in 1992 by the effort of Enome, Kosugi, Nakajima et al.

The latest synthesized picture shown in Fig. 7 provides an example of coronal observations of interest to this symposium.

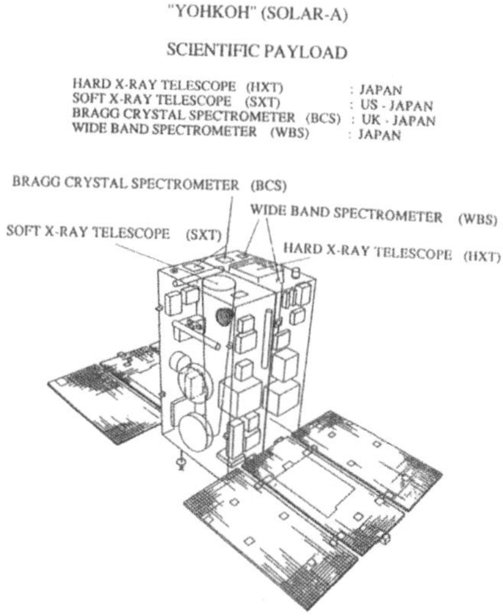

Fig. 8. The YOHKOH payload.

3. X-ray observations

Rocket observations of solar X-rays were pioneered by Friedman, of NRL. The first image of the X-ray sun was obtained with a pin-hole camera on board a sounding rocket during the early 1960s. In 1958, Peterson of UCSD had observed a hard X-ray burst with a balloon flight. An exceptional event happened in a related astronomical field, when, in 1962-3, Giacconi, Gursky and Paolini of AS&E and Rossi of MIT discovered celestial X-rays. This was immediately taken as one of the most intriguing discoveries of Astrophysics of the century. In the following years Giacconi and Vaiana of AS&E developed the grazing incidence telescope as an imaging device in the soft X-ray band, as Riccardo Giacconi has already described in this session.

In the meantime the technology of the modulation collimator for the hard X-ray band, photon energy over 10 keV, was being developed. This is the technique of X-ray tomography, and the software for handling data is known to be essentially similar to that of the medical CT scan. Takakura et al., in 1969, launched a balloon which carried rotating modulation collimators. During this flight luckily a solar flare occurred and its location was precisely determined coincident with the optical flare.

In the late 1960s Vaiana succeeded in making X-ray pictures of the sun using sounding rockets, and, in 1973, with the AS&E telescope on board SKYLAB, discovered the X-ray coronal holes, which is one of the breakthroughs in Solar Physics, among the many interesting features of the X-ray corona.

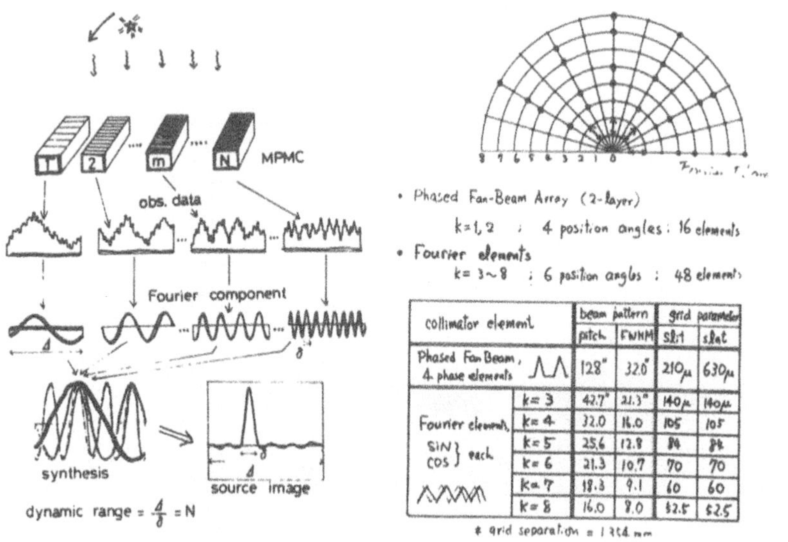

Fig. 9. Principles of operations of a Fourier Transform Telescope.

In 1981 a solar X-ray satellite, "HINOTORI", launched from Japan carried a rotating modulation collimator, which produced a number of images of hard X-ray flares.

For the solar maximum period which was expected to come around 1990 a Japanese solar X-ray satellite "SOLAR-A", renamed "YOHKOH" after the launch, was planned and designed in collaboration among US, UK and Japanese scientists (Fig. 8).

Its soft X-ray telescope (SXT) is a technical extension of Vaiana's telescopes, and the hard X-ray telescope (HXT) is the extension of the modulation collimator:

the name Fourier-transform-telescope, as it is also called, describes its principle, as indicated in Fig. 9. Fig. 10 shows the design of HXT.

"YOHKOH" was launched 3 days before Vaiana's death. Some preliminary results will be reported in this symposium. X-ray pictures are acquired every second or a fraction of second. As of now over 30,000 pictures of soft X-ray image like Fig. 11 have been obtained. We had thought Vaiana should have been the first person to be presented this picture.

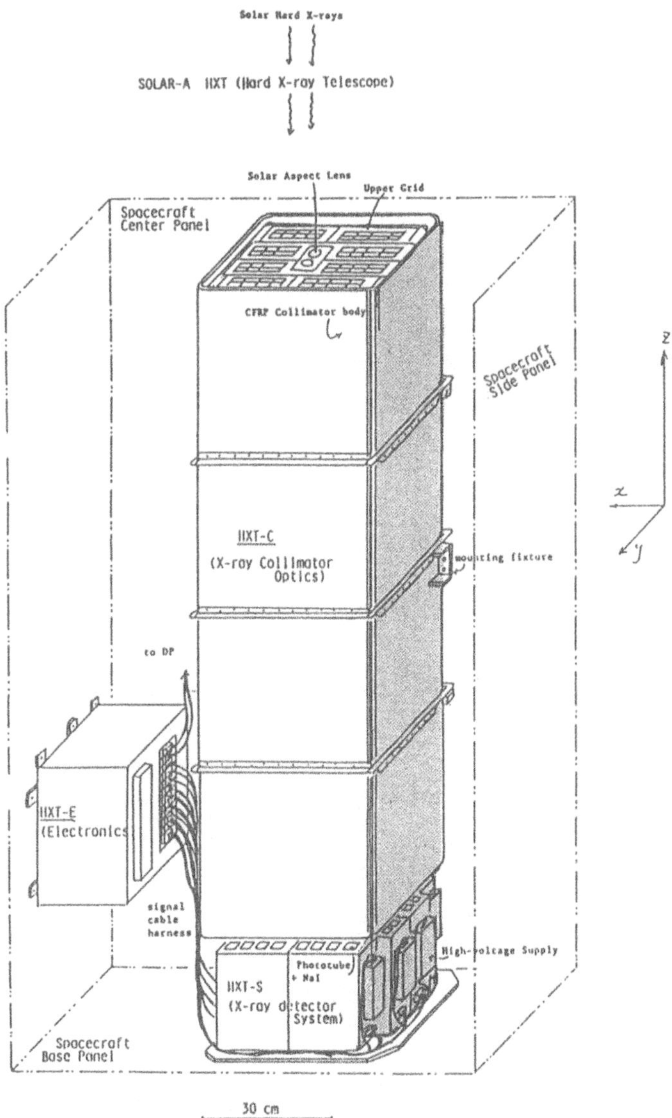

Fig. 10. The YOHKOH Hard X-ray Telescope.

Fig. 12 shows an example of the hard X-ray flare pictures which are obtained typically every ten seconds when the flare appears. This is from a square portion of a soft X-ray picture.

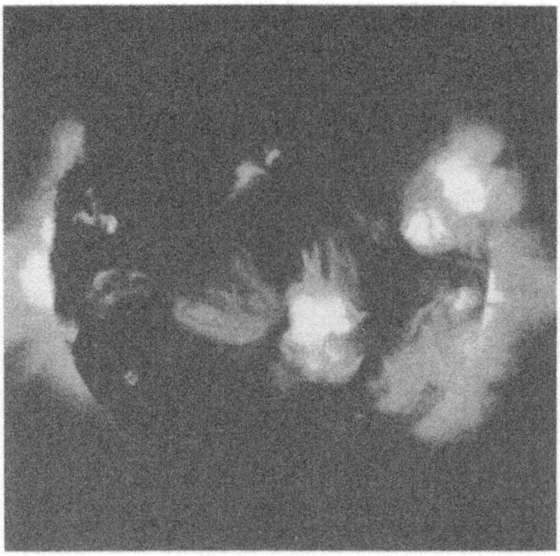

Fig. 11. One of the over 30,000 (up to June 1992) images taken by the YOHKOH SXT.

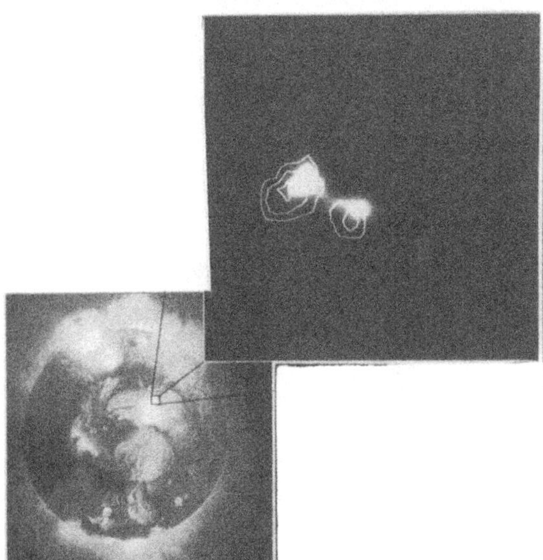

Fig. 12. An example of the HXT flare observations (upper panel).

4. Coronograph

The coronograph of TAO at Mt. Norikura has undergone through two generations and has been well active through several cycles of solar maxima. Fig. 13 shows a coronograph picture of the flare of April 20, 1992.

5. Solar neutrino

Fig. 14 shows the principle of operations of the neutrino detectors of KAMIOKANDE II, which have detected flux from SN 1987a. The experiment conducted by Koshica and Totsuka of the University of Tokyo, was originally intended to detect a proton decay (GUT) (0.1 - 1 per year!), has been continuously monitoring the neutrino flux from the sun.

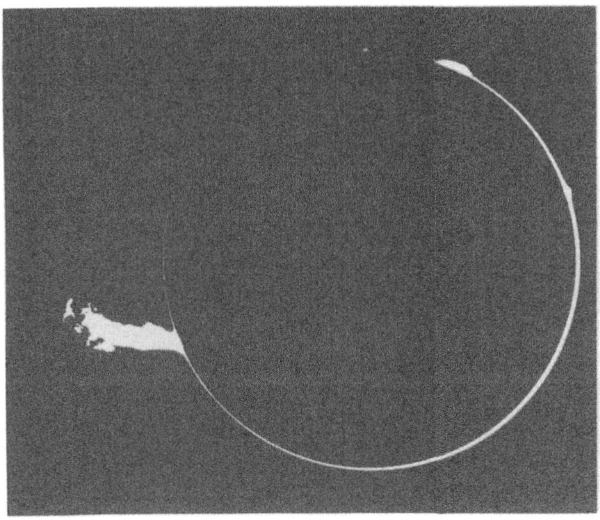

Fig. 13. A picture of the flare of April 20, 1992, taken from the TAO coronograph at Mt. Norikura.

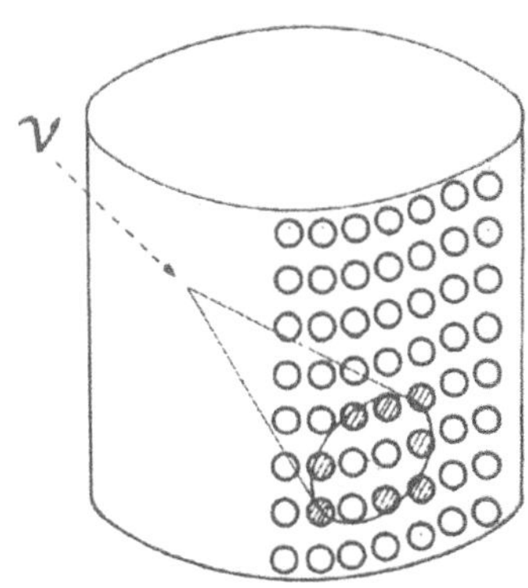

Fig. 14. Principle of operation of the 200 ton water tank KAMIOKA detector.

6. Solar rotation

The angular momentum of the surface layer of the sun has been studied during over four cycles of activity by Yoshimura et al. of TAO.

7. Epilogue

In closing, I congratulate with the colleagues in Palermo for the great role which their Observatory has played in European Astronomy and must say that to have known Pippo and his family and friends was one of my happiest experiences in my life.

PART II

THE SOLAR CORONA

NIXT HIGH RESOLUTION OBSERVATIONS

L. GOLUB

Smithsonian Astrophysical Observatory
60 Garden Street Cambridge, MA 02138

Abstract. Progress in x-ray optics, beginning in the 1960's and continuing to the present day, has been intimately tied to progress in our understanding of the coronae of the Sun and of solar-like stars. Innovative instrumentation, first used on sounding rockets and then flown on satellites, has been the most productive route for progress. This paper presents an overview of the status of coronal x-ray imaging research at the time that *Skylab* was launched in 1973, and of the contributions made by *Skylab* and by the *Einstein* Observatory. We then discuss the new technique of multilayer imaging for soft x-rays and the contributions which the NIXT experiment is making to our understanding of the structuring and heating of the corona.

Key words: Sun – corona – X-rays

1. Coronal Studies ca. 1973

I would like to start by thanking the organizers of this meeting: for the historical session yesterday, Georgia Foderà Serio, and for the local arrangements for the rest of this meeting, Salvatore Sciortino. Of course, we should certainly acknowledge the efforts of the Scientific Organizing Committee, and in particular the head of the committee, Salvatore Serio, who has done an outstanding job in ensuring that this meeting happened and in putting together a superb combination of good talks and excellent location for the meeting.

Nearly twenty years ago, I began working in the field of coronal X-ray astronomy, when I joined 'Pippo' Vaiana's group at A.S. and E. That was just before the launch of Skylab and at a time when coronal imaging studies from space had been going on for about ten years. When I joined the group, the first thing Pippo did was hand me a copy of a paper they had recently completed, so that I could become familiar with the field. Table 1 shows my summary of what was in that paper.

It was a compilation and distillation of results from a decade of sounding rocket work and, as you can see, almost all of the major areas which are now standard in studying the corona were discussed and described and studied in that paper. Riccardo Giacconi, in his excellent summary yesterday of Pippo's career, discussed much of this work, and I will not go through it again. What I will do is essentially add a few footnotes to Riccardo's presentation, seen from my own perspective as a member of Pippo's group.

It was clear in 1973 that the most important finding which had come out of the sounding rocket research was the intimate connection between the magnetic field of the sun and the location of the coronal emission. Although there were some clues to this connection from earlier observations using ground-based methods, the ability to see the corona on disk rather than at the limb made the connection quite clear. Figure 1 is an example, and shows a magnetogram and an X-ray image of several active regions in which we can see the location of magnetic fields at the surface; the black and white areas indicate, respectively, fields going into and out of the sun. Those are just the areas where the X-ray corona is bright and has the appearance of loop-like structures connecting the opposite-polarity fields. The corona should

J.F. Linsky and S. Serio (eds.), Physics of Solar and Stellar Coronae, 71–82.

Table I
Summary: Identification and Analysis of Structures in the Corona.

o Identified 'the importance of the magnetic field in ordering the shapes of all coronal features'

o Studied six types of coronal structure:
- Active Regions
- Active Region Interconnections
- Large Scale Quiet Coronal Structure
- Coronal Holes (and high speed wind streams)
- X-ray Bright Points
- Filament Cavities

o Developed Methods for Quantitative Analysis of X-ray Images

thus be viewed as a three-dimensional structure sitting above the solar surface and, at the very least, controlled by the geometry of the magnetic field.

What was also known in 1973 was that the corona would re-arrange itself in response to the emergence of new magnetic flux by developing interconnections between active regions and, under certain circumstances, by opening up rather than forming bright closed regions. What was seen in the latter case was large areas of reduced X-ray emission which were called coronal holes. One result in particular which is often thought-of as a Skylab result, but which was reported before Skylab (Krieger, Timothy and Roloff 1973) is that these coronal holes very often are the source of recurrent high-speed streams in the solar wind. Figure 2 shows one of the ways in which this result was established. By *in situ* measurements at one AU of the solar wind and its speed, it was possible to trace the solar wind back along Archimedes spirals to the location on the solar surface at which the wind originated. Very often, almost invariably, it was seen that a large coronal hole was situated at that place on the surface.

The small, isolated, compact emission features known as 'coronal X-ray bright points' were identified in the late sixties and the fact that they were magnetic bipoles and that they were found in large numbers, broadly distributed over the entire Sun, was known before Skylab. However, details such as lifetime and variability in spatial distribution of these features had to wait until an extensive time-sequence of data became available.

2. The *Skylab* Era

In May of 1973, the first US space station, called Skylab, was launched, and it was in operation into the beginning of 1974. Three different teams of astronauts were sent up to operate the space station, and in particular, from our point of view, they operated the Apollo Telescope Mount, known as ATM. This was a collection of solar telescopes, including the X-ray telescope known as S-054 built by Pippo.

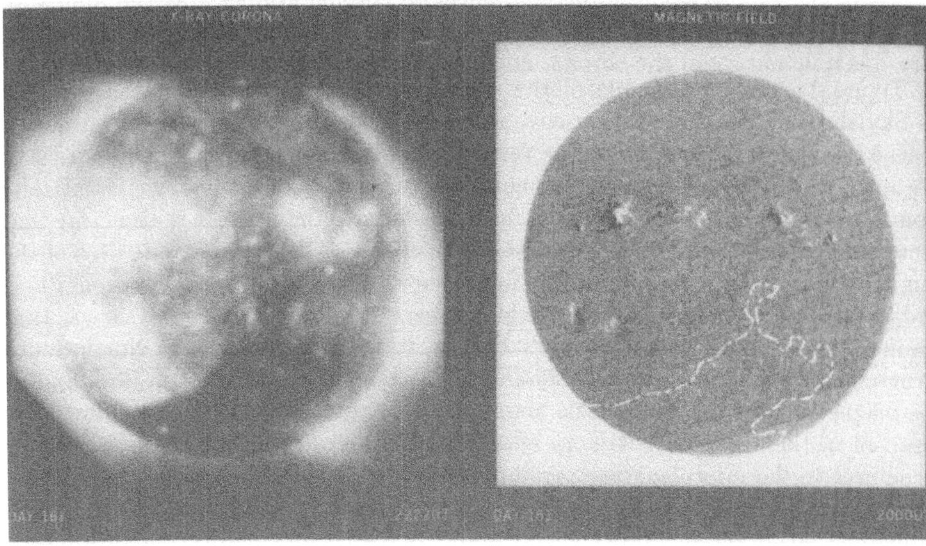

Fig. 1. Comparison of x-ray image with ground-based magnetogram

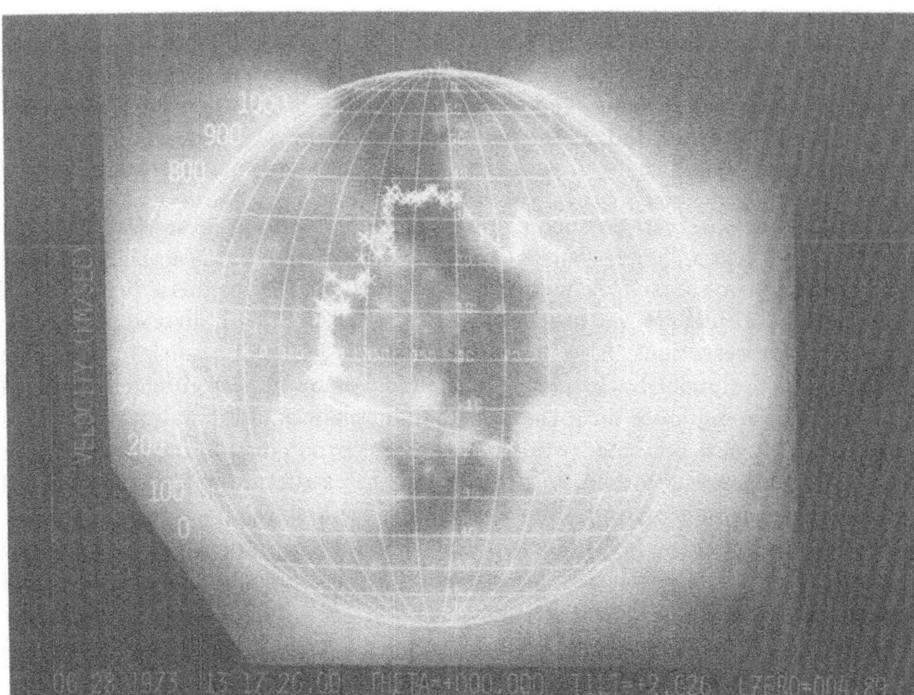

Fig. 2. Superposition of solar wind speed against source location in the corona

This telescope operated continuously for over eight months and recorded 30,000 images of the sun. What this mission added to coronal studies was the ability to see how the coronal features grew and evolved and changed. Also, it revealed the very dynamic nature of the corona: not just in active regions, but everywhere.

Figure 3 gives an example of the changing nature of the corona as revealed by Skylab. It shows three consecutive rotations of the Sun, with horizontal panels spaced one week apart. Thus, in the vertical columns one sees the same quadrant of the sun on successive rotations. It is quite clear that the corona changes drastically from one month to the next and even from one week to the next. It is also clear that these changes occur mainly in response to the evolution of the magnetic field of the sun. The first vertical column in this figure shows three rotations of a coronal hole and its evolution. One of the results which came out of the Skylab analysis was that coronal holes tend to rotate rigidly - that is, to maintain a straight North-South structure, rather than rotating differentially as the surface of the sun does and as the magnetic fields do. The puzzle about the rigid rotation has probably now been resolved in the sense that it is an effect due to the opening up of the boundary structures to the interplanetary medium, which tends to occur rigidly.

Figure 3 also shows on the middle rotation a major outbreak of activity which made that part of the sun have the appearance of a solar maximum corona. However, only one week later in the adjacent panel, we see a very faint, low-lying, diffuse corona, which looks like a solar-minimum corona; and in fact, a few years later in 1976, sounding rocket flights showed that the solar corona at minimum is dominated by X-ray bright points and small low-lying structures (Figure 4).

The next figure illustrates the dynamic nature of active region evolution. Figure 5 shows approximately one and one-half days of observation of an existing active region which was growing, and of newly-emerging regions nearby. Many flares occurred in this location, and, as the figure shows, the complex interconnections between regions grew and changed very rapidly, as did the magnetic structure within each of the active regions.

It was in fact observations such as these which led us to propose that the magnetic field is not just a passive container for coronal material, but is actively involved in the heating-process itself. A heating theory involving the magnetic field directly via dissipation of currents was proposed by Rosner *et al.* (1978), directly as a result of the Skylab observations. Also in 1978, and also as a direct result of the Skylab data analysis, the paper by Rosner, Tucker and Vaiana appeared which set forth the idea that coronal loops form the basic building blocks of the outer atmosphere of the sun. The view presented was that structure is not just an intrusion into an otherwise homogeneous corona, but rather that the structure is fundamental and that the corona is built up out of an aggregation of these relatively isolated 'mini-atmospheres'. One of the by-products of this view was the famous RTV scaling law for coronal loops, which related observable quantities (pressure, temperature and loop-size) within a given loop structure.

3. The *Einstein* Observatory

The development of this view of magnetic field-related coronal heating, combined with the observations from OSO-8 which indicated that waves originating in the

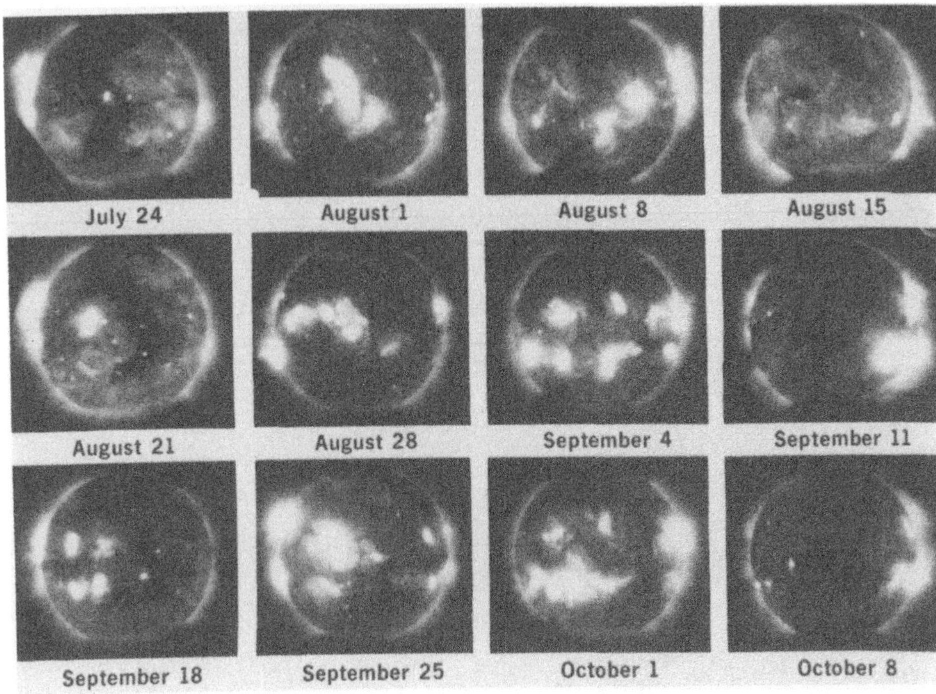

Fig. 3. The x-ray corona from Skylab: three consecutive solar rotations

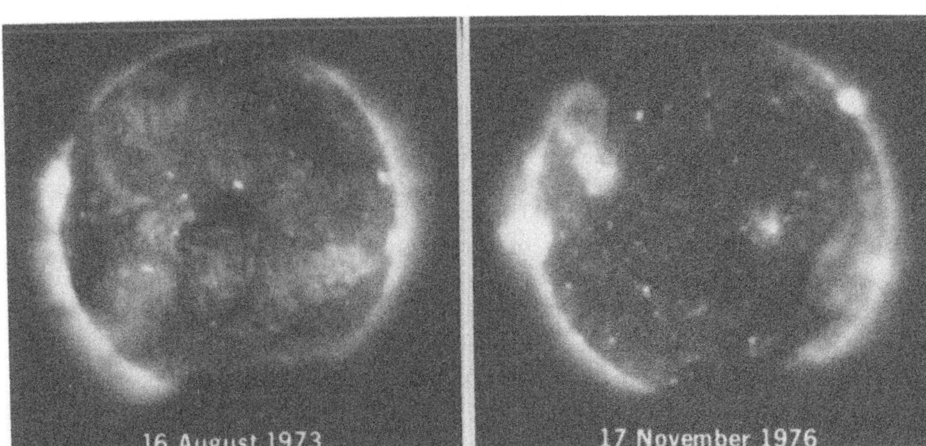

Fig. 4. Solar cycle variation of x-ray bright points: declining phase (1973) *vs.* minimum (1976).

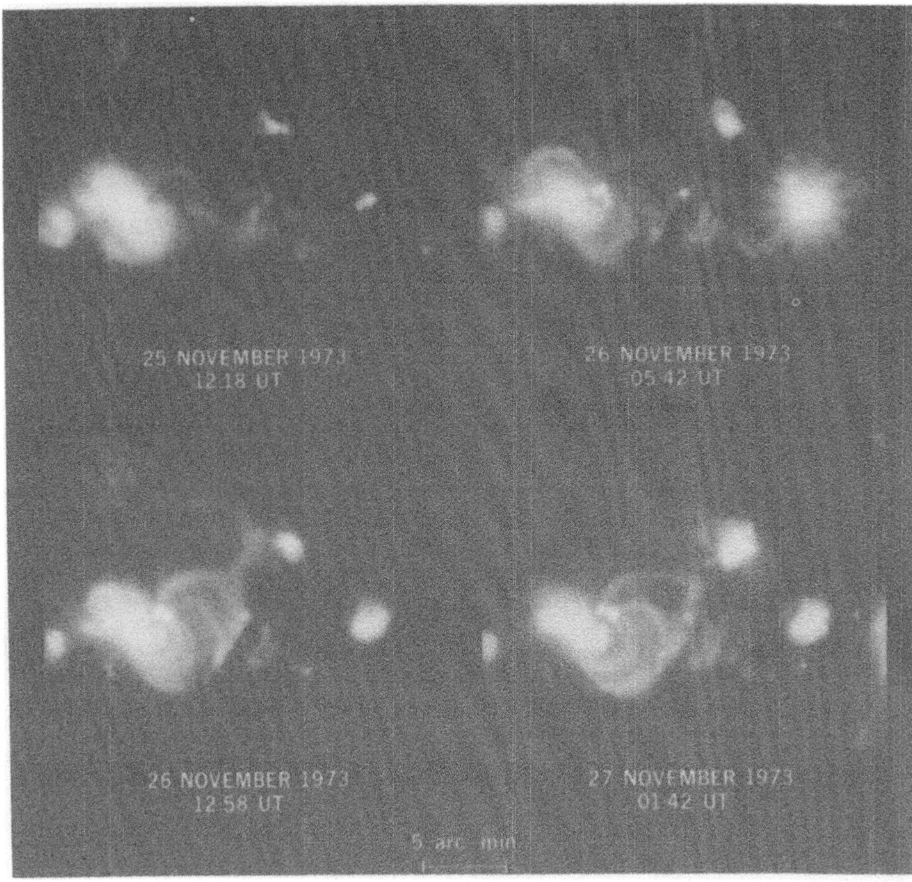

Fig. 5. Rapid evolution of the x-ray corona near active regions

photosphere did not reach the corona (Athay and White 1978, 1979), were both powerful incentives for a change in our views. Moreover, at just this time, the Einstein Obervatory was launched and it also played a key role in the changing of our view on the nature of coronal heating. The Einstein Observatory, for the first time, had sufficient sensitivity to detect ordinary stars - stars like the sun. What was seen was that, contrary to expectations, the sun was not situated near the high end of coronal heating but was in fact a rather weak X-ray emitter. This situation is described in figure 6a. In this figure, we indicate by a dashed line the prediction of X-ray emission from late-type stars as a function of spectral-type, based on accoustic heating theories.

Contrasted to this prediction are the actual observations. What was observed from Einstein was, first, that all spectral types from F through M were seen at about the same level of X-ray luminosity; and within each spectral type, there is observed a large range of X-ray luminosity. This can be summed up by saying that

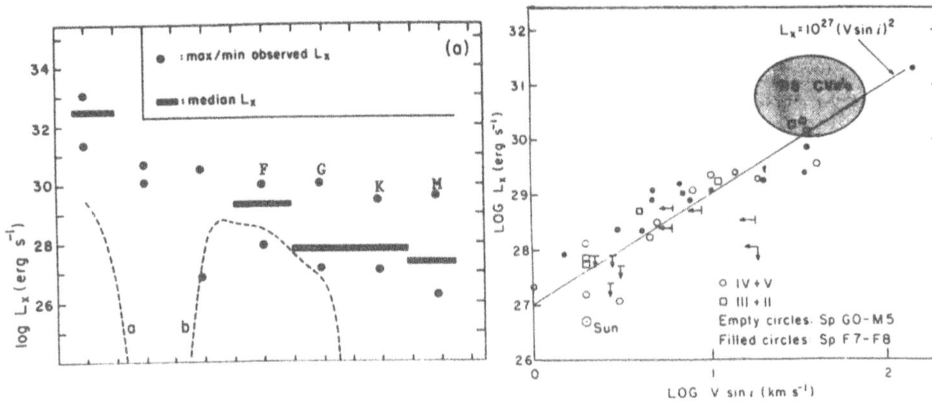

Fig. 6. a) Predicted *vs.* observed stellar x-ray emission levels for solar-type stars; b) correlation between x-ray luminosity and rotation speed for late-type stars.

spectral type is not at all a good indicator of coronal emission. The natural question to ask at this point is: does there exist a parameter which *is* a good predictor of X-ray luminosity? It turns out that, for normal late-type stars at least, there is such a parameter: stellar rotation rate. Figure 6b (from Pallavicini *et al.* 1981) shows a correlation for late-type dwarfs and giants between the X-ray luminosity and the (projected) rotation velocity v sin i. The correlation is quite strong and for dwarfs and giants alike holds over all spectral types F through M.

The correlation with rotation is, in a general sense, an argument for a connection between stellar X-ray emission and magnetic-field strength because the efficiency of stellar dynamos ought to be related to the rotation-rate of the star. As a way of testing this idea, we focused on a particular group of stars, namely those of spectral type late A and early F. The reason for doing this was that arguments based on studies of solar magnetic-field generation (Golub *et al.* 1981) indicated that magnetic-field production ought to be taking place at the bottom of the sun's outer convection zone. It seemed that this idea could be tested by looking at stars earlier than the sun, particularly those in which the convection zones were very shallow or almost non-existent. This work was done by Jurgen Schmitt as part of his thesis work while at Harvard. He examined the group of stars which we called 'those at the onset of convection'. The results of this study are best shown by comparing two figures. Figure 7a shows X-ray luminosity vs. rotation rate for this group of stars. For comparison, we have drawn the L_x vs. $v\ sin^2\ i$ curve onto the figure. It is clear that, for this group of stars, the correlation totally falls apart: the X-ray luminosity is well below the curve, which fits all of the other later-type stars very well.

What Schmitt did next was to model the convection zones of these stars so that the X-ray luminosity could be plotted against a different parameter, namely the Rossby number. This number is dimensionless and it is the ratio between the

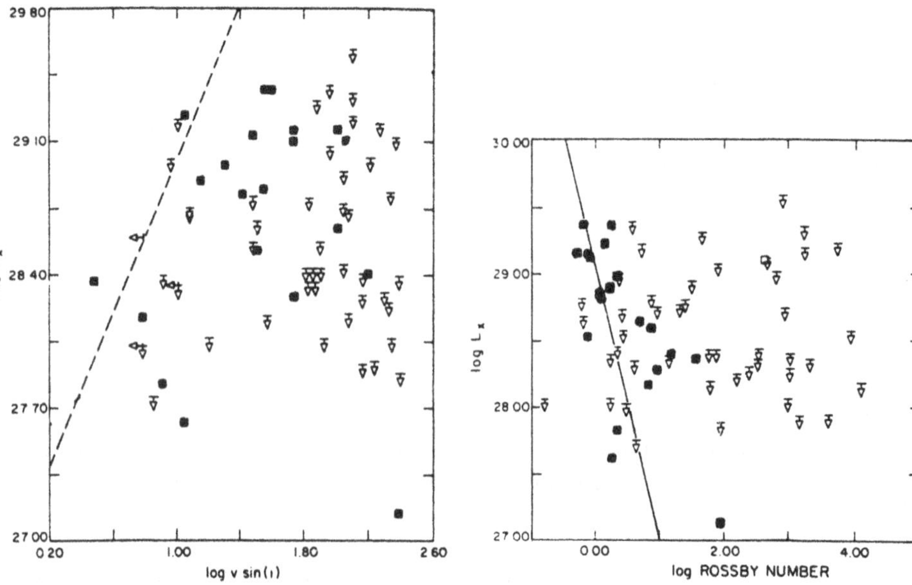

Fig. 7. a) X-ray luminosity *vs.* rotation rate for stars with shallow convection zones; b) Correlation with Rossby number for the same sample.

stellar rotation period and the convective turnover period for the star. This is in turn related to a number called the dynamo number, which, in a crude sense, is used to indicate the efficiency of magnetic field generation via the dynamo. Although the correlation should not really have been as good as it turned out, figure 7b shows quite clearly that plotting X-ray luminosity vs. Rossby number tightens up the correlation enormously for this group of stars. This gives further support to the idea that the X-ray luminosity of late-type stars is in fact closely related to the amount of magnetic flux generated in those stars. (Note that the line in fig. 7b is a simple linear fit between x-ray luminosity and dynamo number.)

Having done all of this work gives us good reason to believe that coronae and X-ray emission on stars and on the sun are intimately connected with the presence of magnetic fields on the surface of these stars. What it does not tell us, however, is *how* the magnetic fields are involved in producing the X-ray emission.

4. The NIXT Experiment

That brings us to the work which is going on now in the Normal Incidence X-ray Telescope (NIXT) program. The NIXT experiment uses multi-layered coated X-ray optics. These have been described in detail in the literature, for instance in Spiller *et al.* 1991. Briefly put, the way these mirrors work is by depositing an artificial Bragg crystal onto a figured surface such that the X-rays diffract back from the surface at normal incidence. This has the effect of focusing the X-rays. Because

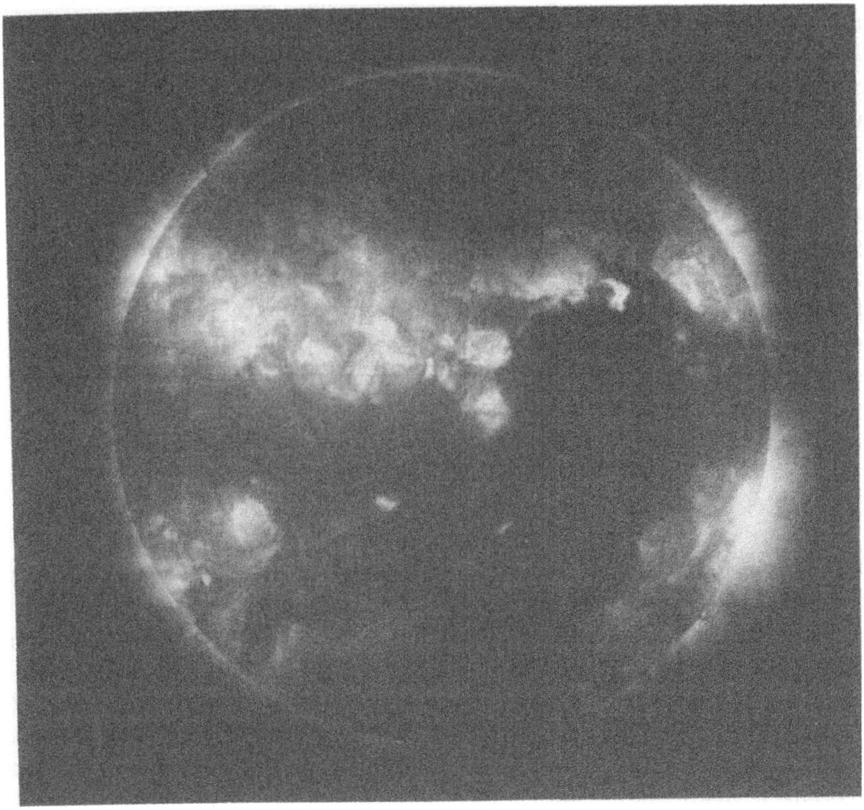

Fig. 8. NIXT high resolution x-ray image, 11 Sept. 1989

such mirrors are made by traditional optical fabrication techniques rather than by the more exotic techniques that are needed to produce grazing-incidence mirrors, the normal-incidence telescopes tend to have better figure quality, and to be more lightweight and less expensive than grazing-incidence devices. In addition, as we describe below, the multi-layer has the property that scattering is greatly reduced in comparison to reflections from a single surface. On the other hand, multi-layer X-ray optics have one major limitation: they do not work yet at short wave-lengths. This severely limits their usefulness in X-ray astronomy and, to some extent, in solar astronomy. The normal-incidence and grazing-incidence techniques should therefore be viewed as complementary, and both of them find their uses in X-ray astronomy. An example of the image quality which is obtained using multi-layered coatings is given in Figure 8, which is an image of the solar X-ray corona taken with the NIXT experiment on 11 September 1989.

Data from this flight have been analysed and discussed in detail in Golub *et al.* 1990. Briefly, what we can say here about this image is that you see fine structure in the corona at all locations and down to the resolution limit of the telescope, which was about 1 arcsecond. It would be surprising if the corona ran out of detail at just our telescope's resolution, so it is extremely likely that the fine structure of the corona extends down to much smaller scale sizes. In this figure, we also see that there is no hot X-ray emission above sunspot umbrae, and that the bright structures emanating from these strong field regions terminate in the sunspot penumbra. I will return to this subject in a moment.

One other feature of this photo is a flare occurring in the NW, about midway between sun center and the limb. The fact that this flare, which was 100 times brighter than any of the active regions in the photo, does not seem at all spectacular, is attributable to the extremely low scattering of the multi-layer coatings.

I will discuss two of the projects on which our group is working at the present time. These are, first, coronal heating, in which we will discuss the way in which the new high-resolution observations are affecting our ideas about the nature of the physical processes responsible for heating the corona; second, we will discuss rapid variability of coronal loop structures as seen during the NIXT rocket flights.

4.1 Heating of the Corona

In 1978, in Rosner *et al.*, we proposed a direct active involvement of the magnetic field in the coronal heating process. Since that time, there have been a number of new observational results which have implications for current-based coronal heating models. These are observations about the non-static coronal loops: Sheely and Golub 1978, Raymond and Foukal 1982 and, as I will discuss below, some of the very recent NIXT observations. The development of vector magnetograph instrumentation has also shown that our 1978 supposition that currents are present in the corona is justifiable by direct measurements of all three components of the magnetic field. Hagyard *et al.* 1984, Gary *et al.* 1987, for instance, argue that field-aligned coronal currents are measured. In addition, spectroscopic observations such as those by the SMM spacecraft and reported by Martens *et al.* 1985 argue that active region loops in the corona have a high temperature component, greater than 10^7 K, but occupying a very small fraction of the coronal volume. Finally, observations of microflares by Lin *et al.* 1984 and of microwave bursts seen at the VLA by Bastian 1991, combined with the above observations, argue that coronal heating proceeds via an intermittent, highly localized process, which, according to the NIXT observations, takes place in numerous long, thin coronal loops.

All of these considerations have been examined recently in a paper by Beaufume, Coppi and Golub 1992, in which we have reexamined our ideas from 1978, taking into account these new observational results. We conclude that coronal loops must be viewed as very dynamic features in which a number of processes, including heating, heat diffusion, cooling, evaporation of material from the chromosphere, all are taking place nearly simultaneously and with approximately equal time scales. Thus the view that coronal loops are static structures is almost certainly incorrect: they must be viewed as only quasi-steady at best, in response to episodic and repeated heating events.

We can further ask what is the lower limit to the transverse scale size which might be observed and which might be relevant to the coronal fine-structuring. This

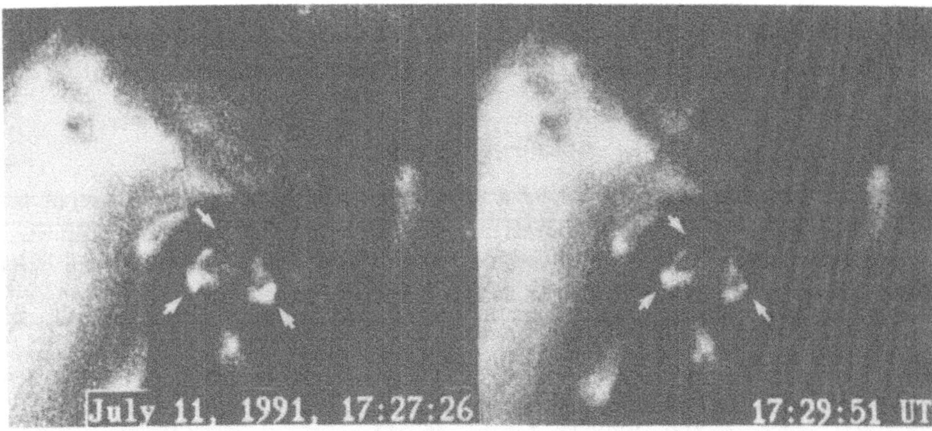

Fig. 9. Short-term variability of x-ray features. NIXT x-ray image taken 11 July 1991. Note that F.O.V. is 2X2 arcmin.

lower limit would be given by the skin-depth of current penetration due to magnetic diffusivity over the lifetime of the loop. Taking a lifetime of 10^4 and a magnetic diffusivity given by either the classical or an anomalous resistivity, we calculate a transverse scale size for coronal loops of less than 10^5 centimeters. This corresponds to .001 arcseconds, which is certainly well below the present resolution limit of X-ray telescopes, although not below the theoretical resolution of such telescopes.

4.2 Rapid Variability of Coronal Structures

As described earlier, the rapid variability of coronal structures was clearly evidenced by the *Skylab* results. Now, the dynamic nature of the corona is further emphasized by our most recent study from the NIXT rocket data, in which we have made blink comparisons between carefully co-aligned images taken at the start and the end of our rocket flights. Typically, this implies a time separation of four or five minutes. Figure 9 shows one such comparison: most of the loops do not change on this time scale, nor are they expected to, as discussed by Golub, Hartquist and Quillen in 1989. However, several of the smaller loops labelled in the figure do show changes and they are of a particularly unexpected type. What we observe are rapid brightening and fading events in small, bright x-ray structures surrounding sunspots. Comparisons between the x-ray data, H-alpha and magnetogram data (Golub and Zirin 1993) show that these intense x-ray emitting regions are located approximately one penumbral radius distance beyond the outer edge of the penumbra, and that these are strong magnetic field regions resembling satellite spots but quite clearly bipolar.

Time sequences of the ground-based data show that the magnetic fields in these regions are moving together, with the appearance of being 'pushed'. Moreover, the x-ray data show that the outer bipoles are connected inward toward the sunpot umbrae, giving the appearance of a continuous loop, with segments alternatively above and below the photosphere. Altogether, we seem to be observing one of the few situations on the Sun where there is the appearance of reconnection driven by an external flow.

Acknowledgements

The NIXT program is supported by a grant from NASA to the Smithsonian Institution (SI). Many of the collaborative programs herein discussed are supported by a grant from the NSF to SI. We would like to thank Anne Davenport and Ripa Rashid for help with the preparation of this manuscript.

References

Athay, R.G. and White, O.R.: 1978, *Ap. J.* **226**, 1135
Athay, R.G. and White, O.R.: 1979, *Ap. J. Suppl.* **39**, 333
Bastian, T.: 1991, *Ap. J. Lett.* **370**, L49
Beafume, P., Coppi, B. and Golub, L.: 1992, *Ap.J.* **393**, 396
Golub, L., Maxson, C.W., Rosner, R., Serio, S., Vaiana, G.S.: 1980, *Ap.J.* **238**, 343
Golub, L., Rosner, R., Vaiana, G.S., Weiss, N.O.: 1981, *Ap.J.* **243**, 309
Golub, L. *et al.*: 1990, *Nature* **344**, 842
Krieger, A.S., Timothy, A.F., Roeloff, E.C.: 1973, *Solar Phys* **29**, 505
Lin, R.P., *et al.*: 1984, *Ap. J.* **283**, 421
Pallavicini, R.P., Golub, L., Rosner, R., Vaiana, G.S., Ayres, T., Linsky, J.L.: 1981, *Ap.J.* **248**, 279
Rosner, R., Tucker, W.H., Vaiana, G.S.: 1978a, *Ap.J* **220**, 643
Rosner, R., Golub, L., Coppi, B. Vaiana, G.: 1978b, *Ap.J.* **222**, 317
Schmitt, J.H.M.M., Golub, L., Harnden, F.R., Jr., Maxson, C.W., Rosner, R., Vaiana, G.S.: 1985, *Ap.J.* **290**, 307
Sheeley, N.R. and Golub, L.: 1979, *Sol. Phys.* **63**, 119
Spiller, E. *et al.*: 1991, *Optical Eng.* **30**, 1109
Vaiana, G.S., Krieger, A.S. and Timothy, A.F.: 1973, *Sol. Phys* **32**, 81

HIGH RESOLUTION THERMALLY DIFFERENTIATED IMAGES OF THE CHROMOSPHERE AND CORONA

ARTHUR B.C. WALKER, JR.
Departments of Physics and of Applied Physics and Center for Space Science and Astrophysics,
Stanford University
Stanford, CA 94305

RICHARD B. HOOVER
Space Science Laboratory, Marshall Space Flight Center
Huntsville, AL 35812

and

TROY W. BARBEE, JR.
Lawrence Livermore National Laboratory
Livermore, CA 94558

Abstract. The development of optical coatings composed of multilayers engineered on atomic scales, which can efficiently and selectively reflect soft X-ray, EUV, and FUV radiation, has significant implications for the study of the solar atmosphere. The first high resolution X-ray images of an astronomical object (the solar corona) formed with normal incidence multilayer optics, were obtained in late 1987. We review the developments which have occurred in multilayer optics technology since 1987, and discuss the advantages that these developments present for solar observations. The most significant advantages of multilayer optics are (i) telescopes with modest apertures ($\sim$ 0.1-0.5 meters) can achieve images with very high ($\sim$ 0.1-0.3 arc seconds) resolution and (ii) the spectral selectivity of multilayers permits the investigation of thermal structures with resolution $T/\Delta T \sim$ 5-10. We describe the analysis of polar plumes observed in 1987 and of small X-ray emitting regions called "bright points" observed in 1991 to illustrate the power of multilayer optics.

Key words: multilayer coatings – X-ray solar images – bright points – solar corona

1. Introduction

The application of the grazing incidence optical configurations conceived by Wolter (1952) to the study of astronomical sources of x-ray emission has had a profound impact on astronomy. It is not yet possible to assess the impact of multilayer optics on astronomy; however, it is clear that these normal incidence optical systems have significant advantages for diagnostic studies of the solar chromosphere and corona. Optical coatings for the efficient reflection of soft x-ray, EUV and FUV radiation at normal incidence were developed in the period from 1971-1981, primarily as a result of the independent work of E. Spiller and T.W. Barbee, Jr. and their collaborators. Spiller (1973; 1992) first described the development of multilayer interference coating for the VUV and FUV in 1971, and fabricated the first structure for use at VUV wavelengths in 1972. Haélbich and Kunz (1976) reported the first experimental results on EUV multilayers in 1976. Techniques for the fabrication of multilayer structures with the

J.F. Linsky and S. Serio (eds.), Physics of Solar and Stellar Coronae, 83–96.

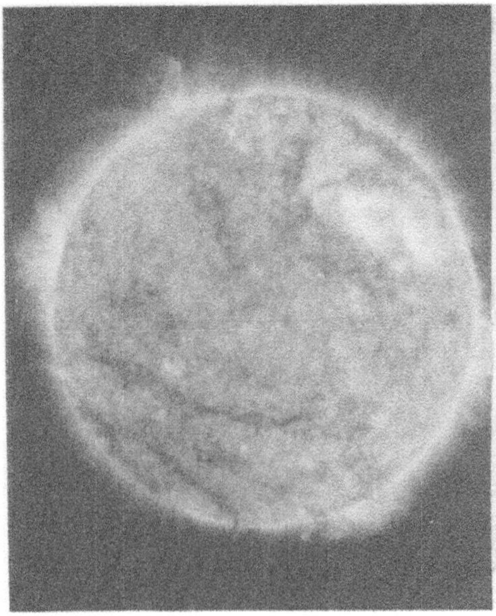

Fig. 1a. The solar corona photographed in the emission of the resonance lines of Fe IX at 171 Å and Fe X at 174.5 Å on 23 October, 1987 at 1809 UT. The image is dominated by material at ~ 1,000,000 K. Note that Figures 1a and 1b are reversed left to right (*i.e.* "mirror images") with respect to one another.

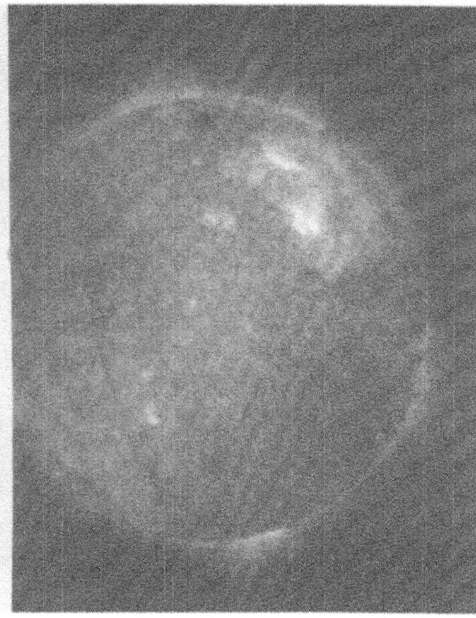

Fig. 1b. The solar corona as photographed on 23 October, 1987 at 1809 UT in a bandpass centered at ~ 256 Å, which contain lines of He II (~ 50,000 K) and Fe XIV (~ 2,500,000 K). The telescope was a single reflection Herschellian. The resolution of the image is limited to ~ 4 arc seconds by the film used.

perfection on atomic scales necessary for efficient reflection of EUV and soft x-ray radiation were first developed by Barbee and Keith (1978), and by Haélbich, Segmuller and Spiller (1979). In 1981, Spiller (1981) and Barbee (1981) demonstrated that multilayer structures were a viable tool for soft x-ray optics. The first successful x-ray/EUV imaging experiments with multilayer optics were performed in the laboratory in 1981 (Underwood and Barbee, 1981; Henry, Spiller and Weisskopf, 1982), and several groups began the development of multilayer astronomical telescopes in the early 1980's (see Walker *et al.*, 1988a,b, and Catura and Golub 1988, for a review of this early work). Underwood *et al.* (1987) made the first astronomical observation with a multilayer telescope, obtaining an image of an active region in the corona with modest resolution (~ 10 - 15 arc seconds) at 44 Å, corresponding to Si XII (T ~ 2,000,000 K) emission. The first high resolution multilayer images of the sun were obtained by Walker *et al.* (1988a), and are shown in Figures 1a and 1b. Figure 1a, was obtained with a pseudo-Cassegrain (*i.e.* spherical rather than conic mirrors) telescope of 62.5 mm aperture which was coated to reflect radiation in a 12 Å bandpass centered at 173 Å; the solar image has a resolution of ~ 1.5 arc seconds. This result, dramatically demonstrated the power of multilayer optical systems for astronomical research, and spurred the application of multilayer optics to other disciplines, such as biology and lithography. In 1989, Golub *et al.* (1990) obtained an image of the solar corona in a bandpass centered at 63.5 Å which contains strong lines of Fe XVI and Mg X, using a large Herschellian mirror of 254 mm aperture, demonstrating the effectiveness of the multilayer technology at soft x-ray wavelengths. At the time of this publication, multilayer optical systems have obtained high resolution images of the solar corona in nine band passes in the wavelength range 44 Å $\leq \lambda \leq$ 304 Å (Walker *et al.*, 1992a).

The advantages of multilayer optics for solar observations can be summarized as follows: *(i)* Compared to grazing incidence optics, normal incidence multilayer optics display lower scatter, suffer less from geometrical aberrations, are less expensive to fabricate and easier to mount and align, and can be made more compact (by folding the optical path). Consequently, it is possible to approach the resolution limit imposed by diffraction for such systems. The combination of filled apertures, and high reflection efficiencies provide high sensitivity optical systems in a compact package. *(ii)* The compact nature and low cost of multilayer telescopes permits the use of multiple telescopes, and therefore the observation of plasma over a broad range of temperatures. *(iii)* The spectral resolving capability of multilayer mirrors (typically $\lambda/\Delta\lambda \sim$ 10-100), sufficient in most cases to isolate a single line multiplet, is a powerful analytical tool, permitting observations with temperature specificity over the entire range of plasma temperatures in the solar atmosphere, 4,500 K < T < 30,000,000 K. *(iv)* Multilayer gratings offer the prospect of attaining high spectral resolution ($\lambda/\Delta\lambda \sim$ 10,000) stigmatic images of the solar atmosphere in the spectral range 10 Å < λ < 400 Å.

2. Multilayer Optics

The concept of multilayer and multilayer interference coated optics is deceptively simple; materials with contrasting optical properties are stacked, with Angstrom scale precision, on a suitably figured substrate so that reflections from successive interfaces are coherently added to form a strong reflected beam. The wavelength of the reflected light is determined by the period of the multilayer structure (typically 20 Å - 1000 Å, permitting radiation of wavelength 40 Å < λ < 2000 Å to be reflected). The properties of the reflected light such as intensity, bandpass and image quality are, however, strongly dependent on a number of factors, including the relative thicknesses and absorption coefficients of the materials, the uniformity of the layers and the nature of the interfaces, the stability of the multilayer structure, and the smoothness of the substrate. Furthermore, the performance of a multilayer optical system is strongly dependent on the use of filters to exclude long wavelength light which is specularly reflected by the multilayer, and high resolution detectors to record the image. Consequently, fundamental advancements in four areas of technology underlie the results that have been achieved and that are anticipated in the near future in imaging XUV [we refer to the soft x-ray and EUV spectral regions (~ 1 A to 1000 Å) as the XUV] and FUV (~ 1000 Å to 2000 Å) radiation; *(i)* optical coatings for the XUV and FUV, *(ii)* ultra smooth mirror substrates, *(iii)* XUV and FUV filters, *(iv)* high resolution XUV/FUV sensitive photographic emulsions.

2.1. Ultrasmooth Mirror Substrates

The surface quality of mirror substrates for XUV and FUV mirrors is of critical importance for two reasons: *(i)* the efficiency of multilayer coatings is dependent on substrate surface quality {this dependence becomes critical for wavelengths below ~ 80 Å, since reflectivity ε decreases as $\varepsilon = \varepsilon_T \exp[-(2\pi\sigma/d)^2]$ where σ is the RMS roughness of the substrate, d is the multilayer lattice constant, and ε_T is the theoretical efficiency for an ideal multilayer}, and *(ii)* scattering from surface inhomogeneities decreases image contrast and degrades image quality. Baker (1989) has shown that figured optical surfaces of very high quality [with figures accurate to λ/100 when tested at ~ 6300 Å (Hoover *et al.*, 1990a,b)] can be polished to have a surface smoothness of less that 2 Å RMS in Zerodur substrates and less that 0.5 Å in sapphire substrates, by using a flow polishing technique. These substrates are of sufficient quality to permit images to be formed at the level of 0.1 arc second resolution.

2.2. Reflective Coatings for Soft X-Rays and the EUV

The development of multilayer coating technology (Barbee, 1990) for the XUV has now progressed to the point that the reflection efficiencies of optical coatings for the wavelength regions ~ $100 < \lambda < 350$ Å are approaching the levels predicted for "perfect" multilayer structures. Measured reflectivities (Hoover *et al.*, 1990a; Barbee *et al.*, 1991), typically ~ 50% (at ~ 150 Å) to ~ 25% (at ~ 300 Å), are ~ 70% - 80% of the reflectivity for ideal multilayer structures. The reflectivity of contemporary multilayers at wavelengths below ~ 70 Å , typically 10% or less, is too low to permit the use of double reflection telescopes for the weak solar lines in this spectral range. The single reflection Herschellian configuration has been used for the multilayer telescopes in this wavelength interval. However, significant progress has been made in perfecting the interfaces between layers to improve reflection efficiency (Barbee, 1990) for these short period multilayers. The fractional bandpass ($\delta\lambda/\lambda$) of a multilayer mirror can range between ~ 1% and 20%, allowing, within limits, a mirror to be specifically designed for a particular application (*i.e.* narrow band imaging, providing a broadband image for a spectrometer, etc.). An important development for astronomical applications is the fabrication of multilayer structures with low contrast between the refractive indices of the materials forming a layer pair, permitting many layers to contribute to the reflected beam, resulting in improved spectral resolution (Barbee, 1990). For the long wavelength EUV (~ 400 Å $< \lambda <$ 1100 Å) Haas and Hunter (1973) and Windt *et al.* (1988a) have shown that heavy metals such as iridium and osmium are reasonably efficient (~ 30%) reflectors. Silicon carbide (Mrowka *et al.*, 1986; Wendt *et al.*, 1988b) becomes an efficient reflector at wavelengths between 600 Å and 2000 Å.

2.3. Reflective Coatings for the FUV and VUV

Aluminum surfaces overcoated with MgF_2 are efficient reflectors in the FUV (Haas and Hunter, 1973). Spiller (1973) and Zukic *et al.* (1990) describe the properties of interference films which can result in enhanced efficiency and spectral selectivity in the reflection of radiation in the FUV (1000 Å $< \lambda <$ 2000 Å) and VUV (2000 Å $< \lambda <$ 3000Å). Interference films can also be used as filters. These films are commercially available, and can be fabricated with bandpasses between ~100 Å and 500 Å wide. For example, we have obtained coated Ritchey-Chrétien Optics from Acton Research Inc. with efficiencies for two reflections of 25% at 1216 Å, and 60% at 1550 Å.

2.4. Filters

For stellar sources of soft x-ray, EUV and FUV radiation such as the sun, the intensity of ultraviolet, visible, and infrared radiation exceeds the intensity of the short wavelength radiations by factors between 100 and 1,000,000. For multilayer optical systems, two multilayer reflections are able to discriminate against nearby offband radiation by factors of ~ 100 - 300, consequently if the objective is to obtain an image of structures as seen in a strong solar line such as He II λ 304 Å , Fe IX λ 171 Å, etc., then we only need to be concerned with "leakage" from distant XUV lines, and from UV and visible radiation.

Unfortunately, multilayer coatings are reasonably efficient (~ 50%) reflectors for UV and visible light. If a polychromatic detector such as film is used, it is necessary to attenuate off-band radiation by the use of filters. Rejection of visible light by a factor of ~ 10^{10} must be achieved. This is generally accomplished by the use of thin (typically ~ 1000 Å - 2000 Å thick) metallic (*i.e.* aluminum or beryllium) or metal coated plastic films supported by nickel mesh; other materials are used for additional attenuation at specific wavelengths. This is a challenging aspect of multilayer imaging technology; Lindblom *et al.* (1991), Powell (1989) and Spiller *et al.* (1990a) discuss filters in detail.

2.5. Photographic Emulsions

All of the high resolution multilayer images obtained so far have been recorded on XUV sensitive photographic emulsions prepared by Kodak. Hoover *et al.* (1990c) have carried out detailed measurements on the 2 emulsions which we have used, XUV 100, which has a resolving power of 200 *lines/mm* (i.e. 2.5 micron "pixels") and the XUV 649 emulsion, which has a resolution of 2000 *lines/mm* (i.e. 0.25 micron "pixels"). We present images obtained with both films in this paper. We have also investigated several other emulsions; Agfa 8E 56, 10E 56, 8E 75, and 10E 75, which appear to have significant advantages for the FUV, and Kodak 2415, which appears to have significant advantages for the soft x-ray region (Hoover *et al.*, 1992).

2.6. Multilayer Gratings

Spectroscopic analysis in the XUV and EUV is now possible using stigmatic instrument configurations previously confined to the visible, ultraviolet and far-ultraviolet, as a result of the development of multilayer gratings. Barbee (1989) successfully fabricated and tested the first multilayer grating; he has shown that these combined microstructure optical elements disperse wavelengths Bragg-diffracted by the multilayer structure so that all constructive interference occurs at essentially constant angles relative to the zero order Bragg-diffracted beam. Bixler, Barbee and Dietrich (1989) have tested multilayer gratings operating at ~ 200 Å, and have demonstrated resolution as high as $\lambda/\Delta\lambda$ ~ 2000 at normal incidence. These very exciting results demonstrate that *it is now possible to fabricate multilayer gratings with properties which are tailored to the needs of high resolution solar XUV spectroscopic observations.*

2.7. Optical Configurations

In a previous review (Walker *et al.*, 1990a) we discussed optical configurations for multilayer telescopes; more recently we analyzed the factors governing the resolution of multilayer telescopes in depth (Walker *et al.*, 1991). The effective use of conventional optical configurations, such as the Ritchey-Chrétien, the Herschellian, and the off-axis parabola has already been demonstrated by the sub-arc second quality images presented in Section 3. We have also demonstrated the use of a multilayer tertiary mirror in conjunction with a Wolter I primary (Walker *et al.*, 1988b). The use of a multilayer mirror as an imaging Bragg crystal in a Johann curved crystal spectrograph has been demonstrated in the laboratory (Walker, *et al.*, 1990a). Walker *et al.* (1990b) have discussed the design of multilayer coronagraphs. Keski-Kuha, Thomas and Davila (1992) have obtained a high resolution solar spectrum using a multilayer coated grating in the Rowland circle geometry, and Walker, Barbee, and Hoover (1992) are planning to fly an objective grating to obtain spectrally dispersed stigmatic solar images. *In summary, all of the optical configurations discussed by Walker et al. (1990a) have been successfully used in solar observations, or have been studied in depth.*

2.8. Image Quality

Wetherell and Rimmer (1972) have pointed out that the use of aplanatic mirror configurations such as the Ritchey-Chrétien, can permit resolutions of 0.05 - 0.1 arc-second to be achieved over fields of several arc minutes. For the sun, this resolving power corresponds to ~ 35-70 kilometers, which approximates the photon meanfree path in the lower chromosphere. To achieve this resolving power at 5000 Å , apertures of 1.25 - 2.50 meters are required, at H Lyman α (1215.6 Å), apertures of 30-60 cm are sufficient, while at He II Lyman α (304 Å) apertures of 7.5 - 15 cm will suffice. Because they permit observations at FUV, EUV and soft x-ray wavelengths, multilayer optical systems offer a distinct advantage for high resolution observations. Walker *et al.* (1991a) have analyzed the factors that determine the quality of the image formed by a 127 mm aperture multilayer telescope, based on the measured performance of the *MSSTA* rocket payload He II (λ 304 Å) Ritchey-Chrétien telescope (Hoover *et al.*, 1990a,b). The factors considered included *(i)* geometrical aberrations, *(ii)* diffraction, *(iii)* aberrations due to mirror distortion, decentration, tilt, and slope errors, and other imperfections in manufacture and assembly, *(iv)* scattering by surface imperfections, *(v)* defocusing due to errors in primary/secondary separation, *(vi)* image blurring due to finite film or detector resolution, and *(vii)* scattering and diffraction due to the various filters (and supporting meshes) used. The net resolution due to these factors is ~ 0.08 - 0.09 arc seconds. If image blurring due to target motion and spacecraft jitter can be held to ~ 0.03 arc seconds, then a net resolution of 0.1 arc second can be achieved. Can this level of resolution be achieved for chromospheric and coronal observations in the near future? Since the indefinite postponement by NASA of two major NASA solar space borne observatory programs, the *Orbiting Solar Laboratory, OSL* (Title, 1989) and the *Ultra High Resolution XUV Spectroheliograph, UHRXS* (Walker *et al.*, 1990c), the prospects for the launch of a free flying observatory with this capability before the year 2000 are dim. Walker *et al.* (1991) and Spiller *et al.* (1990b) have discussed the prospects for high resolution imaging on rockets. Unfortunately, the present rocket aspect system, SPARCS, is likely to limit resolution to ~ 0.3 arc seconds. However, the use of an actively controlled secondary minor may permit the limitations of the SPARCS to be overcome (Walker *et al* 1991). Walker *et al.* (1992b) point out that a balloon borne VUV telescope, designed to observe the chromosphere in the light of Mg II *h* and *k* (λ 2795.4/ 2802.3 Å) could, with the use of an actively controlled secondary mirror, achieve 0.1 arc second resolution. A telescope with ~ 750 mm aperture is required.

3. Solar Observations with Multilayer Optics

Progress on basic problems relating to the structure and dynamics of the solar chromosphere, corona, and corona/solar wind interface has been limited by the quality of the available soft x-ray and EUV data. In the past, instruments designed to obtain such data have been forced to compromise on three or four of the five primary goals of the observational astronomer [spectral resolution, spatial resolution, temporal resolution, field of view, and temperature determination (achieved by simultaneous observations in several lines or wavelength bands)] to concentrate on the remaining one or two goals. This limitation is due in large measure to the fact that the reflection efficiency of conventional mirrors at normal incidence for EUV and soft x-ray radiation is low, therefore precluding the use of the powerful techniques developed for visible and ultraviolet spectroscopy. The techniques that were available, such as grazing-incidence optics and mechanically collimated Bragg spectrometers, have achieved important but limited success. The images in Figure 1 have demonstrated the power of normal incidence multilayer optics to achieve high angular resolution astronomical images with moderate spectral resolution ($\lambda/\Delta\lambda \sim 30$ to 100) for soft x-rays and the EUV.

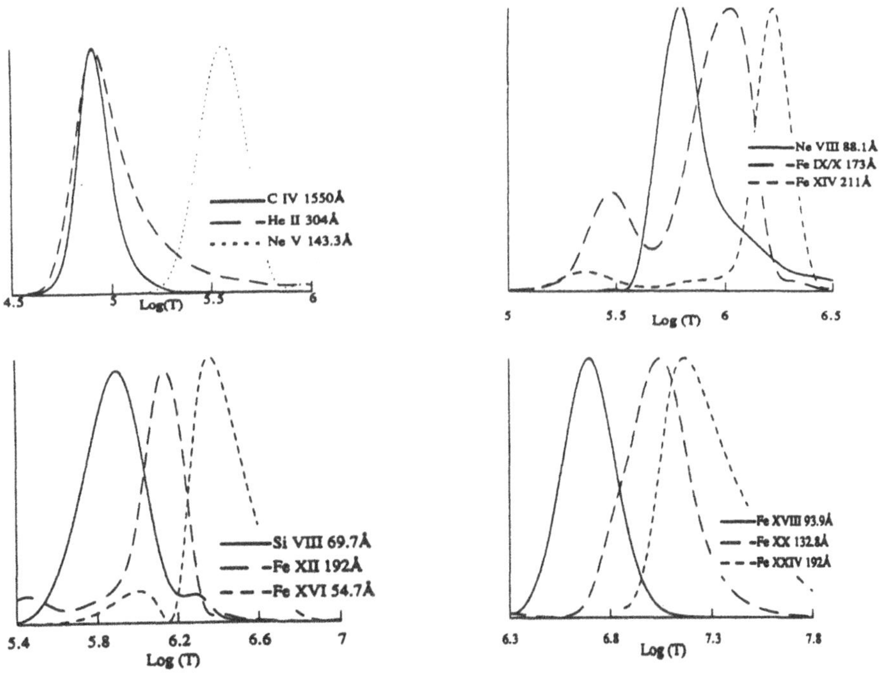

Fig. 2. Calculated temperature response of a set of multilayer telescopes to the solar atmosphere. The response of the λ 1216 Å telescope to the optically think H Lyman α line, which is strongly dependent on plasma geometry, is not shown.

3.1. A Comprehensive Solar Chromospheric/Coronal Observatory

Based on these capabilities we decided to develop a rocket borne solar observatory designed to achieve high angular resolution temperature specific images which would permit us to delineate the structure and dynamics of the solar atmosphere, from the chromosphere (7,000 K), through the hottest structures present in the corona (30,000,000 K). This instrument (Walker *et al.*, 1990d, 1992a), the *Multi Spectral Solar Telescope Array (MSSTA)*, utilizes five 127 mm aperture multilayer Ritchey-Chrétien telescopes and two 63.5 mm Cassegrain telescopes covering the spectral range from ~ 150 Å to ~ 350 Å and as many as six multilayer Herschellian telescopes covering the spectral range 40 Å < λ < 150 Å. Each telescope is able to isolate line multiplets excited over a narrow temperature range (Figure 2), providing full disk images of diagnostic quality arising from structures in the solar atmosphere ranging in temperature (T) from T ~ 50,000 K (He II) to 10,000,000 K (Fe XX). The soft x-ray and EUV images are supplemented by full disk high resolution far-ultraviolet (FUV) images in H I Ly-α (λ ~ 1216 Å) and C IV (λ ~ 1548/1550 Å) that are obtained by two Ritchey-Chrétien telescopes of the same design as the XUV telescopes. The first flight of the *MSSTA* payload occurred on 13 May, 1991. The resulting data sets are intended to address several fundamental problems related to the following solar phenomena: *(i)* the morphology and energetics of the fine structure of the solar chromosphere/corona interface, including the "chromospheric network," spicules, prominences, cool loops, and the magnetic field. *(ii)* the structure, energetics, and evolution of high temperature coronal loops. *(iii)* The large scale structure and dynamics of the corona, including the solar wind interface (represented by phenomena such as polar plumes). *(iv)* The structure and evolution of hot flare generated coronal loops.

3.2. RECENT MULTILAYER OBSERVATIONS OF THE SUN

In Figures 3b-d we present some of the results of the 13 May, 1991 flight of the *MSSTA* payload. Figure 3a represents the configuration of the solar magnetic field. Figure 3b is an image of the chromosphere obtained with the *MSSTA* H Lyman α telescope, and Figure 3c and 3d are images of the corona at ~ 1,500,000 K (Fe XII) and ~ 2,500,000 K (Fe XIV) obtained with two of the *MSSTA* XUV telescopes. Figure 3b was obtained with a 127 mm aperture Ritchey-Chrétien telescope, Figure 3c was obtained with a 75 mm aperture Herschellian telescope, and Figure 3d was obtained with a 62.5 mm aperture pseudo-Cassegrain telescope. Both 3b and 3c display sub-arc-second resolution (~ 0.7 arc seconds); Figure 3d is limited to ~ 1.2 arc seconds resolution. A number of coronal features are visible in the XUV images, including bright points, coronal holes, polar plumes, active region loops, and filaments. The chromospheric network is clearly seen in Figure 3b. The dominant influence of the magnetic field (Figure 3a) is apparent in both the chromospheric and coronal structures.

4. Analysis of Multilayer Observations

4.1. THE ANALYSIS OF THE POLAR PLUMES OBSERVED WITH MULTILAYER OTICS

To demonstrate the power of high resolution multilayer XUV images for the analysis of the coronal plasma, Walker *et al.* (1992c) have analyzed the density and temperature in two of the polar plumes observed in Figure 1a. Polar plumes are believed to be the major source of the high-speed solar wind streams associated with coronal holes. Withbroe Feldman and Ahluwalia (1991) have reviewed models of polar plumes and pointed out the importance of high-resolution (~ 1 arc second) observations of polar plumes to an understanding of the physics of coronal holes. Previous observations of polar plumes (on SKYLAB) in the soft x-ray and EUV have been limited to distances of 0.4 solar radii or less above the limb and resolution ~ 3 - 5 arc seconds. Our observations extend to 0.6 solar radii and have ~ 1.5 arc second resolution. Walker *et al.* (1992c) have solved the following equations to develop a dynamic isothermal model of the plumes and interplume regions observed in Figure 1a.

Conservation of Mass:

$$\frac{d}{dr}\left(r^2 \rho v_r\right) = 0$$

Conservation of Momentum:

$$\frac{dv_r}{dt} = -v_r\frac{dv_r}{dr} - \frac{1}{\rho}\frac{dP}{dr} - \frac{GM_\cdot}{r^2} = 0$$

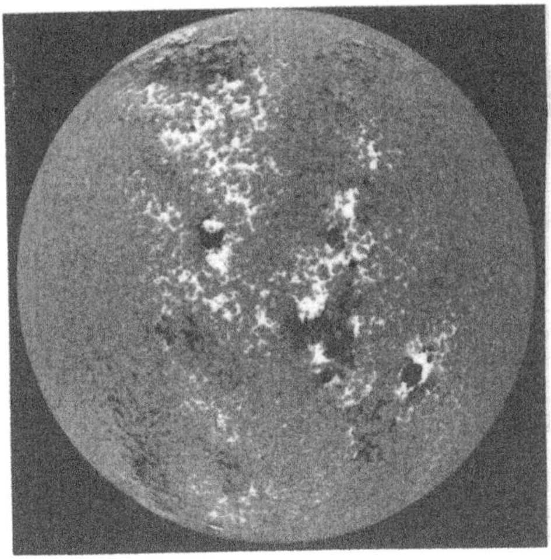

Fig. 3a. National Solar Observatory magnetogram for 13 May, 1991 (Courtesy of Jack Harvey). The north polar axis is oriented toward the left.

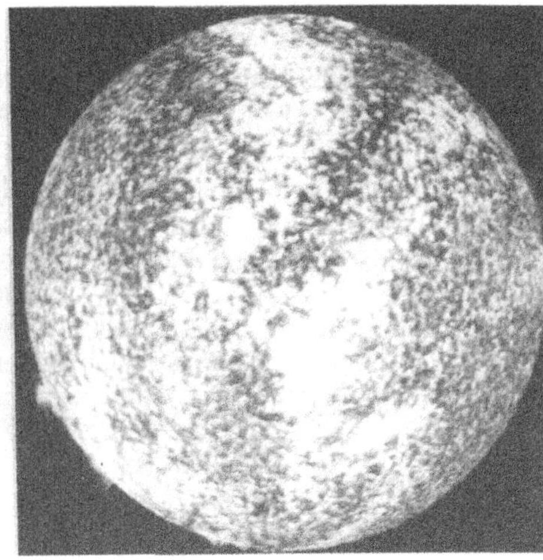

Fig. 3b. Image of the Chromosphere in H Lyman α at 1907 UT on 13 May, 1991. The image is recorded on 649 film.

Fig. 3c. Image of the corona at λ 193 Å (Fe XII) at 1907 UT on 13 May, 1991. The material imaged is at ~ 1,500,000 K. The image is recorded on 649 film.

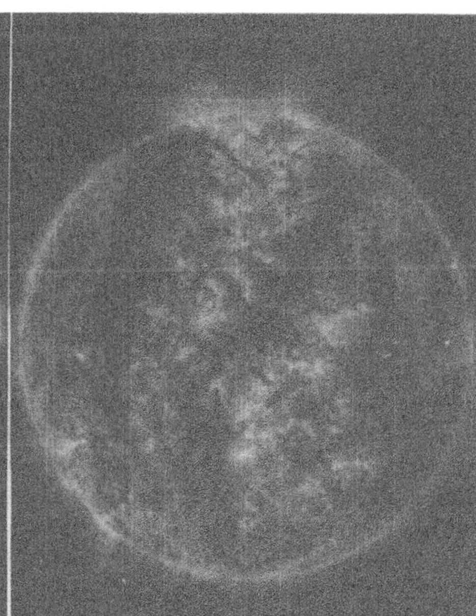

Fig. 3d. Image of the corona at λ 211 Å (Fe XIV) at 1907 UT on 13 May, 1991. The material imaged is at ~ 2,500,000 K. The image is recorded on XUV 100 film.

Ideal Gas Law: $$P = \frac{\rho k T_0}{\mu}$$

The symbols are defined as follows, r is distance from the sun's center, ρ is plasma density, v_r is flow velocity, P is gas pressure, G is the gravitational constant, M is the solar mass, k is Boltzman's constant, T_0 is the temperature of the plasma at the solar limit (by virtue of our assumption of an isothermal atmosphere T_0 is the temperature at all heights), and μ is the mean particle mass. Walker *et al* modeled the plumes as miniature helmet streamers, with a constant cross section embedded in a lower density coronal region of open field geometry, the interplume region. The mean electron density can then be derived from the coronal excitation equation, (Walker, Rugge and Weiss, 1974), which expresses the intensity in any emission line as

$$I_{ij} = \frac{hc}{\lambda_{ij}} a_H A_Z \int_{\Delta x} dx \int_{\Delta y} dy \int_{-\infty}^{\infty} ds\, a_{Zz}(T_e)\, a_{ji}(T_e)\, n_e^2(T_e)$$

where s is the coordinate along the line of sight. This equation is based on the assumption that the observed emission is due entirely to collisional excitation (given by the temperature-averaged excitation function a_{ji}) from the ground level (j) to the excited level (i), and that the ionization temperature is equal to the electron temperature. The quantities A_Z (abundance of element Z), a_{Zz} (population of ionization stage z of element Z), a_H (number of hydrogen atoms per electron), and the excitation functions are taken from Mewe, Gronenschild, and van den Oord (1985). The strongest lines in the bandpass are Fe IX (171 Å), and Fe X (174.5 Å), however, all lines in the bandpass are included in the computation. The film density at position x,y is then given by

$$D(x, y, \langle n_e^2 \rangle) = \frac{1}{4\pi (A.U.)^2} \int_{\Delta x} dx \int_{\Delta y} dy \int_{a/2}^{A/2} dr\, 2\pi r\, V(r,x,y) \times \Sigma_{ij} \int_0^{\infty} dT\, I_{ij}\left[n_e^2(T)\right]$$

where the integral over s has been replaced by an integral over T, and the dependence of the emissivity I_{ij} on the emission measure n_e (T) has been made explicit. For the plume regions, the integral will be dominated by the plumes, rather than by the interplume material in front of and behind them. Walker *et al.* (1992c) assumed that each plume has a constant density over its cross section. The results of the analysis are shown in Figure 4 and Table 1. The densities derived by Walker *et al* (1992c) are in good agreement with previous results (Ahmed and Webb, 1978; Ahmed and Withbroe, 1977, Bohlin, Shultz and Tousey, 1978). Clearly, multilayer images provide a powerful technique for the study of large scale coronal structure.

4.2. Coronal Bright Points and the Chromosphere/Corona Interface

Figures 3b and 3c provide simultaneous high resolution observations of the corona and the chromosphere. The Fe XII emission, which is responsible for the image in Figure 3c, corresponds to coronal material in the temperature range ~ 1,000,000 K to ~ 2,500,000 K. We have identified more than 30 of the small bright features called bright

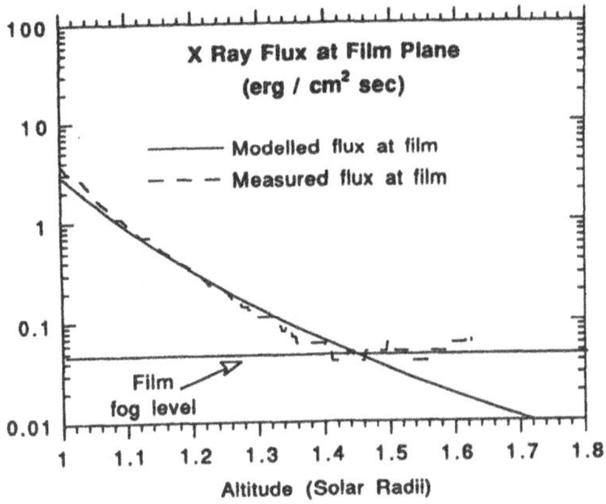

Fig. 4a. The predicted and observed emission from a typical polar plume observed in Figure 1a, based on an isothermal model with solar wind flow included. (From Walker *et al.*, 1992c)

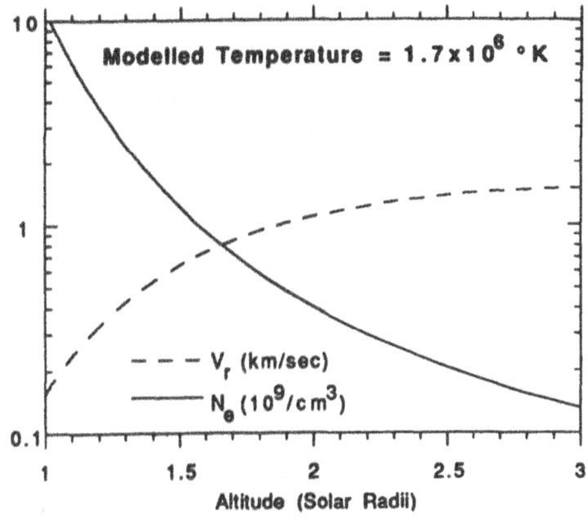

Fig. 4b. Density and solar wind flow velocity versus height for a typical polar plume observed in Figure 1a, derived from an isothermal model with solar wind flow included. (From Walker *et al.*, 1992c)

TABLE 1

Temperature and density of polar plumes and interplume regions.

Plume ID	Temp (K)	ρ at limb (cm^{-3})	ρ at 1.5 R_O
1S	1.7×10^6	11×10^9	12×10^8
2S	1.9×10^6	1.7×10^9	1.7×10^8
3S**	1.5×10^6	1×10^9	0.8×10^8
4S	1.6×10^6	11×10^9	11×10^8
6S	0.949×10^6	1.1×10^9	$.8 \times 10^8$
7S	1.6×10^6	12×10^9	12×10^8

**interplume region

points in this image. It has been widely assumed that these emission features are small loops, however they have not been resolved in previous images. We have identified several bright points which do indeed appear to be resolved into small loops ~ 10 - 20 arc seconds across (Figure 5a). One of these features also appears to coincide with a pair of bright structures in the H Lyman α image which we identify as the foot points of the loop. Walker and Patience (Patience, 1992) have used the numerical models of Vesecky, Antiochos, and Underwood (1979) to model this bright point/loop. The model of Vesecky *et al.* solves the equations of energy conservation, momentum, and hydrostatic equilibrium, assuming a uniform energy input to the loop. The energy loss processes are radiation and conduction. Walker and Patience have compared the emission of the bright point with the loop model of Vesecky *et al* with an energy input of 10^{-3} ergs/cm^3. The model predicts an average energy flux of ~ 0.0445 ergs/cm^2-sec at the film plane for a typical "bright point" loop. Our measured typical bright point loop flux is 0.045 ergs/cm^2-sec, in good agreement with the model. The model of Vesecky *et al* was constrained to have zero conductive flux at a chromospheric temperature of 30,000 K, consequently predicting a high chromospheric luminosity at ~ 30,000 K - 50,000 K. Avrett (1991) has recently carried out a detailed calculation of the structure of the chromosphere, including radiative transport, and particle diffusion, assuming that the chromosphere is heated entirely by conduction from the corona. Avrett's model predicts that most of the chromospheric energy loss (Lyman α is the main radiative loss mechanism for the chromosphere) occurs at T~ 40,000 K-60,000 K. The agreement between the two models may be fortuitous, since Vesecky *et al.* use an optically thin model, however *both* models predict a H Lyman α luminosity which is within a factor of 2 of the luminosity we observe from Figure 5b.

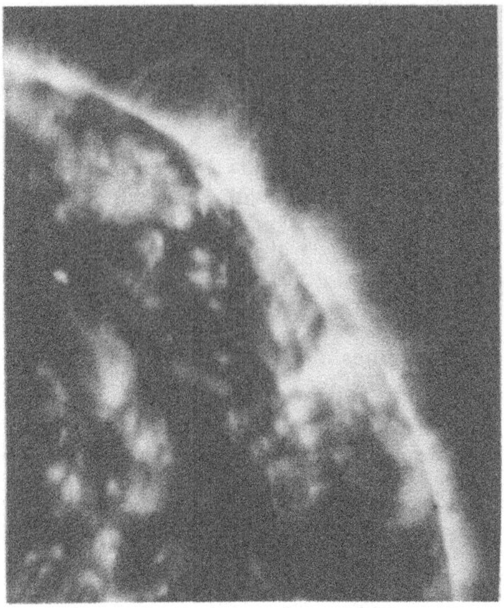

Fig. 5a. An enlargement of Figure 3c showing bright points, including several which exhibit loop like structure. (scale: 10" = 0.75mm)

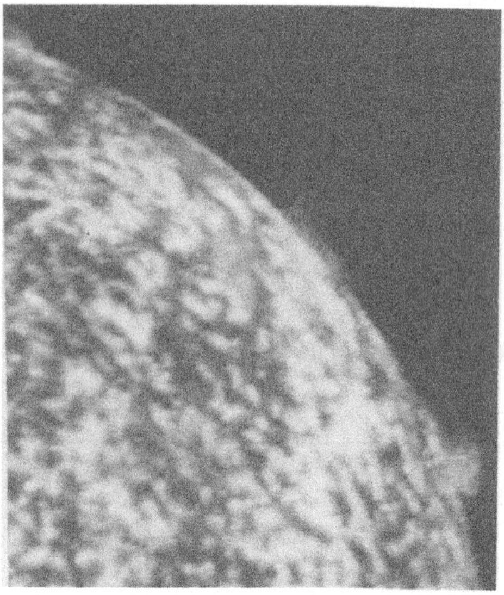

Fig. 5b. An enlargement of Figure 3b, showing bright knots of emission which form the chromospheric network structure. (scale: 10" = 0.94mm)

5. Conclusions

Multilayer optics have been shown to be highly effective for the study of both the large scale structure and the fine scale structure of the solar atmosphere. The *best* soft x-ray and EUV images of the sun obtained to date, with 0.7 arc second resolution, have been achieved using these relatively new techniques.

References

Ahmad, I.A., and Webb, D.F.: 1978, *Solar Phys.* **58**, 323.

Ahmad, I.A., and Withbroe, G.L.: 1977, *Solar Phys.* **53**, 397.

Avrett, E.H.: 1991, in P. Ulmschneider, E.R. Priest, and R. Rosner (eds.) *Mechanisms of Chromospheric Heating*, Springer-Verlag, Berlin, 91.

Baker, P.C.: 1989, in R.B. Hoover (ed.) *X-Ray/EUV Optics for Astronomy and Microscopy, Proc. SPIE* **1160**, 263.

Barbee, T.W., Jr.: 1981, in D.T. Atwood, and B.L. Henke (eds.) *Low Energy X-Ray Diagnostics, Proc AIP* **75**, 124.

Barbee, T.W., Jr.: 1989, *Rev. Sci. Instr.* **60**, 1588.

Barbee, T.W., Jr.: 1990, *Optical Eng.* **29**, 711.

Barbee, T.W., Jr., and Keith, D.C.: 1978, in H. Winick, and G. Brown (eds.) *X-ray Instrumentation for Synchrotron Radiation Research, SSRL Report 78/04* **III-26**.

Barbee, T.W., Jr., Weed, J.W., Hoover, R.B., Allen, M.J., Lindblom, J.F., O'Neal, R.J., Kankelborg, C.C., DeForest, C.E., Paris, E.S., Walker, A.B.C., Jr., Willis, T.D., Gluskin, E., Pianetta, P., and Baker, P.C.: 1991, *Optical Eng.* **30**, 1061.

Bixler, J.V., Barbee, T.W., Jr., and Dietrich, D.D.: 1989, in R.B. Hoover (ed.) *X-Ray/EUV Optics for Astronomy and Microscopy, Proc SPIE* **1160**, 648.

Bohlin, J.D., Shultz, N.R., and Tousey, R.: 1978, *Space Res.* **XV**, 651.

Catura, R.C., and Golub, L.: 1988, *Rev. Phys. Appl.* **23**, 1741.

Golub, L., Herant, M., Kalata, K., Lovos, I., Nystrom, G., Prado, F., Spiller, E., and Wilczynski, J.: 1990, *Nature* **344**, 842.

Haas, G., and Hunter, W.R.: 1973, in B. J. Thompson and R. R. Shannon (eds.) *Space Optics*, National Academy of Sciences, 525.

Haélbich, R.P., and Kunz, C.: 1976, *Optics Comm.* **17**, 287.

Haélbich, R.P., Segmuller, A., and Spiller, E.: 1979, *Appl. Phys. Lett.* **34**, 184.

Henry, J.P., Spiller, E., and Weisskopf, M.: 1982, *Appl.Phys. Lett.* **40**, 25.

Hoover, R.B., Barbee, T.W., Jr., Baker, P.C., Lindblom, J.F., Allen, M.J., DeForest, C.E., Kankelborg, C.C., O'Neal, R.H., Paris, E., and Walker, A.B.C., Jr.: 1990a, *Optical Eng.* **29**, 1281.

Hoover, R.B., Baker, P.C., Hadaway, J.B., Johnson, R.B., Gabardi, D.R., Walker, A.B.C., Jr., Lindblom, J.F., DeForest, C.E., and O'Neal, R.H.: 1990b, in R.B. Hoover and A.B.C. Walker, Jr. (eds.) *X-Ray/EUV Optics for Astronomy, Microscopy, Polarimetry, and Projection Lithography, Proc. SPIE* **1343**, 189.

Hoover, R.B., Walker, A.B.C., Jr., DeForest, C.E., Allen, M.J., and Lindblom, J.F.: 1990c, *Proc. SPIE* **1343**, 175.

Hoover, R.B., Walker, A.B.C., Jr., DeForest, C.E., Watts, R.N., and Tarrio, C.: to be publ. in *Proc. SPIE* **1742**, 1992.

Keski-Kuha, R.A., Thomas, R.J., and Davila, J.M.: 1992, *Proc. SPIE* **1546**, 614.

Lindblom, J.F., O'Neal, R.H., Walker, A.B.C., Jr., Powell, F.R., Barbee, T.W., Jr., Hoover, R.B., and Powell, S.F.: 1991, *Optical Eng.* **30**, 1134.

Mewe, R., Gronenschild, E.H.B.M., and van den Oord, G.H.J.: 1985, *Astron. Astrophys. Suppl. Ser* **62**, 197.

Mrowka, S., Jelinsky, P., Bowyer, B., Sarger, G., and Chòyke, W.J.: 1986, in J. L. Culhane (ed.) *Proc. SPIE* **597**, 160.

Patience, J.L.: 1992, "Modeling coronal bright points as small x-ray emitting loops," Senior honors thesis, Dept. of Physics, Stanford University, Stanford, CA.

Powell, F.: 1989, *Proc. SPIE* **1160**, 37.

Spiller, E.: 1973, in B.J. Thompson and R. R. Shannon (eds.) *Space Optics*, National

Academy of Sciences, Washington, D.C., 570.

Spiller, E.: 1981, in D.T. Atwood, and B.L. Henke (eds.) *Low Energy X-Ray Diagnostics, Proc AIP* **75**, 131.

Spiller, E.: 1992, *Proc. SPIE* **1546**, 489.

Spiller, E., Grebe, K., and Golub, L.: 1990a, *Optical Eng.* **29**, 625.

Spiller, E., McCorher, R., Wilczynski, J.S., Golub, L., Nystrom, G.U., Tukacx, B., and Welch, C.W.: 1990b, *Proc. SPIE* **1343**, 134.

Title, A.: 1989, in *Solar and Stellar Granulation*, R.K. Rutlen, and G. Severino (eds.), Kluwer Academic Publ. NATA ASI **263**, 29.

Underwood, J.H., and Barbee, T.W., Jr.: 1981, *Nature* **294**, 429.

Underwood, J.H., Bruner, M.E., Haisch, B.M., Brown, W.A., and Acton, L.W.: 1987, *Science* **238**, 61.

Vesecky, J.F., Antiochos, S.K., and Underwood, J.H.: 1979, *Astrophys.J.* **233**, 987.

Walker, A.B.C., Jr., Barbee, T.W., Jr., Hoover, R.B., and Lindblom, J.F.: 1988a, *Science* **241**, 1781; 1988b, *J. de Physique Colliques* **C1 49**, C1-175.

Walker, A.B.C., Jr., Lindblom, J.F., O'Neal, R.H., Hoover, R.B., and Barbee, T.W., Jr.: 1990a, *Phys. Scripta* **41**, 1053.

Walker, A.B.C., Jr., Allen, M.J., Barbee, T.W., Jr., and Hoover, R.B.: 1990b, *Proc. SPIE* **1343**, 415.

Walker, A.B.C., Jr., Lindblom, J.F., Timothy, J.G., Barbee, T.W., Jr., Hoover, R.B., and Tandberg-Hanssen, E.: 1990c, *Optical Eng.* **29**, 698.

Walker, A.B.C., Jr., Lindblom, J.F., O'Neal, R.H., Allen, M.J., Barbee, T.W., Jr., and Hoover, R.B.: 1990d, *Optical Eng.* **29**, 581.

Walker, A.B.C., Jr., Lindblom, J.F., Timothy, J.G., Hoover, R.B., Barbee, T.W., Jr., Baker, P.C., and Powell, F.R.: 1991, *Proc. SPIE* **1494**, 320.

Walker, A.B.C., Jr., Barbee, T.W., Jr., and Hoover, R.B.: 1992, private communication.

Walker, A.B.C., Jr., Hoover, R.B., and Barbee, T.W., Jr.: 1992a, "The Multi Spectral Solar Telescope Array: Initial results and future plans," to be publ. in *Proc. SPIE* **1742**

Walker, A.B.C., Jr., Timothy, J.G., Hoover, R.B., and Barbee, T.W., Jr.: 1992b, "Ultra high resolution images of the solar chromosphere and corona using coordinated rocket and balloon observations," to be publ. in *Proc. SPIE* **1742**.

Walker, A.B.C., Jr., DeForest, C.E., Hoover, R.B., and Barbee, T.W., Jr.: 1992c, to be submitted to *Solar Phys.*

Walker, A.B.C., Jr., Rugge, H.R., and Weiss, K.: 1974, *Astrophys. J.* **188**, 423.

Wendt, D.L., Cash, W.C., Jr., Scott, M., Arendt, P., Newnam, B., Fisher, R.F., and Schwartzlander, A.B.: 1988a, *Appl. Optics* **27**, 246.

Wendt, D.L., Cash, W.C., Jr., Scott, M., Arendt, P., Newnam, B., Fisher, R.F., Schwartzlander, A.B., Takes, P.Z., and Pinner, J.M.: 1988b, *Appl. Optics* **27**, 279.

Wetherell, W.B. and Rimmer, M.P., *Appl Optics* **11**, 2817.

Withbroe, G.L., Feldman, W.C., and Ahluwalia, H.S.: 1991, in *Solar Interior and Atmosphere*, A.M. Cox, W.C. Livingston, and M.S. Mathews, eds., Univ. of Arizona Press, Tucson, 1087.

Wolter, H.: 1952, *Ann. Physik* **10**, 94; *ibid.*, 286.

Zukic, M., Torr, D.G., Sparrn, J.F., and Torr, M.R.: 1990, *Appl. Optics* **29**, 4284.

NEW ASPECTS OF SOLAR CORONAL PHYSICS REVEALED BY YOHKOH

YUTAKA UCHIDA

Department of Astronomy, University of Tokyo

Abstract. The *Yohkoh* Satellite has been providing excellent data since its launch on August 30, 1991. In addition to some remarkable new findings about flares, one of the surprises of the mission is the finding of pronounced dynamical behavior of the corona. Full details of this dynamical behavior are revealed by the high and regular cadence, high sensitivity observations by the Soft X-ray Telescope aboard *Yohkoh*. Some of the initial results concerning the coronal dynamical behavior obtained to date are reviewed.

Key words: Soft X-ray observations of the Sun, Corona, Active region corona, Solar activity

1. Introduction

Since its launch on August 30, 1991, the *Yohkoh* satellite has provided us with excellent data. In this paper, we try to provide an overview of the early *Yohkoh* Soft X-ray Telescope (SXT) results about coronal phenomena by citing some of the works of the *Yohkoh* team members as well as our own work. We, however, avoid overlaps with the presentations here by other *Yohkoh* colleagues; Dr. Lemen (see Lemen et al. 1993) has already given an introductory talk covering in-orbit functioning of SXT and general aspects of its data, and Dr. Tsuneta (1993) has talked about flares and some conspicuous events as seen with SXT. To obtain a more complete view of the *Yohkoh* initial results, the readers are referred to their talks and to the report by Dr. Kosugi (1993) in these Proceedings describing the initial results from the hard X-ray telescope (HXT), and to the Special Letter Issue of the *Publications of the Astronomical Society of Japan* (vol 44, October 1992) devoted to the early results from *Yohkoh*. The results from the Bragg Crystal Spectrometers (BCS) and from the Wide-band Spectrometers (WBS) are included in this special issue in addition to those from SXT and HXT.

We use the term "corona" here to refer to the entire outer atmosphere consisting of the general background corona and the corona above active regions. Heretofore, the corona has not been recognized as a dynamic entity. This was probably because our knowledge of it was based on the low cadence observations then available which could only provide us with only sets of pictures at instances short compared with the characteristic time scales of its intrinsic variation.

High sensitivity, wide dynamic range observations by the *Yohkoh*–SXT, obtained with a high and regular cadence, however, have changed this situation drastically, and have revealed the "dynamic" character of the solar corona. (Here the term "dynamic" does not mean that the timescale of the change is comparable to the Alfvén-crossing time or acoustic-crossing time as used in some theoretical treatments. Rather, we use this term to indicate that the various changes occur actively with timescales much shorter than the timescale which was previously believed to be characteristic of the variations of the "averaged" corona, for example a fraction of the solar cycle period in which the corona changes its fundamental shape. We use the term "dynamic" in this sense throughout this paper.) Such variability has

J.F. Linsky and S. Serio (eds.), Physics of Solar and Stellar Coronae, 97–111.

been suggested, for example, by the pioneering results of *Skylab* in a less clear way, but it may be said that such a dynamic behavior has been recognized in its full form for the first time by the *Yohkoh*–SXT images with the use of the video-movie representations.

We now know that the "quiet background corona" is not at all a quiet and isolated entity, but rather, forms a dynamic system with the "active region corona" showing many interesting dynamic interactions appearing at all times. The phenomena described below are typical of the dynamical coronal environment, in stark contrast to the "quiet" impression of the background corona which was based on previous observations.

In §2, we first describe our findings concerning the active region corona. In §3, some of the dynamical behavior of the background corona itself is presented, and then the correlated dynamics of the "active region corona – background corona" system is described in §4. Brief descriptions of some other corona-related information are given in §5, and §6 is devoted to a general discussion of our results.

2. Some Findings about the Active Region Corona

2.1. Transient Loop Brightenings in Active Regions — Supply of Heated Mass to the Active Region Corona

Shimizu et al. (1992) found that injections of already heated mass into the pre-existing magnetic loop patterns in active regions occur very frequently for the most "active" of the active regions but less frequently for the other active regions. This was not clearly noted in the *Skylab* data, probably due to the lower cadence images obtained using film, and partly due to considerably higher scattering of its mirror compared with that of the *Yohkoh*–SXT. This phenomenon of transient brightening of the loops inside active regions was shown to have a mean lifetime of several to tens of minutes, corresponding to GOES B-class tiny brightenings. The brightenings come up, in many cases, from one side of a pre-existing loop pattern, with a velocity of a few hundred km s^{-1} and the density and temperature are $(5 - 12)\times 10^9$ cm^{-3} and $(4 - 7)\times 10^6$ K, respectively, with the estimated thermal energy of the order of 10^{26-29} ergs. In some cases they show evidence for the interaction of two or more loops, but in other cases they are *non-steady injections of already heated mass into magnetic loops from one of the footpoints below.*

The total output in the soft X-ray range and the frequency of such brightenings from an active region were plotted on a log-log diagram for many active regions by Shimizu et al. (1992), who found a slope of about 0.5. In order for the transient brightenings to be the source of heating for the whole active region, there must be a trend of larger brightenings occurring in more "active" active regions. At this time it may be more reasonable to say that we have found a new component of active region heating and mass supply, in addition to some other steady-state heating mechanisms with mass supply due to evaporation process by the heat deposited operating in the background (e.g. Sakurai 1991).

Our preliminary interpretations for this heated mass injection mechanism may include either magnetic reconnection taking place low in the atmosphere near one of the footpoints of such loops, or dynamical relaxation of the magnetic twists which are produced below the photosphere by convective motions and released in the form of non-linear packets of magnetic twists pushing up mass along the loops as a sweeping pinch (Uchida and Shibata 1985; Shibata and Uchida 1986). The latter process was also considered in the loop flare model of Uchida and Shibata (1988). The key observation needed to discriminate among different possible explanations is whether magnetograms show that these brightenings come preferentially from the boundaries of opposite polarity regions or from the area where the flux changes in regions of one and the same polarity inside the active regions. Hα pictures and magnetograms obtained during well-coordinated observations with *Yohkoh* will be examined to test whether these brightenings occur near the polarity reversal line in the active region or not.

2.2. Ubiquitous Expansion of the Active Region Corona

Uchida et al. (1992) found that the active region corona expands occasionally, and in some cases almost continually, rather than being in a magnetohydrostatic equilibrium constrained by the sunspot magnetic field as previously imagined on the basis of the then available observations. This expansion, typically with a velocity of 10 km s^{-1}, is ubiquitous, as if it were a part of the intrinsic nature of active regions. One possibility is that the expansion may have something to do with some trigger such as the loop brightenings in active regions mentioned in §2.1. The expansion is also seen when GOES low-C class flares occur in an active region, and it is not due to an explosive blast, but rather to a retarded slow expansion. These expansions are similiar to the cases in which we do not have any flares of appreciable magnitude, except for the transient brightenings we are discussing.

The expansion is also observable when the active region is on the disk, but it is best seen and it is easiest to measure the velocity of expansion when the active region is on the limb (see Fig. 1). In many cases, the shape of the active region is more or less retained during the expansion, and the velocity of the forefront, as measured from time-lapse differenced images for several events, ranges from several to a few tens of km s^{-1}, with some indication of acceleration.

Does the observed expansion correspond to the outward motion of matter, or are we observing the propagation of some kind of wavefront? We believe that the observed expansion is a real expansion, rather than a wave-like phenomenon, because any wave (MHD fast wave, slow wave, etc.) would propagate in the corona with a velocity one order of magnitude or more faster. Another possible interpretation, i.e. the "opening up – reclosing front" as usually considered in the case of arcade formation type events, can be excluded in this instance because there are many cases where no opening-up of that active region is observed to procede it, and the shape of the front is, in general, not a cusp.

Some evidence from the Japanese interplanetary mission *Sakigake* gives indirect support that there are some interplanetary clouds having a systematic out-of-ecliptic plane magnetic field component (Nakagawa et al. 1989) for which no pronounced solar surface counterparts like flares or disappearance of dark filaments have been found. It is reported that the velocities of such clouds are slower, and the

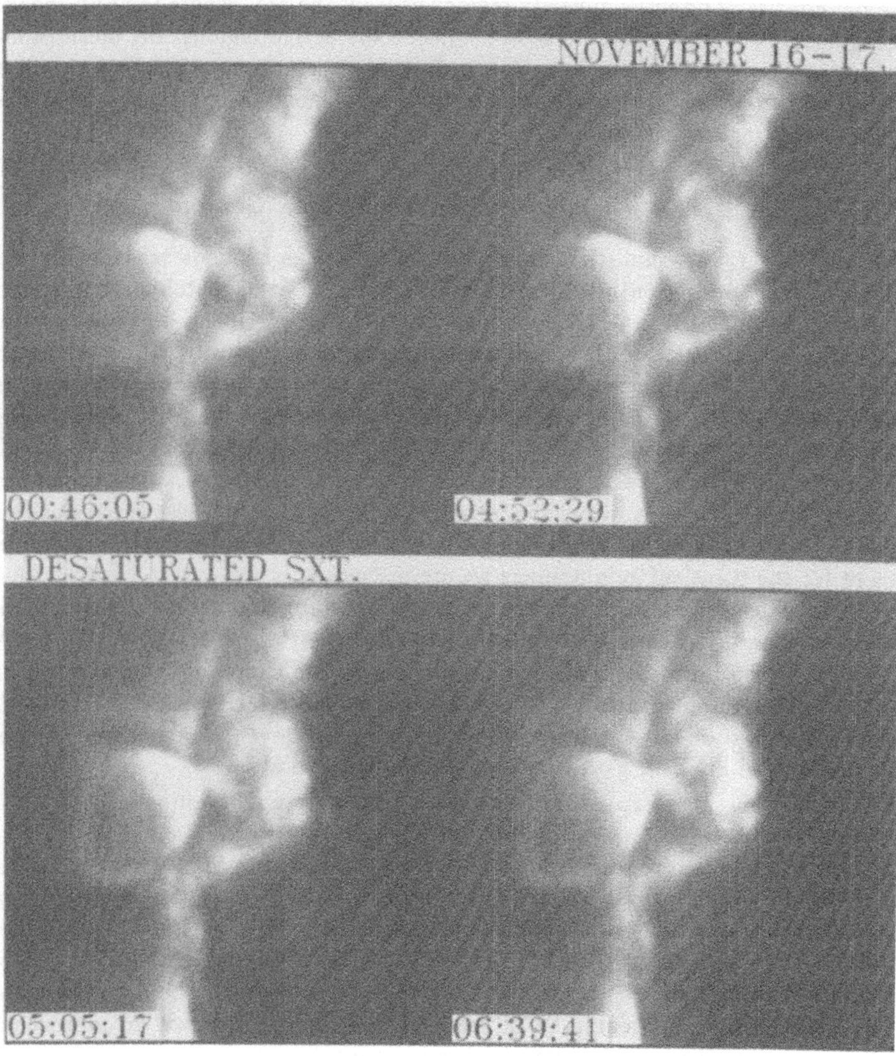

Fig. 1. (A) An active region at the east limb passage on 1991 November 16-17 is shown as an example. The picture is overexposed, and the bright parts are not flares (there are no corresponding GOES enhancements).

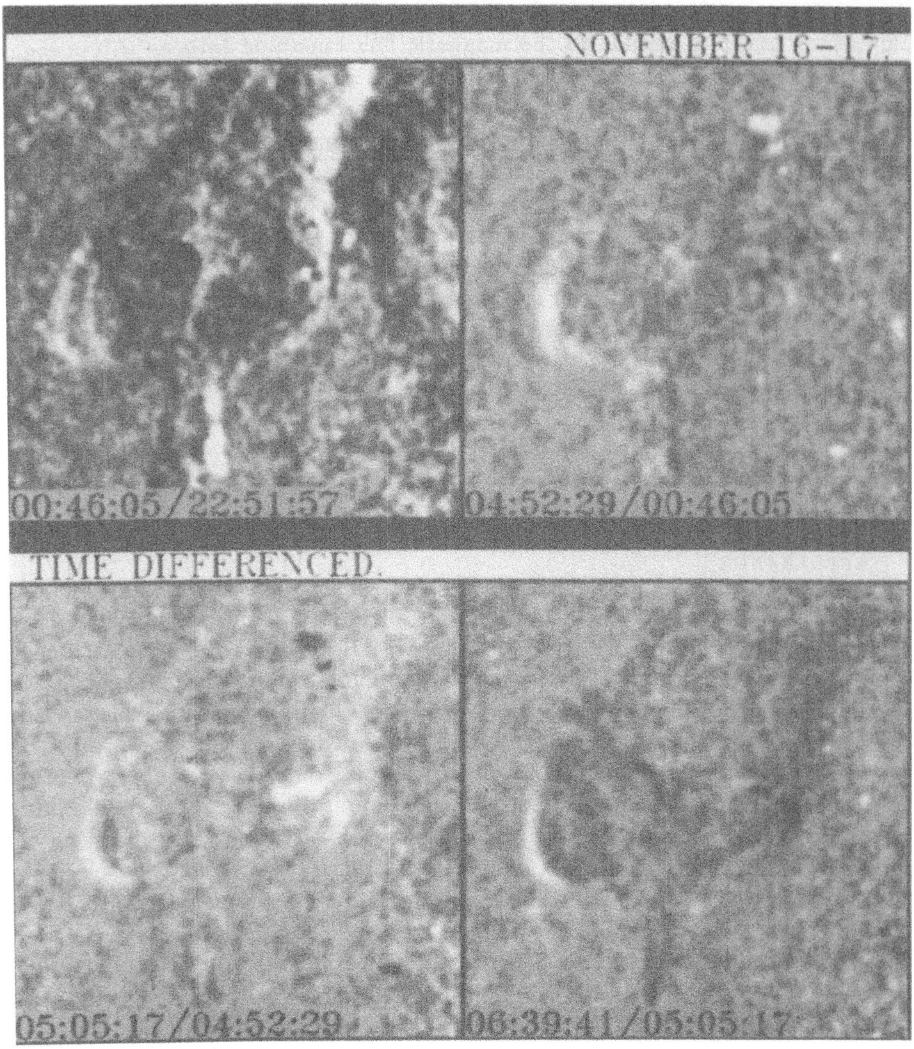

Fig. 1. (B) Time-lapse differenced pictures of the same region as in (A). White part is the newly brightened part during the time-lapse, and it can be seen that the top of the active region corona has expanded during the time-lapse. The velocity is generally of the order of a few to a few tens of km s^{-1}.

densities are systematically higher. Such clouds, which have roughly uni-directional systematic fields of their own and which come one after another with some recurrence, might correspond to the interplanetary extensions of the expanding active region corona discussed here.

If it is established that the expanding active region corona actually moves out into interplanetary space, the estimated mass flux and the magnetic flux lost from the Sun are roughly $\dot{M} \sim 10^{11}$ g s^{-1} and $\dot{\Phi} \sim 10^{16.5}$ Mx s^{-1}, respectively. Then $\dot{M}$ is an appreciable fraction of the total mass loss rate of the Sun through the solar wind, and $\dot{\Phi}$ is similar to the magnetic flux emerging into an active region (Zwaan 1987). The expansion of the active region corona, if established, is a newly recognized componant of the mass loss from the Sun. Contemporary models of the solar wind that are based on the thermal expansion model (Parker 1963), as well as the possible mechanism of solar wind acceleration by Alfvén waves in fast streams (e.g. Hollweg 1981), give a plasma stream parallel to the interplanetary magnetic field, whereas the observed expanding active region corona gives a flow with a magnetic field perpendicular to the motion. Our active region expansion is likely to be a magnetically driven motion, and may show magnetic behavior similar to the clouds observed by *Sakigake*. Thus the expansion of the active region corona that we have found may influence our understanding of the solar wind, and more generally, the winds of other stars.

3. Spontaneous Dynamical Behavior of the Background Corona

3.1. Formation of X-ray Arcades Related to Hα Dark Filament Disappearance

It was already known that the disappearance of Hα dark filaments outside of active regions sometimes causes an extended low-energy Hα double-ribbon flare (Michalitsanos and Kupferman 1974), but without pronounced non-thermal events or the production of high energy particles. On the other hand, Svestka (1976) has reported on *Skylab* observations that show that an X-ray emitting arcade was formed above the region where a dark filament disappeared, connecting the Hα double-ribbons in the corona above, and lasting for more than 20 hours. Wrinkled thready cool material, usually referred to as an erupting prominence, is often seen high in the corona after the disappearance of a dark filament from the solar surface. The relation of the newly formed X-ray arcade to the erupting Hα dark filament and Hα flaring has been discussed in terms of "opening up – reclosing magnetic arcade" type models (Sturrock 1966; Hirayama 1974; Kopp and Pneuman 1976), but a much more detailed examination of the phenomenon than heretofore possible can now be made by using the *Yohkoh* results.

X-ray arcade formation after the disappearance of Hα dark filaments and Hα flaring seems to be a general process, and physically similar phenomena exist on a wide range of scales and magnetic energy densities. The scale ranges from those occurring in active regions as arcade flares (cf. Tsuneta et al. 1992), to X-ray brightenings associated with the disappearances of quieter dark filaments outside active regions (McAllister et al. 1992 to be described in this section), to the formation of fainter X-ray arcades in the circumpolar region with the disappearance of polar filaments but without appreciable Hα enhancements (§3.2).

Yohkoh has already obtained many examples of this phenomenon, and analyses are in progress. Here, we mention one or two cases of special interest, but a more complete account will come later in following papers.

An impressive event of this type occurred in the very initial phase of SXT observations on September 28, 1991. In this event where a beautiful arcade formed after the disappearance of a dark filament, many of the details of the time evolution from the pre-event through the post-event could be analyzed by making use of the fine-structure enhancing techniques usually referred to as the "unsharpened image-masking" method (McAllister et al. 1992). This X-ray flaring event (a brightening in the GOES sub-C class) started in the region where characteristic arches cross with a "tail-like" feature, and then formed an arcade in its later phase. It appeared in the north-east quadrant of the solar disk near the center at about 11:18 UT on September 28. A brightening occurred at the contact point of the two Hα dark filaments which formed separately and merged a few days before the event. After an interruption of observation by satellite night, SXT found that the brightening at the crossing point had developed into an increased brightening of the whole region centered at that location and had expanded towards the north and south along the pre-existing thread structure. Within two hours, a region with a length of 3×10^{10} km brightened, and an arcade-like feature appeared perpendicular to the "spine-like" feature. At 15:56 UT, the "spine-like" feature seemed to have risen, and the arcade was stretched up like a tent. This continued until the satellite changed to another test mode, interrupting the observation before anyone recognized that this event was occurring.

Figure 2 shows fine-structure-enhanced images of this event. Note that the "spine-like" structure along the length of the arcade does not seem to be the locus of the reconnecting points of the opened-up overlying magnetic arcade as suggested in the previous models. The fine-structure-enhanced image clearly shows that the "spine-like" structure has more than two bright threads very similar to the pre-existing field structure below! The "spine-like" structure, which had essentially the same shape as the underlying field structure before the event, stretched the initially round arcade above into a triangular shape like a tent as time progressed, suggesting that the "spine-like" structure is rising from below. This evolution is conspicuously different from the possible previous picture in which the arcade was the locus of the reconnecting points of a once-opened up field reclosing from above.

Another example is the event that occurred on May 7, 1992 (Khan et al. 1992, in preparation). This event occurred in the south-east quadrant near the limb associated with the disappearance of a rather dense dark filament. It occurred at the time of a C2-class X-ray flare with an Hα surge and both type-II and type-IV radio bursts. The arcade had a varying width, wider at its northern-most end and very narrow at its southern-most end. It also had some tail-like structure beyond the limb. The narrow width at the southern end may indicate a difference in the growth, but the arcade kept its shape as it faded.

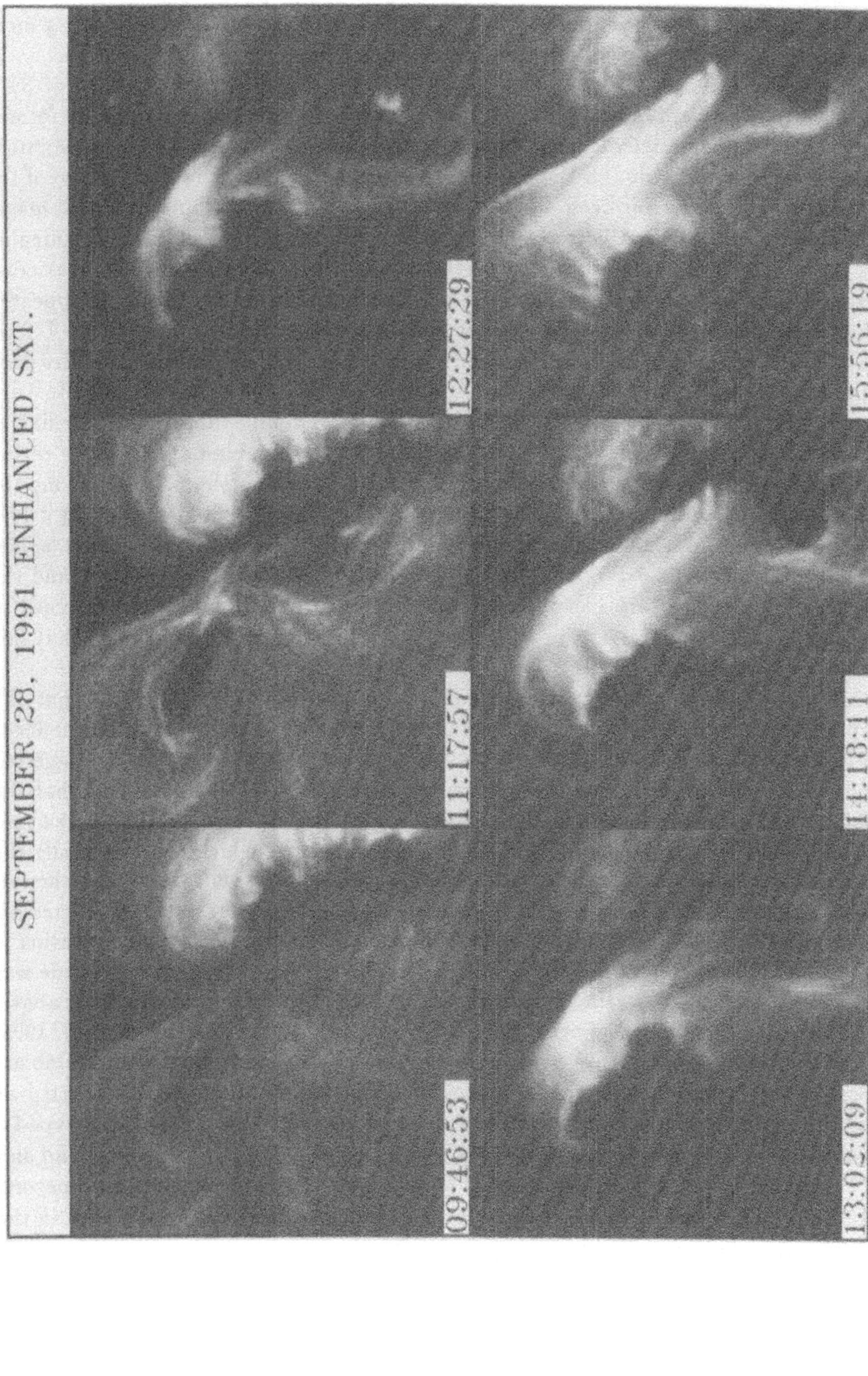
SEPTEMBER 28, 1991 ENHANCED SXT.
09:46:53
11:17:57
12:27:29
13:02:09
14:18:11
15:56:19

Fig. 2. (opposite) Time evolution of the X-ray arcade formation event of September 28, 1991 (from McAllister et al. 1992). The pre-event frame shows that a large scale arch-type structure existed, together with a tail-like structure extending towards the south. At the crossing point, a small brightening with a cross shape started at 11:18, and the region above that point continued to brighten. The bright region began to show "spine plus an arcade" structure at 13:02 onward. In the frame at 14:18, we see that there are two "spines" together with the arcade structure. The arcade structure was pulled up and stretched like a tent. There were brightened threads at that part of the "tail", showing spinning motion during the event. (Since the brightness scale is normalized for each frame, the change in the brightness of the structure on the right-hand side reflects the steep brightening of the arcade in time.)

These are only a few examples out of the many events of this type that were reported by Svestka (1976) and can now be studied in detail as a result of the high cadence observations of *Yohkoh*–SXT. The results mentioned here, and those to be obtained from now on by *Yohkoh*–SXT will contribute greatly to the physical understanding of the process.

3.2. Formation of Large Scale Faint Arcades in the Polar Region

Large scale reconfigurations of the coronal structure leading to the formation of faint large arcades was first observed on November 12, 1991 (Tsuneta et al. 1992), and a few similar events have been observed subsequently. The November 12 event occurred in an area that appeared to be a coronal hole and surprised the observers, but it was confirmed later that the region had a faint polar dark filament and actually was not a unipolar coronal hole region (K. Harvey, private communication). The event started as a weak brightening in a coronal structure on the east side of a "coronal hole" extending towards the north-west quadrant. Gigantic faint loops successively "stood up" from the east end of the "coronal hole" to the west end like a wave propagating across the "coronal hole" over 6×10^5 km with a velocity ≈ 100 km s^{-1}. Tsuneta et al. (1992) concluded, based on both the height and width of the arcade expanding in time and the west end of the arcade showing a cusp structure, that a reclosing process after the opened up of the magnetic field was occuring through quasi-steady magnetic reconnections.

The event indicates that (i) faint large scale versions of the phenomenon mentioned in §3.1 can exist in the polar area, and (ii) there can exist "pseudo-coronal holes", which are dark polar areas but not magnetically unipolar regions.

Another, somewhat different version of a similar phenomenon may be seen (Fig. 3) in the April 27, 1992 event (Khan et al. 1992, in preparation). About 2 hours before the start of this event, activation began in a low-lying filamentary X-ray feature connecting an active region on the south with the region on the west of the area which was to become the "stage" of the event through the region between the footpoints of the high arches. This seemed to "pull down" the high arches toward the region of the "stage". After about ten hours a beautiful new arcade, an array of high semi-circular loops, emerged above the "stage". Although there were small

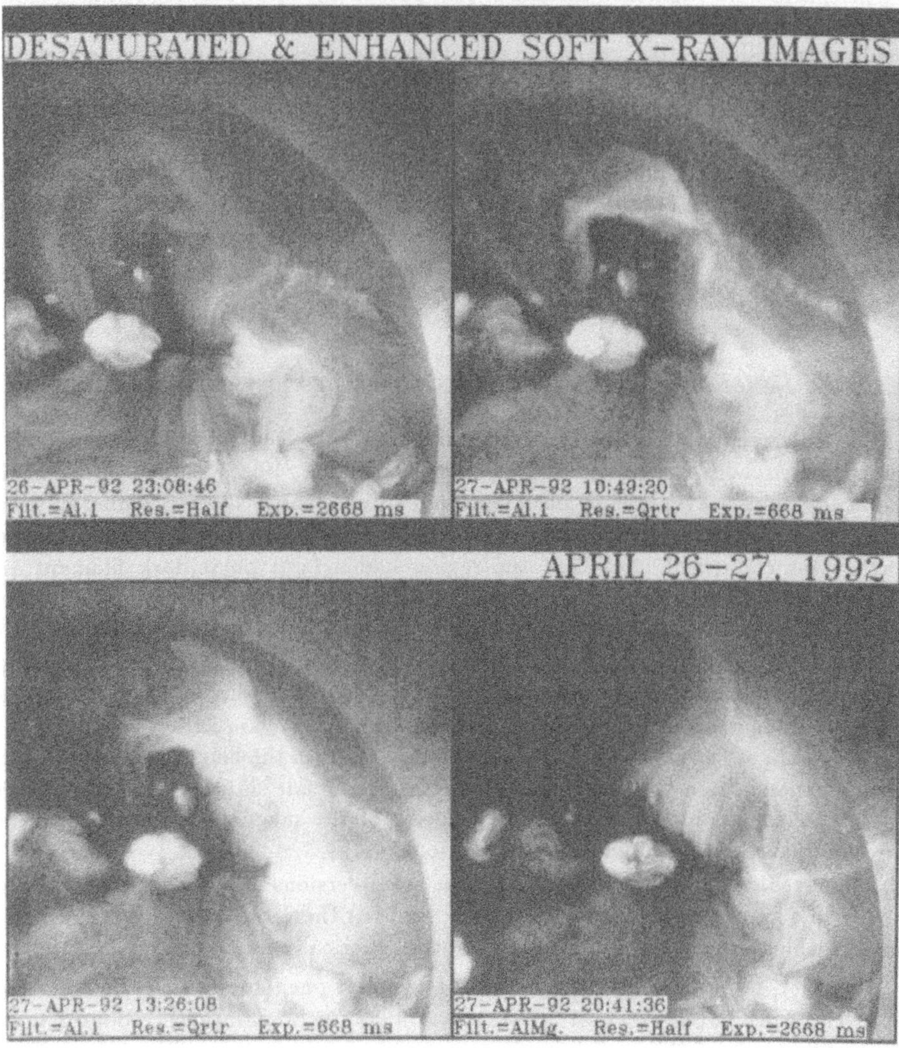

Fig. 3. Faint arcade formation of April 27, 1992. The upper-left frame shows the pre-event situation. The large scale arches in the northern part are deformed later in the upper-right frame through interaction with a low-lying filament, and appear to be "torn down" and changed into a final new arcade structure in about ten hour's time in the last frame.

pieces of dark filaments, indicating that the field polarity reversal line was passing through the general region, there was no dark filament in the part of the future "stage" of the arcade formation. It seems clear that magnetic reconnection occurred and that the final formation of the arcade was the relaxation process to the lowest energy state, a current-free array of arches. We are now analyzing the data obtained throughout the entire process to study the physical processes in detail.

4. Reconfigurations of Coronal Structures due to Small Changes in Active Regions

4.1. Propagating "Reconfiguration Waves" due to Brightenings in Active Regions

The March 27, 1992 event provides an excellent example of the interplay of the active region with the surrounding corona (McAllister et al. 1992, in preparation). Figure 4 shows the evolution of this event. A brightening in the active region led to a brightening of a widespread triangular region to its north, probably via the upper-left channel. This brightening of the triangular region seemed to agitate further the previously inconspicuous highly sheared S-shaped threads on the north-west through its tipped structure. This led to the formation of an arcade from the S-shape, very similar to that of the September 28, 1991 event. The S-shaped threads brightened from the contact point, and the brightening extended toward the south-west along the S-shaped structure. This led to the formation of an arcade as the S-shaped threads changed into a widening bright region. The multiple "spine-like" structures appeared also in this case, with shapes similar to one of the initial S-shaped threads. That part of the final arcade seems to be almost exactly the same as the September 28, 1991 event. In this event, all the processes causing it were seen from the outset. The initial disturbance came from the small brightening in the active region that caused a brightening of the adjacent area, which in turn triggered the formation of the arcade. This type of propagating influence originating in the active region itself is frequently seen. Our tentative interpretation is that we are seeing a propagation of the effect of the magnetic reconnections; a small fraction of the magnetic flux left over in the process of reconnection on each side of the unconnected magnetic regions due to its separation may reconnect with a nearby region of weaker field, and thus affect greatly the surrounding weaker magnetic field regions. The counterpart of the flux absorbed by such a left-over field at a far end again exerts a strong influence on the distant region having a weaker field, and so on. Thus, a small change in a strong-field region can affect the surrounding regions in a cascading manner, thereby affecting quite a large area of the quiet corona.

These results from SXT suggest that the surrounding background corona is constantly affected by changes in the strong field regions that accompany field reconfigurations and heating.

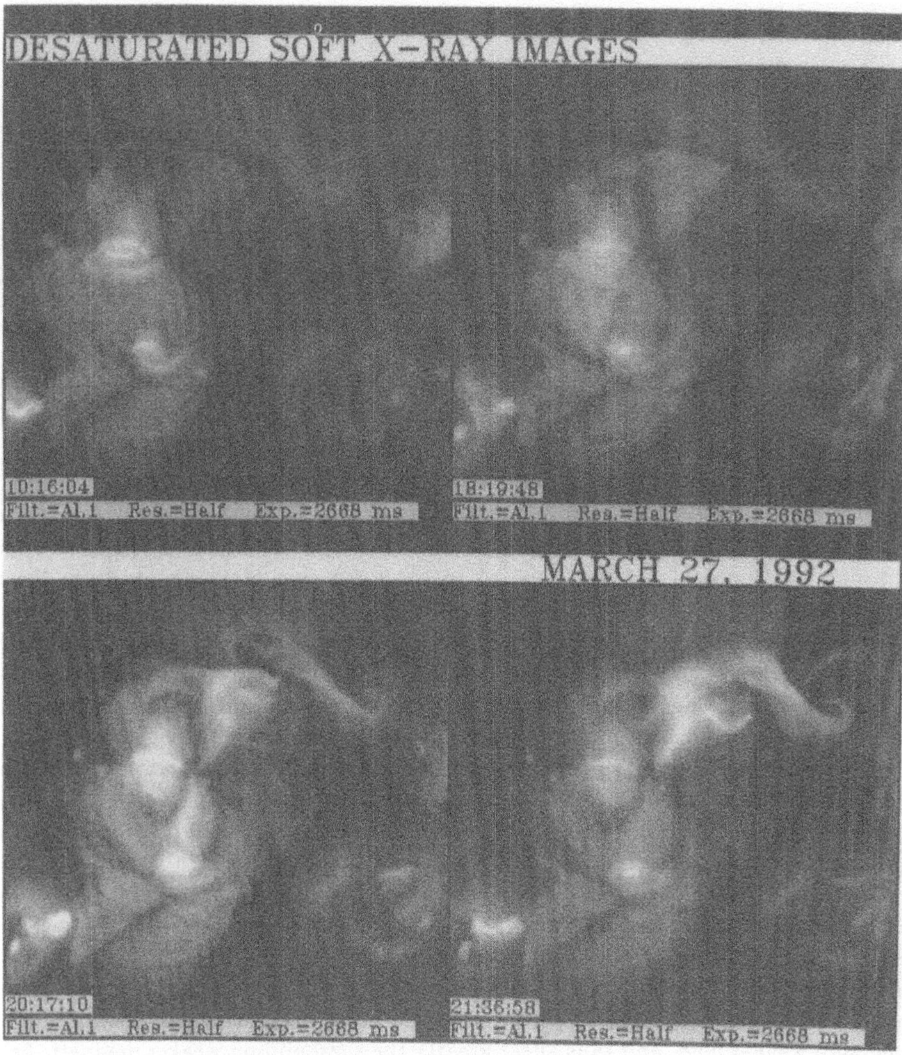

Fig. 4. Propagating "reconfiguration wave" of March 27, 1992. A brightening in the active region on the left influences a triangular region in the upper-middle, and the change of that region stimulates a region further north. Note that an initially rather inconspicuous loop structure began to brighten up with an elongated S-shape at 20:17. This elongated S-shaped structure changed over into an arcade similar to that of September 28,1991.

4.2. Injection of Heated Mass into Adjacent Loop Structures from Active Regions

There are times when hot mass is injected into the surrounding magnetic structure by means of a flow along magnetic flux tubes from an "active" active region (Uchida et al. 1992, in preparation). The active region NOAA 7172, which experienced a flare on May 18, 1992, still retained a considerable activity level even after being rearranged into an "active" active region on the north, together with a helix-looking magnetic structure on the south-west. Typical examples of the injection of heated mass into magnetic loops in the present context are seen in the repetitive events of heated mass injection from May 21, 1992 through May 27, 1992.

The loop structure in the active region in the north participating in the event had one helicity, while the helix-looking loops on the south seemed to have the opposite. Brightenings occurred near the boundary of these two regions, and heated mass was injected repeatedly into the southern helix from the region of brightenings between May 21 through 27. In these events the injected mass did not greatly affect the shape of the helix-looking structure on the south. The velocity of injection was estimated to be $\approx$ 400 km s^{-1}, and the injection occurred at the same time as the brightenings in the boundary region. Fortunately, we have high cadence partial frame images as well as the BCS S XV data for the May 26 event. Therefore, we have line-of-sight velocity information together with the proper motion velocity of the ejected material. We may be able to estimate the rate of magnetic field reconnection if the mass transfer from the hot active region to the helical part occurred through the magnetic reconnection process.

5. Brief Comments on Other Coronal Features: Coronal Holes, Coronal Dark Channels, and Related Phenomena

Coronal holes are one of the signatures of the global magnetic behavior of the Sun. One enigmatic behavior of the coronal holes is that they execute rigid-body rotation while all other surface signatures (e.g. sunspots and dark filaments) show differential rotation. The reason for this is not yet clear, but the *Yohkoh* observations of the faint corona will give information about the boundary of the coronal hole (Tsuneta 1993).

Yohkoh has already provided important data concerning this problem, and has identified "coronal dark channels" connecting polar holes across the equator (Tsuneta et al. 1992). Whether the magnetic field in these channels changes polarity somewhere near the equator or is unipolar down into the other hemisphere may provide critical information concerning the polarity reversal in polar fields near the maximum of the solar cycle. This remains to be confirmed.

As mentioned in §3.2, the faint arcade formation of November 12, 1991 occurred in a region that appeared to be a coronal hole and remained dark for a long time. The presence of such "pseudo coronal holes" in soft X-ray observations suggests that closed coronal loops must have been in that region before the event, if the event is truly a faint version of the arcade formation process. Deep exposures are required (a) to verify the existence of faint coronal loops in what appear to be dark coronal holes, (b) to confirm whether or not really hot and rarefied plasma exists in the open field regions of coronal holes, and (c) to say whether or not the fast streams of the solar wind come from an aggregate of small versions of the X-ray bright points existing in the coronal holes.

6. Summary and Discussion

Even in this early phase of analysis of the *Yohkoh*–SXT results, it is evident that *Yohkoh*–SXT has opened up a new field of solar physics dealing with the dynamical behavior of the background and active region corona.

We can now say that the old magnetohydrostatic, steady heating picture of an active region and the background corona based on the previous low cadence images is not correct. Instead, injections of *already heated mass from below* supply at least a certain fraction, if not all, of the mass and energy of the active regions. Also, the active region corona is almost continually expanding. These expansions may be related to the mass injections mentioned above, although perhaps not directly, because the injected mass itself is *confined* in flux tubes inside the active regions. The mass injections may turn out to be the cause of destabilization of such coronal active regions, and they may in turn bring a new component (magnetically driven mass loss) into the solar mass loss problem.

The background general corona turns out not to be an isolated quiet entity, but is rather a system closely linked with the active region corona. Changes in the active regions mentioned above, for example, can affect the corona in an overwhelming way. In some cases propagating "magnetic reconnection waves", caused by changes in the magnetic connectivity in the stronger magnetic field regions, can propagate further away from an active region in cascade. In some other cases magnetohydrodynamic disturbances may simply perturb distant magnetic structures transiently in the form of simple waves, but in some other cases they may influence distant structures permanently either stimulating some instabilities, or through such "magnetic reconnection waves".

Dynamical changes in the general background corona itself, either spontaneous or by stimulation, often occur at the same time as disappearances of Hα dark filaments. Coronal mass ejections (CME's) may, in this way, either be related to a flare occurring at a distance from the site of the CME, or be unrelated to a flare but related to a much less energetic phenomenon, an active region transient loop brightening that can greatly affect the magnetic structure in the surroundings as in the case described in §4. The required energy may come from the release of the magnetic buoyancy of the structure anchored by the magnetic ties. We feel that the dark filament disappearance itself in these events is merely a passive feature, not a driver of the energy liberation, but one of the results of an instability of the field structure as a whole, which can be triggered by a small disturbance at a distance.

An interesting point is that various phenomena in the solar corona, such as arcade formation in relation to the disappearing dark filament, seem to have examples or analogues with scales ranging from very large in the polar region to small inside active regions. These may correspond to the weak field case and the strong field case, respectively, with similar field configurations. Thus the large and faint arcade above a disappearing polar dark filament described in §3.2, the medium scale hotter arcade discussed in §3.1, and the relatively quiet version of flares with Hα double-ribbon structures in active regions (long-enduring thermal flares, e.g., Tsuneta et al. 1992) may all belong to the same physical class and occur in similar magnetic field configurations but with different magnetic field strengths and different scales.

7. Acknowledgements

The author acknowledges all those who made this very successful satellite with its high quality instruments possible, especially the Japanese, US, and UK core members. Thanks are also due to all the members of the Team for supporting the program through satellite operations. The author is also grateful to Profs. M. Oda, J. Nishimura, and Y. Tanaka for their support of the program from the very initial phase which made this successful satellite possible. Finally, the author acknowledges the collaboration of Drs. J. Khan and A. McAllister in the preparation of this talk, and discussions with Drs. H. Hudson, S. Tsuneta, K. Shibata, A. McAllister, J. Khan, K. Strong, and T. Shimizu.

References

Hirayama, T.: 1974, *Solar Phys.* **34**, 323

Hollweg, J.V.: 1981, *Solar Phys.* **70**, 25

Kopp, R.A. and Pneuman, G.W.: 1976, *Solar Phys.* **50**, 85

Kosugi, T.: 1993, this volume

Lemen, J.R., Acton, L.W., Bruner, M.E., Strong, K.T., Hirayama, T., and Tsuneta, S.: 1993, this volume

McAllister, A.H., Uchida, Y., Tsuneta, S., Strong, K., Acton, L.W., Hiei, E., Bruner, M.E., Watanabe, Ta., and Shibata, K.: 1992, *Publ. Astron. Soc. Japan* **44**, in press

Michalitsanos, A., and Kupferman, P.: 1974, *Solar Phys.* **36**, 403

Nakagawa, T., Nishida, A., and Saito, T.: 1989, *J. Geophys. Res.* **94**, 11761

Parker, E.N.: 1963, *Interplanetary Dynamical Processes*, Interscience: New York

Sakurai, T.: 1991, in Y. Uchida, R.C. Canfield, T. Watanabe, and E. Hiei, ed(s)., *Flare Physics in Solar Activity Maximum 22*, Springer Verlag: Berlin, 245

Shibata, K. and Uchida, Y.: 1986, *Solar Phys.* **103**, 299

Shimizu, T., Tsuneta, S., Acton, L.W, Lemen, J.R. and Uchida, Y.: 1992, *Publ. Astron. Soc. Japan* **44**, in press

Sturrock, P.A.: 1966, *Nature* **221**, 695

Svestka, Z.: 1976, *Solar Flares*, Reidel: Dordrecht

Tsuneta, S.: 1993, this volume

Tsuneta, S., Takahashi, T., Acton, L.W., Harvey, K.: 1992, *Publ. Astron. Soc. Japan* **44**, in press

Uchida, Y. and Shibata, K.: 1985, in M. Kundu and G.D. Holman, ed(s)., *Unstable Current Systems and Plasma Instabilities in Astrophysics*, Reidel: Dordrecht, 287

Uchida, Y. and Shibata, K.: 1988, *Solar Phys.* **116**, 291

Uchida, Y., McAllister, A., Strong, K., Ogawara, Y., Shimizu, T., Matsumoto, R., and Hudson, H.S.: 1992, *Publ. Astron. Soc. Japan* **44**, in press

Webb, D.F., Krieger, A.S., and Rust, D.M.: 1976, *Solar Phys.* **48**, 159

Zwaan, C.: 1987, *Ann. Rev. Astron. Astrophys.* **25**, 83

DYNAMICS OF THE SOLAR CORONA
OBSERVED WITH THE *YOHKOH* SOFT X-RAY TELESCOPE

SAKU TSUNETA

Institute of Astronomy, The University of Tokyo, Mitaka, Tokyo 181, Japan

and

JAMES R. LEMEN

Lockheed Palo Alto Research Lab., Palo Alto, CA 94304, USA

Abstract. The *Yohkoh* Soft X-ray Telescope (SXT) is providing a new perspective on solar flares and the solar corona in general. Some selected topics from the initial results of the *Yohkoh* SXT are presented. Continuous soft X-ray observations from the pre-flare through the post-flare phases indicate that magnetic reconnection of a neutral sheet at the top of the flare loop participates in the flare energy release. The full sun X-ray images show many examples of global changes of magnetic field structure (restructuring) and its activation (heating) with time scale of tens of minutes to hours. Frequent microflares are observed in active regions, especially in flare-producing regions. The broadband filter analysis shows that active region plasmas generally have components with temperatures up to 5 MK in addition to the 2 MK component. A thin long coronal hole along the North-South direction shows rigid rotation with the equatorial rotation rate.

Key words: Yohkoh – Soft X-ray Images – Sun – Corona – Flare – Rotation

1. Introduction

The *Yohkoh* Soft X-ray Telescope (SXT) (Acton *et al.* 1988; Bruner *et al.* 1989; Tsuneta *et al.* 1991) was powered on September 2, 1991, following the successful launch of the spacecraft (Ogawara *et al.* 1991) on August 30. After the in-orbit software and mechanism testing, beautiful first light X-ray and optical images have been taken. Regular observations began in early October after extensive testing of the flight software. The in-orbit performance of the telescope, technically and scientifically, exceeds even the most optimistic expectations prior to launch. *Yohkoh* SXT images are drastically changing our view of the solar corona and the behavior of the magnetized plasma in general (see also Uchida 1993 and Kosugi 1993).

The SXT consists of a grazing incidence X-ray mirror with a co-aligned optical system, two filter wheels housing X-ray and optical analysis filters, a shutter mechanism and the CCD camera. The CCD employed is a 1024 x 1024 virtual phase CCD covering the whole sun. The telescope is controlled by two micro-processors. The spatial resolution of the SXT mirror is as high as 2.5–3 arcsec (FWHM) over the entire solar disk. X-ray scattering due to the micro-roughness of the mirror is far smaller than the previous generation grazing incidence telescopes. Its excellent in-orbit optical performance was verified by the sharp edge of the X-ray limb and the low background around an intense compact flare kernel.

The time resolution of full frame images covering the entire sun ranges from 32 sec to 137 min, depending on the image size and the on-chip summation modes (x1, x2 and x4), while that of the partial frame image covers a 2.6 x 2.6 arcmin^2 area with 64 x 64 pixels in 2 sec. A larger field of view to cover an extended region of interest (the observing region) is available by sacrificing the time resolution. Full frame and partial frame images are taken in an interleaving way. Although there

J.F. Linsky and S. Serio (eds.), Physics of Solar and Stellar Coronae, 113–130.

is a large degree of freedom in these observing modes, the default observing modes currently used are as follows: Typical size of the observing region is 5.2 x 5.2 arcmin^2 with spatial resolution of 2.5 arcsec and time resolution of 32 sec. The cadence is increased to 2 sec for flare observations. The full frame images typically have a time resolution of 128 sec with spatial resolution of 5 arcsec.

Two filter wheels are equipped with five broadband X-ray filters and two optical filters (28 Å bandpass centered at 4310 Å and 184 Å at 4616 Å). High quality optical images co-aligned to X-ray images can be taken simply by positioning the filter wheels at one of the optical filters, permitting accurate alignment of X-ray images with optical images (see Tsuneta *et al.* 1991). The five different X-ray filters provide some temperature diagnostics capability over the wide temperature range from 10^6 K to 10^8 K.

The telescope is controlled with a flexible software system. A nested DO-loop control structure allows observers to change the observing sequence flexibly, depending on the observation purpose and the solar activity. Some of the operation parameters are obtained from analysis of the X-ray images with the onboard computer: For example, an automatic exposure control task adjusts the exposure time over a wide range to follow the rapidly changing intensity of flares, thus maintaining the time cadence of images with proper exposures. An automatic observing region search task tracks (specified) active regions to compensate for the solar rotation and the spacecraft attitude drift, allowing continuous small field observations with high time resolution.

The design of SXT was optimized to address the central problems of solar physics as much as possible under the limited spacecraft resources. As of autumn 1992, about 600,000 X-ray and optical images have been taken. An X-ray movie created from these images reveals a remarkable new world: The solar corona is full of highly transient phenomena, global and local, on many time scales, as opposed to the *Skylab* result. A few of these are: (1) Activation (sudden brightening in X-rays) of large scale coronal structures with sizes comparable to the solar size (Tsuneta *et al.* 1992b). (2) Global restructuring of the magnetic field structure such as successive creation of closed field lines from open field lines (Tsuneta *et al.* 1992b). (3) Small transient brightenings (microflares) frequently seen in energetic active regions (Shimizu *et al.* 1992). (4) Steady high temperature components of active region plasmas reaching 5 MK (Hara *et al.* 1992). (5) Frequent mass and magnetic field ejections (expansions) into the interplanetary space from active regions (Uchida *et al.* 1992). (6) High speed (a few hundreds of km s^{-1}) X-ray jets (Shibata *et al.* 1992). (7) Strong X-ray emission from emerging flux regions. (8) Bright, sometimes irregular-shaped (non-potential) magnetic loops, as thin as a single SXT pixel (2.5 arcsec) in active regions and the quiet corona. (9) Thin long coronal holes along the North-South direction rigidly rotating with the equatorial rate (Hayashi 1992).

In this paper, we will present some of the initial discoveries from the *Yohkoh* SXT. The topics include continuous observations of a flare from the pre-flare through the post-flare phase, global activation and restructuring of the coronal magnetic fields, soft X-ray microflares, steady high temperature components of active region plasmas, and the rotation of coronal "dark channels".

2. Solar Flares as On-going Magnetic Reconnections

The unique features of SXT described above make it an ideal tool for flare studies. Whereas *Skylab* observed only a handful of flares in a limited way, SXT can make continuous observations of a flaring region before the onset of a flare, through its development, and during the recovery of the corona after the end of the flare with high spatial and time resolution. With these data in hand, we can address central questions such as what the magnetic configuration of flares is, and how its elements interact, and then dissipate.

In this paper, we will concentrate on an M3 flare that occurred on the East limb on February 21, 1992 (Tsuneta *et al.* 1992a) because it is too early to summarize all of the *Yohkoh* flare results, and because this flare is a good example to demonstrate the observing capability of SXT. Figure 1 shows the time profile of the flare. The flare continued for about 4 hours, and we have good data coverage from the pre-flare through the post-flare phases. The time profile is very gradual without any spiky component, and the X-ray spectrum is very soft with virtually no X-rays detected above 30 keV. From its appearance, this flare can be classified as a long-duration soft X-ray event.

2.1. X-RAY MORPHOLOGY

Figure 2 shows the time evolution of the flare loop from the pre-flare through the decay phase. Although the flare intensity appears similar in these images, the exposure time was optimized for each image and changed by one order of magnitude during the course of the flare. The top of the loop is the brightest in X-rays, and the height of the loop is about 50,000 km. The soft X-ray flare has a "helmet streamer"-like structure, and the basic arch structure does not change throughout the flare. However, the main loop structure continuously expands in both height and footpoint separation with a speed of about 20 km s^{-1}.

This basic structure can already be seen in the pre-flare images taken 3 hours prior to the flare. Figure 3 covers a wider area in both spatial and time scales from X–9 hours through X+30 hours, where X is the starting time of the hard X-ray flare (3:10 UT). Expansion and restructuring of the active region magnetic field can be seen almost continuously for 8 hours prior to the flare. The speed of the expansions is roughly a few km s^{-1} to 10 km s^{-1}. A small "helmet streamer"-like structure already appears at X–7 hours associated with the global expansion of the active region. This expansion continues until X–3 hours, and a large structure overlying the bright loop, which will flare later, can be seen 3 hours before the flare. This overlying structure will have a major fan-shaped sweeping expansion between X–3 hours and X–30 min, and the clear helmet streamer arch appears at X–30 min. If there were a causal relationship between these series of expansions and the flare, the flare phenomena would not be local, but rather would involve the global field structure of the active region.

Although the basic flaring-arch configuration appears with the first restructuring around X–7 hours, the second expansion between X–3 hours and X–30 min is followed by the gradual start of the flare. A horizontal structure going through the

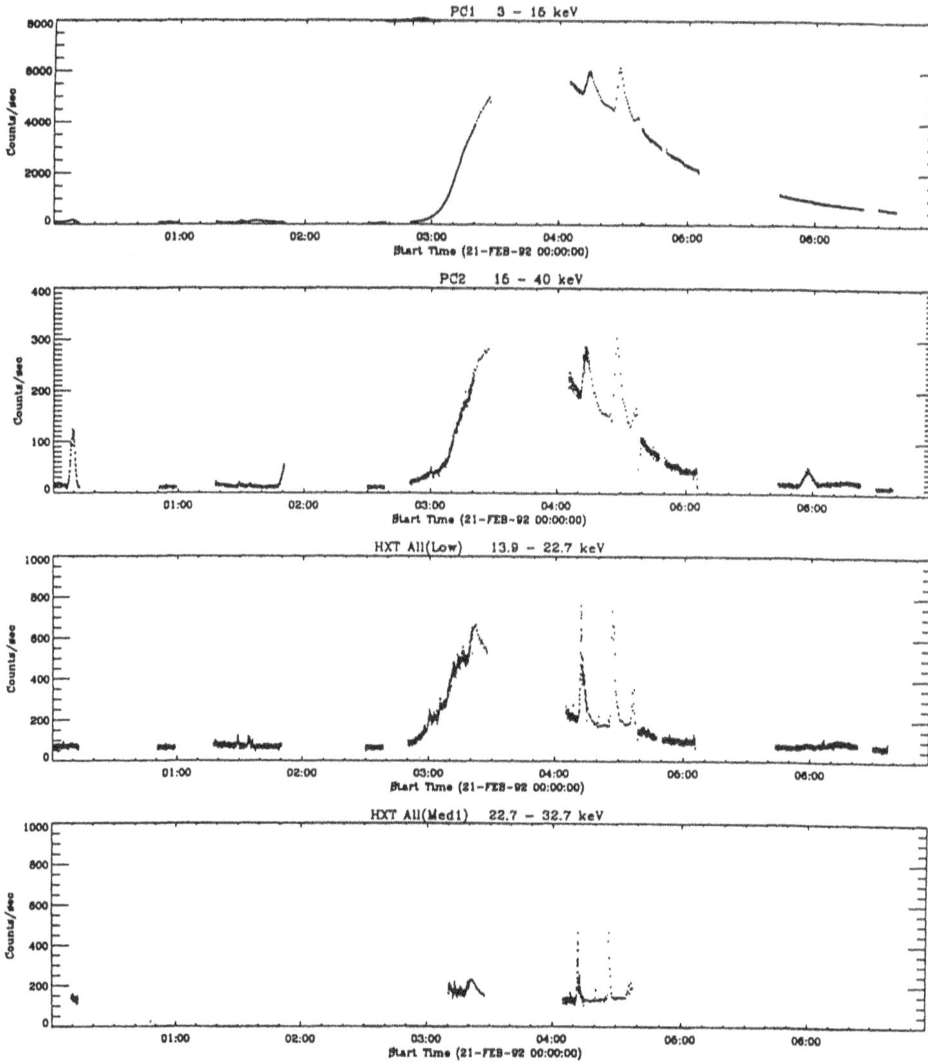

Fig. 1. X-ray time profile of the February 21, 1992 flare from 3 keV to 33 keV. The spiky components seen between 4:10 UT and 4:40 UT are from another flare that simultaneously occurred on the West limb.

neutral sheet section of the helmet streamer is seen around 1 hour to 30 minutes before the flare. This structure does not expand during the global expansion of the active region fields, and is not seen in the flare and post-flare images. We suspect, therefore, that this horizontal component perpendicular to the neutral sheet impedes the start of reconnection. The flare would then start when the horizontal fields rise as a filament eruption and escape from the neutral sheet.

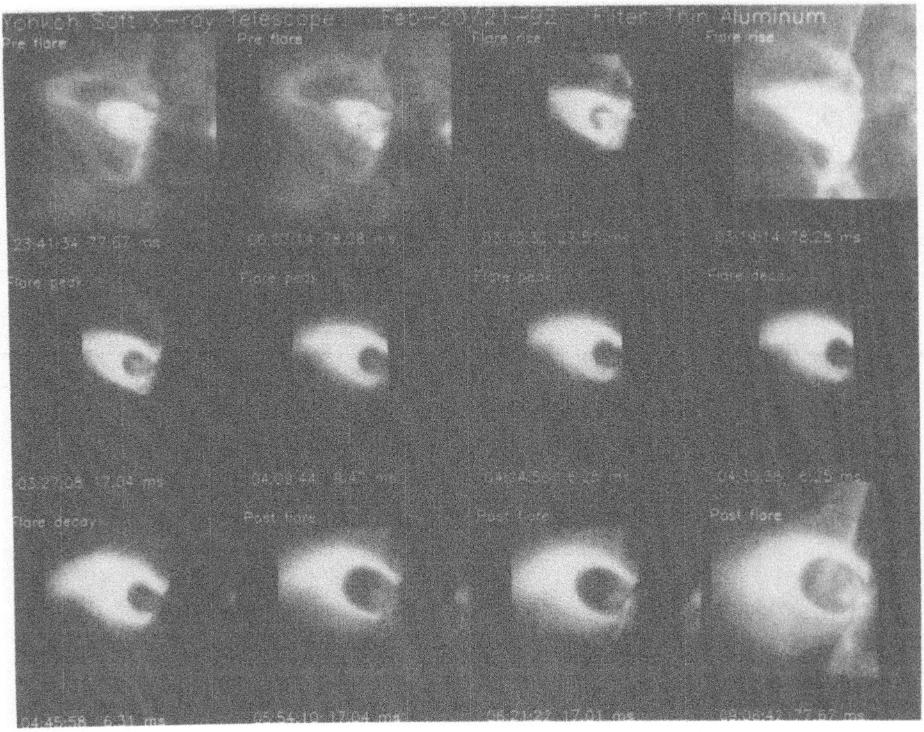

Fig. 2. Evolution of the flare loop from the pre-flare phase through the post-flare phase. The pixel size is 2.5 arcsec. The flare started around 3:10 UT. The smaller field has 2.6 arcmin FOV, while the larger one has 5.2 arcmin FOV.

2.2. Temperature structure

We have derived the temperature structure of the flaring plasma using a pair of broadband filters (Hara 1992). Figure 4 shows the X-ray images, temperatures, emission measures, and pressures obtained from the temperatures and the emission measures from the start of the flare through the decay phase. In the rising and peak phases, there are some time-dependent hot and cool islands. The temperatures of the hot islands reach 13 MK. Note that the temperature derived here is the weighted mean of the actual temperature distribution with the overall wavelength response of the telescope, and does not necessarily represent the (single or mean) temperature of the plasma. The high-density region is formed at the top of the arch with increasing size and emission measure with time.

A more distinct temperature structure appears in the decay phase: The temperature of the outer shell of the arches is higher and decreases toward the inner arches. The peak temperature is about 11 million degrees. The high-density region at the loop top continues to grow vertically as the flare continues. Its temperature

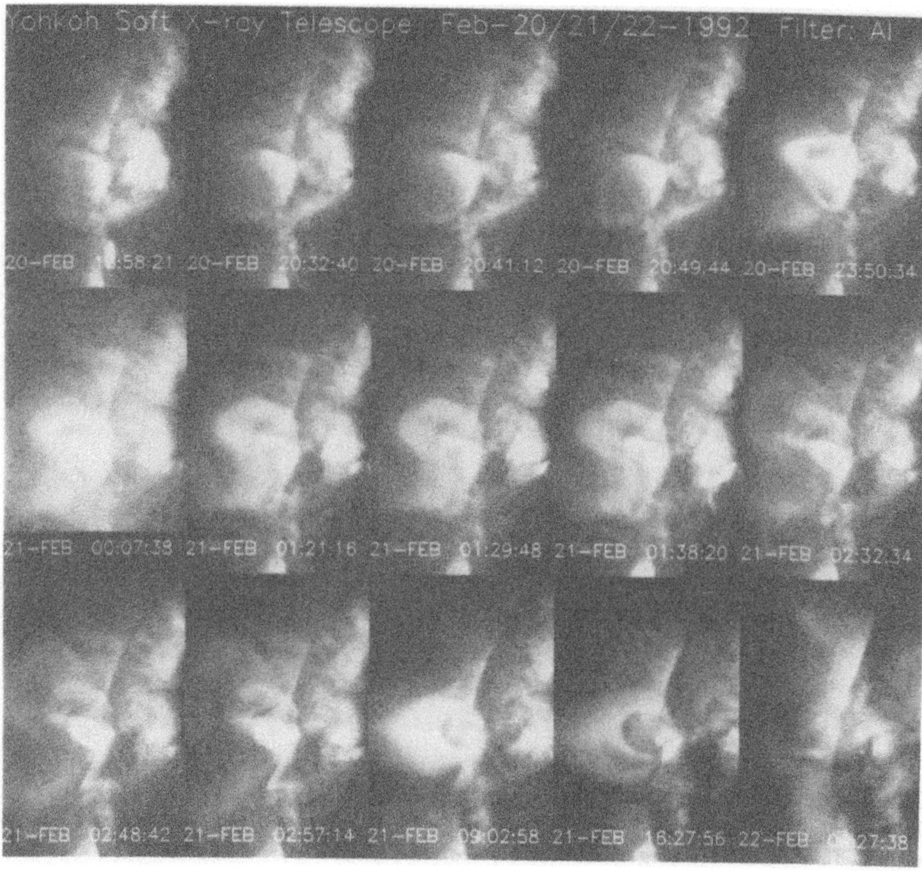

Fig. 3. The flaring active region on the East limb from 9 hours before the flare through 30 hours after the flare. The images do not contain the main part of the flare. Pixel size is 4.9 arcsec, and the field of view is 8.2 x 16.4 arcmin^2. Continuous expansion and restructuring are seen in the pre-flare images.

is, however, as low as 7 million degrees in the peak and decay phases possibly due to radiative cooling. The loop top emission measure at the flare maximum phase is about 1.3×10^{47} cm^{-3} (2.45 arcsec)$^{-2}$, implying a minimum plasma density of $n = 1.9 \times 10^{15}/L^{0.5}$(cm), where L is the line-of-sight thickness of the arcade: $n = 4.3 \times 10^{10}$ cm^{-3} for $L = 2 \times 10^4$ km.

The pressure maps obtained from these temperature and emission measure maps are much smoother than the temperature and emission measure maps. The loop top region generally has higher pressure. The pressure difference between the loop top and the footpoints is, however, smaller than a factor of 2. The peak pressure is about 40 dyn cm^{-2} at the loop top, assuming the line-of-sight thickness $L = 2 \times 10^4$

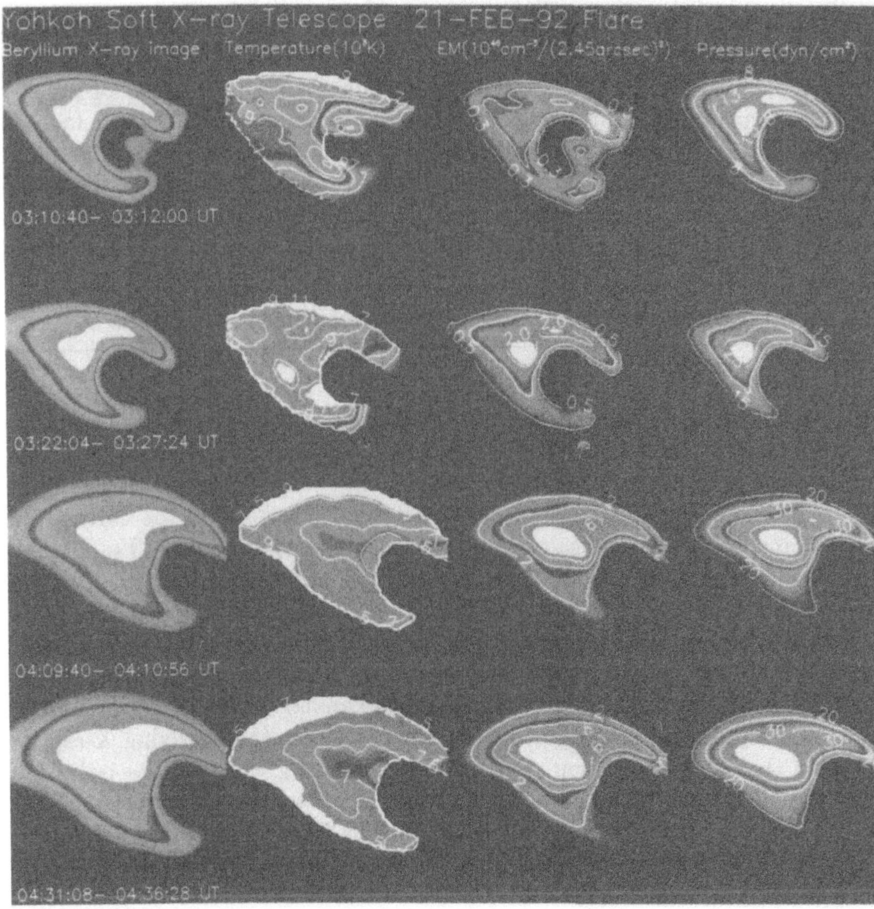

Fig. 4. Temperature–emission measure and pressure maps as well as X-ray images from the rise phase through the decay phase. The temperatures are in units of 10^6 K, the emission measures in units of 10^{46} cm$^{-3}(2''.45)^{-2}$, and the pressures in units of dyn cm^{-2}.

km. This gives a minimum magnetic field of about 30 G to sustain the plasma. The loop top has the highest density and pressure from the rising phase of the flare.

2.3. Solar flares as an on-going magnetic reconnection

The pre-flare observation suggests that the current sheet is created by magnetic restructuring a few hours prior to the flare, and the neutral sheet may be "activated" by the rise of the horizontal structure going through it. An X-type or Y-type reconnection point is then formed at the top of the arch. The gradual increase of the flare arch height and the separation also confirms the idea that reconnection is going on at the loop top (Pneuman 1981). The higher temperature in the outer arch also suggests that the flare energy is supplied by an ongoing reconnection process at the neutral sheet or the slow shock front formed by magnetic reconnection.

If magnetic reconnection is responsible for the flare energy release, there should be other types of energy release involving reconnection, interaction of emerging flux with coronal fields (Shibata, Nozawa and Matsumoto 1992), and loop-loop interaction (Tajima, Brunel and Sakai 1982). Solar flares are diverse phenomena as revealed by *Hinotori* hard X-ray observations (Tsuneta 1987). Since *Yohkoh* has already observed more than 200 flares, a comprehensive understanding of solar flare phenomena is now within our reach with this wonderful data set.

3. Global Restructuring of Coronal Magnetic Fields

3.1. Example of restructuring

The SXT synoptic movie demonstrates that the solar corona is full of highly transient phenomena, both global and local, on many time scales. This highly dynamical behavior of the coronal magnetic fields was not fully revealed by the *Skylab* results. Figure 5 shows an example of the large-scale "restructuring" of the coronal magnetic fields involving spatial scales comparable to one solar radius (Tsuneta *et al.* 1992b). The formation of the closed loop structure propagates westward, reaching the West limb 12 hours after the start of the event at a speed of 20–40 km s^{-1}. The loop structure increases its size and footpoint separation at a speed of 2–4 km s^{-1}. The closed loop observed at the West limb shows a cusp structure. The height of this cusp structure also increases with time at a similar speed. Magnetogram data indicate that the magnetic neutral line is located under the overlying X-ray loop. This shows that the X-ray loop structure is connecting the opposite polarity regions of the global photospheric magnetic field. In order to see the pre-event X-ray structure, many consecutive full sun X-ray images have been integrated, creating images with a few minutes exposure before the event. No large scale structure is, however, visible in those images.

A He I 10830 Å image on the previous day shows a filament structure located on the neutral line, so there must be closed field lines over the neutral line before the event, which are not seen in X-rays. This filament is no longer visible after the event. Therefore, the open loop structure with a neutral sheet would be formed by the eruption of the prominence. Then the closed field structure is formed from the open field due to X-type magnetic reconnection in a non-explosive way. The boundary of the cusp seen in X-rays locates the separatrix of the reconnection field.

3.2. Global restructuring and flares

We suggested in the previous section that magnetic reconnection in a neutral sheet at the loop top, created by pre-flare magnetic restructuring, can explain a long duration flare. The present observations demonstrate that a similar process is responsible for the restructuring of the global magnetic field with a larger spatial dimension and a much smaller energy scale. The only difference between flares and global restructuring would be in magnetic field strength: The magnetic field

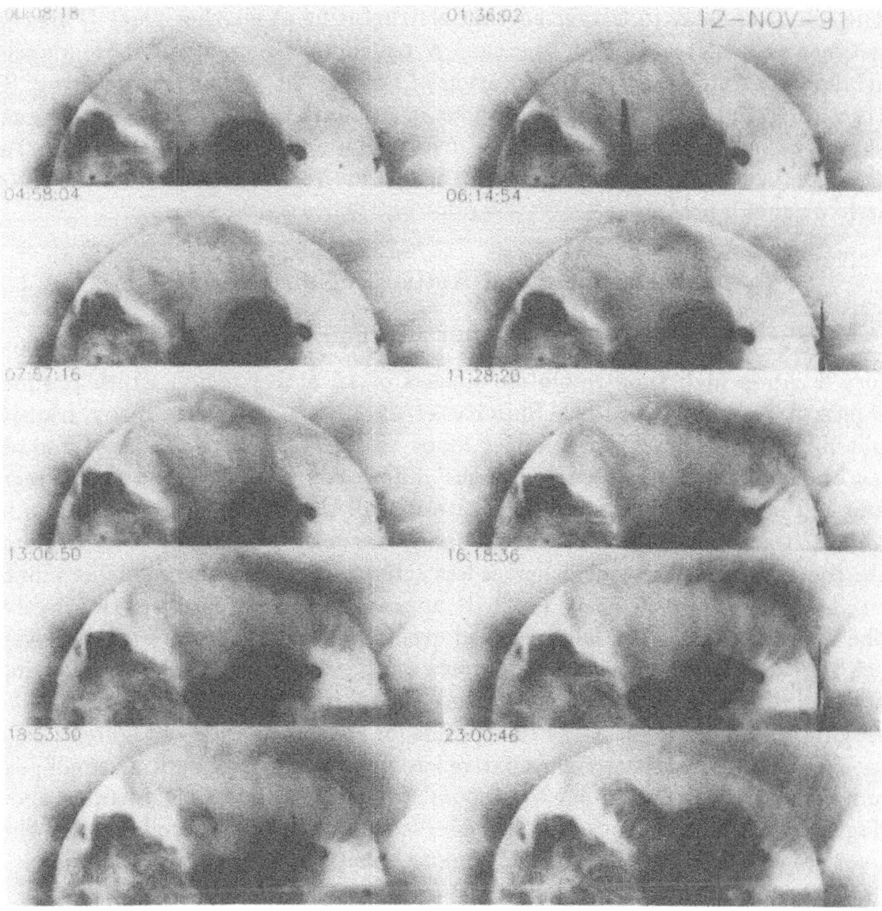

Fig. 5. Large-scale coronal restructuring that occurred on November 12, 1991. The closed loop or cusp structure is formed over 10 hours. The exposure time is 2.6 sec.

strength of this event is much smaller than that of flares, because this event occurs in the general corona with a larger scale size. Nevertheless, magnetic reconnection appears to play a fundamental role both for long duration flares and non-explosive restructuring of the quiet sun.

3.3. Avalanche of global restructuring

The SXT movie impressively shows that the solar corona is full of such global restructuring, and we suggest that non-explosive magnetic reconnection is responsible for the restructuring. Furthermore, the SXT movie shows many examples of successive propagation of the global restructuring, which sometimes appears more complex than this simple example. When the local restructuring occurs somewhere on the sun, it appears to trigger another restructuring at another place. This chain sometimes propagates all over the sun. A restructuring creates a new magnetic neutral sheet, which may be the location of subsequent global restructuring. The restructuring and reconnection go on like an avalanche, and the coronal magnetic field may be in the self-organized critical state (Lu and Hamilton 1991; Bak, Tang and Wiesenfeld 1988) against the disturbance due to the evolution of the photospheric magnetic fields.

4. Microflares and Active Region Heating

4.1. Microflares and Magnetic Loop Interaction

Figure 6 shows high time resolution images of an active region which produced energetic flares (Shimizu 1992; Shimizu *et al.* 1992). There are many transient brightenings (microflares) of compact loops as shown by arrows. These compact loops are faint and are obscured by the diffuse component of the active region before and after the microflares. The durations of the brightenings are from a few minutes to tens of minutes. These brightenings are observed roughly every 3 minutes in the energetic active regions, whereas less active regions produce a smaller number of brightenings (Shimizu *et al.* 1992). It appears that multiple loops are involved in the brightenings for spatially resolved events, although there are smaller events which appear to be a single source (Shimizu *et al.* 1992). Figure 7 shows an example of the evolution of a microflare, suggesting the multiple loop interaction.

Figure 8 is a *GOES* soft X-ray time profile for the same period as Figure 6. *GOES* data can serve as a total coronal radiative loss monitor. Surprisingly, all small peaks seen by *GOES* are in one-to-one correspondence with the small brightenings shown in Figure 6 without exception. SXT observed more transient brightenings which are not seen in *GOES*, because SXT is more sensitive in detecting microflares.

4.2. Relation with Hard X-ray Microflares

In the previous solar maximum, Lin *et al.* (1984) discovered microflares in hard X-rays above 20 keV with non-thermal hard X-ray spectra, and suggested that they could be a contributor to the coronal and/or active region heating.

Figure 9 shows examples of the hard X-ray microflares observed by Lin *et al.* (1984) together with the *GOES* time profile. Small enhancements are seen in the *GOES* data associated with the hard X-ray microflares. These enhancements in the *GOES* profile appear very similar to the peaks corresponding to the SXT brightenings (Fig. 8). This suggests that the brightenings observed by SXT are the soft X-ray counterpart of the the hard X-ray microflares.

If this is true, then the microflares seen by SXT may have essentially a non-thermal nature, and the non-thermal electrons are produced by multiple loop interactions. Figure 7 shows that the multiple footpoints of the loops brighten first, followed by the brightening of the entire loops. The footpoint brightenings may be due to the bombardment of the non-thermal electrons, and the following loop

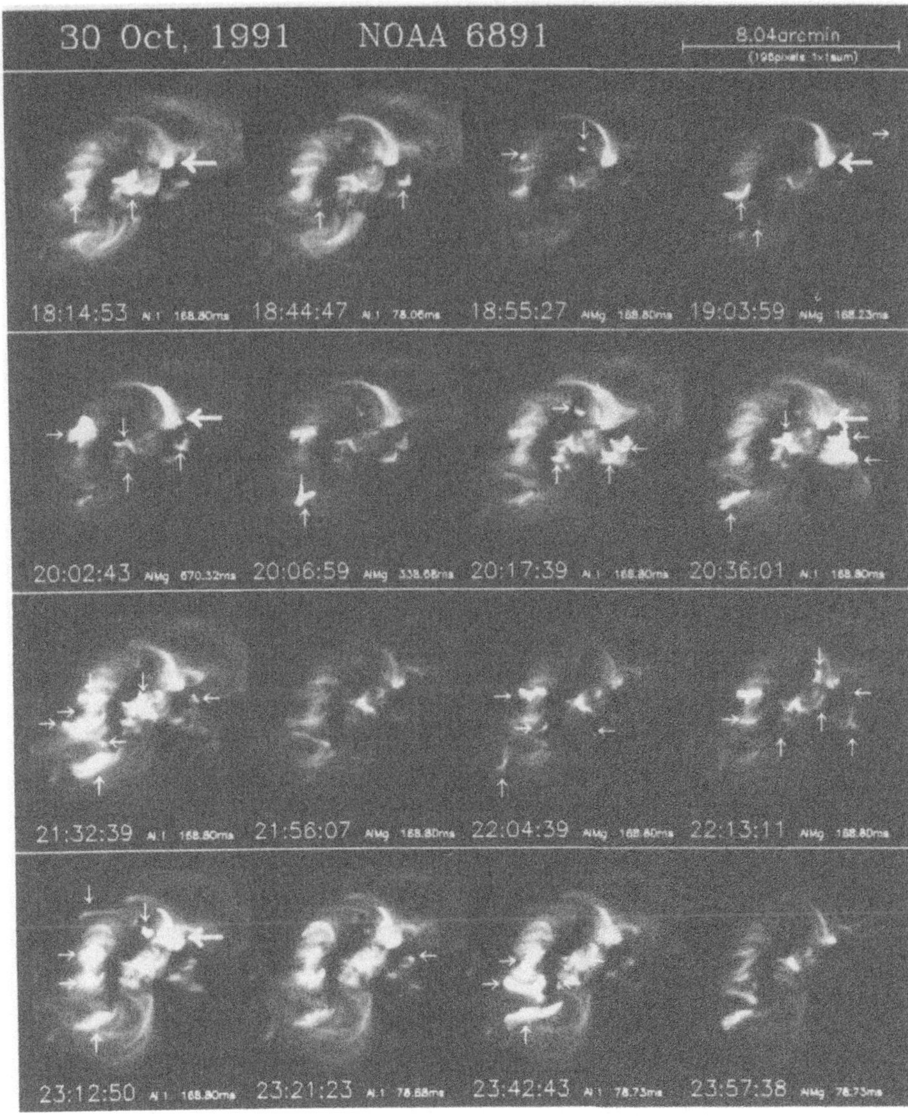

Fig. 6. X-ray images of the active region NOAA 6891. Frequent microflares are seen as indicated by arrows (Shimizu 1992; Shimizu *et al.* 1992). The FOV is 8 arcmin, and the pixel size is 2.5 arcsec.

brightenings may be due to the injection of the heated plasma from the bombarded footpoints. Shimizu *et al.* (1993) are trying to estimate the energy input to the active region plasma by obtaining the integral rate of occurrence of these brightenings.

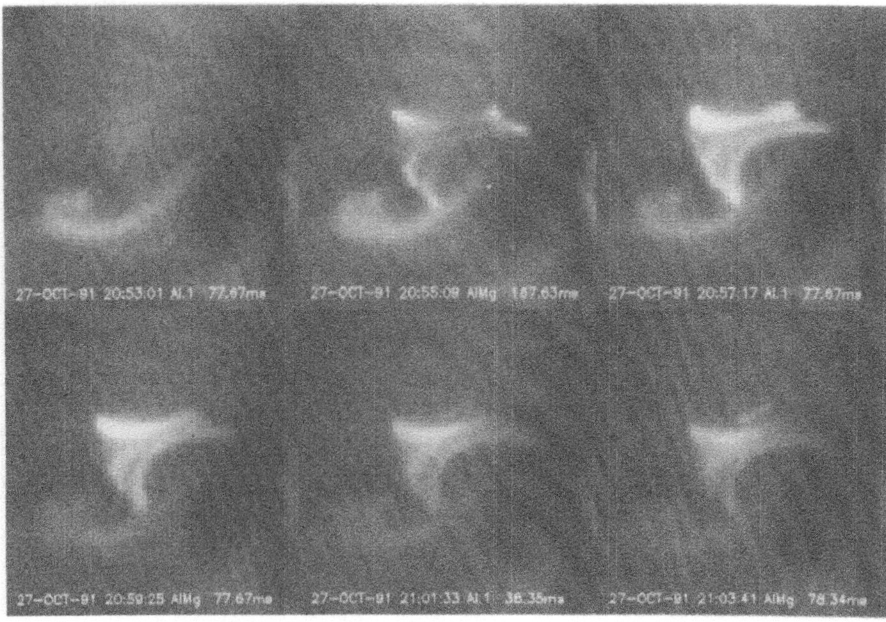

Fig. 7. Evolution of a microflare that occurred on October 27, 1991. The footpoint brightenings are followed by brightening of the entire loops.

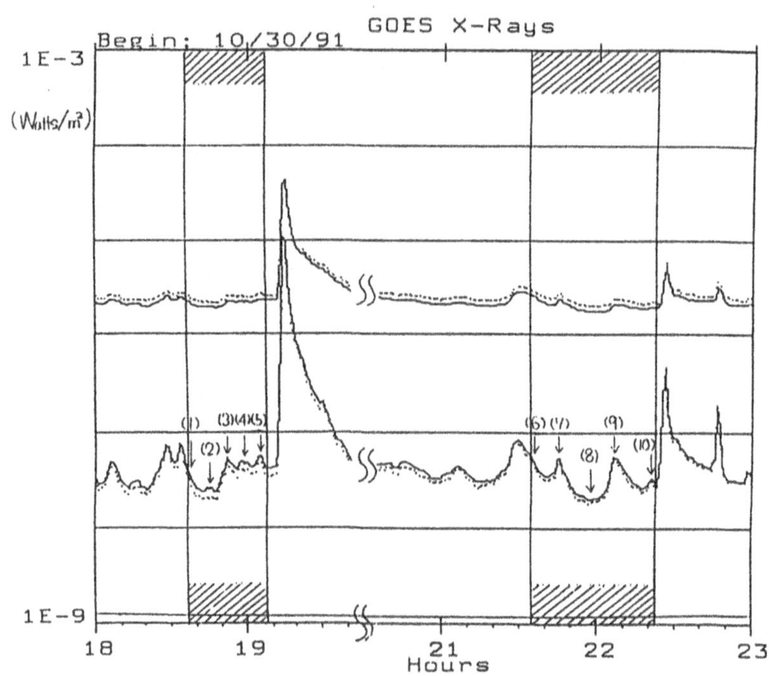

Fig. 8. Small spikes (microflares) have one-to-one correspondence with the brightenings of the compact loops observed by SXT (Shimizu 1992; Shimizu *et al.* 1992).

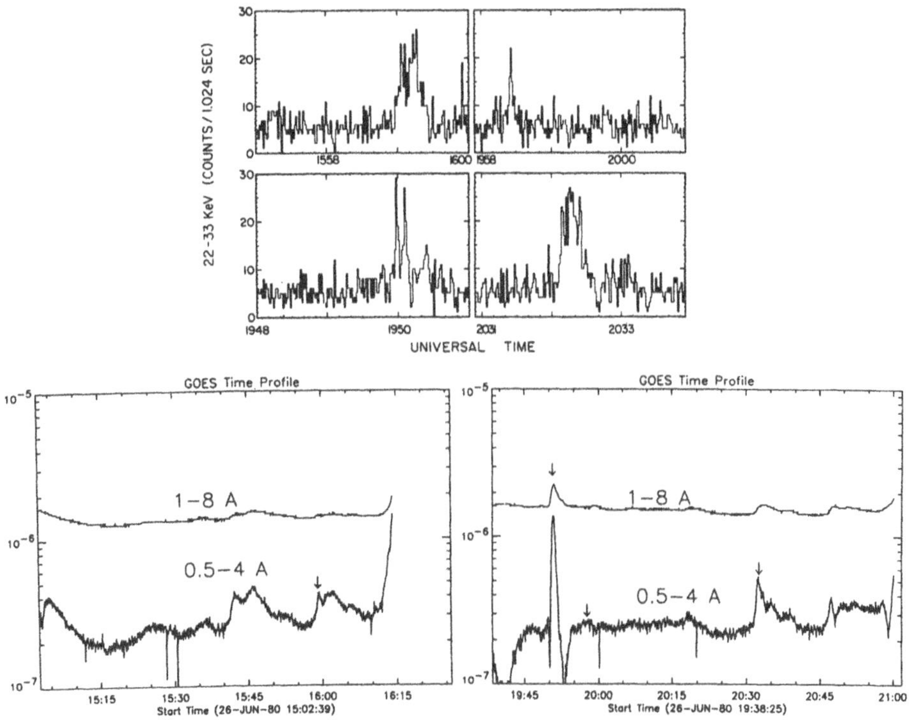

Fig. 9. Hard X-ray microflares observed by Lin *et al.* (1984) and the soft X-ray counterparts seen in the *GOES* plot.

4.3. Magnetic reconnection and non-thermal electrons

Lin (1985) reported that low-energy non-thermal electrons not associated with hard X-ray and optical flares are frequently detected in the interplanetary space, and he interpreted them as due to "coronal flares." Comparison of SXT observations with hard X-ray observations implies that non-thermal particle acceleration is generally associated with magnetic reconnection. Therefore, we suggest that non-explosive magnetic reconnection associated with the global coronal restructuring reported in the previous section is the source of these non-thermal electrons observed in interplanetary space.

5. Temperature Structure of Active Regions

SXT has the capability to obtain pixel-by-pixel plasma temperatures using the broadband filters. Figure 10 shows the temperatures of a bright loop and the diffuse area in an active region obtained from all possible pairs of five analysis filters (Hara *et al.* 1992). The temperatures thus obtained are steadily concentrated around 6

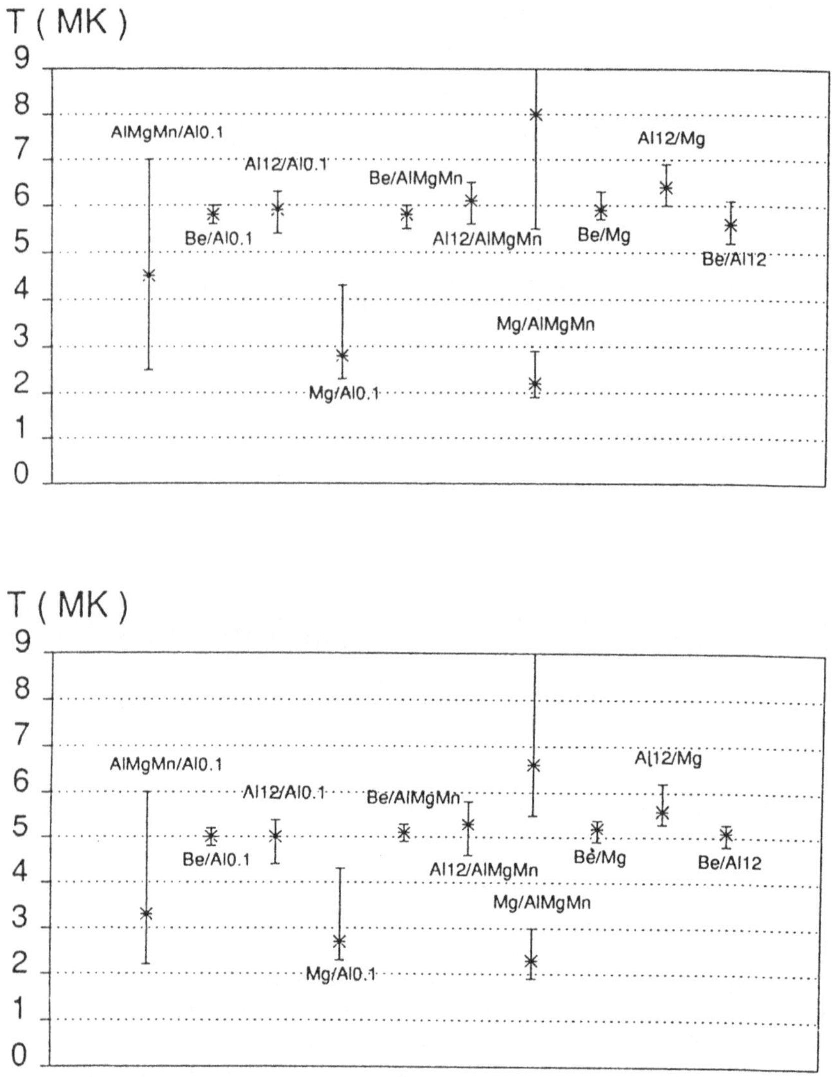

Fig. 10. Temperature and emission measure of an active region obtained from all pairs of filters. Upper panel: bright loop in the active region. Lower panel: diffuse area in the same active region (Hara *et al.* 1992).

MK for the loop, and about 5 MK for the diffuse area. The emission measure is about 10^{27-28} cm^{-5}. Note that the single temperature obtained from the broadband filters is the weighted mean of the possible differential emission measure structure of the active region plasma with the telescope wavelength (temperature) response. It is, therefore, misleading to assume that there is a single temperature plasma with the temperature obtained from the broadband filters.

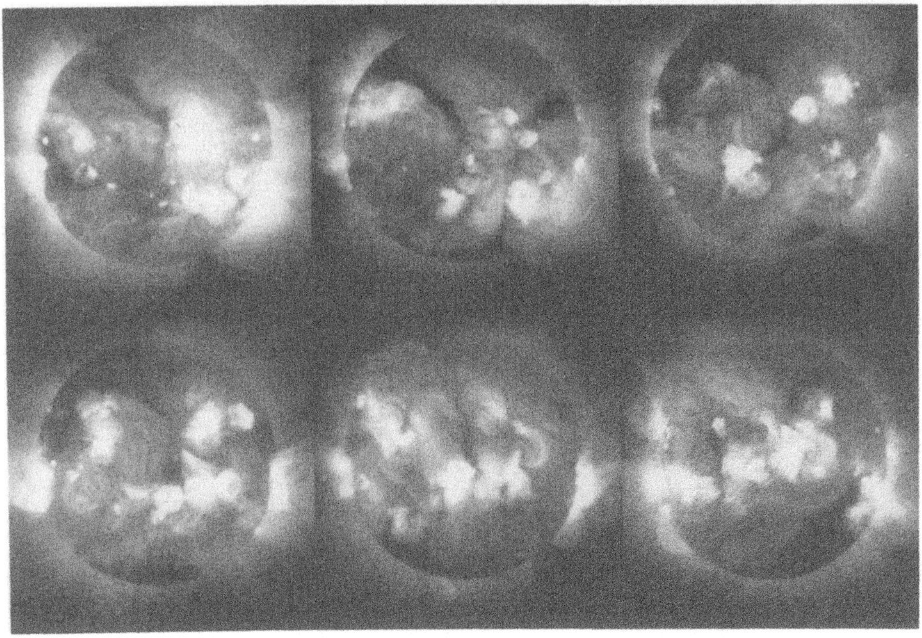

Fig. 11. A coronal dark channel stable over five solar rotations from September 1991 through early February 1992. Exposure dates are (from bottom to top, from left to right) September 20, October 17, November 14, and December 11, 1991, and January 6, and January 31, 1992. Left is North, and up is West.

Figure 10 also shows that some filter pairs give much lower temperatures around 2-3 MK, indicating the presence of a lower temperature component. The emission measure is about 10^{28-29} cm^{-5}. This low temperature component is consistent with the temperature obtained from coronal line observations. This indicates that the differential emission measure distribution of active region plasmas generally extends to around 5-6 MK. This is not inconsistent with *Skylab* results. Levine and Pye (1980) implied the presence of plasmas with temperature higher than 2-3 MK.

6. Rigidly Rotating Coronal Dark Channels

We have not yet seen any large scale coronal holes as observed by *Skylab* (Krieger 1977). Instead, we often see thin long dark areas along the North-South direction (Fig. 11). These areas are generally connected to northern and/or southern polar coronal holes. We call these areas "coronal dark channels."

Although the coronal dark channels are often transient, some of the channels last at least as long as several solar rotations as shown in Figure 11. Figure 12 shows a sketch of the boundary of the coronal dark channel. Dotted lines indicate the location of the channel when it is assumed to have followed the differential rotation derived by sunspot motions (Newton and Nunn 1951). We can clearly see

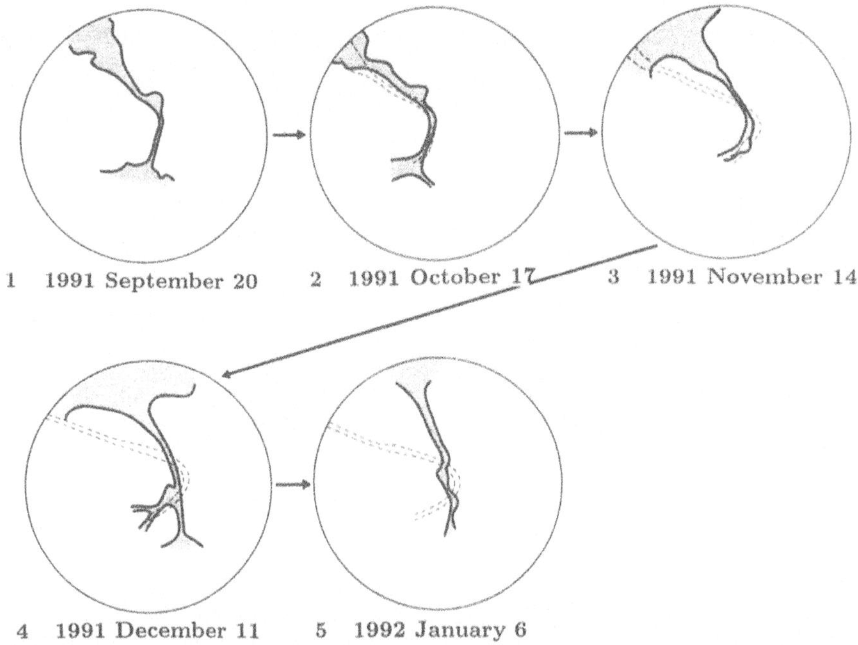

Fig. 12. A sketch of the coronal dark channel seen in Fig. 11. The solid line represents the boundary of the dark channel, and the dotted line is the location of the channel when it is assumed to have followed the differential rotation derived from the sunspot motions (Hayashi 1992).

that the channel is rotating almost rigidly (Hayashi 1992; Ogawara *et al.* 1993).

Figure 13 shows that the dark channel rotates almost rigidly with an angular speed close to the equatorial rotation speed. The angular velocity of the channel is almost constant over the latitude. A similar rigid body rotation is found for the *Skylab* large-scale coronal hole (Timothy, Krieger and Vaiana 1975). The present observations provide the new result that even these very thin structures in the longitudinal direction (smaller than active region size) show this rotation law. We may hypothesize that in general long-lived coronal magnetic structures, including the coronal dark channels, do not differentially rotate as sunspots (Antonucci and Dodero 1978, 1979; Sheeley, Nash and Wang 1987; Wang, Nash and Sheeley 1989). We are producing X-ray synoptic charts to see the rotation nature of the global coronal fields in more detail (Takahashi 1993).

Acknowledgements

The *Yohkoh* SXT is the result of many years of hard work by many scientists and engineers both in Japan and in the United States. We would like to express our sincere thanks to my colleagues, especially to M. Abe, L. Acton, M. Bruner, R. Caravalho, S. Freeland, H. Hara, B. Jurcevich, S. Kubo, M. Morrison, N. Nitta, Y.

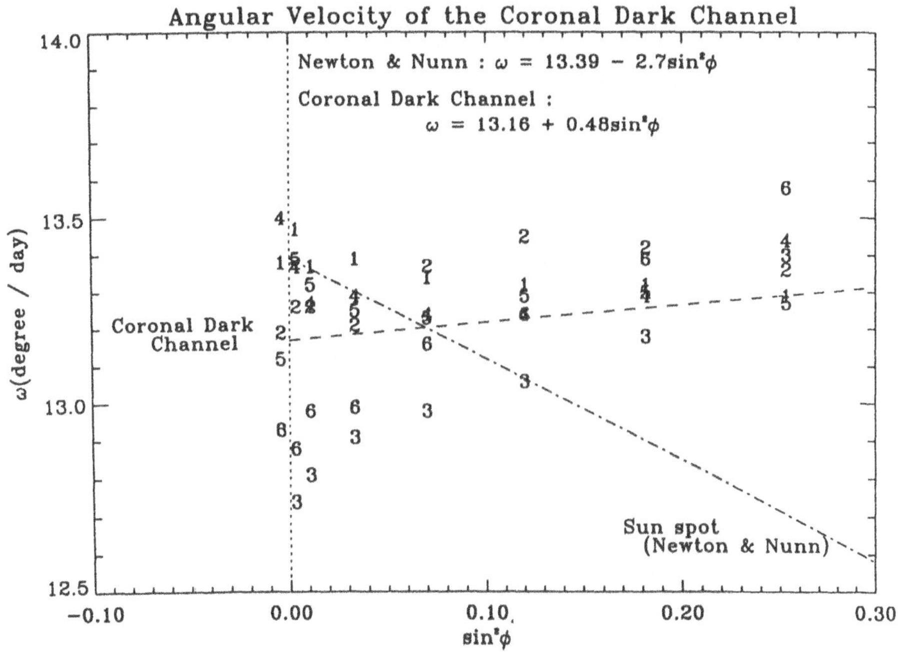

Fig. 13. Rotation speed of a coronal dark channel median and sunspots as a function of latitude (Hayashi 1992). The numbers are the rotation number of the channel starting from September 1991, and indicate the rotation speed of the channel with rotation number.

Ogawara and T. Shimizu, who significantly contributed to the successful fabrication, testing and operation of SXT.

References

Acton, L., Bruner, M., Stern, R., Hirayama, T., Tsuneta, S., Watanabe, T., and Ogawara, Y.: 1988, *Adv. Space. Res.* **8**, 11-93

Antonucci, E. and Dodero, M.: 1977, *Solar Phys.* **53**, 179

Antonucci, E. and Dodero, M.: 1979, *Solar Phys.* **62**, 107

Bak, P., Tang, C. and Wiesenfeld, K.: 1988, *Phys. Rev. A* **38**, 364

Bruner, M., Acton, L., Brown, W., Stern, R., Hirayama, T., Tsuneta, S., Watanabe, T., and Ogawara, Y: 1989, *Yosemite conference on outstanding problems in solar system plasma physics, Am. Geophysical Union Monograph* **54**, 187

Hara, H.: 1992, *Master Thesis*, Department of Astronomy, The University of Tokyo

Hara, H., Tsuneta, S., Acton, L. W., Lemen, J. R., and McTiernan, J. M.: 1992, *Publ. Astron. Soc. Japan* **44**, in press

Hayashi, K.: 1992, *Undergraduate Research Program Report*, Department of Astronomy, The University of Tokyo

Krieger, A.: 1977, *Coronal Holes and High Speed Wind Streams*, Colorado Assoc. Univ. Press, 71

Kosugi, T.: 1993, this volume

Levine, R. and Pye, J.: 1980, *Solar Phys.* **66**, 39

Lin, R.: 1985, *Solar Phys.* **100**, 537

Lin, R., Schwartz, R., Kane, S., Pelling, R., and Hurley, K.: 1984, *ApJ* **283**, 421

Lu, E. and Hamilton, R.: 1991, *ApJ* **380**, L89

Newton, H. and Nunn, M.: 1951, *Monthly Notices Roy. Astron. Soc.* **111**, 413
Ogawara, Y., Takano, T., Kato, T., Kosugi, T., Tsuneta, S., Watanabe, T., Kondo, I., and Uchida, Y.: 1991, *Solar Phys.* **136**, 1
Ogawara, Y., *et al.*: 1993, in preparation.
Pneuman, G: 1981, *Solar Flare Magnetohydrodynamics*, Gordon and Breach, 379
Sheeley, N., Nash, A., and Wang, Y.: 1987, *ApJ* **319**, 481
Shibata, K., Nozawa, S., and Matsumoto, R.: 1992, *Publ. Astron. Soc. Japan* **44**, 265
Shibata, K., *et al.*: 1992, *Publ. Astron. Soc. Japan* **44**, in press
Shimizu, T.: 1992, *Master Thesis*, Department of Astronomy, The University of Tokyo
Shimizu, T., Tsuneta, S., Acton, L., Lemen, J., and Uchida, Y.: 1992, *Publ. Astron. Soc. Japan* **44**, in press
Shimizu, T., *et al.*: 1993, in preparation.
Tajima, T., Brunel, F., and Sakai, J.: 1982, *ApJ (Letters)* **245**, L45
Takahashi, T.: 1993, *Master Thesis*, Dept. of Astronomy, Univ. of Tokyo, in preparation
Timothy, A., Krieger, A., and Vaiana, G.: 1975, *Solar Phys.* **42**, 135
Tsuneta, S.: 1987, *Solar Phys.* **113**, 35
Tsuneta, S., Acton, L., Bruner, M., Lemen, J., Brown, W., Caravalho, R., Catura, R., Freeland, S., Jurcevich, B., Morrison, M., Ogawara, Y., Hirayama, T., and Owens, J.: 1991, *Solar Phys.* **136**, 37
Tsuneta, S., Hara, H., Shimizu, T., Acton, L., Strong, K., Hudson, H., and Ogawara, Y.: 1992a, *Publ. Astron. Soc. Japan* **44**, in press
Tsuneta, S., Takahashi, T., Acton, L., Bruner, M., Harvey, K., and Ogawara, Y.: 1992b, *Publ. Astron. Soc. Japan* **44**, in press
Uchida, Y.: 1993, this volume
Uchida, Y., *et al.*: 1992, *Publ. Astron. Soc. Japan* **44**, in press
Wang, Y., Nash, A., and Sheeley, N.: 1989, *Science* **245**, 712

Hard X-Ray Solar Flares Observed by *Yohkoh* and Particle Acceleration

TAKEO KOSUGI

National Astronomical Observatory, Mitaka, Tokyo 181, Japan

Abstract. A brief review is given of the initial six months (October 1991 – March 1992) of solar flare observations with the Hard X-ray Telescope (HXT) aboard *Yohkoh*. The *Yohkoh* HXT has the following advanced capabilities: (i) four-band, simultaneous imaging in the X-ray energy ranges of 14–23, 23–33, 33–53, and 53–93 keV; (ii) high angular resolution of $\sim$ 5 arcsec with a wide field of view covering the whole Sun; (iii) high time resolution of 0.5 s; and (iv) high sensitivity with a total effective area exceeding 50 cm^2. More than 250 solar flares were observed in the initial six months, including four GOES X-class flares. It was revealed that hard X-ray sources of flares are much more complicated than expected; they sometimes vary quite rapidly, and even at the same time they show different shapes at different X-ray energies. Typical features were derived, however, from observations of several well-analyzed flares. Their implications for the particle acceleration mechanisms of solar flares are discussed.

Key words: Sun: Flares – Sun: Hard X-rays – Particle acceleration

1. Introduction

The Hard X-ray Telescope (HXT; Kosugi *et al.*, 1991) aboard the *Yohkoh* satellite (Ogawara *et al.*, 1991) started routine observations on October 3, 1991 after about one month of adjustment/calibration following the launch (August 30, 1991).

Solar flares produce plenty of high-energy electrons, and these electrons immediately emit hard X-rays while colliding with ambient plasma ions (bremsstrahlung). The total intensity of flares in hard X-rays usually varies rapidly, with time scales of the order of seconds or less, suggesting quite rapid electron accelerations operating impulsively (flare impulsive phase). The energy contained in these high-energy electrons, estimated from hard X-ray emission, is not a negligible part; sometimes it is almost comparable to the total flare energy. Thus the impulsive electron acceleration is one of the key problems of flare theory. Since hard X-ray propagation is almost unaffected by the solar atmosphere, we can expect that imaging observations of flares in this range provide the most direct information concerning the acceleration, transfer, and confinement of high-energy electrons. Based upon this expectation, the first two hard X-ray imaging observations were made during the previous sunspot cycle maximum around 1980–81 by the Solar Maximum Mission and *Hinotori* satellites, and many fruitful results were obtained from these experiments. Nevertheless, they were just preliminary ones; these imaging observations were made in the photon energy range below $\sim$ 30 keV, where thermal emission from a plasma hotter than some 10^7 K may contaminate the hard X-rays emitted from high-energy electrons. Moreover, the angular and temporal resolutions were not sufficient to follow rapidly changing, impulsive phenomena in a complicated structure.

The *Yohkoh* HXT was designed so as to provide a new look at solar flares by imaging hard X-ray sources of purely nonthermal electron origin. Instrumentally it is an imager of the Fourier-synthesis type with 64 modulation subcollimators, each independently measuring the spatially-modulated incident photon count; a set of 32 cosine and sine pairs of photon-count data is telemetered to the ground

J.F. Linsky and S. Serio (eds.), Physics of Solar and Stellar Coronae, 131–138.

TABLE I
Monthly summary of HXT flare observations.

Year	1991			1992			
Month	Oct.	Nov.	Dec.	Jan.	Feb.	Mar.	TOTAL
No. of events	46	22	66	39	69	9	251
in Flare mode*)	32	16	48	17	39	3	155
in Quiet mode**)	14	6	18	22	30	6	96
GOES X class	1	1	1	1	0	0	4
M class	15	6	34	12	26	1	94
C class	24	12	28	25	33	8	130
(not reported)	6	3	3	1	10	0	23

* An observing mode of *Yohkoh*, in which HXT data are taken in all energy bands with temporal resolution 0.5 s.
** Another mode, in which HXT data are taken only in the L-band (14–23 keV) with temporal resolution 2.0 s.

and synthesized there into an image. The adoption of this novel design principle enabled HXT to achieve the advanced capabilities which include the following:

i) Simultaneous imaging in four energy bands, namely, the L-band (14–23 keV), M1-band (23–33 keV), M2-band (33–53 keV), and H-band (53–93 keV).

ii) Angular resolution as fine as $\sim$ 5 arcsec with a wide field of view covering the whole Sun.

iii) Basic temporal resolution of 0.5 s.

iv) High sensitivity with a total effective area exceeding 50 cm^2.

The details of the HXT design have been described elsewhere (Kosugi *et al.*, 1991). Here we simply point out that we have already confirmed the above capabilities in orbit (Kosugi *et al.*, 1992).

Table I shows a monthly summary of HXT flare observations during the initial six months since October 1991. More than 250 solar flares were observed during this period, four of which were GOES X-class flares. Detailed analyses of these flares have been in progress; some initial results have been summarized by Kosugi *et al.* (1992), Sakao *et al.* (1992), and Matsushita *et al.* (1992), all published in the *Yohkoh* Special Letter Issue of the *Publications of the Astronomical Society of Japan* (Vol. 44, No. 5, October 1992). In the following, we will briefly review these initial Letter papers, as well as some recent findings, with particular attention to the electron acceleration in the flare impulsive phase.

2. Characteristics of HXT Flare Images

So far several tens of flares have been well analyzed. It has been revealed that the hard X-ray sources of these flares are much more complicated than expected; each flare shows a different source structure from the others, and even within a flare the structure changes from one moment to another, as well as from one energy band to another. Even so, there seem to exist some common, or typical, characteristics of hard X-ray sources, and they have been deduced from initial analyses of a limited number of flares, although they need be confirmed by further studies.

One of the basic characteristics that we have found is that hard X-ray flares observed in the HXT L-band (14–23 keV) usually show one or more long, thin structures which seem to trace magnetic loops. In the images of a flare occurring just on the limb (December 2, 1991), we observed a bright loop reaching a projected height of about 10,000 km above the photosphere with a knot near the loop top being the brightest feature. The knot seems to be located where a weaker, secondary loop crosses the brighter loop. The hard X-ray images in this band generally resemble the corresponding soft X-ray images taken with the Soft X-ray Telescope (SXT; Tsuneta *et al.*, 1991) aboard *Yohkoh*.

On the other hand, at higher energies the sources are generally more compact and patchy. We have several typical examples in which hard X-rays are emitted from two separate sources (double-source structure) at both ends of the long, thin structure seen in the L-band, suggesting that the electrons (possibly accelerated near the loop top) propagate along the magnetic loops and stream into the lower atmosphere at their footpoints. There are, of course, cases where more than two compact sources brighten almost simultaneously. In contrast, some flares show a simple, single-source structure. The latter may be due to the still insufficient angular resolution of HXT. Anyway, it is clear that we need detailed studies in this regard, which will be discussed later.

Sometimes we observe a sudden shift of the hard X-ray source from one location to another during the impulsive phase of a flare. In the flare of October 29, 1991 at 09:55 UT (Yaji *et al.*, in preparation) whose impulsive phase consists of three spikes which are separated by about one minute in time, the second spike originates from a site some 40,000 km away from the first spike site. It is concluded that the disturbance that triggers the second spike travels at a velocity of about 500 km s^{-1}, implying that it is an MHD disturbance. The third spike is located at the same site as the second. Correspondingly, the third spike is characterized by the softest hard X-ray spectrum among the three, while the second spike is the hardest. Maybe the aftermath of the second spike, i.e., an increased ambient plasma density due to evaporation of chromospheric material, may hinder the efficient acceleration of electrons.

Similarly in cases of intense flares, we observe in the precursor phase hard X-ray sources which appear in an area much wider than those in the impulsive phase. Since the number of flares whose precursor phase has been well analyzed is still small, we cannot conclude that this is the general case. It is suggestive, however, that some restructuring of the active-region coronal magnetic fields commences not from inside but globally from the outer boundary, at least in some cases.

In order to clearly show the essence of these characteristics, we present here as an example the X-class, intense flare on November 15, 1991 at 22:34 UT. This flare occurred near the disk center at S13 W19 in the NOAA active region 6919. The HXT observation of this flare has been reported by Sakao *et al.* (1992), and related observations with *Yohkoh* SXT and at the Mees Solar Observatory by Hudson *et al.* (1992) and Canfield *et al.* (1992), respectively.

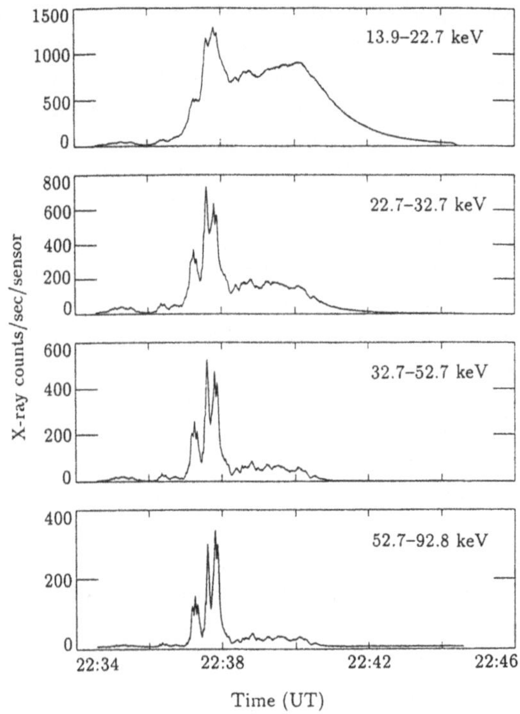

Fig. 1. Hard X-ray time histories of the November 15, 1991 solar flare. The average counting rates of the HXT 64 subcollimators are given (after Sakao *et al.*, 1992).

The hard X-ray flare can be divided into three stages (22:34–22:37, 22:37, 22:38–22:42 UT) from its time history shown in Fig. 1. The first is the precursor stage, where several hard X-ray sources appear in a relatively diffuse area, say about 1 arcmin square, and expand quite rapidly (Figs. 2a and b). During the second stage corresponding to the impulsive phase, hard X-ray emission concentrates in an elongated, compact source which seems to straddle the magnetic neutral line (Fig. 2c). Then during the third stage, namely, the post-impulsive or gradual phase, the brightest portion shifts towards just above the neutral line with a new subsource appearing as designated by an arrow (Fig. 2d).

The impulsive phase consists of three distinct spikes, whose peaks and valleys we call hereafter P1, P2, and P3, and V1 and V2, respectively. The behavior of the hard X-ray sources during these three spikes is very interesting, as compactly summarized in Fig. 3. First, the structure evolves from a single source at P1 to double sources at P2 and P3. Second, the double-source structure is more pronounced at higher X-ray energies (> 30 keV), and also at peaks (P2 and P3) rather than at valleys (V2 and 22:38:00 UT). The double sources can be interpreted as footpoint sources of a single magnetic loop seen as an elongated source in the L-band (14–23 keV; not shown). If this is the case, the second point mentioned above may be evidence

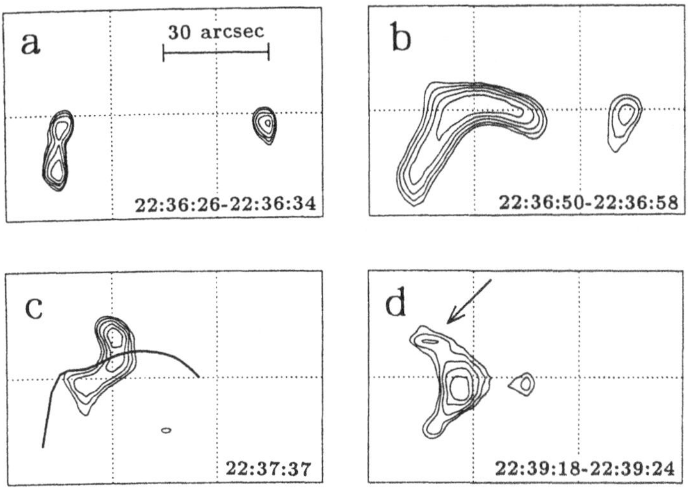

Fig. 2. Evolution of hard X-ray source structure of the November 15, 1991 flare: (a) precursor phase, (b) just before the onset of the impulsive phase, (c) impulsive phase, and (d) gradual phase. All are images in the L-band (14–23 keV) except for the M1-band (23–33 keV) image for (c), in which the magnetic neutral line (thick line) is overlaid. Contour levels are 71, 50, 35, 25, and 18% of the peak brighteness (after Sakao *et al.*, 1992).

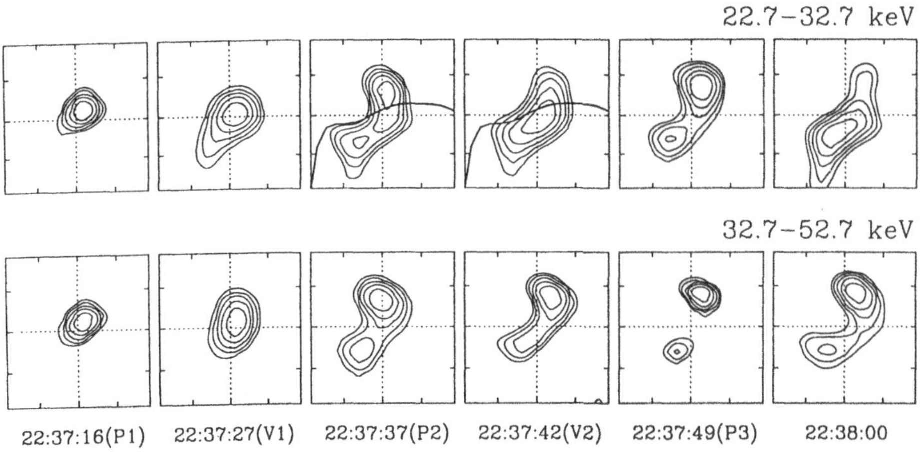

Fig. 3. Comparison of hard X-ray images between the peaks and valleys, as well as between the M1- (upper) and M2-band (lower) images, during the three impulsive spikes of the November 15, 1991 flare. The field of each map is 40 by 40 arcsec. Contour levels are the same as in Fig. 2 (after Sakao *et al.*, 1992).

for high-energy electron precipitation during the spikes with a small fraction of the electrons being reflected and trapped in magnetic mirrors. This interpretation is consistent with the observational fact that the hard X-ray spectrum hardens at the peaks and softens at the valleys (not shown but deducible from Fig. 1), i.e., soft-hard-soft specral evolution of impulsive spikes (Kane and Anderson, 1970; see also Dennis, 1988), because the precipitating electrons yield a harder spectrum than trapped electrons if the electron spectra are the same (e.g., Melrose and Brown, 1976; Kosugi *et al.*, 1988).

Further, perhaps definitive, support for this interpretation was recently obtained by Sakao (private communication). Using a newly-developed, sophisticated data analysis tool, he compared flux time histories of the double sources for the second and third spikes, and found that the double sources vary quite similarly with no time lag (accuracy of less than 0.1 s). This means that the acceleration site must exist between the two sources, i.e., near the loop top.

As pointed out previously, the hard X-ray source structure evolves from a single source at P1 to double sources at P2 and P3. The double-source structures at P2 and P3 are not the same, however. A careful examination of Fig. 3 reveals a slight increase in separation between the double sources from ~13 arcsec at P2 to ~15 arcsec at P3. Sakao (private communication) successfully traced the double-source structure well into the post-impulsive phase after 22:38 UT, and found that this increase in separation continues to ~30 arcsec. Also he found that the angle sustained by the line connecting the double sources and the magnetic neutral line increases as time progresses. Thus we believe that each spike in the impulsive phase, as well as each small hump seen in the time histories of Fig. 1 during the post-impulsive phase, corresponds to the energy release occurring in a different magnetic loop, and that a multiple loop system flares up progressively from underlying, strongly-sheared loops to overlying, less sheared loops. The single source at P1 may be explained by a short length of the loop compared with the HXT angular resolution.

From Fig. 1 it is quite clear that a new, gradual component appears in the post-impulsive phase. This new component is characterized by a very steep spectrum, suggestive of thermal origin. A qualitative analysis of this component, as well as the impulsive spikes, is in progress to clarify the quantitative relationship between the impulsive, precipitating electrons and the gradual, hot thermal plasmas.

3. Average Height of Hard X-ray Sources

As is suggested by different source structures at different X-ray energies, the heights may differ at different X-ray energies. In fact, such height variations were observed in several flares which occurred near the solar limbs; the height is usually lower by a few to several arcsec at higher-energy bands. Probably this height difference is due to a change in the relative intensity between the loop-top and footpoint sources. An alternative explanation is that, when high-energy electrons precipitate, higher-energy X-rays originate from lower and denser layers because higher-energy electrons have smaller collision cross sections and can penetrate deeper. We can evaluate which of the two explanations is the correct one if the measurement is precise enough. Unfortunately the HXT angular resolution (~5 arcsec) is not suit-

TABLE II
Average height of hard X-ray sources.

HXT energy band	Height from photosphere	Height from L-band
L : 14–23 keV	$(9.7 \pm 2.0) \times 10^3$ km	——
M1: 23–33	——	$(-1.0 \pm 0.3) \times 10^3$ km
M2: 33–53	——	(-2.0 ± 0.5)
H : 53–93	——	(-3.2 ± 0.7)

able for making such a precise measurement. A statistical approach, however, may provide a hint to this question.

Matsushita *et al.* (1992) compared the centroid locations of X-ray brightenings with the corresponding Hα flare locations for about a hundred flares, and derived the average height of hard X-ray sources and its variation with X-ray energy. Most of the flares included in this statistic are of the impulsive type. The results are summarized in Table II. Here the heights in the M1-, M2-, and H-bands are compared with the height in the L-band, so that no misalignment between the HXT and Hα observations is involved in this part. (The uncertainty level given in this table is a standard deviation of the statistics, not involving any possible systematic errors.)

There are two important findings included in Table II: one is an unexpectedly large average height of X-ray sources above the photosphere, and the other is the height dispersion between the M1-, M2-, and H-bands. The former might be due to some unknown systematic errors involved in the comparison of locations between hard X-ray and Hα sources, but a much larger error than that given in the table seems unlikely because most of the limb flares we have imaged in the L-band show a loop-like structure whose top reaches an altitude as high as 10 arcsec ($\sim$ 7,000 km) or higher. Probably the great altitude, say higher than some 5,000 km, is real, maybe resulting from predominance of a loop-top source at least in the L-band. Note that it is very difficult to interpret such a high altitude in terms of the electron precipitation model (e.g., Brown and McClymont, 1975). On the other hand, the latter finding, i.e., the lower altitudes towards higher X-ray energy, is most easily explained by the electron precipitation model, as suggested in the first paragraph of this section.

Thus, without doubt, further detailed studies are required both observationally and theoretically. Regarding observational studies, we are conducting detailed analyses of near-the-limb flares, from which we expect to derive some more definitive conclusions in the near future.

4. A Short Summary

We have confirmed the performance of *Yohkoh* HXT in orbit. All the advanced capabilities are now found to be achieved. More than 250 solar flares, including four GOES X-class events, were detected during the initial six months of observation. Detailed HXT data analyses of several interesting flares have revealed several important features concerning the behavior of impulsive hard X-ray sources, and further, the characteristics of the electron acceleration. From these initial studies, we have reconfirmed that *Yohkoh* HXT is a powerful instrument. Nevertheless, solar flares are a very complicated phenomenon involving many processes. We need to develop comprehensive studies, including cooperative data analyses among the four *Yohkoh* instruments, especially between HXT and SXT, as well as between *Yohkoh* and ground-based, optical and radio instruments. Such efforts have just begun and we expect further fruitful results will follow in the near future.

Acknowledgements

The author wishes to express his thanks to the late Prof. Keizo Kai, former Principal Investigator of HXT, for his leadership in preparing this successful telescope. The staff of ISAS, the staff of NASA Deep Space Network, and all members of the *Yohkoh* team are acknowledged for their earnest support in the *Yohkoh* operation, without which we could not have detected so many flares in this short period. This work is a group effort of the HXT core members, namely, Y. Ogawara, K. Makishima, T. Murakami, T. Sakao, S. Masuda, M. Inda, K. Yaji, and K. Matsushita. This work is partially supported by the Scientific Research Fund of the Ministry of Education, Science and Culture under Grant No. 02452011. The Yamada Science Foundation provided the author with travel funds to attend the Symposium held at Palermo, Italy.

References

Brown, J.C. and McClymont, A.N.: 1975, *Solar Phys.* **41**, 135.

Canfield, R.C., Hudson, H.S., Leka, K.D., Mickey, D.L., Metcalf, T.R., Wuelser, J.-P., Acton, L.W., Strong, K.T., Kosugi, T., Sakao, T., Tsuneta, S., Culhane, J.L., Phillips, A., and Fludra, A.: 1992, *Publ. Astron. Soc. Japan* **44**, in press.

Dennis, B.R.: 1988, *Solar Phys.* **118**, 49.

Hudson, H., Acton, L., Hirayama, T., and Uchida, Y.: 1992, *Publ. Astron. Soc. Japan* **44**, in press.

Kane, S.R. and Anderson, K.A.: 1970, *Astrophys. J.* **162**, 1003.

Kosugi, T., Dennis, B.R., and Kai, K.: 1988, *Astrophys. J.* **324**, 1118.

Kosugi, T., Makishima, K., Murakami, T., Sakao, T., Dotani, T., Inda, M., Kai, K., Masuda, S., Nakajima, H., Ogawara, Y., Sawa, M., and Shibasaki, K.: 1991, *Solar Phys.* **136**, 17.

Kosugi, T., Sakao, T., Masuda, S., Makishima, K., Inda, M., Murakami, T., Ogawara, Y., Yaji, K., and Matsushita, K.: 1992, *Publ. Astron. Soc. Japan* **44**, in press.

Matsushita, K., Masuda, S., Kosugi, T., Inda, M., and Yaji, K.: 1992, *Publ. Astron. Soc. Japan* **44**, in press.

Melrose, D.B. and Brown, J.C.: 1976, *M.N.R.A.S.* **176**, 15.

Ogawara, Y., Takano, T., Kato, T., Kosugi, T., Tsuneta, S., Watanabe, T., Kondo, I., and Uchida, Y.: 1991, *Solar Phys.* **136**, 1.

Sakao, T., Kosugi, T., Masuda, S., Inda, M., Makishima, K., Canfield, R.C., Hudson, H.S., Metcalf, T.R., Wuelser, J.-P., Acton, L.W., and Ogawara, Y.: 1992, *Publ. Astron. Soc. Japan* **44**, in press.

Tsuneta, S., Acton, L., Bruner, M., Lemen, J., Brown, W., Caravalho, R., Catura, R., Freeland, S., Jurcevich, B., Morrison, M., Ogawara, Y., Hirayama, T., and Owens, J.: 1991, *Solar Phys.* **136**, 37.

PROPERTIES OF SMM FLARES

J. SYLWESTER and B. SYLWESTER
Space Research Centre of Polish Academy of Sciences
Wroclaw, Poland

Abstract. It is known that X-ray observations of the solar corona contain information on the flare energy release. In this paper we shall describe the use of density-temperature and emission measure-temperature diagrams for the analysis of the flare heating function. Simple considerations of flare behaviour in the diagrams are supported by results of calculations of hydrodynamic flare models. The model calculations have been performed using the PALERMO-HARVARD 1D code. We start our considerations from a description of the so called quasi-steady-state type of changes in a coronal solar loop. Next we discuss the flare decay phase and the dependence of the pattern of evolution in the diagrams on the time variations of flare heating rates. Finally we discuss a simplified energetics model for the flare rise phase and the dependence of the form of global flare thermal energy rise on the equivalent geometrical flare characteristics, i.e. loop semilength, L, and cross-sectional area, A. We illustrate our considerations by examples of flares observed in soft X-rays by the Solar Maximum Mission Bent Crystal Spectrometer (BCS).

Key words: flares – heating processes – diagnostics – X-rays

1. Introduction

According to the picture derived from EUV and X-ray observations (SKYLAB, NIXT, *YOHKOH*) the plasma of the solar corona is usually confined in loop-like magnetic flux tubes anchored in the photosphere. In addition to providing a channel for heat deposition, the magnetic field acts both by confining plasma motions along its direction and by suppressing thermal conduction across the field. For the purpose of this paper, we shall assume that the loops are insulated from one another and uniform in the direction perpendicular to the magnetic field.

2. The Steady-State

There are four parameters describing the plasma conditions in the loop: the loop semilength, L, the base pressure, p, the maximum temperature, T (usually at the summit), and the maximum volumetric heating rate, E_{Hmax}. If the heating inside the loop is stationary over sufficiently long time and is symmetric with respect to the summit, only two of these parameters are independent. There exist two relationships between them which for loops smaller than the pressure scale height and uniform heating along the coronal portion of the loop are known as steady-state (S-S) scaling laws (Rosner, Tucker and Vaiana, 1978, hereafter RTV); $T^3 \propto pL$, $E_{\mathrm{H}} \propto p^{7/6} L^{-5/6}$ (hot loop solution discussed by Antiochos and Noci 1986). Instead of using the pressure, one can use the summit density, N, as a parameter. This leads to a modified equivalent pair of relationships:

$$T^{3.5} = 1.2\,10^6 E_{\mathrm{H}} L^2, \tag{1}$$

$$N\,L = 1.4\,10^6\,T^2. \tag{2}$$

These equations relate plasma conditions at the loop summit to the volumetric heating rate under the assumption that the heating is constant in time and uniform ($E_{\mathrm{H}} = E_{\mathrm{Hmax}}$ everywhere along the loop of semilength L). The values of

J.F. Linsky and S. Serio (eds.), Physics of Solar and Stellar Coronae, 139–146.

constants in Eqs. (1) and (2) have been derived based on results of hydrostatic loop modelling (the PALERMO-HARVARD code) under the assumption that the heating is uniformly distributed along the coronal portion of the loop. In this case the loop heating power is $\mathcal{E}_{\rm tot} = E_{\rm H} V$, where V is the loop volume. If only a smaller portion of the loop length is heated or the heating function is non-uniform along the loop, we can introduce a parameter characterizing the equivalent heated portion of the loop ρ. For the loop in which the maximum energy release occurs in the corona, somewhere close to the loop summit, it can be defined as $\rho = \int_0^L E_{\rm H}(l)dl/LE_{\rm Hmax}$. In the case of non-uniform heating, $\mathcal{E}_{\rm tot}$ is ρ times smaller than for uniform heating and the plasma density is scaled down by a factor $\sqrt{\rho}$ (since the power of the loop heating must be balanced by radiative losses which scale as N^2). For a given ρ, the scaling relation (2) transforms to

$$N\,L = 1.4\,10^6 \sqrt{\rho}\,T^2. \tag{3}$$

The summit temperature is approximately the same as in the case of uniform heating (until $\rho \ll 1$), since the maximum rate of heating ($E_{\rm Hmax}$) does not change and the "conductive" coupling relation (Eq. 1) is weakly dependent on plasma density.

3. The Quasi-Steady-State

Let us now consider the evolution of thermodynamic characteristics of the plasma at the loop summit under conditions of slowly changing heating along the loop (gentle variations). Under such conditions Eqs.(1) and (3) are valid and the role of motions in carrying the energy can be neglected. We can distinguish three specific cases of evolution:

(a) the spatial distribution of the heating is constant (ρ = const) and the power of the loop heating ($\mathcal{E}_{\rm tot}$) is changing,

(b) the maximum volumetric rate of energy deposition ($E_{\rm Hmax}$) is constant and $\mathcal{E}_{\rm tot}$ is changing (in proportion to ρ),

(c) The $\mathcal{E}_{\rm tot}$ is constant and the spatial distribution of the energy deposition along the loop is changing (i.e. $E_{\rm Hmax} \propto \rho^{-1}$).

Each of these simple cases of "gentle" evolution of the heating has a corresponding characteristic pattern of behaviour in the density-temperature ($N-T$) diagram in which the summit plasma parameters are plotted (Fig. 1). In case (a), the evolution proceeds along the lines with the slope $\xi = dlogT/dlogN = 1/2$. The slopes of these lines are characteristic for the steady-state scaling law (Eq. 2). In the rest of this paper we shall refer to this case as steady-state (S-S) (if $\rho = 1$) or quasi-steady-state (QSS) (if $\rho < 1$). In case (b), Eq. (1) predicts that the maximum temperature of the plasma should be approximately constant, but the density should depend on $E_{\rm Hmax}$. When the distribution of heat in the loop tends to be uniform, the plasma conditions at the summit tend to reach the S-S values along a horizontal trajectory. In case (c), $\mathcal{E}_{\rm tot}$ is constant and the density changes reflect changes of $E_{\rm Hmax}$ which are followed by related variation of the maximum temperature (Eq. 1). The density variations tend to accommodate the temperature dependence of the total radiative losses ($P_{\rm r} \propto T^{-1/2}$, RTV). This case is characterized by a steep slope of

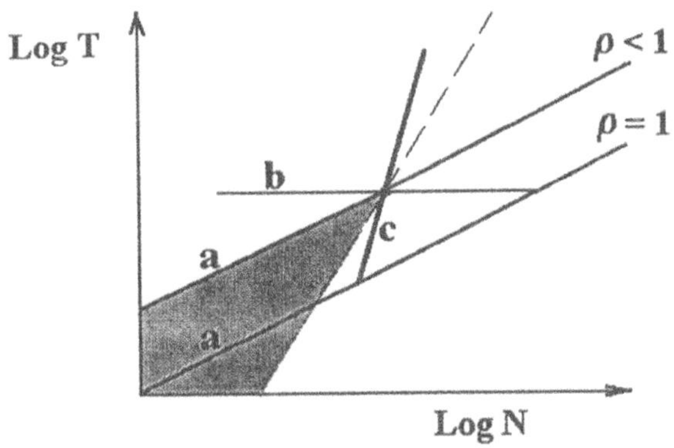

Fig. 1. The scheme of characteristic trajectories for the three cases discussed in the text: a - lines with slopes $\xi = dlogT/dlogN = 1/2$, b - line with slope $\xi \approx 0$, c - line with slope $\xi \approx 4$. The dashed line in the figure ($\xi \approx 2$) corresponds to the so-called OFF case discussed in Section 4. The marked area corresponds to all of these flare decay phase trajectories for which the decrement of flare heating is of the order of the thermodynamic time given by Eq. (4).

the evolutionary path in the $(N-T)$ diagram. Evolutionary lines for cases (b) and (c) are limited by the S-S line.

Two characteristic time scales for restoring the steady situation (after a small perturbation of the heating rate value or heating rate distribution along the loop) appear to be important. The shorter one characterizes the conductive coupling (few to several tens of seconds for the solar coronal loops). The longer one represents the thermodynamic time scale:

$$\tau_{\rm th} = 3.7\,10^{-4}\,\frac{L}{\sqrt{T}}, \tag{4}$$

introduced by Serio et al. (1991). This time scale characterizes the rate of decrease of the entropy per particle at the loop summit after an abrupt switch-off of the heating. It is shown later that the time $\tau_{\rm th}$ corresponds also to the increment of the thermal energy of the plasma during the rise phase of flares. Provided that the characteristic time of flare heating variations is longer than a few thermodynamic times, the distribution of parameters along the loop is expected to be close to QSS. As an example of this type of slow variation of the heating, one may consider evolution of the non-flaring active region loops or some flare loops during the decay phase (as will be shown in the following section).

4. Decay Phase Analysis

A frequent subject of debate is whether the heating operates during the decay phase of flares. Jakimiec et al. (1992) considered this problem in detail, showing that the slope of the decay phase trajectory in the (N-T) diagram is directly related to the value of the decrement time τ of the flare heating rate.

To allow for a direct comparison with observations, instead of using the $(N-T)$ diagrams, we shall use their observational counterparts, the $(\sqrt{\epsilon}-T)$ diagrams. The summit density N is usually a parameter inaccessible from the analysis of soft X-ray spectra. The square root of the emission measure, $\sqrt{\epsilon}$, characterizes well the plasma density under the assumption of constant emitting volume. The plasma temperature can be determined (in the isothermal approximation) using appropriate line ratio diagnostics. Therefore, the $(\sqrt{\epsilon}-T)$ diagrams can be constructed based on the analysis of the soft X-ray spectra. It should be noted, however, that the value of the characteristic slope $\zeta = dlogT/dlog\sqrt{\epsilon}$ in the $(\sqrt{\epsilon}-T)$ diagrams differs from the corresponding value of ξ in the $(N-T)$ diagrams, but the general pattern of behaviour does not change. We should derive characteristic values of the slope for lines corresponding to cases (a), (b) and (c) before using the $(\sqrt{\epsilon}-T)$ diagrams for diagnostic purposes.

B. Sylwester et al. (1992) verified that T_{Ca} is a good measure of the summit plasma temperature and that $\sqrt{\epsilon_{\mathrm{Ca}}}$ is a good measure of the average density of the plasma contained in the coronal portion of the loop. In Fig. 2 we present (after B. Sylwester et al. 1992) a plot that summarizes results of many hydrodynamic calculations of different flare models related to scenario (a) from the previous section. We remind the reader that in this scenario the value of $\mathcal{E}_{\mathrm{tot}}$ decreases exponentially during the decay phase without changing the shape of the spatial distribution of heating along the loop. The line plotted in Fig. 2 shows the dependence of the slope $\zeta = dlogT_{\mathrm{Ca}}/dlog\sqrt{\epsilon_{\mathrm{Ca}}}$ on the decrement time τ of the heating decay. The value of τ is expressed in units of the thermodynamic time τ_{th}. Inspection of Fig. 2 confirms that for slowly decreasing heating (large τ/τ_{th}), the slope of the decay trajectory corresponds to the S-S slope already discussed (the value of the slope is $\zeta = 0.37$). For models characterized by a rapid switch-off of flare heating the slope during the decay is $\zeta = 1.46$ (Fig. 2). These two characteristic limiting branches represent slow (QSS) and fast (OFF) cooling regimes (a and dashed lines in Fig. 1). It was found that slopes for these two branches do not depend on the behaviour of the heating prior to decay, the initial temperature, or the loop size (in the solar case).

For times τ of the order of τ_{th} the intermediate slopes have been obtained (Fig. 2). For these intermediate cases the dependence of the slope ζ on L and/or T is substantial but cancels out when plotted in terms of τ_{th}. The dependence shown in Fig. 2 may be directly used as a diagnostic of flare heating based on the $(\sqrt{\epsilon_{Ca}}-T_{Ca})$ diagrams obtained from BCS observations. For details of T_{Ca} and ϵ_{Ca} determinations see Fludra et al. (1991).

For those observed flares with sensible estimates of the characteristic flare lengths, the "absolute" value of the characteristic decay time of the heating (in s) and the value of the "absolute" heating rate (in ergs $\mathrm{cm}^{-3}\ \mathrm{s}^{-1}$) can be derived from Eqs. (1) and (4).

B. Sylwester et al. (1992) analyzed the decay phase evolution for 66, mostly large, flares observed by the BCS. Several characteristic patterns of flare evolution during the decay phase have been distinguished:

(*I*) OFF–like, which have decay phase slopes close to those obtained for models with an abrupt switch-off or with rapid decay of the heating rate ($\tau/\tau_{\mathrm{th}} < 0.2$).

(*II*) QSS–like, with slopes corresponding to flare models with quasi-steady-state

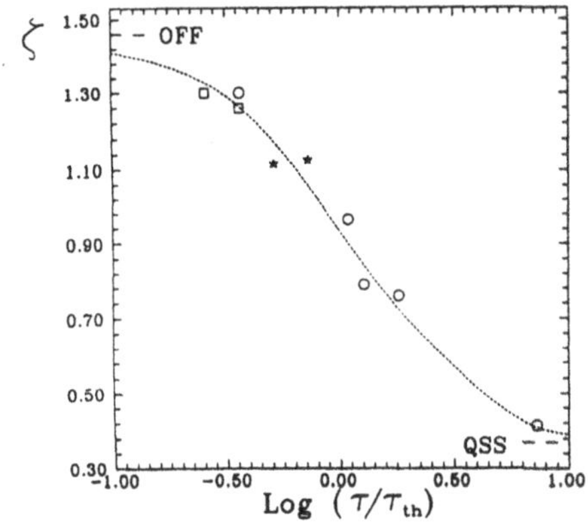

Fig. 2. The dependence of the slope ζ of the decay phase trajectories in the emission measure-temperature diagram on the decrement time τ of flare heating expressed in terms of the thermodynamic time τ_{th}. For details see B. Sylwester et al. (1992).

evolution (very slow heating rate decay, $\tau/\tau_{th} > 5$).

(III) Intermediate, having slopes between those for OFF-like and QSS-like cases (see corresponding marked area in Fig. 1).

(IV) Those characterized by decay phase slopes not predicted by the considered simple flare models, for instance those related to rearrangement of the heating function profile inside the loop (cf. case (b) and (c) in Fig. 1).

(V) Those for which the decay phase of the evolutionary path is composed of two or more branches corresponding most probably to the decay of different loop structures.

Examples of the observed flare evolution for each of the cases discussed above are shown in Fig. 3. For about 50% of the analysed flares the emission measure-temperature diagrams resemble those obtained for simple hydrodynamic flare models (cases I, II and III). This suggests that the assumptions made in the Palermo-Harvard modelling are appropriate for real conditions in many solar flares. For at least 70% of flares it is necessary to assume that the flare heating persisted into the decay phase. The flares for which emission measure-temperature diagrams belong to groups (IV) and (V) seem to be much more complicated. They might involve a set of distinct loop structures within the same flaring active region. Each of the loops may successively dominate the soft X-ray emission during the flare evolution and be responsible for part of the evolutionary path in the diagram. For some flares from categories (IV) or (V) the possible explanation for the observed behaviour in the diagram may be the rearrangement of the heat distribution along the loop. Forthcoming *YOHKOH* high resolution X-ray images of the flaring loops may substantially help us understand the temporal and spatial variations of the heat release in solar flaring loops.

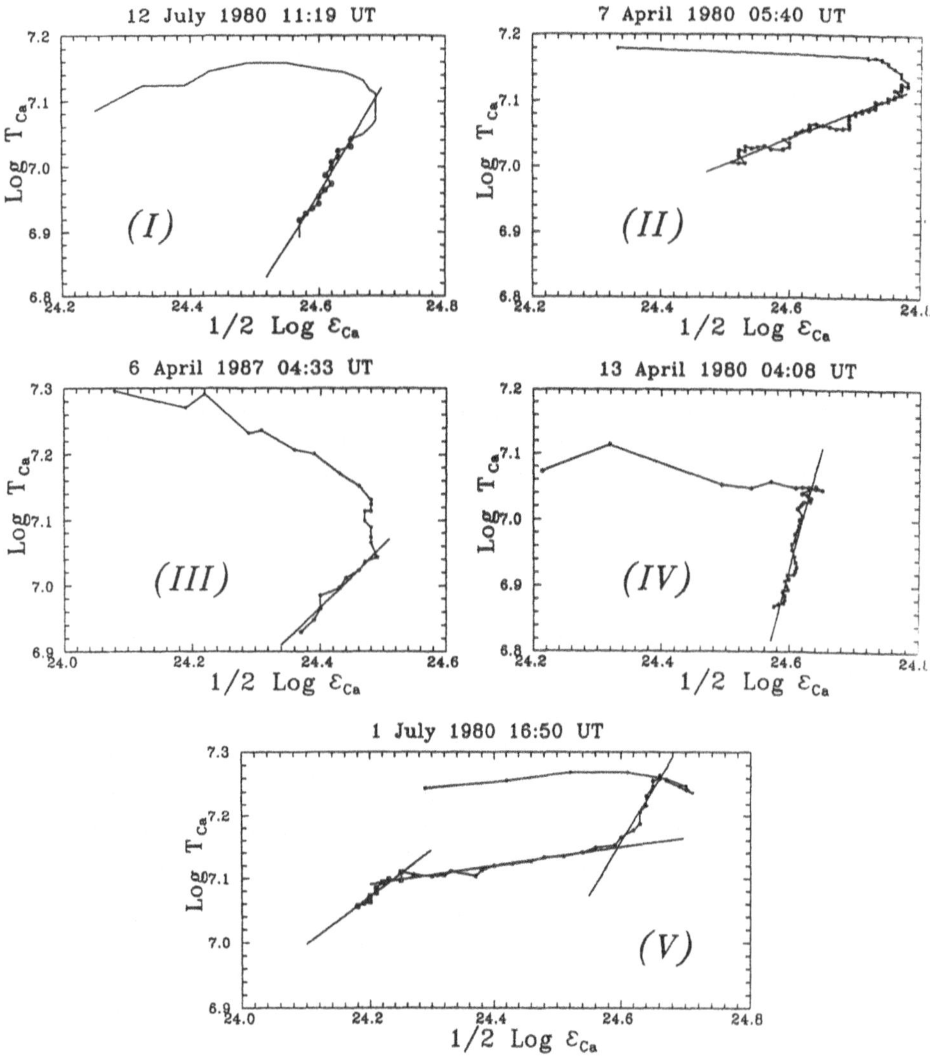

Fig. 3. Examples of the characteristic patterns of the observed flare evolution for five categories discussed in the text: (*I*) - the heating has been switched-off abruptly; (*II*) - the heating changes so slowly that the evolution occurs along the S-S line during the decay; (*III*) - the decrement of flare heating is about equal to the thermodynamic time (cf. Eq. 4); (*IV*) - the rearrangement of the heating during the decay most probably occurred (observed slope $\zeta = 3.6$ is steeper than that characteristic for the OFF case); (*V*) - the decay evolutionary path is composed of three branches with different slopes. The first branch corresponds to the OFF case, the second resembles branch b in Fig. 1 and the last may be interpreted by assuming $\tau/\tau_{th} \approx 2$. It is possible that different loop structures are responsible for branches with different slopes. Diagrams are plotted in cgs units.

5. Rise Phase Energetics

For some simple flares it is possible to investigate equivalent flare loop parameters, provided that reliable estimates of the time variations of the flare thermal energy ($E_{\rm th} = 3kNTV$) are known (J. Sylwester et al. 1993). However, it is not possible to determine directly the flare thermal energy based on temperatures and emission measures which are available from the analysis of the soft X-ray spectra (the volume V and/or the density of the flaring plasma are usually not known). Instead, it is possible to follow the time variations of the so-called thermodynamic measure η ($\eta = T\sqrt{\epsilon}$; $E_{\rm th} = \eta\sqrt{V}$). The temperature T and emission measure ϵ estimates should refer to that part of the flaring loop where the bulk of the thermal energy is stored. As discussed earlier, the temperatures and emission measures obtained from analysis of the BCS spectra provide information about the flare heating variations during flares and thus are very useful parameters for constructing diagrams. These parameters can also be used to determine the value of the thermodynamic measure.

Results of the hydrodynamic modelling of the flaring plasma show that under the assumption of a constant heating rate (the value and profile along the loop), the time behaviour of the flare total thermal energy (or thermodynamic measure) can be very well approximated (see Fig. 4) by the following simple expression:

$$\eta(t) = \eta_0 \left[1 - exp\left(-\frac{t\sqrt{\rho}}{\tau_{\rm th}}\right)\right] . \qquad (5)$$

Here η_0 represents the value of the thermodynamic measure characteristic for the corresponding S-S situation. It would be reached when the heating operates for a sufficiently long time, i.e. when $\tau/\tau_{\rm th} \gg 1$. The value of η_0 depends on the flaring loop geometry, i.e. the cross-sectional area, A, the loop semilength, L, and the parameter ρ ($\eta_0 \propto \sqrt{\rho}\sqrt{A/L}$). It was found that the characteristic increment time of the rise of thermal energy (or thermodynamic measure) is equal to the thermodynamic time (Eq. 4). In Fig. 4 we show both the calculated and the fitted values of $E_{\rm H}(t)$. The calculated values correspond to one of the models discussed in detail by Jakimiec et al. (1992). This model is characterized by a long (600 s) phase of constant flare heating. One can therefore conclude that the thermodynamic time characterizes the fastest rate of change of density-dependent thermodynamic characteristics for both the rise and decay phases of flares.

Values of η_0 and $\tau_{\rm th}/\sqrt{\rho}$ can be obtained from the best fit of Eq. (5) to the observed time variations of $\eta(t)$ during rise phase. Empirical determination of $\tau_{\rm th}$ for both flare rise and decay allows one to estimate the upper limit for the effective loop semilength and the lower limit for the cross-sectional area of the flaring loop, provided that the summit plasma temperature is known.

In Table I, we present values of the geometrical characteristics derived from analysis of the BCS observations for the rise phase of three flares described in more detail by B. Sylwester et al. (this volume). Since ρ is not known a priori, values in the second and third columns in the table set upper limits for the flaring loop semilengths and lower limits for the areas. The estimated value for the volume does not depend on ρ and therefore can be used to estimate the plasma density $N_{\rm max}$ based on the observed maximum value of the emission measure. These values of

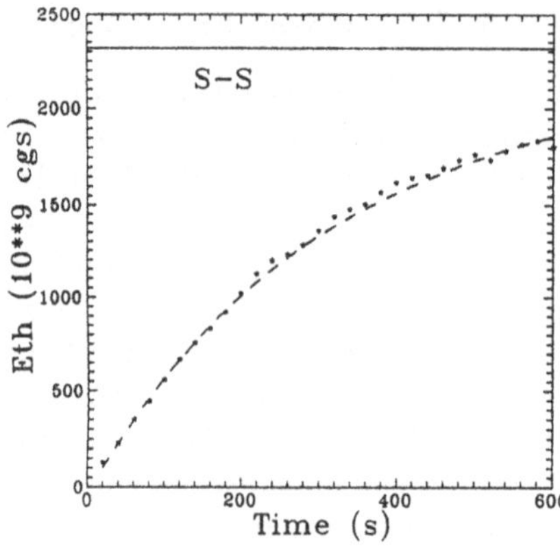

Fig. 4. The fit of values predicted by Eq. (5) to the rise time variations of the total thermal flare energy. Full points - values calculated from hydrodynamic modelling, dashed line - values of η (multiplied by $V^{1/2}$) obtained from the approximation given in Eq. (5).

TABLE I
Characteristics of Three Flares Observed on June 29, 1980

BCS peak time (UT)	$L/\sqrt{\rho}$ (10^9 cm)	$A\sqrt{\rho}$ (10^{16} cm^2)	Volume (10^{26} cm^3)	N_{max} (10^{11} cm^{-3})	$d\,\rho^{1/4}$ (arc sec)
02:37	2.4	3.4	1.6	0.6	2.9
10:43	1.3	1.2	0.32	3.1	1.7
18:25	2.5	1.2	0.58	2.6	1.7

maximum density are given in the fifth column. Assuming further that the footpoint cross-sectional area is of a circular form, we are able to estimate lower limits for the apparent effective diameter of the flaring loop (last column).

The application of this procedure to the analysis of the flare observed by the GOES broad-band detectors has been made by J. Sylwester et al. (1993). They found that the values of the effective geometrical parameters determined from the rise of the thermodynamic measure compare well with those estimated from direct imaging (NIXT observations, Herrant et al. 1991).

References

Antiochos, S.K. and Noci, G.: 1986, *ApJ.* **301**, 440

Fludra, A., Lemen, J.R., Jakimiec, J., Bentley, R.D. and Sylwester, J.: 1989, *ApJ* **344**, 991

Herrant, M., Pardo, F., Spiller, E. and Golub, L.: 1991, *ApJ* **376**, 797

Jakimiec, J., Sylwester B., Sylwester, J., Serio, S., Peres, G. and Reale, F.: 1992, *A&A* **253**, 269

Rosner, R., Tucker, W.M. and Vaiana, G.S.: 1978, *ApJ* **220**, 643

Serio, S., Reale, F., Jakimiec, J., Sylwester, B. and Sylwester, J.: 1991, *A&A* **241**, 197

Sylwester, B., Sylwester, J., Serio, S., Reale, F., Bentley, R.D. and Fludra, A.: 1992, *A&A* , in press

Sylwester, J., Sylwester, B., Jakimiec, J., Garcia, H.A., Serio, S. and Reale, F.: 1993, *Advances in Space Res.* , submitted

ANALYSIS OF FLARE EVOLUTION IN THE EMISSION MEASURE-TEMPERATURE DIAGRAM FOR THREE EVENTS OBSERVED BY SMM

B. SYLWESTER, J. SYLWESTER, M. SIARKOWSKI
Space Research Centre of Polish Academy of Sciences Wroclaw, Poland

and

S. SERIO, F. REALE
Istituto di Astronomia and Osservatorio Astronomico, Palermo, Italy

Abstract. We have examined the time profile of the flare heating function based on an analysis of evolutionary paths in the emission measure-temperature diagram. Three limb flares on 29 June 1980 have been selected for the analysis. The observations consist of the soft X-ray spectra and images obtained by instruments placed aboard the Solar Maximum Mission (SMM) satellite. The emission measures and the temperatures have been derived from the CA XIX soft X-ray spectra obtained by the Bent Crystal Spectrometer (BCS). The images obtained by the Flat Crystal Spectrometer (FCS) have been used to estimate the size of one of the flares. We illustrate how the behaviour of the emission measure-temperature diagram allows one to study the rate of energy deposition during both the rise and decay phases of the flare evolution.

Key words: Flares – X-rays – Solar Physics – Hydrodynamics

1. Introduction

The aim of this contribution is the investigation of the flare heating function $E_H(t)$ for selected events observed by the BCS and FCS instruments aboard SMM. The heating function is an important physical quantity because it describes the rate of thermal energy release (per unit volume) in solar flares. In a previous paper by Jakimiec et al. (1992) (P II) it was shown, based on hydrodynamic calculations, that the most efficient diagnostic of the heating function is the analysis of the flare evolution in the (N-T) diagram, where N and T are the summit density and temperature of the plasma contained in the flaring loop. During the heating phase, the temperature of the summit plasma adjusts quickly to the actual value of the energy deposition rate. Thus the temperature variations can be used as a trace of the heating rate variations. During the decay phase the slope of the evolutionary heating path in the (N-T) diagram is strictly related to the character of the decrease of the heating rate. In particular, two characteristic branches of the evolution have been distinguished: one has a steep slope, where $E_H(t)$ decreases rapidly, or is switched-off abruptly (the so-called OFF branch), and the other has a less steep slope where $E_H(t)$ decreases slowly enough to allow the evolution to proceed along the steady-state line (the so-called quasi-steady-state (QSS) branch). The slopes of these two limiting branches have been found to be independent of the loop length and/or initial conditions.

The intermediate (between OFF and QSS) slopes in the (N-T) diagram can be accommodated by models with the flare heating decay constant τ of the order of the thermodynamic decay time τ_{th} (Serio et al. 1991) (P I) and, therefore, depend on the loop semilength and the temperature (Sylwester et al. 1992) (P III). If the

J.F. Linsky and S. Serio (eds.), Physics of Solar and Stellar Coronae, 147–150.

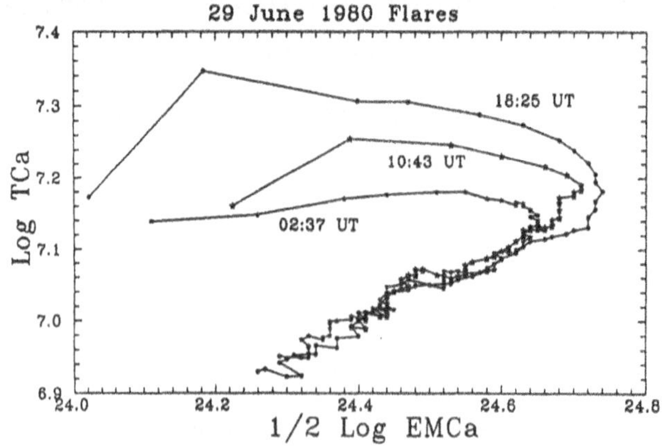

Fig. 1. The emission measure-temperature diagram for three analysed flares.

slope of the decay evolutionary path is determined from the observations and the loop length and initial temperatures are known, the absolute value of the e-folding decay time for the heating rate can be estimated.

The observational counterpart of the (N-T) diagram is the emission measure-temperature diagram ($\sqrt{\epsilon} - T$). As was shown in P III, T_{Ca} and ϵ_{Ca}, as derived from Ca XIX spectra, are related to the summit temperature and density of the flaring structure so they can be used as their characteristics. Therefore, the ($\sqrt{\epsilon} - T$) diagram is in fact the diagnostic diagram for the flare energy release $E_H(t)$. In this research we have used BCS Ca XIX spectra to infer the information about the mean plasma temperature T_{Ca} and emission measure ϵ_{Ca}.

2. Data Analysis

We have selected three large flares observed by the BCS. They occurred on June 29, 1980 in AR NOAA 2522 situated at the west solar limb (S29 W90). They were cospatial in X-rays, gave rise to coronal transients, and had similar appearance in the SMM-observed characteristics as shown Table I.

TABLE I
Basic characteristics of analysed flares.

BCS peak time (UT)	GOES import.	Δt (s)	T_{max} (MK)	ϵ_{max} (10^{49} cm^{-3})	E_H (ergs cm^{-3} s^{-1})	τ (s)
02:37	M3.6	110	18	1.8	4.3[a]	640[b]
10:43	M4.2	180	22	2.5	25	325
18:25	M4.2	230	26	3.0	14[a]	550[b]

[a] - the lower limit (see text for details)
[b] - the upper limit

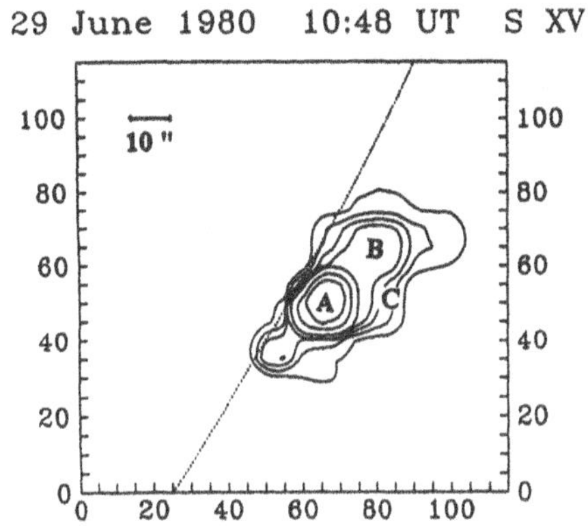

Fig. 2. The deconvolved FCS flare image obtained in the S XV line. The dotted line marks the position of the solar limb as seen by the FCS white light sensor. The contours are at fractions of 0.05, 0.1, 0.15, 0.2, 0.3, 0.5 and 0.7 of the maximum intensity. The locations of three patches which dominate successively are labelled as A, B and C.

The plasma temperatures and emission measures have been derived in the isothermal approximation from the analysis of the Ca XIX spectra (Fludra et al. 1989). The evolution of these parameters in the $(\sqrt{\epsilon} - T)$ diagram is presented in Fig. 1. The duration, Δt, of the rise phase heating increases for the successive events (third column in Table I), and the maximum temperature and emission measure get systematically higher values (fourth and fifth columns). During the decay phase, the pattern of behaviour in the $(\sqrt{\epsilon} - T)$ diagram is similar for the analysed flares. Two evolutionary branches can be distinguished with average slopes $\zeta_1 \approx 0.60$ and $\zeta_2 \approx 0.76$. However the evolution along the first branch lasted for 13.5, 7.0 and 3.5 minutes, respectively, for successive flares.

We have chosen the flare at 10:43 UT for a detailed analysis because this event is well imaged by the FCS. To increase the spatial resolution to about 5", the FCS images in high temperature line S XV (T≈ 16 MK) have been deconvolved. The isophotes of the deconvolved image (10:48 UT) are shown in Fig. 2.

Inspection of such images indicates that during more than 10 minutes from flare start the dominant structure in the S XV line emission is a compact region of the characteristic dimension ≈ 13" which does not change in size or position (A in Fig. 2). The patches B and C twice change their positions by more than 10" (after 10:51 UT). The relative importance of the S XV emissions coming from patches A, B and C is illustrated in Fig. 3.

The light curves presented were obtained by summing the pixels which compose individual patches. Patch A is the main emitting source during the rise phase and during the early decay. Its emission is responsible for the first branch of the $(\sqrt{\epsilon} - T)$ diagram until 10:51 UT. The second branch is associated with the decay of the B and C patches. Analysis of the S XV images of this flare indicates that the location of the initial energy release (source A) was low in the corona, most probably in the flare loop seen in the edge-on position above the limb. The derived semilength of the loop is L = 14 Mm (comparable to the upper limit obtained by J. Sylwester,

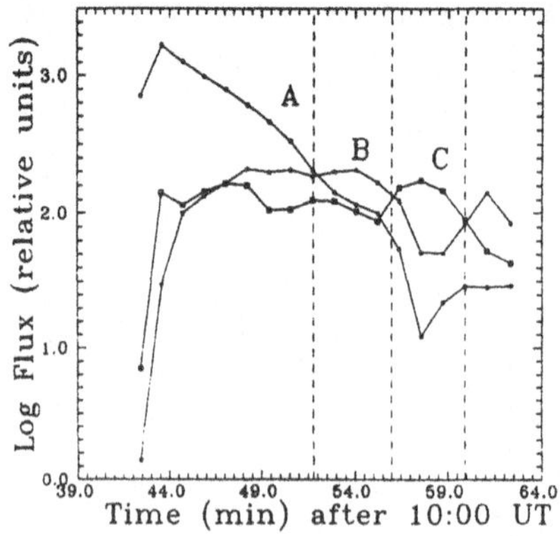

Fig. 3. The S XV line light curves for three patches of the middle flare.

this volume). We concentrated our attention on the analysis of the source A. The behaviour of the evolutionary path in the $(\sqrt{\epsilon} - T)$ diagram closely resembles that obtained for one of the calculated flare models (see P II and P III). During the rise phase the model is characterized by constant heating lasting for 180 s. However a heating rate $E_H(t) \approx 25$ ergs cm^{-3} s^{-1} (2.5 times higher than in the model) had to be assumed to explain the derived rise phase temperatures. During the early decay phase the slope is $\zeta = 0.61$ in the $(\sqrt{\epsilon} - T)$ diagram. This points towards $\tau/\tau_{th} = 2.88$, which leads to an e-folding time of the heating decay $\tau = 325$ s.

3. Summary and Conclusions

Our analysis gives insight into the time behaviour of the heating during the entire flare. For the analysed flares the main conclusions from this study are:

1) The rise phase heating appears to be constant in time. Estimated durations of the heating phase and values of the heating rates are given in columns 3 and 6 of Table I, respectively. The value of the heating rate for the middle flare has been estimated based on the measured size of patch A. For other flares the values given have been derived based on the upper limits of loop semilengths (J. Sylwester, this volume). Successive flares achieve higher emission measures.

2) The early decay phase evolutions resemble each other in the $(\sqrt{\epsilon} - T)$ diagram, for three flares (slopes ζ=0.59, 0.61 and 0.61, respectively) which indicates a value $\tau/\tau_{th} = 2.9$. The corresponding values τ of the decrements are given in the last column of Table I.

References

Fludra, A., Lemen, J.R., Jakimiec, J., Bentley, R.D. and Sylwester, J.: 1989, *ApJ* **344**, 991

Jakimiec, J., Sylwester, B., Sylwester, J., Serio, S. and Reale, F.: 1992, *A&A* **253**, 269 - Paper II

Serio, S. Reale, F., Jakimiec, J., Sylwester, B. and Sylwester, J.: 1991, *A&A* **241**, 197 - Paper I

Sylwester, B., Sylwester, J., Serio., S., Reale, F., Bentley, R.D. and Fludra, A.: 1992, *A&A* in press, - Paper III

HYDRODYNAMICS AND DIAGNOSTICS OF CORONAL LOOPS SUBJECT TO DYNAMIC HEATING

G. PERES

Osservatorio Astrofisico di Catania
V.le A. Doria 6, 95125 Catania Italy

F. REALE and S. SERIO[1]

Istituto ed Osservatorio Astronomico
Piazza del Parlamento 1, 90134 Palermo Italy

[1] *Also at IAIF - CNR*
Via Archirafi 36, 90100 Palermo, Italy

Abstract. We present preliminary results on hydrodynamic simulations of coronal loops maintained close to stationary conditions by a sequence of microflares. We simulate microflares as small (either randomic or periodic) heating episodes which account for all the "stationary heating". We synthesize the emission in some lines and bands observable by X-ray instruments. In particular we report the emission in the bands of the Soft X-ray Telescope on board Yohkoh, in the band detected by the Normal Incidence X-ray Telescope and in the lines detected by the X-ray Polychromator on board the Solar Maximum Mission. Our scope is both to explore feasibility of directly detecting microvariability, and to devise possible diagnostics of variable vs. steady heating.

Key words: Hydrodynamics - Flares - Solar Physics - Spectroscopy

1. Introduction

Although the mechanism of coronal heating is still largely unknown, it is obvious that the energy release which maintains the plasma in coronal loops at a few million degrees can hardly proceed as a constant process, even outside evident flares. Quite likely the "steady" coronal heating endows a time structure, especially if, as proposed by several authors, it is due to the release of energy stored in the magnetic field via some kind of instability (cf. for instance, Galeev et al. 1981).

In this respect it has been proposed that a significant fraction – if not the totality – of the heating in coronal loops is provided by a sequence of very small flare-like episodes of energy release: microflares and nano-flares (Parker 1988, 1989). One of the facts quoted in support of this idea is the energy distribution of frequency of hard X-ray flares: a power law with index -1.8 (Datlowe et al 1974, Dennis 1985). The distribution is poorly known at low energies, i.e. approximately below 10^{27} erg, but if such distribution, reasonably enough, continues unchanged at lower energies, microflares could provide a significant fraction of the coronal heating.

Further support for the presence of microflares comes from a few observational facts. Lin et al. (1984) detected very small flares, with total energy below 10^{27} erg, using a balloon-borne hard X-ray detector; the events distribution was consistent with the power law mentioned above. Schadee, De Jager and Svestka (1983) report the presence of a rather hot plasma component even outside flares, as well as detection of a few small events observed with the Hard X-ray Imaging Spectrometer on board the Solar Maximum Mission (SMM). As for soft X-ray observations, data taken with the S-054 telescope on board Skylab (Vaiana 1978; Sheeley and

J.F. Linsky and S. Serio (eds.), Physics of Solar and Stellar Coronae, 151–158.

Golub 1979) did show that X-ray coronal loops vary or flicker. The data taken, at a higher sensitivity and time resolution, with the Soft X-ray Telescope (SXT) on board Yohkoh (Tsuneta 1992, Uchida 1992, Lemen 1992) seem to confirm largely such a scenario. Haisch et al. (1988) also report on variability seen in X-ray lines with the X-ray Polychromator (XRP) on board the SMM satellite. Therefore a certain amount of low-level variability is present and strongly supports the idea that a number of events contributes to coronal heating.

A unified scenario of heating ranging from large to small events is quite appealing, but the presence of microflares to such an extent to heat significantly the corona is far from being proved. In order to assess their presence from observations, it is important to characterize a corona heated by microflares so as to provide an effective diagnostics.

2. The Model

Our approach is to compute, by means of a detailed hydrodynamic code, the evolution of plasma in a loop maintained close to stationary coronal conditions by a sequence of heating episodes; the latter ones providing the only heating mechanism. For a direct comparison with observations, we synthesize the emission from the calculated evolution of temperature, density and velocity. In particular we have concentrated our attention on the emission in the band where NIXT, the normal incidence X-ray telescope based on multilayer mirrors (Golub et al., 1990), is sensitive, in the bands to which SXT/Yohkoh is sensitive and in the lines (O VIII, Ne IX, Mg XI, Si XIII, S XV, Ca XIX, Fe XXV) observed by the XRP/SMM. The last three lines are also detected by the Bragg Crystal Spectrometer on board Yohkoh at a sensitivity ten times higher.

On one hand we aim at understanding the characteristics of coronal loops subject to a pulsed heating; on the other hand we aim at devising a diagnostics of the heat variability, particularly in soft X-rays.

We have used the Palermo-Harvard hydrodynamic model of the coronal plasma confined inside the loops (Peres et al. 1982; Peres and Serio, 1984; Peres et al 1987; Peres 1989). We have considered several simulations of loop evolution, starting from different initial conditions, and for different characteristics of the heating.

The initial configuration is a loop kept in static conditions by an appropriate constant power density; at *time* = 0 we switch off this term replacing it with pulsed heating of constant average power and we let the loop plasma evolve under the influence of the pulsed heating. We have considered two alternative formulations of heating: either a parameterized function of time and space which provides some freedom in modeling different locations of heat deposition inside the loop, or according to a model of heating by non-thermal electron beams, accelerated at the loop apex, precipitating along the loop legs and heating the ambient plasma by Coulomb collisions (for this last formulation cf. Nagai and Emslie 1984; Reale, Peres and Serio, 1985). The parameterized formulation of input power per unit volume is

$$Q(s,t) = A \times f(t) \times g(s) \tag{1}$$

$$g(s) = exp[-(s - s_*)^2/2\sigma^2] \tag{2}$$

where s, s_* and σ are, respectively, the field line coordinate, the center and the spread of the gaussian heating distribution. The power is deposited inside the coronal loops either as diffused in the highest part of the loop or concentrated at the base of the loop. $f(t)$ is a time modulation factor; we use an analogous modulation also for the electron beam evolution.

For $f(t)$ we have so far considered two possibilities: either a periodic sequence of identical and short pulses, or a sequence of random pulses of variable intensity and duration. In both cases the power is constant during each pulse and its time average is such to maintain the loop around the initial static conditions. In the randomic case both the time sequence of heating events and their duration is random with a Poissonian distribution; the average event duration is 1/10 of the average repetition time. As for the intensity of each pulse, we have considered the mechanism which Rosner and Vaiana (1978) proposed for flares and other cosmic bursting events, and based on a "load and release" mechanism, i.e. a randomic release of energy stored at a constant rate. The total energy released at any random pulse therefore equals that delivered in the analogous stationary loop, *over the time between the end of the previous pulse and the end of the on-going one.*

3. Results

We find that a loop can be maintained close to coronal static conditions by a pulsed heating as long as it provides enough energy and as long as the characteristic time of repetition of the heating events is of the order or less of the characteristic cooling time as defined by Serio *et al.* (1991):

$$\tau_c = 120\ L_9\ T_7{}^{-0.5} \tag{3}$$

where L_9 is the loop semilength in units of 10^9 cm and T_7 is the loop apex temperature in units of 10^7 K. A large set of simulations, of which we report just a small subset of results, shows that time intervals between heating episodes larger and larger than the loop cooling time bring to larger and larger variations in plasma characteristics and in loop emission. For any specific loop model and a given characteristic time of pulses repetitions, the largest variability tends to occur in the hardest bands and hottest lines.

Figure 1 reports an example of the evolution we obtain in a typical active region loop with semilength of 2×10^9 cm and initially with plasma pressure of 6 dyne cm^{-2} at the base of the corona. The heating is a prescribed function of time and space and is delivered in the highest part of the loop with total energy in each impulse of the sequence as shown on the right axis. The characteristic repetition time of the sequence is 600 s, to be compared with the characteristic thermal decay time of 424 s according to the formulation of Serio et al. (1991). The figure shows that the temperature, like other plasma characteristics, is maintained close to the value of the initial static condition.

We find large variations in the light curves of this loop model synthesized in the NIXT band and in some of the lines detected by the XRP/SMM. However,

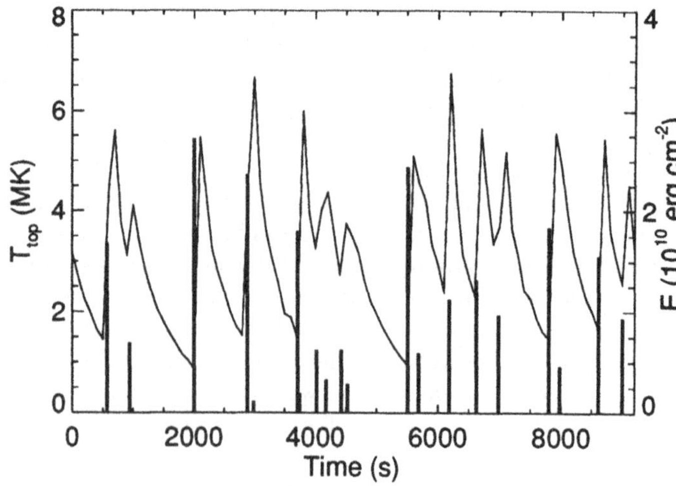

Fig. 1. Evolution of temperature at the apex of a symmetrical loop subject to a random sequence of heat pulses. The vertical solid lines mark the time when the heat pulse begins and their height yields the total energy delivered by each pulse.

in order to evaluate the detectability of such variations, one should consider how they would stand against the background of the active region loops, taking into account the angular and time integration of the possible instruments to be used. In fact spatial integration over large pixels, which might include many loops and considerable background, tends to reduce the detectable variability. In this respect it is worth noting that the two instruments which make the XRP, the Flat Crystal Spectrometer (FCS) and the Bent Crystal Spectrometer (BCS), integrate over a field of view of, respectively, 13" × 13" and 6′ × 6′; the BCS on board Yohkoh views the entire Sun simultaneously. On the other hand NIXT, thanks to its sub-arcsecond angular resolution, should detect variations more easily. Fig. 2 reports the distribution of emission in the band of NIXT along the field line coordinates and vs. time. The heating is delivered in the coronal (topmost) part of the loop; the variations are indeed very large in the corona, and they are very large at the base of the loop despite the heating occurs at the loop apex, because NIXT is very sensitive to plasma at the temperature range close to the base of such loop. We have found an even larger variability at the loop base when the heating occurs by non-thermal electron beams penetrating in the lower corona and chromosphere or is delivered at the base of the loop. NIXT present time resolution of a few tens of seconds and time coverage of, at most ten minutes, should be improved in order to detect easily variations like.

Analogously, we have computed the emission of the loops subject to variable heating, as detected by SXT/Yohkoh. Its spatial resolution around 2 arcsec, time resolution in standard observing mode of 2 second - which can be improved signifi-

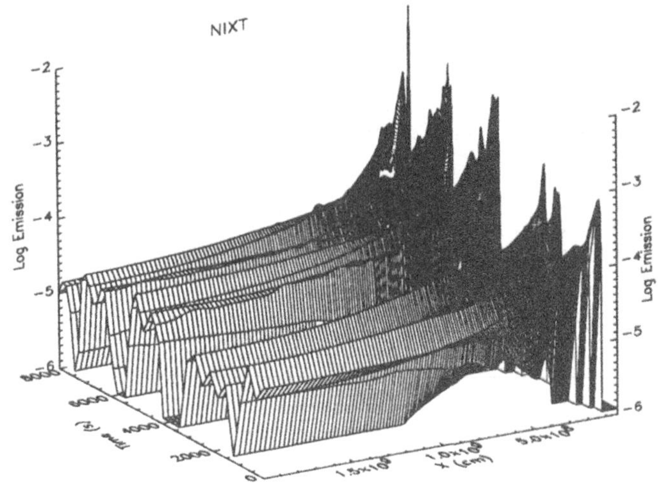

Fig. 2. Distribution of emission per unit volume ($erg\ cm^{-3}\ sec^{-1}$) in the NIXT band along the field line coordinates and vs. time for a microflaring loop model. The origin of spatial coordinates is located on the photosphere and the other extremum is at the loop apex; the model is spatially symmetric with respect to the loop apex so we show only half of it. The average repetition time of the random heating is 1200 sec.

cantly - and good time and space coverage, would be rather appropriate for catching microflare events. In Fig. 3 we report the emission in the E band of SXT of first loop model discussed above, the loop of semilength 2×10^9 cm but subject to a periodic pulsed heating with characteristic time 600 sec. It is immediately evident that the emission detected should vary rather significantly. In the light of these results it is plausible that the marked variability of some loops observed with Yohkoh (Tsuneta 1992; Uchida 1992, Lemen 1992) might be caused by variable heating.

Now we address the question: Since it should be hard to detect the possible microflares with XRP/SMM, how can we infer their presence and their features from the loop plasma characteristics and, in particular, from the differences with respect to typical stationary loop?

In this perspective, in alternative to the direct detection of variability caused by the heat impulses, one might aim at detecting the effect they have on other features such as spectra. In the following we concentrate on spectral characteristics and, in particular, we describe a diagnostics of microflares based on line ratios measured with XRP/SMM.

In Fig. 4 we report dependence of the ratio between the photon count of SXV and OVIII line observed by XRP/SMM, on pulse repetiton time for several simulations of pulsed heating localized at the top of a loop, both for a loop 2×10^9cm long and with base pressure 6 dyne cm^{-2} and for a loop four times shorter and with a coronal pressure of 24 dyne cm^2. The photon counts are properly averaged in time and along the loop so as to reproduce closely the conditions of typical observations

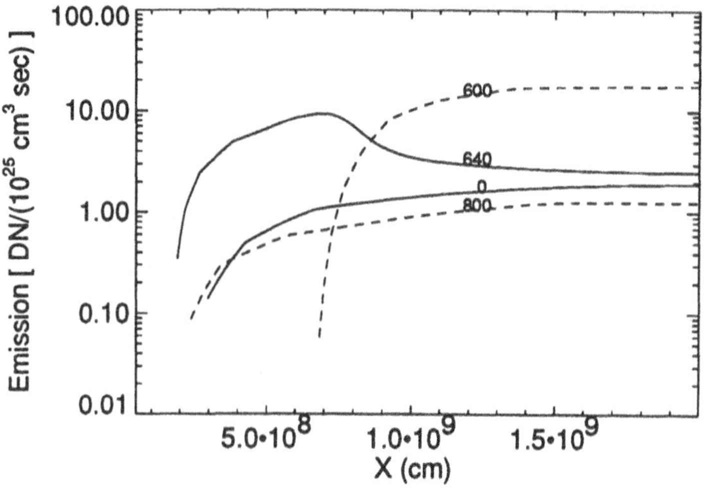

Fig. 3. Emission in the E band of SXT/Yohkoh synthesized from a model loop with a semilenght of 2×10^9 cm and subject to a periodic sequence of heating pulses occurring every 600 seconds since the beginning of the simulation. The time in seconds since the beginning of the simulation is reported on each curve. One DN (Digital Number unit of Yohkoh's CCD) is equivalent to, approximately, 365 eV (100 electrons).

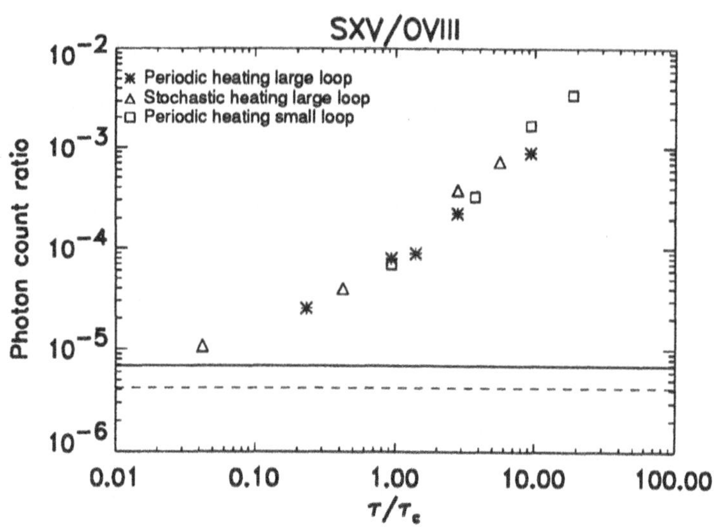

Fig. 4. Ratio between the count rates in the lines of SXV and O VIII vs. the ratio of heat pulses characteristic time and loop cooling time. We report the results for two loops with semilength of 2×10^9 cm and 5×10^8 cm respectively. The horizontal lines yield the value of the ratio for the corresponding static loops: dashed line shorter loop, solid line longer loop.

of small loops by means of the XRP/SMM. The ratio can change by many orders of magnitude with increasing repetition time. Results of simulations both with stochastic and with periodic sequences are reported in this figure and the results align along a curve irrespective of the periodicity or randomicity of the heating time scale. Analogous results are found for other XRP line ratios; in particular the ratio of hot over cooler lines count rates which we compute from the loops heated by microflares can increase up to a few orders of magnitude, with the ratio of repetition time over loop cooling time, with respect to what one expects for a loop in static conditions.

On the basis of these results we might be able to devise a diagnostics of variable loop heating: even if we do not detect variations in line intensities, ratios very different from those expected in static conditions speak for variable heating. Furthermore, if one also detects line variations, the characteristic time of variations and the average line ratio, obtained independently, could provide the cooling time of the structures.

4. Conclusions

In this paper we have discussed some of the physical characteristics one can predict for loops maintained around stationary conditions by a sequence of heat pulses. We have also presented some possible diagnostics.

We have seen that, leaving other characteristics unchanged, there is no basic significant difference between cases with periodic and cases with randomic heating sequence, the important parameter being the ratio of average pulse repetition time to loop cooling time.

Direct detection of variability caused by microflares is possible in observations with high spatial resolution in the case of well isolated loops or of loops significantly brighter than the background. NIXT and Yohkoh should play a significant role in the task of detecting microflares, although one cannot rule out the possibility of detecting interesting microvariability also in the data obtained by other instruments with angular resolution such as the XRP/SMM which, instead, allows good spectral resolution and coverage.

With a different diagnostic approach, we have shown that the ratio of count rates in the X-ray lines detected by the XRP/SMM appear to provide an interesting diagnostics, which depends strongly on the ratio τ/τ_c and very little on the loops average static conditions. It is conceivable that analogous diagnostics could be derived from other lines and also in other spectral regions.

Acknowledgements

We acknowledge partial support from the Agenzia Spaziale Italiana and from the Italian Ministero dell'Università e della Ricerca Scientifica e Tecnologica. We also acknowledge support from Italian Consiglio Nazionale delle Ricerche (CNR) for calculations performed on the CRAY-YMP at CINECA (Bologna, Italy).

References

Datlowe, D. W., Elean, M. J., and Hudson, H. S. 1974, *Solar Phys.*, **39**, 155.
Dennis, B. R. 1985, *Solar Phys.*, **100**, 465.
Galeev, A. A., Rosner, R., Serio, S., Vaiana, G. S. 1981, *ApJ.*, **243**, 301.
Golub, L., Herant, M., Kalata, K., Lovas, I., Nystrom, G., Pardo, F., Spiller, E., and Wilczynski, J. 1990, *Nature*, **344**, 842
Haisch, B. M., Strong, K. T., Harrison, R. A., Gary, G. A. 1988, *ApJS.*, **68**, 371.
Lemen, J. 1992, In *This volume.*
Lin, R. P., Schwartz, R. A., Kane, S. R, Pelling, R. M., and Hurley, K. C. 1984, *ApJ.*, **283**, 421.
Nagai, F. and Emslie, E. G. 1984, *ApJ.*, **279**, 896.
Parker, E. N. 1988, *ApJ.*, **330**, 474.
Parker, E. N. 1989, *Solar Phys.*, **121**, 271.
Peres, G., Rosner, R., Serio, S., and Vaiana, G. S. 1982, *ApJ*, **252**, 791.
Peres, G., and Serio, S. 1984, *Mem. Soc. Ast. It.*, **55**, 749.
Peres, G., Reale, F., Serio, S., and Pallavicini, R. 1987, *ApJ*, **312**, 895.
Peres, G. 1989, *Solar Phys.*, **121**, 289.
Reale, F., Peres, G., and Serio, S. 1985, *A&A*, **152**, L5.
Rosner, R., and Vaiana, G. S. 1978, *ApJ*, **222**, 1104.
Schadee, A., De Jager, C., and Svestka, Z. 1983, *Solar Phys.*, **89**, 287.
Serio, S., Reale, F., Jakimiec, J., Sylwester, B., Sylwester, J. 1991, *A&A*, **241**, 197.
Sheeley, N. R. Jr., and Golub, L. 1979, *Solar Phys.*, **63**, 119.
Tsuneta, S. 1992, In *This volume.*
Uchida, Y. 1992, In *This volume.*
Vaiana, G. S. 1978, *Space Research*, **18**, 331.

SOFT X-RAY LINE SHIFTS AS SIGNATURE OF THE FLARE HEATING PROCESS

E. ANTONUCCI, M.A. DODERO
University of Torino, Torino, Italy
G. PERES
Osservatorio Astrofisico di Catania, Catania, Italy
F. REALE, S. SERIO
Istituto ed Osservatorio Astronomico, Palermo, Italy
and
B.V. SOMOV
Sternberg Astronomical Institute, Moscow State University, Moscow

Abstract. The observed blue-shifted emission in soft x-ray lines typical of energetic solar flares, discovered by analyzing the spectroscopic data obtained with the Bent Crystal Spectrometer of the Solar Maximum Mission, is interpreted at the light of the results of numerical simulations of both thermal and non-thermal flares. The soft x-ray line profiles with pronounced blue wings observed during the impulsive phase are found to be consistent with a flare model with a hot thermal source located near the footpoints of the magnetic fluxtube. These same profiles, however cannot be reproduced with models of flares heated by electron beams.

Key words: Flares – X-rays – Activity of the Sun

1. Introduction

Hydrodynamic models of solar flares can reproduce with accuracy the main physical properties of the thermal source (T $\sim 10^7$ K), emitting soft x-rays (1 - 3 Å) during flares. This conclusion is generally valid for both thermal flares, characterized by bulk energization in a small volume where temperatures can be as high as 10^8 K, and non-thermal flares, where energy is converted mainly in acceleration of particles to energies above 10 -20 keV.

The same hydrodynamic models however do not generally reproduce the soft x-ray line profiles with pronounced blue-wings observed (e.g. in Ca XIX and Fe XXV spectra) during the impulsive phase of energetic flares (class M and X). In the present paper we suggest that the property of reproducing the line profile can be used in an attempt to discriminate among different models and infer the nature of the energy transport process in flares.

2. Simulations of the Properties of the Soft X-Ray Thermal Source

We report here, as an example, the results of numerical simulations of the emission in the Ca XIX spectral region (3.165 - 3.231 Å) for non-thermal flares, where the atmosphere is heated by collisions with the ambient plasma of accelerated electrons injected in a coronal loop along the field lines. The numerical results for thermal flares can be found in Antonucci et al. (1987).

Different cases of non-thermal flares are considered with beam parameters varying within the typical range of the values inferred from the observations (Table I).

J.F. Linsky and S. Serio (eds.), Physics of Solar and Stellar Coronae, 159–162.

TABLE I
Parameters of the Electron Beam.

Energy Flux	ϕ (erg cm^{-2} s^{-1})	-	5 x 10^{10}	1 x 10^{11}
Lower Energy Cutoff	E_c (keV)	10	15	20
Spectral Index	γ	4	6	8

Power is released in the loop in accord with a gaussian time profile with FWHM $\simeq$ 30 s. The coronal loop is semicircular with length 4 x 10^9 cm and cross-section 2.5 x 10^{17} cm^2.

The numerical code used to simulate the Ca XIX spectral emission consists of the hydrodynamic code by Peres et al. (1982), adapted to the non-thermal case by Reale, Peres and Serio (1985) and the spectral synthesis code developed by Antonucci et al. (1982) and updated by Antonucci, Gabriel and Dennis (1984).

The simulated Ca XIX spectral emission integrated over the entire coronal loop can be analyzed with the same techniques used for the observed, spatially unresolved Ca XIX spectra. We can derive the emission measure, EM, the average temperature of the soft x-ray source, T, and the average evaporation velocity, V, of the plasma convected from the impulsively heated chromosphere to the corona (Fig.1).

A comparison of the maximum values, attained during the events, of the simulated and observed physical parameters (Antonucci, Gabriel and Dennis, 1984) of the soft x-ray source confirms the general consistency of models with the observations (Table II). (Observed emission measures are spread over a wider range of values, indicating that some events would require either beams with a higher energy flux than those indicated in Table I, in order to induce a higher amount of evaporation in a single loop, or the presence of a system of impulsively heated loops.) In first approximation, therefore, hydrodynamic models can be considered sufficient to explain the formation of the soft x-ray thermal source observed in flares through convection of heated chromospheric plasma into the corona: process known as chromospheric evaporation.

3. Simulations of Ca XIX Spectral Line Profiles

Ca XIX synthetic spectra derived for a non-thermal flare taking place in a coronal loop of density 6 x 10^9 cm^{-3}, with typical beam parameters (e.g., energy flux 5 x 10^{10} ergs cm^{-2}, lower energy cutoff 15 keV, spectral index 6) do not show in any case pronounced blue-wings as in the observed spectra. In fact, in non-thermal flares, as far as energy is injected in the loop the emission measure of the upflows always exceeds that of the chromospheric plasma already accumulated in the loop: the result being that the entire Ca XIX line is blue-shifted (Fig. 2).

We have also considered thermal models with a very hot source at the top or, in alternative, at the base of a loop. Only in this second case we find the typical

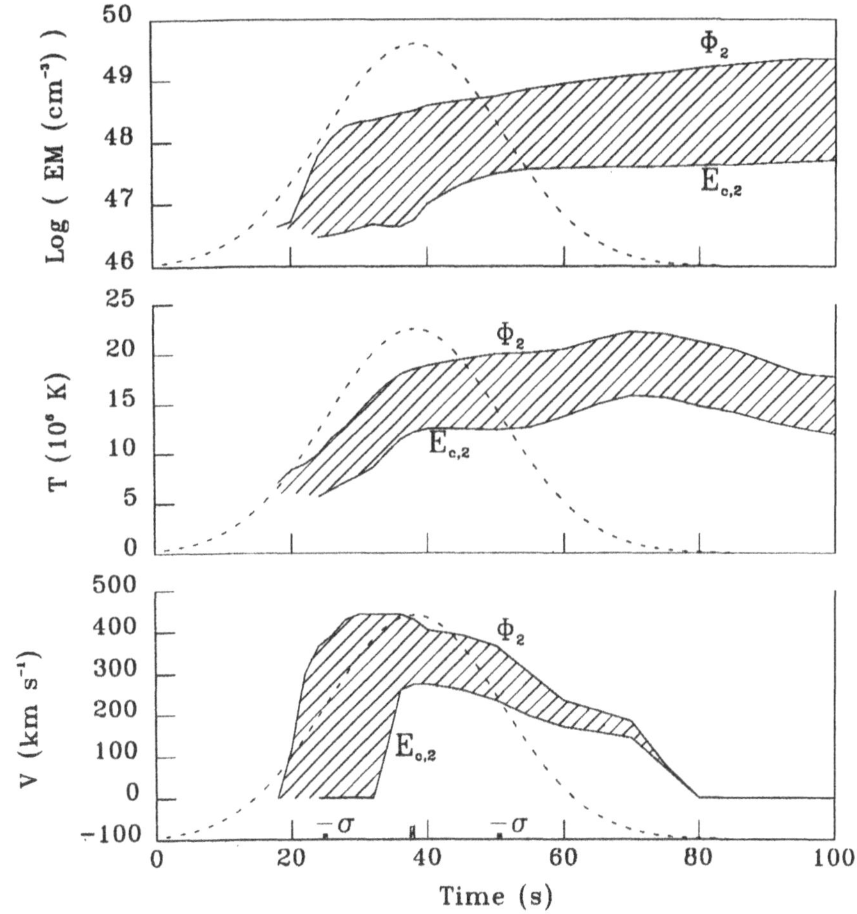

Fig. 1. The dashed areas define the regions where the simulated parameters: emission measure, EM, temperature, T, and evaporation velocity, V, are found. The limiting cases correspond to the highest input energy flux, Φ_2, and energy cutoff, $E_{c,2}$, of Table I. The gaussian profile of the rate of energy release is indicated in arbitrary units.

Ca XIX line profile observed in energetic solar flares with strong blue wings formed in a relatively short time.

In fact, for thermal flares with a very hot region, with input power density $\geq 10^2$ ergs cm^{-3} s^{-1}, near the footpoint of the loop, at 2 x 10^8 cm, the simulated soft x-ray line profiles show blue wings consistent with those typically observed in energetic flares (Fig.2). This is a consequence of the fact that the emission measure of the soft x-ray source, resulting from accumulation of hot ($\simeq 10^7$ K) plasma convected from the impulsively heated chromosphere to the corona, becomes in relatively short time ($\geq$ 35 s) larger than the emission measure of the upflowing chromospheric plasma itself.

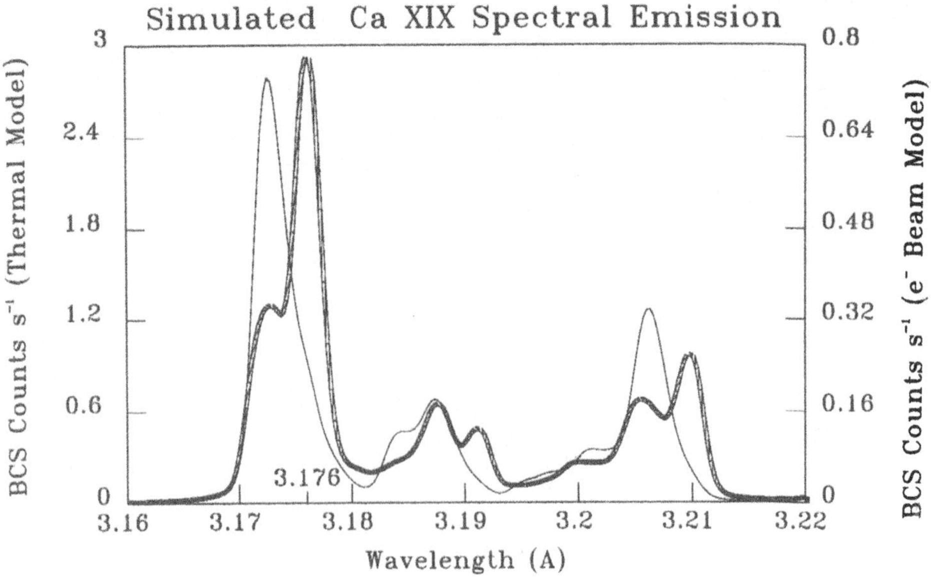

Fig. 2. Comparison of Ca XIX spectra simulated in the cases of a thermal flare with energy source near the loop base (thick line) and a non-thermal flare (thin line).

TABLE II
Physical Parameters of Flaring Loops

	Emission Measure (10^{48} cm^{-3})	Average Temperature (10^7 K)	Evaporation Velocity (10^7 cm s^{-1})
Observations	8 - 200	1.2 - 2.3	1.5 - 4.9
Simulations	0.6 - 25	1.6 - 2.2	2.7 - 4.9

We can therefore suggest that the strong blue-wings in the Ca XIX spectral lines can be interpreted as signature of thermal flares with a very hot region, localized in the low atmosphere, near the loop footpoints. This very hot region responsible for the flare might result from local conversion of magnetic into kinetic energy or trapping and thermalization of non-thermal particles in the lower corona.

References

Antonucci, E. et al.: 1982, *Solar Phys.* **78**, 107

Antonucci, E., Gabriel, A.H. and Dennis, B.R.: 1984, *Astrophys. J.* **287**, 917

Antonucci, E., Dodero, M.A., Peres, G., Serio, S. and Rosner, R.: 1987, *Astrophys. J.* **322**, 522

Peres, G., Rosner, R., Serio, S. and Vaiana, G.: 1982, *Astrophys. J.* **252**, 791

Reale, F., Peres, G. and Serio, S.: 1985, *Astron. and Astrophys.* **152**, L5

SIPHON FLOW MODELS OF CORONAL LOOPS

S. ORLANDO and S. SERIO*
Istituto ed Osservatorio Astronomico, Palermo
and
G. PERES
Osservatorio Astrofisico di Catania

Abstract. We report on the development of a siphon flow model of coronal loops. In order to achieve high reliability, especially close to critical points, we solve the hydrodynamic differential equations with two independent algorithms and different criteria for the evaluation of the precision. We present preliminary results obtained with a complete model which takes into account simultaneously gravity effects, thermal conduction, radiative losses from an optically thin plasma and a parameterized input power function. For the special case without gravity we have obtained scaling laws that extend the Rosner, Tucker and Vaiana (1978) scaling laws.

Key words: Hydrodynamics – Loop Models – Solar Physics

1. Introduction

Many authors have studied models of steady state flows in coronal loops (Cargill and Priest 1980, Noci 1981, Antiochos 1984, Thomas 1988). These models provide a powerful and interesting generalization of hydrostatic models of coronal loops (Rosner, Tucker and Vaiana - RTV - 1978, Vesecky, Antiochos and Underwood 1979, Serio et al. 1981). Although the latter ones can reproduce reasonably well EUV and X-ray observations (Pallavicini et al. 1981), recently it has been demonstrated that siphon flows loop models might provide a fit to detailed EUV solar observations (Peres, Spadaro and Noci 1992) significantly better than static models. On the other hand present flow models are limited because they do not take into consideration at the same time relevant physical effects such as, for example, gravity and thermal conduction. A full exploration of the parameters space has not yet been tackled. In this paper, we have developed a siphon flow loop model which takes into proper account a wide set of physical effects.

2. Model equations

Our model describes loop siphon flows in steady state. We have assumed that the magnetic flux tube is rigid and unperturbed by the flow within (low β plasma). The loop is assumed to be semicircular with constant cross section. The fluid is assumed to be a completely ionized hydrogen plasma, with a small fraction of heavy ions which do not influence its motion. We include the effects of gravity, thermal conduction, radiative losses and uniform heating along the loop. The fluid equations in this case are:

$$\rho v = j \tag{1}$$

$$\rho v \frac{\partial v}{\partial s} = -\frac{\partial P}{\partial s} + \rho g \tag{2}$$

* also IAIF, CNR Palermo

J.F. Linsky and S. Serio (eds.), Physics of Solar and Stellar Coronae, 163–166.

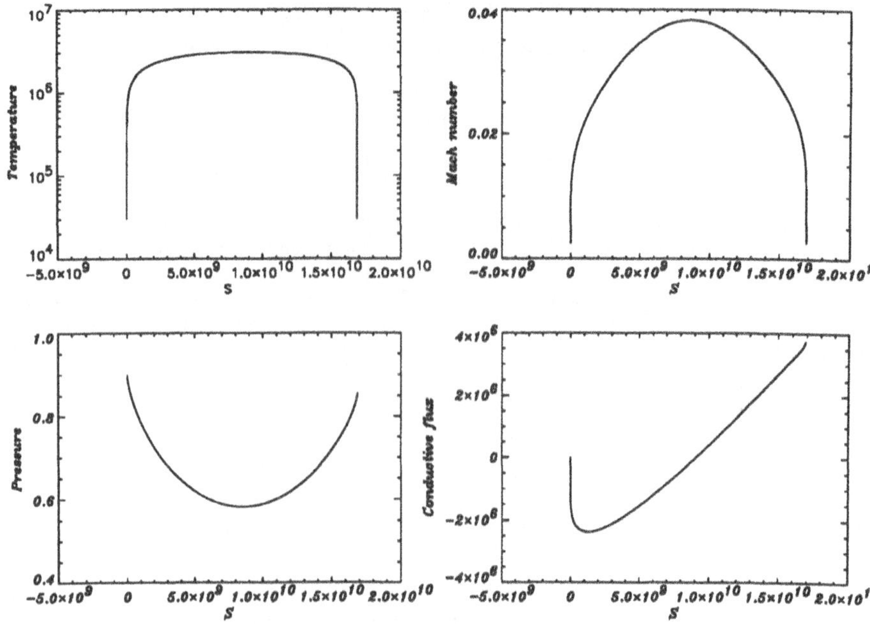

Fig. 1. Typical distribution of main physical quantities along the loop

$$\frac{1}{2}\left(5P\frac{\partial v}{\partial s}+3v\frac{\partial P}{\partial s}\right)+\frac{\partial F_c}{\partial s}=E_H-\frac{P^2}{4k_b^2T^2}\Lambda(T) \quad (3)$$

$$F_c=-\kappa T^{5/2}\frac{\partial T}{\partial s};\quad \rho=m_Hn;\quad P=2nk_BT;\quad g=-g_0\cos\left(\frac{\pi s}{2L}\right);\quad \kappa=10^{-6}$$

where m_H is the hydrogen mass, n the hydrogen number density, v the velocity, s the field line coordinate, P the pressure, g the gravity component along field lines, F_c the conductive flux reckoned from the lower boundary of the ascending branch, E_H the input power per unit volume, T the temperature, k_B the Boltzmann constant, $\Lambda(T)$ the radiative losses (RTV 1978), κ the thermal conductivity, L the semilength of the loop. We have limited our preliminary study to the case of subsonic flows. We have fixed the temperature at the lower boundary to be $3\cdot 10^4K$; there we set the conductive flux to be zero as in Antiochos (1984). In order to obtain a highly reliable solution, even close to critical points, we use two independent codes to solve our system of differential equations. The first is based on a fourth-order Runge Kutta method with adaptive stepsize control and the second on a Bulirsch Stoer method (Press et al. 1986, Stoer and Bulirsch 1980). Our codes have been cross checked by direct comparison of results (the difference between results obtained with the two codes are typically at the 6th significant decimal place) as well as by comparison with results of hydrostatic models (setting $v=0$) and of siphon flow models in the literature (Cargill and Priest 1980, Thomas 1988, Montesinos and Thomas 1989, Antiochos 1984, Noci et al. 1989, Peres et al. 1992). The tests have been highly satisfactory. In figure 1 we present an example of the solutions that we find.

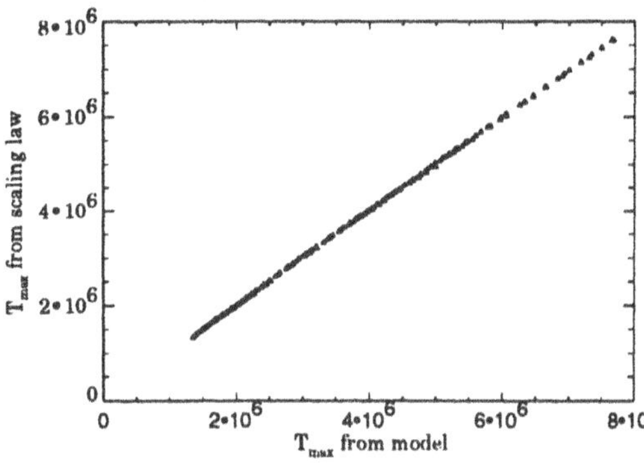

Fig. 2. Correlation between T_{max} value given by model and scaling law

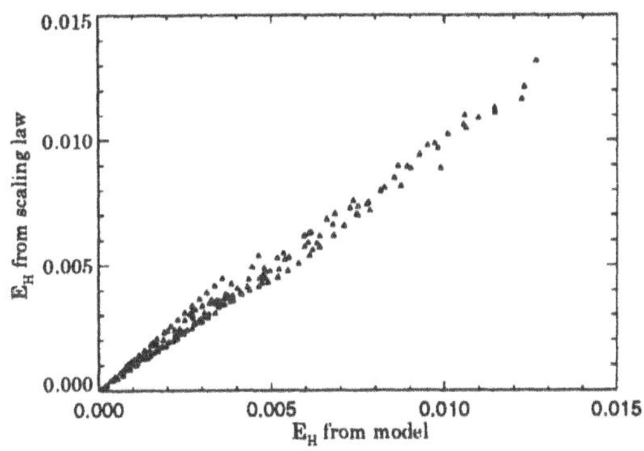

Fig. 3. Correlation between E_H value given by model and scaling law

3. Special case: no gravity

We have first applied our model to the special case without gravity in order to generalize the RTV scaling laws but also to open the path to the derivation of scaling laws for steady state siphon flow models with gravity. With these goals in mind the model has been run to compute a set of solar coronal loops with values of length, base pressure and base flow ranging over a grid of values from $6 \cdot 10^8$ to $2 \cdot 10^{10}$ cm, from 0.17 to 3.15 dyne cm^{-2} and from 10^4 to 10^5 cm s^{-1}, respectively. Our results are summarized in the following scaling laws relating the maximum temperature and input power density in the loop to base dynamic pressure $P_0 + \rho_0 v_0^2$ and flow velocity v_0 (physical quantities with the suffix 0 refer to the loop lower boundary at the ascending branch):

$$T_{max} = \nu(v_0)[(P_0 + \rho_0 v_0^2)L]^{\alpha(v_0)} \tag{4}$$

$$E_H = \mu(v_0)(P_0 + \rho_0 v_0^2)^{\gamma(v_0)} L^{\delta(v_0)} \tag{5}$$

$$\nu(v_0) = 1.35 \cdot 10^3 (1 - \frac{11}{8}\chi + \frac{1}{2}\chi^2 + \frac{11}{9}\chi^4) \tag{6}$$

$$\alpha(v_0) = \frac{1}{3} + \frac{2}{23}\chi + \frac{2}{19}\chi^2 - \frac{2}{9}\chi^4 \tag{7}$$

$$\mu(v_0) = 6 \cdot 10^4(1 - \frac{24}{5}\chi + 7\chi^2 - \frac{13}{3}\chi^4) \tag{8}$$

$$\gamma(v_0) = \frac{7}{6} + \frac{1}{2}\chi \tag{9}$$

$$\delta(v_0) = -\frac{5}{6} + \frac{4}{9}\chi - \frac{3}{8}\chi^4 \tag{10}$$

where

$$\chi = (\tfrac{10v_0}{c_b})^{1/2} \qquad c_b = (\tfrac{2k_BT_0}{m_H})^{1/2}$$

In figures 2 and 3 we report the predictions of the scaling laws vs. the model results, respectively for the maximum temperature and for the input power density. The correlation is particularly good for T_{max}. The evident scatter present in the E_H plot is to be attributed to the more complicated fitting to determine the coefficients of the scaling law.

4. Conclusions

The scaling laws that we have found generalize to siphon flows the RTV scaling laws to which they reduce setting $v = 0$. Our siphon flow model will be used in the future: a) to explore the parameter space of solutions also close to possible critical points, b) to study loops with length comparable to, or larger than, the typical pressure scale height, c) as a starting point to compute emission in spectral lines and bands of interest.

References

Antiochos, S.K. 1984, *Ap. J.* **280**, 416.
Cargill, P.J., and Priest, E.R. 1980, *Solar Phys.* **65**, 251.
Montesinos, B., and Thomas, J.H. 1989, *Ap. J.* **337**, 977.
Noci, G. 1981, *Solar Phys.* **69**, 63.
Noci, G., Spadaro, D., Zappala', R.A., and Antiochos, S.K. 1989, *Ap. J.* **338**, 1131.
Pallavicini, R., Peres, G., Serio, S., Vaiana, G.S., Golub, L., and Rosner, R. 1981, *Ap. J.* **247**, 692.
Peres, G., Spadaro, D., and Noci, G. 1992, *Ap. J.* **389**, 777.
Press, W.H., Flannery, B.P., Teukolsky, S.A., and Vetterling, W.T., 1986, Numerical Recipes, (Cambridge University Press).
Rosner, R., Tucker, W.H., and Vaiana, G.S. 1978, *Ap. J.* **220**, 643.
Serio, S., Peres, G., Vaiana, G.S., Golub, L., and Rosner, R. 1981 *Ap. J.* **243**, 288.
Stoer, J., and Bulirsch, R., 1980, Introduction to numerical analysis, (Springer-Verlag New York Inc.).
Thomas, J.H. 1988, *Ap. J.* **333**, 407.
Vesecky, J.F., Antiochos, S.K., and Underwood, J.H. 1979, *Ap. J.* **233**, 987.

SPECTRAL PREDICTIONS BY A THERMAL HYDRODYNAMICAL MODEL OF A SOLAR FLARE LOOP

GAN,W.Q.* and RIEGER,E.

Max-Planck-Institute of Extraterrestrial Physics, 8046 Garching, Germany

Abstract. We have computed CaXIX spectra (3.163-3.22 Å) as well as Hα and CaII K line profiles based on a thermal hydrodynamical model of solar flare. The results show an overall agreement with the observations. We therefore conclude that the thermal hydrodynamical model can rather well describe the observed spectral characteristics of solar flares.

Key words: Sun: flares — hydrodynamics — Sun: chromosphere of — Sun: corona of

1. Introduction

Theoretical spectral predictions have been investigated by a number of authors (e.g., Cheng et al. 1984; Peres et al. 1987; Antonucci et al. 1987; Canfield & Geyley 1987; Emslie & Alexander 1987; Li et al. 1989). Emslie & Alexander (1987) calculated the CaXIX w line profile based on a non-thermal electron heating hydrodynamical model. They found that the simulated line profiles agree with observations except for the greater blueshift of the line (3~5 mÅ). They attributed this difference to the calibration of the instrument (BCS) in SMM. But McClyments & Alexander (1989) investigated in detail the calibration of BCS and showed that the blueshifts of CaXIX w lines obsered for 9 disk flares are all smaller than 1 mÅ. This discrepancy between the theoretical prediction and the observations was recently explained to be due to the limited temporal resolution of the instrument (Antonucci 1989; Li et al. 1989). In the optical waveband, only Canfield & Gayley (1987) made a calculation for Hα line profiles based on the non-thermal electron heating hydrodynamical model of Fisher et al. (1985). By using the theoretical Hα line profiles, Wuelser & Marti(1989) and Graeter (1990) successfully explained the observed Hα line profiles with respect to explosive chromospheric heating by non-thermal electrons. Some Hα kernels which could not be due to non-thermal origin were assumed by them as of thermal origin. But because of the lack in studying the Hα line response to the heating by a thermal conduction front, they could not confirm their conjectures.

In this paper, we compute CaXIX spectra as well as Hα and CaII K line profiles based on the thermal hydrodynamical model of Gan et al. (1991) and discuss their implications to the observations.

2. CaXIX Spectra

The method to calculate the CaXIX spectra is similar to that of Antonucci et al. (1987) but we consider the effect of the loop orientations. Figure 1 shows the results for the loop located in the center of solar disk and on the limb, respectively. From the figure, we see that the stationary component of the w line predominates at the initial stage of the energy release. However, with the occurance of chromospheric evaporation, the blue-shifted component becomes gradually dominant until

* Humboldt Research Fellow, on leave from Purple Mountain Observatory, Nanjing, China

J.F. Linsky and S. Serio (eds.), Physics of Solar and Stellar Coronae, 167–170.

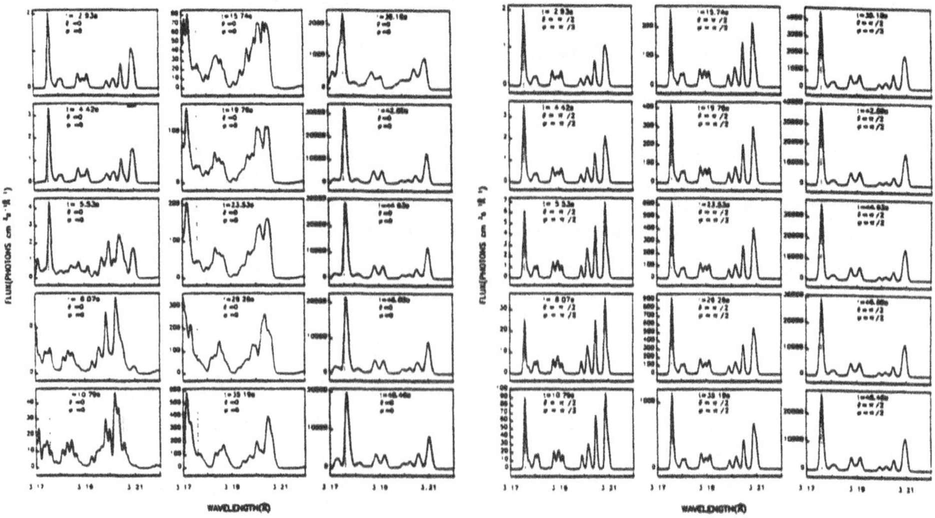

Fig. 1. CaXIX spectra for the thermal hydrodynamical model of Gan et al. (1991), assuming the loop located in the center of solar disk ($\theta = 0$) and on the limb, respectively

the evaporated material gets to the top of the loop (t=39.18 s). The maximum blueshift may be as large as 6 mÅ. Another characteristic at the early stage is that the intensity of the z, j lines may be greater than that of the w line. It is also noticed that with increasing θ, the blueshift of the line decreases. At the limb, the blueshift or blue-symmetry disappears completely even when $\varphi \neq \pi/2$.

3. Hα and CaII K Line Profiles

The method is based on the previous static code (Gan & Fang 1987). The main improvement is to increase greatly the frequency points for each of transitions so as to keep enough accuracy in calculating the transition probabilities at high velocities.

Figure 2 and 3 show respectively the evolutions of Hα and CaII K line profiles. The first impression from the figures is that under the thermal conductively heating model, the intensities of Hα and CaII K lines may be strong enough. At the initial phase (within ~10 s), the Hα line profile shows to have an additional redshifted emission component. Then the basic characteristic is the reversed red asymmetry. The maximum intensity of Hα does not take place at the initial phase of the energy release but in a later phase. The variations of CaII K line are also of some interest. The intensity of K_1 at the red side is greater than that at blue one. This phenomenon is quanlitatively consistent with the observations of Gan & Fang (1987b) and Fang et al. (1992). This means that the usually observed K_1 characteristic can be explained by the chromospheric condensation under the thermal model of solar flares.

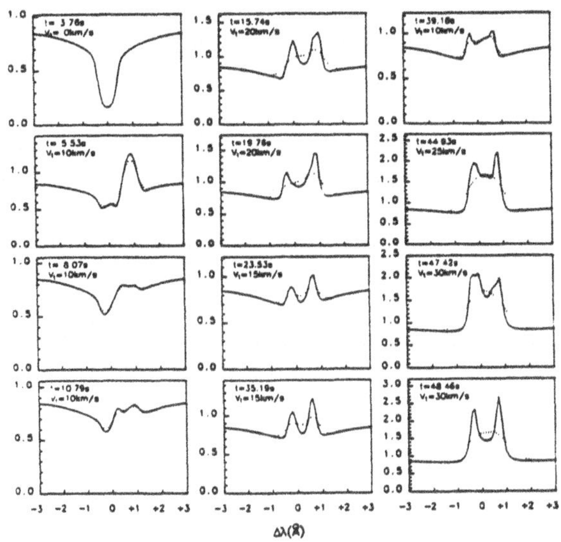

Fig. 2. Hα line profiles for the thermal hydrodynamical model of Gan et al. (1991). The dashed-line is the profile convolved with a Gaussian macro-turbulence velocity

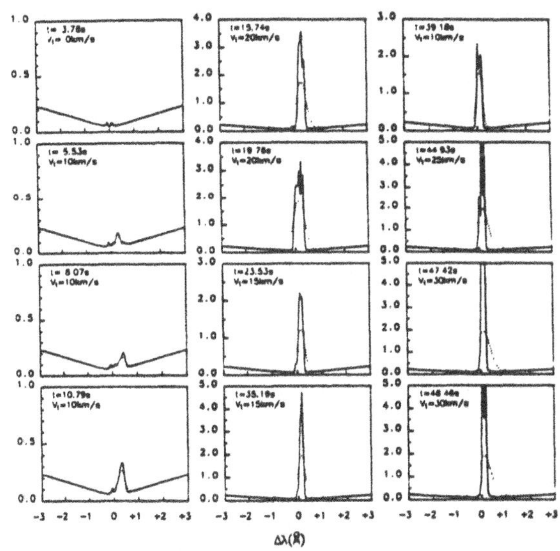

Fig. 3. The same as Figure 4 but for CaII K line profiles

4. Discussion

It should be pointed out that some of the characteristics of CaXIX w lines calculated here are qualitatively consistent with observations in general. The only discrepancy is the dominant blueshifted component in the early stage of the impulsive phase. Just as shown by Antonucci et al. (1987) and Li et al. (1989), when we integrate the flux within the first 42.89 s, the stationary component turns out to be dominant. This supports the view that the long integration times used in the published observations to date tend to wash out the impulsive phase blueshifted profiles. Obviously, this view needs to be checked in the future by using higher temporal resolution observations. On the other hand, in the field of hydrodynamical models, how to decrease the blueshift of the line is indeed a topic. We find that the initial coronal density plays an important role in determining the velocity of the chromospheric evaporation. Another possibility to decrease the blueshift is that if we assume the coronal loop is not semi-circular, but a low-lying arc in the early stage of the impulsive phase, then the velocity along the line of sight will undoubtedly be small. We hope that YOHKOH SXT can provide observations to test this conjecture.

The intensities of Hα and CaII K lines, as well as their asymmetric characteristics shown in Section 3 turn out to be comparable to the observations. This means that there is no difficulty for a thermal model to produce a chromospheric flare. We compare our results here with those of Canfield & Gayley (1987). It is noticed that the Hα lines in both models are all reversed red asymmetry, but in the non-thermal case, the red peak is weaker than the blue one, while in the thermal case, the situation is just contrary. Another difference is that the intensity of the Hα line given by Canfield & Gayley (1987) seems to be too great compared with the observations. Gan et al. (1992) studied empirically the Hα line profiles affected by the chromospheric condensation. In comparison of the empirical results with our self-consistent study here, we may find largely coincidence. This demostrates that the empirical study can be taken as a very useful preliminary diagnosing method, if we take into account that constructing a hydrodynamical model is complicated and time-consuming.

References

Antonucci E.: 1989, *Solar Phys.* **121**, 31
Antonucci E., Dodero M., Peres G., Serio S., Rosner R.: 1987, *ApJ* **322**, 522
Canfield R.C., Gayley K.G.: 1987, *ApJ* **322**, 999
Cheng C.C., Karpen J.T., Doschek G.A.: 1984, *ApJ* **286**, 787
Emslie A.G., Alexander D.: 1987, *Solar Phys* **110**, 295
Fang C., Hiei E., Yin C.Y., Gan W.Q.: 1992, *PASJ* **44**, 63
Fisher G.H., Canfield R.C., and McClymont,A.N.: 1985, *ApJ* **289**, 414
Gan W.Q., Fang C.: 1987, *Solar Phys.* **107**, 311
Gan W.Q., Fang C., Zhang H.Q.: 1991, *A&A* **241**, 618
Gan W.Q., Rieger E., Fang C., Zhang H.Q.: 1992, *submitted to Solar Phys.* ,
Graeter M.: 1990, *Solar Phys.* **130**, 337
Li P., Emslie A.G., Mariska J.T.: 1989, *ApJ* **341**, 1075
McClyments K.G., Alexander D.: 1989, *Solar Phys.* **123**, 161
Peres G., Reale F., Serio S. Pallavicini R.: 1987, *ApJ* **312**, 895
Wuelser J.P., Marti H.: 1989, *ApJ* **341**, 1088

ENERGY RELEASE TOPOLOGY IN SOLAR FLARES: AN XUV PERSPECTIVE

CHUNG-CHIEH CHENG
E.O. Hulburt Center for Space Research
Naval Research Laboratory, Washington, D.C. 20375, U.S.A.

Abstract. The *Skylab* XUV observations demonstrate the importance of mutiloop structures and global activation of an active region in the flare energy release topology. In multiple loop systems, different loops may play different roles during the flare evolution. The existence of simple loop flares indicates that loop interaction is not a necessary condition of flare occurrence. For some flares, sudden flare energy release is not preceeded by a slow energy buildup phase.

Key words: solar flares – XUV

1. Introduction

In this paper, I will discuss some aspects of XUV flare observations from *Skylab* that have bearings on the magnetic configuration and energy release topology in solar flares. The NRL XUV slitless spectrograph (SO 82A) on *Skylab* recorded discrete images of solar flares with a spatial resolution of $\sim 2''$ in line emissions (170-630 Å) which sample the full range of flare temperatures between 2×10^4 and 2×10^7 K. While the *Skylab* XUV observations lack time resolution and coverage, the spectrally separated flare images permit us to study the spatial and thermal structures in flares. These "old" XUV observations could provide useful insights for the interpretation of the recent X-ray observations of flares from the Japanese solar satellite *Yohkoh*.

2. Flare Loop Structures: Simple and Multiple

The XUV images of flares obtained from *Skylab* show unambiguously that the cool and the hot flare plasmas have related loop structures (Widing and Cheng 1974, Cheng and Widing 1975). One of the simplest cases is the 5 September 1973 flare. Figure 1 shows the XUV images of the flare observed simultaneously in the chromospheric He II lines and the high temperature Fe XXIV lines (2×10^7 K) together with the photospheric magnetogram. At the flare maximum (1831 UT), the two bright He II emission patches are located directly over regions of strong vertical fields with opposite magnetic polarity, while the hot 10-20 million degree Fe XXIV plasma is concentrated in a loop-like structure located in between the He II patches. As the flare cools (at 1837 UT), the hot plasma disappears and a loop appeares in Fe XIV-XVI at coronal temperatures, which clearly shows that the magnetic configuration of the flare is a simple loop.

Most flares, however, are much more complicated, involving many loops. Figure 2 shows the XUV images of the 15 June 1973 flare. The complex loop structure of the active region changes drastically during the flare evolution (Cheng 1977a); the flare loops or bundle of loops are often activated simultaneously or in succession. Observations in X-ray and UV from *SMM* have shown that a multiloop interacting

J.F. Linsky and S. Serio (eds.), Physics of Solar and Stellar Coronae, 171–174.

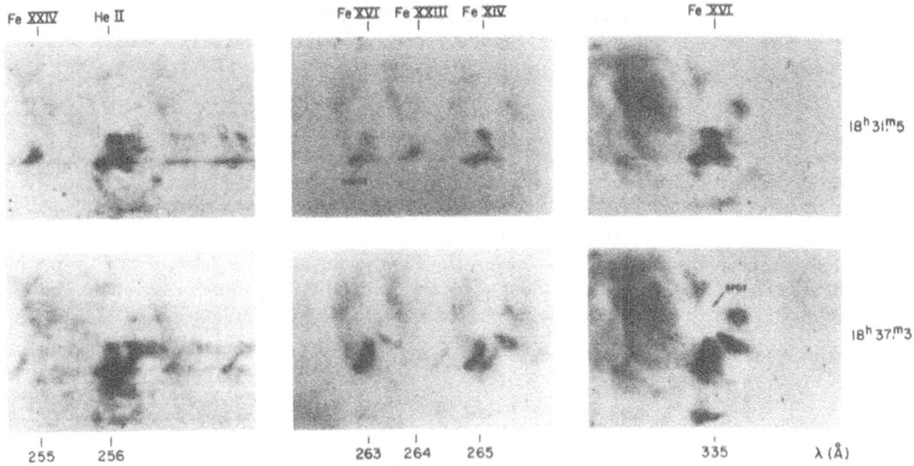

Figure 1

system is the dominating configuration of flare energy release (Machado et al. 1988, Cheng 1982, Cheng et al. 1985, Cheng and Pallavicini 1987, 1988).

The flare plasmas are highly dynamic. Figures 1 and 2 show that, during the initial phase, material is ejected and is most conspicuous in the He II, Fe XV-XVI lines. For the 15 June flare, the ejecta originated from the flare loop top with an upward velocity of about 400 km s^{-1}, while for the 5 September flare, the ejecta came from the footpoint with a velocity of about 200 km s^{-1}. The hot flare plasma in the 15 June flare also shows explosive mass motion, as indicated by a streak-like feature in the Fe XXIV images at the beginning of the flare (Figure 2).

3. Loop Interactions: Heating and Instabilities

In a multiloop system, it is not obvious what roles each loop plays and what are their interactions in the flare energy release process. The 21 January 1974 flare is especially revealing in this respect. Figure 3 shows the time development of the flare in various XUV lines. At 2316 UT, during the rise phase of the soft X-ray burst, we see a cool loop in He II and a small hot loop containing the Fe XXIII-XXIV emissions located just beneath the lower footpoint of the He II loop (Widing and Hiei 1984, Cheng and Widing 1990). The two loops shared a common footpoint, and the interaction between them probably triggered the energy release in the two loops. In the small loop, this is manifested in heating, while in the cool He II loop, MHD instabilities. At 2318 UT, the He II loop exhibited dramatic morphological changes that suggested wriggling of the loop, and eventually the originally smooth loop was disrupted. Cheng (1977b) has shown that the structural evolution in the cool loop is consistent with the interpretation of a kink instability caused by a large current of the order of 7×10^{10} amperes in the loop. The observed kinking behavior in the He II loop provides strong evidence for the existence of large current or twist in flaring loops. Depending on the degree of stress in the loops involved, energy release can take different forms. In the present case, one loop is heated to high temperature while the other suffers plasma instabilitites without appreciable heating. X-ray observations from *SMM* have corroborated the XUV results and

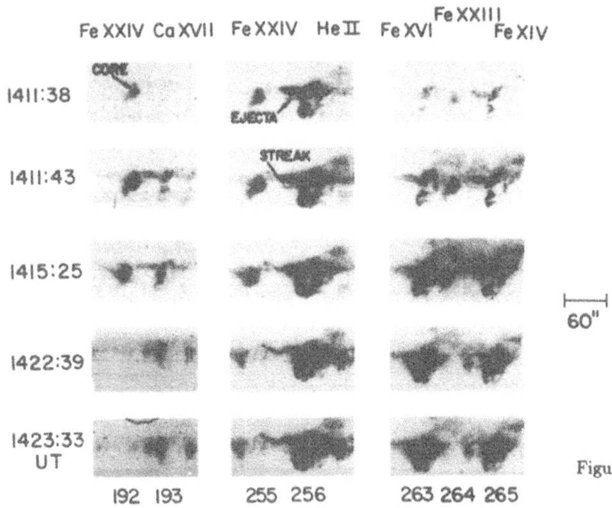

Figure 2

provided further examples of how flare energy release topology depends on the state of internal stress of the particpating loops (Machado et at. 1988).

4. Global Aspect of Flare Energy Release

How the evolution of an active region as a whole affects the flare energy release is not well known. One clue is provided by the preflare observations of the 15 June 1973 flare. Figure 4 compares the preflare images at 1218:46 UT and 1352:39 UT with those near the flare maximum at 1412:43 UT and later (Cheng 1977). The active region at $\sim$ 2 hours before the flare onset consisted of loops of various sizes and temperatures with the western-most loop most enhanced. When the flare occurred at $\sim$ 1412 UT, the originally enhanced loops in Fe XVI and Si XII disappeared and emissions in He II, Ne VII, Mg IX, and Fe XV-XVI appeared as two ribbons on each side of the magnetic neutral line. In between the ribbons a hot plasma in Fe XXIII-XXIV appeared. As the flare cooled, the active region gradually returned to its preflare morphology. It appears that, at the time of the flare, the energy that powered the active region was halted and diverted to the flaring loops that caused their sudden heating. These flaring loops were not the loops originally seen before the flare. All of the images in Figure 4, except at 1352 UT, have the same exposure time, so the disappearance of the preflare loop seen at 1218 UT is real.

5. Discussion

The *Skylab* XUV observations show the prevalence of loop interactions in the flare energy release process, in which participating loops play different roles. Although a majority of flares involve multiloop systems, there are flares that occur in a simple loop; the 5 Sept. 1975 flare described earlier being a primary example. This indicates that loop interaction is not a necessary condition for flare occurrence. A highly twisted loop with a large current is susceptible to plasma or MHD instabilities that release the stored magnetic energy. For flares with multiloop systems, the

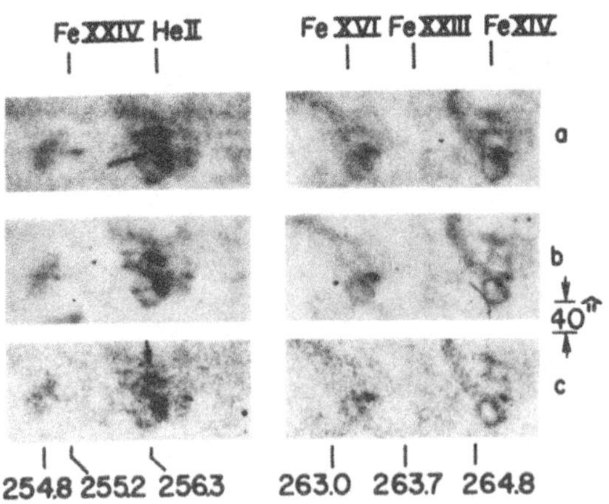

Figure 3 (a) 2315:56 (b) 2317:50 (c) 2318:02 UT

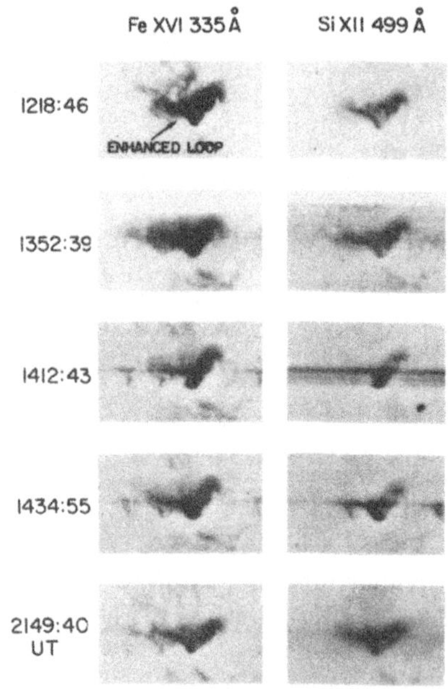

Figure 4

interaction between loops can occur through mechanical impact or electromagnetic coupling. The interaction between loops provides a convenient way of triggering the release of stored energy in individual loops. Since the individual loop plays a pivotal role, the study of simple loop flares is important to the understanding of flares with more complicated morphology.

As imporant as an individual magnetic loop is to the flare physics, the global aspect of the energy release process, involving the activation of the entire active region, can not be neglected. An example is the 15 June 1973 event, for which energization of the flare loops occurs at the time of the flare. That is, energy is suddenly injected into the flare loops perhaps due to some large-scale disturbance of subphotospheric origin. For these flares, slow energy buildup in stressed loops during the preflare phase is not a prerequisite of energy release.

References

Cheng, C.C.: 1977a, *Solar Phy.* **55**, 413.
Cheng, C.C.: 1977b, *Ap.J* **213**, 558.
Cheng, C.C., and Widing, K.G.: 1975, *Ap.J.* **201**, 735.
Cheng, C.C. et al.: 1981, *Ap.J. Letter* **248**, L39.
Cheng, C.C. et al.: 1982, *ApJ* **253**, 353.
Cheng, C.C., Pallavicini, R., Acton, A., and Tandberg-Hanssen, E.: 1985, *Ap.J.* **298**, 887.
Cheng, C.C., and Pallavicini, R.: 1987, *Ap.J.* **318**, 459.
Cheng, C.C., and Pallavicini, R.: 1988, *Ap.J.* **324**, 1138.
Cheng, C.C., and Widing, K.G.: 1990, *Adv. Space Res.* **10(9)**, 97.
Machado, M.E. et al.: 1988, *Ap.J.* **326**, 425.
Widing, G.S., and Cheng, C.C.: 1974, *Ap.J.Letter* **194**, L111.
Widing, G.S., and Hiei, E.: 1984, *Ap.J.* **281**, 426.

PLASMA TEMPERATURE DISTRIBUTION DURING THE IMPULSIVE PHASE OF SOLAR FLARES

R. MARTIN, E. ANTONUCCI

University of Torino, Istituto di Fisica, Torino, Italy

and

B.V. SOMOV

Sternberg Astronomical Institute, Moscow State University, Moscow, Russia

Abstract. The plasma temperature distribution during flares is derived from an analysis of Ca XIX (3.176 Å) and Fe XXV (1.850 Å) spectra observed with the Bent Crystal Spectrometer (SMM), using the fact that their emissivity curves peak at different temperatures. The temperature $T_{Ca,Fe}$, derived from the Ca XIX to the Fe XXV resonance line ratio, is compared with the temperatures T_{Ca} and T_{Fe} obtained from the ratio of the dielectronic recombination satellites to the resonance line intensity. During some very impulsive flares the temperature T_{Fe} exceeds $T_{Ca,Fe}$ and T_{Ca}. This effect can be explained with the existence of a "hot" component in the coronal plasma, with a temperature within 4 - 10 10^7 K.

Key words: Flares – X-rays – Corona of the Sun

1. Introduction

The temperature distribution in the coronal plasma during the impulsive phase of solar flares is investigated for a set of events of class M and X, observed with the Bent Crystal Spectrometer (BCS) onboard SMM. This study is based on the analysis of the soft X-ray spectra emitted from the Ca XIX and Fe XXV ions at 3.176 Å and 1.850 Å respectively, whose emissivity curves peak at different temperatures, 3 10^7 K and 5 10^7 K, respectively.

The temperature distribution in a coronal plasma can in principle be investigated by deriving the plasma differential emission measure $\phi(T)\,dT = n^2{}_e\,dV$, starting from a set of observed line intensities:

$$I_i = \frac{0.8\,h\nu}{4\pi\,L^2}\,A_i \int_{T_{min}}^{T_{max}} \frac{N_i}{N_E}(T)\,C(T)\,\phi(T)\,dT \qquad erg\,cm^{-2}\,s^{-1} \qquad (i = 1,...N),$$

where n_e is the electron density, L the Earth–Sun distance, A_i the element abundance, N_i/N_E(T) the ionization balance fractional term and C(T) the collisional excitation coefficient. This method of inversion, used by several authors (e.g., Sylwester, Schrijver and Mewe 1980, Jakimiec et al. 1984, Fludra and Sylwester 1986), gives reliable results only when a large number of spectral lines is available.

Since the information we have from BCS spectra is derived from few lines, we use a different approach to the problem. We investigate the thermal structure of the coronal plasma by deriving: the calcium temperature, T_{Ca}, and the iron temperature, T_{Fe}, from intensity ratios of the dielectronic recombination satellite lines (e.g., k and j lines, respectively) to the resonance line w (Gabriel 1972, Antonucci et al. 1982), and a temperature, $T_{Ca,Fe}$, derived from the ratio of the Ca XIX to the Fe XXV resonance line (Antonucci, Dodero and Martin 1990). All these quantities define an average electron temperature of the source, obtained by weighting the

J.F. Linsky and S. Serio (eds.), Physics of Solar and Stellar Coronae, 175–178.

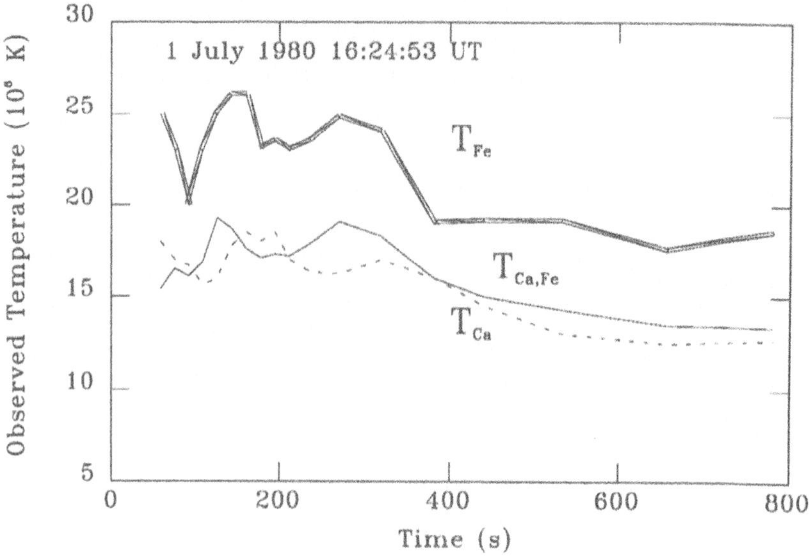

Fig. 1. Observed plasma temperatures during the July 1, 1980 flare.

plasma thermal distribution with the ion emissivity curves. If the coronal plasma is truly isothermal, the same values for all temperatures are obtained; while, temperature differences are a signature of the existence of a multithermal plasma.

2. Modeling of the Observational Results

We have studied the temporal evolution of the plasma temperatures T_{Ca}, T_{Fe} and $T_{Ca,Fe}$ in a set of solar flares observed during 1980. While in some flares no significant temperature differences are observed, in other cases such differences are evident. For example, in the flares of May 9, 1980 and July 1, 1980 the iron temperature T_{Fe} significantly exceeds both $T_{Ca,Fe}$ and T_{Ca} (Fig. 1). Flares with this behaviour are very impulsive events. For such flares, T_{Fe} lies in the range between 15 - 26 10^6 K, with $\Delta T_{max} = [T_{Fe} - T_{Ca,Fe}]_{max} \simeq 9\ 10^6$ K.

The observed temperature differences can be related to the characteristics of the plasma temperature distribution. We assume a reliable model of the plasma temperature distribution and calculate the various temperatures as a function of the model parameters. As a first step, we considered a singly peaked differential emission measure $\phi(T)$, with a gaussian profile. The parameters characterizing $\phi(T)$ are the total emission measure $\mathrm{EM} = \int \phi(T)\, dT$, the full width at half maximum, FWHM, and the temperature of the peak, T_p. The value of the temperature T_p is varied within 8 10^6 to 30 10^6 K in steps of 2 10^6 K. For each position, we calculate the theoretical line intensities $I_w(Ca)$, $I_k(Ca)$, $I_w(Fe)$ and $I_j(Fe)$. Ionization balance data used in these calculations are from Jacobs et al. (1977, 1980), and element

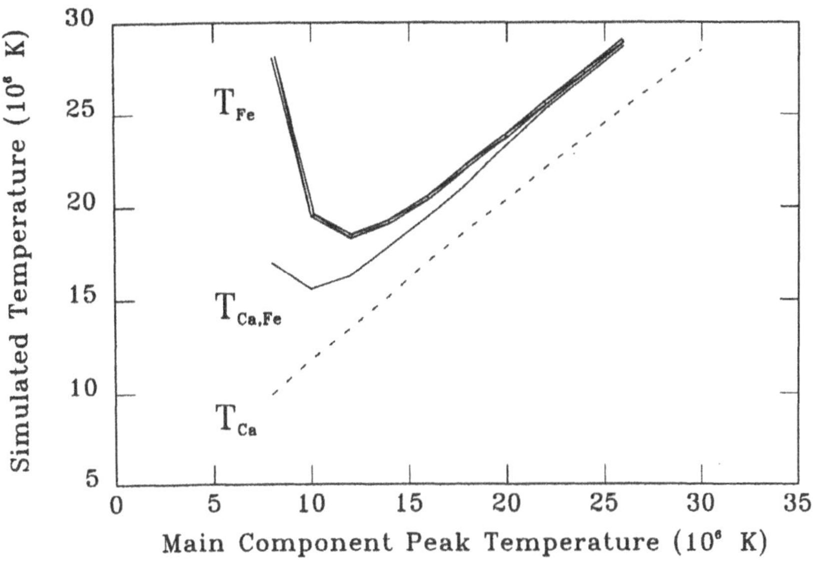

Fig. 2. Plasma temperatures calculated for a $\phi(T)$ profile with a main component ($n_e^2 V = 10^{49}$ cm^{-3}) with variable peak temperature and a "hot" component ($n_e^2 V = 3\ 10^{46}$ cm^{-3}) at a temperature of 10^8 K.

abundances are from Veck and Parkinson (1981). From the corresponding intensity ratios, the expected plasma temperatures T_{Ca}, T_{Fe} and $T_{Ca.Fe}$ are obtained. With typical total emission measures within 10^{49} - 10^{50} cm^{-3}, by simply varying the $\phi(T)$ parameters, we cannot reproduce any significant difference between temperatures.

Therefore we assume a differential emission measure $\phi(T)$ with a double structure, by adding to the main component a "hot" component with a gaussian profile, and a peak temperature, T_p, between 40 - 100 10^6 K. During each simulation, this "hot" component is maintained at a fixed temperature, while the principal one is shifted in the same way as previously described. The parameters used in the analysis are varied within the limits reported in Table I. By properly choosing the parameters of the two gaussian components, it is possible to reproduce the observed temperature differences (Fig. 2). Results consistent with the observational data are found by coupling a main component with total EM in the range 10^{49} - 10^{50} cm^{-3}, centered at $T_p \sim 8$ - 20 10^6 K, with a high-temperature component with total EM between 10^{46} - 10^{47} cm^{-3}, centered at T_p between 40 - 100 10^6 K.

3. Interpretation of the Observational Results and Conclusions

The need of an additional "hot" component in the model differential emission measure $\phi(T)$ to reproduce the observations is consistent with the results by Lin et al. (1981) and by Tanaka (1986).

TABLE I
Model of Plasma Temperature Distribution

	Main Component	"Hot" Component
Emission Measure (cm^{-3})	10^{49} - 10^{50}	10^{46} - 10^{47}
Peak Temperature (10^6 K)	8 - 30	40 - 100
Width (FWHM) d(log T)	0.4 - 0.6	0.2 - 0.4

The existence of a "hot" plasma component in flares can be explained by considering the theoretical predictions on MHD plasma flows from a reconnecting high-temperature turbulent current sheet (HTCS). Current sheets are proposed as sites where the primary energy release occurs during solar flares (Somov 1992). From a current sheet, high-temperature plasma is expelled at the Alfvén velocity. Given the temperature of the HTCS as an input parameter, it is possible to compute the differential emission measure of the plasma ejected from the current sheet and accumulated in the surrounding region in a typical time scale $\tau \sim 10^2$ s (Antonucci, Dodero and Somov 1992). Attributing to the HTCS a temperature $T_{e,CS}$ in the range between 30 - 100 10^6 K, total EM values within 10^{46} - 10^{47} cm^{-3} are obtained, which are indeed consistent with the emission measure of the "hot" component used to reproduce accurately the observed temperature values.

As a general conclusion, this work supports the existence, in very impulsive solar flares, of a significant contribution to the soft X-ray spectral emission from high-temperature coronal plasma at temperatures within 40 - 100 10^6 K, whose presence can be directly related to the primary energy release processes in a reconnecting current sheet.

References

Antonucci, E. et al.: 1982, *Solar Phys.* **78**, 107
Antonucci, E., Dodero, M.A. and Martin, R.: 1990, *Astrophys. J. Suppl.* **73**, 147
Antonucci, E., Dodero, M.A. and Somov, B.V.: 1992, *this same issue* ,
Fludra, A. and Sylwester, J.: 1986, *Solar Phys.* **105**, 323
Gabriel, A.H.: 1972, *Monthly Notices Roy. Astron. Soc.* **160**, 99
Jacobs, V.L. et al.: 1977, *Astrophys. J.* **211**, 605
Jacobs, V.L. et al.: 1980, *Astrophys. J.* **239**, 1119
Jakimiec, J. et al.: 1984, *Adv. Space Res.* **7**, 203
Lin, R.P. et al.: 1981, *Astrophys. J. Letters.* **251**, L109
Somov, B.V.: 1992, *Physical Processes in Solar Flares*, Kluwer Academic Publishers, Dordrecht, London
Sylwester, J., Schrijver, J. and Mewe, R.: 1980, *Solar Phys.* **67**, 285
Tanaka, K.: 1986, *Publ. Astron. Soc. Japan* **38**, 225
Veck, N. and Parkinson, J.: 1981, *Monthly Notices Roy. Astron. Soc.* **197**, 41

NON-THERMAL LINE PROFILES IN SOLAR FLARES

E. ANTONUCCI, M.A. DODERO

Istituto di Fisica, University of Torino, Torino, Italy

and

B.V. SOMOV

Sternberg Astronomical Institute, Moscow State University, Moscow, Russia

Abstract. We have analyzed the profiles of soft x-ray lines observed with the Flat Crystal Spectrometer onboard SMM during solar flares. Soft x-ray lines are predominantly emitted from two distinct sources, with characteristic temperatures within 5 - 8 10^6 K and 16 - 25 10^6 K, respectively. Both sources emit lines characterized by non-thermal profiles. The lower temperature source is characterized by a non-thermal excess in line width, corresponding to an average velocity of 70 km s^{-1}, independent of temperature. Higher velocities are observed for the high-temperature flare source. In this second case, non-thermal velocities increase with plasma temperature. The observed dependence of velocity on temperature is discussed in relation to the process of conversion of magnetic into kinetic energy in a high-temperature turbulent current sheet and it is found to be consistent with the velocity-temperature distribution expected for a plasma produced by a current sheet during reconnection.

Key words: Flares – X-rays – Corona of the Sun

1. Introduction

During the impulsive phase of solar flares, the width of soft x-ray lines is larger than expected from thermal broadening (Doschek et al., 1980, Antonucci et al. 1982, Tanaka et al. 1982). The velocity associated with the non-thermal excess in line broadening is observed to increase from active region values of 60 km s^{-1} to flare values of 100 - 200 km s^{-1}, one or two minutes before the hard x-ray burst (Antonucci, Gabriel and Dennis 1984). This is a clear evidence of the fact that non-thermal mass motions in flare plasmas, at least early in the flare, are not a signature of secondary processes related to the response of the ambient atmosphere to impulsive deposition of energy, but they are more likely related to the primary energy release process. One of the processes proposed for the conversion of magnetic into kinetic energy during flares, magnetic reconnection in a high-temperature turbulent current sheet (HTCS) has indeed the property of producing at the same time MHD plasma flows, local heating and particle acceleration (Somov, 1992).

2. Non-Thermal Velocities in Flare Plasmas

We have analyzed the profiles of a number of soft x-ray lines, O VIII, Ne IX, Mg XI, Si XIII, S XV, Ca XIX, Fe XXV, emitted during solar flares in the temperature range from 3 10^6 K to 5 10^7 K, which have been observed with the Flat Crystal Spectrometer (FCS) during SMM. The line width can be characterized by a Doppler temperature T_{D}. The thermal broadening expected for each line can be derived in the following way. We derive the differential emission measure, $\Phi(T)$, from the intensities of the lines observed in each FCS spectral scan obtained during the flares of Table I. The emission lines, computed for each plasma element within the temperature interval dT, with $n_{\mathrm{e}}^2\, dV = \Phi(T)\, dT$, and convolved with a gaussian

J.F. Linsky and S. Serio (eds.), Physics of Solar and Stellar Coronae, 179–182.

TABLE I
Solar Flare List

Date	Time (UT) FCS Scan	SMM Pointing Coordinates	A.R.	GOES Classification
25/8/80	13 07 19	N18 W57	2629	M1
5/11/80	22 29 05	N10 E08	2776	M1
5/11/80	22 33 25	N10 E08	2776	M4
23/1/85	07 33 12	S11 W57	4617	M1
24/4/85	09 41 31	N05 E25	4647	X2
30/4/85	23 48 04	N03 W64	4647	C2

profile to allow for thermal broadening at the temperature T, are added to construct the thermal line expected from a plasma with differential emission measure $\Phi(T)$. The average plasma temperature, T_e, is derived from the broadening of this line. The observed non-thermal excess in line width is then expressed in terms of the velocity, $V_t = \sqrt{2k(T_D - T_e)/m}$, where m is the ion mass.

The temperature distribution of the flare plasma, in agreement with previous studies (e.g. Bornmann and Strong, 1988), consists of two plasma sources at different temperatures, 5 - 8 10^6 K and 16 - 25 10^6 K, respectively. The dependence of the non-thermal velocity, V_t, on the temperature T_e is analyzed separately for lines formed predominantly (80 %) in the low and high-temperature regions. The average velocity in the low-temperature region has a value of 70$\pm$ 10 km s^{-1}, independent of temperature. In the high-temperature flare plasma region non-thermal velocities are higher and increase with temperature (Fig. 1).

3. A Model for Non-Thermal Velocities in Flare Plasmas

We suggest that the observed non-thermal velocities are related to the existence of one or several current sheets in the flare region (see also Antonucci, Tsinganos and Rosner, 1986), and we attempt to relate the observed dependence of non-thermal velocity on plasma temperature to the process of conversion of magnetic into kinetic energy in a reconnecting HTCS (Somov, 1992).

Plasma outflows at the Alfvén velocity, V_A, are expected from a reconnecting current sheet. The average outflow velocity would consist of two components, $V = \pm(1/2)\, V_A \times cos\alpha$, where α is the angle between the current sheet and the line-of-sight. Therefore the lines emitted from the outflowing plasma would be characterized by a non-thermal broadening.

In the HTCS model (Somov, 1992): (i) the Alfvén velocity is much larger than the drift, or reconnection, velocity: $V_A \gg v_d$, (ii) the temperature inside the HTCS is much higher than the temperature of the inflowing plasma.

The Alfvén velocity is: $V_A = B_o\,(4\pi\, M\, n_s)^{-1/2}$ where, B_o is the reconnecting magnetic field, n_s is the plasma density inside the HTCS, M is the average 'molecular' weight.

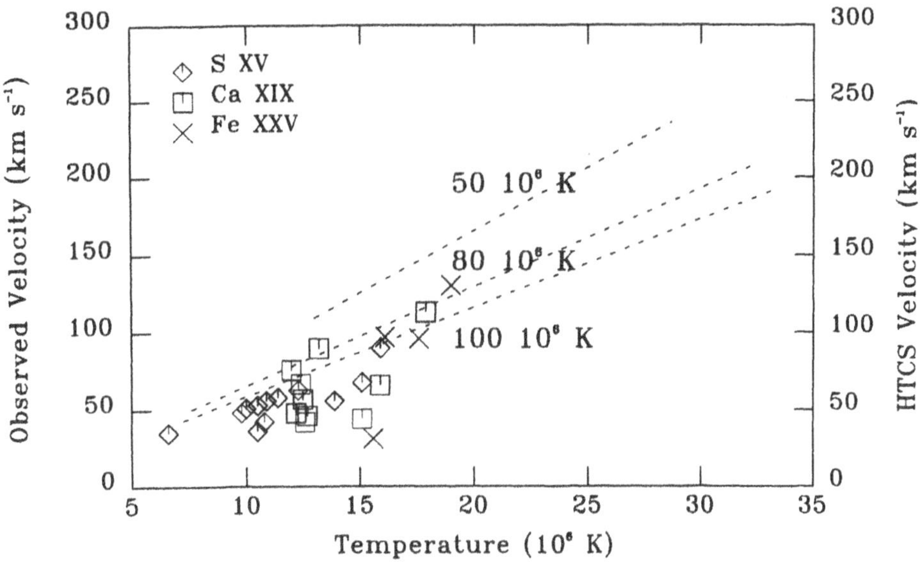

Fig. 1. Observed non-thermal plasma velocity derived from the profiles of S XV, Ca XIX and Fe XXV lines, compared with the functions $V_t(T)$ predicted for the plasma produced by a current sheet at temperature 50, 80 and 100 x 10^6 K.

Let us assume a balance of magnetic and gas pressure across the HTCS: $B_o^2\,/\,(8\pi) = n_s\,k\,(T_{e.CS} + T_{i.CS})$ where $T_{e.CS}$ and $T_{i.CS}$ are the current sheet electron and ion temperatures, then the non-thermal velocity is related to the Alfvén velocity as follows:

$$V_t\,(T_{CS}) = \frac{1}{2}\,V_A\,(T_{CS})\,cos\,\alpha = \frac{1}{2}\sqrt{\frac{2\,k}{M}}\,(1+\theta^{-1})^{1/2}\,\sqrt{T_{CS}}\,cos\,\alpha\,.$$

Here θ is the electron-to-ion temperature ratio, $T_{CS} = T_{e.CS}$ and $T_{i,CS} = \theta^{-1}T_{CS}$. The value of θ depends on the state of plasma turbulence in the HTCS. For example, in a HTCS with a small transverse component of the magnetic field in the marginal regime of ion-acoustic turbulence, $\theta = 6.5$; then, $V_t\,(T_{CS}) \simeq 6.0\ \ 10^3\,\sqrt{T_{CS}}\,cos\,\alpha$ cm s^{-1}. This case can be related to the 'hot' phase of flares (Somov, 1992).

The function $V_t\,(T_{CS})$ cannot be easily derived, since we in general do not have a direct measure of the temperature, T_{CS}, of the current sheet itself. However, we suggest to measure the velocity-temperature distribution of the plasma produced by the current sheet in the nearby region. We therefore compute the differential emission measure, $\phi\,(T)$, and the velocity distribution, $v\,(T)$, for the plasma produced in the flare region by a HTCS at temperature $T_{e.CS}$ and velocity V_A, in a time $\tau \simeq 10^2$ s, for different HTCS temperatures (Table II). We derive the profile expected for a spectral line, assuming that each element of the plasma with emission measure, $\phi\,(T)\,d(T)$, is characterized by a velocity $v(T)$. The width of the lines

TABLE II
Energy and Plasma Production by a Reconnecting HTCS

Temperature	T_e , 10^6 K	50	80	100
Energy Release Rate	P_s , 10^{28} $erg\, s^{-1}$	2.01	4.38	7.80
Plasma Production Rate	dN/dt , 10^{35} s^{-1}	2.53	3.75	5.00
Emission Measure	EM , 10^{46} cm^{-3}	2.38	5.21	9.25
HTCS Outflow Velocity	$V_t\,(T_{CS})$, 10^7 $cm\, s^{-1}$	4.2	5.4	6.0

computed in such a way, compared with that of the thermal line computed for the same plasma with the method described in section 2, allows us to determine the non-thermal broadening and in turn the velocity, $V_t\,(T)$, as a function of temperature. That is, we can predict the temperature-velocity dependence expected for the plasma directly produced by a current sheet. The observed velocities V_t, are found to be consistent with the functions, $V_t\,(T)$, derived for the current sheets at higher temperatures: T_{CS}= 80 and 100 x 10^6 K (Fig. 1).

4. Conclusions

The non-thermal profiles of soft x-ray lines emitted from flare plasmas are an indication of the presence of macroscopic plasma motions with a symmetric velocity distribution along the line-of sight. An analysis of the profiles of flare lines observed with the FCS has shown that while there is no dependence of non-thermal velocity on temperature in the region at 5 - 8 x 10^6 K, in the high-temperature source, at 16 - 25 x 10^6 K, non-thermal velocities are increasing with temperature.

This effect can be interpreted as evidence for mass motions related to the plasma ejected, at the Alfvén velocity, from a reconnecting high-temperature turbulent current sheet. The distributions of temperature and velocity for the plasma produced by a current sheet, in a time comparable with the time scale for radiative cooling, can be easily derived. The soft x-ray lines emitted from this plasma are characterized by non-thermal broadenings consistent with a velocity dependence on temperature as derived experimentally from the FCS lines, at least for current sheets with T_{CS}= 80 - 100 10^6 K. Therefore non-thermal broadenings in flare lines and their dependence on temperature are consistent with the velocity-temperature distribution of the plasma produced from one or more reconnecting current sheets.

References

Antonucci, E. et al.: 1982, *Solar Phys.* **78**, 107
Antonucci, E., Gabriel, A.H. and Dennis, B.R.: 1984, *Ap. J.* **287**, 917
Antonucci, E., Rosner, R. and Tsinganos, K.: 1986, *Ap. J.* **301**, 975
Bornmann, P.L. and Strong, K.T.: 1988, *Ap. J.* **333**, 1014
Doschek, G.A., Feldman, U., Kreplin, R.W. and Cohen, L.: 1980, *Ap. J.* **239**, 275
Somov, B.V.: 1992, *Physical Processes in Solar Flares*, Kluwer Academic Publ.: Dordrecht, London
Tanaka, K., Watanabe, T., Nishi, K. and Akita, K.: 1982, *Ap. J.* **254**, L59

CORONAL LOOP INTERACTION

RAYMOND N. SMARTT
National Solar Observatory/Sacramento Peak
*National Optical Astronomy Observatories**
Sunspot, NM 88349 USA

and

ZHENDA ZHANG
Department of Astronomy
Nanjing University, Nanjing, China

Abstract. High-resolution images of post-flare loop systems in Fe XIV (5303Å) and FeX (6374Å) display transient enhancements at the projected intersections of some loops. The brightness of such enhancements gradually increases to a marked maximum and then fades with a typical lifetime of about thirty minutes. The maximum in the red line lags that of the green line by about ten minutes, while $H\alpha$ reaches a maximum about ten minutes later; this sequence gives a measure of the cooling time, and hence electron density. The observed phenomenon is interpreted as localized loop coalescence and partial magnetic reconnection, with the possibility of an increase in the current density, and heating of the plasma in the immediate vicinity of the X-point.

Key words: Post-flare loops – Magnetic reconnection – Red line – Green line

1. Introduction

The coalescence of current loops leading to magnetic reconnection has been widely studied in relation to solar flares (see, for example, Tajima *et al*, 1987 and Tajima *et al*, 1982). In such treatments, internally- and externally-driven magnetic reconnection of the plasma is considered over different time scales and energies. Sakai and de Jager (1991) treat the case of 3-dimensional X-type current-loop coalescence, where two flux-tubes interact over a localized region. They show that, following a strong plasma collapse due to the pinch effect, a point-like plasma explosion can be driven while fast magnetosonic shock waves can also be excited. While such studies are concerned with the relatively high energies that characterize commonly-observed flares, it appears likely that the underlying physical processes in the high-energy case are relevant to the phenomena described here, that of interacting post-flare coronal loops. The interactions are treated as a situation of externally-driven reconnection in the low-energy case of post-flare coronal loops.

2. Observations

The observational data are from the 20-cm aperture NSO/SP emission-line coronagraph that records photographically sequential images in Fe XIV (5303Å), FeX (6374Å) and Hα, with a normal cadence of one minute. High-resolution post-flare loop images in the green line show occasional marked enhancements in the vicinity of the projected intersection of two loops (Smartt and Zhang, 1987; Zhang and Smartt, 1991). From the data it is evident that such enhancements occur as one loop develops or moves to intersect a neighbor, producing an X-point of coalescence.

* Operated by the Association of Universities for Research in Astronomy, Inc. (AURA) under cooperative agreement with the National Science Foundation.

J.F. Linsky and S. Serio (eds.), Physics of Solar and Stellar Coronae, 183–186.

Fig. 1. Flare-associated coronal loops, recorded in Fe XIV (5303Å), on April 28, 1980, at 1932 UT, showing three enhancements at locations where adjacent loops intersect.

The morphology of the post-flare loop events as seen at their maximum phase, especially clear in cases where the plane of the loops are not strongly inclined to the line-of-sight, is a brightness at the point of intersection far greater than the simple sum of the brightnesses of the individual loops, and some enhancement extending partially along the loops away from the intersection point, illustrated in Figure 1. Further, there is some partial "filling in" of the space between the loops, marked by a sharp boundary (see Figure 1), evidently a process of redistribution of magnetic flux between non-aligned loops, with partial magnetic reconnection. More than 90 events have been analyzed. It is found that typical enhancements increase to a maximum over a period on average of about 15 minutes and then gradually fade with a lifetime of the order of 30 minutes. Usually at least several such events are observed in a post-flare loop development. The maximum brightness varies between different events, with a trend that events occurring at low heights are brighter than events occurring much higher, and hence later in the post-flare loop development. It should be pointed out that much weaker events, if they occur, would not be detected in data of this quality. Moreover, events that might occur over only a very small spatial scale would also not be detected except in the case of images that were recorded with exceptionally-high angular resolution. These points are emphasized because much of the data are recorded close to the resolution and/or exposure limits of the photographic film. Corresponding enhancements at the same location appear in the FeX (6374Å) images. The overall red-line emission images are generally much fainter than those of the green line, and the enhancements in the red-line are correspondingly weaker, but nevertheless are usually well-defined in the form of a highly localized bright patch. Each red line maximum lags the corresponding maximum in the green line by at least a few minutes, with a mean of about 10 minutes, a measure of the cooling time. About 10 minutes later again, Hα reaches a maximum at the site of the reconnection. In some cases it is completely absent just minutes before. It should be pointed out that a post-flare loop system is itself a highly dynamical system in which the loops, typically in a complex configuration, gradually increase in height. During the development of such a system, which usually lasts fifteen hours or more, new loops can appear, while others disappear, with a systematic trend to simpler configurations, a decreasing number of loops and fainter emission. Apparently it is this dynamical property of post-flare loops that increases strongly the probability of loop coalescence, as compared with coronal loops in general – but there is some observational evidence that loop interactions can also occur in "quiet" coronal loop systems. The dynamical property of post-flare loops also results in some uncertainty in tracing the morphology of a loop coalescence event through its life. Nevertheless, the characteristics of the enhancements as described above represent a very well-defined phenomenon.

3. Discussion

On the assumption that radiative cooling dominates, the electron density has been estimated to be $n_e \leq 10^{11}$ cm^{-3} (Smartt and Zhang, 1978). It is noted that this estimate of the electron density neglects the possibility that cooling could also occur through plasma expansion (Tamano, 1991). The processes involved in heating the plasma at the X-point following coalescence remain uncertain without spectral information and data from other wavelength regimes, especially since the current density might be enhanced also in this region due to non-linear processes. For example, Dungey (1953) has pointed out that in the vicinity of a magnetic neutral point, a small disturbance can cause the current density to become large in the neighborhood of the neutral point.

It is interesting to note that the third case treated by Sakai and de Jager (1991) of a particular two-loop interaction in which the planes of the loops are mutually inclined such that the region of interaction is relatively small, that is, that the length of interaction, $L \approx a$, where a is the radius of the current loop, appears to have similar geometry to that observed in the post-flare loop systems discussed here and illustrated in Figure 1. However, we have made a first estimate of the energy, E, involved in the reconnection via the coalescence instability (Tajima et al, 1982), in which $E = (LB^2a^2/2)\ln(L/a)$, where $L \sim 7 \times 10^8$cm, B is the loop magnetic field ($\sim 100G$), and $a \sim 10^8$cm, which gives $E = 7 \times 10^{28}$ ergs. The numerical values applied here are somewhat uncertain, especially the value of B; nevertheless, the value for E is thought to be a reasonable order-of-magnitude estimate, corresponding to that of a very small flare.

4. Conclusion

Coronal loop interactions that produce transient enhancements are described. These data constitute an extremely consistent observational feature. However, interpretation of the enhancements in terms of the plasma processes involved remains uncertain. But the observational clues suggest two-loop coalescence and partial magnetic reconnection. Such scenarios have been investigated theoretically in relation to flare studies. Hence, it appears likely that the enhancements involve plasma processes similar to those in small flares. Spectral data and observations at other wavelengths, such as are now becoming available from YOHKOH, are extremely desirable to guide analysis and interpretation.

References

Dungey, J.W.: 1953, *Phil. Mag.*, **44**, 725.

Sakai, J-I., de Jager, C.: 1991, *Solar Phys.* **134**, 329.

Smartt, R. N. and Zhang, Z. 1987: in G. Athay and D. S. Spicer (eds), *Theoretical Problems in High-Resolution Solar Physics II:* NASA Conference Publication **2483**, p. 129.

Tajima, T., Sakai, J., Nakajima, H., Kosugi, T., Brunel, F. and Kundu, M.R., 1987, *Ap. J.* **321**, 1031.

Tajima, T., Brunel, F. and Sakai, J. 1982, *Ap. J.* **258**, L45.

Tamano, T.: 1991, *Solar Phys.* **134**, 187.

Zhang, Z., and Smartt, R. N.: 1991, *Acta Astronomica Sinica* **31**, 233.

QUIESCENT SOLAR PROMINENCES: A TWO-DIMENSIONAL MODEL

R. OLIVER and J. L. BALLESTER

Departament de Física, Universitat de les Illes Balears, 07071 Palma de Mallorca, Spain

and

E. R. PRIEST

Mathematical and Computational Sciences Department, University of St. Andrews, St. Andrews KY16 9SS, Scotland

Abstract. Using analytical approximations we have studied the effects of different external magnetic configurations on the half-width, mass, and internal magnetic structure of a quiescent solar prominence, modelled as a thin vertical sheet of cool plasma. Firstly, we build up a zeroth-order model and analyse the effects produced by a potential or linear force-free coronal field. Secondly, the effects of these external magnetic configurations on a two-dimensional model proposed by Ballester and Priest (1987) are studied. The main effects are a change of the half-width with height, an increase of the mass, a decrease of the magnetic field strength with height and a change in the shape of the magnetic field lines.

Key words: Prominences — magnetohydrodynamics

1. Introduction

Quiescent solar prominences are seen at the solar limb as vertical sheets of cool plasma which lie above and along photospheric neutral lines and last from days to months. It has been suggested (Priest, 1989) that prominences and models in which the magnetic field goes through the prominence like a normal magnetic arcade be said to possess Normal Polarity, whereas those in which it goes through in the opposite direction be said to possess Inverse Polarity. Reviews of observations and theories on prominences can be found in Poland (1986), Ballester and Priest (1988), Priest (1989), and Ruzdjak and Tandberg-Hanssen (1990).

Ballester and Priest (1987) proposed a two-dimensional model for a vertical prominence sheet by allowing slow variations of the magnetic field and plasma properties with height. They studied the behaviour of the magnetic field and the width of the prominence with height. Also, the shapes of the magnetic field lines were obtained.

Here, we use this model to study how the external magnetic configurations (potential and linear force-free) affect the half-width, mass and internal magnetic configuration. Fig. 1 shows the location of the prominence together with the coronal arcade, being x_0 the position of top and $L - x_0$ the arcade half-width. For additional details see Oliver et al. (1991).

2. Zeroth-Order Model

We have modified the classical one-dimensional model due to Kippenhahn-Schlüter (1957) by imposing coronal boundary conditions at a finite distance. We take as external magnetic field configuration either a current-free (potential) or a linear

J.F. Linsky and S. Serio (eds.), Physics of Solar and Stellar Coronae, 187–189.

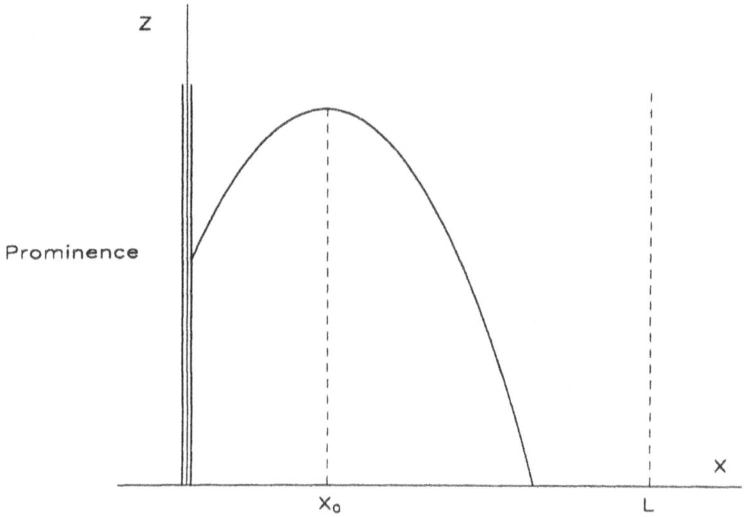

Fig. 1. Sketch of the coronal magnetic configuration with the prominence located in the YZ plane and an external arcade

force-free arcade, assuming that the range of altitudes is much smaller than the coronal pressure and the magnetic scale height. The external magnetic field and pressure will not vary with height, so that the edge between the prominence and the corona is vertical. Then, the matching of the two structures provides us with an equation for the position of the boundary in terms of the prominence scale height, coronal pressure and the external magnetic field strength. When the external magnetic field strength is increased, both the half-width and the mass of the prominence are increased, while increasing the coronal pressure produces a decrease in the half-width. Moreover, the larger the shear of the coronal field, the smaller the half-width and supported mass of the sheet.

3. First-Order Model

We perturb the one-dimensional zeroth-order model, making it two-dimensional, but considering as boundary conditions those given by a two-dimensional structure (potential or linear force-free). Assuming that $kz << 1$ and $z/\Lambda_c << 1$, imposing the equality of the magnetic field components and the gas pressure and linearizing the equations, we obtain the shape of the new boundary.

There are two basic ways the width of the prominence changes with height: In some cases the sheet broadens at the base with respect to its zeroth-order size and narrows as height increases, whereas in other cases the opposite behaviour is found. On the other hand, the internal magnetic field lines are not equally spaced as in the zeroth-order model: A solution in which the sheet narrows with height possesses field lines that bend more as height increases, and visa-versa. In all the cases, the magnetic field is found to decrease with height.

4. Mass of the Prominence

Since now the magnetic field is z dependent, the supported mass has to be obtained from an integration of the Lorentz force over the height of the prominence. The calculation of the perturbed supported mass always yields larger values than the zeroth-order ones. Furthermore, the greater the external magnetic field perturbation the larger the supported mass, as it is expected.

5. Conclusions

Taking analytical approximations we have studied the effects that two different different external magnetic configurations (potential and linear force-free) produce on a two-dimensional prominence modelled as a thin vertical cool sheet of matter. The main conclusions are:

(1) The width of the prominence is either found to increase or to decrease with height. This behaviour depends on the coronal plasma β and the importance of the perturbation.

(2) The prominence mass is always found to grow with the perturbation.

(3) The magnetic field strength inside the prominence always decreases with heigth. This is an inmediate consequence of the matching of the internal structure to the form of coronal field we have taken.

(4) When the width of the prominence grows with height, the internal magnetic field lines flatten with height and visa-versa.

The first-order model gives variations of the prominence characteristics with height (such as width) and suggests that it would be fruitful in the future to obtain more observational results on these variations.

References

Ballester, J. L. and Priest, E. R.: 1987, *Sol.Phys.* **109**, 335

Ballester, J. L. and Priest, E. R., eds.: 1988, *Dynamics and Structure of Solar Prominences*, Universitat de les Illes Balears: Palma de Mallorca

Kippenhahn, R. and Schlüter, A.: 1957, *Z.Astrophys.* **43**, 36

Oliver, R., Ballester, J. L., and Priest, E. R.: 1991, *Sol.Phys.* **134**, 123

Poland, A., ed.: 1986, *Coronal and Prominence Plasmas*, NASA Conf. Publ. 2442: Greenbelt

Priest, E. R., ed.: 1989, *Dynamics and Structure of Quiescent Solar Prominences*, Kluwer Academic Publishers: Dordrecht

Ruzdjak, V. and Tandberg-Hanssen, E., eds.: 1990, *Dynamics of Solar Prominences*, Springer-Verlag: Berlin

MHD WAVES IN A SOLAR PROMINENCE

R. OLIVER and J. L. BALLESTER

Departament de Física, Universitat de les Illes Balears, 07071 Palma de Mallorca, Spain

and

A. W. HOOD and E. R. PRIEST

Mathematical and Computational Sciences Department, University of St. Andrews, St. Andrews KY16 9SS, Scotland

Abstract. The presence of short- and long-period oscillations in quiescent solar prominences has been abundantly reported during recent years. In this paper, we use the Kippenhahn-Schlüter model to investigate the magneto-acoustic-gravity fast and slow modes of vibration of a prominence. First of all, we obtain analytically the MHD modes in the absence of gravity for a prominence with a purely horizontal magnetic field and, later, we solve the full problem numerically. Our results suggest that short-period horizontal oscillations detected in limb prominences could be due to the fundamental slow mode and its first harmonics, while the fast modes could produce vertical oscillations with periods around 3 minutes or periods ranging from 10 minutes to several hours, although the latter are suspicious to be unphysical.

Key words: MHD waves — prominences

1. Introduction

The oscillations in solar prominences have been mainly detected in the velocity field and the line intensity, and their periods have been classified in two categories: oscillations of short period, between 3 and 10 minutes, and oscillations of long period, mostly between 40 and 80 minutes (see the review by Tsubaki, 1988). Short-period oscillations have been detected in prominences located at the limb. However, in filaments observed near the disk centre these oscillations are hard to detect and only recent observations have provided the evidence of their existence. Long-period oscillations have only been detected in limb prominences, so that they correspond to horizontal plasma oscillations. In this paper, we investigate the MHD modes of oscillation of a prominence whose equilibrium configuration is given by a Kippenhahn and Schlüter (1957) model —for additional details, see Oliver et al. (1992).

2. Basic Equations

2.1. The Equilibrium State

To describe the equilibrium state of a solar prominence we use the Kippenhahn and Schlüter (1957) model, where the temperature of the plasma is uniform and the magnetic field and the density only vary in the direction of the x-axis (across the prominence). The y- and z-axis are placed along the prominence and in the vertical direction, repectively. The configuration is symmetric about $x = 0$ and the boundary of the prominence is located at $x = \pm x_p$.

J.F. Linsky and S. Serio (eds.), Physics of Solar and Stellar Coronae, 191–194.

2.2. Equations of Waves

We now assume small perturbations of the variables about the equilibrium state and consider the basic MHD equations for an ideal plasma. Moreover, we only take into account waves with motions in the xz-plane which propagate in the vertical direction with a wave number k_z. We obtain two coupled, second-order ordinary differential equations for the velocity components in the horizontal and in the vertical direction (i.e., v_x and v_z). Due to the symmetry of the equilibrium configuration, the solutions must also be symmetric; we have found that either v_x is even and v_z is odd (which corresponds to a kink mode), or v_x is odd and v_z is even (which corresponds to a sausage mode). Moreover, we have considered four different sets of boundary conditions to be imposed at $x = \pm x_p$, namely

$$v_x(\pm x_p) = v_z(\pm x_p) = 0, \tag{1}$$

$$v_x'(\pm x_p) = v_z(\pm x_p) = 0, \tag{2}$$

$$v_x(\pm x_p) = v_z'(\pm x_p) = 0, \tag{3}$$

$$v_x'(\pm x_p) = v_z'(\pm x_p) = 0. \tag{4}$$

Expression (1) corresponds to the rigid plate condition, while expressions (2) to (4) are similar to the so-called "flow through" conditions used at the corona-photosphere interface. We have considered a variety of boundary conditions in order to allow for a wide range of modes that may exist. The real state will probably be a combination of them.

3. Horizontal Field Limit

Since it is not possible to obtain a simple analytical solution to our differential equations as they stand, we have considered the simpler case of a medium with no gravity and horizontal magnetic field. Then, the pair of differential equations can be solved analytically. After imposing any of the boundary conditions (1) to (4) and the parity of the mode (kink or sausage) we obtain the dispersion relation, which allows us to calculate the frequency of oscillation ω in terms of the vertical wave number k_z. Fig. 1 shows the dimensionless frequency squared $\bar{\omega}^2$ versus the dimensionless wave number $\alpha \equiv \bar{k}_z$ for the kink modes when rigid plate boundary conditions are imposed (similar diagrams are obtained for the other sets of boundary conditions). The plasma velocity of the slow modes (plotted as solid) is almost horizontal, whereas the fast modes (plotted as dashed) are characterized by a plasma velocity which is almost vertical. Thus, horizontal oscillations, detected in limb prominences, are likely to be produced by slow modes, while those detected in filaments can be associated with fast modes.

We have considered parameter values typical of a prominence, though the half-width must be very small since this horizontal magnetic field limit is only valid for very thin prominences. Taking x_p= 300 km, the periods of oscillation are smaller than 2 minutes for slow modes and smaller than one minute for fast modes.

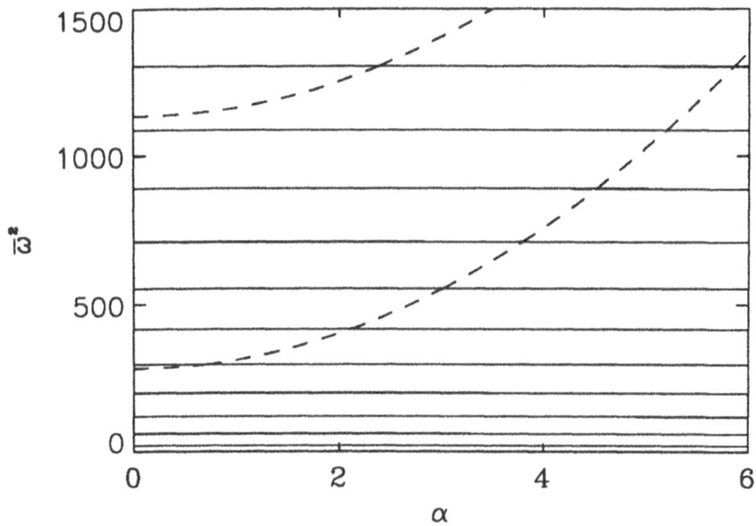

Fig. 1. Magneto-acoustic ($g = 0$) kink modes in a uniform medium permeated by a uniform and horizontal magnetic field. *Solid lines*: slow modes; *dashed lines*: fast modes.

4. Numerical Computations

The full coupled differential equations have been solved numerically by means of a fourth-order Runge-Kutta algorithm. Since we want ω to be real, the presence of gravity implies that the vertical wave number k_z must be complex (i.e., given by $\bar{k}_z = \alpha + i\beta$). This means that waves propagating vertically have an amplitude which can either grow ($\beta > 0$) or decay ($\beta < 0$) with height. We have imposed a set of boundary conditions selected from Equations (1) to (4) and the parity of the mode (kink or sausage). Then, after fixing the value of α, we have obtained the corresponding ω and β, together with the eigenfunctions v_x and v_z.

The numerical results obtained with rigid plate boundary conditions are quite similar to those obtained in the horizontal field case (Fig. 1), although the fast modes are shifted towards higher frequencies (smaller periods). On the other hand, the value of the imaginary part of k_z is very small, so that the change of velocity with height is almost negligible. In the numerical computations we have considered a half-width of 3000 km, so that the fundamental slow modes and some of their harmonics give periods between 16 and 3 minutes, in good agreement with the observed short-period horizontal oscillations. The fundamental fast mode produces vertical oscillations with periods around 3 minutes.

When integrating the system of ODEs with boundary conditions given by Equations (2) to (4), the resulting dispersion diagrams are quite different to the previous ones. Moreover, two fast modes producing long-period oscillations (between 10 and 120 minutes for boundary conditions given by Equation (3) and several hours for boundary conditions given by Equation (4)) could be responsible for long-period vertical oscillations, although no observations of such type have been reported at present. For these sets of boundary conditions the imaginary part of k_z is positive for most of the modes found, giving a wave amplitude which increases with height. Therefore, most of these modes are likely to be unphysical.

5. Conclusions

In this work we have modelled the propagation of MHD waves in a solar prominence described by the model of Kippenhahn and Schlüter (1957), restricting ourselves to the prominence region without taking into account the coronal medium.

First of all, we have used an analytical approximation, the limit of horizontal magnetic field and no gravity, which has allowed us to obtain the magneto-acoustic modes in a very thin prominence. The resulting theoretical periods obtained for the fundamental even and odd slow modes (100 and 50 seconds) are shorter than the observed ones (180 – 300 seconds), due to the fact that only narrow prominences can be considered in this analytical limit. Long periods cannot be obtained in this approximation.

Next, we have considered the full problem and have solved it numerically. We have obtained the magneto-acoustic-gravity fast and slow modes finding that, for a very thin prominence, the frequencies display a very good agreement with those obtained in the analytical case. For the half-width used here (3000 Km) and rigid plate boundary conditions, the periods of interest to our study are between 3 and 16 minutes for the fundamental slow modes and some of their harmonics, which is in excellent agreement with observations of horizontal short-period oscillations. A fast mode with a period around 3 minutes could be responsible for some of the reported shot-period oscillations in filaments. The other sets of boundary conditions yield quite different results: the dispersion diagrams are different and two fast modes producing long-period vertical oscillations appear, although most of the modes found are suspicious to be unphysical.

References

Kippenhahn, R. and Schlüter, A.: 1957, *Z.Astrophys.* **43**, 36

Oliver, R., Ballester, J. L., Hood, A. W., and Priest, E. R.: 1992, *ApJ* , in press

Tsubaki, T.: 1988, in Solar and Stellar Coronal Structures and Dynamics, ed(s)., *R. C. Altrock*, National Solar Observatory, 140

CALIBRATED SOLAR EUV SPECTRUM FROM SERTS

ROGER J. THOMAS and WERNER M. NEUPERT
Laboratory for Astronomy and Solar Physics, Code 680, NASA-Goddard Space Flight Center, Greenbelt MD 20771 USA

and

WILLIAM T. THOMPSON
Applied Research Corporation, Landover MD 20785 USA

Abstract. The Solar EUV Rocket Telescope and Spectrograph (SERTS) provides spatially imaged high-resolution spectra at wavelengths of 235 – 450 Å in first order, including many emission lines formed over the temperature range of $4.7 \leq \log T \leq 7.5$ characteristic of the corona and upper transition region. The instrument utilizes a grazing-incidence Wolter-2 telescope feeding a normal-incidence toroidal grating spectrograph which is quasi-stigmatic throughout its broad bandpass. Photometric calibrations have now been performed for each of the optical components carried on the 1989 and 1991 flights of this experiment. Calculations have also been carried out to determine the residual atmospheric extinction in the EUV at rocket altitudes during each flight, resulting in relative intensity measurements good to ± 25% over the first-order spectral range. An absolute photometric scale accurate to within a factor of 2 was then derived by fitting our solar observations to reported values for the average He II quiet sun flux at 304 Å. In addition, an absolute wavelength scale for the spectrograph was established by fitting postflight laboratory measurements to precisely known EUV standards. Solar wavelengths were then determined to better than 5 mÅ for most lines, often representing the highest accuracy presently available in this spectral range. A catalog is being developed from these calibrated observations that will list the absolute wavelength, intensity, and width for all lines detected in the averaged EUV spectrum of a solar active region. For each identified line, the catalog will also indicate the responsible ion and atomic transition, as well as the temperature of formation.

Key words: Solar – EUV – Spectrum

1. Introduction

The SERTS solar rocket experiment combines good spatial and spectral resolutions ($\approx$ 5 *arcsec* and 50 *m*Å) over a wavelength range of 235 – 450 Å and a field of view nearly 5 *arcmin* long, while simultaneously providing EUV spectroheliograms over an even larger FOV (Neupert *et al.* 1992a). The spectrograph has a dispersion of 2.2 Å mm^{-1} and plate scale of 11 $\mu m\ arcsec^{-1}$. Its flight on 5 May 1989 produced observations of a large solar active region (NOAA AR5464 at S18W49). The 7 May 1992 flight obtained spectra of primarily quiet areas, extending across the solar limb, and featured the first use of a multilayer-coated toroidal grating (Thomas *et al.* 1991, Davila *et al.* 1992). Before and after each flight, the spectrograph recorded laboratory spectra of He II, Ne II, and Ne III in order to verify focus and alignment, and to establish instrumental parameters. At present, special EUV-sensitive photographic film (Kodak 101-07) is used to record the near-stigmatic spectral images, which are then digitized with a Perkin-Elmer PDS Scanning Microdensitometer. The film's D – log E curve is derived independently for each flight or laboratory run by comparing EUV images of the same source region made with different exposure times.

For the initial spectral catalog, we selected the longest available exposure (246 *s*) of the solar active region taken on the 1989 flight, and spatially averaged it along the

J.F. Linsky and S. Serio (eds.), Physics of Solar and Stellar Coronae, 195–198.

entire slit length (first correcting for a slight tilt of the slit relative to the dispersion plane). The instrument's full spectral resolution is preserved by using 5-μm pixels in wavelength, corresponding to 11 mÅ each. After background subtraction, a wavelength varying noise level was estimated from fluctuations in the remaining baseline over several spectral regions apparently devoid of emission lines. Any feature whose peak exceeded this noise level by at least a factor of three was fit by either a single or double Gaussian profile, with each Gaussian providing a measured position, amplitude, and width. A few such features were rejected as plate flaws since their fitted profiles are narrower than the minimum possible instrumental width. The rest form the basis of the EUV spectral catalog described below.

2. Wavelength Scale

The wavelength scale is derived from laboratory EUV spectra of a hollow-cathode gas-discharge lamp, recorded with the SERTS instrument shortly after its flight and reduced by procedures similar to those outlined in the previous section. Positions are determined by Gaussian fits for eleven Ne II lines and for the He II line at 304 Å, all of which have absolute wavelengths that are known to better than 0.3 mÅ (Kaufman and Edlen 1974). The derived scale is given by a third-order polynomial (in pixel position relative to the He II centroid) fitted to the known laboratory wavelengths. Residuals from this fit are $\leq$ 0.5 mÅ for all twelve lines, clearly demonstrating the intrinsic accuracy of the SERTS spectrograph. The laboratory wavelength scale is then used for flight data by referencing all measured solar line positions to that of the solar He II 304 Å line.

One complication arises from the fact that the 304 Å line is saturated in the 1989 long-exposure flight frame over most of the slit length, and so we could use only the ends of the slit where the intensity was sufficiently low to set the wavelength scale reference position. Thus, the accuracy of the absolute scale depends on the assumption that the wavelength of the laboratory He II line is the same as that of the solar line from locations on two edges of the observed solar active region. Wavelength shifts due to line-of-sight velocity differences in the rocket's motion during an exposure or due to solar rotation variations along the slit would lead to apparent profile broadening of less than 2 mÅ and are neglected here. The derived wavelength scale is obviously most accurate over the range that was fitted to the measured laboratory standards, namely 304 – 448 Å. It becomes less certain as it is extrapolated toward the spectrograph's short wavelength limit of 235 Å. Considering all sources of error, the absolute EUV wavelengths determined from the averaged solar active region spectrum should be accurate to better than 5 mÅ for most lines in the catalog. In many cases, these are the most accurate solar wavelengths yet obtained in this spectral region.

3. Photometric Calibration

Direct photometric calibrations in the EUV have been carried out for each of the SERTS optical components, including the Wolter-2 grazing-incidence telescope, aluminum blocking filter, and normal-incidence toroidal grating. The telescope reflectance was measured at 304 Å, and its gradual variation with wavelength was modeled by theoretical calculation. The absolute efficiencies of the filter and grating were both measured over the full wavelength range at the Synchrotron Ultraviolet Radiation Facility (SURF-2) of the National Institute of Standards and Technology. Combined with the geometric collecting area of the telescope, these efficiencies give the instrument's effective area as a function of wavelength, which was found to range from 0.01 cm^2 at 250 Å to about 0.1 cm^2 beyond 380 Å.

In flight, the response of the instrument is also affected by EUV extinction along the line of sight due to the Earth's atmosphere at rocket altitudes. Calculations of this effect averaged over each film exposure were made from the detailed MSIS-86 thermospheric model of Hedin (1987), using laboratory cross sections for O (Angel and Samson 1988), N_2 (Samson *et al.* 1987), and O_2 (Samson *et al.* 1982). We find that atmospheric extinction typically varies by around $\pm$ 8% of its mean value over the SERTS spectral range for any given exposure.

Wavelength differences in the photometric response of the flight film were examined by using the above results to compare predicted intensity ratios of line pairs that are known to be density-*insensitive* with those measured by SERTS, a technique described by Neupert and Kastner (1983). This comparison shows that the photon response of our film does not vary significantly with wavelength over the observed range (*i.e.*, a given number of incident photons will cause the same film darkening at all wavelengths). The comparison also provides a check on the overall relative calibration of the instrument, indicating an accuracy well within ±25% from 250 to 450 Å, and even better over shorter spectral intervals.

Finally, the relative calibration obtained in this way was put onto an absolute scale by forcing our observations of the quiet sun He II emission at 304 Å to agree with an average of reported values for that flux available in the literature, namely 6730 ± 650 $erg\ cm^{-2}s^{-1}sr^{-1}$. As a check, we then used the resulting absolute calibration to determine a value for the quiet sun He II emission at 256 Å as observed by SERTS, which turned out to be within 4% of the average of other reports. An inversion of this scale can also be used to derive the absolute response of our flight film, with the result that a film density of 1.0 corresponds to an illumination rate of 1.3×10^9 $ph\ cm^{-2}$ at any EUV wavelength in the first-order SERTS range. Based on this calibration, the 1989 version of the instrument has a 2σ-sensitivity level of 5.0 $erg\ cm^{-2}s^{-1}sr^{-1}$ at 304 Å, improving to around 1.4 $erg\ cm^{-2}s^{-1}sr^{-1}$ above 380 Å. We estimate that absolute intensities determined for most of the lines in the catalog should be good to within a factor of 2.

4. EUV Spectral Catalog

There are 193 emission lines that meet our selection criteria in the averaged SERTS spectrum of a solar active region. Of these, 167 have at least preliminary identifications based on comparisons with previous observations (Behring *et al.* 1976, Dere 1978, Feldman *et al.* 1987, Kelly 1987). Almost all of the identified lines come from first order. However, there are also a number of very strong lines of Fe IX–XIV at wavelengths between 170 and 225 Å that are seen in second order, although with considerably reduced photometric accuracy. In addition, there are 26 lines in the catalog for which no identifications are presently available. Some of these may be marginal detections, but at least 15 were measured with signals of more than 3σ. Furthermore, nearly half of the unidentified lines were also seen in the flare spectra described by Dere (1978), indicating that they are valid solar features.

For the identified lines, those formed at the lowest temperatures are of He II (formed near log T = 4.7), C IV (5.0), and Ne III (5.1). Above this, there is complete temperature coverage between Ne V (at log T = 5.5) and Ca XVIII (6.8). Four elements have sequences of at least 4 consecutive stages of ionization: Mg V–IX, Si VII–XI, S XI–XIV, and Fe IX–XVII. There are six isoelectronic sequences represented by at least 4 members: Li-like, Be-like, B-like, C-like, Na-like, and Mg-like. And there are nine ions which each have more than 6 lines in the catalog: Mg VII (10 lines), Mg VIII (7), Si IX (7), Si X (8), Fe XI (7), Fe XII (10), Fe XIII (17), Fe XIV (12), and Fe XV (8). Thus, this calibrated EUV spectrum will be ideal for numerous detailed studies of solar plasma characteristics, such as differential emission measure, spectroscopic density or temperature diagnostics, line formation mechanisms, and abundance variations. The results of several such studies utilizing SERTS observations have recently been reported (*e.g.* Jordan *et al.* 1992, Keenan *et al.* 1992, Neupert *et al.* 1992b).

We are presently investigating the proposed identification of each line by comparing its relative intensity against predictions. A final version of the catalog is being prepared for publication. It will list absolute wavelength, intensity, and width, along with all fitting uncertainties. For each identified line, the catalog will also indicate the responsible ion and atomic transition, as well as the temperature of formation.

Acknowledgements

This work has been supported under RTOP grant 879-11-38 from the Solar Physics Office of NASA's Space Physics Division.

References

Angel, G. C., and Samson, J. A. R. (1988), *Phys. Rev. A* **38**, 5578.

Behring, W. E., Cohen, L., Feldman, U., and Doschek, G. (1976), *Ap. J.* **203**, 521.

Davila, J. M., Thomas, R. J., Thompson, W. T., Keski-Kuha, R. A. M., and Neupert, W. M. (1992), *Proc. 10th Colloq. on UV and X-ray Spectroscopy*, Berkeley.

Dere, K. P. (1978), *Ap. J.* **221**, 1062.

Feldman, U., Purcell, J. D., and Dohne, B. (1987), *Atlas of EUV Spectroheliograms from 170 to 625 Å*, NRL, Washington DC.

Hedin, A. E. (1987), *J. Geophys. Res.* **92**, 4649.

Jordan, S. D., Thompson, W. T., Thomas, R. J., and Neupert, W.M. (1992), *Ap. J.* in press.

Kaufman, V., and Edlen, B. (1974), *J. Physical and Chemical Reference Data* **3**, 825.

Keenan, F. P., Thomas, R. J., Neupert, W. M., Conlon, E.S., and Burke, V. M. (1992), *Solar Phys.* in press.

Kelly, R. L. (1987), *J. Physical and Chemical Reference Data* **16**, Suppl. 1, 1371.

Neupert, W. M., and Kastner, S. O. (1983), *Astron. Astrophys.* **128**, 181.

Neupert, W. M., Epstein, G. L., Thomas, R. J., Thompson, W. T. (1992a),*Solar Phys.* **137**, 87.

Neupert, W. M., Brosius, J. W., Thomas, R. J., Thompson, W. T. (1992b),*Ap. J. Lett.* **392**, L95.

Samson, J. A. R., Rayborn, G. H., and Pareek, P. N. (1982), *J. Chem. Phys.* **76**(1), 393.

Samson, J. A. R., Masuoka, T., Pareek, P. N., Angel, G. C. (1987), *J. Chem. Phys.* **86**(11), 6128.

Thomas, R. J., Keski-Kuha, R. A. M., Neupert, W. M., Condor, C. E., and Gum, J. S. (1991), *Applied Optics* **30**, 2245.

DISCUSSION FOLLOWING PAPERS ON THE SOLAR CORONA

Discussion following paper by L. Golub

F. Walter: How much improvement in the NIXT spatial resolution comes from observing a single line with a narrow temperature range, whereas Skylab imaged over a broad spectral band with a wider range of coronal temperatures?
L. Golub: I believe that most of the improvement is due to the multilayer mirrors, which have very good figure quality and greatly reduced scattering compared with a single reflecting surface. It is also the case that in other spectral lines, such as FeX, different structures are seen so that there may also be some lack of sharpness in broadband observations.
R. Rosner: The function which tells us the emission efficiency of a given ion as a function of temperature, $g(T)$, has a width ΔT which is roughly proportional to T. At temperatures of several million degrees, this width is also of the same order, 10^6 K. Hence, everything else being equal, it is hard to see how temperature sensitivity can have much to do with the sharpness of the NIXT images.
J. Schmitt: What are typical values of the coronal filling factor derived from your NIXT observations? This would be very valuable to compare to stellar results.
L. Golub: Unfortunately, there is not a simple answer to your question. As I said earlier, the structures seen at 1×10^6 K are very different from those seen at 3×10^6 K and in some cases, such as active regions, the different loops seem to be intermixed within the coronal volume. Possibly the best answer is obtained by studying density-sensitive line pairs, and comparing the true density with the average obtained from an EM analysis. Such measurements were done from Skylab by Vernazza *et al.* and more recently by R. Thomas using SERTS rocket data. The results typically give filling factors in the range 1-10% at coronal temperatures.
A. Maggio: Can you comment on the available observational evidence for variable cross sections along the coronal loops?
L. Golub: Typically we find that loops have nearly constant cross section. However, these loops are seen to have widths of about an arcsecond, which is at our resolution limit. Therefore, I believe that the observed constancy is misleading and that we cannot yet determine the true variability in individual loop cross sections.
B. Foing: Is the rigid rotation of coronal holes linked with the magnetic field being deeply anchored in the convection zone? Does the rotation period of coronal holes equal the internal rotation period derived from helioseismology?
L. Golub: This was a suggestion made in the Skylab days by Frankenthal and others in our group. I believe from more recent work by the NRL group that the answer may be directed outward rather than inward, with the opening of the boundary structures controlling to some extent the shape of the coronal hole.
Y.-M. Wang: Our group at NRL has done simulations of coronal holes which indicate that their rigid rotation is due to the approximately current-free nature of the large-scale coronal field. The coronal holes represent patterns of open field lines rather than material regions tied to the photosphere. The open-field footpoints

J.F. Linsky and S. Serio (eds.), Physics of Solar and Stellar Coronae, 199–207.

shadow the rigidly rotating outer coronal field, which in turn tends to track the low-latitude photospheric flux.
S. Serio: Do your data imply that most of the coronal structures remain unchanged during the $\sim$ 5 minutes of a NIXT flight, or vice-versa?
L. Golub: The 5-minute duration of a rocket flight is too short for observing most of the changes. The loops which we see varying during the flight are typically rather small ones. Most loops do not show obvious changes during the flight.
R. Thomas: I would like to clarify Leon's description of my "filling factor" result. What I actually determine is the product of the filling factor and the source thickness along the line of sight. My very preliminary result for solar active region loops at temperature $\log T \simeq 6.2$ is that this product is on the order of $10^7 - 10^8$ cm. Thus, a filling factor of 1-10% is implied if one uses the commonly assumed value of 10^9 cm for the source thickness. An alternate interpretation is that the AR filling factor is near unity and the loop thickness is relatively small, on the order of 0.1 - 1 arcsecond. The latter interpretation seems more consistent with the small loop thickness that Leon inferred from his analysis of NIXT images.

Discussion following paper by Y. Uchida

R. Rosner: Could you clarify the numbers you quoted earlier on the dynamic nature of AR (active region) loops: Did the 1/3 min (active AR) and 1/2 hr (quiescent AR) timescales refer to the brightening rate of a single loop, or to the loop brightening frequency for the entire AR? If it is the latter, then these AR loops should be reasonably well-described by quasi-static models for most of their lifetime (since their evolution timescales are for the most part long compared to sound crossing times, much less Alfvénic crossing times). Could you comment?
Y. Uchida: The numbers are for a typical active region, and the continuous flickering of "active" regions seen in the movie indicates that more "active" ones should be variable. As to the term "dynamic" and "quasi-static," I agree that the expansion velocity is smaller than the Alfvén velocity, and therefore, the expansion timescale is longer than the Alfvén crossing timescale. You are right that this case is "static," but would you call the active region "static" when looking at it in the movie? What is an appropriate word for this phenomenon?
G. Field: My question is related to Bob Rosner's. Your work shows that the corona is "dynamic" in the everyday use of the word, but it seems to me that it is evolving more slowly than the Alfvén crossing timescale. If that is true, at any one time, the corona must be in magnetohydrostatic equilibrium. Is that correct?
Y. Uchida: Yes, I agree that the term "dynamic" I used is not valid by the definition that the timescale is smaller than the Alfvén or acoustic crossing times, but if we must call all sub-Alfvénic or subsonic velocity phenomena "static," the definition of the term "dynamic" is too narrow, and misses drastically changing phenomena. The term "dynamic" I used here is not in the everyday use of the word, but surely not according to the narrow definition either. I suggest that a proper definition may be to compare the timescale of the change with the lifetime of the object.
L. Golub: Perhaps I can help to clarify this discussion about loop brightenings.

Some active regions, as seen by Skylab and the NIXT rocket, seem to emit X-rays in a relatively steady manner with very little variability. Other active regions show, in addition to the steady emission, some occasional loop brightenings which give these regions a more dynamic appearance. "Rapid" variability in this context means that many loops per hour brighten, and this is still long compared with Alfvén crossing times so it is not rapid in that sense. Also, because the steady X-ray emission is always present in both "active" and "inactive" regions, it would seem to be separate from any regular heating mechanism.

Y. Uchida: Thank you. I think that *Yohkoh* could study this phenomenon with its high and regular cadence, and its dark filter combinations could suppress the general active region loops which almost saturated the Skylab films.

B. Foing: Why do you interpret the transient brightenings in active regions as associated with mass motions that can significantly contribute to the solar wind? This matter cannot leave the Sun, since it is constrained by closed loops.

Y. Uchida: As can be seen near the start of the video, we always find such tiny brightenings inside very active regions. There are many such brightenings and the expansion seems to be continuous. We attributed the phenomenon of active region brightenings to a mass outflow, because we can argue that it is neither a wave-like densification (which is too slow) nor "re-closing loops" (because they were never opened up)! The dome-shaped expansion keeps its brightness, indicating that the mass in the loop is actually conserved.

G. Peres: You mentioned that sometimes loops appear to brighten from the base. Could this be the evaporation of chromospheric plasma conductively heated from the top of the loop?

Y. Uchida: This possibility cannot yet be excluded. However, I do not think that hot regions from which heat is conducted are seen before the hot-jet comes up, nor are any special events, which can provide high energy electrons, observed! I think that jets are likely due to a different mechanism as I mentioned in my talk.

S. Serio: I want to call your attention to another timescale, the loop cooling time, in addition to Alfvén and sound crossing times. If heating is unstable over times comparable to the cooling time, this will modify the Rosner-Tucker-Vaiana (RTV) $T^3 \sim pL$ scaling law, as is observed in many cases.

Y. Uchida: There is at least one class of nonsteady heating, and there are loops that are supplied hot matter dynamically which, therefore, do <u>not</u> follow the RTV scaling law.

E. Priest: (1) When two loops reconnect, the total magnetic helicity is conserved, which may explain the appearance of twist in one of your new loops. Have you been able to test for magnetic helicity conservation? (2) The simplest ways that a coronal arcade can form are by new flux emergence, close-down of open field, or filling of preexisting field lines. Which, if any, of these are happening in the two examples you showed? (3) An expansion of 10 km sec^{-1} of an active region represents 1 solar radius per day and so obviously cannot continue indefinitely. What is the long-term evolution of an active region? Does it expand, erupt and expand from below again, or does it expand, fade and expand from below?

Y. Uchida: (1) We are thinking of performing such an analysis, but have not yet done so. There are indeed events suggesting such a decrease of helicity. (2) We be-

lieve that we see some cases of heating of pre-existing loops, but we may also see the reclosing cases. We do not yet have an explanation of why a magnetic field that is kept open for a long time suddenly closed as in the November 12, 1991 event which Tsuneta showed. (3) I think that the active region expansion is caused by a destabilization of the magnetic configuration by the addition of new flux. The active region itself is replenished by the new flux, while the expanding outer part may continue to be accelerated somewhat like the case of CME's.
M. Livshits: We observe many steady-state loops in the coronal green line. The dynamic events are rare, except at maximum activity. What do you think about the frequency of dynamic events? Is it different in the X-ray and optical ranges?
Y. Uchida: The active region expansion I talked about has a long timescale of hours to a day. It is slow, but the point is that it is directed only outward, and was not noticed until now. I suggest that the expansion of active region loops could also be seen in the optical if regular cadence long-time observations were obtained and made into a movie. The phenomenon is ubiquitous and almost all active regions will show this type of "dynamic" behavior even in the optical range.

Discussion following paper by S. Tsuneta and J. Lemen

Y.-M. Wang: Why do you believe that coronal "dark channels" differ from coronal holes? Have you compared the dark channels with He I 10830 Å images from Kitt Peak?
S. Tsuneta: The coronal dark channels appear to be much thinner than Skylab coronal holes. We have not yet verified that the coronal dark channels we observed have the same properties as the Skylab holes. It seems that the dark channels do not have a clear counterpart in the He I 10830 Å images.
E. Priest: How much energy is released in your microflares, and how does it compare with the heating rate required for quiet and active regions?
S. Tsuneta: Much work is needed to answer this subtle and important question. From the temperature, emission measure and loop size, we can estimate the conductive and radiative fluxes, which are 10^{26-27} and 10^{23-25} ergs s^{-1}, respectively. Since the total energy loss is 10^{28-29} ergs, the microflares are small flares. To obtain the total energy output of microflares, we need to know the intensity versus occurrence rate. This is being done [see the answer to Rosner's question below]. Since we do not see such brightenings outside of active regions, the microflares do not contribute to the general coronal heating outside of active regions.
B. Somov: What is a typical value of the emission measure for the compact high temperature component in active regions?
S. Tsuneta: The emission measure is about 10^{27-28} cm^{-5} for the hot component (temperature 4-5 MK) in active regions and is 10^{28-29} cm^{-5} for the cool component (1-2 MK). Thus the higher temperature component has a smaller emission measure. It is not clear whether active region plasmas consist of two distinct temperature components or have continuous differential emission measure distributions.
R. Rosner: Have you constructed plots of the number of transient loop events versus their intensity? In other words, have you been able to connect your loop

transients in a quantitative way to Lin's observations?

S. Tsuneta: My student, Toshi Shimizu is constructing this plot. The integral rate increases with a slope of about -1 down to a certain X-ray luminosity as in Lin's plot [see Figure 4 in Ap.J. **283**, 421 (1984)]. Then, the curve becomes less steep. The combination of short exposure times (tens of msec) and long separations between exposures ($1 \sim 2$ min) causes a serious "dead time" for short duration microflares. The saturation of the plot towards the fainter events could be due partly or entirely to this dead time effect. We are trying to correct the plot for this effect. It is not easy to connect our plot to Lin's plot, because his integral rate is versus the 20 keV photon flux, whereas our plot is the integral rate versus the much softer total flux observed by SXT (5-40 Å). Thus both plots must be converted to a common energy scale which introduces some assumptions. I believe, however, that we have found with SXT the soft X-ray counterpart to Lin's microflares in hard X-rays.

J. Schmitt: I have two questions concerning the soft X-ray analogs to the hard X-ray microflares: (1) What are the shortest timescales on which you have observed these events? (2) What leads you to think that this is thermal emission?

S. Tsuneta: (1) The duration of the soft X-ray microflares varies from 1 to about 10 minutes. Since the exposure interval is about 1 minute, shorter events could exist but not be detected by the SXT. We are planning higher time resolution observations. (2) There is no clear reason to believe that the soft X-ray emission from the microflares is totally thermal. As I speculated from the comparison with Lin's hard X-ray microflares, the energy supply may be nonthermal. If the soft X-ray emission is due to nonthermal electrons, however, these electrons will produce a substantial amount of high temperature plasma which also emits soft X-rays.

G. Peres: Microflares on timescales of a few seconds are unlikely to be thermal. How large are the intensity variations that you see in microflaring loops?

S. Tsuneta: The brightness distribution along the microflaring loops is highly time dependent. Initially the contact point of different loops brightens and then the entire loop gradually brightens.

F. Chiuderi Drago: If coronal holes rotate rigidly and active regions do not, one should see active regions overtaking coronal holes. Have you ever seen that?

S. Tsuneta: Yes, the coronal holes should overtake the active regions at high latitudes, but we have not found one such case in the SXT full Sun movie. My student, Takahashi, is trying to create X-ray "synoptic charts" to more clearly identify the global nature of the coronal dark channels and large scale coronal structure. I expect that we will be able to answer your question with the X-ray synoptic charts.

L. Golub: We did indeed see active regions moving through coronal holes during Skylab. It thus appears that the photospheric field does show differential rotation, while the coronal hole structures rotate rigidly.

S. Tsuneta: Your observations are very interesting.

E.R. Priest: I was fascinated by your beautiful examples of loop interaction and restructuring. When two loops interact, important theoretical questions are: (1) How fast does reconnection take place? (2) How much energy is released? Do your observations provide any answers to these questions?

S. Tsuneta: The formation of closed loop structures continues over a few hours to tens of hours. It is a much longer process than solar flares. Since we do not see any clear enhancement in the full Sun soft X-ray flux associated with the restructuring, the energy involved in the process is much smaller than the total solar coronal radiative energy. Some of the restructuring is associated with prominence eruptions, and there is one clear example of a prominence eruption that "drives" reconnection. The process is very much consistent with the series of models proposed by Hirayama, Kopp, Pneuman, and Sturrock. I strongly believe that we will be able to answer all of these important questions with SXT observations soon.

Y. Uchida: It is better not to think of the rigid-body rotation of the coronal dark channel as indicating some deep-rooted phenomenon. The weak magnetic field there cannot oppose the differential rotation of the subphotospheric layers. I have suggested that the dark channel may be a "shadow" of an "active longitude belt" on the opposite hemisphere which "absorbs" the opposite polarity flux of neighboring regions, and the stationary behavior of the opened-up flux region simply reflects the stationary behavior of such "active longitude belts".

S. Tsuneta: That is an interesting idea, but I have an impression from the SXT movie that there is no active longitude belt to "absorb" the flux in neighboring regions. Instead, some of the coronal dark channels are stable over many rotations against the appearance and disappearance of active regions in the neighboring area.

N. Weiss: The rotation rate in the radiative zone, as inferred from helioseismology, corresponds to the angular velocity at intermediate latitudes, while coronal holes rotate at the equitorial rate. It seems more likely to me that they correspond to a nonaxisymmetric dynamo wave that drifts azimuthally, relative to the plasma.

Y.-M. Wang: To understand why coronal holes rotate rigidly, one should read Nash et al. (Solar Physics **117**, 359) or Wang & Sheeley (Ap.J. **365**, 372).

G. Peres: In the video you have shown that a significant fraction of loops seem to turn on and off without changing shape. How common are these occurrences?

J. Lemen: We have not yet performed a systematic study, but it appears that loop geometries are preserved throughout active region transient brightenings, and even some flares, i.e., the loops appear to brighten and cool without much change in geometry. More studies are planned.

M. Livshits: We saw loop structures on the quiet Sun disk. Do you observe loop arcades in quiet Sun regions on or near the limb?

J. Lemen: Large loop arcades have been seen at high solar latitudes both on the disk and at the limb, and they are not associated with particular active regions.

B. Byrne: (1) What are the temperatures of the extended loop structures which compare in morphology with the white-light corona? (2) What is the spatial scale of the white light flare footpoint separation?

J. Lemen: (1) A detailed temperature analysis of the filter ratio data for the extended corona has not been performed, to my knowledge. I would guess that a typical temperature is about 1×10^6 K. (2) About 20 arcseconds separation.

J. Schmitt: You showed the good spatial correlation between continuum and hard X-rays for the November 15 flare. Yet, the hard X-ray emission seemed to start before the white light emission. Is this real or just a question of sensitivity?

J. Lemen: It is probably the sensitivity of the white light measurements. At the peak of the flare, the white light excess was only 35% over the normal photosphere.

R. Pallavicini: How many of those active region brightenings in *Yohkoh* pictures would be visible in integrated X-ray emission, i.e. have you compared the *Yohkoh* data with full-disk GOES data?

J. Lemen: It is possible to identify individual brightenings observed in the GOES X-ray light curves. When there was only one active region on the disk, we can often make a one-to-one correspondence for each transient brightening observed.

S. Serio: During the poster session I will show a movie obtained by Vaiana from Skylab data. The comparison with your video will clearly show 18 years of progress in X-ray mirrors and detectors and also that many of the features of the X-ray corona had already been seen in Skylab data.

G. Field: Are there examples of new phenomena you have found that required the specific capabilities of your instrument?

J. Lemen: I think that the three orders of magnitude reduction in the scattering wings of the point spread function for the *Yohkoh*/SXT compared to the Skylab telescope will make the study of faint features much more fruitful. Furthermore, the SXT has much better time resolution and the exposure control is automatically adjusted to prevent saturation during flares. Finally, the SXT is expected to observe over a larger portion of the solar cycle.

Y.-M. Wang: To what extent does your result that loop magnetic fields do not diverge depend on the assumed density and velocity profiles across the loop?

J. Lemen: This is simply an observational comment that makes no hydrodynamic assumptions about the loop.

C.-C. Cheng: In your video that shows the formation of a streamer there is a brightening at one of the legs. Is this a flare? What is the relationship between this brightening and the large-scale loop preceding the streamer formation?

J. Lemen: It may be a sub-flare which acts to destabilize the magnetic field geometry. I think the analysis has not been done adequately, however, to demonstrate the connection between this brightening and the subsequent formation of the streamer.

Discussion following paper by T. Kosugi

G. Field: How do you interpret the sequence of events that occurred during the flare of November 15? How do you know that these loops are involved?

T. Kosugi: Our speculation is that the three spikes in the impulsive phase of this flare are due to successive flaring or "chain-reaction" of three adjacent loops. Judging from the increasing separation with time of the double sources seen in hard X-ray images as well as in white light, the chain reaction starts from below and develops to overlying loops. Because the hard X-ray and white light double sources are located low in the solar atmosphere, it is hard to interpret this increasing separation as due to an expanding loop.

Discussion following paper by J. Sylwester and B. Sylwester

R. Pallavicini: If my memory is correct, the flare observed by NIXT to which you apply your model is a two-ribbon flare, while your model refers to a closed magnetically confined loop. Could this affect your analysis and conclusions?

J. Sylwester: As pointed out by Leon Golub in his talk, the flare in soft X-rays is seen as a single constant cross-section area loop. Emission of this loop dominates the emission of the entire solar disc by a factor of ~ 100. Therefore it appears safe to apply our method of analysis to the GOES fluxes.
S. Serio: This technique is very useful also for obtaining insight into the distribution of volumes in stellar flares, where we lack spatial resolution.

Discussion following paper by G. Peres et al.

Y. Uchida: The *Yohkoh* observations clearly show an intermittent component of the heating and mass supply for the coronal loops of active regions, giving observational support for your treatment. What is the physical process providing the heat source in your model? Did you simply assume a heat source?
G. Peres: Yes I did.
G. Field: Perhaps the impulsive heating is due to shock waves, generated when magnetic instabilities drive supersonic motions. The characteristic timescale is the Alfvén crossing time for the unstable portion of the loops, perhaps a few seconds.

Discussion following paper by W.-Q. Gan and E. Rieger

S. Serio: Does your code include viscosity? Peres and Reale have found that viscosity affects the velocity distributions in the loop.
W. Gan: Viscosity is not included, and it is valuable to check its influence further.
F. Reale: Antonucci et al. have shown that the stationary component of the Ca XIX line during the impulsive phase of flares may be explained by the presence of impulsive heating located at the base of the loop. This possibility could not be ruled out by observations before *Yohkoh* because they do not have sufficient time resolution.
W. Gan: In our model, the stationary component (Ca XIXw) dominates after 40 sec. I noticed that in Antonucci et al. (1987, and poster paper at this meeting) the time required is 60 sec. Among the more than 200 flares observed by *Yohkoh* (time resolution is 5 sec or better), only about 10 flares showed dominant blueshifted emissions in the impulsive phase (Cheng, private discussion). Most flares showed stationary-dominated blue asymmetry in Ca XIXw, but nearly all of the current hydrodynamical models predict blueshifted Ca XIXw emission. In my talk, I suggested two ways to decrease the velocity in theoretical models. One is to increase the initial coronal density. Another is to assume that the loop is a low-lying arc, rather than a semi-circular loop in the impulsive phase. Certainly other changes could improve the theoretical model, such as the location of the heating source.

Discussion following paper by C.-C. Cheng

R. Pallavicini: I wonder about the disappearance of active region structures during a flare. Is this effect real, or is the flare so bright that it swamps the contribution of other structures? How common are these loop "disappearances?" Could this be a temperature effect?

C. Cheng: With about the same exposure times, the enhanced loops observed before the flare should also appear in the flare exposure, if they are still there. We have not done any photometry on the S082A plates, and it would be interesting to compare the total energy content in the pre-flare active region, the flare, and the post-flare active region. Previous whole-Sun X-ray observations have shown that, for some flares, the soft X-ray light curves exhibit a noticeable drop in intensity just before the flares take off. No, it could not be a temperature effect.
L. Golub: If the loops became hot during flares then they could disappear, because the observed spectral line is formed only at a lower temperature.
C. Cheng: No. The XUV spectrograph covers the entire 170-630 Å spectral range, which includes many emission lines formed in plasmas with temperatures from 10^4 to 2×10^7 K. An initially bright Fe XVI ($\sim 3 \times 10^6$ K) loop, when flaring and becoming hotter, will certainly appear in lines found at higher temperatures, such as Fe XXIII-XXIV.
B. Foing: For the high velocity events and flares, how does the kinetic energy content compare with the thermal energy content?
C. Cheng: For XUV streaks, associated with the flares, the kinetic energy content is somewhat greater than the thermal energy.
G. Field: You stated that no change in the morphology of the magnetic field is observed during a flare. On the other hand, you attributed the energy release to a kink instability, which does involve a change of morphology. How do you reconcile this apparent contradiction?
C. Cheng: I said that the global magnetic field morphology of the active region is not altered by the flare as evidenced in the images of the post-flare loop system whose morphology differs very little from that before the flare. The 15 June flare occurred in a new set of loops whose magnetic morphology did change during the flare. One of the loops suffered kink instabilities and was completely disrupted, while the others were heated to the high temperature of 2×10^7 K.

Discussion following paper by R. Smartt and Z. Zheng

B. Somov: Is it possible to estimate the plasma density where the optical brightening is observed?
R. Smartt: The plasma density at the time of the observed interaction can be estimated from the observed cooling time, assuming that radiative cooling dominates, but the estimate is temperature dependent. When accurate microdensitometry of the enhancement observed in the green line and the red line is done, we should provide reasonably accurate estimates of the temperature during the interaction.
N. Weiss: What observations would be needed to provide direct evidence of the change in topology associated with reconnection of interacting loops?
R. Smartt: Some loops are observed at the limb with their planes nearly normal to the line-of-sight. In such cases, the initial and final topologies are mostly self-evident. Otherwise, disk-centered observations of these interactions, as recorded in high-angular-resolution X-ray images, for example, would be required to establish typical topological changes along a radial direction.

PART III

STELLAR CORONAE: THE PRE-ROSAT PICTURE

STELLAR CORONAL EMISSION: WHAT WE HAVE LEARNED FROM PRE-ROSAT OBSERVATIONS

S. SCIORTINO

Istituto ed Osservatorio Astronomico
Piazza del Parlamento 1, 90134 Palermo Italy

Abstract. Thanks to the success of *Einstein* and, later, of *Exosat*, the past decade has seen the unexpected flowering of the *Study of X-ray Emission from Normal Stars.* At the opening of the "ROSAT-era", when the availability of new data promises exciting advances, I review major accomplishments in the field, such as: i) the unpredicted discovery of coronal emission from nearly all types of stars, ii) the influence of stellar evolution on X-ray emission, iii) the role played by both optically-selected and X-ray-selected stellar samples in our present understanding of coronal emission, and iv) the contribution of coronal emission from late-type stars to the X-ray make-up of the Galaxy.

Key words: Stars: coronae - X-rays: stars -X-rays: diffuse background

1. The Years of Discoveries

The past 15 years have seen the early phase and the mature development of a new frontier in the realm of astrophysical research: *The study of X-ray emission from normal stars.* G.S. Vaiana has been one of the pioneers of this new frontier and was for me a great honor to work with him in the past decade, rich of pleasure for many exciting discoveries. I will try in this review to describe the advancement in this field to which he has given fundamental contributions.

Before 1979 only few stellar coronal sources were known (Figure 1). The coronae of Capella (Catura et al., 1975) and α Cen (Nugent & Garmire, 1978) could still be explained by an acoustic heating mechanism, while the evidence of high temperature components in the coronae of several RS CVn's seen with HEAO-1 (Walter et al., 1980) indicated the need for a non-gravitational confinement. For part of the astrophysical community this last finding was not surprising. In fact, mainly as a result of the program driven by G. S. Vaiana leading to the Skylab S-054 X-ray telescope (see the reviews of Vaiana & Rosner, 1978 and also of Golub, in this volume), the study of the solar corona had clearly shown how complex and structured was the corona of the Sun, pointing toward the need for magnetic confinement of coronal plasma.

With the launch of the *Einstein* Observatory, and since its early observations, our perception of the pervasiveness of stellar X-ray emission has changed drastically. In his memorable paper (Vaiana et al., 1981), based on results of a first round of *Einstein* observations, G.S. Vaiana with his group and several collaborators presented an X-ray survey of about 150 stars, concluding that, with the exception of late-B/early-A and late giants and supergiants, all stars are X-ray emitters. This discovery was a breakthrough in stellar coronal physics. It ruled out, as essentially incorrect, all those models that tried to predict/explain stellar X-ray emission without taking into account the critical role played by the magnetic field both in structuring and heating the stellar coronae.

Since these early results, the Palermo-Cfa stellar group has been engaged in the homogeneous analysis of all stellar data gathered with the Imaging Proportional Counter (Giacconi et al., 1979; Gorenstein et al., 1981) on board the *Einstein* Observatory, which covered approximately 10% of the sky with $\sim$ 4000 images. The task of extracting quantitative information for more than 35000 stars from IPC images has been made possible by the final highly-sophisticated IPC data processing

J.F. Linsky and S. Serio (eds.), Physics of Solar and Stellar Coronae, 211–224.

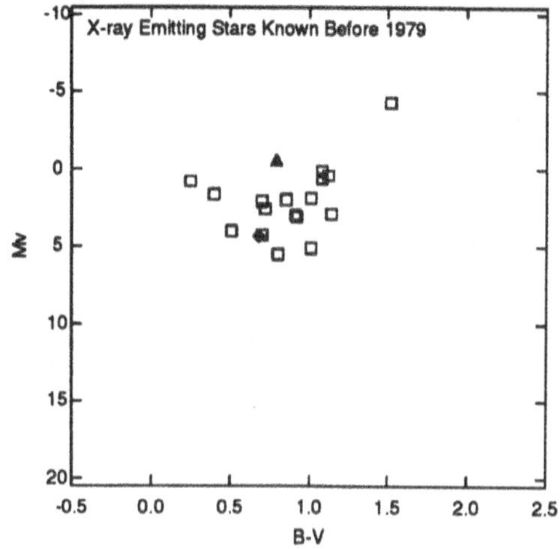

Fig. 1. HR diagram of the stellar coronal sources known up to 1979. Except for Capella (filled triangle) and α Cen (filled diamond), all other stars were RS CVn-like systems (squares). Based on these spare data no clear answer could have been given about the mechanism responsible of X-ray emission. However, already at those early days RS CVn's coronal temperatures gave indication for a non-gravitational confinement of emitting plasma.

system (Harnden et al., 1984), while handling all the available information has called for the realization of the Einstein Stellar X-ray Database (Sciortino et al., 1988; Harnden et al., 1990). This effort has resulted in the *Einstein* Stellar Survey (Vaiana, 1990; Vaiana et al., 1990).

The result of this effort is illustrated in figure 2, showing the HR diagram of most (*but not all !*) of the stars with, at least, an X-ray flux (or upper limit) measured with the *Einstein* Observatory.

Among the stars surveyed but not shown in figure 2 are those listed in the SAO catalog with magnitude fainter than 6.5 (the limiting magnitude of the Bright Stars Catalog). This sample is largely composed of yellow main sequence and giant stars up to distances of 100-200 pc, and likely contains a population of young intense X-ray sources (Micela et al., 1992a; Favata et al., 1992a). The space density of this population is quite small and their discovery is virtually impossible exploring only the volume inside the sphere of 25 pc radius made accessible by available catalogs of nearby stars. This explains why their existence has not been recognized in the *Einstein* dG survey of Maggio et al. (1987).

If all stars would emit X-rays at the same level of the Sun, then the study of X-ray emission from normal stars would have been a very meager affair. As shown in figure 3, only a handful of stars, all inside 10 pc, have been detected at the solar X-ray emission level: $L_X \sim 10^{27}$erg s^{-1}, while the majority of stars have been detected at significantly higher (up to three orders of magnitude) L_X. In other words, it has been only the surprising discovery of the wide spread of L_X for each spectral type and luminosity class, evident since early days (Vaiana et al., 1981), that has allowed the rapid growth of our knowledge. **This large spread shows that "classical" stellar parameters cannot explain the X-ray emission level**, indicating that coronal emission is a manifestation of "stellar activity" determined by "non-classical" stellar parameters. Since 1979, it has been

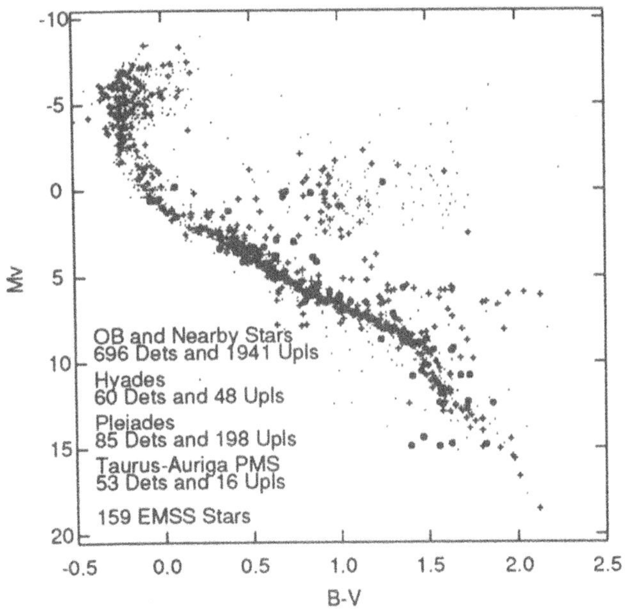

Fig. 2. HR diagram of stars surveyed with the *Einstein* Observatory, based on X-ray surveys of optically selected samples of O stars (Chlebowsky et al., 1989; Sciortino et al., 1990), B stars from SAO (Grillo et al., 1992), nearby field dA (Schmitt et al., 1985), dG (Maggio et al., 1987), dK and dM (Barbera et al., 1992b) stars, Pleiades (Micela et al., 1990), Hyades (Micela et al., 1988), and optically and X-ray selected Taurus-Auriga PMS stars (Damiani & Micela, 1992). Detections are indicated by pluses, while stars with measured upper limits are indicated by dots. The X-ray selected stars (bullets) are taken from the Extended Medium Sensitivity Survey stars (Fleming, 1988; Stocke et al., 1991).

clear that the magnetic field is a crucial ingredient (Rosner & Vaiana, 1980). Its interplay with stellar rotation rate, spin-down with increasing stellar age, depth of convection zone would result in stellar dynamos of various efficiency. In this interpretational framework, the spread in the observed X-ray luminosity will mainly reflect that of the efficiency of the stellar coronal heating. It is hypothesized that the heating mechanism is mainly due to the interaction between stellar magnetic fields and coronal plasma, a mechanism clearly at work on the Sun. In the past ten years this picture has passed many observational tests, as I will briefly discuss in the following. I will not discuss evidence coming from temporal variations of coronal emission level, since this subject has been reviewed by Pallavicini, in this volume.

2. What Parameters does L_X Depend upon in Late-type Stars ?

2.1. L_X AND STELLAR DYNAMOS

Stellar magnetic fields have to be continuously regenerated by dynamo action (see Weiss, in this volume and references therein), and a more intense dynamo action will result generally in a more intense X-ray emission (but see the caveat in the following). If this scenario is correct, there are various predicted trends and relations that should occur between the X-ray emission level and other stellar parameters,

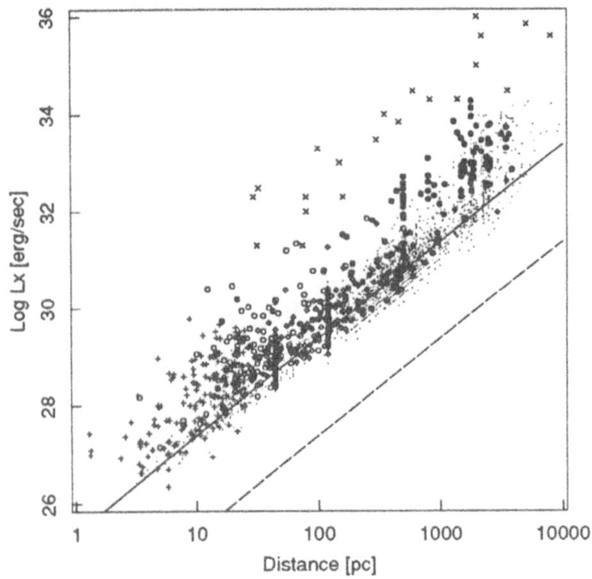

Fig. 3. L_X vs. distance for various stellar samples. Detected sources are indicated with different symbols as follows: Main sequence G, K, M stars (plus), main sequence A-F stars and yellow giants (circles), Orion, Pleiades, Hyades and EMSS stars (diamonds), OB stars (bullets), HEAO-1 compact sources (crosses). Small dots indicate upper limits to the X-ray luminosity. The lines represent limiting sensitivities of 2 10^{-13}erg cm^{-2} s^{-1} (solid), typical of Einstein observations, and of 2 10^{-15}erg cm^{-2} s^{-1} (dashed), expected for the majority of AXAF-I observations.

such as: (i) stellar rotation rate; (ii) dynamo efficiency at A/F type boundary, as a result of the depth of the convection zone; (iii) ineffectiveness of $\alpha - \omega$ dynamo action at the low-mass end, where stellar models predict a completely convective structure.

2.1.1. L_X *and Rotation Rate*

Pallavicini et al. (1981) have shown that the X-ray luminosity of late-type stars, independently of spectral type and luminosity class, correlates with stellar rotation rate, as $L_X \propto (v \sin i)^2$ (cf. figure 5 in Pallavicini et al. paper). Adopting a somewhat larger sample of dwarf G stars, and considering both detections and upper bounds, Maggio et al. (1987) have confirmed this finding. Micela et al. (1985a,1990) and Caillault & Helfand (1985) have pointed out that this relation breaks down for the fast rotating, intense X-ray emitting Pleiades dK stars (see figure 4).

A similar break is clearly evident also among the stellar coronal sources discovered with the Extended Medium Sensitivity Survey (Fleming et al., 1989), the majority of which have rotational velocities much higher than those of nearby field stars (see Figure 4). These results can be naturally interpreted (cf. Rosner, 1989) as the existence of a saturation level in stellar coronal emission occurring when the entire stellar surface is covered with solar-like active regions. In such a case a further increase of the stellar rotation rate cannot result in an increase of the stellar X-ray emission level.

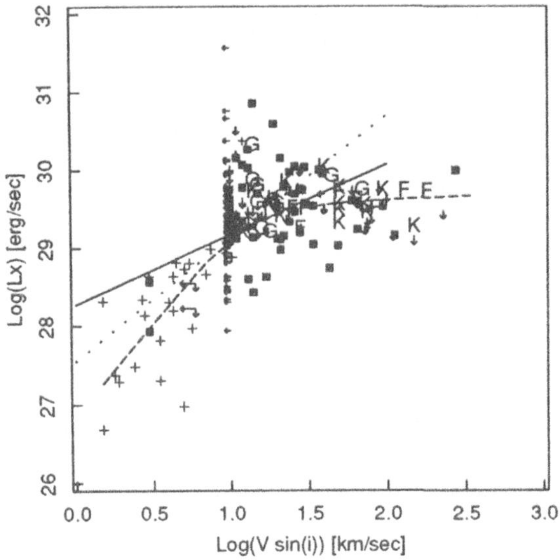

Fig. 4. L_X vs. rotation rate for (i) a sample of late type stars (pluses) from Maggio et al. (1987), (ii) a sample of Pleiades stars from Micela et al. (1990), each letter indicates the corresponding spectral type, (iii) a sample of EMSS stellar sources (filled squares) from Fleming (1988). The dotted line is the $L_X \propto (v \sin i)^{2.1\pm0.5}$ relation, found for the late-type stars (Pallavicini et al. 1981; Maggio et al. 1987) and holding also for the entire class of RS CVn's (not shown in the figure). The solid line is the regression fit, $L_X \propto (v \sin i)^{1.21\pm0.13}$ to the Pleiades data (Micela et al. 1990), while the dashed line is a robust local fit to both the field and Pleiades star samples. It is evident that the local fit provides also a good description of the coronal emission of the EMSS stellar sources.

2.1.2. L_X *and Depth of Convection Zone*

The two principal ingredients determining the efficiency of an $\alpha - \omega$ stellar dynamo are the rotation rate and the structure of the convection zone. Dynamo efficiency is usually expressed with a single parameter, the dynamo number, $N_D \propto \alpha(\nabla_r\Omega)d^4/\eta^2$, where α is the mean elicity of convective velocity times convective turnover time (τ_c) at the base of convection zone, $\nabla_r\Omega$ is the radial gradient of angular velocity, d is the scale length of convection, and η is the turbulent magnetic diffusivity. Assuming $\alpha \sim \Omega d$ and $\eta \sim d^2\tau_c$ (Durney & Latour, 1978), and making the further naive assumption $\nabla_r\Omega \sim \Omega/d$ then $N_D \sim (\Omega\tau_c)^2 \sim R_o^{-2}$. The so-called Rossby number, R_o, is a parameter that gives a measure of dynamo efficiency in terms of measured rotational rate and of τ_c as derived from models. Noyes et al. (1984) have shown that in a wide range of spectral types the intensity of Ca h&k lines is well correlated with R_o. A similar results has been obtained correlating X-ray luminosity with R_o (Micela et al., 1985b; Schmitt et al., 1985; Maggio et al., 1987), **indicating a common magnetic origin for many manifestations of stellar activity.**

Since convection is an essential ingredient for stellar dynamos, and stellar interior models predict the growth of the convection zone at late-A spectral types, no coronae should develop on normal late-B/early-A stars (but see Linsky, in this volume, for a discussion of peculiar B/A stars) while X-ray emission should smoothly increase together with the depth of convection zone in the late-A/early-F spectral type range. Schmitt et al. (1985) have performed a detailed study of the X-ray emission level among A and early-F stars showing (i) that early A stars as a class are not X-ray emitters, and (ii) that the X-ray emission level of late-A/early-F stars correlate better with the Rossby number than with the rotational velocity, as expected in the region of rapid change of the depth of convection zone with spectral type.

2.1.3. L_X *at the Low-Mass End*

The expectation of a lower X-ray emission level in low-mass stars (later than M5) has been recently addressed by Barbera et al. (1992a,b,c), based on the final data processing of the full set of *Einstein* observations and on a detailed analysis to investigate and to remove at the best the biases of the optically-selected stellar sample. This work shows with high statistical significance the existence of a drop in X- ray emission level (as indicated both by X-ray luminosity and surface flux) at spectral type later than $\sim$ M5. This drop was previously indicated by Bookbinder (1985) on the basis of a smaller, less optically characterized stellar sample and an early version of the *Einstein* data analysis system.

2.2. L_X and Stellar Age

One of the most important results of stellar X-ray astronomy has been the discovery that the X-ray luminosity level decreases with increasing stellar age (Vaiana, 1981, 1983; Caillault & Helfand, 1985; Micela et al., 1985a, 1988, 1990).

Young pre-main sequence stars where recognized as a class of intense X-ray emitters ($L_X \sim 10^{29-30}$erg s^{-1}) since early observations (Ku & Chanan, 1979; Ku et al., 1982; Gahm, 1980; Feigelson & Decampli, 1981; Walter & Kuhi, 1981; Feigelson & Kriss, 1981), but only the systematic analysis of various coeval stellar samples (for a recent review see Vaiana et al.,1992 and references therein), with the adoption of survival analysis techniques (Avni et al., 1980; Schmitt, 1985; Feigelson & Nelson, 1985) to retain upper limits as well as detections information, has clearly shown the decrease of the X-ray emission level from the PMS stage and along the entire MS lifetime. This can be explained as an effect of stellar spin-down with age (for recent reviews, see Soderblom 1991, Bovier 1991, Stauffer 1991) resulting from the magnetic braking (Collier-Cameron et al. 1991) due to intense stellar magnetic activity associated with high rotational velocity.

This ageing effect is so relevant to be one of the crucial ingredients for a realistic modeling of the stellar coronal content of the Galaxy. It influences substantially stellar X-ray counts (Favata, Micela and Sciortino 1992a,b), is a sensible diagnostics of stellar birthrate in the Galaxy in the last billion years (Micela et al., 1992a), and gives a natural explanation of the origin of the diffuse soft X-ray background feature called the Galactic Ridge (cf. §5.1).

2.3. L_X and Evolution Toward The Red Giant Branch

Changes of stellar structure with evolution toward the red giant branch, especially stellar angular momentum redistribution and development of the convection zone, influence the level of "stellar activity" (for a review of some of the theoretical aspects see Rosner, in this volume). Based on IUE data of 22 stars, Linsky and Haisch (1979) have suggested the existence of an UV dividing line going from the K2III to the G8Ib spectral types, separating earlier stars having line emission coming from transition regions from later ones showing mainly evidence of strong winds.

The final, detailed analysis of the entire sample of late giants and supergiants listed in the BSC and surveyed with the *Einstein* Observatory (Maggio et al., 1990) has demonstrated on firm statistical grounds the existence of an X-ray dividing line, originally suggested on more sparse data (Ayres et al., 1981), occurring for stars of spectral type later than K3III. The analysis of a substantially larger sample has confirmed the existence of the UV dividing line likely at the same position of the X-ray one (Haisch et al., 1990).

Using both X-ray and IUE data, our group has recently studied the evolution of stellar activity in early post-main-sequence phases for stars with 1.1-1.9 $M_{\odot}$. Available data are consistent with a picture in which the onset of convection when stars leave the main sequence accounts for the raise of X-ray luminosity. However, the rotational velocity of stars with $M < 1.6\ M_{\odot}$, already declined during the main sequence life-time, due to intense magnetic activity, keeps declining until the dynamo ceases to be effective, and the X-ray emission consequently drops (for more details see Maggio et al., in this volume).

3. Low-Resolution Spectroscopy of Stellar Coronae

Medium/high resolution spectra are available only for a handful of stars (cf. Mewe, in this volume). Hence, most of the knowledge on the physical conditions of stellar coronal plasma is based on low resolution spectra, such as those gathered with the *Einstein* IPC, or on broad band measurements, such as those of the *Exosat* CMA. The analysis of about 170 IPC spectra of nearby stars (Schmitt et al., 1990) has shown that: (i) coronal temperatures are in the range $10^6 - 10^{7.5}$ K with dF-dK stars being cooler than dM stars; RS CVn systems have very hot coronae (see figure 2 of Schmitt et al. 1990), (ii) often a two-temperature model, as well as a continuous emission measure distribution model, gives a better fit to high S/N spectra (see figures 7, 8 and 9 of the cited paper). A comparison of *Einstein* IPC and *Exosat* fluxes for a sample of $\sim$ 15 stars detected with both instruments has also indicated the existence of multi-temperature components in coronal emitting plasmas (Pallavicini et al., 1988).

4. X-ray Emission from Early-type Stars

One of the most unpredicted discovery of the *Einstein* Observatory has certainly been the recognition that OB stars are intense X-ray emitters with X-ray and bolometric luminosities being roughly proportional (Harnden et al., 1979). Based on a small sample of OB stars, Pallavicini et al. (1981) have shown that X-ray luminosity do not correlate with rotation rate. Hence, X-ray emission from massive stars is not associated with dynamo action. Given the lack of a convection zone in OB stars, this is not really surprising.

With the availability of the final catalog of O stars surveyed with *Einstein* imaging instruments (Chlebowsky et al., 1989) we have carried out a detailed study of X-ray emission from O stars (Sciortino et al., 1990) concluding that: (i) the majority of O stars have X-ray luminosity between 10^{31-32}erg s^{-1}, (ii) $\log L_X/L_{Bol}$ ranges between -6 and -7, (iii) L_X/L_{Bol} seems to correlate with $\dot{M}V_{\infty}^2$, (iv) none of the stellar and wind parameters considered, including L_{Bol}, accounts for the large spread of L_X, (v) there is no evident relation between L_X and association membership and/or position in the Galaxy.

Many models have been proposed to explain the intense X-ray emission from OB stars. They belong mainly to two classes: those where a hot corona lies near the stellar surface, at the base of a cooler high-speed wind (Cassinelli & Olson, 1979; Waldron, 1984), and those where shocked X-ray emitting matter is distributed throughout the wind, and the shocks are due to the instability of a radiatively-driven wind to density perturbations (Lucy & White, 1980; Lucy, 1982).

A very crucial test of all those models is the comparison between predicted and observed spectra (Cassinelli & Swank, 1983). Only Lucy's model, with an ad-hoc choice of shock intensity distribution, seems to match the observations (cf. figure 5 of Cassinelli & Swank, 1983), while all models assuming the existence of a corona at the base of the wind are ruled out by the lack of the predicted strong absorption feature below 0.5 keV (cf. figures 3, and 4 of the cited paper). S. Owoki and collaborators, and Pauldrak and collaborators have undertaken long term projects for studying the physics of wind-induced instability. This study should make possible to predict the resulting X-ray emission. For further details the reader is referred to the paper by Kudritzki, in this volume.

5. Stellar Coronal Emission and the Origin of Diffuse Soft X-ray Background (DXRB)

Since the seminal paper of Rosner et al. (1981), several authors have evaluated the contribution of stellar coronal sources to the DXRB (Kahn & Caillault, 1986; Caillault et al., 1986; Caillault, 1990). In the following I will limit myself to consider only the most recent results (see Table 1) based on a detailed stellar spatial distribution taken from the Bachall & Soneira Galaxy model (Bahcall & Soneira, 1980; Bahcall, 1986) and a detailed description of interstellar absorption at X-ray wavelenghts.

Schmitt & Snowden (1990) have computed the contribution of M stars assuming that: (i) their coronal spectra can be modeled with a continuous emission measure distribution with $\log T_{\rm max} = 7.1$ and $\alpha = 0.9$ (Schmitt et al., 1990); (ii) their X-ray luminosity function can be parametrically described with a log-normal distribution of given $< L_X >$. With a similar approach Ottman & Schmitt (1992) have evaluated the contribution of RS CVn's whose spectra have been modeled with $\log T_{\rm max} = 7.5$ and $\alpha = 3$.

Kashyap et al. (1992) have evaluated the contribution of all spectral types (including RS CVn's), giving also an estimate of the contribution of halo population stars. In their calculations (i) they have modeled coronal spectra both in terms of the same continuous emission measure distribution model adopted by Schmitt and Snowden (1990), and in terms of a two-temperature description with $T_{\rm low} = 1.4\ 10^6$ K and $T_{\rm high} = 1.74\ 10^7$ K, derived from the stellar coronal temperature survey of Schmitt et al. (1990), where the emission measure of the high-T component, $EM_{\rm high}$, increases with increasing total L_X, in analogy with the solar case; (ii) they have adopted the full set of maximum likelihood X-ray luminosity functions derived from the *Einstein* Stellar Survey as available at the end of 1990. Due to the limited knowledge on X-ray luminosity functions at high L_X, from which the major part of stellar contribution to DXRB derives, those authors have obtained confidence intervals for the predicted stellar contribution through a bootstrap analysis.

A comparison of the predictions of the above two models is shown in table 1 for a limiting sensitivity of 10^{-10}erg s^{-1} cm^{-2}, namely that of the all sky survey of McCammon et al. (1983). The discrepancies can be explained (i) by the tails of X-ray luminosity functions retained by Kashyap et al. (1992) in their calculation, and

TABLE 1
Comparison of Recent Predictions of Stellar Coronal Source Contribution to Soft DXRB for a Limiting Sensitivity of 10^{-10} erg sec^{-1} cm^{-2}

Energy	Band	High Galactic Latitude dM	All MS	RS CVn[(a)]	Low Galactic Latitude dM	All MS	RS CVn[(a)]
Schmitt and Snowden (1990) and Ottman and Schmitt (1992)							
< 0.3	C	...	-	...	...	-	...
0.45-0.65	M_1	5%	-	...	9%-18%	-	-
0.65-0.85	M_2	10%	...	1%	20%-40%	...	5%
0.85-1.15	I	8%	...	1%	20%-40%	...	14%
1.15-2.0	J	< 3%	...	1%	7%-14%	...	16%
2.0 - 6.0	HEAO1[(b)]	...	...	-	...	...	6%
Kashyap et al. (1992)[(c)]							
< 0.3	C	...	< 3%	-	...	< 3%	...
0.45-0.65	M_1	...	4%-8%	~ 1%	...	7%-16%	...
0.65-0.85	M_2	...	7%-17%	~ 1%	...	14%-34%	...
0.85-1.15	I	...	12%-30%	~ 1%	...	29%-71%	...
1.15-2.0	J	...	9%-22%	~ 1%	...	27%-66%	...

(a) Notice that the limited knowledge on space density, scale height and X- ray luminosity function of RS CVn's makes the calculation of their contribution to DXRB much more uncertain than that of main sequence stars (cf. Ottman and Schmitt, 1992).
(b) The 6% contribution in the 2-6 keV band accounts for ~ 27% of the Galactic Ridge excess, however this prediction leaves unexplained the strong Fe 6.7 Kev line seen in the Ridge and in stars formation regions with Ginga (Koyama 1988a,b).
(c) M stars account for ~ 60%-75% of the overall stellar contribution, the exact amount depending on energy band and assumed stellar spectra. The intensity of the stellar contribution to DXRB decreases by a factor 2 when the limiting sensitivity becomes 10^{-14}erg cm^{-2} s^{-1}.

(ii) by differences in the assumed model stellar spectrum. Future 100 eV resolution observations, such as those which will be possible with CCD instruments, should better constrain stellar spectra, and allow a further advancement in the field.

5.1. The Origin of the Galactic Ridge

Our group in Palermo has re-evaluated the stellar contribution to DXRB to investigate the origin of the Galactic ridge (Micela, 1991; Micela et al., 1992b).

The predictions shown in figure 5 have been evaluated in the [0.16 - 3.5] keV energy band and for a limiting flux, $f_x \sim 10^{-13}$ erg cm^{-2} s^{-1}, typical of *Einstein* observations, assuming that:

- The nearby dA-dF field stars are representative of the dA-dF stars in the entire Galaxy and have coronal temperatures of 0.3 keV.
- The dG, dK, dM stars have been divided into 3 age ranges with scale heights, X-ray luminosity functions, and coronal temperatures as listed below, and

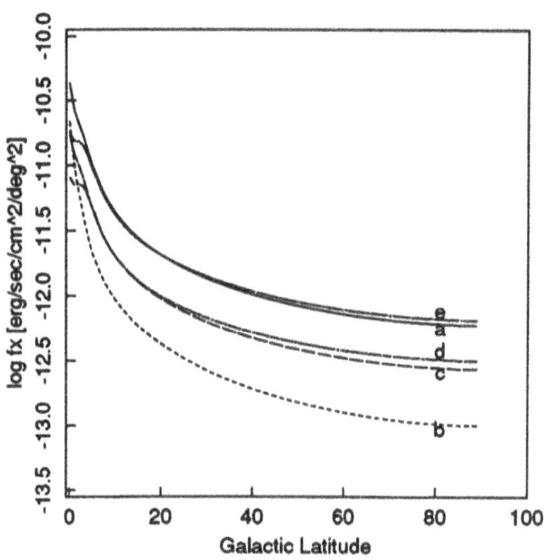

Fig. 5. Predicted contribution of disk population stars to DXRB as a function of galactic latitude for a limiting X-ray flux of 10^{-13}erg cm^{-2} s^{-1} (line *a*). The contributions from stars of the three distinct age ranges considered are also shown: "Pleiades-age" (line *b*), "Hyades-age" (line *c*), and "Old-Disk age" (line *d*) stars. For comparison, we shows also the predicted contribution assuming that all stars belong to the "Old-Disk age" group (line *e*).

with densities derived assuming a constant stellar birth rate:

Age (yr)	Scale Height (pc)	Prototype X-ray Lum. Fun.	Coronal Temp. (keV)
$10^7 - 10^8$	140	Pleiades	1.1
$10^8 - 10^9$	240	Hyades/Young Disk	0.5
$10^9 - 10^{10}$	400	Old Disk	0.3

This model seems capable to explain the Galactic Ridge in terms of unresolved coronal emission from young disk population stars (line *a*), and predicts a stellar contribution toward galactic poles of only few percent of the observed diffuse soft X-ray background in the [0.16 - 3.5] keV energy band. For comparison we show in figure 5 (line *e*) a calculation performed assuming a single stellar age, and a coronal temperature equal to 0.3 keV. At high galactic latitude this single age, single T model is essentially indistinguishable from the other (line *a* of Fig. 5), while at low galactic latitude the predicted emission accounts only for $\sim$ 25% of the observed excess X-ray emission in the Galactic ridge.

In figure 6 we compare as a function of galactic longitude a model prediction and the actual IPC background measurements in the Galactic ridge region. The points are measuraments taken from the *Einstein* IPC survey of soft X-ray background (Micela et al., 1991), and the lines are the predicted profiles computed for $b = 0°$ (lower) and $b = 5°$ (upper) including the average level of non-astrophysical background (dashed line).

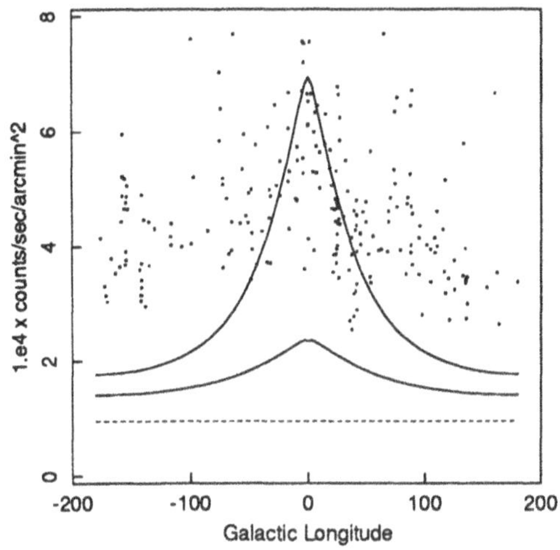

Fig. 6. Comparison of a model prediction of the [.16 - 3.5] keV X-ray background with *Einstein* Observatory IPC measurements, in the region of sky with $|b| < 5°$, namely toward the Galactic ridge. The dashed line is the average level of non-astrophysical contamination. The lower and upper solid lines are the predicted profiles computed for $b = 0°$ and $b = 5°$, respectively.

6. A Glance into the Future

Since many of the new exciting results from ongoing ROSAT stellar surveys are reported in many other papers of this volume (Bromage, 1992; Casanova et al., 1992; Caillault et al., 1992; Fleming, 1992; Harnden, 1992; Pye, 1992; Schmitt, 1992), I will not even attempt to discuss them. Looking at the stellar X-ray astronomy of the next decade I want to notice that future more sensitive, high resolution observations covering the energy range up to $\sim$ 10 keV, such as those that will be available with ASTRO-D, Jet-X, AXAF and XMM, will allow to better probe the origin of the Galactic Ridge. In particular, the origin of the iron 6.7 keV emission line, seen both in Galactic Ridge and in star formation regions with the few degrees spatial resolution instrument of Ginga (Koyama, 1988a; Koyama, 1988b), is an open and intriguing research problem.

As a result of stellar density scale heights, and shape of X-ray luminosity functions, if we had a very high X-ray sensitivity, we would expect a *clustering of stars* in the f_x/f_v vs. m_v plane for each pointing direction. Based on the Palermo-ESA X-ray Galaxy model (Favata et al. 1992c), and taking into account the effect of stellar age as discussed by Micela, Sciortino and Favata(1992a) and Favata, Micela and Sciortino (1992a,b), one can calculate the predicted positions of the centroids of the distributions in the f_x/f_v, m_v plane.

Figure 7 shows the contour plot of number density (number of star per unit m_v, and unit f_x/f_v) for the dA, dG, dM0-dM5 stars at $b = 20°$ and $l = 90°$. Centroid positions depend on latitude: at the poles they move towards smaller m_v (apparently more luminous stars), and on the plane they move towards larger m_v (less luminous stars). The limiting flux required to observe the centroids is very low: for instance, at $b = 20°$, we predict $f_x \sim 10^{-(17\div18)}$ erg cm^{-2} s^{-1}, that becomes $f_x \sim 10^{-(16\div17)}$ erg cm^{-2} s^{-1} at the pole. At this sensitivity, the number of stars per square degree is very high and obviously there will be source confusion at any

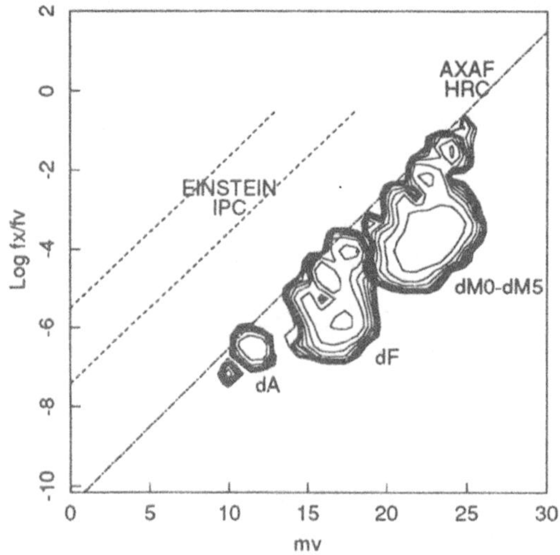

Fig. 7. Contours of model-predicted stellar number density in the log f_x/f_v - m_v plane, for dA, dG and dM0-dM5 stars at $b = 20°$, and $l = 90°$. The number density of each group has been normalized to its maximum value. Contours cover the range between 1 and 1/10 of the maximum value. The range of limiting sensitivities achieved by the *Einstein* IPC observations, and the foreseeable limiting sensitivities for the AXAF HRC observations are shown. Even a mission with sensitivity such as that of AXAF will not be able to sample the centroid of the distribution of M stars.

foreseeable sensitivity and resolution that instruments will reach in near future. As shown in figure 7, **AXAF will detect only the tail of the stellar distribution of M stars. Hence, even AXAF (or AXAF-like missions) will not be able to resolve the stellar soft X-ray background.**

Acknowledgements

I thank my colleagues M. Barbera, F. Favata, and G. Micela for having provided material in advance of publication, and A. Maggio, F. Reale, and S. Serio for their critical reading of this paper. I acknowledge support from Ministero della Università e della Ricerca Scientifica e Tecnologica, from Agenzia Spaziale Italiana, and from GNA-CNR.

References

Avni, Y., Soltan, A., Tananbaum, H., and Zamorani, G. 1980, ApJ, 238, 1800.

Ayres, T. R., Linsky, J. L., Vaiana, G. S., Golub, L., and Rosner, R. 1981, ApJ, 250, 293.

Bahcall, J. N. 1986, ARA&A, 24, 577.

Bahcall, J. N. and Soneira, R. M. 1980, ApJS., 44, 73.

Barbera, M., Micela, G., Sciortino, S., Vaiana, G. S., F. R. Harnden, J., and Rosner, R. 1992a, Mem. Soc. Astr. It., in press.

Barbera, M., Micela, G., Sciortino, S., Harnden, F. R., Jr., and Rosner, R. 1992b, ApJ, submitted.

Barbera, M., Micela, G., Sciortino, S., Harnden, F. R., Jr., and Rosner, R. 1992c, In *This volume.*

Bookbinder, J. 1985, PhD Thesis, Harvard University.

Bouvier, J. 1991, In *Angular Momentum Evolution of Young Stars*, S. Catalano and J. Stauffer, editors, page 41, Kluwer Academic Publisher.

Bromage, G. 1992, In *This volume.*
Caillault, J. P. 1990, PASP, 102, 989.
Caillault, J. P. and Helfand, D. J. 1985, ApJ, 289, 279.
Caillault, J. P., Helfand, D. J., Nousek, J. A., and Takalo, L. O. 1986, ApJ, 304, 318.
Caillault, J. P., Gagne, M., Stauffer, J. 1992, In *This volume.*
Casanova, S., Feigelson, E. D., Montmerle, T. 1992, In *This volume.*
Cassinelli, J. P. and Olson, G. L. 1979, ApJ, 229, 304.
Cassinelli, J. P. and Swank, J. H. 1983, ApJ, 271, 681.
Catura, R. C., Acton, L. W., and Johnson, H. M. 1975, ApJL, 196, 47.
Chlebowsky, T., Harnden, F. R., Jr., and Sciortino, S. 1989, ApJ, 341, 427.
Collier-Cameron, A., Jianke, L., and Mestel, L. 1991, In *Angular Momentum Evolution of Young Stars*, S. Catalano and J. Stauffer, editors, page 297, Kluwer Academic Publisher.
Damiani, F. and Micela, G. 1992, In *This volume.*
Durney, B. R. and Latour, J. 1978, Geophys. Ap. Fluid. Dyn., 9, 241.
Favata, F., Micela, G., and Sciortino, S. 1992a, In *This volume.*
Favata, F., Micela, G., and Sciortino, S. 1992b, A&A, submitted
Favata, F., Micela, G., and Sciortino, S., Vaiana, G. S. 1992c, A&A, 256, 86.
Feigelson, E. D. and Decampli, W. M. 1981, ApJL, 243, 89.
Feigelson, E. D. and Kriss, J. A. 1981, ApJL, 248, 35.
Feigelson, E. and Nelson, P. L. 1985, ApJ, 257, 695.
Fleming, T. H.. 1988, PhD Thesis, University of Arizona.
Fleming, T. H. 1992, In *This volume.*
Fleming, T. H., Gioia, I., and Maccacaro, T. 1989, ApJ, 340, 1011.
Gahm, G. F. 1980, ApJL, 242, 163.
Giacconi, R. et al. 1979, ApJ, 230, 540.
Gioia, I. M., Maccacaro, T., Schild, R. E., Wolter, A., Stocke, J., Morris, S., and Henry, J. P. 1990, ApJS., 72, 567.
Golub, L. 1992, In *This volume.*
Gorenstein, P., Harnden, F. R., Jr., and Fabricant, D. G. 1981, page 869.
Grillo, F., Sciortino, S., Micela, G., Vaiana, G. S., and Jr., F. R. H. 1992, ApJS, in press.
Haisch, B., Bookbinder, J., Maggio, A., Vaiana, G. S., and Bennett, J. 1990, ApJ, 361, 570.
Harnden, F. R., Jr. et al. 1979, ApJL, 234, 51.
Harnden, F. R., Jr., Fabricant, D. G., Harris, D. E., and Schwarz, J. 1984, SAO Special Report 393, Smithsonian Astrophysical Observatory.
Harnden, F. R., Jr., Sciortino, S., Micela, G., Maggio, A., Vaiana, G. S., and Schmitt, J.H.M.M. 1990, In Elvis, M., editor, *Imaging X-ray Astronomy*, page 313. Cambridge University Press.
Harnden, F. R., Jr. 1992, In *This volume*
Kahn, S. M. and Caillault, J. P. 1986, ApJ, 305, 526.
Kashyap, R., Rosner, R., Micela, G., Sciortino, S., Vaiana, G. S., and Harnden, F. R. 1992, ApJ, 391, 667.
Koyama, K. 1988a, Comments Astrophys., 12, 287.
Koyama, K. 1988b, ISAS Res. Note, 391.
Ku, W. H.-M. and Chanan, G. A. 1979, ApJL, 234, 59.
Ku, W. H.-M., Righini-Cohen, J., and Simon, M. 1982, Science, 215, 61.
Kudritzki, R. P. 1992, In *This volume.*
Linsky, J. 1992, In *This volume.*
Linsky, J. L. and Haisch, B. 1979, ApJL, 229, 27.
Lucy, L. B. 1982, ApJ, 255, 286.
Lucy, L. B. and White, R. L. 1980, ApJ, 241, 300.
Maggio, A., Sciortino, S., Vaiana, G. S., Majer, P., Golub, L., Bookbinder, J., Harnden, F. R. J., and Rosner, R. 1987, ApJ, 315, 687.
Maggio, A., Sciortino, S., Vaiana, G. S., Majer, P., Golub, L., Bookbinder, J., Harnden, F. R. J., and Rosner, R. 1987, ApJ, 315, 687.
Maggio, A., Vaiana, G. S., Haisch, B. M., Stern, R. A., Bookbinder, J., Harnden, F. R., and Rosner, R. 1990, ApJ, 348, 253.
Maggio, A., Sciortino, S., Bianchi, L., Harnden, F. R., Jr., and Rosner, R. 1992b, In *This volume.*
McCammon, D., Burrows, D. N., Sanders, W. T., and Kraushaar, W. L. 1983, ApJ, 269, 107.
Mewe, R. 1992, In *This volume.*

Micela, G. 1991, PhD Thesis, Universita' di Palermo.

Micela, G., Sciortino, S., Serio, S., Vaiana, G. S., Bookbinder, J., Golub, L., Harnden, F. R. J., and Rosner, R. 1985a, ApJ, 292, 172.

Micela, G., Sciortino, S., and Serio, S. 1985b, In Oda, M. and Giacconi, R., editors, *X-Ray Astronomy '84*, page 43, Bologna.

Micela, G., Sciortino, S., Vaiana, G. S., Schmitt, J. H. M. M., Stern, R. A., Harnden, F. R., and Rosner, R. 1988, ApJ, 325, 798.

Micela, G., Sciortino, S., Vaiana, G. S., Harnden, F. R., Rosner, R., and Schmitt, J. H. H. M. 1990, ApJ, 348, 557.

Micela, G., Harnden, F. R., Rosner, R., Sciortino, S., and Vaiana, G. S. 1991, ApJ, 380, 495.

Micela, G., Sciortino, S., and Favata, F. 1992a, ApJ, submitted.

Micela, G., Sciortino, S., and Favata, F. 1992b, in preparation.

Noyes, R. W., Hartmann, L. W., Baliunas, S., Duncan, D. K., and Vaughan A. H 1984, ApJ, 279, 763.

Nugent, J. and Garmire, G. 1978, ApJL, 226, 38.

Ottman, R. and Schmitt, J. 1992, A&A, 256, 421.

Pallavicini, R. 1992, In *This volume.*

Pallavicini, R., Golub, L., Rosner, R., Vaiana, G. S., Ayres, T., and Linsky, J. L. 1981, ApJ, 248, 279.

Pallavicini, R., Monsignori-Fossi, B. C., Landini, M., and Schmitt, J. H. M. M. 1988, A&AS, 191, 109.

Pye, J. 1992, In *This volume*

Rosner, R. 1989, In *Highlights of Astronomy*, Vol. 8, page 521

Rosner, R. 1992, In *This volume.*

Rosner, R. and Vaiana, G. S. 1980, In Giacconi, R. and Setti, G., editors, *X-Ray Astronomy*, page 129.

Rosner, R. et al. 1981, ApJL, 249, L5.

Schmitt, J. H. M. M. 1985, ApJ, 293, 178.

Schmitt, J. H. M. M. 1992, In *This volume.*

Schmitt, J. H. M. M., Golub, L., Harnden, F. R. J., Maxson, C. W., Rosner, R., and Vaiana, G. S. 1985, ApJ, 290, 307.

Schmitt, J. H. M. M., Collura, A., Sciortino, S., Vaiana, G. S., Harnden, F. R., Jr., and Rosner, R. 1990, ApJ, 365, 704.

Schmitt, J. and Snowden, S. L. 1990, ApJ, 361, 207.

Sciortino, S., Harnden, F. R., Jr., Maggio, A., Micela, G., Vaiana, G. S., Schmitt, J., and Rosner, R. 1988, In Murtagh, F. and Heck, A., editors, *Astronomy from Large Databases.*

Sciortino, S., Vaiana, G. S., Morossi, C., Ramella, M., Harnden, F. R., Jr., Schmitt, J.H.M.M, and Rosner, R. 1990, ApJ, 220, 361.

Soderblom, D. R. 1991, In *Angular Momentum Evolution of Young Stars*, S. Catalano and J. Stauffer, editors, page 151, Kluwer Academic Publisher.

Stauffer, J. R. 1991, In *Angular Momentum Evolution of Young Stars*, S. Catalano and J. Stauffer, editors, page 117, Kluwer Academic Publisher.

Stocke, J. T. et al. 1991, ApJS, 76, 813.

Vaiana, G. S. 1981, The Institute of Space and Astronautical Science, Report N. 597

Vaiana, G. S. 1983, In *Solar and Stellar Magnetic Fields: Origins and Coronal Effects*, Stenflo, J. O., editor, p. 165, Dordrecht: Reidel.

Vaiana, G. S. 1990, In *Imaging X-ray Astronomy*, Elvis, M., editor, p. 61, Cambridge University Press.

Vaiana, G. S. and Rosner, R. 1978, ARA&A, 16, 393.

Vaiana, G. S. et al. 1981, ApJ, 245, 163.

Vaiana, G. S. et al. 1990, in *The Einstein Observatory Astrophysical Databases*, Vol. 3 (on floppy disk).

Vaiana, G. S., Maggio, A., Micela, G., and Sciortino, S. 1992, Mem. Soc. Astr. It., page in press.

Waldron, W. 1984, ApJ, 282, 256.

Walter, F. M. and Kuhi, L. V. 1981, ApJ, 250, 573.

Walter, F. M., Cash, W., Charles, P. A., and Bowyer, C. S. 1980, ApJ, 236, 212.

Weiss, N. 1992, In *This volume.*

STELLAR CORONAL X-RAY SPECTROSCOPY

R. MEWE
SRON-Laboratory for Space Research, Sorbonnelaan 2,
3584 CA Utrecht, The Netherlands

Abstract. Since the Solar corona was detected as an X-ray source, X-ray spectroscopy has proven to be an invaluable tool in studying the tenuous hot (1–10 MK) coronae surrounding the Sun and many other stars throughout a large part of the HR diagram, with the probable exception of single very cool giant and supergiant stars. This paper reviews our present observational knowledge of stellar X-ray spectroscopy, focussing on the understanding of the structure of the coronae around cool solar-type stars, as has emerged from the medium-resolution spectral X-ray observations carried out with the *EINSTEIN* and *EXOSAT* observatories. We show some examples of differential emission measure modelling for simulated spectra to be observed by the spectrometer on board the recently launched Extreme Ultraviolet Explorer (*EUVE*). Finally, we consider the perspectives offered by high-resolution dispersive spectroscopy with future missions like NASA's Advanced X-ray Astrophysics Facility *AXAF* and ESA's X-ray Multi-Mirror Mission *XMM* and the expected performance of non-dispersive spectroscopy with superconductor technology currently in development. A few spectral simulations are given to demonstrate the diagnostic capabilities.

Key words: X-ray spectroscopy – Coronal modeling – EUVE spectroscopy

1. Introduction

Because of its proximity the Sun is the only star whose surface can be observed in detail. Among all celestial X-ray sources the Sun is the only one that can be studied in sufficient detail to establish the nature of the physical processes involved so that it serves as a benchmark for the studies of all other stars. High-resolution spectroscopy in a broad wavelength region ranging from radio to hard X-rays is a powerful tool to study a variety of phenomena associated with magnetic fields in the photosphere and the overlying outer atmosphere, e.g., active regions, emerging flux regions, plages, filaments, magnetic loop structures, X-ray bright points, etc. X-ray pictures from space-based experiments such as on *SKYLAB* have exhibited that the corona of the Sun is highly structured (e.g. atlas by Zombeck, Vaiana *et al.* (1978)), i.e. it consists of a variety of discrete structures - involving a range of time and length scales - which appear to outline the magnetic field lines emerging from the convection zone below the photosphere. Among the brightest in X-rays and EUV are those regions in which the field lines close back to the surface and which confine the hot emitting plasma in *closed magnetic loop structures*. In this context we should mention the pioneering work of Giuseppe Vaiana and co-workers concerning the studies of the Solar corona, especially the physical description of such loop structures. For review articles, e.g. Vaiana and Tucker (1974), Vaiana and Rosner (1978), Rosner, Golub, and Vaiana (1985), for *SKYLAB* results on the description of structures in general, e.g. Vaiana *et al.* (1973a,b, 1976), of active regions and magnetic fields, e.g. Poletto, Vaiana *et al.* (1975), of coronal holes, e.g. Timothy, Krieger, and Vaiana (1975), of flares, e.g. Pallavicini, Vaiana *et al.* (1975), and for the work on loop scaling laws, e.g. Rosner, Tucker, and Vaiana (1978) (hereafter referred to as RTV).

Since the detection in 1948 of X-rays from the Solar corona, many other cosmic sources have been identified as X-ray emitters containing hot gas at temperatures exceeding one million Kelvin. Observations with sensitive imaging X-ray telescopes onboard the *EINSTEIN* and *EXOSAT* satellites have demonstrated the ubiquity of X-ray emission of "normal" stars (i.e. stars for which the X-rays do not originate from accretion onto a compact object but are produced by thermal emission in a hot tenuous envelope, the corona) throughout a large part of the HR diagram (e.g. Vaiana *et al.* 1981).

J.F. Linsky and S. Serio (eds.), Physics of Solar and Stellar Coronae, 225–236.

X-ray spectroscopy has proven to be an invaluable tool in modelling astrophysical plasmas. Based on assumed plasma models one can infer from the observations relevant physical parameters such as electron temperature, density, emission measure, ion and elemental abundances, mass motions, orbital velocities, and redshifts. Though the existence of many types of cosmic X-ray sources has been well established, only poor spectral information is available because the majority of the X-ray observations was done with only limited or no spectral resolution. Early space instruments such as proportional counters allowed to determine merely the overall shape of the X-ray spectra of stellar coronae and it was not until the 1980s that the introduction of solid state, Bragg crystal, and transmission grating spectrometers on the *EINSTEIN* and *EXOSAT* satellites with modest spectral resolution ($\lambda/\Delta\lambda \simeq 10$–$100$) provided for the first time a taste of more detailed spectroscopy of stellar coronae and motivated the use of more sophisticated emission models of hot optically thin plasmas (e.g. Raymond and Smith 1977; Mewe *et al.* 1985, 1986a).

The next generation of spectroscopic space missions near the end of the century, viz. NASA's Advanced X-ray Astrophysics Facility (*AXAF*) (Weisskopf 1987) and ESA's X-ray Multi-Mirror Mission (*XMM*) (Barr *et al.* 1988) will give us the opportunity to measure emission lines with a resolution of better than 0.05 Å in the wavelength range $\sim$ 1–140 Å, allowing much better to resolve temperature and density structure of stellar coronae and to interpret the origins of different spectral components.

This paper concentrates on the coronae of late-type stars from the point of view of X-ray spectroscopic diagnostics. For recent reviews on coronal spectroscopy see e.g. Mewe (1991a) and Schmitt (1990). For reviews on stellar coronae from the point of view of the physical processes, cf. e.g. Rosner *et al.* (1985), Serio (1985), Linsky (1985), Pallavicini (1988, 1989), Schmitt (1988), Mewe (1991b), or from the point of view of the Solar-stellar connection of Solar-type stars, e.g., Schrijver (1990, 1991). I briefly discuss the spectral model of an optically thin plasma, spectral modelling in terms of a differential emission measure, velocities in binaries and some results of medium-resolution X-ray spectroscopy with *EXOSAT*. Finally, I briefly consider the capability for plasma diagnostics of high-resolution dispersive spectroscopy with future missions like *AXAF* and *XMM* and the expected performance of non-dispersive spectroscopy with cryogenic superconductors (e.g. bolometer and SIS-junction) currently in development.

2. Optically thin thermal plasmas

The X-ray emission from a stellar corona is typically optically thin thermal plasma emission from a very hot, tenuous, and highly ionized gas in which the atoms have all, or almost all, of their electrons stripped due to collisions with energetic electrons. Temperatures are typically above one million degrees Kelvin.

This is the standard *coronal model* that was first applied by Elwert (1952) to the Solar corona. It assumes that the radiative power loss is compensated by heating so that the gas is in a steady state of statistical equilibrium both for the bound atomic states and for the ionization balance. Electron collisions control the ionization state and emissivity of the gas and the plasma electrons (and ions) are relaxed to Maxwellian energy distributions with a common temperature T (for details, e.g. Mewe 1990a, 1991a). Deviations from this model due to photo-ionization, optical depth, high density, non-Maxwellian electron distributions, and transient ionization are discussed by Raymond (1988) and Mewe (1990a). Examples of "coronal" plasmas are: stellar coronae, supernova remnants, the hot gas in the interstellar medium and in galaxies and clusters (for spectra, e.g. Mewe 1990b).

Line emission dominates the soft X-ray spectrum and the cooling of coronae at temperatures up to $\sim$ 10 MK and comes mainly from electron impact excitation of bound levels within a highly ionized atom, followed by spontaneous radiative decay (cf. Mewe *et al.* 1985). There is an additional (generally smaller) contribution from continuum emission due to free-free ("bremsstrahlung"), free-bound ("recombination") and two-photon radiation (cf. Mewe *et al.* 1986a).

TABLE I
Prototype late-type stars for *AXAF* and *XMM* observations
Listed are source distance d (pc), interstellar column density N_H (10^{18} cm^{-2}), temperature T_i (MK) and reduced emission measure $\varepsilon_i' = \varepsilon/d^2$ (10^{50} cm^{-3} pc^{-2}) of component i.

Object	Spectral type	d	N_H	T_1	ε_1'	T_2	ε_2'	Ref.[1]
Capella	(G5III+)F9III	13	5	5	3	25	9	[1]
σ^2 CrB	F8V(+G1V)	23	5	6	1	25	7	[1]
λ And	G8III-IV(+?)	23	5	8	1	20	6	[1]
HR 1099	G5IV+K1IV	35	5	7	2	50	5	[2]
AR Lac[2]	K0IV	50	10	3	0.4	14	1.2	[3]
AR Lac	G2IV	50	10	3	0.4	14	0.8	[3]
κ^1 Cet	G5V	9	3	3	0.6			[1]
Algol[3]	(B8V+)KIV	27	5	3	1.5	25	3.5	[4]
Algol[4]		27	5	69	10.2			[4]
Procyon	F5IV-V(+DF)	3.5	2	0.6	1	2	1.5	[1]
α Cen A	G2V	1.3	1	2	0.15			[5]
α Cen B	K1V	1.3	1	2	0.35			[5]
BY Dra	M0Ve+M2Ve	16	3	20	2			[1]

[1] [1] Mewe et al. (1990); [2] Swank *et al.* (1981); [3] Ottmann *et al.* (1992); [4] Van den Oord, Mewe (1989); [5] Golub et al. (1982); [2] eclipsing binary (P=1.983^d); [3] quiescence; [4] flare peak.

2.1. Simulated High-Resolution Spectra of Stellar Coronal Sources

The overall appearance of the X-ray spectra will be dominated by the ionization structure which is controlled by electron impact plus auto-ionization and by radiative plus dielectronic recombination and varies dramatically with temperature throughout the range 0.1–100 MK. In extremely hot ($T > 100$ MK) plasmas all abundant elements are nearly fully ionized and the X-ray emission is dominated by the free-free continuum from hydrogen and helium, but in a wide temperature range (0.01–10 MK) the X-ray spectra of optically thin sources are rich in emission lines from many ions, which are powerful diagnostics of plasma parameters such as electron temperature, density, and elemental abundances. A combined determination of emission measure $\varepsilon = \int n_e^2 dV$ and density n_e yields the emitting volume V, hence the product Lf of loop length and filling factor, i.e. the fraction of the apparent volume that actually emits the X-rays (f may be much less than unity if the emitting volume viewed suffers a high degree of filamentation). This will allow us to constrain the coronal dimensions and structures in single stars and non-eclipsing binaries, while this is currently only rarely possible for eclipsing binaries. For various methods to diagnose plasmas, e.g. Mewe (1991a) with references therein.

Figure 1 illustrates the possibilities of high-resolution X-ray spectroscopy. For a number of prototype stars we have calculated spectra between 3–140 Å convolved

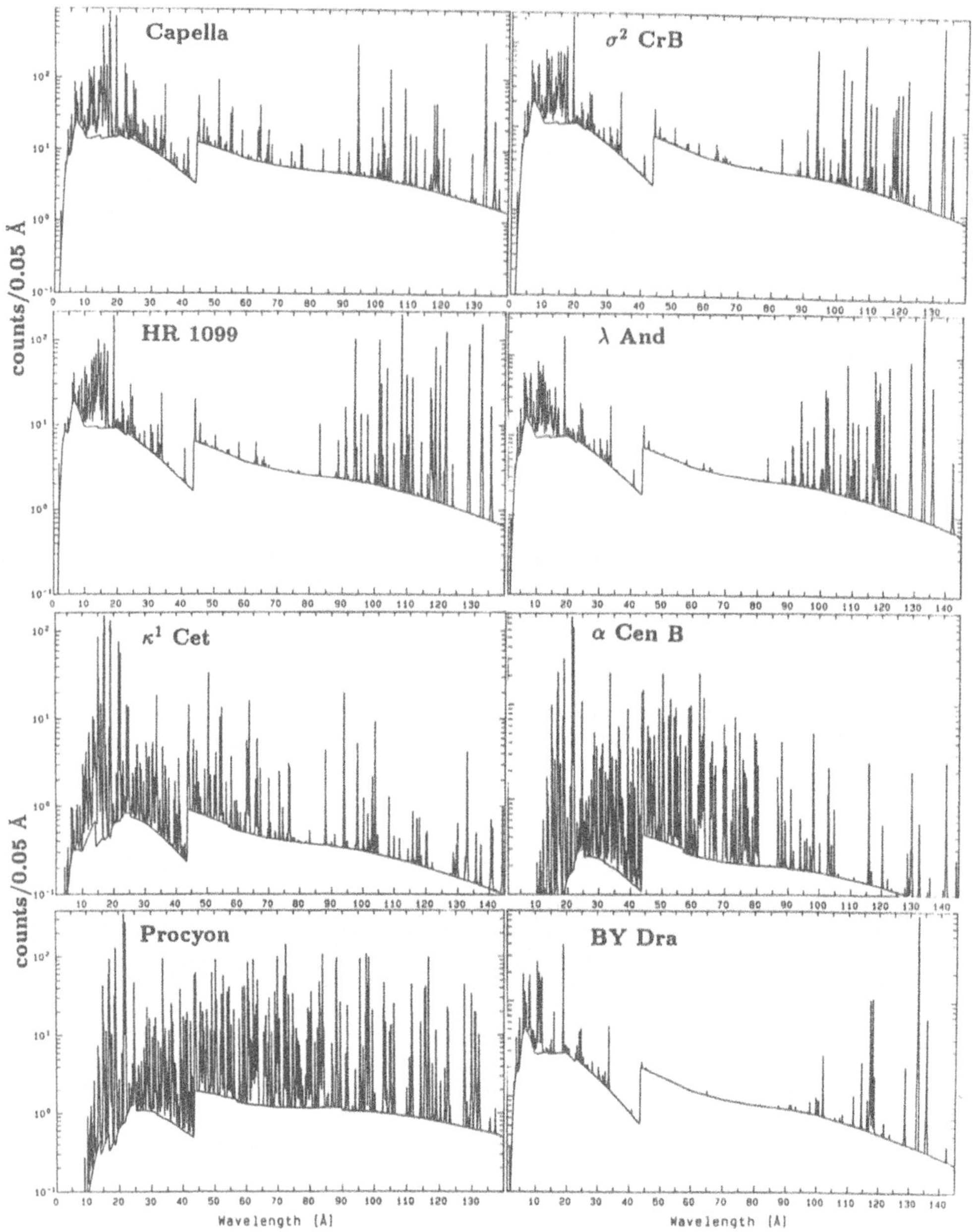

Fig. 1. X-ray spectra of various stellar coronal sources simulated for the *AXAF* Low Energy Transmission Grating Spectrometer (LETGS). Source parameters from Table 1. Exposure time is 10^4 s.

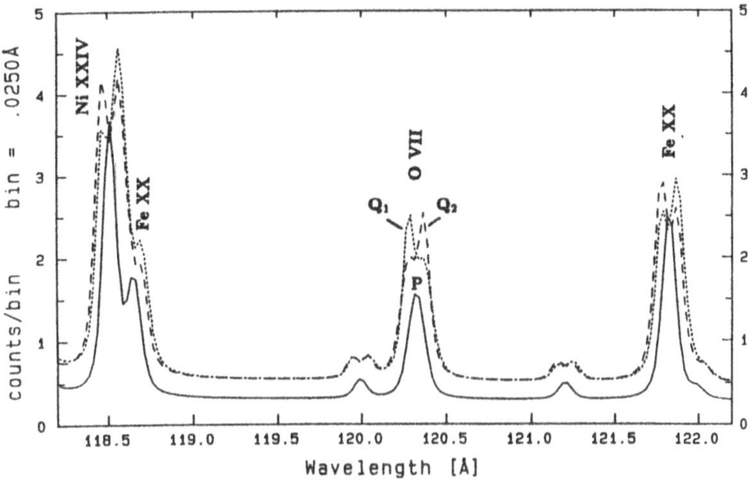

Fig. 2. Simulated *AXAF*-LETGS spectra of the eclipsing binary AR Lac. Dotted (or dashed) line shows the spectrum at quadrature (Q_1 (or Q_2,, resp.)) when both stars are observed and the G-star is approaching (or receding) with v = -116 km/s (or v = +116 km/s), solid line shows spectrum during totality of the primary eclipse (P) when the K-star is only observed because it occults completely the smaller G-star. For parameters see text.

with the response of the high-resolution Low Energy Transmission Grating Spectrometer (LETGS) on *AXAF* (Brinkman *et al.* 1987) and for an exposure time of 10^4 s. The source parameters used for these simulations are derived from *EXOSAT*, *EINSTEIN*, and *ROSAT* observations and listed in Table 1. It is clearly seen that the improved spectral resolution ($\Delta\lambda \simeq 0.05$ Å) of the *AXAF* spectrometer allows to resolve virtually all important spectral lines yielding the possibility to resolve the coronal temperature and ionization structure in more detail and to determine abundances from line/line and line/continuum ratios.

Spectroscopy with high spectral resolution ($\Delta\lambda \simeq 0.05$ Å) at longer wavelengths ($\gtrsim$ 100 Å) with the LETGS on *AXAF* will permit for the first time detailed studies of spectral line *profiles* in the X-ray region. This may provide wind expansion velocities of stellar coronae, flow velocities along active-region loops, orbital velocities in X-ray binaries, upflow velocities in stellar flares, etc. The Doppler shift caused by the orbital motions in close binaries may suffice in some cases to determine the relative activity level of the binary components. Observations at the two quadratures yield the intensities of the two components separately, hence indicate which star is the stronger X-ray source or whether the X-rays are emitted by the corona of both stars or perhaps by a common corona.

As an illustrative example of Doppler imaging in the X-ray region we consider the eclipsing binary AR Lacertae which was observed by *EINSTEIN* (Walter *et al.* 1983), *EXOSAT* (White *et al.* 1990), and *ROSAT* (Ottmann *et al.* 1992). The system consists of a G2 IV primary star and a K0 IV secondary star revolving around each other with a period of 1.98 day (orbital velocity 116 km/s, whereas the surface equatorial velocities of the two synchronized components are v_1 = 39 km/s (G) and v_2 = 72 km/s (K)). Based on results of *ROSAT* eclipse measurements (cf. Table 1) we assume the following input model for the simulations: a compact 2-T corona around the G-star with $T_{1(2)} = 3(14)$ MK, $\varepsilon_{1(2)}/d^2 = 0.4(0.8) \times 10^{50}$ cm^{-3} pc^{-2} and height $H \ll$ stellar radius R_* and an extended, hot 2-T corona around the K-star with $T_{1(2)} = 3(14)$ MK, $\varepsilon_{1(2)}/d^2 = 0.4(1.2) \times 10^{50}$ cm^{-3} pc^{-2} and

$H \sim R_*$; $N_H = 10^{19}$ cm^{-2}, exposure time $t_{exp} = 2 \times 10^4$ s. The lines are broadened thermally and in addition by the stellar rotation which we approximate by velocity profiles with halfwidths $\Delta v_{1/2} \simeq v_1$ (G-corona) or $\simeq 2v_2$ (K-corona). Resulting spectra with the O VII line at 120.83 Å (formed by the cooler components), and the Fe XX line at 121.83 Å and the Ni XXIV line (blended by an Fe XX line) at 118.52 Å (from the hot components) are presented in Figure 2 for various phases of the binary. The differences observed in the spectra can be used to disentangle the contributions from the two binary components.

2.2. Spectral modelling in terms of a Differential Emission Measure

In many sources the X-ray spectrum will not be a unique function of one single temperature, but instead will be determined by a distribution in temperature across the emission region. Then the observed spectra can be described in terms of a differential emission measure (DEM) distribution, which is defined as follows. The line and continuum spectral intensity in wavelength bin λ_j can be expressed as $\propto \int F(\lambda_j,T)\varphi(T)\,dT \equiv \int F(\lambda_j,T)T\varphi(T)\,d(\ln T)$, where $F(\lambda_j,T)$ is the spectral emissivity for the line plus continuum emission as a function of temperature T integrated over wavelength bin λ_j, convolved with the instrument response function, and the DEM is given by $\varphi(T)dT = n_e^2 dV$ (n_e electron density, V plasma volume). N.B. total emission measure is: $\varepsilon = \int n_e^2 dV = \int T\varphi(T)d(lnT)$. We use here $T\varphi(T)\Delta\ln(T)$ in the graphic representation of the DEM. The differential emission measure distribution is derived from the observed spectrum by deconvolving $\varphi(T)$ from the observed spectral intensities, using theoretical emission functions. For this deconvolution we apply an iterative technique, originally proposed by Withbroe (1975), and modified by Sylwester *et al.* (1980). The method places no restrictions on the functional form of the differential emission measure distribution, although the final result is subject to an implicit smoothing because spectra with neighbouring values of T are more or less alike. From an analysis of the library of spectra in terms of the mutual correlations it can be shown that the widths in temperature of the correlation functions determine the "temperature resolution", i.e. the scale on which we can extract detail of the DEM distribution. From the derived DEM distribution we can assess contributions to the coronal energy balance. A detailed DEM analysis has been performed on the *EXOSAT* transmission grating spectra (TGS) of late-type stars by Lemen *et al.* (1989) and Schrijver *et al.* (1989) to constrain the basic properties of stellar magnetic loops, and by Mewe *et al.* (1991) to extract the DEM for simulated spectra for the *AXAF*-LETGS. In the case of the *EXOSAT* observations Lemen *et al.* (1989) have derived the DEM distributions of the cool stars Capella, σ^2 CrB, and Procyon. These are dominated by coronal plasma emission in two relatively narrow temperature intervals: 5 MK and 25 MK for Capella and σ^2 CrB, and 0.6 MK and 2–3 MK for Procyon.

The existence of three dominant temperatures in the coronae of cool stars may contain information on the stability of the corresponding loops. Lemen *et al.* (1989) note that these temperatures correspond with temperature intervals where the radiative loss $\Lambda(T)$ increases with T. If radiation is the dominant cooling mechanism, the loop apex temperatures may tend to reach values for which the loop plasma is relatively stable against heating perturbations.

The doubly peaked DEM distributions suggest a model corona comprising two distinct ensembles of quasi-static magnetic loops with maximum temperatures near the dominant temperatures given above. Mewe *et al.* (1986b) and Schrijver and Mewe (1986) attempted to derive from the *EXOSAT* spectra constraints for the loop geometry by comparing the observed spectra with computed spectra using the RTV loop model. They show that the spectra observed for Capella and σ^2 CrB are

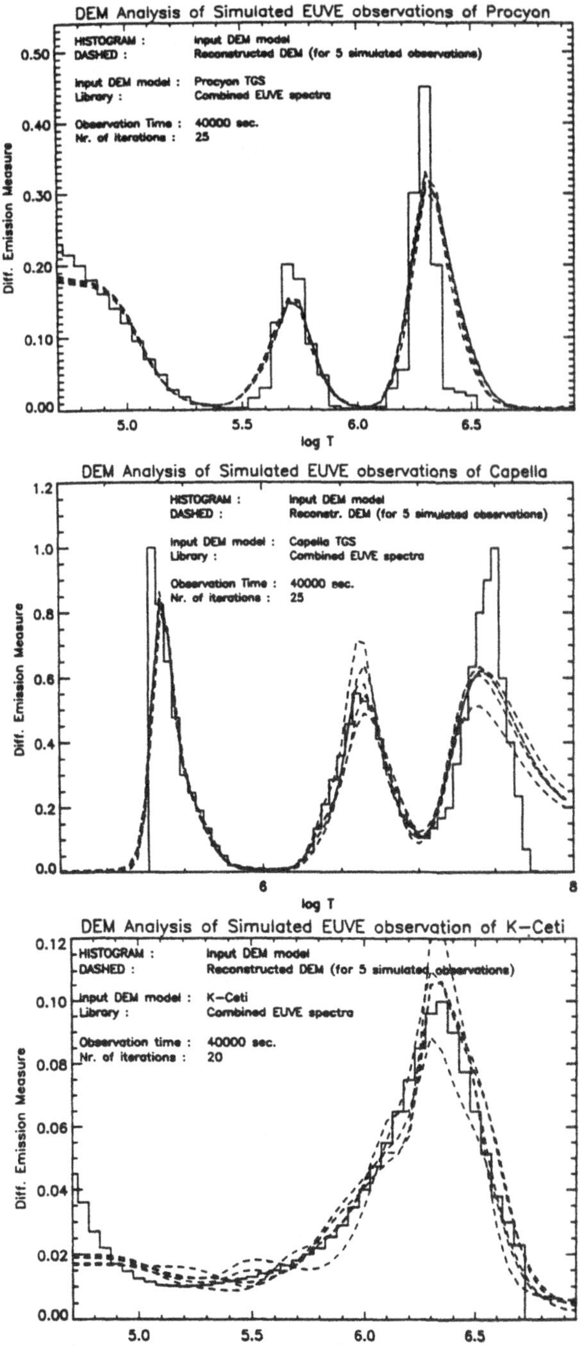

Fig. 3. Results of EUVE DEM modelling for the models of Procyon (panel a), Capella (b), and κ^1 Cet (c). Plotted vs. temperature T are the values of the differential emission measure DEM $\equiv T\varphi(T)\Delta\ln(T)$ with $\Delta\ln(T[\mathrm{K}])$=0.115. The input model is shown as a solid histogram and the simulated observations as dashed curves.

incompatible with the RTV model for static loops of constant cross section: The RTV model predicts much too strong radiation beyond 140 Å caused by the emission from Fe VIII-XI formed around 1 MK. One possibility is that the cross-section of the loop increases with height so that most of the matter in the loop is at a high temperature, and thus the spectrum appears isothermal. To test this hypothesis Schrijver *et al.* (1989) performed a two-component analysis of the *EXOSAT* spectra using computed loop spectra covering a range of apex temperatures T_m and expansion factors Γ (= ratio of loop cross section at apex to cross section at footpoint). The model spectra were generated with a revised computer code originally developed by Vesecky *et al.* (1979), using the spectral model of Mewe *et al.* (1985). The fits for Capella and σ^2 CrB suggest the cool (5 MK) component originating in loops expanding significantly with height ($\Gamma \gtrsim 5$) and indicate also $\Gamma > 1$ for the hot (25 MK) component. This is indirectly observed on the Sun: coronal condensations over magnetic bipolar regions have a projected area roughly an order of magnitude larger than the area of the underlying photospheric plage (e.g. Schrijver 1987). However, recent results with the soft X-ray telescope on *YOHKOH* seem to contradict this because it was found that the majority of loops have nearly constant cross sections along their lengths, rather than expanding with height (J.R. Lemen *et al.*, this conference).

We have applied a variant of the Withbroe-Sylwester method in a study of DEM models of three prototype cool stars proposed to be observed by the Extreme Ultraviolet Explorer (EUVE) (e.g. Bowyer *et al.* 1992): (i) Procyon, which has a very cool coronal component with a temperature below 1 MK), model based on IUE observations (Jordan *et al.* 1986) and on *EXOSAT*-TGS observations (Lemen *et al.* 1989); (ii) Capella, which possesses a very hot corona with temperatures around 5 and 25 MK, model based on *EXOSAT*-TGS observations (Lemen *et al.* 1989) and IUE observations (Doschek and Cowan 1984); (iii) κ^1 Cet as prototype for a Solar-type corona with temperature 2–3 MK, model based on that proposed by Doschek and Cowan (1984) for the quiet Sun, but scaled with *EXOSAT* observations (cf. Table 1) and extended below 1 MK towards lower temperatures using *IUE* data for χ^1 Ori (Jordan *et al.* 1987). For interstellar density N_H cf. Table 1. We have simulated spectra for a standard observing time of 4 10^4 s, taking into account the sky background and the statistical Poisson noise. In Figure 3 we have plotted the differential emission measure DEM $\equiv T\varphi(T)\Delta\ln(T)$ per interval $\Delta\ln(T[\mathrm{K}]) = 0.115$ (or $\Delta^{10}\log(T[\mathrm{K}]) = 0.05$). The results show that by combining the three wavelengths bands (70–200 Å, 100-400 Å and 280–760 Å) we are able to constrain the DEM distribution between $\sim$ 0.05 MK and 10 MK. Outside this temperature region apparently no strong spectral lines are available in the three wavelength bands to constrain the DEM.

3. Applications of non-dispersive spectroscopy

Though spectroscopic techniques with dispersive elements such as gratings and Bragg crystals offer the possibility of measuring with relatively high spectral resolution ($\lambda/\Delta\lambda \gtrsim 500$) they have also certain disadvantages. For example: relatively low efficiency ($\sim$ 10% for gratings, 1% for Bragg crystals), relatively small band width ($\sim$ 0.1–2 keV), incapacity for observing non-pointlike sources whose finite extent will degrade the spectral resolution. The development of cryogenic superconductors will probably open a new era of observing X-ray sources through the combination of high efficiency ($\gtrsim$ 80%), large band width (0.1–10 keV) and high spectral resolution (ΔE a few eV). Moreover, the application of series of detectors in an array will allow also imaging of extended sources.

The spectral resolving power R for such detectors can be written (assuming a

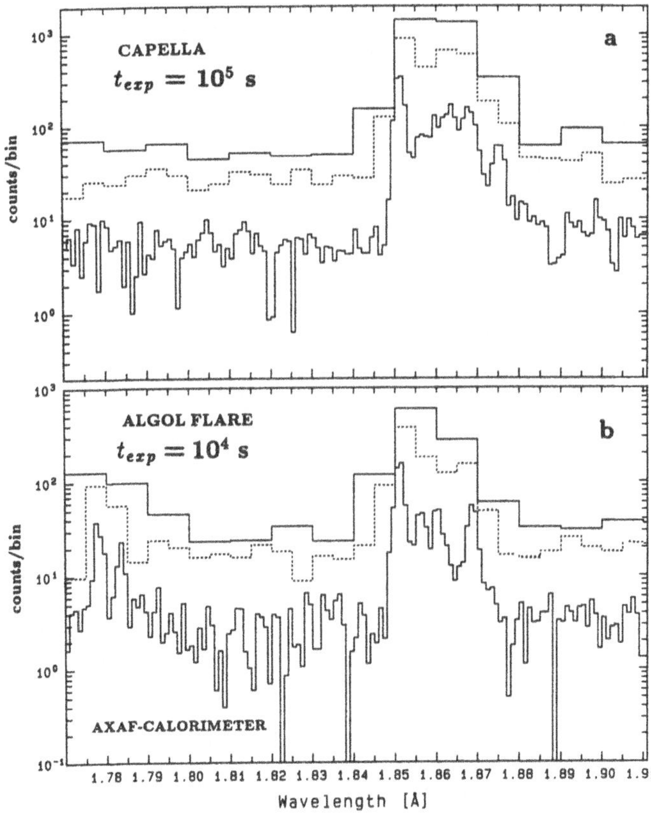

Fig. 4. Simulated *AXAF* calorimeter spectra with random noise added of Capella (panel a) and of a flare on Algol (b) for an exposure time of $t_{exp} = 10^5$ s and of 10^4 s, respectively, and with wavelength bins of 0.001 Å, 0.005 Å, and 0.01 Å. Source parameters from Table 1.

gaussian response profile with $\Delta E/E$ (FWHM) $\simeq 2\sqrt{\ln 2}(\sigma_N/N) = 2.36(\sigma_N/N)$ (e.g. Fraser 1989): $R \equiv E/\Delta E \simeq 0.4(N/\sigma_N) = 0.4(E/FW)^{1/2}$. Here σ_N is the square root of the variance (corrected by the Fano factor F which is typically ~ 0.2) of the number N $(=E/W)$ of secondary electrons, electron-hole pairs, or quasi-particles, created by the absorption of an X-ray photon of energy E, in the case of a gas discharge detector like a gas scintillation proportional counter (GSPC), a semi-conductor like a charge-coupled device (CCD), or a cryogenic superconductor like bolometer or superconductor-isolator-superconductor(SIS)-junction, respectively. Finally, W is the energy to create the secondary particles (typically ~ 30 eV, 3 eV, or 3 meV, for the three cases, resp.). So the resolving powers for GSPCs, CCDs, and SISs are approximately in the ratio 1:4:100. With superconductors a resolving power $\gtrsim 1000$ may be obtainable at $E \gtrsim 1$ keV.

For dispersive spectrometers the spectral resolving power will be largely determined by the optical aberrations of the dispersive element and telescope and of the pixel resolution of the detector, whereas for non-dispersive detectors the statistical fluctuations in the photon-charge conversion process in the detector will mainly contribute to the resolution. The important property of non-dispersive spectrometers is the combination of high resolution and efficiency together with the possibility of imaging which is especially in advantage for spectroscopy at high energy ($E \gtrsim 1$ keV) and for extended sources.

4. Simulations of high-resolution Fe K and L spectra

The few observations with the spectrometers aboard the *EINSTEIN* (Giacconi *et al.* 1979; for references see Mewe 1991a) and *EXOSAT* observatories (Brinkman *et al.* 1980) have shown that spectra of even modest spectral resolution ($\lambda/\Delta\lambda = 10$–$100$) allow the identification of coronal material at different temperatures whose existence may relate to a range of possible magnetic loop structures in the outer atmospheres of these stars. These low- and medium-resolution spectral X-ray results require at least a two-temperature interpretation, although differences in the derived temperatures may suggest a wide range of temperatures. The improved spectral resolution ($\gtrsim 1000$) of the next generation of dispersive and non-dispersive spectrometers is needed to fully resolve the temperature structure of stellar coronae. I consider a few cases.

Mewe (1991a, Fig. 4) compares *EXOSAT*-TGS spectra of Capella with spectra simulated for the LETGS on *AXAF* (Brinkman *et al.* 1987) and the reflection grating spectrometer (RGS) on *XMM* (Brinkman *et al.* 1989). The simulated spectra of the Fe L-shell blend around 15 Å ($2s$–$3p$ lines) and of the strong $2s$–$2p$ lines between 90–140 Å clearly demonstrate the diagnostic capabilities of the instruments and show that spectrometers with a resolving power of $\gtrsim 500$ up to ~ 1000 or more resolve the L-blend in its many individual components, allowing more refined temperature, density, and velocity diagnostics.

The high-resolution X-ray K-shell spectra from highly ionized iron and calcium in the wavelength regions 1.7–2 Å and 3–3.3 Å provide superior diagnostics of very hot plasmas with temperatures in the range 10–100 MK. The availability of such spectra of the flaring Sun as obtained with the Bragg crystal spectrometers aboard the spacecrafts *P78-1*, *SMM*, and *Hinotori* (for detailed references cf. Mewe (1991a)) has stimulated the diagnostics of hot thin plasmas. Up to now the iron K-shell blend at 6.7 keV has been detected from *non-solar* stellar coronae only at rare occasions: e.g. in *EXOSAT* ME observations on Algol (White *et al.* 1986, van den Oord and Mewe 1989) and UX Ari (Pasquini *et al.* 1988) and *GINGA* LAC observations on UX Ari (Tsuru *et al.* 1989).

In high-resolution iron spectra of very hot plasmas the temperature can be diagnosed from the He-like singlet/triplet ratio, from the intensity ratio of Fe XXVI Ly α and the Fe XXV w He-like resonance line, or from the ratio of a resonance line and a nearby dielectronic recombination satellite line (e.g. He-like triplets with their satellites). However, an extremely high spectral resolution is required in order to avoid too much blending of the stronger line features that are most important for the diagnostics of the plasma parameters by the hundreds of weaker unresolved satellite lines. Mewe (1991a, Fig. 8, cf also Schmitt 1990) has demonstrated the effects of spectral resolution on the K-shell spectrum of helium-like iron and its satellites in the interval 1.84–1.92 Å for a series of resolving powers R from 10^4 (typically for the Bragg spectrometers used for Solar observations) down to 10^3 (as expected for the bolometer proposed for *AXAF* (Zhang *et al.* 1990)). It is shown that in the latter case, though the important temperature diagnostic lines become blended, one can still distinguish between all ionization stages from Fe XX through Fe XXV. Figure 4 shows simulations of observations on Capella and Algol (peak flare) (parameters from Table 1) with the bolometer on *AXAF* (with resolution $\Delta E = 7$ eV and effective area ~ 170 cm^2, Linsky 1990). With the bolometer we begin to approach the resolution of a Bragg crystal spectrometer but with a four times bigger area.

5. Conclusions

The results from broad-band and medium-resolution spectral observations with *EINSTEIN* and *EXOSAT* already permitted the identification of coronal gas at different temperatures to be associated with various possible loop structures, but the application of high-resolution X-ray spectroscopy as can be performed at the end of our century with spectrometers onboard *AXAF* and *XMM* will provide promising perspectives for the study of the physics of coronal sources as has been illustrated in this paper. Line profile studies will be important in the studies of the activity of single and binary late-type stars and of the dynamics of stellar flares. The wavelength region around the Fe L-shell spectrum (7–18 Å) and in particular the extreme ultraviolet regions at 90–140 Å (Fe $2s$–$2p$) and 170–300 Å (Fe 3ℓ–$3\ell'$) are very promising for dispersive spectroscopy with $R \gtrsim 500$–1000 which is attainable with present-day technology of transmission and reflection gratings and position-sensitive detectors. The helium-like triplets with their satellites in the K-shell spectra of highly ionized iron and calcium at short wavelengths ($\lesssim 3$ Å) provide a very valuable diagnostics for electron temperature, emission measure, ionization balance, and (in limited cases also) electron density. However, to accomplish this goal, an extremely high spectral resolution ($\gtrsim 2000$) will be required which with currently available technology can only be obtained with crystals and perhaps in the future with bolometers and Nb-junctions if a resolution can be reached within a factor of two from the theoretical limit.

Acknowledgements

This work has been supported by Space Research Organization Netherlands (SRON).

References

Barr, P. *et al.*: 1988, *The High-Throughput X-Ray Spectroscopy Mission: The Mission Science Report*, ESA SP-1097

Bowyer, S., Jelinsky, P., Christian, C.A., Malina, R.F.: 1992, in: *The Extreme Ultraviolet Explorer All-Sky Survey and Guest Investigator Spectroscopy Mission*, Proc. Cambridge Workshop on Cool Stars, Stellar Systems, and the Sun (Oct. 1991, Tucson, Arizona)

Brinkman, A.C., Aarts, H.J.M., Branduardi-Raymont, G. *et al.*: 1989, in: *EUV, X-Ray, and Gamma-Ray Instrumentation for Astronomy and Atomic Physics* **1159**, eds. C. J. Hailey, O. H. W. Siegmund, Soc. Photo Opt. Instr. Eng., Washington, p. 495

Brinkman, A.C., Dijkstra, J.H., Geerlings, W.F.P.A.L. *et al.*: 1980, *Appl. Optics* **19**, 1601

Brinkman, A.C., van Rooijen, J.J., Bleeker, J.A.M. *et al.*: 1987, *Astrophys. Lett. and Communications* **26**, 73

Doschek, G.A., Cowan, R.D.: 1984, *Astrophys. J. Suppl. Ser.* **56**, 67

Elwert, G.: 1952, *Z. Naturf.* **7a**, 432

Fraser, G.W.: 1989 in: *X-Ray Detectors in Astronomy* Cambridge Univ. Press, Cambridge

Giacconi, R. *et al.*: 1979, *Astrophys. J.* **230**, 540

Golub, L. *et al.*: 1982, *Astrophys. J.* **253**, 242

Jordan, C., Ayres, T.R., Brown, A. *et al.*: 1987, *Monthly Notices Roy. Astron. Soc.* **225**, 903

Jordan, C., Brown, A., Walter, F.M., Linsky, J.L.: 1986, *Monthly Notices Roy. Astron. Soc.* **218**, 465

Lemen, J.R., Mewe, R., Schrijver, C.J., Fludra, A.: 1989, *Astrophys. J.* **341**, 474

Linsky, J.L.: 1985, *Solar Phys.* **100**, 333

Linsky, J.L.: 1990, in: *High Resolution X-ray Spectroscopy of Cosmic Plasmas*, eds. P. Gorenstein, M.V. Zombeck, Proc. IAU Coll. 115, Cambridge Univ. Press, Cambridge, p. 94

Mewe, R.: 1990a, in: *Physical Processes in Hot Cosmic Plasmas*, eds. W. Brinkmann, A.C. Fabian, F. Giovanelli, Kluwer Acad. Publ., Dordrecht, p. 39

Mewe, R.: 1990b, in: *Atomic Spectra and Oscillator Strengths for Astrophysics and Fusion*, ed. J.E. Hansen, North-Holland, Amsterdam, p. 67
Mewe, R.: 1991a, *Astron. Astrophys. Rev.* **3**, 127
Mewe, R.: 1991b, *Adv. Space Sci. Rev.* **11**, (1)127
Mewe, R.: 1992, in: *The physics of chromospheres, coronae and winds*, ed. C.S. Jeffery, Cambridge
Mewe, R., Gronenschild, E.H.B.M., van den Oord, G.H.J.: 1985, *Astron. Astrophys. Suppl. Ser.* **62**, 197
Mewe, R., Lemen, J.R., van den Oord, G.H.J.: 1986a, *Astron. Astrophys. Suppl. Ser.* **65**, 511
Mewe, R., Lemen, J.R., Schrijver, C.J.: 1990,, *Adv. Space Sci. Rev.* **10**, (2)129
Mewe, R., Lemen, J.R., Schrijver, C.J.: 1991, *Astrophys. Space Sci.* **182**, 35
Mewe, R., Schrijver, C.J., Lemen, J.R., Bentley, R.D.: 1986b, *Adv. Space Sci. Rev.* **6**, No. 8, 133
Ottmann, R., Schmitt, J.H.M.M., Kürster, M.: 1992, *Astrophys. J.* , in press
Pallavicini, R.: 1988, in: *Hot Thin Plasmas in Astrophysics*, ed. R. Pallavicini, Kluwer Acad. Publ., Dordrecht, p. 121
Pallavicini, R.: 1989, *Astron. Astrophys. Rev.* **1**, 177
Pallavicini, R., Vaiana, G.S., Kahler, S.W., Krieger, A.S.: 1975, *Solar Phys.* **45**, 411
Pasquini, L., Schmitt, J.H.M.M., Pallavicini, R.: 1988, in: *Activity in Cool Star Envelopes*, eds. O. Havnes, B.R. Pettersen, J.H.M.M. Schmitt, J.E. Solheim, p. 241
Poletto, G., Vaiana, G.S., Zombeck, M.V., Krieger, A.S., Timothy, A.F.: 1975, *Solar Phys.* **44**, 83
Raymond, J.C.: 1988, in: *Hot Thin Plasmas in Astrophysics*, ed. R. Pallavicini, Kluwer Acad. Publ., Dordrecht, p. 3
Raymond, J.C., Smith, B.W.: 1977, *Astrophys. J.* Suppl. **35**, 419
Rosner, R., Golub, L., Vaiana, G.S.: 1985, *Ann. Rev. Astron. Astrophys.* **23**, 413
Rosner, R., Tucker, W.M., Vaiana, G.S.: 1978, *Astrophys. J.* **220**, 643 (RTV)
Schmitt, J.H.M.M.: 1988, in: *Hot Thin Plasmas in Astrophysics*, ed. R. Pallavicini, Kluwer Acad. Publ., Dordrecht, p. 109
Schmitt, J.H.M.M.: 1990, in: *High Resolution X-ray Spectroscopy of Cosmic Plasmas*, eds. P. Gorenstein, M.V. Zombeck, Proc. IAU Coll. 115, Cambridge Univ. Press, Cambridge, p. 110
Schrijver, C.J.: 1987, *Astron. Astrophys.* **180**, 241
Schrijver, C.J.: 1990, in: *New windows to the uiniverse* **1**, eds. F. Sanchez, M. Vazquez, Cambridge Univ. Press, Cambridge, p. 233
Schrijver, C.J.: 1991, *Reviews in modern Astronomy* **4**, 1
Schrijver, C.J., Lemen, J.R., Mewe, R.: 1989, *Astrophys. J.* **341**, 484
Schrijver, C.J., Mewe, R.: 1986, in: *Cool Stars, Stellar Systems, and the Sun*, eds. M. Zelik, D.M. Gibson, Springer, New York, p. 300
Serio, S.: 1985, in: *Proc. ESA Workshop, Cosmic X-Ray Spectroscopy Mission*, Lyngby, Denmark, ESA SP-239, p. 59
Swank, J.H., White, N.E., Holt, S.S., Becker, R.H.: 1981, *Astrophys. J.* **246**, 208
Sylwester, J., Schrijver, J., Mewe, R.: 1980, *Solar Phys.* **67**, 285
Timothy, A.F., Krieger, A.S., Vaiana, G.S.: 1975, *Solar Phys.* **42**, 135
Tsuru, T. *et al.*: 1989, *Publ. Astron. Soc. Japan* **41**, 679
Vaiana, G.S. *et al.*: 1973a, *Astrophys. J.* **185**, L47
Vaiana, G.S. *et al.*: 1981, *Astrophys. J.* **245**, 163
Vaiana, G.S., Krieger, A.S., Timothy, A.F.: 1973b, *Solar Phys.* **32**, 81
Vaiana, G.S., Krieger, A.S., Timothy, A.F., Zombeck, M.: 1976, *Astrophys. Space Sci.* **39**, 75
Vaiana, G.S., Rosner, R.: 1978, *Ann. Rev. Astron. Astrophys.* **16**, 393
Vaiana, G.S., Tucker, W.M.: 1974, in: *X-ray Astronomy*, eds. R. Giaconni, H. Gursky, D. Reidel, Dordrecht, p. 169
Van den Oord, G.H.J., Mewe, R.: 1989, *Astron. Astrophys.* **213**, 245
Vesecky, J.F., Antiochos, S.K., Underwood, J.H.: 1979, *Astrophys. J.* **233**, 987
Walter, F.M., Gibson, D., Basri, G.: 1983, *Astrophys. J.* **267**, 665
Weisskopf, M.C.: 1987, *Astrophys. Lett. and Communications* **26**, 1
White, N.E., Culhane, J.L., Parmar, A.N. *et al.*: 1986, *Astrophys. J.* **301**, 262
White, N.E., Shafer, R.A., Horne, K.*et al.*: 1990, *Astrophys. J.* **350**, 776
Withbroe, G.L.: 1975, *Solar Phys.* **45**, 301
Zhang, J. *et al.*: 1990, in: *High Resolution X-ray Spectroscopy of Cosmic Plasmas*, eds. P. Gorenstein, M.V. Zombeck, Proc. IAU Coll. 115, Cambridge Univ. Press, Cambridge, p. 361
Zombeck, M.V., Vaiana, G.S., Haggerty, R. *et al.*: 1978, *Astrophys. J. Suppl.* **38**, 69

TIME VARIABILITY OF STELLAR X-RAY EMISSION

R. PALLAVICINI

Osservatorio Astrofisico di Arcetri
Largo Enrico Fermi 5
I-50125 Florence, Italy

Abstract. Time variability of stellar coronal emission is reviewed using results from the *Einstein*, *EXOSAT*, *GINGA* and *ROSAT* satellites. Discussed topics include the search for rotational modulation of coronal emission, the monitoring of eclipsing binary systems, the detection of flares (and possibly microflares) at X-ray wavelengths. A detailed comparison is made between solar and stellar X-ray variability on time scales ranging from minutes to years. It is shown that the amplitude of solar X-ray variability due to rotation and the activity cycle is much larger than typically observed in other late-type stars. This suggests that the currently available samples of detected coronal sources are strongly biased towards active stars with large fractional areas covered by magnetically confined active regions.

Key words: X-rays – coronae – variability – eclipses – flares

1. Introduction

It may be appropriate in a meeting in memory of the late Giuseppe Vaiana to recall that "Pippo" was probably the first to discuss in some detail time variability of stellar X-ray emission. He did so in a review paper presented at the IAU Symposium No. 102 held in Zurich in August 1982 (Vaiana 1983). In that paper, Pippo reported on early results from the *Einstein* Observatory and discussed the variability of stellar coronal emission on time scales ranging from several minutes to a couple of years. In the ten years elapsed since then, our understanding of stellar X-ray emission and its time behaviour has increased enormously; however, as far as time variability is concerned, the progress has not been as significant as in other areas of stellar coronal research. The reasons can easily be understood.

First of all, the vast majority of the *Einstein* stellar observations were of short duration (a few thousand seconds each), and many sources were observed only once. Moreover, even when repeated observations of the same source had been obtained, often using different satellites, there was the problem of cross-calibrating instruments that have different responses and are sensitive to different passbands (this is the case for instance of *EXOSAT* vs. *Einstein* observations). More importantly, even using a single satellite, it is difficult to disentangle the short-term variations due to flares from the more gradual variations caused by the rotation of the star and/or by the existence of stellar activity cycles. Although the continuous observations made by the *EXOSAT* satellite over intervals of up to three days have somewhat alleviated this problem (see reviews by Pallavicini 1988, 1989), our knowledge of stellar X-ray variability remains very limited. For only one source, the Sun, we have indeed a good knowledge of X-ray variability on all time scales, from less than a second to several decades.

In this paper, I will summarize our present understanding of stellar X-ray variability as has emerged mainly from *Einstein* and *EXOSAT* observations. I will also mention briefly the more recent *ROSAT* results though the analysis of such data in search of time variability is still at an early stage. Nevertheless, I will stress several

J.F. Linsky and S. Serio (eds.), Physics of Solar and Stellar Coronae, 237–248.

times the important contribution expected from *ROSAT* in this area too. Before discussing other stars, I will present some results about the Sun, since the solar data can, at least to some extent, be used as a guideline in interpreting the spatially unresolved, and much more fragmentary, data available at present for other late-type stars.

2. Solar X-ray variability

When observed in soft X-rays, the Sun is an extremely variable star. This is well known to solar physicists who are familiar with the plots of the integrated X-ray emission of the Sun published regularly in *Solar Geophysical Data.* To stellar physicists, however, it may come somewhat as a surprise to hear that in X-rays the Sun has been observed continuously as a star for nearly three decades. This has been possible through a series of space missions, such as the *SOLRAD* and *GOES* satellites, that have recorded the integrated emission of the Sun in selected soft X-ray bands (typically 0.5-3 Å, 1-8 Å, 8-20 Å, 44-60 Å). Cross-calibration of different detectors has allowed the long term variations of the Sun to be monitored over the last three activity cycles (Kreplin 1970, Kreplin et al. 1977, Wagner 1988).

A simple inspection of some recent issues of *Solar Geophysical Data* will show daily plots of solar X-ray emission over the spectral bands 0.5-4 Å and 1-8 Å recorded by one of the geostationary satellites of the *GOES* series (developed by NOAA). The integrated flux (averaged over 5 min intervals) appears to vary in a virtually continuous way, with amplitudes that range from the smallest detectable variations (a few tens percents) to two or three orders of magnitude. Most of the variability that appears in the daily plots is due to flares (over time scales from several minutes to a few hours). However, some of the variations in the integrated solar X-ray flux are also due to more subtle changes that occur in active regions as a consequence of fluctuations in the heating rate and/or the emergence of new magnetic flux (e.g. Haisch et al. 1988). Although prominent in soft X-rays, these variations are negligible when compared to the bolometric luminosity of the Sun. In the 1-8 Å band the X-ray flux is typically at the level of $\sim 10^{-3}$ erg cm^{-2} s^{-1}, while the Solar Constant is 1360 watt m^{-2}. Hence, soft X-ray emission in this passband contributes only $\sim 10^{-9}$ of the solar luminosity. In spite of this, the X-ray variations are quite conspicuous, and may reach several orders of magnitudes in a large solar flare. As a rule, the amplitude of the variations is larger in the harder spectral bands.

In addition to short term variability, the solar flux changes on longer time scales, that are associated with the rotation of the Sun, the birth and decay of individual active regions and the 11-year activity cycle (with the subsequent varying numbers of active regions present on the solar surface at any one time). We know this because we can relate directly the variations observed in the spatially integrated solar X-ray flux to the spatial distribution of such emission as revealed by high-resolution X-ray images. Obviously, this link can be made only in the case of the Sun and this is the reason why solar observations are of such paramount importance for interpreting time variations observed in spatially unresolved stellar observations.

A good example is the photographic atlas of the Sun compiled by Zombeck et al. (1978) using images obtained by the AS&E soft X-ray telescope on board *SKYLAB* (a similar atlas can now be made using the new higher quality *YOHKOH* images). For each day during the first six months of the *SKYLAB* mission (May 29 to November 27, 1973), the Zombeck et al. atlas shows a full disk image of the Sun in two different soft X-ray bands: 2-32 + 44-54 Å and 2-17 Å. The spatially resolved X-ray images are compared with the 8-20 Å flux obtained over the same period by the *SOLRAD* 9 satellite (developed by NRL). We encourage the reader to browse through the atlas to have a vivid impression of coronal variability in a star like the Sun over time scales ranging from days to months. We also suggest to look at Fig. 18 of the Zombeck et al. paper to see how the full-disk X-ray flux and other indices of solar activity changed during the six months covered by the atlas.

Some interesting considerations can be made. First, the integrated 8-20 Å flux (obtained by making daily averages and subtracting the contribution of flares) showed a well-defined modulation with a period of slightly less than a month, that is clearly due to the rotation of the Sun. The amplitude of the modulation in the 8-20 Å passband is quite large (typically a factor 10), but varies from cycle to cycle; also the general shape of the light curve varies from one rotation to the next, but in all cases the X-ray daily flux follows quite strictly other activity indices, such as the sunspot number, the Ca II plage index and the 10.7 cm radio flux. The modulation is caused by the non-uniform distribution of active regions across the solar surface, while the different amplitudes and shapes from one cycle to the next are caused by the intrinsic variations in the number, extension and intensity of the active regions, which grow and decay on time scales that are comparable to a few solar rotations or less. Direct comparison with spatially resolved X-ray images confirms that the Sun was almost free of bright active regions at times of low integrated X-ray flux, while being covered by many active regions at times of maximum integrated X-ray flux (compare for instance the appearance of the Sun on September 4 and November 11, 1973, two days in which the integrated 8-20 Å fluxes differed by a factor 20).

There are also large variations of the average full-disk X-ray flux of the Sun throughout the solar cycle. Wagner (1988) have used the 1-8 Å *GOES* data to estimate the variations throughout the cycle of the background X-ray flux (i.e. the flux due principally to active regions, with the contribution of flares and coronal mass ejections subtracted). He found that the annually smoothed daily background 1-8 Å flux (which takes out the contribution of transient events and smooth the effect of the solar rotation) varied by at least a factor 85 from the solar minimum in May 1975 to the solar maximum in September 1981 and back again to the minimum in October 1986. Similar results were obtained previously for softer X-ray spectral bands observed with the *SOLRAD* satellites. For instance, Kreplin (1970) estimated that from the minimum in 1964 to the maximum in 1969 the 8-20 Å flux varied by a factor $\sim$ 200, and that the 44-60 Å flux varied by a factor $\sim$ 20. From *SOLRAD* 9 data published in *Solar Geophysical Data* I estimated that the average 8-20 Å flux varied by at least a factor 60 from the solar maximum in 1969 to early 1974 (when *SOLRAD* 9 ceased to operate). These data are all consistent, and show that the amplitude of the variation due to the activity cycle is quite large in the solar case, probably as large as a factor $\sim$ 100 in the spectral bands 1-8 and 8-20 Å and a factor $\sim$ 10 in the band 44-60 Å.

3. Stellar X-ray variability

Two important conclusions can be derived from the previous discussion: a) the variability of solar X-ray emission due to flares, rotational modulation and the activity cycle is large; b) the amplitude of such variability depends on the spectral band, and is larger at shorter wavelengths. Thus, before we can extrapolate the solar results to other stars we must determine which of the most commonly used solar passbands are more similar to those that have been used in obtaining stellar observations. In order to do so, I have used the line + continuum model of Mewe et al. (1985) for an optically-thin thermal plasma in ionization equilibrium and I have computed the ratio of the 1-8, 8-20 and 44-60 Å fluxes to the flux in the *Einstein* IPC passband. These ratios are plotted as a function of temperature in Fig. 1. For comparison, I have also plotted the flux ratios of the *EXOSAT* LE and of the *ROSAT* PSPC passbands to the *Einstein* IPC passband.

As shown by Fig. 1, over a large range of coronal temperatures (from $\sim 3 \times 10^6$ K to more than 10 million K) the largest contribution to the IPC passband comes from the 8-20 Å flux. Only at very low coronal temperatures ($\leq 2 \times 10^6$ K), the contribution of the 44-60 Å flux becomes prominent, while the 1-8 Å flux is the major contributor at temperatures larger than $\sim$ 20 million K. Thus, for most stellar coronal observations, the 8-20 Å solar data should be the most relevant for estimating the amplitude of the expected variability. Very soft sources may be expected to show a lower level of X-ray variability, while the amplitude of the observed variations are likely to be larger in intrinsically harder sources such as RS CVn binaries and active flare stars.

If the solar 8-20 Å data were really indicative of the expected level of variability of stellar coronal sources observed with either *Einstein*, *EXOSAT* or *ROSAT*, we should conclude that these sources must be highly variable, even outside flares. By extrapolating the solar results to other stars, we may expect variations of a factor $\sim$ 10 due to rotational modulation and a factor $\sim$ 100 due to stellar cycles. Such large variations of the *quiescent* emission of X-ray coronae, if they exist, would have profound implications for our understanding of stellar coronae. For instance, they could affect severely the correlations that have been found in the past between X-ray emission and other stellar parameters, such as rotation and chromospheric emission (Pallavicini et al. 1981, Schrijver 1987). Note that most of these correlations were based on stars observed only on one occasion and/or on data obtained at different times at the various wavelengths.

The X-ray observations obtained so far do *not* support the existence of such large variations in the quiescent emission of stars. Rather, the observations suggest that long-term variations (due either to rotational modulation or activity cycles) are small, at most a factor 2 or 3 (Ambruster et al. 1987, Pallavicini et al. 1990b). There is no evidence of larger variations than could not be explained as due to the sporadic occurrence of flares. Unfortunately, our monitoring of coronal sources in X-rays has been so far very fragmentary, and it is often difficult to decide which fraction of the observed variability is due to transient flare-like activity and which is intrinsic to quiescent emission. However, if variations as large as a factor 10 to 100 were present, they should have been detected, at least occasionally.

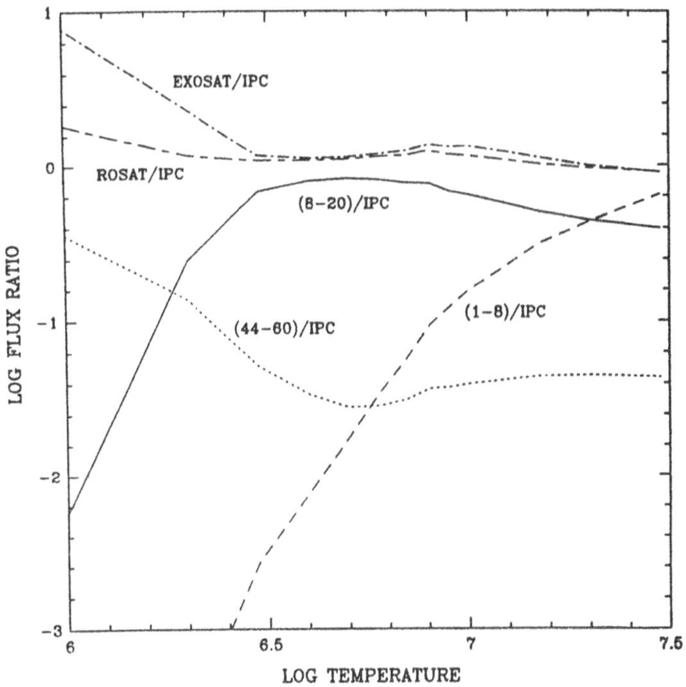

Fig. 1. Flux ratios of different X-ray spectral bands to the *Einstein* IPC passband. Also shown is the ratio of the *EXOSAT* LE and of the *ROSAT* PSPC passbands to the IPC passband.

Since it is unlikely that the Sun is unique at X-ray wavelenghts, the most plausible explanation is that our sample of X-ray detected coronal sources is highly biased towards active stars. The sensitivity of *Einstein* was such that only $\lesssim 1\%$ of the stars detected by *Einstein* emit at levels as low as the Sun. A similar problem is likely to occur for the *ROSAT* All-Sky Survey, which, being a flux-limited survey, is strongly biased towards the high luminosity tail of the stellar luminosity function. Clearly, the amplitude of the variations that can be detected in the quiescent emission of stars depends on the fraction of the stellar surface covered by active regions. A larger coverage factor will produce a smaller modulation due to the rotation of the star, as well as a smaller variation during the activity cycle. Moreover, only the largest flares can be detected in stars that have a quiescent background flux that is much larger than the solar one. To test this hypothesis, it would be interesting to monitor systematically for a long period of time a nearby inactive star, in much the same way that has been done for the Sun over the past thirty years.

Having somewhat clarified the relationship between solar and stellar variability data, we can now turn to the stellar observations themselves. I must say from the very beginning that sufficient data have been gathered so far only for a few specific classes of stellar objects (mainly RS CVn binaries and dMe flare stars) and the discussion therefore will be limited by necessity to these classes of stars. We know

(Collura et al. 1989) that at least some early-type stars are variable on time scales of hours to days (and possibly even minutes). However, the physical reasons for this variability are still poorly understood and early-type stars will be excluded from the rest of this paper. I will also not discuss long-term variability that could be produced by activity cycles in late-type stars: the time coverage is too limited, and the contamination by short-term variability too large, to allow determination of X-ray stellar cycles, at least for the time being. Extreme variability is certainly present in pre-main sequence objects and this variability is commonly interpreted as originating from magnetic processes similar to those operating in other late-type stars (Walter and Kuhi 1984, Montmerle 1985, 1992, Feigelson 1987, Gahm 1988, 1989). However, the data are too sparse to decide whether this variability originates entirely from flares or, more likely, is a combination of flaring and non-flaring activity, as in other late-type stars. PMS objects will also not be discussed in detail in this review.

A thorough investigation of time variability of coronal sources observed with *Einstein* was carried out by Ambruster et al. (1987) who analyzed a sample of 19 late-type stars (mostly flare stars) using a optimized χ^2 test. They found that short-term variability (on time scales from several minutes to hours) was common to all stars in the sample, with typical amplitudes of $\sim$ 30-50%. The variability appears rather stochastic (as is the solar one), and the same source, observed at different times, may be variable in one observation and constant in another. A problem with the data analyzed by Ambruster et al. is the presence of large data gaps due to the low orbit of the *Einstein* Observatory. Such gaps make extremely difficult to determine the physical nature of the observed variability and to distinguish for instance between slow evolving flares (on time scales of a few hours) and the repeated occurrence of shorter lived events (tens of minutes) overimposed on a more constant quiescent background. The *EXOSAT* data which allow continuous coverage of a source for up to three days represent a significant improvement in this respect and allow a better separation between flaring and quiescent emission. Pallavicini et al. (1990b) have analyzed the entire sample of dMe flare stars observed by *EXOSAT* (25 sources for a total monitoring time of nearly 300 hours), confirming the high degree of variability of the quiescent emission of late-type stars and the caotic nature of such variability. They also found, in agreement with Ambruster et al. (1987), that long-term variations (on times scales of months to years) were small (a factor 2 or 3, at most), at least for the active stars in the sample.

4. Some selected results on X-ray variability

In this section I will discuss in more detail a few selected topics on the X-ray variability of stellar coronal sources.

4.1. Rotational modulation

There have been several attempts in the literature to determine the rotational modulation of stellar coronal emission. In most cases the results have been inconclusive, owing to the superposition of flaring and non-flaring variability and the lack of observations extending over more than one rotation period (e.g. Collier Cameron 1988, Vilhu 1992). Probably the best case is that reported by Agrawal and Vaidya (1988) who detected a 20% variation in the *Einstein* IPC light curve of HR 1099 in antiphase with the photometric light curve (i.e. maximum X-ray emission at the time of maximum spot coverage). Agrawal (1988) claimed to have detected a similar signature of rotational modulation in *Einstein* observations of the binary flare star Gliese 867, which showed a 40% decrease over three days. However, Pollock et al. (1991) have shown on the basis of higher resolution *EXOSAT* data and a more sophisticated analysis of the *Einstein* data that both components of the visual binary were contributing to the observed emission: apparently, the observed variations were due to flaring activity on both stars, rather than to rotational modulation of component A (as was supposed by Agrawal).

The *ROSAT* All-Sky Survey has provided a novel method for detecting rotational modulation at X-ray wavelengths. Since the Survey data consist of snapshots (typically $\sim$ 20 sec long) regularly spaced at intervals of 96 min for periods of two days or longer, we can construct light curves for bright sources whose rotational period is of the order of a few days or less. An application of this technique to the rapidly rotating active star AB Dor (P = 0.514 days) has been discussed by Künster et al. (1992), who have combined more than one month of observations of this star and have analysed them as a function of orbital phase. The data show a substantial scatter (about a factor of 2) that largely masks the presence of a possible rotational modulation. On the other hand, *ROSAT* pointed observations of the same source presented by Künster at this meeting show better evidence of rotational modulation, in addition to erratic variations. It is interesting to note in passing that the variations shown by this active star, including rotational modulation, are less than a factor of 2, much less than what could have been guessed on the basis of observations on an inactive star like the Sun. This supports our previous conclusion that a large coverage factor may drastically reduce the amplitude of long-term variability of active stars with respect to the solar case.

A systematic analysis of the *ROSAT* All-Sky Survey data in search of rotational modulation in short-period RS CVn binaries is being carried out at present by Dempsey and coworkers: preliminary results have been reported at this conference. When fully reduced, these data are expected to provide important insights into the spatial distribution of X-ray emitting structures and on the fraction of the star covered by active regions.

4.2. Eclipse observations

Eclipsing binary systems provide a unique opportunity to infer the spatial distribution of hot coronal material around stars. The changes that are produced in the light curve as one star passes in front of the other permit, at least in principle, to determine the location, horizontal extent and height of the bright emitting structures. Unfortunately, the solution is often not unique and there are large ambiguities in the interpretation of the light curve. Rotational modulation is a form of self-eclipse, whith the active regions being eclipsed by the star itself as they move behind the stellar limb.

Several eclipsing binary systems have been observed in recent years using instruments on board *Einstein*, *EXOSAT* and *ROSAT*. These include Algol, the RS CVn systems AR Lac, TY Pyx and ER Vul, and the dMe flare star YY Gem (Walter et al. 1983, White et al. 1986, 1987, 1990, Culhane et al. 1990, Pallavicini et al. 1990a). Being the closest and brightest eclipsing binary among RS CVn stars, AR Lac has a special importance and has been observed for eclipses by all three satellites.

AR Lac consists of a G2 IV primary (with $R_* = 1.5\ R_\odot$) and a K0 IV secondary (with $R_* = 2.8\ R_\odot$) separated by 9.1 $R_\odot$ and orbiting each other with a period of 1.98 days. Using the *Einstein* Observatory, Walter et al. (1983) observed 17% of the orbital period of AR Lac, including a deep primary eclipse (when the G star was occulted by the K star) and a shallow secondary eclipse (when the K star was behind the G star). From this they concluded that compact coronal structures were present on both stars and, in addition, that an extended corona, highly inhomogeneous in longitude, was around the K0 IV component. Although we now know, from the subsequent analysis of *EXOSAT* data, that the solution found by Walter et al. was probably not unique, the interesting point is the inferred presence of very large coronal structures that were tentatively identified by Walter et al. with the high-temperature component ($T \sim 20$ million K) known to be present in RS CVn binaries (Swank et al. 1981).

Complete coverage of one orbital period of AR Lac was obtained by White et al. (1990) using both the Low Energy (LE) detector (sensitive to the spectral band 0.04 - 2 keV) and the Medium Energy (ME) detector (1 - 6 keV) on board *EXOSAT*. The primary eclipse was observed by both detectors, but there was no obvious eclipse in the ME data. This indicates that there must be a high-temperature component sufficiently extended to avoid eclipses by the two stars, while at the same time compact structures must also exist close to the surface of at least one of the components. Similar results were obtained in the *EXOSAT* observation of the eclipsing binary TY Pyx (Culhane et al. 1990).

White et al. (1990) have carried out extensive model simulations of the *EXOSAT* LE light curve of AR Lac, using both a χ^2 fitting and maximum entropy techniques. While the absence of a deep eclipse in the ME data clearly indicates that high-temperature plasma (at $T \geq 10^7$ K), probably filling the entire space between the stars, must exist, there are large ambiguities with regard to the location and extent of the X-ray bright regions on the two stars. Acceptable solutions are obtained both by locating the structures on each of the two components, or by assuming that only one of the two stars is X-ray bright. There are also large uncertainties with regard to the height of the various structures.

New observations of AR Lac obtained recently by *ROSAT* have further complicated the matter (Schmitt 1992; see also the results presented by Ottmann at this conference). The *ROSAT* PSPC has observed AR Lac both in the Full-Sky Survey and in the pointed mode. The primary eclipse was seen in all cases, while there is only marginal evidence for a shallow dip preceding secondary eclipse. This is consistent with previous *Einstein* and *EXOSAT* results. However, in contrast with *EXOSAT*, the primary eclipse is seen in all *ROSAT* X-ray energy bands, including the hardest one (1.1-2.4 keV). Spectral analysis of the PSPC data indicates that the major contribution to the *ROSAT* hard band comes from plasma that is virtually at the same temperature as that contributing to the *EXOSAT* ME band (where the eclipse was not seen). This could indicate either that the eclipse was too shallow to be detectable in the less sensitive and more noisy *EXOSAT* ME data or that the structure of the corona has sustantially changed between the two observations. Detailed analysis of the *ROSAT* data is required to discriminate between the two possibilities.

4.3. Flares and microflares

Flares are the most commonly observed and best understood form of variability in late-type stars. A large variety of X-ray flares have been observed from M dwarf flare stars, from RS CVn and Algol-type binaries and from pre-main sequence objects (Kahler et al. 1982, Haisch et al. 1983, de Jager et al. 1986, 1989, Doyle et al. 1988a,b, White et al. 1986, 1990, Tagliaferri et al. 1988, Tsuru et al. 1989, Pallavicini et al. 1990b, Culhane et al. 1990, Stern 1992, Stern et al. 1992).

A flare has also been observed from the A-type visual binary Castor (Pallavicini et al. 1990a), although in this case it is not yet clear whether the emission originated in one of the A-type primaries of the system (both spectroscopic binaries), or in an unseen late-type companion. Surprisingly, prior to the *ROSAT* mission there was only one reported case of an X-ray flare from a "normal" solar-type star (the G0 dwarf π^1 UMa; Landini et al. 1986). This was most likely due to insufficient time devoted in the past to X-ray monitoring of otherwise unconspicuous stars. In the *ROSAT* All-Sky Survey there are in fact many examples of flares from "normal" solar-type stars (Schmitt, this conference).

As shown by extensive analysis of *Einstein* and *EXOSAT* data (see reviews by Haisch 1983 and Pallavicini et al. 1990b), the light curves, time scales and flare temperatures are all similar to those typically observed in solar flares, but the released energies are orders of magnitude larger. The peak temperatures are usually in the range $2 - 4 \times 10^7$ K, though higher temperatures have occasionally been reported for some flares on RS CVn and Algol-type systems (see review by Linsky 1991). The time scales range from several minutes to a few hours and there is some indication of the existence of different classes of stellar X-ray flares, similar to solar compact and 2-ribbon events (Pallavicini et al. 1977, 1990b). The total energies released in the X-ray passband range from $\sim 10^{30}$ to 10^{34} erg for flares on dMe stars, and from $\sim 10^{35}$ to 10^{36} erg for flares on RS CVn binaries and PMS objects. For comparison, the typical energies of solar X-ray flares (integrated over the X-ray band and throughout the flare lifetime) are $\sim 10^{28}$ to 10^{31} erg. The events observed on RS CVn binaries and PMS objects are usually more energetic and longer-lived than those observed on dMe stars (Montmerle et al. 1985, White et al. 1986, Tsuru et al. 1989, Stern et al. 1992). We cannot exclude, however, that smaller and shorter-lived events also exist on these stars, but are missing due to the much higher quiescent X-ray emission of these stars.

The observations obtained so far show that at least for flares on dMe stars there is a strong analogy between solar and stellar events, except for the much larger energies involved in the stellar case. It is expected therefore that models

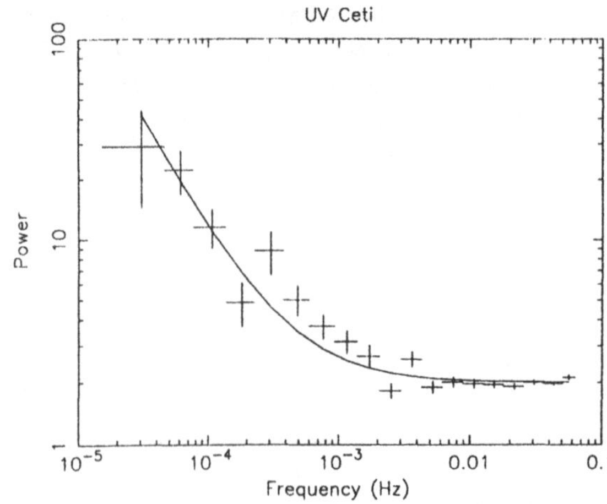

Fig. 2. Power spectrum of UV Ceti obtained by combining four different *EXOSAT* observations of the star.

similar to those developed in the solar case should also be able to reproduce the observations of stellar flares. Reale et al. (1988) and Cheng and Pallavicini (1991) have applied hydrodynamic models of transient events in a magnetically confined loop structure to stellar flares observed from *Einstein* and *EXOSAT*, and have found a good agreement between observations and theoretical predictions. Poletto et al. (1988) have applied magnetic reconnection models developed for solar two-ribbon flares to long-duration events on M dwarf stars. Although formal fitting of the data is not a proof that these models represent a correct description of stellar events, they are at least consistent with the available data.

Spatially resolved observations of the Sun at a variety of different wavelengths have revealed that in addition to flares there are also less energetic shorter-lived events ("microflares") that occur randomly in space and time (Lin et al. 1984, Porter et al. 1987). Since the number of flares and microflares increases rapidly towards lower energies (approximately as $dN/dE \sim E^{-\beta}$ where $\beta \sim 1.8$; Hudson 1991), it has been speculated that these events (or events of still lower energy such as the "nanoflares" proposed on theoretical grounds by Parker 1988, 1989) may have enough energy to heat the solar corona. While this suggestion is still a matter of debate, the existence of "microflares" is a well established observational fact in the solar case. Much more difficult, if at all possible, would be to identify the presence of such events in disk-integrated soft X-ray observations of other late-type stars.

Butler et al. (1986) have claimed to have found a signature of stellar microflares in *EXOSAT* observations of some dMe flare stars. By comparing the X-ray data (binned over time intervals of 30 and 60 sec) with simultaneous optical observations in the Hγ line, they have noticed the simultaneous occurrence of many short-lived events at both wavelengths. From this, and other indirect evidences, they concluded that the quiescent X-ray emission of dMe stars, and possibly of all late-type stars, results from the continuous occurrence of "microflares" lasting from tens of seconds to several minutes and with characteristic energies of $\sim 2\times10^{30}$ erg. This suggestion, though attractive, has raised some controversy, since the correlation between X-ray and optical events is far from being perfect, the binning of the X-ray data was too short to ensure a good statistics, and finally it is unclear whether these short-lived fluctuations occur continuously as claimed.

In order to check the above suggestion, Collura et al. (1988) and Pallavicini et al. (1990b) have independently analyzed the *EXOSAT* data using a variety of statistical tests, including χ^2 tests, autocorrelation techniques and power spectrum analysis. The conclusion was that there is no statistically significant evidence in the *EXOSAT* data for the presence of *continuous* short-term variability of the type suggested by Butler at al. Variability is present, but it occurs on longer time scales and larger energies than those suggested by Butler et al. for their "microflares". Moreover, if microflares are important for heating stellar coronae, their frequency distribution must be much steeper than a simple extrapolation of the observed flare frequency distribution to lower energies (see also similar conclusions by Ambruster et al. 1987 based on *Einstein* data). As an example, I show in Fig. 2 a power spectrum of UV Ceti obtained by combining all *EXOSAT* observations of this star, taking out only a major flare that occurred on December 23, 1985 (see Pallavicini et al. 1990b for details). The normalization is such that the counting statistics noise corresponds to a power of 2. Clearly, there is no evidence of power in excess of the noise level for frequencies higher than $\sim 2 \times 10^{-3}$ Hz. If variability on shorter time scale exists, it is below the detection threshold of *EXOSAT* data.

References

Agrawal, P.C.: 1988, *Astron. Astrophys.* **204**, 235

Agrawal, P.C., and Vaidya, J.: 1988, *Mon. Not. R. Astron. Soc.* **235**, 239

Ambruster, C.W., Sciortino, S., and Golub, L.: 1987, *Ap. J. Suppl.* **65**, 273

Butler, C.J., Rodonò, M., Foing, B.H., and Haisch, B.M.: 1986, *Nature* **321**, 679

Cheng, C.-C., and Pallavicini, R.: 1991, *Ap. J.* **381**, 234

Collier Cameron, A., Bedford, D.K., Rucinski, S.M., Vilhu, O., and White, N.E.: 1988, *Mon. Not. R. Astron. Soc.* **231**, 131

Collura, A., Pasquini, L., and Schmitt, J.H.M.M.: 1988, *Astron. Astrophys.* **205**, 197

Collura, A., Sciortino, S., Serio, S., Vaiana, G.S., Harnden, F.R.Jr., and Rosner, R.: 1989, *Ap. J.* **338**, 296

Culhane, J.L., White, N.E., Shafer, R.A., and Parmar, A.N.: 1990, *Mon. Not. R. Astron. Soc.* **243**, 424

de Jager, C., and 19 other authors: 1986, *Astron. Astrophys.* **156**, 95

de Jager, C., and 19 other authors: 1989, *Astron. Astrophys.* **211**, 157

Doyle, J.G., Butler, C.J., Byrne, P.B., and van den Oord, G.H.J.: 1988a, *Astron. Astrophys.* **193**, 229

Doyle, J.C., Butler, C.J., Callanan, P.J., Tagliaferri, R., de la Reza, R., White. N.E., Torres, C.A., and Quast, G.: 1988b, *Astron. Astrophys.* **191**, 79

Feigelson, E.D.: 1987, in T. Montmerle and C. Bertout, ed(s)., *Protostars and Molecular Clouds*, CEN Saclay: Gif-sur-Yvette, 123

Gahm, G.F.: 1988, in A.K. Dupree and M.T.V.T. Lago, ed(s)., *Formation and Evolution of Low Mass Stars*, Kluwer: Dordrecht, 295

Gahm, G.F.: 1990, in L.V. Mirzoyan, B.R. Pettersen and M.K. Tsvetkov, ed(s)., *Flare Stars in Star Clusters, Associations and the Solar Vicinity*, Kluwer: Dordrecht, 193

Haisch, B.M.: 1983, in P.B. Byrne and M. Rodonò, ed(s)., *Activity in Red Dwarf Stars*, Reidel: Dordrecht, 255

Haisch, B.M., Linsky, J.L., Bornmann, P.L., Stencel, R.E., Antiochos, S.K., Golub, L., and Vaiana, G.S.: 1983, *Ap. J.* **267**, 280

Haisch, B.M., Strong, K.T., Harrison, R.AA., and Gary, G.A.: 1988, *Ap. J. Suppl.* **68**, 371

Hudson, H.S.: 1991, *Solar Phys.* **133**, 357

Kahler, S., and 30 other authors: 1982, *Ap. J.* **252**, 239

Kreplin, R.W.: 1970, *Ann. Geophys.* **26**, 567

Kreplin, R.W., Dere, K.P., Horan, D.M., and Meekins, J.F.: 1977, in O.R. White, ed(s)., *The Solar Output and its Variations*, Colorado University Press: Boulder, 287

Kürster, M., Schmitt, J.H.M.M., and Fleming, T.A.: 1992, in G.A. Bookbinder and M.S. Giampapa, ed(s)., *Cool Stars, Stellar Systems, and the Sun*, Astron. Soc. Pacific: San Francisco, in press

Landini, M., Monsignori-Fossi, B.C., Pallavicini, R., and Piro, L.: 1986, *Astron. Astrophys.* **157**, 217

Lin, R.P., Schwartz, R.A., Kane, S.R., Pelling, R.M., and Hurley, K.C.: 1984, *Ap. J.* **283**, 421

Linsky, J.L.: 1991, *Mem. Soc. Astron. Ital.* **62**, 307

Mewe, R., Gronenschild, E.H.B.M., and van den Oord, G.H.: 1985, *Astron. Astrophys. Suppl.* **62**, 197

Montmerle, T., Feigelson, E.D., Bouvier, J., and André, P.: 1992, in E.H. Levy and J.I. Lunine, ed(s)., *Protostars and Planets III*, University of Arizona Press: Tucson, in press

Montmerle, T., Kock-Miramond, L., Falgarone, E., and Grindlay, J.: 1985, *Ap. J.* **269**, 182

Pallavicini, R.: 1988, *Mem. Soc. Astron. Ital.* **59**, 71

Pallavicini, R.: 1989, *Astron. Astrophys. Reviews* **1**, 177

Pallavicini, R., Golub, L., Rosner, R., Vaiana, G.S., and Linsky, J.L.: 1981, *Ap. J.* **248**, 279

Pallavicini, R., Serio, S., and Vaiana, G.S.: 1977, *Ap. J.* **216**, 108

Pallavicini, R., Tagliaferri, G., Pollock, A.M.T., Schmitt, J.H.M.M., and Rosso, C.: 1990a, *Astron. Astrophys.* **227**, 483

Pallavicini, R., Tagliaferri, G., and Stella, L.: 1990b, *Astron. Astrophys.* **228**, 403

Parker, E.N.: 1988, *Ap. J.* **330**, 474

Parker, E.N.: 1989, *Solar Phys.* **121**, 271

Poletto, G., Pallavicini, R., and Kopp, R.A.: 1988, *Astron. Astrophys.* **201**, 93

Pollock, A.M.T., Tagliaferri, G., and Pallavicini, R.: 1991, *Astron. Astrophys.* **241**, 451

Porter, J.G., Moore, R.L., Reichmann, E.J., Engvold, E., and Harvey, K.L.: 1987, *Ap. J.* **323**, 380

Reale, F., Peres, G., Serio, S., Rosner, R., and Schmitt, J.H.M.M.: 1988, *Ap. J.* **328**, 256

Schrijver, C.J.: 1987, *Astron. Astrophys.* **172**, 111

Schmitt, J.H.M.M.: 1992, in R. Pallavicini, ed(s)., *Solar and Stellar Coronae*, Soc. Astron. Ital.: Roma, in press

Stern, R.A.: 1992, in Y. Tanaka and K. Koyama, ed(s)., *Frontiers of X-ray Astronomy*, Universal Academy Press: Tokyo, 259

Stern, R.A., Uchida, Y., Tsuneta, S., and Nagase, F.: 1992, *Ap. J.* , in press

Swank, J.H., White, N.E., Holt, S.S., and Becker, R.H.: 1981, *Ap. J.* **246**, 214

Vaiana, G.S.: 1983, in J.O. Stenflo, ed(s)., *Solar and Stellar Magnetic Fields: Origins and Coronal Effects*, Reidel: Dordrecht, 174

Tagliaferri, G., Giommi, P., Angelini, L., Osborne, J.P., and Pallavicini, R.: 1988, *Ap. J. Letters* **331**, L113

Tsuru, T., and 12 other authors: 1989, *Publ. Astron. Soc. Japan* **41**, 679

Vilhu, O.: 1991, in A. Treves, G.C. Perola and L. Stella, ed(s)., *Iron Line Diagnostics in X-ray Sources*, Sringer-Verlag: Heidelberg, 30

Wagner, W.J.: 1988, *Adv. Space Res.* **8**, (7)67

Walter, F.M., Gibson, D.M., and Basri, G.S.: 1983, *Ap. J.* **267**, 665

Walter, F.M., and Kuhi, L.V.: 1984, *Ap. J.* **284**, 194

White, N.E., Culhane, J.L., Parmar, A.N., Kellett, B.J., Kahn, S., van den Oord, G.H.J., and Kuijpers, J.: 1986, *Ap. J.* **301**, 262

White, N.E., Culhane, J.L., Parmar, A.N., and Sweeney, M.A.: 1987, *Mon. Not. R. Astron. Soc.* **227**, 545

White, N.E., Shafer, R.A., Horne, K., Parmar, A.N., and Culhane, J.L.: 1990, *Ap. J.* **350**, 776

Zombeck, M.V., Vaiana, G.S., Haggerty, R., Krieger, A.S., Silk, J.K., and Timothy, A.: 1978, *Ap. J. Suppl.* **38**, 69

X-RAYS AND ACTIVITY IN PRE-MAIN SEQUENCE STARS

FREDERICK M. WALTER

State University of New York at Stony Brook

Abstract. Low mass pre-main sequence stars are rapidly rotating and fully convective, the perfect recipe for copious magnetic activity and coronal X-ray emission. *EINSTEIN* observations revealed a large number of serendipitous X-ray sources in regions of star formation – the low mass pre-main sequence naked T Tauri stars. These stars are coeval and cospatial with, and outnumber, the classical T Tauri stars. I summarize the properties of this X-ray discovered population, and the characteristics of stellar activity at young ages.

Key words: Stars: pre-main sequence – Stars: activity – Stars: evolution

1. The Low Mass Pre-Main Sequence Stars

Joy (1945) noted that the T Tauri stars, irregular variables associated with nebulosity, exhibited "emission lines resembling the solar chromosphere". The T Tauri stars are low mass stars in their pre-main sequence (PMS) phase of evolution (Herbig 1962; see Bertout 1989 and Appenzeller & Mundt 1989 for recent reviews). The most prominent spectral characteristic of the classical T Tauri stars (CTTS) is the Hα emission line, with equivalent widths $W_\lambda(H\alpha)\geq 10$Å, and occasionally exceeding 100Å. The emission measures n_e^2V of the CTTS "chromospheres" exceed that of the Sun by factors of $\sim 10^6$. If indeed attributable to a solar-like chromosphere, then the CTTS should have coronal X-ray luminosities $L_X \sim 10^{33}$ erg s^{-1} and would be among the brightest X-ray sources in the sky, with intensities about $\frac{1}{5}$ that of the Crab Nebula in the 1-2 keV range. Obviously, they are not.

We now know that the strong "chromospheric" emission from the CTTS arises not in a compact solar-like chromosphere, but rather in an extended volume related to the putative circumstellar accretion disk. The CTTS is a composite system with a low mass PMS star at its heart. The star is surrounded by circumstellar material, the source of the infrared excess, which is likely in a flattened, disk-like configuration (Adams *et al.* 1987). The source of the ultraviolet excess is the boundary layer, the interface between the disk, in Keplerian rotation, and the much more slowly rotating photosphere. Here the accreting material must decelerate and lose most of its kinetic energy. The boundary layer has the spectrum of an 8-10$\times 10^3$ K black body (Herbig & Goodrich 1986). The gross spectrum of a CTTS is the sum of the photospheric spectrum and these two continuum excesses. Spatially extended emission with a large volume emission measure, perhaps associated with the boundary layer and inner disk, produces the "chromospheric" emission spectrum.

2. X-Ray Observations of Star Formation Regions

Early X-ray observations of star forming regions (SFRs) were carried out with non-imaging experiments. Observations with *UHURU*, *ANS*, and *Ariel 5* revealed a source with $L_X \sim 3\times 10^{33}$erg s^{-1} (2-11 keV) coincident with the Orion nebula. *SAS-3* observed a weaker source with a 2-11 keV luminosity of $\sim 10^{33}$ erg s^{-1}. Bradt & Kelley (1979) inferred that the source of the emission was extended, and argued that the X-rays were produced in the colliding stellar winds of the Trapezium stars.

J.F. Linsky and S. Serio (eds.), Physics of Solar and Stellar Coronae, 249–256.

The situation clarified greatly once Ku & Chanan (1979) reported the initial *EINSTEIN* observation of the Trapezium. The IPC *image* revealed 22 unresolved X-ray sources. The OB stars of the Trapezium were the brightest X-ray source, but most of the other sources were associated with nebular variables (i.e., classical T Tauri stars). The X-rays were thought to arise in shocks, due to either accretion onto the star or the winds shocking the ISM. But it was now clear that young low mass stars could be bright X-ray sources ($L_X \sim 10^{30}$ erg s^{-1}).

EINSTEIN observations of CTTS by Gahm (1980), Feigelson & DeCampli (1981), and Walter & Kuhi (1981, 1984), showed that about $\frac{1}{3}$ of the CTTS were detected in X-rays, with typical $L_X \sim 10^{30}$ erg s^{-1}. Gahm argued that the CTTS detected by *EINSTEIN* were those which flare the most in the optical, and that the source of the X-ray emission was flares. Walter & Kuhi (1981) argued that the CTTS with the largest veiling and the strongest Hα emission were the least likely to be X-ray sources. They suggested that the coronae were geometrically thin and that the X-ray emission was "smothered", or absorbed, by the geometrically extended envelope. Montmerle *et al.* (1983) reported observations of 47 X-ray sources associated with the ρ Oph cloud. Because the log N-log S plot for these sources had a slope of -1.4, similar to that of solar flares, they surmised that the X-rays arise from a superposition of flares. They observed a large X-ray flare in ROX-20; X-ray flares have also been seen in DG Tau (Feigelson & DeCampli 1981) and AS 205 (Walter & Kuhi 1984). Walter & Kuhi showed that the X-ray temperatures and surface fluxes were comparable to those of the RS CVn systems, and suggested that the source of the X-rays was a hot (T$\sim 10^7$K) solar-like corona.

3. The Naked T Tauri stars

The biggest surprise in the *EINSTEIN* images of SFRs was the sheer number of X-ray sources detected. The space density of stellar X-ray sources in active SFRs exceeded that anywhere else in the sky, with the exception of the core of the Pleiades.

3.1. The Taurus-Auriga Region

Feigelson & Kriss (1981) and Walter & Kuhi (1981) studied 5 serendipitous X-ray sources in the fields of CTTS in the Tau-Aur region. These proved to be late-K stars with relatively weak Hα emission (W_λ(Hα) a few Å), and prominent Li absorption. The strong Li absorption line, W_λ(Li)$\sim$0.5Å, is suggestive of stellar youth, because lithium is destroyed in nuclear reactions at temperatures prevalent at the base of the convection zone. Depletion of the surface abundance of Li in low mass, convective stars is both expected (e.g., Bodenheimer 1965), and observed (e.g., Skumanich 1972, Duncan 1981). The CTTS have Li abundances log $N_{Li} \sim 3$ (Zappala 1972; Basri *et al.* 1991), where the hydrogen abundance log N_H=12. Log N_{Li}=3 is the present galactic Li abundance; the current solar surface abundance is log $N_{Li} \sim 1$. The combination of abundant Li, Hα emission, and projection near dark clouds indicated strongly that these stars were PMS, though they clearly had anemic levels of "chromospheric" activity, at least as compared to the CTTS. These were the naked T Tauri stars (NTTS).

Mundt *et al.* (1983) studied these 5 stars in more detail. Radial velocities showed them to be members of the T association. V826 Tau proved to be a double-lined spectroscopic binary (SB2) with a circular 4.8 day orbit. Unlike the CTTS, which have easily measurable near-IR excesses, four of the stars exhibited no near-IR excesses (V836 Tau has a small 2μm excess). Herbig *et al.* (1986) discussed 11 PMS stars discovered in an objective prism CaII H&K survey. Feigelson *et al.* (1987) discussed the X-ray sources in the region of the Tau-Aur clouds; Walter *et al.* (1988; W88) studied the 33 PMS stars among these X-ray sources. In all, these studies added 44 previously unknown PMS stars to the 84 known CTTS (Cohen & Kuhi 1979). Extrapolating from the incompletely sampled X-ray observations, W88 estimated that the Tau-Aur complex contained of order 1000 NTTS.

3.2. Other Star Forming Regions

While at present Tau-Aur is arguably the best studied SFR at all wavelengths, other SFRs have been well observed in X-rays. The basic picture, that NTTS are common, holds for *all* regions observed. Other SFRs well observed with *EINSTEIN* include ρ Oph and the Sco OB2 association, Chamaeleon I, CrA, and much of the constellation of Orion.

Feigelson & Kriss (1989) presented *EINSTEIN* IPC observations of the Cha I cloud, a compact, isolated cloud forming mostly low mass stars. They found 22 X-ray sources, 17 of which which they identified with previously catalogued PMS stars. They identified another 4 with new NTTS. Walter (1992) confirmed 4 NTTS and found 2 new NTTS from high-resolution spectra of 11 stars.

Bouvier & Appenzeller (1992) identified 30 PMS stars as likely optical counterparts to 24 X-ray sources of Montmerle *et al.* (1983). Most of these stars are weak emission stars with inferred ages between 10^6 and 10^7 years.

Walter (1986) discussed 6 X-ray discovered PMS stars in the Sco OB2 association and CrA cloud. More complete studies of 5 IPC fields in Sco OB2, with an age of 5×10^6 years, revealed 28 NTTS and one new CTTS (Walter *et al.* , in preparation). The NTTS are coeval with or younger than the B stars, and the slope of the initial mass function (IMF) is consistent with the field star IMF.

Strom *et al.* (1990) studied the X-ray sources in the L1641 cloud, and discuss 83 optical counterparts to 65 X-ray sources. Most have $W_\lambda(H\alpha)<5$Å; 78% of the optical counterparts *do not* have significant IR excesses, and are NTTS. Inferred ages range from 3×10^5 to 10^7 years.

3.3. The True Low Mass PMS Population

All these observational studies point in the same direction: there is a large population of previously unknown and largely unanticipated (but see Herbig 1978) PMS stars in SFRs. These stars appear very normal - they have neither strong emission lines nor continuum excesses. The *observed* ratio of NTTS to CTTS in all regions observed ranges from about 1 (Tau-Aur; W88) to 3 (L1641; Strom *et al.* 1990) to 10 (Sco OB2; Walter *et al.* in preparation). After corrections for various sources of incompleteness, the ratio can only increase; W88 estimated that the NTTS out-

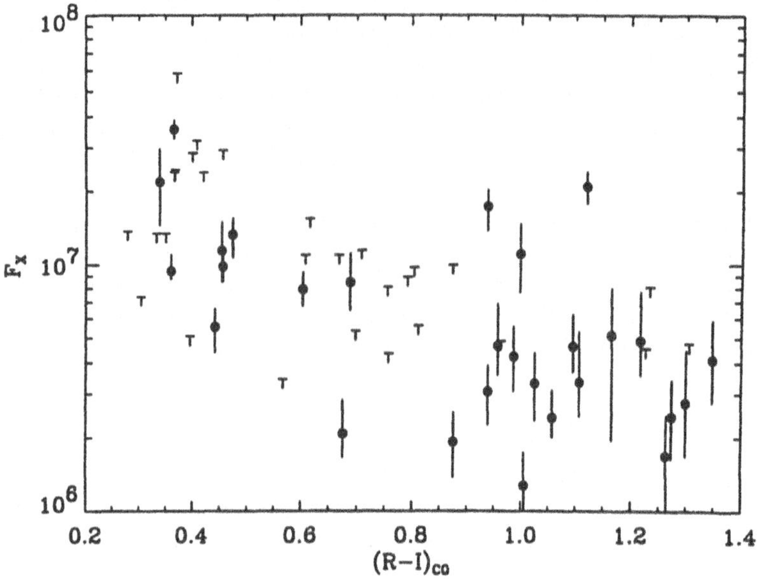

Fig. 1. the X-ray surface flux as a function of the Cousins R-I color for the NTTS in the Sco OB2 association. T represents the fluxes of the Tau-Aur sample (Walter *et al.* 1988).

number the CTTS by 10:1 in Tau-Aur, when integrated over the entire complex and over the last 3×10^7 years. Low mass PMS stars are much more common than was thought a mere 10 years ago, and we have X-ray imaging satellites to thank for this revelation.

4. Stellar Activity and the NTTS

While the "chromospheric" activity of the CTTS is largely not of stellar origin, the X-rays must arise either in a stellar corona or in flaring associated with the disk. The boundary layer cannot generate X-rays. The emission measures and temperatures of the quiescent (non-flaring) coronae of CTTS (Walter & Kuhi 1984) are similar to those of the RS CVn binaries. This suggests a coronal source for the X-rays. The large X-ray flares on DG Tau and AS 205 occurred on stars which did not have quiescent X-ray emission detectable with the *EINSTEIN* IPC (although AS 205 is a steady source in a ROSAT PSPC image); the source of these flares is not known.

The NTTS do not suffer from the ambiguities inherent in the composite CTTS systems. The X-rays give us a picture of the young solar corona, at ages of a few $\times10^6$ years. Because we see the stellar photospheres, we can use the Barnes-Evans relation to derive stellar radii and emission surface fluxes. Walter & Barry (1991) discuss the mean observed coronal and chromospheric flux levels of the NTTS. The fluxes are comparable to those seen in the most active late-type stars, including young stars on the main sequence. They show that the emission surface flux for single stars decays exponentially in the square root of the stellar age.

Figure 1 shows the X-ray surface flux, corrected for extinction, for a sample of NTTS. The mean F_X is about 10^7 erg cm^{-2} s^{-1}, but there is a tendency for the bluer stars to have higher X-ray surface fluxes ($\log(F_X) \propto M/2$, where M is the stellar mass). A similar relation is seen in chromospheric emission in clusters. The surface fluxes are comparable to those of RS CVn systems. There is a tight relation between the Hα emission flux (measured by subtracting the spectrum of an inactive star of the same spectral type) and the Ca II K line surface flux, with $F_{H\alpha} \propto \sqrt{F_{CaII}}$. The flux-flux relations are considerably shallower than seen on the main sequence. There is no discernable rotation-activity relation. These differences between the NTTS and main sequence stars may be attributable to saturation of the activity.

The mean X-ray to bolometric flux ratio for the 28 NTTS in the Sco OB2 association is $\frac{f_X}{f_{bol}}=10^{-3.5}$. This is consistent with the empirical upper limit $\frac{f_X}{f_{bol}} \sim 10^{-3}$ (Vilhu & Walter 1987). Three stars lie above the limit, but show discrepant ratios of the X-ray to chromospheric fluxes, implying likely X-ray variability or flaring.

5. The Importance of the NTTS for Astrophysics

In addition to providing an opportunity to study the properties of PMS *stars* , the NTTS afford an opportunity to explore other significant astrophysical problems. I summarize some of them for completeness.

Calibration of PMS stellar models. The masses and ages of PMS stars are estimated by placing them in an H-R diagram, and comparing their locations with theoretical evolutionary tracks and isochrones. These evolutionary models have not been subject to observational scrutiny, because we have no independently determined masses and ages for PMS stars. To measure stellar masses we need spectroscopic binary systems. Few CTTS are confirmed binaries (Mathieu 1992), in part because of selection effects, but perhaps largely because large disks are dynamically unstable in a close binary system. The NTTS, on the other hand, seem to have a normal frequency of binary systems (Mathieu *et al.* 1989). So far we have identified 4 SB2s among the NTTS, and over a dozen SB1s. Eventually we will find an eclipsing system which we can use to pin down the mass of a PMS star.

The Initial Mass Function. The large number of low mass PMS stars has global implications for how stars form in the first place. Is star formation bi-modal? After applying admittedly large correction factors for incompleteness, we find that the slope of the IMF between about 0.5 and 20$M_\odot$ in Tau-Aur and in Sco OB2 is close to that in the field (Walter & Boyd 1991). There is no evidence for truncated IMFs in loose OB associations.

Timescales for Disk Dissipation and Planetesimal Formation. Unless our solar system is a fluke, planetary systems form out of the circumstellar disks which surround the CTTS. The large NTTS population, many with ages comparable to or younger than the CTTS, shows that the remnant circumstellar material generally dissipates on timescales of order 3×10^5 years (W88; Strom *et al.* 1989). If one is confident that planets actually form, then this time is an upper limit to the time it takes the warm dust to accumulate into large inefficient radiators. These observed disk dissipation and survival (Strom *et al.* 1989) timescales constrain the planetary formation models. On the other hand, if planets require a long incubation

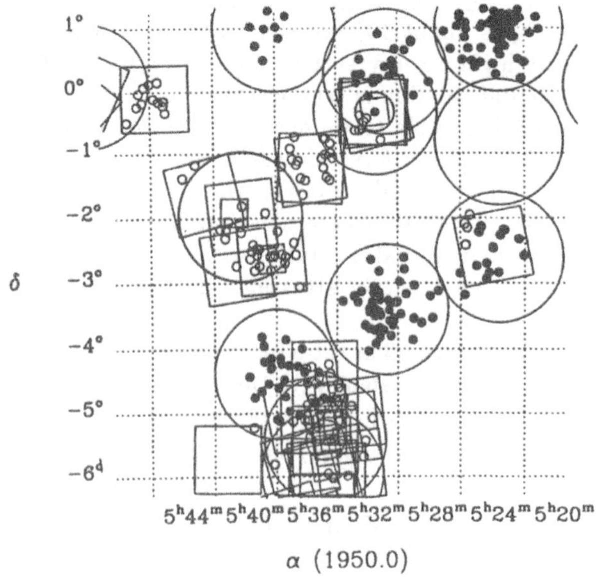

Fig. 2. NTTS and ROSAT PSPC sources in and near the belt of Orion. The squares represent the *EINSTEIN* pointings. Open circles represent optically identified PMS stars associated with IPC sources. The filled circles represent ROSAT PSPC sources.

period, then perhaps only the CTTS, the few PMS stars which retain their disks for a long period, will form planets, and planetary systems will be rare.

6. The View From ROSAT

Other contributions to this symposium have highlighted recent results from ROSAT. With its full sky coverage, increased sensitivity, and deep pointings in selected areas, the number of X-ray sources in SFRs is climbing rapidly. I have programs to obtain PSPC observations of selected regions in the CrA and Orion SFRs. The density of X-ray sources is 2-3 times larger than even the deep (10 ksec) *EINSTEIN* IPC images. Figure 2 illustrates the situation in the belt of Orion. In a recent observing run we obtained low dispersion spectra of 31 stellar counterparts to PSPC sources in Orion; 25 proved to be PMS stars. If this ratio holds, there are thousands of new low mass PMS stars in the Ori OB1b association alone.

It is clear that the ROSAT observations will provide an unprecedented global outlook at SFRs. Eventually we will be able to map the distribution of low mass PMS stars, discovered by virtue of their X-ray emission, on the sky. We will learn the IMF of the SFRs, both those forming and those not forming high mass stars. We will find eclipsing SB2s among the NTTS, and will be able to measure PMS stellar masses and radii directly. We will study the activity of these fully convective stars in some detail. This is a monumental undertaking, one that will take many years, and many months of observing time. But it is an undertaking that will yield important results in areas of galactic, stellar, and even planetary astronomy.

Acknowledgements

This research has been supported by grants from the National Science Foundation and the National Aeronautics and Space Administration to SUNY Stony Brook.

Appendix

A. A Note on Nomenclature

Walter (1986) called these X-ray selected low mass PMS stars naked T Tauri stars. The activity and spectral energy distributions of the NTTS are different from those of the CTTS. However, the kinematics and the distributions of masses and ages of the NTTS and the CTTS are indistinguishable (with the caveat that there does exist in Tau-Aur an older population of NTTS). The NTTS do *not* have detectable near-IR excesses, and their Hα emission levels are well below those of the CTTS. Mundt *et al.* (1983) called them post-T Tauri stars, following Herbig (1978), but the ages of the NTTS are generally comparable to those of the CTTS. The NTTS may be post-T Tauri in an evolutionary sense (they probably are), but this requires the assumption that all low mass PMS stars begin as CTTS, and then become NTTS when their circumstellar material dissipates. Walter (1986), and later W88, argued that the photospheres of the CTTS and the NTTS were identical, and that the essential differences,the IR excesses and strong Hα emission of the CTTS, arose in the circumstellar environments. The NTTS are what the CTTS would be were one to strip them of their circumstellar gas and dust - in other words, naked T Tauri stars[1].

It has become fashionable to refer not to NTTS, but rather to a class of WTTS, or *weak-lined* T Tauri stars[2]. These stars are classified solely using $W_\lambda(H\alpha)$; stars with $W_\lambda(H\alpha)<10Å$ are considered WTTS (Herbig & Bell 1988). The draw to this class is the ease of classification – one needs only a low dispersion spectrum.

All NTTS are WTTS: a solar-like chromosphere cannot produce $W_\lambda(H\alpha)>10Å$. The converse, that all WTTS are NTTS, is not true, because a small $W_\lambda(H\alpha)$ need not imply a lack of circumstellar material. I describe two types of WTTS which are not necessarily NTTS:

1. **G stars.** W_λ is the ratio of the emission to the continuum fluxes. For stars with equal Hα surface fluxes, $\frac{W_{\lambda 1}}{W_{\lambda 2}} \propto \frac{T_2^4}{T_1^4}$, where T is the photospheric temperature. The G2 star SU Aur is one of the most active low mass PMS stars. Although the Hα surface flux exceeds that of normal stellar chromospheres by nearly 2 orders of magnitude, $W_\lambda(H\alpha)$ is only 5Å, and it is WTTS.
2. **Stars with large passive disks.** Stars with passive disks are not actively accreting, have no significant boundary layer, and have small Hα equivalent

[1] If NTTS are naked, the CTTS could be considered clothed.

[2] WTTS is somewhat of misnomer: the WTTS really are not weak-lined. Their photospheric lines are of normal strength – it is the veiled CTTS which have weak absorption lines because of veiling (Hartigan *et al.* 1991). Their emission lines are as strong as "normal" stellar emission lines get – it is only through comparison with the CTTS, whose emission lines are partly of circumstellar origin, that the lines can be considered weak. To avoid having to remove "WTTS" from the lexicon, I suggest that the WTTS be redefined as the *wimpy* T Tauri stars.

widths. They may well have large near-IR excesses, from reprocessing of the stellar flux in the CS disk. An example may be HDE283447.

Lada (1987) classified PMS stars from their broadband spectral energy distributions. Class 3 sources are essentially black bodies; class 1 sources have flat λF_λ spectra. Class 2 sources have near-IR spectral energy distributions flatter than black bodies. That the NTTS do not have measurable near-IR excesses places them in class 3 with normal stars; WTTS may fall in either of classes 3 or 2.

The NTTS and WTTS are defined using different physical criteria, and are not identical populations, It is important to keep this distinction in mind when generalizing about the properties of PMS stars.

References

Adams, F.C., Lada, C.J., & Shu, F.H.: 1987, *ApJ* **312**, 788.
Appenzeller, I. & Mundt, R.: 1989, *A&ARev* **1**, 291.
Basri, G., Martin, E.L., & Bertout, C.: 1991, *A&A* **252**, 625.
Bertout, C.: 1989, *ARAA* **27**, 351.
Bodenheimer, P.: 1965, *ApJ* **142**, 451.
Bouvier, J. & Appenzeller, I.: 1992, *A&AS* **92**, 481.
Bradt, H.V., & Kelley, R.L.: 1979, *ApJL* **228**, L33.
Cohen, M. & Kuhi, L.V.: 1979, *ApJS* **41**, 743.
Duncan, D.: 1981, *ApJ* **248**, 651.
Feigelson, E.D. & DeCampli, W.M.: 1981, *ApJL* **243**, L89.
Feigelson, E.D. & Kriss, G.A.: 1981, *ApJL* **248**, L35.
Feigelson, E.D. & Kriss, G.A.: 1989, *ApJ* **338**, 262.
Feigelson, E.D., Jackson, J.M., Mathieu, R.D., Myers, P.C., & Walter, F.M.: 1987, *AJ* **94**, 1251.
Gahm, G.F.: 1980, *ApJL* **242**, L163.
Hartigan, P. *et al.* : 1991, *ApJ* **382**, 617.
Herbig, G.H.: 1962, *Adv.Astron.Astr* **1**, 47.
Herbig, G.H.: 1978., in Problems of Physics and and Evolution of the Universe, ed(s)., , Academy of Sciences of the Armenian SSR: Yervan, 171.
Herbig, G.H. & Bell, K.R.: 1988, Lick Observatory Contribution 1111.
Herbig, G.H. & Goodrich, R.W.: 1986, *ApJ* **309**, 294.
Herbig, G.H., Vrba, F.J., & Rydgren, A.E.: 1986, *AJ* **91**, 575.
Joy, A.H.: 1945, *ApJ* **102**, 168.
Ku, W.H.-M. & Chanan, G.A.: 1979, *ApJL* **234**, L59.
Lada, C.: 1987, in Star Forming Regions, ed(s)., *M. Peimbert and J. Jugaku*, Reidel:Dordrecht, 1.
Mathieu, R.D.: 1992, in Binary Stars as Tracers of Stellar Formation, ed(s)., *A. Duquennoy and M. Mayor*, Campridge Press:Cambridge, in press.
Mathieu, R.D., Walter, F.M., & Myers, P.C.: 1989, *AJ* **98**, 987.
Montmerle, T., Koch-Miramond, L., Falgarone, E., & Grindlay, J: 1983, *ApJ* **269**, 182.
Mundt, R., *et al.* : 1983, *ApJ* **269**, 229.
Skumanich, A.: 1972, *ApJ* **171**, 565.
Strom, K.M., Strom, S.E., Edwards, S., Cabrit, S., and Skrutskie, M.F.: 1989, *AJ* **97**, 1451.
Strom, K.M., *et al.* : 1990, *ApJ* **362**, 168.
Vilhu, O., & Walter, F.M.: 1987, *ApJ* **321**, 958.
Walter, F.M.: 1986, *ApJ* **306**, 573.
Walter, F.M.: 1992, *AJ* **104**, 758.
Walter, F.M., & Barry, D.C.: 1991, in The Sun in Time, ed(s)., *C. Sonnett, M.S. Giampapa, and M.S. Matthews*, University of Arizona:Tucson, 633.
Walter, F.M., and Boyd, W.T.: 1991, *ApJ* **370**, 318.
Walter, F.M., Brown, A., Mathieu, R.D., Myers, P.C., and Vrba, F.J.: 1988, *AJ* **96**, 297 (W88).
Walter, F.M. & Kuhi, L.V.: 1981, *ApJ* **250**, 254.
Walter, F.M. & Kuhi, L.V.: 1984, *ApJ* **284**, 194.
Zappala, R.R.: 1972, *ApJ* **172**, 57.

A-TYPE AND CHEMICALLY PECULIAR STARS

JEFFREY L. LINSKY*

Joint Institute for Laboratory Astrophysics, National Institute of Standards and Technology and University of Colorado, Boulder, CO 80309-0440

Abstract. Conventional wisdom holds that early-type and late-type stars should have very different outer atmospheres, because the early-type stars lack deep convective zones and thus a mechanical energy source to heat their outer layers. I will show that the magnetic chemically peculiar (CP) stars hotter than about spectral type A2 display many of the phenomena seen in the most active late-type stars. In particular, many CP stars are luminous nonthermal radio sources and ROSAT confirms that many are also luminous coronal x-ray sources like the RS CVn systems. A wind-fed magnetosphere model has been proposed to explain both the nonthermal radio and the x-ray emission. In this model the stellar wind plays the role of a mechanical energy source analogous to the role played by convection in the active late-type stars. By comparison, the chemically normal single stars in the spectral range B2–A5 typically show neither nonthermal radio emission nor x-ray emission, and thus appear to be very different from the CP stars of similar spectral type and from the active late-type stars.

Key words: chemically peculiar stars – A-type stars – X-ray emission – radio emission

1. Is Conventional Wisdom Correct for CP Stars?

Since the first convincing discovery of a flare on a star (UV Ceti) other than the Sun by Luyten (1949), astronomers have studied solar-like phenomena on late-type stars in increasing detail. Included among these phenomena are chromospheres, multimillion degree coronae, nonthermal radio emission, dark starspots, active regions, flares, and other transient or nonclassical phenomena observed across the electromagnetic spectrum. The apparent absence of these phenomena in the A- and B-type stars has been the basis for a widely held belief that the outer atmospheres of early-type and late-type stars are fundamentally different: early-type stars lack the deep convective zones that presumably are required for the dynamo generation of turbulent magnetic fields, nonradiative heating of the outer layers, and the acceleration of nonthermal electrons.

I have long felt that this "conventional wisdom" was at minimum too simplistic and perhaps completely wrong. My opinion was based on the presence of strong magnetic fields in the photospheres of many slowly rotating chemically peculiar (CP) A- and B-type stars measured by Babcock (1958) and subsequent workers. These fields are easily measured because they are ordered (simple dipoles or quadrupoles), unlike the complex fields observed on the Sun and presumed to be present on active late-type stars. Since large fields (up to 34 kG for the mean surface field in the case of Babcock's star) are observed on some stars, might not magnetic fields also be present but not detected on other A- and B-type stars? This could occur because conventional magnetometers that measure Zeeman lines in polarized light have great difficulty measuring fields of complex geometry like the Sun due to field cancellation or if there is rapid stellar rotation.

The presence of strong magnetic fields is a necessary but not sufficient condition for solar-like activity. All that would be required for active phenomena on the

* Staff member, Quantum Physics Division, National Institute of Standards and Technology. Official contribution of N.I.S.T.; not subject to copyright in the United States.

J.F. Linsky and S. Serio (eds.), Physics of Solar and Stellar Coronae, 257–266.

magnetic A- and B-type stars is a source of mechanical energy, a role played by convection in the late-type stars. The presence of chemical peculiarities is generally explained by diffusion, which is suppressed by convection and, presumably, other sources of mechanical energy (see review by Vauclair and Vauclair 1982). This logic serves as a theoretical explanation for the apparent absence of "active" phenomena in these stars. Table I summarizes the conventional wisdom concerning magnetic CP stars prior to 1985. X-ray and 6-cm radio luminosities are given for the Sun and the RS CVn systems, which are among the most active of the late-type stars as indicated by their large x-ray and radio luminosities and their energetic flares. The more active star in a RS CVn system is typically an early K0 subgiant with a deep convective zone and large rotational velocity due to tidally-induced synchronism of rotation and revolution. For more information concerning RS CVn systems and their radio and x-ray emission see, for example, Strassmeier *et al.* (1988), Drake, Simon, and Linsky (1989), and Dempsey *et al.* (1993).

TABLE I
Summary of Conventional Wisdom before 1985.

Stellar Property	CP Stars	Active Cool Stars
1. Convective zone	No	Yes
2. Stable atmosphere	Yes	No
3. Magnetic field	Ordered	Complex, flux tubes, plages
4. Rotation rate	Slow for spectral type	Fast for spectral type
5. Winds	Indirect evidence (polar jets)	Yes in open field regions
6. Hot coronae (x-ray emission)	No evidence before ROSAT	$\log L_x^{\odot}=27.0$–28.0 $\log L_x^{RS}=29.5$–32.0
7. Nonthermal radio emission	No evidence before 1985	$\log L_6^{\odot}=10.5$–12.0 $\log L_6^{RS}=14.5$–18.0
8. Coronal heating mechanism	Not required	MHD Processes (recombination or shocks)
9. Electron accel. mechanism	Not required	Magnetic field reconnection produces electric fields

In this talk I will attempt to answer the heretical question: **Are the magnetic chemically peculiar stars high-temperature analogs of the active cool stars?** I will summarize what we have been learning about the nonthermal radio and high-temperature x-ray emission from these stars, and describe a magnetosphere model that we have developed for these stars. In this model, weak radiation-driven winds provide the mechanical energy source, which is required with the strong magnetic fields to explain the observed active phenomena. I conclude that the winds in the A- and B-type CP stars play a role analogous to convection in the late-type stars.

2. First Test of Conventional Wisdom: The Search for Nonthermal Radio Emission

Stimulated perhaps by the correlation of strong magnetic fields with radio emission on the Sun, several groups have searched unsuccessfully for radio emission from CP stars (see Drake *et al.* 1987 for a summary). Drake *et al.* (1985) reported the first radio detections of the CP star σ Ori E (B2Vp He–S), discovered serendipitously, and another helium-strong (He–S) star HR 1890 (B1.5V He–S), together with 12 nondetections. This minisurvey revealed two properties of these stars that were confirmed in subsequent surveys: the radio spectra are typically flat between 2 and 6 cm, which is consistent with gyrosynchrotron emission from a nonthermal distribution of electrons and the high radio luminosity of some sources. For example, $\log L_6$=17.8 for σ Ori E, which is nearly 10^7 times more luminous than the Sun.

In their 6-cm VLA survey of 34 CP stars, Drake *et al.* (1987) increased the list of detections from two to five stars. They detected three He–S stars with early-B spectral types (δ Ori C was added to the previous detections of HR 1890 and σ Ori E), and for the first time detected two helium weak (He–W) and silicon strong stars with late-B or A0 spectral types (IQ Aur and GL Lac). The latter is often called Babcock's star. The He–S stars generally lie within the temperature range $T_{\rm eff}$ = 18,000–23,000 K, while the He–W and Si strong stars lie within the range $T_{\rm eff}$ = 11,000–16,000 K.

Drake *et al.* (1987) noted that no classic Ap stars with enhanced abundances of Sr, Cr and Eu were detected, even though the survey sensitivity extended down to values of L_6 200 times smaller than the luminosity of σ Ori E. The SrCrEu stars are cooler than the He–W stars, which in turn are cooler than the He–S stars. I will show later that this sequence of effective temperatures plays a critical role in explaining the origin of the radio emission. Drake *et al.* (1987) provided further support for the idea that gyrosynchrotron emission from mildly relativistic electrons is the emission mechanism by showing that the radio spectra are flat between 2 and 20 cm, the emission from σ Ori E and HR 1890 is variable on time scales of hours, and the radio luminosities of all detected stars are very large and, incidentally, similar to the most radio luminous of the RS CVn binaries. Additional support for the gyrosynchrotron emission process is provided by the measurement of mild polarization in three sources by Linsky, Drake, and Bastian (1992), and the determination of high brightness temperatures ($T_B > 10^9$ K) for σ Ori E and HD 37017 by Phillips and Lestrade (1988) using VLBI techniques. See, however, Leone (1991) for arguments in favor of gyroresonance emission.

The detection of 11 additional He–W stars at 6 cm allowed Linsky *et al.* to identify statistical properties of the sample and to develop a model for the magnetospheres of these stars. They noted that on average the He–S stars are 20 times more radio luminous than the cooler He–W stars ($<\log L_6>$=17.7 compared with 16.4 for the He–W stars) and 1000 times more luminous than θ Aur (A0p Si), which is the coolest and least radio luminous star detected. This range in L_6 is much too large to be explained by changes in the stellar radii or surface areas. With 16 detections and 46 upper limits, they were able to find meaningful correlations with effective temperature ($L_6 \sim T_{\rm eff}^{6.6}$) and with the root mean square value of the photospheric magnetic field ($L_6 \sim B_{rms}^{1.36}$). They found no correlation with stellar rotational velocity, but Leone (1991) presented evidence that the radio emission from σ Ori E and HD 37017 vary in phase with their rotational and magnetic periods.

Although mass loss rates have not been measured for these stars, radiation-driven wind theory predicts that the mass loss rate ($\dot{M}$) should depend on a high power of T_{eff}. Also, IUE spectra of the C IV and Si IV lines in these stars indicate the presence of both polar jets and circumstellar plasma located near the magnetic equator (e.g. Shore, Brown, and Sonneborn 1987). Linsky *et al.* conclude that $L_6 \sim \dot{M}^{0.49} B_{rms}^{0.95}$, with a correlation coefficient of 0.90 for 16 sources. This indicates that mass loss can play a critical role in the radio emission process. They also conclude that the **magnetic CP stars must be a new class of radio stars** as the radio emission from the active cool stars does not depend on the mass loss rate, and the radio emission from the O-type stars depends on the mass loss rate but not, apparently, on the magnetic field strength.

Radio emission has not been observed from any of the chemically normal B-type stars (later than about spectral type B1) or the A-type stars. The most stringent upper limits for 6-cm emission from nearby chemically normal A-type stars (see Brown *et al.* 1990) are $\log L_6 < 12.73$ for α PsA (A3 V) and $\log L_6 < 12.48$ for α Aql (Altair, A7 IV-V). These radio luminosity upper limits are, respectively, factors of 100 and 175 times smaller than the observed value of L_6 from the least radio luminous CP star θ Aur (A0p Si). The nature of the companion to HD 28867, which is a radio source and may be a young B star (White, Pallavicini, and Kundu 1992), however, requires further study. Thus some aspect of the chemical peculiarity phenomenon, presumably the presence of a strong magnetic field, is critical for the creation of strong radio emission in the B- and A-type stars.

2.1. A Wind-fed Magnetosphere Model

Linsky *et al.* (1992) proposed that the observed properties of radio emission from the magnetic CP stars may be understood as optically thick gyrosynchrotron emission from a nonthermal distribution of electrons produced in a current sheet far from the star. Their model (see Fig. 1) consists of a rotating dipolar magnetosphere in which stellar wind plasma flows out from near the magnetic poles along the field lines. Far from the star ($r \gtrsim 10R_\star$), the gas pressure exceeds the magnetic pressure and the field lines near the equator are drawn out into current sheet configurations (see Havnes and Goertz 1984 and André *et al.* 1988) analogous to planetary magnetotails. These current sheets are likely locations of plasma heating and particle acceleration, as strong magnetic field gradients produce electric fields and tearing mode instabilities may be important. In the complex geometry provided by the usual displacement of the rotational and magnetic poles, there are many possible scenarios for particle acceleration and heating. In the "direct injection" scenario proposed by Linsky *et al.* (1992), the electrons return to near the star by traveling along magnetic fields, where they are reflected and radiate in the strong fields near the magnetic poles. Uchida (1985) has discussed a similar model and Usov and Melrose (1992) have described in detail the plasma heating mechanisms that may

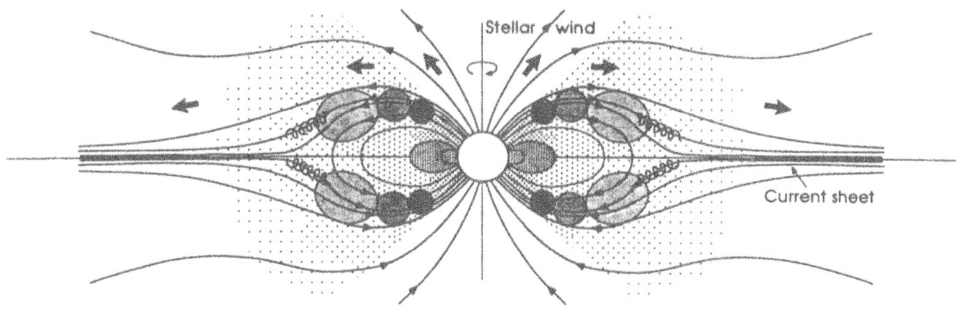

Fig. 1. A cartoon showing a cross section through a CP star and its magnetosphere. For simplicity, an aligned rotator is shown. A dense stipple indicates the presence of dense, trapped thermal plasma. A light stipple indicates the possible presence of tenuous thermal plasma. The small dark circles show schematically where high frequencies originate; larger, lighter circles designate the location of the source with decreasing frequency.

operate in the current sheets surrounding early-type magnetic stars.

The model proposed by Linsky *et al.* accounts for time variability in terms of intrinsic variability in the acceleration process and rotational modulation of the projected area of the emitting region. The flat microwave spectra result from two factors: (i) a decrease in source size with increasing frequency and (ii) a decrease in effective temperature of the emitting electrons with increasing frequency. The circular polarization observed on some sources may be an opacity effect resulting from the increasing magnetic field strength with decreasing radius. Finally, the decreasing radio luminosity with decreasing stellar $T_{\rm eff}$ along the main sequence is explained by the decreasing mass loss rate for radiation driven winds.

While both the CP stars and the RS CVn systems radiate by the gyrosynchrotron process with similar radio luminosities (see Table II and Fig. 2), the physical processes leading to electron acceleration are likely different. For CP stars, the radio emission requires a wind to distend the field lines, create current sheets, and accelerate the electrons far from the star. In the RS CVn model proposed by Morris, Mutel and Su (1990), on the other hand, plasma is trapped in the strong field region of the corona near the star and some of the trapped electrons are accelerated by an unspecified process. The model proposed by Linsky *et al.* for the nonthermal radio emission from magnetic CP stars may also be relevant to certain of the weak T Tauri stars and other PMS objects (e.g., André *et al.* 1988).

3. Second Test of Conventional Wisdom: The Search for Coronal X-ray Emission

3.1. What Has Been Learned From *Einstein*

In their initial survey of stellar x-ray emission using the *Einstein* satellite, Vaiana *et al.* (1981) first called attention to a significant gap in the x-ray H-R diagram;

TABLE II
Comparison of Radio Parameters of Different Types of Stars.

Parameter	Sun	dMe	RS CVn	He–S	He–W
Spectral type	G2 V	dM4e	K0 IV	B2 V	B8 V
$R_\star/R_\odot$	1.0	0.4	3	4	3
B_{photo}(kG)	1.5	3–5	≈1	2.0–10.0	≤ 1.0–35.0
f_{photo}	0.02	0.2–0.7	0.3?	≈1	≈1
Topology	complex	complex	complex	dipole/quad	dipole
$< fB_{photo} >$	0.03	0.6–3.5	≈0.3	2.0–10.0	≤ 1.0–35.0
$< fB_{photo} > R_\star^2$	0.03	0.1–0.6	≈3	32–160	≤ 9.0–315
log L_6	10.5–12.0	13.0–14.6	14.5–18.0	17.4–17.9	15.7–17.9
log L_x	27.0–28.0	27.7–29.8	29.5–32.0	30.2	30.0
log L_x/L_{bol}	-6	-4 to -2	-6 to -3	-7	-6
flaring	yes	yes	yes	yes?	yes?
active regions	yes	yes	yes	—	—
$(R/R_\star)^2 T_B$	8×10^4	1×10^8	1×10^{10}	2×10^{10}	$0.04–6 \times 10^{10}$
$R/R_\star$	1	—	3	<4	—
T_B	8×10^4	—	1×10^9	$\geq 1 \times 10^9$	—

none of the early-B stars were detected as x-ray sources. They explained this gap between the O-type stars, where the x-rays may be produced by shocks in their winds, and the solar-like cool stars with turbulent magnetic fields as due to the absence of convection in the B-type stars stars. They did list a few late-B and A-type stars as x-ray detections, however, including both Sirius and Vega. Shortly thereafter, Cash and Snow (1982) used *Einstein* to observe 17 A-type CP stars and three B7–A1 chemically normal stars. Of the CP stars that were not known to be binaries, only two Ap stars (the SrCr star ω Oph and the HgMn star β Scl) were detected with x-ray luminosities of logL_x = 28.4 and 28.8, respectively. Cash and Snow were concerned, however, that the x-ray emission from the two CP and three chemically normal stars might come from previously unknown cool dwarf companions that are optically faint but x-ray bright. As shown in Table II, the range in logL_x for dMe stars is 27.7–29.8.

This concern, as well as the problem of ultraviolet light leaks in the *Einstein* HRI detector, have frustrated attempts to determine whether any CP or chemically normal A- and B-type main sequence stars are coronal x-ray sources. For example, Golub *et al.* (1983) found no additional single CP stars as x-ray sources, but identified Sirius A (A1 V), Vega (A0 V), and Altair (A7 IV-V) as x-ray sources. Schmitt *et al.* (1985) showed that the x-ray detections of the first two stars were actually detections of stellar ultraviolet light by the *Einstein* HRI, but they argued that the x-ray detection of Altair by the IPC was not due to a UV light leak and very unlikely to be due to a cooler companion star because the inferred temperature of the x-ray emitting plasma was much lower than is found for active late-type stars.

In their study of the x-ray emission from a large (125 objects) and properly selected sample of chemically normal late-A and early F-type field stars, Schmitt *et al.* (1985) concluded that the few detections of A-type stars (except for Altair) with the *Einstein* IPC can be explained by x-ray emission from previously unknown M dwarf companions, since the measured $\log L_x \sim 29.0$ is consistent with active M dwarf coronae. Also, the cumulative distribution function (CDF) of L_x for the A stars ($0.1 \leq B-V \leq 0.3$) is qualitatively different from the CDF for the F stars ($0.3 \leq B-V \leq 0.5$), confirming that coronal x-ray emission begins near $B-V = 0.3$. They also found that the CDF for the x-ray emission from binary systems containing an F dwarf star is similar to that of a mixture of F and M dwarfs, implying that the M dwarf companions of F stars are not overly luminous compared to single M dwarfs. The observed quiescent and flare x-ray emission from Castor (A1 V + A5 Vm) can also be explained as coming from an unseen late-type companions to one or both of these single-lined spectroscopic binary systems (Pallavicini *et al.* 1990).

In the young Ursa Major cluster, Schmitt *et al.* (1990) identified one single A-type star, β Eri (A3 III) as an x-ray source, but unlike Altair, its x-ray temperature is similar to that found in a typical F- or G-type giant and thus a companion of this type could be the emission source. All five detected A-type stars in the Hyades Cluster are known to be members of multiple systems (Micela *et al.* 1988) so their x-ray properties are in doubt. As one proceeds to younger clusters such as the Pleiades (Micela *et al.* 1990) and the Orion Nebula (Caillault and Zoonematkermani 1989), the detected A-type stars are far more x-ray luminous than field A-type stars and thus more likely than a putative late-type companion to be the x-ray source. However, x-ray luminous weak-line T Tauri stars are possible companions for stars as young as those in Orion.

The early B-type stars present a different situation. Caillault and Zoonematkermani (1989) identified seven early-B stars in the Orion Nebula as luminous x-ray sources, but their L_x/L_{bol} ratios lie in the range 10^{-6} – 10^{-8} that characterizes the O-type stars for which the x-ray emission is thought to be formed in shocks in their radiation driven winds rather than in magnetically heated coronae. Grillo *et al.* (1992) searched the whole *Einstein* archive to identify 74 B-type x-ray sources. Most of these have spectral types B0-B3 with L_x/L_{bol} ratios in the range 10^{-6} – 10^{-8}. These are likely to be stars with x-rays produced in their winds. On the other hand, 34 of 1058 stars in the spectral range B5-B9 were detected as x-ray sources. Grillo *et al.* conclude that x-ray emission from the late-B stars is much less common than in the hotter stars, but whether the emission comes from late-type companions of these late-B stars is not yet known.

3.2. What is being learned from ROSAT

The previously described investigations with the *Einstein* satellite left unanswered the question of whether the CP stars are x-ray sources. Several investigations with the ROSAT satellite are attempting to answer this question. The first of these to be completed is the search for x-ray emission from all 102 CP stars with radio detections or upper limits. Drake *et al.* (1992) searched the ROSAT All-Sky Survey, which has a detection threshhold of 3×10^{-13} ergs cm^{-2} s^{-1} in about 600 s integration, detecting x-ray sources at the locations of 11 CP stars. Four of these

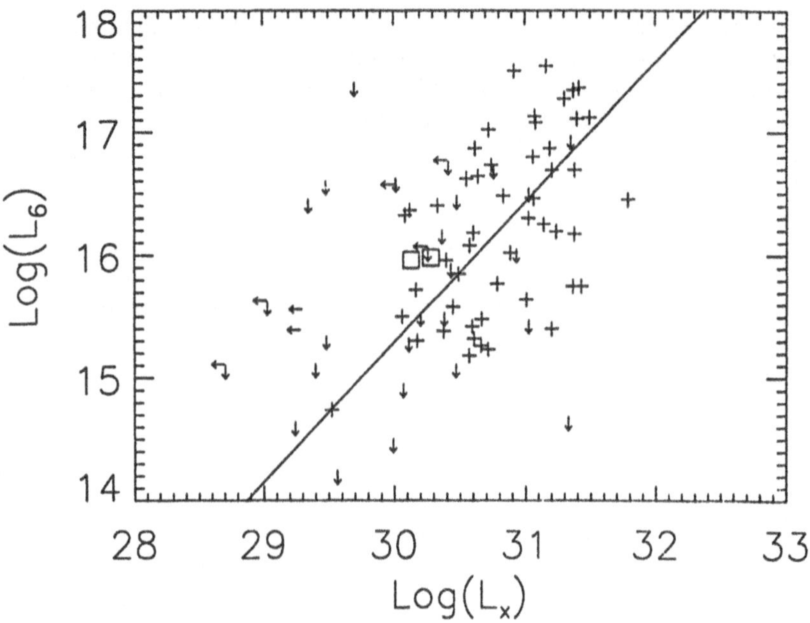

Fig. 2. Comparison of radio and ROSAT x-ray emission for CP stars (squares) and RS CVn systems (pluses). Upper limits are indicated by arrows, and the solid line is a maximum likelihood fit to the RS CVn data from Dempsey *et al.* (1993).

stars have hotter companions that are likely wind x-ray sources like many of the O-type stars. Of the remaining seven stars, HR 3089 and θ Car are early-B He–S stars. The x-ray emission from these two stars could be produced in their winds as the L_x/L_{bol} ratios are near 10^{-7} like that of the O-type stars. The star 52 Her (A2Vp SrCrEu) is the least x-ray luminous ($\log L_x = 28.8$) of the group. Since this value of L_x is consistent with that of an active M dwarf star, one cannot rule out a previously unknown companion star as the source of emission.

The remaining four stars in the ROSAT All-Sky Survey sample are He–W and Si rich stars: ν For (B9.5p Si), HR 5624 (Ap Si), HR 5988 (B8p He–W), and HR 6054 (B7 IV He–W). They are characterized by $\log L_x \geq 29.5$, which is more luminous than essentially all single M dwarfs stars, and $L_x/L_{bol} \geq 10^{-6.3}$, which makes it unlikely that the x-rays are produced in stellar winds as for the O-type stars. Drake *et al.* therefore concluded that these four stars are the first detected members of a **new class of coronal x-ray CP stars**. Nonthermal radio emission has also been detected from two of these stars (HR 5988 and HR 6054). Their radio and x-ray luminosities are plotted in Fig. 2 along with those of the RS CVn systems for which Dempsey *et al.* (1993) have obtained ROSAT All-Sky Survey data and for which radio data are available. The radio fluxes of the other two stars are now being reduced. The two CP stars with detected radio and x-ray emission have luminosities consistent with the distribution of RS CVn systems (see Fig. 2) but not the M dwarfs, which typically have $\log L_6 \leq 14.6$ and $\log L_x \leq 29.5$. This reinforces the conclusion that the x-ray and radio emission are intrinsic to the CP stars rather than to unknown M dwarf companions, and that the x-ray and radio emission from the CP stars have properties similar to those of the brightest coronal sources among the late-type stars, the RS CVn systems.

The ROSAT All-Sky Survey study of CP stars is being followed by pointed observations of individual stars. For example, Drake (private communication) finds that HR 5624 (Ap Si) is a strong but variable x-ray source with a mean count rate about 60% larger than that found during the All-Sky Survey. On the other hand, 3 Sco (B8IIIp HeSi) and HR 5942 (B5IV He) were not detected as x-ray sources, despite their being nonthermal radio sources. These two stars would lie far to the left of the solid line in Fig. 2.

The positive identification of four He–W stars as x-ray sources strengthens the hypothesis advanced by Mitskevich and Tsymbal (1992) that the observed weakness of He I lines among the late-B CP stars is not due to a true deficiency of helium, but rather to the "superionization" of He I by x-ray and XUV photons shining down on the photosphere from a hot corona. They find that the He I lines are most affected for stars in the effective temperature range 14,000 K $< T_{\rm eff} <$ 20,000 K, which is appropriate for the late-B and middle-B stars, but that the effect becomes unimportant for stars with $T_{\rm eff} \geq$ 25,000 K, corresponding to spectral type B0 V and earlier. The measurement of x-ray fluxes from a larger sample of CP stars is needed to test this interesting hypothesis.

4. Conclusions

My story is not complete as additional pointed x-ray observations are needed to study the x-ray time variability and coronal plasma temperatures of CP stars, and additional radio observations, including VLBI observations, are needed to better understand the sizes and nonthermal electron energies in the magnetospheres of these stars. Nevertheless, I believe that the case is now made that at least some CP stars display phenomena very similar to those found in active late-type stars. We have developed a wind-fed magnetosphere model that can account for both the nonthermal radio and coronal x-ray emission by using the wind as a mechanical energy source analogous to the convective motions in the late-type stars. This model should be modified to include oblique magnetic geometries and to make quantitative predictions of the x-ray emission. I encourage CP star theorists to accept this challenge. By comparison, the chemically normal single stars in the spectral range B2–A5 typically show neither nonthermal radio emission with very small upper limits nor x-ray emission, and thus appear to be very different from the CP stars of similar spectral type and from the active late-type stars.

5. Acknowledgments

I thank NASA for support through Interagency Transfers W-17,772 and H-04630D to the N.I.S.T. and grant NAG5-1797 to the University of Colorado. I thank Dr. Stephen Drake and Robert Dempsey for permission to cite our ROSAT results prior to publication. I also thank Dr. Drake for comments on the manuscript.

References

André, P., Montmerle, T., Feigelson, E.D., Stine, P.C., and Klein, K.-L.: 1988, *ApJ* **335**, 940

Babcock, H.W.: 1958, *ApJS* **3**, 141

Brown, A., Vealé, A., Judge, P., Bookbinder, J.A., and Hubeny, I.: 1990, *ApJ* **361**, 220

Caillault, J.-P. and Zoonematkermani, S.: 1989, *ApJL* **338**, L57

Cash, W. and Snow, T.P.Jr.: 1982, *ApJL* **263**, L59

Dempsey, R.C., Linsky, J.L., Schmitt, J.H.M.M., and Fleming, T.A.: 1993, *ApJS* , in press

Drake, S.A., Abbott, D.C., Bastian, T.S., Bieging, J.H., Churchwell, E., Dulk, G., and Linsky, J.L.: 1987, *ApJ* **322**, 902

Drake, S.A., Abbott, D.C., Bieging, J.H., Churchwell, E., and Linsky, J.L.: 1985, in R.M. Hjellming and D.M. Gibson, eds., *Radio Stars*, D. Reidel: Dordrecht, 247

Drake, S.A., Linsky, J.L., Schmitt, J.H.M.M., and Rosso, C.: 1992, *ApJL* , submitted

Drake, S.A., Simon, T., and Linsky, J.L.: 1989, *ApJS* **71**, 905

Golub, L., Harnden, F.R.Jr., Maxson, C.W., Rosner, R., Vaiana, G.S., Cash, W.Jr., and Snow, T.P.Jr.: 1983, *ApJ* **271**, 264

Grillo, F., Sciortino, S., Micela, G., Vaiana, G.S., and Harnden, F.R.Jr.: 1992, *ApJS* **81**, 795

Havnes, O. and Goertz, C.K.: 1984, *A&A* **138**, 421

Leone, F.: 1991, *A&A* **252**, 198

Linsky, J.L., Drake, S.A., and Bastian, T.S.: 1992, *ApJ* **393**, 341

Luyten, W.J.: 1949, *ApJ* **109**, 532

Micela, G., Sciortino, S., Vaiana, G.S., Harnden, F.R.Jr., Rosner, R., and Schmitt, J.H.M.M.: 1990, *ApJ* **348**, 557

Micela, G., Sciortino, S., Vaiana, G.S., Schmitt, J.H.M.M., Stern, R.A., Harnden, F.R.Jr., and Rosner, R.: 1988, *ApJ* **325**, 798

Mitskevich, A.S. and Tsymbal, V.V.: 1992, *A&A* , in press

Morris, D.H., Mutel, R.L., and Su, B.: 1990, *ApJ* **362**, 299

Pallavicini, R., Tagliaferri, G., Pollock, A.M.T., Schmitt, J.H.M.M., and Rosso, C.: 1990, *A&A* **227**, 483

Phillips, R.B. and Lestrade, J.-F.: 1988, *Nature* **334**, 329

Schmitt, J.H.M.M., Golub, L., Harnden, F.R.Jr., Maxson, C.W., Rosner, R., and Vaiana, G.S.: 1985, *ApJ* **290**, 307

Schmitt, J.H.M.M., Micela, G., Scortino, S., Vaiana, G.S., Harnden, F.R.Jr., and Rosner, R.: 1990, *ApJ* **351**, 492

Shore, S.N., Brown, D.N., and Sonneborn, G.: 1987, *AJ* **94**, 737

Strassmeier, K.G., Hall, D.S., Zeilik, M., Eker, Z., and Fekel, F.C.: 1988, *A&AS* **72**, 291

Uchida, Y.: 1985, in A.B. Underhill and A.G. Michalitsianos, eds., *The Origin of Nonradiative Heating/Momentum in Hot Stars*, NASA Conf. Publ. No. 2358, 199

Usov, V.V. and Melrose, D.B.: 1992, *ApJ* **395**, 575

Vaiana, G.S. *et al.* : 1981, *ApJ* **244**, 163

Vauclair, S. and Vauclair, G.: 1982, *ARAA* **20**, 37

White, S.M., Pallavicini, R. and Kundu, M.R.: 1992, *A&A* **257**, 557

CORONAL STELLAR EMISSION IN GALAXIES

G. FABBIANO

Harvard-Smithsonian Center for Astrophysics

Abstract. The X-ray luminosity of galaxies is not dominated by their stellar coronal emission. However, stars are there and their presence must account for a fraction of the X-ray emission. Stellar coronal emission in the Galaxy may explain the Galactic Ridge. In nearby spirals, a fraction of the unresolved emission must be ascribed to stars. Even in more distant X-ray faint elliptical galaxies, there may be evidence for stellar emission.

Key words: Galaxies – Galactic X-ray emission

1. Introduction

In the optical band, studying galaxies is also studying stars. This is not true in X-rays, where the luminosity is dominated by different sources. In spiral galaxies a population of hard-spectrum discrete X-ray sources dominates the emission in the *Einstein* band ∼(0.2–4 keV). These are likely to be X-ray binaries and young supernova remnants. A hot gaseous interstellar medium (ISM) may be also present, and has been convincingly detected in starburst galaxies. In X-ray-bright elliptical galaxies the X-ray luminosity is dominated by the emission of a hot ISM. However, emission from a population of low-mass binaries, similar to those in the bulge of M31, may be the dominant source of X-rays in X-ray-faint ellipticals (see Fabbiano 1989 and references therein; Fabbiano, Kim and Trinchieri 1992; Kim, Fabbiano and Trinchieri 1992a and b).

What is the role of stellar coronal emission in the X-ray luminosity of galaxies? Clearly it does not dominate the flux detected in the *Einstein* band: the X-ray to optical luminosity ratio expected from a 'normal' stellar population in spiral galaxies is smaller than the average measured value for the sample of galaxies observed with *Einstein* (see Fabbiano 1989). Moreover, the average X-ray spectrum of spiral galaxies is harder (kT>3 keV; Kim, Fabbiano, and Trinchieri 1992) than that expected from pure stellar emission for a measureable mix of stellar types (see Vaiana 1990). In elliptical galaxies this contribution would be even harder to discern.

However, the stellar contribution must be there at a certain level. As Pippo Vaiana was fond of saying, if there are $\sim 10^{11}$ suns in a spiral galaxy emitting 10^{28}ergs s^{-1} each, we have $\sim 10^{39}$ ergs s^{-1}, which is not far off the observed range of X-ray luminosity of spirals ($\sim 10^{39-40.5}$ergs s^{-1}).

In this talk I will review the evidence of stellar coronal emission in the Galaxy and in nearby spiral galaxies; I will then discuss the UV and X-ray observations of a starburst galaxy, where the X-ray emission of the young stellar population may be very significant; I will conclude with some recent puzzling results on X-ray faint elliptical galaxies that may suggest a detectable amount of stellar emission.

2. The Galaxy

Stars in the Galaxy could be detected in large numbers and studied with the *Einstein Observatory* (e.g. Linsky 1990; Vaiana 1990 and references therein). Although stars are individually faint X-ray sources, given to their large numbers they can be

J.F. Linsky and S. Serio (eds.), Physics of Solar and Stellar Coronae, 267–273.

TABLE I
X-ray Components of the Galaxy
(Watson 1990)

Source	log $< L_x >$	log L_x(component)
X-ray Binaries	37–38	39.3
SNR	34.5	37.5
CVs	31.5	36.5–37.5
Stars: #		
0 ($\sim$ 5000)	33	38.7
B-K (5×10^9)	28	37.7
M (10^{10})	28	38
RSCVn (10^6)	31	37

responsible for $\sim$ 1/4 of the Galactic X-ray emission (see Table 1, adapted from Watson 1990).

A diffuse hard emission is present in the Galactic plane, the 'Galactic Ridge', whose nature is not well understood. Both unresolved low luminosity sources and diffuse processed have been suggested (see Watson 1990 and references therein). As part of her Ph. D. thesis, Micela (1991) explored the possibility that this Ridge emission may be explained with the integrated emission of the stellar population of the Galactic plane. The results are surprisingly encouraging. Both the flux and the hard X-ray spectrum of the Ridge as detected in the *Einstein* IPC (fig. 1) can be entirely explained with the integrated emission of young stellar clusters in the Galactic plane (see figs 2 and 3). This is because there is a spectral dependence on age of the coronal stellar spectrum, with a marked hardening in younger stars (see Vaiana 1990).

3. M31 and M33

Let us now move to the two spiral galaxies closest to us: M31 and M33. Coronal emission from individual stars would not be detectable in the *Einstein* observations, which are dominated by a number of individual bright sources ($L_X \sim 10^{36-38}$ergs s^{-1}). In the bulge of M31, *Einstein* and *ROSAT* high resolution HRI observations have led to the detection of close to 80 point-like sources (Van Speybroeck et al 1979; Trinchieri and Fabbiano 1990; Primini et al 1992). The spectrum of the integrated emission of the bulge of M31 shows that the majority of these sources are likely to be low-mass binaries (Fabbiano, Trinchieri and Van Speybroeck 1987). When all the detected sources are removed from the HRI images of the bulge, one finds a residual diffuse emission, which cannot entirely be explained with the integrated emission of fainter sources belonging to the same population as

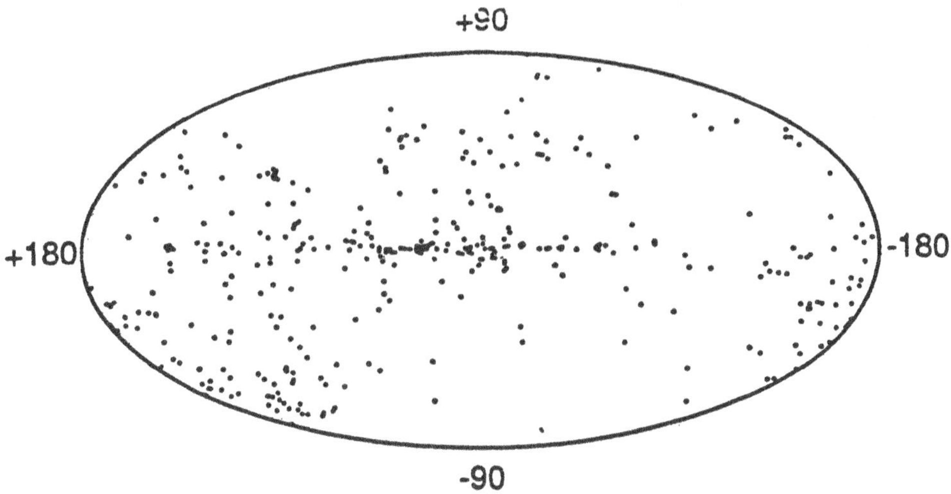

Fig. 1. Hardness ratio map of the IPC background: the hardest regions. The Galactic Ridge can be clearly seen (Micela 1991).

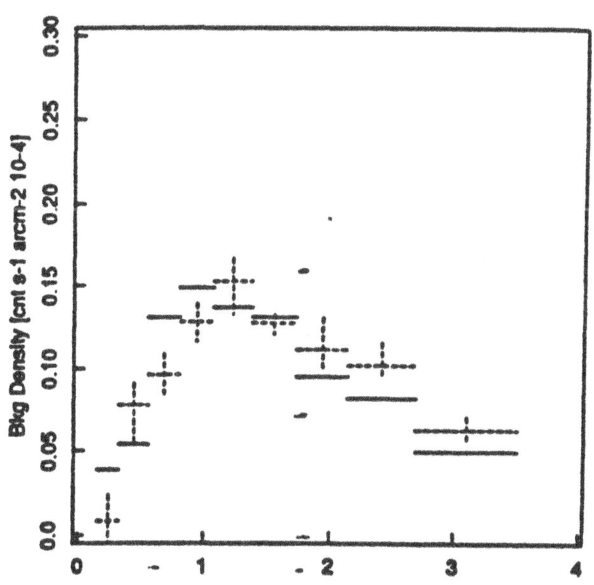

Fig. 2. IPC spectrum of Galactic Ridge (dotted lines) compared with the average spectrum of young stellar clusters in the galactic plane (solid lines); Micela 1991.

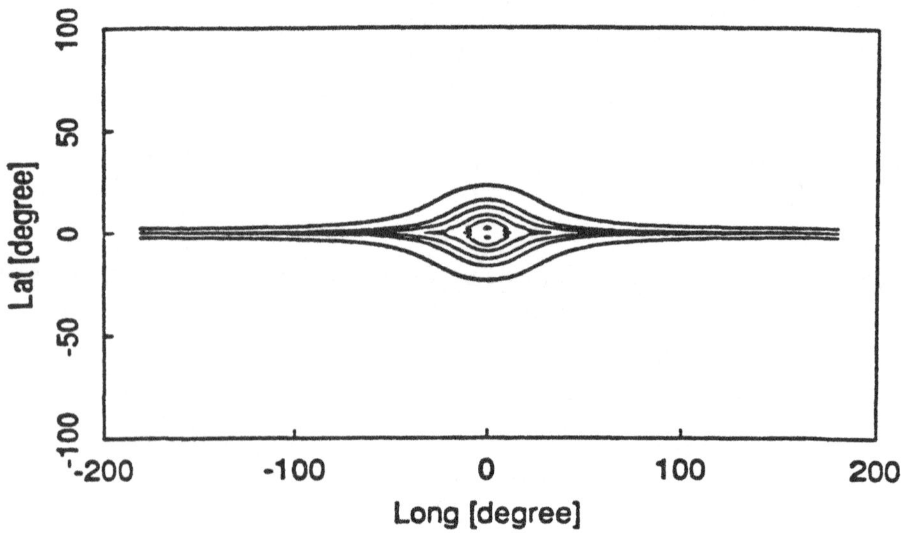

Fig. 3. Contour map of the predicted stellar contribution to the x-ray emission of the Galaxy. Notice the ridge and bulge. The flux values are consistent with the observed diffuse emission (Micela 1991)

the detected point sources (Trinchieri and Fabbiano 1991; Primini et al 1992). This residual emission could be due all or in part to stellar coronae. Given the integrated luminosity of $\sim 4 \times 10^{38}$ergs s^{-1}, $\sim 10^{10}$ stars could be responsible for most of the emission.

The *Einstein* images of M33 show that the emission is dominated by bright point-like sources. Based on their spectra and variability, these sources are most likely accretion binaries (Long et al. 1981; Markert and Rallis 1983; Trinchieri, Fabbiano, and Peres 1988). One of them exhibits a 1.8 day binary period, reminiscent of a massive X-ray binary (Peres et al. 1989). To find any evidence of stellar coronal emission in this galaxy, one can subtract from the images all these individual sources, and then look at the residual diffuse emission. A fraction of this emission must be due to stars. Trinchieri, Fabbiano, and Peres (1988) found that there is an emission component that cannot be resolved in individual sources above the detection threshold of the *Einstein* observation. This emission follows the exponential shape of the optical light, and it can be decomposed into two spectral components: a harder component (kT $>$ 2 keV) with $L_X \sim 2 \times 10^{38}$ergs s^{-1}, and a softer component (kT$<$ 1 keV) with $L_X \sim 0.4 - -3 \times 10^{38}$ ergs s^{-1}. The coronal emission of the estimated stellar content of M33 (see Trinchieri, Fabbiano, and Peres 1988) could amount to a few$\times 10^{37}$ ergs s^{-1} and thus be responsible for a substantial fraction of at least the softer emission component.

4. NGC5204, a blue starburst galaxy

In all the cases discussed above, stellar coronal emission cannot play a dominant role in the X-ray emission detected in the *Einstein* band. But there is an instance in which stellar emission may dominate the X-ray luminosity. This is NGC 5204, a blue star-forming galaxy, dwarf companion to M101. This galaxy was observed both with *Einstein* and with IUE (Fabbiano and Panagia 1983). The *Einstein* observation led to the detection of an $L_X \sim 4.2 \times 10^{39}$ergs s^{-1}, for a distance of 7.2 Mpc. The IUE data could be explained with the integrated emission $\sim 2-20 \times 10^4$ OB supergiants and up to $7-70 \times 10^5$ OB stars.

Since the X-ray luminosity of an OB star is $\sim 10^{30-32}$ergs s^{-1}, the total X-ray luminosity of the stellar component detected with IUE could approximate the detected L_X in the most favorable case.

5. X-ray-faint ellipticals: do we see the emission of M stars?

Although everybody agrees that the X-ray emission of X-ray bright E and S0 galaxies is dominated by the hot ISM, there has been a certain amount of controversy on the nature of the emission of X-ray faint early-type galaxies: do these galaxies retain their ISM, or is their X-ray emission totally dominated by a population of low-mass binaries, as in the bulge of M31? (see Fabbiano 1989 and references therein; Canizares, Fabbiano and Trinchieri 1987; Fabbiano, Gioia, and Trinchieri 1989).

To address this problem, Kim, Fabbiano and Trinchieri (1992) derived the average spectral parameters of elliptical galaxies in bins of different X-ray-to-optical ratio. While this analysis showed that the spectrum tends to harden in X-ray fainter ellipticals, thus confirming the dominance of the X-ray binary emission in the latter, a totally unexpected result is given by the ellipticals with the lowest L_X/L_B ratio (see fig. 4). In these galaxies, the distribution of the spectral counts suggests the presence of a *very soft* component of the X-ray emission. If the emission of these galaxies consists of two components, a hard -binary- component, plus a second component, the *Einstein* data suggest a temperature for the very soft component of 0.16-0.22 keV (at 90% confidence). The luminosity of this component would be $\sim$1/3-1/2 ($\sim 10^{40}$ergs s^{-1}) of the total X-ray emission. These results are confirmed by the *ROSAT* observations of two of these galaxies: NGC 4365 and NGC 4382 (Fabbiano et al. 1992, in preparation). The *ROSAT* PSPC data cannot be fitted with a single emission component, while a two-component thermal model gives an acceptable χ^2. The confidence contours for the two temperatures are shown in fig. 5 for NGC 4365.

What is the origin of this very soft emission? Could we been detecting the integrated coronal emission of M stars in these galaxies? The temperature of the soft component is compatible with this hypothesis (see Figure 11 in Vaiana 1990), however the number of M stars may be a bit on the high side. If a typical M star emits $\sim 10^{27}-10^{28}$ erg s^{-1}, $\sim 10^{12}-10^{13}$ M stars would be needed. Other explainations are possible, including the emission of a cool ISM (Pellegrini et al 1992, in preparation) or ultra-soft binaries, similar to those detected with *ROSAT*

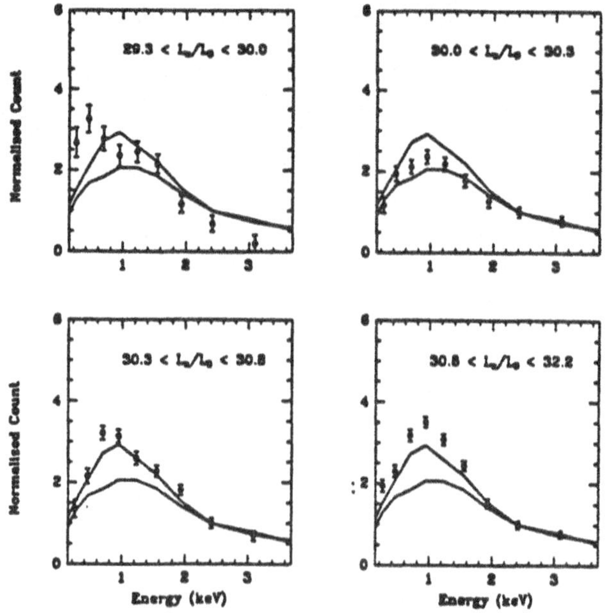

Fig. 4. Average IPC observed spectra of elliptical galaxies in four ranges of L_x/L_B. Note the excess emission in the lowest channel in the group with lowest L_x/L_B (Kim, Fabbiano, and Trinchieri 1992).

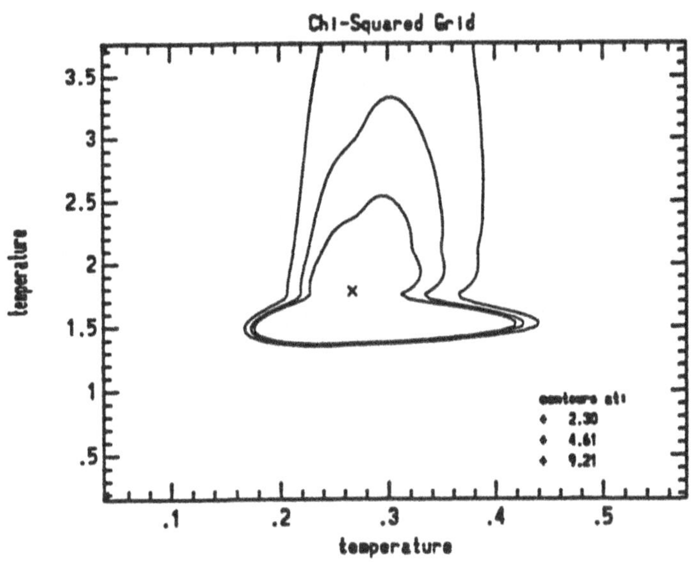

Fig. 5. Confidence contours for a two-temperature fit to the ROSAT PSPC data of NGC 4365 (Fabbiano et al. 1992, in preparation).

in the LMC (Greiner, Hasinger, and Kahabka 1991). If this very soft emission is of stellar origin, then we would expect it to be present in all early-type galaxies, and to scale approximately with the optical luminosity. If this were the case, it would have been undetectable in more luminous ellipticals. The origin of this very soft emission is an open problem, which will require future spectral and spatially resolved spectral observations to solve.

6. Conclusions

Although the X-ray luminosity of galaxies is not dominated by the integrated contribution of stellar coronae, stellar coronae can give a sizeable contribution. There is a tendency in the high-energy non-stellar community to forget about stars, and to try to explain unresolved emission components both in the Galaxy and in external galaxies with other than stellar coronal emission. But the stars are there, and their baseline contribution should always be considered before looking for more exotic phenomena.

This work was partly supported under the NASA ROSAT Guest Observer Program.

References

Canizares, C.R., Fabbiano, G., and Trinchieri, G. 1987, *Ap.J.*, **312**, 503.
Fabbiano, G. 1989, *Ann.Rev.Ast.Ap.*, **27**, 87.
Fabbiano, G., Gioia, I.M., and Trinchieri, G. 1989, *Ap.J.*, **324**, 749.
Fabbiano, G., Kim, D.-W., and Trinchieri, G. 1992, *Ap.J.Suppl.*, **80**, 531.
Fabbiano, G. and Panagia, N. 1983, *Ap.J.*, **266**, 568.
Fabbiano, G., Trinchieri, G., and Van Speybroeck, L.S. 1987, *Ap.J.*, **316**, 127.
Greiner, J., Hasinger, G., and Kahabka, P. 1991, *Ast.Ap.*, **246**, L17.
Kim, D.-W., Fabbiano, G., and Trinchieri, G. 1992a, *Ap.J.Suppl.*, **80**, 645.
Kim, D.-W., Fabbiano, G., and Trinchieri, G. 1992b, *Ap.J.*, **393**, 134.
Linsky, J.L. 1990, in *Imaging X-ray Astronomy*, ed. M. Elvis, (Cambridge: University Press), p.39.
Long, K.S., D'Odorico, S., Charles, P.A., Dopita, M.A. 1981, *Ap.J.(Lett.)*, **246**, L61.
Markert, T.H. and Rallis, A.D. 1983, *Ap.J.*, **275**, 571.
Micela, G. 1991 Ph.D. thesis, Univ. di Palermo (Italy).
Peres, G., Reale, F., Collura, A., and Fabbiano, G. 1989, *Ap.J.*, **336**, 140.
Primini et al. 1992, preprint.
Trinchieri. G. and Fabbiano. G. 1991, *Ap.J.*, **382**, 82.
Trinchieri, G., Fabbiano, G., and Peres, G. 1988, *Ap.J.*, **325**, 531.
Vaiana, G.S. 1990, in *Imaging X-ray Astronomy*, ed. M. Elvis (Cambridge: University Press), p.61.
Van Speybroeck, L., Epstein, A., Forman, W., Giacconi, R., Jones, C., Liller, W., and Smarr, L. 1979, *Ap.J.(Lett.)*, **234**, L45.
Watson, M.G. 1990, in *Windows on Galaxies*, eds. G. Fabbiano, J. Gallagher, and A. Renzini, (Dordrecht: Kluver), p.177.

THE G.S. VAIANA X-RAY ASTRONOMY CALIBRATION AND TESTING (XACT) FACILITY

A. COLLURA

IAIF-CNR Via Archirafi 36, 90100 Palermo, Italy

M. BARBERA, G. INZERILLO, A. MAGGIO, G. MICELA, F. MIRABELLO, S. SCIORTINO and S. SERIO

Istituto ed Osservatorio Astronomico di Palermo, Palazzo dei Normanni, 90134 Palermo, Italy

and

G. PERES

Osservatorio Astrofisico di Catania, Cittá Universitaria, Catania, Italy

Abstract. We describe a calibration and testing facility dedicated to the development of instrumentation for X-ray Astronomy. It is installed at the Osservatorio Astronomico di Palermo. The main component of the facility is a vacuum system consisting of a ~ 16 meter pipe, and of a 1 meter long cylindrical test chamber with a 0.5 meter radius. It is evacuated by means of cryogenic pumps and of an oil-free assembly of mechanical pumps, ensuring an extremely clean and contaminant-free vacuum.

The X-ray generation system includes a X-ray source, designed to privilege efficiency and versatility, a gas flow monitoring proportional counter and a set of filters and anodes that can be interchanged maintaining vacuum condition.

X-ray detection is presently performed by means of two gas flow proportional counters. A gas scintillation proportional counter is under development, while imaging detectors will be added later.

The facility is controlled and operated by a data acquisition and control system including two CPU's and a microprocessor controller, allowing remote running of most experiments.

We envisage to use the facility in the development and calibration of filters and detectors for X-ray Astronomy. It will start operations by the end of 1992. The first project in schedule is the calibration and development of the filters for the High Resolution Camera of the AXAF mission.

Key words: X-rays - Instrumentation

1. Introduction

The Osservatorio Astronomico di Palermo (OAP) is involved in several space astrophysics projects. The most far-reaching projects include:

- The European Space Agency (ESA) X-ray Multi Mirror European Photon Imaging Camera (XMM-EPIC) (Peacock and Ellwood 1988).
- The Wide Field X-ray Telescope (WFXT), a satellite project from an Italian-NASA collaboration.
- The NASA Advanced X-ray Astrophysics Facility (AXAF) mission (Tananbaum 1990) a sophisticated X-ray astronomy mission featuring an advanced X-ray telescope.

In the framework of these collaborations, the OAP has built a X-ray test facility for the development and testing of hardware for space astrophysics missions.

The facility, that will start operating by the end of 1992, has been designed to be easy to operate and versatile; its applications range from the calibration and development of X-ray detectors to measurements of transmissivity and reflectivity

J.F. Linsky and S. Serio (eds.), Physics of Solar and Stellar Coronae, 275–278.

of materials. Its size is intermediate between that of small facilities available in several laboratories and that of big facilities like the 130 meter Panther facility at MPE (Garching, Germany) (Aschenbach et al., 1979, Stephan et al. 1981), or the $\sim$ 600 meter MSFC facility (Huntsville, Al, USA). This, and its efficient design, make it very suitable for a variety of tests.

The first project to be carried on will be the development and testing of filters for the AXAF High Resolution Camera.

In this paper we describe the facility, discuss the technological choices, and present the upgrades planned for the next future.

2. Description

The facility includes, in addition to a number of tools and accessories which are quite standard and will not be described here, a stainless steel (AISI 304 L) vacuum chamber, a set of cryopumps to produce the vacuum in the chamber, a X-ray source with an independent vacuum system based on a magnetic levitation turbo-molecular pump, three gas detectors, and support electronics for data acquisition and control. In the following we will describe the facility in detail.

2.1. The Vacuum Chamber

The vacuum chamber was manufactured by the CINEL of Padua (Italy). It includes a 16 meter pipe connecting the X-ray source head to the test section, and the test section itself.

The pipe consists of 11 tubes, each either 1 or 2 meter long, assembled together to reach the overall length of the pipe. The diameter of the pipe increases from the source end to the test section end, following the divergence of the X-ray beam, from a minimum of 150 *mm* to a maximum of 630 *mm*, thus limiting the volume (and the internal surface of the pipe) to that just necessary to obtain, at the test end, a beam with the same diameter of the end of the pipe. The advantages of such a configuration are obvious and can be summarized in a reduction of the pumping capacity - and/or of the time - necessary to reach the desired level of vacuum, an increase in versatility and operability, a reduction of the weight of the chamber and of the overall realization cost. Each section has, in addition to the ports for pumping and instrumentation, a number of ports for inspection or feedtroughs. Internally, a number of baffles prevent grazing incidence of the radiation on the surface of the pipe. The flanges of the ports and those for the connection of the tubes are Con Flat type for diameters up to 250 *mm*, and ISO type for bigger diameters. The pumping ports are equipped with ISO flanges.

The test section is a cylinder of 1 *m* length and 1 *m* diameter, with several ports for inspection, instruments and feedtroughs, a pumping port and a door in the back having the same diameter of the chamber. The test section can be isolated from the pipe by means of a sliding gate valve with the same diameter of the pipe (630 *mm*). The problem of anchorage inside the test chamber has been solved placing in the bottom of the test section two rows of 38 *mm* Con Flat flanges. Such flanges are closed with blank flanges to which bolts have been welded. The bolts penetrate in the chamber and are the anchorage points, offering the advantage of being removable by simply substituting the flanges with normal blank flanges if the whole diameter of the chamber were needed. The internal surface of the test section has been treated, like that of the pipe, with quartz microspheres. The chamber (pipe plus test section) is rated for a minimum pressure of 2×10^{-7} $mbar$, though normal operation will be in the 10^{-6} $mbar$ scale. The whole chamber is mounted on a stainless steel bakeable modular frame.

2.2. The Vacuum System

The vacuum system is based on cryogenic pumps for the chamber and on a magnetic levitation turbo molecular pump for the X-ray source.

The cryogenic pumps (Edwards Coolstar series), have a total nominal pumping capacity of 3900 l/sec, of which 2400 l/sec are allocated to the pipe, distributed over 4 units, and 1500 l/sec are allocated to the test section. The 5 units are driven by 2 independent compressors. Pre-vacuum is obtained by means of an oil-free rotative dry-pump of advanced technology (Edwards DP80). The vacuum obtained by such a system is extremely clean and with no carbon contamination, therefore particularly suitable for the development of contamination-sensitive devices like microchannel plate detectors.

The magnetic levitation turbo molecular pump (Edwards MAGLEV STP 300 model) allocated to the X-ray source chamber is able to generate a vacuum in the 10^{-8} $mbar$ scale. Its characteristics of cleanness are comparable to those of the cryopumps. Its nominal pumping speed is 300 l/sec. The pre-vacuum system, the only oil-lubricated part, is equipped with a special oil vapor trap located between the pre-vacuum system and the turbo molecular pump so as to ensure, for such a small system, complete blockage of contaminating vapors.

2.3. The X-ray Source System

The principal element of the system is a Manson multi-anode microfocus X-ray source. Electrons produced by a hot filament are accelerated by an electric field and impinge on an anode, producing X-rays. Six different anodes can be selected, without breaking the vacuum in the source chamber. The source also contains selectable filters that can be used to further define the beam spectrum. It can produce X-rays in the 0.1 - 10 keV band, with a spot size of ~ 1 mm. The design of this X-ray source privileges efficiency, which is one of the most interesting features of the X-ray generation system. It produces a flux of $\sim 5 \times 10^5$ $photon\ cm^{-2}\ s^{-1}$ at 16 meters, with only 15 W of power supply. An immediate advantage of such a large efficiency is that no cooling system is required. The X-ray source produces two matched orthogonal beams, one of which can be used for monitoring the flux or, alternatively, to run a different experiment. The quite strong X-ray flux of the source allows experiments of shorter duration and the possibility of using devices that reduce the efficiency of the system, like a monochromator.

2.4. The detectors

Two commercial Gas Flow Proportional Counters produced by LND, with $\sim$ 18% energy resolution are presently available. The gas supply system is directly controlled by the Data Acquisition and Control System (see 2.5) which ensures a constant gas density, checks the existence of the proper operation conditions, and allows remote operation. A built-in-house Gas Scintillation Proportional Counter is in the phase of development and will be described elsewhere.

2.5. The Data Acquisition and Control System

The data acquisition and control system, partly built by SCIENCE.wks and partly provided by Edwards, is based on three CPU's, the first one controlling the low and high vacuum pumping system, the second one controlling the X-ray source, the gas system and the detectors, and the third one dedicated to data acquisition and analysis. It controls the vacuum level and the status of the valves, and allows automatic running of the pumping cycles, like start pumping from the atmosphere, regeneration of the cryopumps, or operation of gate valves. It establishes the high voltage level for the X-ray source and the proportional counter high voltage supplies, selects the gas for each proportional counter, receives temperature and pressure data and operates the Gas Control System to maintain the gas density in the proportional counters. It also monitors the system status to protect the facility and the counters. It collects X-ray spectra, i.e. the integrated number of counts and rate data from the three counters. It is also capable of commanding and logging data from the three counters.

3. Future Upgrades

Some further developments, improvements and extensions are already planned for the immediate future.

A monochromator based on a transmission grating will be installed in the pipe in order to have the possibility of selecting a narrow passband in the X-rays produced by the source. The combination of the monochromator with the energy resolution of the detectors, along with the strong X-ray flux produced by the source, will allow an almost continuous energy band selection in the X-rays used for calibration.

A source of UV radiation will be added to the facility to extend its calibrating possibilities to lower energies. This is particularly useful also for the search and calibration of UV leaks in filters for X-ray detectors.

A micro-channel plate imaging system will be added to the available detectors. It will include a workstation dedicated to image processing.

References

Aschenbach, B., Brauninger, H., Trumper, J., 1979, *SPIE* **184**, 243.

Peacock, A. and Ellwood, J., 1988 *Space Science Reviews* **48**, 343.

Stephan, K.H., Predehl, P., Aschenbach, B., Brauninger, H., Ondrusch, A., 1981, *SPIE* **316**, 203.

Tananbaum, H., 1990, *Imaging X-ray Astronomy* 15, M. Elvis eds., Cambridge University Press.

CORONAL EMISSION AT THE LOW MASS END: RESULTS FROM AN OPTICAL-SELECTED SAMPLE OF NEARBY K AND M STARS SURVEYED WITH THE *EINSTEIN* OBSERVATORY

M. BARBERA, G. MICELA and S. SCIORTINO
Istituto ed Osservatorio Astronomico, Palermo
F.R. HARNDEN JR.
Harvard-Smithsonian Center for Astrophysics
and
R. ROSNER
Department of Astronomy and Astrophysics, University of Chicago

Abstract. We have carried out a statistical analysis of the X-ray emission of K and M dwarfs, starting with an optical volume-limited sample (1651 stars), and using all the available *Einstein* Observatory Imaging Proportional Counter (IPC) data homogeneously reduced with the latest processing. We have measured the $0.16 - 3.5$ *keV* X-ray emission of all 257 stars of the optical sample surveyed with the IPC, and have identified a subsample of them statistically *representative* of the population of nearby K and M dwarfs. On the base of this *representative* sample we have confirmed the occurrence of a drop in X-ray emission level around spectral type M5 ($M_V \sim 13.4$), have constructed *unbiased* maximum likelihood X-ray luminosity functions for K, early M ($M_V <$ 13.4) and late M stars ($M_V > 13.4$), and have confirmed with high significance level ($> 3\sigma$) the decrease in X-ray luminosity with age for K and early M stars and with lower statistical significance ($\sim 1\sigma$) for late M stars.

Key words: Late-type stars – Coronae

1. Sample Selection

1.1. Optical Parent Sample

We have compiled a volume-limited sample of essentially all optically known low mass dwarfs (spectral types K and M and luminosity classes IV, V and VI) within 25 parsecs of the Sun (1681 stars) merging data from three optical catalogues (Gliese 1969; Gliese and Jahreiss 1979; Woolley *et al.* 1970). We have investigated the spatial incompleteness of this sample, as a function of M_V, and have found that K stars may be considered spatially complete within 25 pc from the Sun, M stars earlier than $M_V \sim 13$ are incomplete by a factor 2-3 in the same volume, and later M stars are even 10 times less numerous than expected in the same volume. We have subdivided the entire optical sample in young disk, old disk and halo population stars according to a kinematical method (Eggen 1973a, 1973b, 1973c).

1.2. X-ray Sample

We have extracted an X-ray sample including all the stars of the optical parent sample which fall, either serendipitously or as targets, in at least one of the IPC fields (257 stars). We have computed the X-ray luminosity, in the *Einstein Broad* energy band $0.16 - 3.5$ *keV*, from the final processing of the *Einstein* IPC data (REV-1, Harnden *et al.* 1984). We have subdivided the X-ray sample in targeted

J.F. Linsky and S. Serio (eds.), Physics of Solar and Stellar Coronae, 279–282.

(offaxis < 3 arcminutes), and serendipitous stars (offaxis > 3 arcminutes). With this definition the subsample of serendipitous K stars presented a statistically significative excess of young stars respect to the optical parent sample. Following this hint we have found that 23 out of the 131 stars of the serendipitous X-ray sample were indeed associated to targeted stars and for this reason should be considered targeted too. The excess of young K stars of the new serendipitous subsample is not present anymore and the little discrepancy with respect to the optical parent sample is not statistically significant. After comparing the cumulative distance distribution of stars in the optical and X-ray samples in proper subvolumes and after comparing the maximum likelihood X-ray luminosity functions of different X-ray subsamples, we have identified our best available *representative* X-ray sample of nearby low mass stars. This sample consists of all the K stars within 12 pc from the Sun, all the M stars within 8 pc, and only the serendipitous stars at greater distances (66 K and 127 M stars).

2. Results and Conclusions

2.1. X-ray Luminosity Dependence on Spectral Type

Using the X-ray *representative* sample, we have reconsidered the statistical significance of the drop in X-ray emission level previously suggested for stars later than spectral type M5 (Golub *et al.* 1983; Rosner, Golub, and Vaiana 1985; Bookbinder 1985). The data suggest the occurrence of a steep drop in X-ray luminosity at $M_V \sim 13.4$. In order to evaluate if this drop is statistically significative we have proceeded as follows. We have assumed that the late M stars ($M_V > 13.4$) have the same X-ray luminosity function as the earlier M stars, and, under this hypothesis, have evaluated the expected number of detections of late M stars above three given thresholds of $log(L_X)$, namely: 27.5, 28, and 28.5. We have found that the number of observed detections of late M stars is lower than expected, with high statistical significance. To investigate whether this drop reflects simply the decreasing stellar surface of late M stars, we have repeated our analysis considering X-ray surface flux (F_X) instead of X-ray luminosity (L_X). The drop of F_X at $M_V \sim 13.4$, even though less pronounced than in the case of X-ray luminosity, is still evident (figure 1) and the results of an analogous analysis as for L_X show that also adopting X-ray surface flux the drop in X-ray emission level for $M_V > 13.4$ is statistically significative.

2.2. Maximum Likelihood X-ray Luminosity Functions

Since a single X-ray luminosity function is not adequate to describe the entire sample of M stars, we have subdivided them into: early M ($8.5 < M_V \leq 13.4$) and late M ($M_V > 13.4$) stars. In figure 2 we have plotted the cumulative maximum likelihood X-ray luminosity functions of K, early M and late M. We notice that early M stars present a very high luminosity tail, and thus, being the largest population of stars in the galaxy, they are the principal contributors to the soft X-ray stellar background (Rosner *et al.* 1981). We believe that these X-ray luminosity functions of K, early M and late M stars are the less biased that can be built with available X-ray data, up until ROSAT full sky-survey will improve the status of our knowledge.

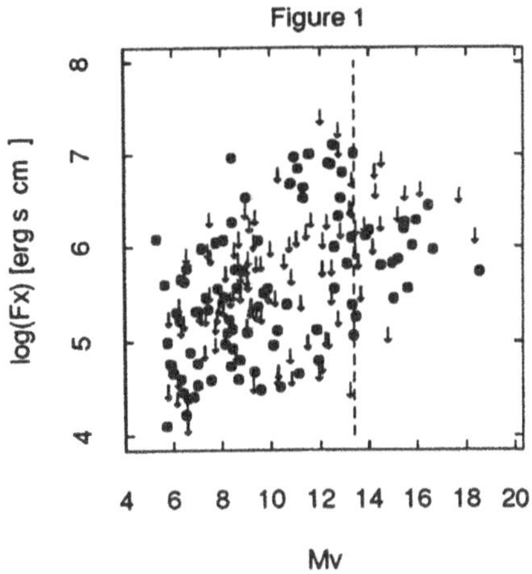

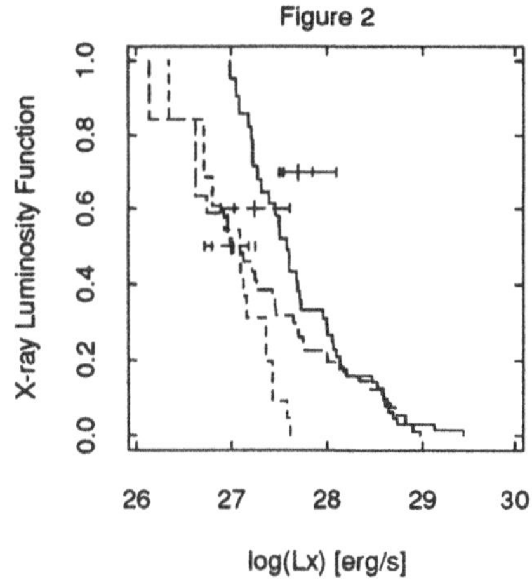

Fig. 1: X-ray surface flux vs. M_v for the *representative* X-ray sample. Dots indicate detections, arrows indicate upper limits, and the vertical line marks $M_v = 13.4$.

Fig. 2: X-ray cumulative luminosity functions for the *representative* samples of K (solid line), early M (long dashed line), and late M stars (short dashed line). The corresponding means are shown with 2 and 3 σ confidence level uncertainties.

2.3. X-RAY LUMINOSITY DEPENDENCE ON AGE

We have built maximum likelihood X-ray cumulative luminosity functions of young and old disk stars of the *representative* X-ray sample separately for K, early M, and late M stars. The derived mean values of $log(L_x)$ with associated 1σ confidence level uncertainties are reported in table 1. The mean of $log(L_x)$ is systematically higher for the young stars, and the Wilcoxon two-sample test we have adopted to compare the X-ray cumulative luminosity functions of young and old stars in each of the three spectral type shows that, both for K and early M stars, the hypothesis that the X-ray luminosity functions of young and old stars are extracted from the same parent population can be rejected with a statistical significance of 3σ (confidence level > 0.998), while for late M stars the statistical significance is only 1σ (confidence level $= 0.799$).

2.4. L_x DEPENDENCE ON L_{bol}

In figure 3 $log(L_x)$ is plotted vs. $log(L_{bol})$ for the *representative* X-ray sample. Squares indicate the flare stars of Pallavicini *et al.* (1990) which are present in our entire X-ray sample. We notice that the good correlation found by Pallavicini *et al.* (1990) in the case of flare stars is not extendable to the entire sample of M stars; indeed flaring stars seem to mark a saturation in the X-ray luminosity level. The oblique solid line is obtained covering the entire stellar surface with regions of $log(F_x) = 7.3$ (typical of active solar regions). This line seems to well reproduce the saturation of X-ray luminosity for K and early M stars; while for late M stars the "actual" detections are about a factor 5 lower. This fact can be interpreted in two

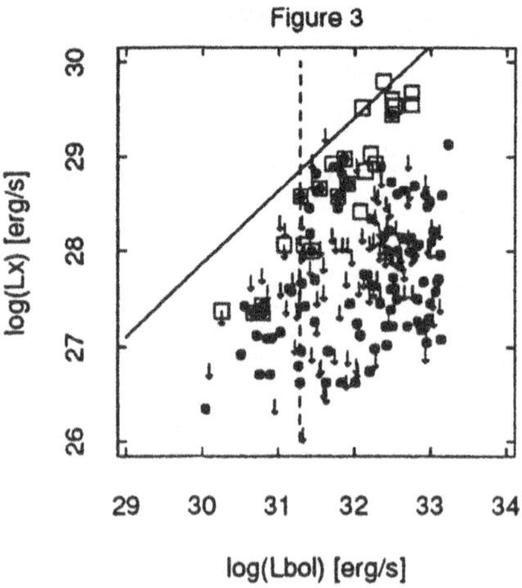

TABLE I

	$< log(L_x) >$ young disk	$< log(L_x) >$ old disk
K	28.0 ± 0.1	27.5 ± 0.1
early M	27.7 ± 0.1	27.0 ± 0.2
late M	27.3 ± 0.1	26.9 ± 0.2

Fig. 3: L_x vs. L_{bol} for the *representative* X-ray sample. Dots indicate detections, arrows indicate upper limits, the vertical dashed line marks $M_v = 13.4$, and the solid oblique line is obtained covering the entire stellar surface with regions of $log(F_x) = 7.3$. Squares indicate flare stars of our entire X-ray sample observed with EXOSAT (Pallavicini *et al.* 1990).

Tab. 1: Mean $log(L_x)$ of young and old disk stars for the K, early M, and late M stars, respectively.

simple ways: the active regions of late M stars have lower X-ray flux ($log(F_x) \sim 6.6$) or the maximum filling factor for late M stars is about 1/5 of that of the K and early M stars. In any case the coronal activity level of late M stars is significantly lower than that observed in K and early M stars.

Acknowledgements

We acknowledge support by Agenzia Spaziale Italiana and Ministero della Università e della Ricerca Scientifica e Tecnologica, (MB, GM, SS), IAIF-CNR (MB), NASA (FRH, RR).

References

Bookbinder, J.A.: 1985, *Ph.D. Thesis*, Harvard University
Eggen, O.J.: 1973a, *P.A.S.P.* **85**, 289
Eggen, O.J.: 1973b, *P.A.S.P.* **85**, 379
Eggen, O.J.: 1973c, *P.A.S.P.* **85**, 542
Gliese, W.: 1969, *Veroffentl. Astron., Rechen-Institut: Heidelberg* **No. 22**,
Gliese, W., Jahreiss, H.: 1979, *Astr. Astrophys. Suppl.* **38**, 423
Golub, L.: 1983, in Activity in Red-Dwarf Stars, ed(s)., *P.B. Byrne & M. Rodonò*, Reidel Publishing Co., 83
Harnden, F.R., Jr., Fabricant, D.G., Harris, D.E., Schwarz,J.: 1984, *SAO. Spec. Rept.* **393**,
Pallavicini, R., Tagliaferri, G., Stella, L.: 1990, *Astron. Astrophys.* **282**, 403
Rosner, R. *et al.*: 1981, *Ap. J. Letters* **249**, L5
Rosner, R., Golub, L., Vaiana, G.S.: 1985, *Ann. Rev. Astron. Astrophys.* **23**, 413
Woolley, R.E.E.A., Penstom, M.J. & Pocock, S.B.: 1970, *Royal Observ. Ann., No. 5* **281**, 815

THE X-RAY EMISSION OF T TAURI STARS IN TAURUS-AURIGA AS SEEN FROM THE EINSTEIN OBSERVATORY IPC *

F. DAMIANI and G. MICELA

Istituto ed Osservatorio Astronomico di Palermo

Abstract. We present a systematic study of the X-ray emission of pre-main-sequence stars in the Taurus-Auriga cluster. The sample consists of the 69 catalogued stars in this region observed with the *Einstein* Observatory, 53 of them have been identified with X-ray sources. The whole sample of the X-ray data, as given by the final processing, have been uniformly analyzed, and have been studied using X-ray luminosity functions, which take into account also non-detections. Our stars follow the trend of decreasing X-ray emission with age, well known for young late-type stars. Available data suggest that the average L_x of stars with companions is less than that of single stars; we discuss this in connection with the dependence of the coronal X-ray emission on rotation. Last, for strong emission-line stars, the (wind) Hα emission is inversely related to L_x, suggesting inhibition of the wind in the stars with bright coronae; in the weak-line stars, instead, a positive correlation of L_x with the emission lines supports their chromospheric origin.

Key words: Stars: coronae – Stars: pre-main-sequence – X-rays: stars

1. Introduction

In the early 80's many papers have appeared concerning X-ray emission of classical T Tauri stars (CTTS) in the Taurus-Auriga star-forming region, observed with the IPC detector (0.16-3.5 keV) of the *Einstein* Observatory; a large fraction of these stars turned out to be strong X-ray sources, as compared to more evolved main sequence stars, with X-ray luminosities around $L_x = 10^{30}$ ergs/sec or larger (e.g. Walter and Kuhi 1984 and references cited therein).

Later, intense X-ray emission proved to be an useful diagnostics to discover new PMS stars with little optical emission-line activity: these were called weak-line T Tauri stars (WTTS) (e.g. Feigelson *et al.* 1987, Walter *et al.* 1988).

Most of these X-ray sources have a low S/N, and others show a non-negligible absorption, so the correct determination of the observed flux depends on the accuracy of the data reduction. Therefore, a systematic reanalysis of the X-ray emission of both CTTS and WTTS, taken together, seems worthwhile.

2. Observations and Data Analysis

The optical parent sample we used is composed by all stars listed by Herbig and Bell (1988) in the Taurus-Auriga region. These are 161 stars, but many more should be present, especially in the latest PMS stages (Herbig and Bell 1988). The X-ray observed subsample consists of 69 stars, falling into 34 IPC fields; 53 stars are identified with X-ray sources, and we computed upper limits for the 16 undetected stars.

The X-ray data of the whole sample have been homogeneously analyzed with the final data processing (Harnden *et al.* 1984). Although the parent sample itself is incomplete, we tried to follow a statistical approach, using maximum-likelihood

* This work was inspired by the late G.S. Vaiana, who has taken part to its early development. Without his contribution this paper would not have been in the present form.

J.F. Linsky and S. Serio (eds.), Physics of Solar and Stellar Coronae, 283–286.

X-ray luminosity functions (Feigelson and Nelson 1985), which take into account the upper limits to X-ray emission of undetected stars.

We have critically examined the distribution of L_X we find. The presence of X-ray selected stars in the parent sample biases the resulting luminosity function towards high values of L_X, as also does the IPC detection threshold (about $L_X \sim 10^{30}$ ergs/sec in our case). Therefore, to evaluate reliably the median L_X, we have first excluded all X-ray selected stars; we have then account for the number of undetected stars, and the number of X-ray unresolved T Tauri pairs, which make uncertain the true number of X-ray emitters in the sample.

In sum, we detect at least 15, and at most 21, out of 37 CTTS in the observed sample. Therefore, the observed median $\log L_X = 29.7$ ergs/sec should be close to the "true" value, but some chance remains that this is an overestimate.

3. Results

3.1. Age Dependence of X-ray Activity

The decrease of the X-ray activity with the stellar age for late-type stars is a known result, for stars of the Pleiades age and older. To check if this trend continues back to the T Tauri age we compare (fig.1) the Tau-Aur median X-ray luminosity with that of the Pleiades (Micela *et al.* 1990), of the Hyades (Micela *et al.* 1988), and the young and old disk stars (Barbera *et al.* 1992).

A similar analysis has been made by Feigelson and Kriss (1989) for the Chamaeleon PMS stars; our median L_X compares well with their data, and strengthens the extension of the known age dependence of X-ray activity to the PMS phase.

3.2. The Effect of Binarity

In a number of cases pairs of stars are found so close to be unresolved by the IPC. Were these bound binary systems or not, these stars likely formed together, so a comparison of "close pairs" with more isolated stars easily reflects the effect of different evolutionary histories.

To study the effect of binarity on X-ray emission, we compare the distribution of (unsplitted) L_X for the pairs with the *convolution* of the L_X-distribution of single stars with itself (fig.2); the latter, if X-ray emission is not affected by binarity, should be identical to the L_X distribution of the pairs. Both these luminosity functions include the L_X upper limits. As a result, the convolved luminosity function of the single stars turns out to be *higher* than the other; the two distributions are different at the 99.4% level according to a Wilcoxon test.

We conclude that if our maximum-likelihood estimate is similar to the "true" luminosity function, binarity is a quenching factor for the X-ray emission of these stars. This effect, if real, might be related to the very early rotational evolution of these systems. In fact, the L_X is strongly dependent on rotation for these stars (Bouvier 1990, Damiani, Micela and Vaiana 1991, 1992).

Storing most of the initial angular momentum in the orbital motion of a multiple system may help to solve the angular momentum problem of star formation

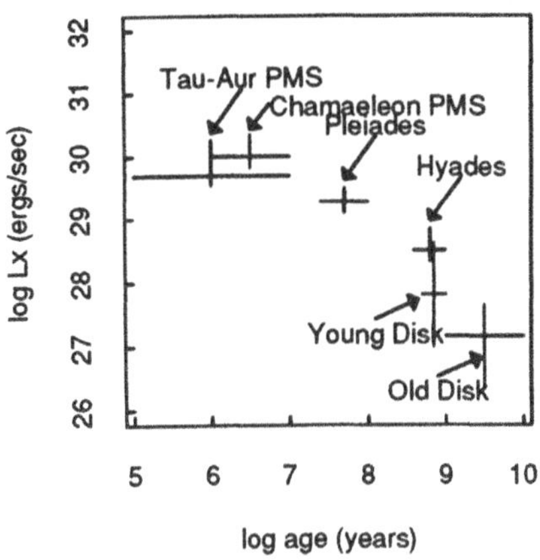

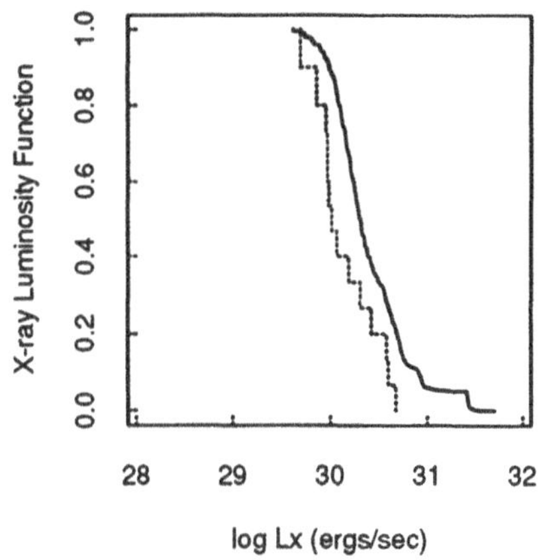

Fig. 1. (Left) Median $\log L_x$ vs. age. Vertical bars show the width ($\pm 1\sigma$) of the L_x distributions.
Fig. 2. (Right) X-ray luminosity function for the unresolved pairs (dotted), and convolution of the single-stars luminosity function with itself (solid).

process (Bodenheimer 1991). It is then possible that single and multiple stars have a different rotational evolution, which may result into their different X-ray behavior.

3.3. X-ray Activity and Hα Emission

a) Classical T Tauri Stars

A strong Hα emission line has been the most useful diagnostics to identify classical T Tauri stars; the line profiles are broad and indicate complex outflows or winds. Given the correlation found between the luminosity in the lines and that in the IR continuum, some physical connection between disk and wind is suggested (Cabrit *et al.* 1990). We note however that some scatter is present in this correlation: thus, for a disk of given accretion rate the efficiency of wind generation is variable. We succeeded in finding a sort of relation between this "wind production efficiency" (as given by the ratio $L_{H\alpha}/F_{25\mu}$) and L_x for these stars; in fact, stars having an efficiency above and below the median have X-ray luminosity functions different above the 3σ level according to a Wilcoxon test (fig. 3).

We interpret this finding as evidence of the competition occurring on stellar surface between different magnetic topologies; namely, while a solar-like X-ray corona, for which there is evidence (e.g. Damiani, Micela and Vaiana 1991, 1992), requires closed magnetic fields to confine and heat the plasma, a stellar wind requires surface regions with open-field topology, to let matter escape. So, the observed anticorrelation between L_x and wind production efficiency may indicate that in some stars closed-field regions dominate, and the wind is inhibited, while in others the opposite situation prevails.

b) Weak-line T Tauri Stars

In the WTTS, instead, the Hα emission is thought to be of chromospheric origin (Walter *et al.* 1988); our data support this hypothesis, as we find a positive corre-

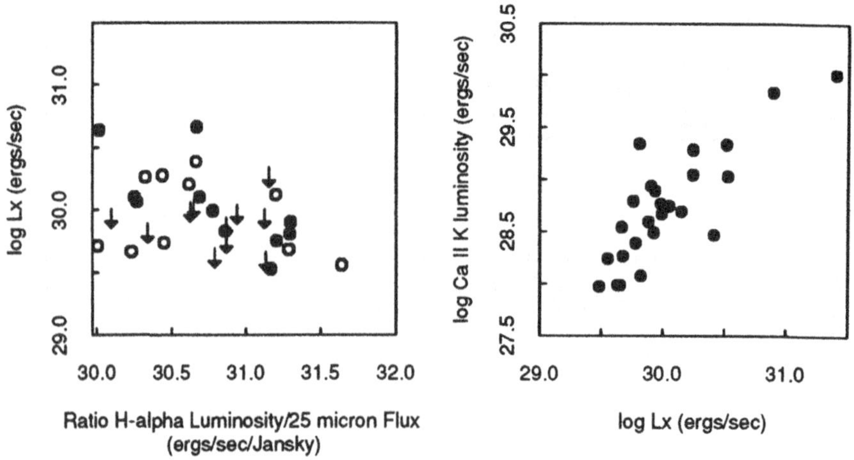

Fig. 3. (Left) L_x vs. ratio $L_{H\alpha}/F_{25\mu}$ (wind production efficiency). Filled dots are isolated stars. Empty symbols are stars in pairs, unresolved by both the IPC and IRAS. Arrows are X-ray upper limits.

Fig. 4. (Right) The Ca II K line luminosity vs. L_x. A much similar correlation is present using the Hα line luminosity, in the case of WTTS alone, as the luminosities of these lines are all correlated.

lation between the L_x and line luminosities for these stars (considering the excess line emission with respect to standard stars). The same relation is found between X-ray- and Ca II K luminosity, another typical indicator of chromospheric activity (fig. 4).

References

Barbera,M., Micela,G., Sciortino,S., Vaiana,G.S., Harnden,F.R.,Jr., Rosner,R. 1992, Mem. SAIt, in press.

Bodenheimer,P. 1991, in *Angular Momentum Evolution of Young Stars*, S.Catalano and J.Stauffer eds., p. 1.

Bouvier,J. 1990, *Astron. J.*, **99**, 946.

Cabrit,S., Edwards,S., Strom,S.E., Strom,K.M. 1990, *Ap. J.*, **354**, 687.

Damiani,F., Micela,G., Vaiana,G.S. 1991, in *Angular Momentum Evolution of Young Stars*, S.Catalano and J.Stauffer eds., p. 89.

Damiani,F., Micela,G., Vaiana,G.S. 1992, in preparation.

Feigelson,E.D., Nelson,P.I. 1985, *Ap. J.*, **293**, 192.

Feigelson,E.D., Jackson,J.M., Mathieu,R.D., Myers,P.C., Walter,F.M. 1987, *Astron. J.*, **94**, 1251.

Feigelson,E.D., Kriss,G.A. 1989, *Ap. J.*, **338**, 262.

Harnden,F.R., Jr., Fabricant,D.G., Harris,D.E., Schwartz,J.1984, *Smithsonian Ap. Obs. Spec. Rept.*, No. 393.

Herbig,G.H., Robbin-Bell,K. 1988, *Third Catalog of Emission-Line Stars of the Orion Population*, Lick Obs. Bull. n. 1111.

Micela,G., Sciortino,S., Vaiana,G.S., Schmitt,J.H.M.M., Stern,R.A., Harnden,F.R., Rosner,R. 1988, *Ap. J.*, **325**, 798.

Micela,G., Sciortino,S., Vaiana,G.S., Harnden,F.R., Rosner,R., Schmitt,J.H.M.M. 1990, *Ap. J.*, **348**, 557.

Walter,F.M., Kuhi,L.V. 1984, *Ap. J.*, **284**, 194.

Walter,F.M., Brown,A., Mathieu,R.D., Myers,P.C., Vrba,F.J. 1988, *Astron. J.*, **96**, 297.

AN ANALYSIS OF THE STELLAR CONTENT OF THE EINSTEIN EXTENDED MEDIUM SENSITIVITY SURVEY USING AGE-HOMOGENEOUS X-RAY LUMINOSITY FUNCTIONS

F. FAVATA

European Space Agency, Astrophysics Division, P.O. Box 299, 2200 AG Noordwijk, The Netherlands

and

G. MICELA and S. SCIORTINO

Istituto e Osservatorio Astronomico, Palazzo dei Normanni, 90134 Palermo, Italy

Abstract. We present an analysis of the stellar component of the *Einstein* Extended Medium Sensitivity Survey (EMSS) in which we compare the observed source counts in the EMSS with a model based on age-homogeneous X-ray luminosity functions, computed from samples of stars in the solar neighborhood and in nearby open clusters. We show that the variation of X-ray luminosity with age is a key parameter in determining the number of coronal sources in an X-ray based survey. Additionally, we discuss the importance of properly accounting for the presence of unresolved binaries.

Key words: Stars: coronae - X-rays: stars - Galaxy: stellar content

1. Introduction

Star counts are a classical tool for the study of the structure of the stellar component of the Galaxy, and they are the observation against which Galaxy models are tested (Bahcall and Soneira 1980, Bahcall 1986, Robin and Crezé 1986). X-ray based surveys of serendipitous sources are the conceptual equivalent of star counts in the X-ray domain. Therefore, serendipitous X-ray surveys can be used to test whether our (local) knowledge on stellar X-ray emission (such as volume-limited X-ray luminosity functions, XLF) is in agreement with the observed characteristics of global galactic stellar X-ray emission. We have developed a numerical model (XCOUNT, Favata *et al.* 1992) to compute the expected number of stellar coronal X-ray sources at a given limiting flux in a given sky direction. The model is based on a modified version of the Bahcall and Soneira star distribution, on the Lockman model of neutral H distribution (Lockman 1984), and on X-ray luminosity functions evaluated for stellar samples subdivided into homogeneous age groups (see Vaiana *et al.* 1992 and references therein). Given the strong variation of X-ray emission with stellar age, this is necessary to properly account for the contribution of both young (scarce, but X-ray bright) and old (abundant, but X-ray faint) stars. Finally, the effect of stellar multiplicity must be properly taken into account.

2. Why is using age-homogeneous XLFs important?

Einstein observations of stars has shown that the X-ray luminosity of stellar coronal sources decreases strongly with age (Micela *et al.* 1988, 1990). Therefore in this present work we have used XLFs computed on age-homogeneous sample of stars, extracted from volume limited samples and from observations of open clusters of known age (Hyades and Pleiades). The reader is referred to Vaiana *et al.* (1992) and references therein for more information on the XLFs.

J.F. Linsky and S. Serio (eds.), Physics of Solar and Stellar Coronae, 287–290.

TABLE I

age	G stars	K stars	M blue	M red
$10^7 - 10^8$ yr	22	13	24	-
$10^8 - 10^9$ yr	16	10	25	-
$> 10^9$ yr	11	2	5	< 1
Total	49	25	54	< 1
"average" XLF	13	11	48	< 1

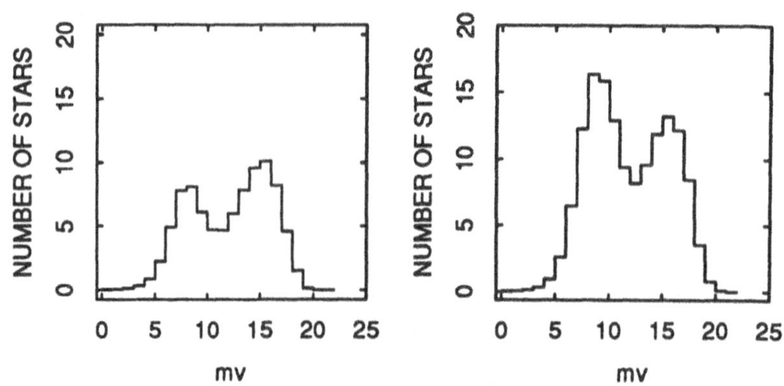

Fig. 1. Predicted EMSS source distributions in m_v, computed with age-homogeneous XLFs (left) and with age-independent XLFs (right)

Given the strong variation of X-ray luminosity with age and the difference in spatial distribution between old and young stars, the usage of an "average" XLF in computing expected source counts or fluxes (such as in computations of stellar contributions to the soft X-ray background) will lead to an under-estimate of the expected number of sources. We have computed the expected star counts by using "average" XLFs, and compared them with the ones computed using age-homogeneous XLFs assuming a constant stellar birthrate, as shown in Table 1. As it can be seen, the usage of average XLFs leads to an under-estimate of the total number of sources.

3. The effect of multiplicity

Most optical Galaxy models will predict only the number density of stars, regardless of their being part of binary system. Multiplicity, therefore, has to be considered *a posteriori*. In the optical domain the change in the characteristics of the unresolved binaries can be small: a G plus M binary will appear as a slightly redder and brighter G star. In the X-ray domain the changes can be dramatic: for example, given that M stars can be as X-ray luminous as G stars (or more), the same G plus M binary can appear twice (or more) over-luminous in the X-ray.

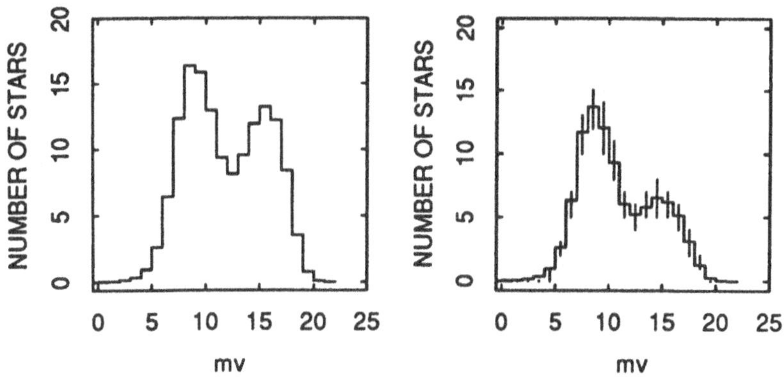

Fig. 2. The expected m_v distribution for EMSS coronal sources generated by the XCOUNT code (left), and the same m_v distribution, after the bootstrap process that takes into account the expected frequency of binaries. (right). The error bars show the standard deviation of the bootstrapped distributions

To take into account the effect of multiplicity in our computation, we have performed a bootstrap operation on the distributions computed with XCOUNT. Assuming a standard fraction of stars in binary systems of 50% (Mihalas and Binney 1981), the appropriate number of sources are extracted at random from the computed distributions (in m_v, B-V and f_x/f_v) to form binaries, whose magnitude, color and X-ray flux are computed appropriately. This operation is repeated several times, generating a large number of possible distributions. The average distribution is computed, together with the formal standard deviation on each bin.

4. Comparison between theory and observation

As it can be seen from the previous graphs, the agreement between the XCOUNT predictions and the EMSS observations is, for the range in apparent magnitudes brighter than 9 and fainter than 14, satisfactory. The brighter magnitude range corresponds roughly to main sequence stars of spectral types A and early F, while the fainter magnitude range contains almost exclusively M stars.

The agreement for the M stars can only be obtained by the proper usage of age dependent luminosity functions and assuming a constant stellar birth rate. If only an *average*, age independent, luminosity function is used, the observation and the expectations disagree by roughly a factor of two. The comparison between computed distributions and EMSS observations shows an excess of observed sources at apparent magnitudes comprised between 9 and 13. This same excess can be seen in the B-V distributions around a color index value of 1.2. These values of apparent magnitudes and color correspond to G and K main sequence stars.

We had shown such an excess to be present in the *Einstein* Medium Sensitivity Survey data (Favata *et al.* 1988), using a less sophisticated numerical model and relying on a much smaller source sample and therefore on smaller statistics. We show that the excess is confirmed on the larger scale of the EMSS and using the more sophisticated XCOUNT model. Note that the EMSS subsample against which we are comparing the predictions has been cleaned from all the known active binary sources, and should therefore only contain *normal* coronal sources.

We believe the excess is to be interpreted in terms of the existence of a source population with optical characteristics similar to the ones of normal main sequence stars, characterized by a low spatial density and high X-ray luminosity. Such a population would easily have been missed in the construction of the locally determined X-ray luminosity functions. The 25 pc solar neighborhood contains about 6.5×10^4 pc^3, therefore a population with a spatial density somewhat lower than 10^{-4} pc^{-3} would easily go undetected.

5. What is the excess yellow population?

Considering that the EMSS subsample considered here has been studied in detail for multiplicity by Fleming (1988), whose careful work should have detected most (if not all) of the active binaries in the sample, we think that few, if any, of the *excess* sources should be active binaries. To understand what their nature might be we have started a program of optical spectroscopy of yellow EMSS stars (Favata et al., 1993). Referring to the observations of the Li I 6707 Å line described in the above paper, it can be seen that an high fraction of the sources show a Li abundance much higher than what would be expected for a random sample of field stars. In particular, some sources show very high Li abundance, comparable to the values commonly observed in pre-main sequence objects (such as the so called naked or weak-lined T-Tauri stars, WTTS). Given that it is well known that X-ray selection is one of the most efficient methods of selecting WTTS objects (Walter *et al.* 1988), we deem it likely that the yellow EMSS excess is composed, at least in part, of pre-main sequence objects. In summary, given the characteristics of the source sample, and on the basis of ongoing optical observations, we consider it likely that a large fraction of these excess yellow sources is composed of very young, active stars, part of which perhaps still in their pre-main sequence stage.

Acknowledgements

Two of us (G.M, S.S) acknowledge support from Ministero della Università e della Ricerca Scientifica e Tecnologica, from ASI, and from GNA-CNR.

References

Bahcall, J.N. and Soneira, R.: 1980, *ApJ Supp.* **44**, 73
Bahcall, J.N.: 1986, *ARA&A* **24**, 577
Favata, F., Rosner, R., Sciortino, S. and Vaiana, G.S.: 1988, *ApJ* **324**, 1010
Favata, F., Micela, G., Sciortino, S. and Vaiana, G.S.: 1992, *A&A* **256**, 86
Favata, F., Barbera, M., Micela, G., Sciortino, S.: 1993, this volume
Fleming, T.H. : 1988, *Ph.D. Thesis*, Univ. of Arizona:Tucson
Lockman, F.J. : 1984, *ApJ* **283**, 90
Micela, G. et al: 1988, *ApJ* **325**, 798
Micela, G. et al: 1990, *ApJ* **348**, 557
Mihalas, D. and Binney, J.: 1981, *Galactic Astronomy*, W. H. Freeman: San Francisco
Robin, A. and Crezé, M. : 1986, *A&A* **157**, 71
Vaiana, G.S. Maggio, A., Micela, G., Sciortino, S.: 1992, *Mem. SAIt*, in press
Walter, F.M. *et al.*: 1988, *AJ* **96**, 297

X-RAY AND UV EMISSION FROM POST-MAIN-SEQUENCE STARS: THE CONNECTION BETWEEN SURFACE ACTIVITY AND EVOLUTION

A. MAGGIO and S. SCIORTINO
Istituto e Osservatorio Astronomico di Palermo
Palazzo dei Normanni, 90134 Palermo, Italy

L. BIANCHI
Osservatorio Astronomico di Torino
10025 Pino Torinese, Italy

F.R. HARNDEN, JR.
Harvard-Smithsonian Center for Astrophysics
60 Garnden Str., Cambridge, MA 02138

and

R. ROSNER
Department of Astronomy and Astrophysics
University of Chicago, Chicago, IL 60637

Abstract. We have studied a sample of 51 nearby stars with estimated masses in the range 1.1 - 1.9 $M_{\odot}$, to monitor the coronal and transition region emission level, during the evolutionary phases from the main sequence to the base of the red giant branch. The available data suggest a mass-dependent variation of the X-ray (0.16-4 keV) luminosity and the C IV (1550) line luminosity, linked to changes of the stellar internal structure, and rotational regime. We discuss a possible interpretation in terms of the changing efficiency of a stellar magnetic dynamo, and the relative importance of mechanical and magnetic heating of the stellar outer atmospheres.

Key words: stars: activity – stars: evolution – stars: X-rays – UV lines

1. Data Sample and Analysis

We have selected a sample of 51 stars (Fig. 1) within a distance of 50 pc, lying mostly along evolutionary tracks between 1.1 and 1.7 solar masses, which have been observed in soft X-rays with the *Einstein* Observatory, and in UV with IUE (Maggio *et al.* 1990; Haisch *et al.* 1990). Two ROSAT targets, and four new IUE observations are also included.

We find that the radiative emission from the outer atmospheres of stars with $M > 1.6 M_{\odot}$ seems to behave differently than for stars with lower mass.

On the main sequence, the X-ray luminosity of most stars with B-V < 0.42 (spectral type F3) is relatively low, at $L_x \sim 3 \times 10^{28}$ $erg\ s^{-1}$. In the early evolutionary phases beyond the main sequence, the X-ray luminosity of the higher mass stars tend to increase sistematically up to $\sim 10^{30}$ $erg\ s^{-1}$, while the lower mass stars show an initial moderate increase followed by a drop, at B-V $\sim$ 0.6, below our sensitivity threshold (Fig. 2a).

A mass-dependent behavior is shown also by the C IV (1550 Å) emission line strength, a transition region diagnostic (Fig 2b). For $M > 1.6\ M_{\odot}$ stars near the main sequence, the C IV luminosity is $L_{CIV} \sim 10^{29}$ $erg\ s^{-1}$, corresponding to a surface flux of $few \times 10^5$ $erg\ cm^{-2}\ s^{-1}$. While evolution proceeds, both the

J.F. Linsky and S. Serio (eds.), Physics of Solar and Stellar Coronae, 291–294.

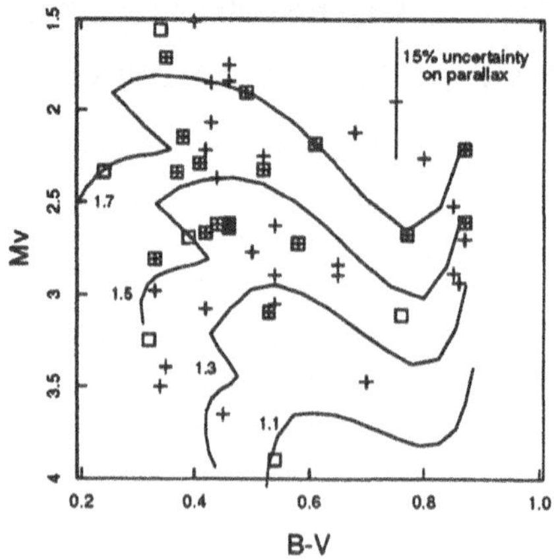

Fig. 1. H-R diagram of the stars in the selected sample. Only stars within a distance of 50 pc, and B-V color less than 0.9 have been included. Squares indicate stars observed in X-rays, while plus symbols are used for IUE observations. The theoretical evolutionary tracks are from Vandenberg (1985), and refer to stars with masses in the range 1.1 - 1.7 $M_{\odot}$. These tracks have been used to estimate the star mass by interpolation. Given the uncertainty on the parallaxes, this estimate may be in error by about $\pm\ 0.2 M_{\odot}$.

luminosity and the surface flux appear to decrease steadly, reaching our detection limit near the base of the red giant branch. On the other hand, for the $M\ \leq\ 1.6\ M_{\odot}$ stars, the C IV emission shows a behavior similar to what found in X-rays, although with a fairly large scatter in the data.

2. Discussion

To understand the above observational picture, at least qualitatively, our working hypothesis is the following: On the main sequence, the earlier, more massive stars in our sample have too shallow convection zones for a magnetic dynamo mechanism to be efficient. The chromospheres and even the transition regions of these stars are dominated by mechanical heating, but not the coronae (see also Simon and Drake 1989; Simon and Landsman 1991). This explains the relatively high UV emission with respect to the X-ray emission. These stars retain an high angular momentum during the main sequence lifetime, since they are not affected by magnetic braking. While evolving, these stars develop a convection zone which triggers the onset of an efficient magnetic dynamo, which in turn provides the required coronal heating. At the same time, the mechanical heating is theoretically expected to decline (Narain and Ulmschneider 1990). On the other hand, for the lower mass stars the magnetic dynamo is already at work on the main sequence, and their rotational velocity is known to decrease with age, together with their activity level (Pallavicini *et al.* 1981; Maggio *et al.* 1987). Upon leaving the main sequence, the depth of the convection zone start to increase, but the rotation ıate keeps declining, and eventually drops below some threshold value where the dynamo ceases to be effective.

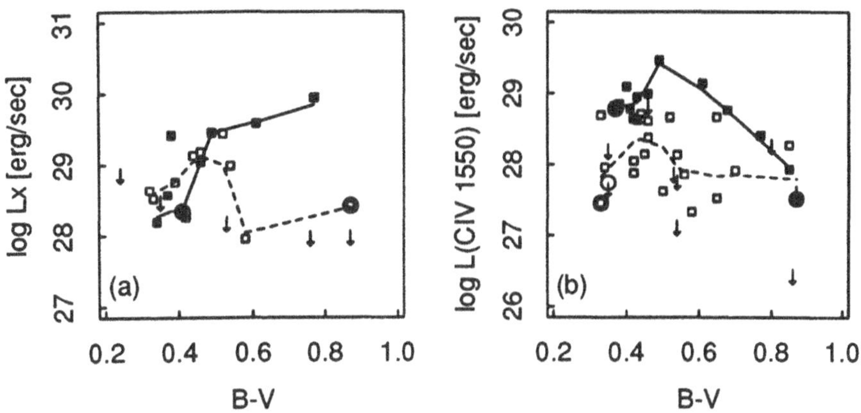

Fig. 2. Scatter plots of X-ray and C IV (1550) luminosity vs. B-V color. Squares are used for detections, and arrows for upper limits; filled symbols identify $M > 1.6\ M_{\odot}$ stars; circles indicate the new ROSAT observations. The lines are the result of a robust locally weighted regression, which takes into account both detections and upper limits, applied separately to the high-mass (solid line) and low-mass (dashed line) subsamples. We remark that, given the uncertainty in the estimated star masses, some of the stars near the chosen mass boundary may have been misassigned.

To test this hypothesis we have searched for correlations between the activity level, and dynamo-related parameters, such as the rotational velocity and the Rossby number (Durney, Mihalas and Robinson 1981). The latter is equal to the ratio P_{rot}/τ_c, where the rotation period P_{rot} has been computed from measured values of the surface velocity and stellar radii obtained using the Barnes-Evans relation, while the convective turnover time τ_c has been properly estimated from theoretical evolutionary models (Rucinsky and Vandenberg 1986).

We find that the X-ray luminosity is not correlated with the rotation rate for the whole sample (Fig. 3a). A significant correlation is found excluding the $M > 1.6\ M_{\odot}$ stars on the main sequence, the $M \leq 1.6\ M_{\odot}$ more evolved stars with $L_x < 10^{28.5}\ erg\ s^{-1}$, and also the most intense X-ray source (the reddest high-mass star in Fig.2a). According to our hypothesis, the first two classes do not correlate because the dynamo mechanism is not effective, while for the latter object the influence of the increased convection zone depth on the efficency of the $\alpha - \omega$ dynamo cannot be neglected. In fact, a significant correlation is found between the X-ray luminosity and the Rossby number, by excluding only the sources with the lowest X-ray emission level (Fig. 3b).

The above conclusions could be certainly strengthened if a larger star sample would be available. Moreover, it would be very useful to have computations of the convective turnover times tested by comparing results from different evolutionary models. Finally, in our analysis we have not taken into account the evolution of the rotational regime in the star interior (Endal and Sofia 1979; Rutten and Pylyser 1988, Pinsonneault *et al.* 1989), and the radial gradient of the rotation rate which is theoretically expected to play an important role in $\alpha - \omega$ dynamo models (Durney

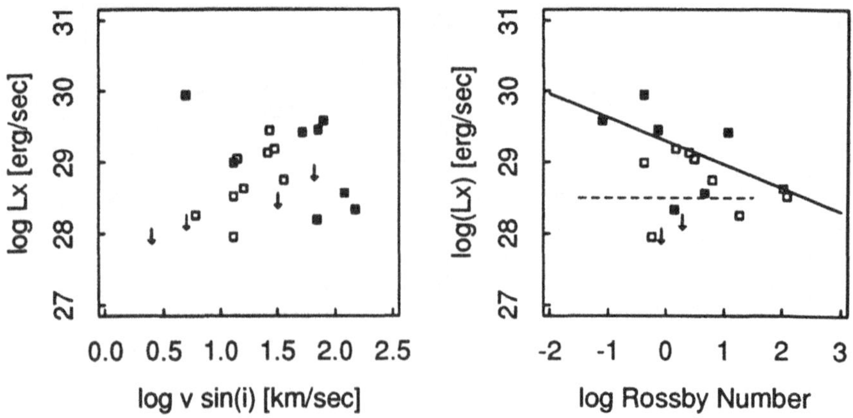

Fig. 3. (a) X-ray luminosity vs. projected rotational velocity. The symbols are as in Fig. 2. Note the existence of fast rotators with low X-ray luminosity, indentified with M > 1.6 $M_\odot$ stars near the main-sequence. (b) X-ray luminosity vs. Rossby number. The best fit power law for all stars with $\log(L_x) > 28.5\ erg\ s^{-1}$ (those below the dashed segment have been excluded), irrespective of their mass or color, yields a slope of $-0.34^{+0.06}_{-0.06}$ with a correlation coefficient of $-0.74^{+0.10}_{-0.16}$. We judge the fit quality reasonably good, given the uncertainties in the evaluation of the Rossby number.

et al. 1982). Nonetheless, our study suggests that evolution exercises its influence on stellar activity by acting on the stellar rotation and convection. In particular, the subphotosperic convection is expected to be strongly influenced during the post-main sequence evolution by the readjustment of the internal stellar structure. The intensity and spectra of the X-ray emission are presumed to vary during the stellar evolution as an indirect consequence of modifications of the internal star structure, and rotational regime.

References

Durney, B.R., Mihalas, D., and Robinson, R.D. 1982, *PASP*, **93**, 537.

Endal, A.S., and Sofia, S. 1979, *Ap. J.*, **232**, 531.

Haisch, B.M., Bookbinder, J., Maggio, A., Vaiana, G.S., and Bennett, J.O. 1990, *Ap. J.*, **361**, 570.

Maggio, A., Sciortino, S., Vaiana, G.S., Majer, P., Bookbinder, J., Golub, L., Harnden, F.R., Jr., and Rosner, R. 1987, *Ap. J.*, **315**, 687.

Maggio, A., Vaiana, G.S., Haisch, B.M., Stern, R.A., Bookbinder, J., Harnden, F.R., Jr., and Rosner, R. 1990, *Ap. J.*, **348**, 253.

Narain, U., and Ulmschneider, P. 1990, *Space Sci. Rev.*, **54**, 377.

Pallavicini, R., Golub, L., Rosner, R., Vaiana, G.S., Ayres, T., Linsky, J.L. 1981, *Ap. J.*, **248**, 279.

Pinsonneault, M.H., Kawaler, S.D., Sofia, S., and Demarque, P. 1989, *Ap. J.*, **338**, 424.

Rucinsky, S.M., and Vandenberg, D.A. 1986, *PASP*, **98**, 669.

Rutten, R.G.M., and Pylyser, E. 1988, *Astr. Ap.*, **191**, 227.

Simon, T., and Drake, S.A. 1989, *Ap. J.*, **346**, 303.

Simon, T., and Landsman, W. 1991, *Ap. J.*, **380**, 200.

Vandenberg, D.A. 1985, *Ap. J. Suppl.*, **58**, 711.

PHOTOMETRIC AND SPECTROSCOPIC OBSERVATIONS OF COOL STARS SERENDIPITOUSLY DISCOVERED BY EXOSAT*

M. CAVALLARO, M. BOSSI and G. GUERRERO
Osservatorio Astronomico di Brera, Via E.Bianchi 46, 22055 Merate (Como) Italy

and

G. TAGLIAFERRI**
ISO Observatory, Astrophysics division, ESA-SSD, ESTEC, Keplerlaan 1, 2200AG Noordwijk, The Netherlands

Abstract. We present photometric and spectroscopic results on a sample of 22 northern cool stars X-ray selected by EXOSAT. Based on $uvbyH_\beta$ Strömgren photometry, new spectral classifications and parallaxes for 16 cool stars were derived, from which we estimate the corresponding X-ray luminosities. Six of the 12 sources tested for light changes showed indications of variability.

Spectroscopy was also carried out in the spectral range 6400Å- 6800Å, with a resolution of 0.5 Å/pixel. The spectral range was chosen to include the H_α and Li 6708Å lines. The H_α line showed very often a complex profile and, in a few cases, was filled in or in emission. In some of these stars, the Li 6708 line has an equivalent width of the order of hundreds of mÅ, confirming the results already obtained on the EXOSAT sample in the Southern hemisphere (Tagliaferri *et al.* 1992).

The present data, though in a preliminary and still incomplete form, show the presence of interesting objects, the nature of which needs to be better determined with further observations.

Key words: X-ray sources – Stars: late type.

1. Introduction

Sizable samples of X-ray selected cool stars have been obtained from the *Einstein* and EXOSAT X-ray surveys (Gioia *et al.* 1990; Giommi *et al.* 1991). Several investigations have been carried out in the past using *Einstein* samples (cf. Caillault *et al.* 1986; Silva *et al.* 1987; Favata *et al.* 1988; Fleming *et al.* 1988, 1989). In particular, the analysis of a sample of 137 X-ray selected cool stars (Fleming *et al.* 1988) led to the discovery of 47 binary systems, many of them belonging to the class of "active" RS CVn and W UMa types. Using a smaller data sample, Favata *et al.* (1988) reported evidence for a new class of X-ray luminous stellar objects, which contribute significantly to the stellar $\log N - \log S$ distribution. They showed that these stars could be active RS CVn, W UMa-like binaries. From previous work it is apparent therefore that X-ray surveys are a very efficient way to discover new active binary systems.

The Low Energy Telescope (LE) and CMA detector on board the EXOSAT satellite (White and Peacock 1988) proved to be very good for detecting serendip-

* Based on data collected with the European satellite EXOSAT and at the S.Pedro Martir Observatory (Baja California) of the Instituto de Astronomia de la UNAM, Mexico, and supported by the Agenzia Spaziale Italiana (ASI)

** On leave from Osservatorio Astronomico di Brera, Milano

J.F. Linsky and S. Serio (eds.), Physics of Solar and Stellar Coronae, 295–298.

itous sources in the 0.05-2 keV X-ray energy band; about one hundred late-type stars have been detected (Giommi *et al.* 1991).

Most of the brightest EXOSAT serendipitous late-type stars ($m_v \leq 10$) are included in the HD, SAO or other catalogues which, however, do not provide information on luminosity class or variability. In order to determine the nature of X-ray selected cool stars in the EXOSAT sample, we started an optical observing campaign both in the southern and northern hemisphere. In the south, observations with the ESO telescopes have shown that many of the stars in the sample are variable (Cutispoto *et al.* 1992), and many are very young stars, with some being Post T-Tauri stars (PTTS) candidates (Tagliaferri *et al.* 1992).

Here we present photometric and spectroscopic results on a sample of northern EXOSAT cool stars. The data were obtained with the telescopes based at S.Pedro Martir, B.C., Mexico (Instituto de Astronomia, UNAM, Mexico).

2. The Observations

From the EXOSAT sample of serendipitous cool stars (Giommi *et al.* 1991) we selected 22 sources with magnitudes and coordinates suitable for observations with the S.Pedro Martir telescopes. For each X-ray position, we have checked that no other plausible optical counterpart is present in the X-ray error box. We then performed $uvbyH_\beta$ Strömgren photometry using the Multichannel Danish Photometer applied to the 150 cm telescope at S.Pedro Martir. Spectroscopy was carried out with the 212 cm telescope at the same Observatory using a grating spectrograph and a CCD detector. Table 1 summarizes the data collected.

The transformation from the instrumental to the standard *uvby* system has been made by means of standard stars suitably selected in Olsen's catalogue (1983), while the physical parameters were obtained using a calibration program kindly supplied by Moon and Dworetsky (1985). The X-ray luminosities given in column 6, have been obtained from the parallaxes calculated from the calibrated absolute visual magnitudes. To this end, the EXOSAT thin lexan filter count rates have been converted to X-ray fluxes, in the spectral band 0.05-2 keV, by using a constant conversion factor of 1×10^{-10} $erg\ cm^{-2}\ count^{-1}$. However the value of the conversion factor from LE counts to energy flux depends on source temperature and interstellar column density, which are both poorly known for our stars. The adopted constant value for the conversion factor is only an appropriate mean for nearby sources with coronal temperatures in the range log T $\approx 6.5 - 7.3$ (see Fig. 2 in Giommi *et al.* 1991). Individual values for the derived luminosities can have errors of up to $\pm 40\%$.

For 12 stars of our sample, extended photometric observations were performed and the variability flag is given in column (9). Six of them showed indications of light variability with *u* amplitude ranging from 0.02 to 0.10 mag. In one case, SAO 80493, a flare seems to be present. The number of measurements and their time resolution did not allow us to reliably assess possible periodicities.

The spectra were obtained in the spectral range 6400Å- 6800Å, so as to include the H_α and Li 6708Å lines. The resolution was of 0.5 Å/pixel. The H_α line showed very often a complex profile and, in a few cases, was filled in or in emission. In

TABLE I

Summary of the observational data for the 22 stars of our sample. *Columns (2) and (3)*: technique of the observations; *columns (4) and (5)*: physical parameters obtained by means of Strömgren photometry; *column (6)*: X-ray luminosity (*erg* s^{-1}); *column (7)*: Li6708Å line intensity compared with the Ca6718Å line; *column (8)*: description of the H_α line: (a) totally in absorption, (b) weakly filled-in, (c) strongly filled-in, (e) in emission; *column (9)*: photometric variability.

Star (1)	Spect. (2)	Phot. (3)	T_{eff} (4)	Sp.T. (5)	$logL_x$ (6)	Li6708 (7)	H_α (8)	Var. (9)
SAO 11215	yes	no	-	-	-	weak	(b)	-
EXO0146+0608	yes	no	-	-	-	weak	(e)	-
BD+30397	no	yes	4280	K5V	29.6	-	-	-
SAO 130113	yes	yes	-	-	-	weak	(b)	yes(?)
HD 22670	yes	no	-	-	-	strong	(a)	-
EXO0422+1709	no	yes	3420	M3V	29.3	-	-	-
EXO0425+1735	no	yes	3500	M5IV	29.4	-	-	-
HD 32199	no	yes	5510	G8V	29.2	-	-	-
EXO0730+3156	no	yes	5270	K0V	30.4	-	-	-
BD+261682	no	yes	4650	K4V	29.0	-	-	-
SAO 80493	no	yes	6340	F6V	29.2	-	-	flare (?)
HD 95976	no	yes	6440	F5V	29.4	-	-	no
HD 99900	no	yes	7070	F1V	30.3	-	-	yes
HD 104753	no	yes	5960	G1V	29.8	-	-	no
HD 143271	no	yes	5520	G8V	29.0	-	-	yes
HD 167389	no	yes	5970	G2IV	28.9	-	-	no
SAO 30858	yes	yes	5950	G1V	29.3	weak	(a)	yes(?)
HD 170527	yes	no	-	-	-	strong	(b)	-
HD 189733	yes	yes	4850	K2V	28.4	weak	(b)	no
BD+164908	yes	yes	5540	G8V	29.9	strong	(c)	yes
HD 220091	yes	yes	-	-	-	weak	(a)	no
HD 222143	yes	yes	5690	G2V	29.3	weak	(a)	no

some stars, the Li line was very strong, with an equivalent width of the order of hundreds of mÅ. This fact can be considered as a strong indication of stellar youth and confirms the results already obtained, based on the EXOSAT sample in the Southern hemisphere (Tagliaferri *et al.* 1992). The distinctive features of Li and H_α lines are given in columns 7 and 8 of Table 1.

3. Conclusion

A clear result that emerges from Table 1 is the high level of coronal X-ray emission for most of our stars. All sources but one have X-ray luminosities in excess of 10^{29} $erg\ s^{-1}$. These values are very high for late type stars (see review by Rosner *et al.* 1985; Pallavicini, 1989), but this is not surprising, since any X-ray selected sample will pick up the brightest objects of any given class of stars (e.g. Fleming *et al.* 1988). X-ray luminosities as high as $10^{30} - 10^{31}$ $erg\ s^{-1}$ are typically found among late type stars only for very active sources such as RS CVn binaries and PMS.

50% of the twelve stars with an extended photometric coverage are variable or suspected variable. The observed color variations are consistent with the hypothesis that the photometric variations were produced by cool spots, whose visibility was modulated by the star rotation. However further observations are needed to better determine the variability's characteristics of these stars.

Finally, although our moderate resolution spectra do not allow us to performe a detailed analysis of the Li 6708Å line, this was so strong in some of our stars to be clearly detected. This is an indication that some of them are very young, possibly PTTS candidate, like those recently detected for a similar EXOSAT sample in the Southern hemisphere (Tagliaferri *et al.* 1992).

In conclusion, the present data, though in a preliminary and still incomplete form, show the presence of interesting objects, the nature of which needs to be better determined with further observations.

References

Cutispoto, G., Tagliaferri, G., Rodonò, M., Giommi, P., Pallavicini, R., Pasquini L.: 1992, *7th meeting on Cool Stars, Stellar System and the Sun*, M.Giampapa and J.Bookbinder eds, Tucson, USA, *in press*

Caillault, J.-P., Helfand, D.J., Nousek, J.A., Takalo, L.O.: 1986, *Ap.J.* **304**, 318.

Favata, F., Rosner, R., Sciortino, S., Vaiana, G.S.: 1988, *Ap.J.* **324**, 1010.

Fleming, T.A., Liebert, J., Gioia, I.M., Maccacaro, T.: 1988, *Ap.J.* **331**, 958.

Fleming, T.A., Gioia, I.M., Maccacaro, T.: 1989, *Ap.J.* **340**, 1011.

Gioia, I.M., Maccacaro, T., Schild, R.E., Wolter, A., Stocke, J.T., Morris, S.L., Henry, J.P.: 1990 *Ap.J. Supp.* **72**, 567.

Giommi, P., Tagliaferri, G., *et al.*: 1991, *Ap.J.* **378**, 77.

Moon, T.T., Dworetsky, M.M.: 1985, *MNRAS* **217**, 305.

Olsen, E.H.: 1983, *Astr. Ap. Suppl.* **54**, 55.

Pallavicini, R.: 1989, *Astr. Ap. Rev.* **1**, 177.

Rosner, R., Golub, L., Vaiana, G.S.: 1985, *Ann. Rev. Astr. Ap.* **23**, 413.

Silva, D.R., Gioia, I.M., Maccacaro, T., Mereghetti, S., Stocke, J.T.: 1987, *Astr. J.* **93**, 869.

Tagliaferri, G., Randich, S., Pallavicini, R., Cutispoto, G.: 1992, *Astr. Ap.* submitted.

White, N.E., Peacock, A.: 1988, *X-ray Astronomy with EXOSAT*, R. Pallavicini and N.E. White eds, *Mem. Soc. Astr. Ital.* **59**, 7.

LITHIUM ABUNDANCE IN NORTHERN-HEMISPHERE ACTIVE BINARY STARS

D. BARRADO, M.J. FERNÁNDEZ-FIGUEROA, E. DE CASTRO and M. CORNIDE
Dpto. Astrofísica. Facultad de Físicas.
Universidad Complutense. E-28040 Madrid (Spain)

Abstract.
Observations of the 6707.8 Å Li I line in a sample of active binary systems have been carried out at the Calar Alto Observatory with the 2.2m telescope. In addition, some single stars with different chromospheric activity levels have also been observed mainly for comparison purposes.

Gaussian fits have been performed in order to separate the Fe I line at 6707.4 Å which is usually blended with the Li I line.

A curve-of-growth method has been used to derive Li abundances. The results are discussed according to the activity levels.

Key words: Abundances – Chromospheres – RS CVn

1. Introduction

Lithium is a light element characterized by its easy destruction by (p, α) nuclear reactions in the stellar interiors. The photospheric Li abundance in late-type stars is strongly related to spectral type and age. Different authors have proposed empirical expressions for the Li abundance as a function of the age by observing galactic open clusters (Boesgaard, 1991; Rebolo, 1989).

RS CVn and BY Dra active binary systems are characterized by synchronized rotational and orbital periods. These systems show high chromospheric activity levels in radio and optical. The components of BY Dra binaries are main sequence stars later than G8, whereas the RS CVn systems have at least a cool evolved component. The presence of the Li I line in these stars (Spite, Maillard and Spite 1984; Pallavicini, Randich and Giampapa 1992) is difficult to explain with the above-mentioned theories which predict a large depletion in stars with deep convective zones.

2. Observations and reduction

The observations were carried out in 1990 June 1-7 at the Coudé focus of the 2.2m telescope at the Calar Alto Observatory using a CCD detector. The resolution limit of the spectra was 0.12 Å per pixel. The total spectral range covered was $\sim$ 100 Å. Typical signal-to-noise ratios were $\sim$ 100. Each CCD exposure was reduced using standard procedures. The package MIDAS was used throughout.

The sample includes RS CVn and BY Dra binary systems (selected from Strassmeier et al. 1988) and single stars (selected from Noyes et al. 1984) with different chromospheric activity levels. In Figure 1a the normalized spectrum of the SB2 σ^2 CrB is plotted. Lines originated in both components can be seen (SB2 spectrum).

The Li I unresolved doublet at 6707.81 Å is very close to a Fe I line at 6707.41 Å. Since most part of the stars in our sample are fast rotators the Li I doublet is blended with the Fe I line. In order to obtain the Li I equivalent width, the contribution of the Fe I line was substracted using an interactive routine. Gaussian

J.F. Linsky and S. Serio (eds.), Physics of Solar and Stellar Coronae, 299–302.

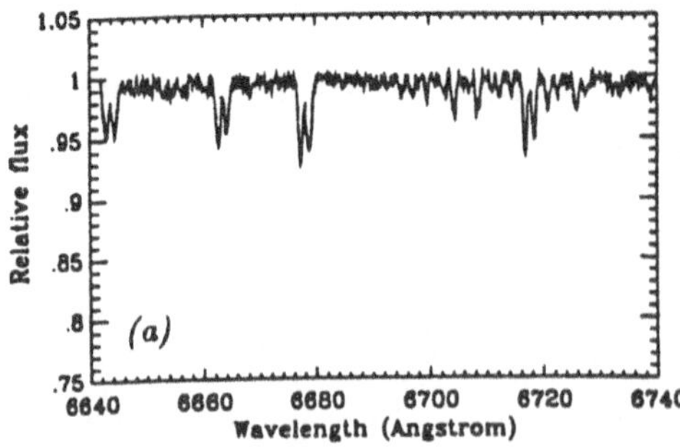

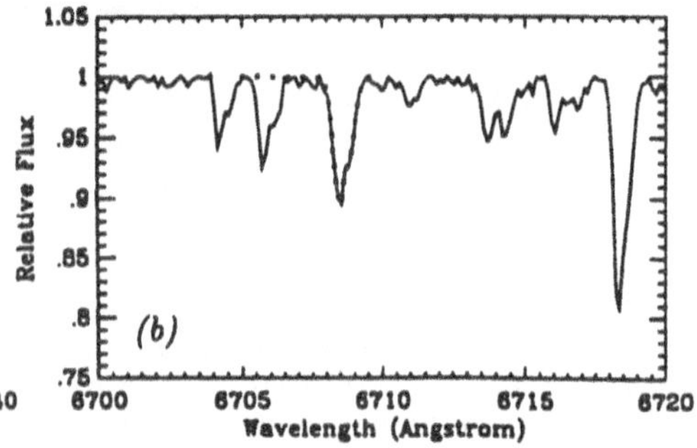

Fig. 1. (a) The normalized spectrum of σ^2 CrB. The Li I lines at 6707 Å are clearly present in both stars. (b) Spectrum of β Com around 6707 Å (solid line) and the synthetic profile obtained for the Li I and Fe I blend (dashed line).

functions were fitted to the line profiles. We took into account different cases: simple spectra of single stars and SB1 and complex SB2 spectra. Figure 1b shows the Li I and Fe I profiles around 6707 Å for β Com.

Once the fit was achieved, the equivalent width was measured for each line. It was assumed that the equivalent width of the whole feature contained no significant contribution from ^{6}Li. For SB2 the contributions from both components to the continuum were taken into account, following the work by Pallavicini *et al.* (1992). When gaussian fits were not feasible the total equivalent width was measured and then the contribution from the blended lines was estimated from spectra of stars with similar spectral type and luminosity class.

The equivalent width of the Li lines was converted into abundances by using the curves of growth of Pallavicini *et al.* (1987). The T_{eff} of non-eclipsing binaries was assigned from spectral type. For single stars the calibration of Böhm-Vitense (1981) was used. A common value of gravity, namely log $g_* = 4.5$, was used for main sequence stars and log $g_* = 3.75$ was used for giants and subgiants. The accuracy of the derived Li abundances (on a scale where log N(H)=12.00) is $\sim$ 0.30 dex, the main source for this uncertainty being the errors in the determination of effective temperatures. The relevant parameters and the results for each star are shown in Table I.

3. Results and discussion

In Figure 2a we plot the Li abundance versus the effective temperature. As it can be seen the active binaries have a higher abundance, in particular the K type stars. An age effect cannot explain the Li abundance excess of RS CVn cool components. The fast rotation of binaries sinchronized with the orbital periods in our sample could be the cause of lower Li depletion .

It is well known that chromospheric activity is correlated with stellar rotation. We have plotted in Figure 2b the Li abundance versus the flux in the Ca II H and K lines. The chromospheric data of single and binaries are from Noyes *et al.*(1984) and Fernández-Figueroa *et al.*(1992) respectively. There is a correlation betweeen

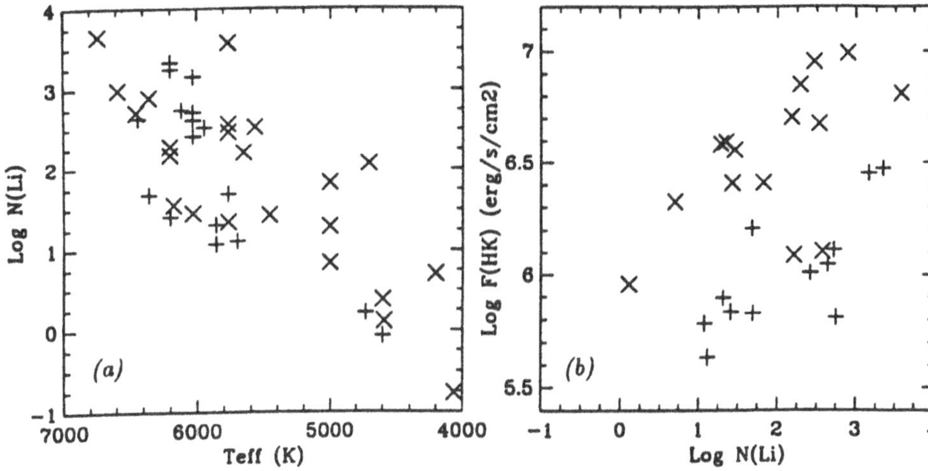

Fig. 2. (a) Li abundance versus effective temperature. The two groups are shown: Rs CVn and BY Dra on one hand (represented by the symbol ×) and the active single stars on the other (represented by the symbol +). (b) Ca II H and K fluxes versus Li abundances (symbols are the same as in panel (a)).

both parameters. The two activity groups are clearly distinguished with slopes very similar and a power law $F_{HK} \propto (N_{Li}/N_H)^{1/3}$ could be fit. For the same Li abundance RS CVn and BY Dra are, on average, 0.5 dex more active in Ca II than the single stars. In each group the spread on Li abundances is mainly a temperature effect as can be infered from Hyades cluster data.

The results presented in this work do not provide a quantitative model for the role of rotation in the Li depletion. But the suggested connection between reduced Li depletion and fast rotation of tidally interactive binaries is in agreement with the evolutionary model of Pinsonneault *et al.* (1989).

Acknowledgements

This work has been partially support by the DGICYT project number PB91-0348.

References

Boesgaard, A. M.: 1991, *Astrophy. J.* **370**, L95

Böhm-Vitense, E.: 1981, *Ann. Rev.* **19**, 295

Fernández-Figueroa, M. J., Montes, D., De Castro, E., Cornide, M.: 1992, *(to be published)*,

Noyes, R.W., Hartmann, L., Baliunas, S., Duncan, D., and Vaugham, A.H.: 1984, *Astrophys. J.* **279**, 763

Pallavicini, R., Cerruti-Sola, M., and Duncan, D. K.: 1987, *Astron. Astrophys.* **174**, 116

Pallavicini, R., Randich, S., and Giampapa, M.S.: 1992, *Astron. Astrophys.* **253**, 185

Pinsonneault, M.H., Kawaler, S.D., Sofia, S., Demarque, P. : 1989, *Astrophys. J.* **338**, 424

Rebolo, R.: 1989, *Astrophys. Space Science* **157**, 47

Spite, M., Maillarsd, J. P., and Spite, F.: 1984, *Astron. Astrophys.* **141**, 56

Strassmeier, K.G., Hall, D.S., Zeilik, M., Nelson, E., Eker, Z., and Fekel, F.C.: 1988, *Astron. Astrophys. Suppl. Ser.* **72**, 291

TABLE I

Relevant stellar parameters and the Li abundance and flux in Ca II H and K lines.

Name	Sp. Type	Duplicity	Eq.width ($m\AA$)	$LogN_{Li}$	F_{HK} ($10^6 erg/s/cm^2$
HD 107067	F8 V	S	104	3.25	
HD 108102	F8 V/F8 V	SB2	22/28	2.18/2.29	5.066/7.098
RS CVn	F5 IV/F0 IV	SB2	35/15	2.71/083	
β Com	F9.5 V	S	74	2.72	1.288
59 Vir	G0 Vs	S	113	3.34	2.979
HD 115404	K2 V	S	9	0.21	
HR 5110	F2 IV/K2 IV	SB2	6/146	1.57/2.06	
τ Boo	F6 IV	S	5	1.68	1.616
ι Vir	F7 IV	S			1.902
α Boo	K1.5 III	S	7	-0.07	
HR 5384	G1 V	S	7	1.31	0.78
5 Ser	F8 III-IV	S	4	1.41	0.682
GX Lib	[G-K V]/K1 III	SB1	18	0.37	
λ Ser	G0 V	S	19	1.69	0.672
χ Her	F8 Ve	S	67	2.75	0.646
ρ CrB	G0 Va	S	4	1.07	0.608
σ^2 CrB	F6 V/G0 V	SB2	63/70	2.90/2.47	9.854/9.016
σ^1 CrB	G1 V	S	63	2.53	
WW Dra	G2 IV/K0 IV	SB2	53/35	2.21/1.28	1.218/3.865
V792 Her	F3 V/K0 III	SB2	79/84	3.66/0.70	/2.125
HR 6469	F2 V/[G0V]? G5 IV	SB1	19	1.43	2.570
26 Dra	G0 Va	S	64	2.62	
HR 6608	G2 IIIb	S	6	1.1	0.430
Z Her	F4 V-IV/K0 IV	SB2	43/78	2.99/1.82	/2.57
V772 Her	G0 V/[M1 V] G5 V	SB1	/ 79	/ 2.57	/ 1.268
V815 Her	G5 V/[M1-2 V]	SB1	178	3.58	6.497
α Lyr	A0 V	S	35		
HR 7162	F9 V	S	46	2.42	1.023
V478 Lyr	G8 V/[dK-dM]	SB1	90	2.53	4.754
17 Cyg	F7 V	S	32	2.64	1.112
ER Vul	G0 V/G5 V	SB2	7/9	1.46/1.35	3.619/3.929
HR 8314	G0 V	S	119	3.16	2.83
BY Dra	K4 V/K7.5 V	SB2	11/8	0.11/-0.77	0.907/

LITHIUM ABUNDANCE DETERMINATION IN EMSS STELLAR SOURCES

F. FAVATA

European Space Agency, Astrophysics Division, P.O. Box 299, 2200 AG Noordwijk, The Netherlands

and

M. BARBERA, G. MICELA and S. SCIORTINO

Istituto e Osservatorio Astronomico, Palazzo dei Normanni, 90134 Palermo, Italy

Abstract. We present high resolution spectroscopical observations of a sample of F, G and K stars from the *Einstein* Extended Medium Sensitivity Survey (EMSS) in the Li I 6708 Å region. Preliminary values for the Li abundances are derived using the curve-of-growth method.

Key words: Stars: abundances - Stars: late-type

1. Introduction

We have been studying the composition of the stellar component of the Einstein Extended Medium Sensitivity Survey (EMSS, Gioia *et al.* 1990), as described in a companion paper in these proceedings (*An Analysis of the Stellar Content of the Einstein Extended Medium Sensitivity Survey Using Age-Homogeneous X-ray Luminosity Functions*), determining that a significant fraction of the yellow star population detected in the EMSS must be somewhat peculiar. To determine the true nature of this yellow star population we have started a campaign of optical observations of yellow EMSS stellar source, conducted at the European Southern Observatory (ESO) in La Silla, Chile and at the Dominion Astrophysical Observatory (DAO) in Victoria, British Columbia. Lithium has played a major role in our observational campaign, due to its value as an age indicator. In this paper we present an overview of the high resolution spectroscopic observations obtained so far of the Li I 6708 Å line.

2. The observed sample

The sample observed so far has been limited in magnitude on the faint side by the size of available telescopes, and in sky distribution from the available time. Although not entirely random, it is representative of the EMSS in the magnitude range covered, and it is being extended to cover all of the EMSS in this magnitude range. The stars observed so far are listed in Table 1.

3. Observations, instruments and data reduction

The sample presented here has been obtained at ESO for the southern sample, using the CAT 1.4 m telescope with the Coudé Echelle Spectrograph (CES) with the short camera and an RCA CCD as detector, at roughly 50,000 resolving power, and at DAO for the northern sample, using the 48" (1.2 m) telescope, using the coudé spectrograph with the 96" camera, image slicer and a Ford CCD as detector. Resolving power was roughly 30,000.

J.F. Linsky and S. Serio (eds.), Physics of Solar and Stellar Coronae, 303–306.

TABLE I

EMSS Name	Name	m_v	Sp. Type	EMSS Name	Name	m_v	Sp. Type
MS0002.8+1602	HD 42	8.57	F5V	MS1109.8+3605	SAO 62451	6.41	G0V
MS0003.3-4201	SAO 214961	7.41	G0V	MS1256.2+3833		9.35	K3V
MS0009.9+1417	SAO 91772	8.5	G5V	MS1330.5-0811	HD 117860	7.2	G0V
MS0011.6+0840		11.54	K0Ve	MS1436.8-2628	SAO 182743	9.73	K4V
MS0031.9-0646	SAO 128830	6.83	F4V	MS1441.7+5208	SAO 29254	8.2	G2V
MS0134.4+2027	HD 9902	8.67	F9V	MS1520.7-0625	HD 136905	7.31	K0III
MS0138.0-5627	HD 10361	5.82	K5V/K0V	MS1521.1+3027	SAO 64673	5.61	F8V
MS0206.2-1019		8.91	G5V	MS1552.0-2338	HD 142361	8.79	G3IV
MS0214.9+1813	HD 14147	7.29	F0V	MS1558.4-2232		11.42	K3e
MS0234.2-0321	HD 16287	8.09	K2V	MS1634.7+2638	HD 149931	7.98	F5V
MS0236.4-0148	SAO 130032	8.99	F8V	MS1704.3+5432	SAO 30239	5.83	F6V
MS0239.9+0704	SAO 110699	8.93	K5V	MS1737.2+6847	SAO 17576	4.8	F5V
MS0300.1-1528	HD 18955	8.44	G8V	MS1751.0+7046		9.63	K5IV
MS0303.8+1717	SAO 93280	8.69	F7V	MS1753.5+1830	SAO 103221	9.21	K4
MS0327.2-2416	SAO 168581	9.16	K7V	MS1810.3+6940	SAO 17800	8.58	K2V
MS0337.6-0202	SAO 130647	8.0	G9V	MS2125.5-1503		11.45	F8V
MS0413.7-6235	SAO 248969	3.35	G8III/I	MS2148.2+1420		11.35	K3IV/V
MS0448.4+1058	HD 30810	6.76	F8V	MS2254.2+0219		10.25	K2V
MS0457.5+0312	SAO 112298	7.47	K2V	MS2302.4-4427	SAO 231427	9.69	G8
MS0535.7-2839	SAO 170610	7.17	F4V	MS2335.2+0305	SAO 128293	7.29	F5V
MS1100.2+6155	SAO 15379	7.11	F8				

In both cases the observations covered a short spectrum stretch (about 40 Å) centered on the Li I resonance line at 6707.81 Å. The observations were reduced using the IRAF package. The reduction consisted of the usual bias removal and flat fielding, followed by spectrum extraction. To ensure that the flat field spectrograph illumination would be the same as the illumination from program objects, flat field exposures were obtained on a bright B star with high rotational velocity (having essentially a flat spectrum on the small spectral stretch observed here). The analysis consisted in the measurement of the equivalent width of the Li I 6707.81 Å line, with deblending from the nearby Fe I 6707.44 Å line when the two lines were only partially blended in the spectrum. For program objects with relatively high rotational velocity the two lines are fully blended. In these cases the relative contribution on the Fe I line to the blend was estimated from the equivalent width of the same line in stars of the same spectral type with no visible Li I line. Finally, the lithium abundance was estimated using published curves of growth (Pallavicini *et al.* 1987).

4. Discussion and implications for the stellar component of EMSS

It can be readily seen from the abundance distribution of Li that the EMSS yellow stars have a Li distribution different from the one of field stars: some EMSS sources (some of the late F and of the K stars) show Li abundances typical of pre-main sequence stars, and several other sources have a lithium abundance more typical of stars in young open clusters. It has been shown that for active stars, the lithium abundance deduced from the Li I 6707 Å line is not always in agreement with other indicators of stellar age. In particular, active stars often show a Li abundance higher than what would be expected on the basis, for example, of the age attributed to them on the basis of stellar evolution theory (Pallavicini *et al.* 1992)

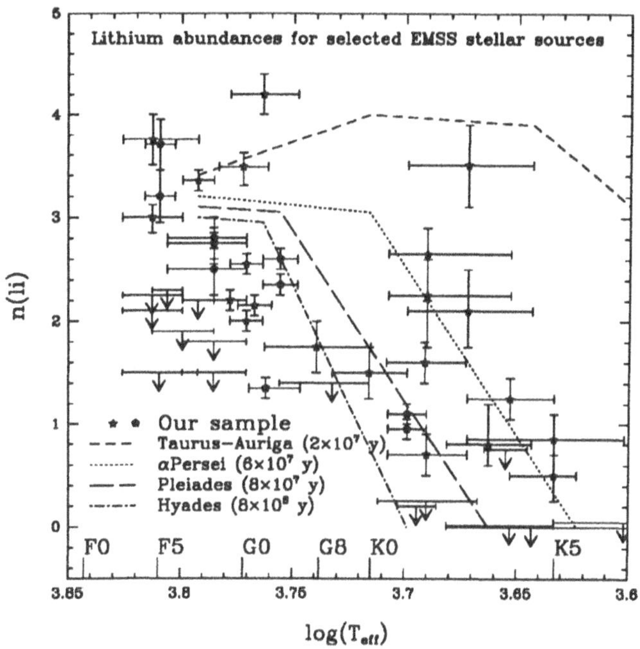

Fig. 1. The calculated Li abundance (expressed as log n(Li), relative to a log hydrogen abundance of 12) is plotted against the log effective temperature for the sample stars. Superimposed for comparison are the approximate upper envelopes of Li abundances for stars observed in open clusters (and therefore of known age), taken from Basri *et al.* (1991). Detections as well as upper limits are plotted. The error bars shown are due to the uncertainty on the effective temperature.

Therefore while the stars with the highest Li abundance in the figure are likely to actually be pre-main sequence objects (it has been shown that X-ray selection is a very efficient method of finding PMS stars, Walter et al. 1988), the age deduced on the basis of Li abundance for the objects with moderately high Li abundance should be treated with caution. We have shown in the past (Favata *et al.* 1988, 1992, 1993) that X-ray selected samples of stars (in particular the EMSS) contain an excess of yellow stars with respect to what can be accounted for in terms of the X-ray emission characteristics of nearby stars and of stars in nearby open clusters. We believe that the data shown here point toward this excess being at least in part composed by very young, active sources, some still at a pre-main sequence stage (perhaps the X-ray selected WTTS of Walter *et al.* 1988)

5. Planned future work

As stated above, the EMSS sample presented here is far from complete. Additional observing runs are already allocated at ESO, and at DAO on the same instruments used so far, to provide a more complete and representative sample. Additionally, we are planning to apply for time at larger telescope, to be able to extend the coverage of our sample to fainter magnitudes. The analysis presented here is preliminary in that is performed using published curves of growth. We plan to refine it by using synthetic spectra computed form model atmospheres, which should give has more precise abundance values. Finally, some of the sources presented here have a large uncertainty in the effective temperature, which is reflected in a large uncertainty in the Li abundance. We plan to perform UBV photometry on these object to determine a more precise effective temperature. Although the refinements described above should improve the accuracy of the results, we believe that the main conclusions, namely that the yellow stars in the EMSS contain a large fraction of object with high Li abundances, are bound not to change.

Lithium abundances deduced from our observations are often much higher than the abundances typical in inactive field stars of the same spectral types. These high Li abundances have been observed in active stars (Pallavicini *et al.* 1992) and similar works on X-ray selected stellar sources from EXOSAT observatory surveys produce similar results (Tagliaferri *et al.* 1993). Although it is clear that Li deduces ages cannot be taken at nominal value for active stars, we believe that these Li observations supply an useful hint to the solution of the problem of excess yellow stars in X-ray selected samples.

Acknowledgements

Three of us (M.B, G.M, S.S) acknowledge support from Ministero della Università e della Ricerca Scientifica e Tecnologica, from ASI, and from GNA-CNR.

References

Basri, G., Martin, E., Bertout, C.: 1991, *A&A* **252**, 625
Favata, F., Rosner, R., Sciortino, S. and Vaiana, G.S.: 1988, *ApJ* **324**, 1010
Favata, F., Micela, G., Sciortino, S. and Vaiana, G.S.: 1992, in Tanaka, Y. and Koyama, K., ed(s)., *Frontiers of X-ray Astronomy*, Universal Academy Press, in press
Favata, F., Micela, G., Sciortino, S.: 1993, this volume
Pallavicini, R., Cerruti-Sola, M. and Duncan, D.K.: 1987, *A&A* **174**, 116
Pallavicini, R., Randich, S. and Giampapa, M.S.: 1992, *A&A* **253**, 185
Tagliaferri, G. *et al.*: 1993, this volume
Walter, F.M. *et al.*: 1988, *AJ* **96**, 297

LITHIUM IN STARS X-RAY SELECTED BY EXOSAT*

G. TAGLIAFERRI **

Osservatorio Astronomico di Brera, Via Brera 28, I-20121 Milano, Italy

S. RANDICH

Dipartimento di Astronomia, University of Florence,
Largo Fermi 5, I-50125 Firenze, Italy

R. PALLAVICINI

Osservatorio Astrofisico di Arcetri, Florence,
Largo Fermi 5, I-50125 Firenze, Italy

and

G. CUTISPOTO

Osservatorio Astrofisico di Catania, Catania,
v.le A.Doria 6, I95125 Catania, Italy

Abstract. We present the results from high resolution spectra in the Li 6708 Å region's line for a sample of 26 X-ray selected stars, all of spectral type later than F. UBV(RI)c photometric data obtained from our observations are also used. Synthetic spectra were computed and Li abundances and metallicities derived for all stars. Although there is a big scatter in the derived Li abundances from star to star, the majority of them have high Li abundances, much higher than what expected in field stars of similar T_{eff}. This suggests that a large fraction of our stars may be constituted by young objects, with possibly some post-T Tauri candidates among them.

Key words: X-ray sources – Stars: late type – Stars: abundances.

1. Introduction

In the past few years we have studied extensively the sample of cool stars detected in the EXOSAT High Galactic Latitude Survey (Giommi *et al.* 1991). To this end, spectroscopic and photometric observations were obtained by us both in the southern and northern hemisphere. In particular, at the European Southern Observatory we have performed medium to high resolution spectroscopy in various spectral ranges, including the Li 6708 Å line region.

Li abundances can give information on the internal mixing of a star and its evolutionary history. Lithium is easily destroyed by convective mixing in stellar interiors when the temperature at the bottom of the convective zone reaches about $2\ 10^6$ K, so the efficiency of Li destruction should depend both on the age of the star and on the depth of the convective zone. Li abundances, as determined by the 6708 Å line, have been extensively used in the past as an age indicator for late type stars. However, even if the mass of a star is well known, it is not straightforward to convert Li abundances into ages. Other parameters, like metallicity and rotation, are likely to affect the abundance of lithium (Soderblom 1991). For instance, in recent years, the presence of Li absorption has been detected in the spectra of stars that are evolved away from the main sequence, such as some red giants and RS CVn binaries (Brown et al. 1989, Gratton and D'Antona 1989, Pallavicini et al.

* Based on observations collected with the European satellite EXOSAT and at the ESO observatory, La Silla, Chile.

** also ISO Observatory, ESA-SSD, ESTEC, Noordwijk

J.F. Linsky and S. Serio (eds.), Physics of Solar and Stellar Coronae, 307–310.

1992). Various mechanisms have been proposed to explain this apparent anomaly, such as the presence of large starspots that enhance the Li absorption line, the production of Li in stellar flares by spallation, an anti correlation between rotation and Li depletion, and the evolution from late A or early F-type progenitors with shallow convective zones.

None of these mechanisms has been confirmed observationally; however given that they are all somewhat related to stellar activity in a broad sense, the study of the Li 6708 Å line should be a useful tool for understanding the nature of our X-ray selected sample of cool stars. In fact this sample is likely to contain both young stars and RS CVn binaries and/or related active stars, that are all known to be strong X-ray emitters.

Here we present the results from high resolution spectra of 26 objects, all of spectral type later than F, obtained in the region of the Li 6708 Å line. UBV(RI)c photometric data obtained by us are also used in the analysis.

2. The Observations

From the EXOSAT list of serendipitous sources we selected all known stars of spectral type later than F and for 26 of these stars we obtained high resolution spectra of the Li I 6708 Å line. The observations in the Li line were carried out in two different runs, March 1990 and October 1990, using the Coudé Echelle Spectrometer fed by the CAT telescope, with the short camera and a CCD detector. The nominal resolving power was R=60,000.

For all of them, but two, we have also performed extensive and repeated photometry, in order to derive accurate magnitude and colors, and search for possible variability. Here we will use these data to determine the physical properties of our stars. The data were collected in 1989 and 1990 using the 50cm and 1m ESO telescopes. The typical accuracy of these data is of the order of about 0.01 magnitudes, with somewhat higher values for the U-B color due to the low photon counting level.

3. Results

Synthetic spectra were computed, and Li abundances and metallicites derived, for all stars in the sample. The computed spectra covered a range of $\approx$20 Å around the Li 6708 Å line and were obtained using a code described in detail by Gratton and Sneden (1990). Appropriate model atmospheres were extracted from the grid of Bell et al. (1976). The synthetic spectra were then convolved either with an instrumental gaussian profile or with the rotation broadening function. That allowed us to derive also accurate rotational velocities. The derived Li abundances depend critically on the assumed value of the star effective temperature ($T_{\rm eff}$) used in the fit. In the case of single stars we obtained $T_{\rm eff}$ directly from the (B-V) color using the scale of Böhm-Vitense (1981). Otherwise, in the case of SB2 binaries, we used the $T_{\rm eff}$ derived from the spectral types which, in turn, were determined from the photometry. Also the gravity, the microturbulence velocity and, when needed, the radii of the stars were derived from the adopted spectral types.

In Fig. 1 we plot the Li abundances as a function of the effective temperature for our stars, together with those of stars in several young clusters and in one star forming region, namely Tau-Aur (Basri et al. 1991), α Per (Balachandran et al. 1989), the Pleiades (Duncan et al. 1983) and the Hyades (Cayrel et al. 1984). From this figure it is apparent that there is a large scatter in the Li abundances from star to star. About one third of our sources have Li abundances that are intermediate between those of stars in Tau-Aur and stars in young open clusters such as the Pleiades and α Per. Several have Li abundances comparable to or larger than the primordial value for Pop I stars (*i.e.* $log\ N(Li) > 3.1$). All remaining objects, but two, have Li abundances comparable to or higher than those of stars in the Hyades. From these data we conclude that the majority of our stars have high Li abundances, much higher than typically observed in field stars of similar $T_{\rm eff}$.

The comparison with cluster data and with PMS objects in Tau-Aur suggests that a large fraction of our X-ray selected stars may be constituted by young objects, with possibly some pre-main sequence candidates among them (those with $log\ N(Li) > 3.1$). We cannot exclude, however, that some objects may have an "anomalous" Li abundance for reasons other than age. We note that only one star of our sample is close to a star forming region (Orion). Thus, those stars that appear to be PMS candidates are more likely Post-T Tauri rather than Weak-lined or Classical T Tauri stars in regions of star formation.

In conclusion, the present data show the existence of interesting objects among the EXOSAT serendipitous sources, the nature of which needs to be better determined by further observations. To this end, we have already obtained high resolution spectra in Hα and in the Ca II H&K lines. Moreover, in order to measure radial and rotational velocities and ascertain whether some of these stars are binaries, we have observed most of them using CORAVEL. Extensive and repeated UBV(RI)c photometric observations have also been obtained. When the full analysis of these data will be completed, we should have enough information to isolate truly young objects from stars that are not young, despite their higher than normal Li abundances.

References

Balachandran, S., Lambert, D., Stauffer, J.: 1988, *Ap.J* **333**, 267.
Basri, G., Martin, E.L., Bertout, C.: 1991, *Astr. Ap.* **252**, 625.
Bell, R.A., Eriksson, K., Gustafsson, B., Nordlund, Å.: 1976, *Astr. As. Suppl.* **23**, 37.
Böhm-Vitense, E.: 1981, *Ann. Rev. Astr. Ap.* **19**, 295.
Brown, J.A., Sneden, C., Lambert, D.L., Dutchover, E.Jr.: 1989, *Astr. Ap.* **71**, 293.
Cayrel, R., Cayrel de Strobel, G., Campbell, B., Dappen, W.: 1984 *Ap. J.* **283**, 205.
Duncan, D.K., Jones, B.F.: 1983, *Ap.J.* **271**, 663.
Giommi, P., *et al.*: 1991 *Ap.J.* **378**, 77
Gratton, R.G., D'Antona, F.: 1989, *Astr. Ap.* **215**, 66.
Gratton, R.G., Sneden, C.: 1990, *Astr. Ap.* **234**, 366.
Pallavicini, R., Randich, S., Giampapa, M.: 1992, *Astr. Ap.* **253**, 185.
Soderblom, D.R.: 1991 *Mem. Soc. Astr. It.* **62**, 43.

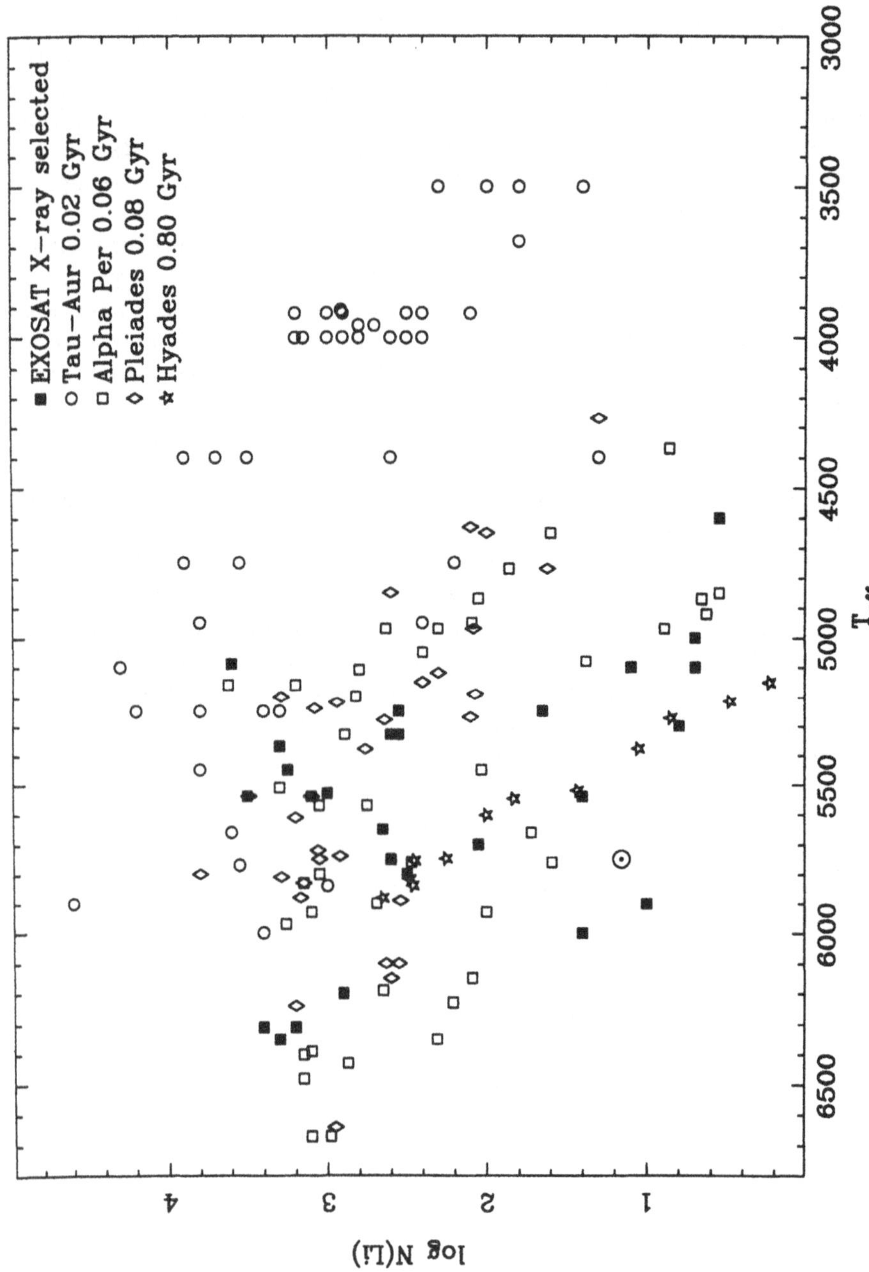

Fig. 1. Li abundances vs effective temperature for our X-ray selected stars, PMS stars in the Tau Aur complex and stars in young open clusters.

EMISSION UP TO 18 KEV IN THE QUIESCENT X-RAY SPECTRUM OF II PEG

J.G. DOYLE
Armagh Observatory
G.H.J. VAN DEN OORD
Sterrenkundig Insituut
and
B.J. KELLETT
Rutherford Appleton Lab.

Abstract. The quiescent X-ray emission shows a power-law tail up to 18 keV. We show that an interpretation in terms of free-free emission by non-thermal particles requires unrealistic large values for the number of particles involved and for the energy input required to maintain the population. The data can however be satisfactorily explained by a differential emission measure distribution indicating the presence of substantial amounts of plasma with temperatures of $10^8 K$.

Key words: II Peg – quiescent X-ray spectrum – thermal – differential emission measure.

1. Quiescent X-ray emission

In mid-August 1989 we obtained two days satellite time on the Japanese X-ray satellite Ginga to monitor the RS CVn star II Peg for flares. Here we present results for the quiescent emission. Briefly, the quiescent X-ray spectrum can not be satisfactory fitted with a single temperature Raymond-Smith model. Instead the best results were obtained for a combination of a single temperature plasma together with a power-law. The reason for this second high energy/temperature component is the tail in the photon distribution, which extended up to 18 keV. Extensive efforts were made to check the validity of this high energy emission. It was not due to poor background subtraction (as the background determination can be a problem in calibrating GINGA data), since this would require an excess of $\sim$ 10% over the 'average' diffuse hard X-ray background. The presence of a second source in the field-of-view was also ruled out, since neither the EINSTEIN Observatory nor EXOSAT reported any other point sources in the field of II Peg. For the single temperature model, the thermal component implied a temperature of 12.4 10^6 K with an emission measure of 1.5 $10^{53} cm^{-3}$. The photon index of the non-thermal component was 2.05. Below we discuss in detail the nature of this additional component.

2. An interpretation

There are various ways to produce the observed power-law emission by both thermal and non-thermal particle distributions. Suppose for example that the power-law component is caused by collisional Bremsstrahlung of energetic electrons interacting with the thermal plasma ($T = 12.4\,10^6 K$) in the coronal volume. The observed spectrum can be inverted to obtain the energy distribution function $f(E)$

J.F. Linsky and S. Serio (eds.), Physics of Solar and Stellar Coronae, 311–314.

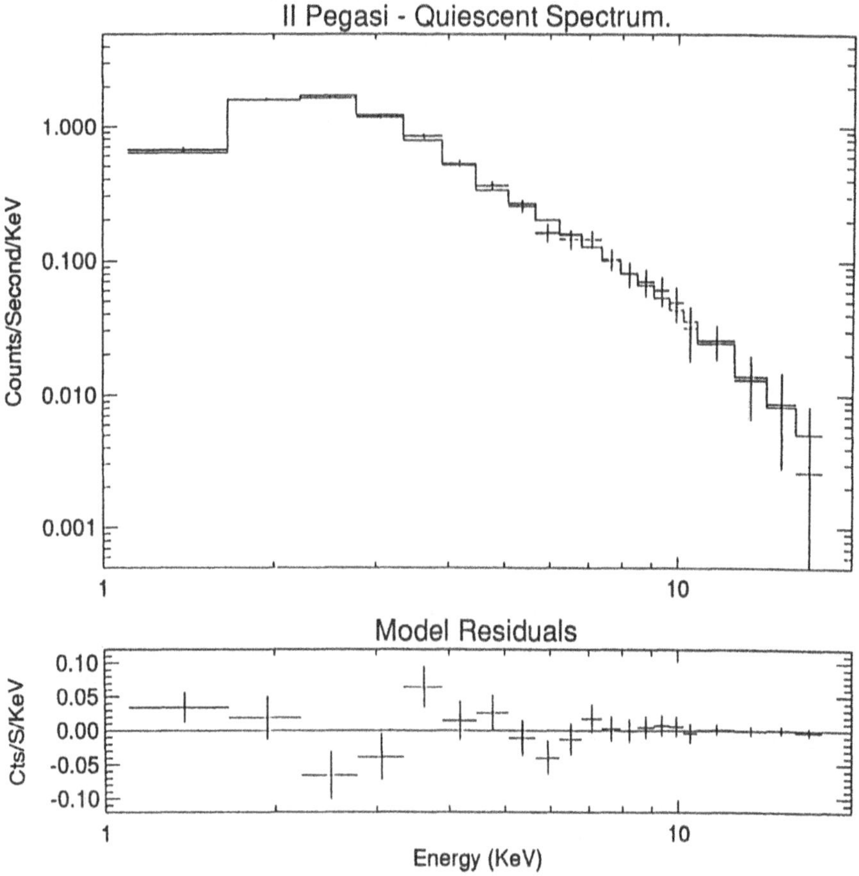

Fig. 1. The quiescent data for II Peg obtained with GINGA in August 1989 (+), plus a model fit of a single temperature Raymond & Smith model ($T = 12.4\,10^6 K$) plus a power-law of $\gamma = 2.05$ for the high energy points. The power-law tail starts becoming prominent at about 6 keV and extends up to 18keV

of the energetic electrons (see e.g. Brown et al. 1991, their Eq. 9). For a photon spectrum given by $J(\epsilon) = A\epsilon^{-\gamma}$ with $A = 1.86\,10^{-3}$ and $\gamma \approx 2$ (i.e. the derived power-law parameters), the electron distribution is given by $f(E) = 6.5\,10^{52}E^{-3/2}/(nV)$ *electrons* cm^{-3} keV^{-1}, where E is in keV. In this expression n is the density of the thermal plasma and V the coronal volume. This electron distribution will loose its energy by Coulomb interactions with the ambient thermal plasma while producing the power-law spectrum. As a consequence continuous injection of electrons has to take place to compensate for the collisional energy losses. The injection rate q(E) [*electrons* s^{-1} keV^{-1}] is given by

$$\frac{\partial}{\partial E}[Vf(E)\frac{dE}{dt}] = q(E) \tag{1}$$

with

$$\frac{dE}{dt} = -2\pi e^4 \Lambda n (\frac{2}{mE})^{1/2} \quad erg\, s^{-1} \tag{2}$$

the energy loss rate by Coulomb interactions in a cold target (Λ is the Coulomb logarithm). From the inferred electron distribution and Eq. (1), we find for the electron injection rate $q(E) = 6.3\,10^{44} E^{-3}\, electrons\, s^{-1}\, keV^{-1}$ with E in keV. Integration of $q(E)$ between 1.07 keV (thermal energy plasma) and 18 keV (detection limit power-law) we find that per second $3.2\,10^{44}$ electrons have to be injected per second in order to maintain the energetic electron population against collisional losses. This represents an energy input between 1.07 and 18 keV of $\int q(E)EdE = 9.6\ 10^{35}\, erg\, s^{-1}$. Because the radiative losses between 1 and 18 keV equal only ($9\,10^{29}\, erg\, s^{-1}$), practically all of the injected energy will go into (fast) heating of the thermal plasma which is not observed.

An alternative explanation for the power-law emission is that we are dealing with energetic particles which are only interacting amongst themselves and not with the thermal ($T_e = 12.4\,10^6 K$) plasma. Self-interaction of energetic particles will on a relatively short timescale result in the formation of a Maxwellian distribution (Trubnikov, 1965). A Maxwellian distribution will however not result in a power-law shape of the photon spectrum. Therefore we need to invoke the presence of the distribution of Maxwellians by means of a differential emission measure distribution $\xi(T)$. For details we refer to Craig & Brown (1976). For thermal Bremsstrahlung from a differential emission measure distribution $\xi(T)$, the photon flux at the Earth is given by

$$J(\epsilon) = \frac{\alpha c}{4\pi d^2}\{\frac{8}{3\pi}\}^{1/2}\sigma_T \int_0^\infty \{\frac{mc^2}{kT}\}^{1/2}\xi(T)\frac{e^{-\epsilon/kT}}{\epsilon}dT \quad ph\, s^{-1}\, cm^{-2}\, keV^{-1} \tag{3}$$

with α the fine structure constant and σ_T the Thomson cross-section. By inverting this expression and using the observed spectrum, we find for the differential emission measure

$$\xi(T) = 6.8\,10^{45}\, T_7^{-3/2} \quad cm^{-3}\, K^{-1} \tag{4}$$

where $T_7 = T/(10^7 K)$. The total emission measure of the plasma with temperatures in excess of temperature T is given by

$$\int_T^\infty \xi(T')dT' = 1.4\ 10^{53} T_7^{-1/2} \quad cm^{-3} \tag{5}$$

This expression shows the interesting fact that it is not unrealistic to identify the thermal component resulting from the spectral fitting ($T = 12.4\,10^6 K, EM = 1.5\,10^{53}\, cm^{-3}$) as the low temperature part of a continuous plasma distribution covering a large range of temperatures. In that case the distinction between a thermal population and a power-law emitting population is only an artefact of our fitting procedure. A substantial amount of coronal plasma has therefore temperatures of the order of $10^8 K$. This hot plasma could not be observed with the previously used instrumentation (e.g. EXOSAT, EINSTEIN) because of the low sensitivity above $\sim 6keV$. Although fits of EXOSAT data implied temperatures of tens of million Kelvin, the instrument could not be used to determine accurately temperatures above $60\,10^6 K$.

3. Conclusions

One of the most surprising aspects of the GINGA observation of II Peg is the X-ray spectrum obtained during the quiescent state. The fact that the spectrum shows a power-law tail up to $\sim$ 18 keV requires the presence of extremely hot ($T > 10^8$ K) plasma in the corona of II Peg. Non-thermal particle populations fail in explaining the spectrum because of the large energy input required. This would result in a preferential collisional heating of the ambient plasma which is not observed.

If indeed a very hot plasma is present in the corona of II Peg then we could expect high temperature plasma also in other stellar coronae. Because only GINGA can be used to trace the presence of such plasmas, the number of observations is limited. We note however that Tsuru et al. (1989) found a power-law like tail in the quiescent spectrum of the RS CVn system UX Ari. Although these authors did not perform a differential emission measure analyses, they found that the spectrum could be explained by a single-temperature fit with a temperature of 46 $10^6 K$ and an emission measure of 2 10^{54} cm^{-3}. This temperature is indeed higher than the temperatures found with EXOSAT. A systematic study of "high-energy quiescent" emission from magnetically active stars remains however necessary.

Acknowledgements

We would like to thank the GINGA Observatory staff for their help in obtaining this data. Research at Armagh Observatory is grant-aided by the Dept. of Education for N. Ireland. We also acknowledge the support provided in terms of both software and hardware by the STARLINK Project which is funded by the UK SERC. G.H.J. van den Oord acknowledges financial support from the Royal Dutch Academy of Sciences (KNAW).

References

Brown, J.C., MacKinnon, A.L., van den Oord, G.H.J., Trottet, G., 1991, A&A 242,L13

Collier Cameron, A., 1988, MNRAS 233,235

Craig, I.J.D., Brown, J.C., 1976, A&A 49,239

Trubnikov, B.A., 1965, *Reviews of Plasma Physics, ed. M.A. Leontovich, Consultants Bureau Enterprises, New York*

Tsuru, T., Makishima, K., Ohashi, T., Inoue, H., Koyama, K., Turner, M.J.L., Barstow, M.A., McHardy, I.M., Pye, J.P., Tsunemi, H., Kitamoto, S., Taylor, A.R. & Nelson, R.F.: 1989, PASJ 41,679

THE LARGE CORONAL STRUCTURE AND MASS LOSS OF ACTIVE LATE-TYPE DWARFS

O.G. BADALYAN and M.A. LIVSHITS

Institute of Terrestrial Magnetism, Ionosphere and Radio Wave Propagation, 142092 Troitsk, Moscow Region, RUSSIA

Abstract. The mass loss rate of late-type dwarfs is estimated under the assumption that structures such as streamers exist in their coronae. Consideration of a Parker-type flow in slowly diverging coronal rays shows that the mass loss $\dot{M} \geq 10^{-11} M_{\odot}$ yr^{-1} in active dwarfs and the values of $\dot{M}$ are still larger for the K-subgiants — members of RS CVn-type binaries. Magnetic "braking" of such a wind should limit the time when the star is active to a duration of $\sim 2 \times 10^8$ years.

Key words: Stars: late-type — Stars: coronae — Stars: mass loss

1. Introduction

G. S. Vaiana was one of the first to show that X-ray emission originates from coronal arches. He noted (Vaiana 1983) that arches can occupy almost the complete volume of coronae of most active late-type dwarfs. Subsequently, it became clear that the soft X-ray spectra are well described by two-temperature models with temperatures $T_1 \sim 3 \times 10^6$ K and $T_2 \sim 20 \times 10^6$ K, and emission measures $EM(T_1) \approx EM(T_2)$ for active late-type stars. Setting aside the arch structure, we have interpreted the low-temperature component of coronal plasma in the framework of a hydrostatic isothermal corona. This allows us to determine in the same manner the densities at the base of the coronae of more than 30 active late-type stars (Katsova *et al.* 1987).

The hydrodynamic expansion of the corona develops at large distances from the stellar surface. On the Sun, the coronal structures at $r > 2R_{\odot}$ are predominantly extended in the radial direction. Moreover, the polarized intensity begins to decrease more slowly at $r \geq 2 - 2.5R_{\odot}$ for all position angles, and this behavior of the intensity is correlated with the onset of gas-dynamic expansion (Badalyan, 1988).

The magnetic force lines of the large-scale field are stretched out by the plasma flow, and this leads to the formation of coronal streamers. Physical processes in streamers are analogous to those in comet or magnetosphere tails (see review by Koutchmy & Livshits 1992). If such structures as streamers exist in the coronae of active late-type dwarfs, they can lead to some of the effects considered below.

2. Plasma Outflow in Streamers of Stellar Coronae

The calculation of mass loss rate from a star $\dot{M}$ is reduced to finding the flux (nvs), where the values of density, velocity and area are attributed to the moving plasma. The theoretical analysis of plasma flow in slowly diverging structures has been done by Badalyan (1991). A streamer is approximated by a cone, and the "beam" divergence is less than purely radial. The radius of the structure is proportional to $(r + x_0)$, where x_0 is the distance between the solar center and the top of the cone, which is located on the back side (behind the center). The velocity of the flow in such a structure for an isothermal plasma is described by the equation (Badalyan

J.F. Linsky and S. Serio (eds.), Physics of Solar and Stellar Coronae, 315–317.

1991)

$$(v/v_c)^2 - ln(v/v_c)^2 = 4(r_c^0/r_c)(r_c/r - 1) + 4ln[(r + x_0)/(r_c + x_0)] + 1, \quad (1)$$

where the critical velocity v_c is reached in a spherically symmetric atmosphere at distance r_c^0, and in the structure at distance r_c.

The densities in stellar coronae have been obtained from X-ray data. In our paper (Badelyan & Livshits 1992) we used the values $EM(T_1)$ and T_1 from the two-temperature models by Lemen *et al.* (1989), Pallavicini *et al.* (1988), and Swank *et al.* (1981).

It is obvious that in low-lying layers where velocities of the outflow are small, the density distribution is satisfactorily described by the hydrostatic approximation. The low-temperature component of the corona in these layers can be considered without taking into account the large-scale structures. At significantly greater heights, the plasma is flowing in isolated streamers. To make definite the description of the transition from nearly stationary ($v << v_c$) plasma to moving plasma in developed structures, we assume that at the level where $v \simeq 0.4v_c$, the density can be considered as hydrostatically distributed.

Thus, the flow (nvs) is defined at the level where $v = 0.4v_c$. The distance r_x corresponding to this condition is calculated from (1); the density n_x at distance r_x is obtained in accordance with the hydrostatic law. For the stars under consideration it has been assumed that the total cross section area of the streamers is equal to 1/4 of the surface of the sphere with radius r_x.

Finally, we have for the rate of mass loss from the star

$$\dot{M} = 3.15 \times 10^{-50} n_x(0.4v_c)\pi r_x^2 \quad (2)$$

in $M_\odot$ yr^{-1}, where the mean mass of the particle $m = 2 \times 10^{-24}$ gram has been adopted to calculate the constant $mt_0/M_\odot$ (t_0 is the number of seconds in one year).

Results of calculations of $\dot{M}$ for the typical active stars are given in Table I. We include the following stars: 111 Tau — the young F star of age 3×10^8 years, the active G star κ Cet, and three well-known spotted stars EQ Vir, CC Eri, and BY Dra. Also calculations have been carried out for two RS CVn-type binaries with periods of 6.4 days (UX Ari) and 104 days (α Aur). The following values are shown in Table I: the name of the star and its spectrum (for three binary stars only the active component is indicated), the stellar radius R and mass M, the temperature $T = T_1$, and emission measure $EM = EM(T_1)$ of the low-temperature component of the coronal emission, the density at the base of corona n_0 (a parameter of the hydrostatic law), the characteristics of the plasma flow $0.4v_c$ and r_c^0, r_c, r_x expressed in the radii of the stars, and the final value of the mass loss $\dot{M}$ in terms of $M_\odot$ yr^{-1}.

3. Conclusion

If stellar coronal streamers exist, the mass loss rate will be 2 or 3 orders of magnitude larger than in the solar case. These values are close to those for a spherically symmetric corona with $T = T_1$. The increase in $\dot{M}$ as compared with the solar case is caused by the high densities in the coronae and by the slow decrease of these densities up to the level where stellar wind is formed. If the developed model is valid, the magnetic "braking" mechanism should limit the time when the star is active to a duration of $\approx 2 \times 10^8$ years.

TABLE I

star name	spectral type	radius in $R_\odot$	mass in $M_\odot$	T_1 $10^6 K$	EM $10^{52}\mathrm{cm}^{-3}$
111 Tau	F8 V	1.2	0.84	3.8	1.1
κ Cet	G5 Ve	1.2	1.84	3.9	0.43
EQ Vir	K5 Ve	0.65	0.46	4.1	0.79
CC Eri	K7 Ve	0.80	0.74	4.7	1.7
BY Dra	M0 Ve	0.75	0.60	4.0	2.1
UX Ari	K0 IV	3.1	1.7	8.0	60
α Aur Ab	F9 III	7.1	2.7	4.7	5.6

star name	n_0 $10^9\mathrm{cm}^{-3}$	$0.4v_c$ km s^{-1}	r_c^0	r_c	r_x	n_x $10^9\mathrm{cm}^{-3}$	$\dot{M}$ $10^{-11} M_\odot\ \mathrm{yr}^{-1}$
111 Tau	2.4	91.6	1.28	6.88	2.26	0.58	1.9
κ Cet	1.6	92.8	1.55	7.64	2.61	0.24	1.0
EQ Vir	4.9	95.1	1.22	6.69	2.17	1.3	1.2
CC Eri	5.7	101	1.37	7.13	2.37	1.2	2.1
BY Dra	7.0	94.0	1.40	7.23	2.41	1.3	2.0
UX Ari	2.1	133	0.49	7.24	1.31	1.6	16.4
α Aur Ab	0.21	102	0.57	7.81	1.48	0.12	6.5

References

Badalyan, O.G.: 1988, *Sov. Astron.* **32(2)**,

Badalyan, O.G.: 1991, *Sov. Astron.* **35(4)**,

Badalyan, O.G., Livshits, M.A.: 1992, *Sov. Astron.* **36(1)**,

Katsova M.M., Badalyan O.G, Livshits M.A.: 1987, *Sov. Astron.* **31(6)**, 652

Koutchmy, S., Livshits, M.: 1992, *Sp. Sci. Rev.*, in press

Lemen, J.R., Mewe, R., Schrijver, C.J., Fludra, A.: 1989, *Astrophys.J.* **341**, 474

Pallavicini, R., Monsigniori-Fossi, B.C., Landini, M., Schmitt, J.H.M.M.: 1988, *Astron & Astrophys.* **191**, 109

Swank, J.H., White, N.E., Holt, S.S., Becker, R.H.: 1981, *Astrophys.J.* **246**, 208

Vaiana, G.S.: 1983, in Stenflo, J.O., ed(s)., *Solar and Stellar Magnetic Fields: Origin and Coronal Effects*, Dordrecht: Reidel, 165

DISCUSSION ON STELLAR CORONAE: THE PRE-ROSAT PICTURE

Discussion following paper by S. Sciortino

T. Fleming: Later at this meeting J. Schmitt will present results from a paper which we have just completed on late-M dwarfs detected in the ROSAT All-Sky Survey. We claim that there is no drop in coronal heating efficiency at spectral type M6. I think the difference in our results stems from the paucity of late-M dwarfs ($\sim$36 later than M5). Since *Einstein* covered only 10% of the sky, there are no data on most of them, whereas ROSAT covered almost 100% of the sky.
S. Sciortino: (1) The representative sample on which Barbera et al. (1992) base their conclusion is composed of $\sim$ 25 stars with $M_v > 13.4$. In this range of M_v *Einstein* observed much more than 10% of the known stellar sample. (2) If $L_x/L_{\rm bol}$ were constant then young very-late type stars should have $L_x \sim 10^{27.5} - 10^{28.5}$ ergs s^{-1}. Only very few such stars should have been seen with *Einstein* observatory in the Hyades cluster (limiting $L_x \sim 10^{28.4}$) and could be hidden in the unidentified objects. ROSAT pointed observations of the Hyades and the Pleiades should definitely clarify the reality of a drop in $L_x/L_{\rm bol}$.
Y. Uchida: Is there any possibility that the scatter in the $L_x/L_{\rm bol}$ diagram can be attributed to magnetic effects? I once suggested a possible modification of radiatively driven winds by magnetic fields in hot stars with energy stored in magnetic fields and released occasionally. Although there is no convection in these stars and therefore no dynamo process in the usual sense, it is possible that the fossil field brought into the star during its formation is strengthened and does the job.
S. Sciortino: The critical issue regarding O stars is that none of the "coronal-like" models proposed to explain X-ray emission does a good job, because the predicted spectra always have a clear absorption feature that has not been observed. The problem is the location of the X-ray emitting regions.
R. Rosner: The primary reason solar-like coronal models for OB stars (which were invented before *Einstein*) are no longer considered valid is that with a few exceptions the absorption of photons near $E \simeq 0.78$ keV is predicted but not observed.
J. Schmitt: I would like to comment on the calculation of the contribution of stars to the DXRB. The essential quality required is $L_{x,\rm mean}$. For a skewed distribution function (like the dM stars) $L_{x,\rm mean}$ and $L_{x,\rm median}$ may differ by almost one order of magnitude. Therefore, in a sample with, say, 100 objects only a very small number in the tail actually determines the mean (with correspondingly large errors). The advantage of using parametric methods is that the whole population contributes in the calculation of the mean. The disadvantage is of course that one must assume some parametric model (which can however be tested).
S. Sciortino: (1) While it is true that $L_{x,\rm mean}$ is the essential quantity, it is also true that our present knowledge is limited and only a full resampling via bootstrap of the "original" ML X-ray luminosity function is the correct way to explore and quantify the range of acceptable computed contributions to the DXRB. The shape of the distribution is also an essential feature. (2) The parametric models make the

J.F. Linsky and S. Serio (eds.), Physics of Solar and Stellar Coronae, 319–323.

assumption that all the stars we observe belong to the same class, so that you can assign them an $L_{x,\mathrm{mean}}$. Adopting the full ML X-ray luminosity function is safer if the tail is indeed due to a somewhat different class of objects.

R. Pallavicini: The existence of a saturation effect is a good way to explain the discrepancies in the X-ray/rotation relationship, but does it imply that the physical conditions (n,T,p) in the coronae of very active stars are the same as in the Sun, i.e. they depend on dynamo and magnetic fields. Do you think that typical active region structures in active stars are similar to those on the Sun?

S. Sciortino: I agree with your suggestion that the physical conditions in very strong X-ray coronal sources can be more extreme than in the Sun. I do not know if this can explain the saturation effect, nor do I feel that we have enough data to properly address the problem. One should keep in mind that the description in terms of a surface area filling factor is really naive, given that the X-ray luminosity is indeed a volume effect as shown clearly by solar observations.

Discussion following the paper by R. Mewe

S. Serio: A comment on the loop expansion factor you require to fit the data. If you use $\Gamma = 1$, probably the data can be fitted equally well using two different loops.

R. Mewe: The EXOSAT transmission grating spectra of the three late-type stars were fitted by a two-component loop model. We obtained for both components an expansion factor which is significantly ($\geq$ 90% confidence) larger than unity, so that in these cases the Rosner-Tucker-Vaiana $\Gamma = 1$ model could be excluded.

J. Linsky: For more than 10 years there has been a controversy concerning whether stellar emission measure distributions truly have peaks at two temperatures, or whether the two peaks are instrumental artifacts. In the next decade high-resolution spectra from XMM and AXAF will provide the answer to this question. In your opinion which of these two possibilities is correct?

R. Mewe: All medium resolution and often also broadband data from *Einstein*, EXOSAT and ROSAT require a two-temperature best fit. The few observations of eclipsing binaries reveal the existence of distinct separate emission regions with usually different temperatures. I think the 2-T structure is real and not an instrumental artifact, but the different locations of the temperature peaks for different instruments is an artifact presumably due to the different passbands of the instruments. In our differential emission measure analysis of our EXOSAT grating spectra we checked very carefully the reality of the apparent minimum in between the two different temperature regimes. I am convinced that AXAF and XMM will confirm this (and may even find a more complicated temperature structure).

Discussion following paper by R. Pallavicini

B. Somov: What is the highest energy at which stellar flares are observable in X-rays?

R. Pallavicini: So far, stellar flares have been observed only in soft X-rays, at energies less than $\simeq$ 10 keV. There are no observations of stellar flares in hard X-rays and there is little hope for such observations in the near future.

A. Maggio: Is there any observational evidence for a relationship between the quiescent X-ray emission and variability levels, either in terms of flare frequency or in terms of percentage modulation on medium/long time scales?

R. Pallavicini: For stars observed at comparable S/N levels, the variability is larger for intrinsically weaker sources, probably because we can detect only the largest variations in intrinsically brighter sources and/or because very active stars are more uniformly covered by X-ray bright sources.

F. Walter: (1) Two out of three AR Lac eclipses were detected by the *Einstein* MPC, which has a $2-6$ keV response, and is thus sensitive to "high" energies. (2) The modulation amplitudes you quoted for the Sun were with flares removed. What would the modulation amplitudes be if one included a mean flare contribution? Would the flares have larger contrast when there are lower mean flux levels, thereby decreasing the modulation amplitudes?

R. Pallavicini: (1) The EXOSAT observation of AR Lac is unique in this sense (but one case against two!). The EXOSAT result for AR Lac, however, is consistent with similar results for other eclipsing binary systems (Algol, TY Pyx). (2) The rotational modulation effect is present also for X-Y9 flares alone, with amplitude comparable to or larger than that of active regions. So I expect that the modulation would in fact be larger if you include flares (and the data would be more noisy).

G. Peres: Solar microflares are events below 10^{27} ergs. When we talk of stellar microflares we mean something much more energetic.

R. Pallavicini: Yes, of course. The stellar "microflares" one can hope to detect at present have luminosities comparable to the star's quiescent luminosity. But we cannot exclude that stellar "microflares" (if they exist) are more energetic than solar "microflares" in the same way as stellar flares are typically stronger than solar flares. The crucial issue is whether these low-amplitude "fluctuations" (independently of how you call them) are continuous or not.

Discussion following paper by F. Walter

R. Pallavicini: (1) Your statement that the NTTS greatly outnumber the CTTS seems to be contradicted by most X-ray surveys that show a ratio 1:2. (2) Do you see any evolutionary effects among different NTT stars in your sample?

F. Walter: (1) It depends when you look. Over time, CTTS presumably evolve into NTTS. In Tau-Aur, the ratio is $\sim$ 1:1 when only stars with ages $\lesssim 10^6$ yr are considered. When extrapolating over the entire Tau-Aur complex, we integrate over $\sim 3 \times 10^7$ yr and get a much larger ratio of NTTS/CTTS. In the SCO OB2 and Ori OB1b associations, with ages $5-10 \times 10^6$ yr, the ratio NTTS/CTTS $\sim$10. In younger, more co-eval associations, such as ρ Oph or Cha I, the ratio is 2:3. It is clearly a time-dependent quantity. (2) Not yet. We will be looking for differences in the IMF's between the T and OB associations, as well as differences in stellar properties (activity levels, rotation) with age.

Discussion following paper by J. Linsky

R. Rosner: Jon Arons published a paper in the early 70's in which he pointed out that if one loaded the magnetosphere of an Ap star with plasma, then such a star would become "active." This is in fact exactly what Jupiter has managed to do.
G. Field: Jupiter's activity is driven by the solar wind, but there is no companion star whose wind hits the A star magnetosphere.
Y. Uchida: The mechanism you are proposing was discussed by me (Y. Uchida 1985, in the NASA symposium, "The Origin of Non-Radiative Heating/Momentum in Hot Stars," p. 199), but not for the A-type stars. I proposed this mechanism for hotter stars in which the fossil field can interact with the radiation-driven winds, which then energize the magnetic field.
S. Sciortino: Can you be sure that the high L_x of the two late B type stars you mentioned are not due to X-rays from a PMS late-type companion? Could you comment on this possibility?
J. Linsky: The two late-B stars with detected X-ray and radio emission have x-ray luminosities $\log L_x = 30.13$ and 30.28, respectively. These X-ray luminosities are much larger than typically found in field dwarfs with spectral types F to late-M. However the late-B stars are relatively young and putative companions would also be young and likely to be X-ray luminous. We have requested ROSAT pointed observations of a sample of chemically peculiar stars to provide a more definitive answer to your question.
F. Damiani: (1) Does your model for the radio emission of Ap-Bp stars predict a $\dot{M} - L_x$ correlation for these stars? (2) Even if this correlation is at present only an empirical one, do you think that a similar "magnetospheric" picture may hold even in the case of the Herbig Ae-Be PMS stars, which do have strong winds and are sometimes (and surprisingly) detected in the radio and X-rays?
J. Linsky: (1) We have not developed this model sufficiently far yet to predict this correlation. (2) I would expect that the Herbig Ae-Be stars would also follow this relation if they have strong magnetic fields.
J. Schmitt: I agree with Jeff that the book on X-ray emission from A-type stars was closed too early. The ROSAT results on A-type stars are still being analyzed but are nevertheless confusing. The nondetection of Vega (A0V) has been confirmed, but a number of other cases have been found (with the ROSAT HRI), where the X-ray emission certainly comes from the A-type star. I would like to encourage observers at other wavelengths to take a close look at the X-ray emitting A stars.

Discussion following paper by F. Favata

G. Tagliaferri: We have performed CORAVEL observations of the EXOSAT sample of stars in order to determine the orbital motions and to check if they are running away from some star associations.

Discussion following paper by G. Tagliaferri

F. Walter: How do you get log $N(\mathrm{Li}) \gtrsim 4$ for Tau-Aur? Most estimates for these stars, e.g. work of Boesgaard and her students, get values closer to the cosmic abundance of ~ 3.

G. Tagliaferri: The values that I plotted for Tau-Aur are not mine but are taken from a recent paper of Basri, Martin, & Bertout (A.&A. **252**, 625 (1992)). These high values are probably due to saturation effects in the Li 6708 Å line, as noted by these authors.

D. Duncan: High lithium abundances ($\gg 3$) are found by others in T-Tauri stars, but when the line is this strong, NLTE effects are likely, and the absolute abundances are less reliable. A logarithmic abundance which appears to be ≈ 4 may correspond to an actual Li abundance significantly < 4.

G. Tagliaferri: This is a remark with which I fully agree.

S. Drake: How are the Li-strong stars that you believe are pre-main sequence in nature distributed spatially?

G. Tagliaferri: With one exception, all of our stars were selected not to be members of any known cluster or association.

M. Barbera: Why should a serendipitiously selected X-ray sample have such a high fraction (1/4 to 1/3) of pre-main sequence (PMS) stars?

G. Tagliaferri: It is well known that PMS stars have very high X-ray luminosities, so it is not unexpected to have such a high percentage of PMS candidates. For instance, many could be T-Tauri stars which have been found recently in optical studies of a binary system with an early-type primary star and a late-type secondary (e.g. Pallavicini et al. 1992; Martin et al. 1992).

R. Rosner: I am struck by your comment that so many of the strong Li stars in your sample are PMS stars. Since these are field stars, it would seem that explaining where these stars come from is at least as difficult a puzzle as explaining why they are very active with strong Li. Any comments?

G. Tagliaferri: They could be PTTS (post T-Tauri stars) that are more evolved than classical T-Tauri stars and are no longer associated with star forming regions, for example, the sample of PTTS proposed by Lindros (1986) and studied by Pallavicini et al. (1992) and Martin et al. (1992). We have CORAVEL observations to see whether our PMS candidates are running away from star forming regions.

F. Walter: The more we observe Li, the more confusing it becomes. There are clear cases of stars with cosmic abundance, e.g., AB Dor, which are not PMS, and should have depleted at least some of their Li. Some PMS stars may be underabundant in Li. Soderblom and collaborators have presented evidence (at a recent AAS meeting) that, in the Pleiades, the more active stars have larger Li abundances. Does activity inhibit Li depletion? Does activity create Li via spallation reactions? A cosmic abundance of Li in a star not associated with a star forming region is interesting, but not necessarily an indication that the star is PMS.

G. Tagliaferri: At the moment the best (or most accepted) classification for AB Dor is that it is a PTTS, somewhat more evolved than a classical T-Tauri star. I agree with your remark, but for our stars that have Li abundances comparable with Tau-Aur PMS (and higher than the Pleiades) I think it is very unlikely that the high Li abundances can be due to the high activity levels. This is probably the case, instead, for the other Aur stars whose Li abundances are like that of the Pleiades.

PART IV

ROSAT OBSERVATIONS OF STELLAR CORONAE

ROSAT OBSERVATIONS OF LATE-TYPE STARS

J.H.M.M. SCHMITT
Max-Planck-Institut für extraterrestrische Physik
8046 Garching, F.R.G.

Abstract. I will review results from X-ray observations of late-type stars in the ROSAT all-sky survey as well as in the ROSAT pointing program. The various background components and the resulting sensitivity of the all-sky survey will be discussed. Among the scientific issues covered are coronal dividing lines, hybrid stars and X-ray microflares.

Key words: stars: X-ray radiation – stars: coronae

1. Introduction

Observations with the *Einstein Observatory* and EXOSAT have shown the ubiquity of stellar X-ray emission (cf. Vaiana *et al.*, 1981). While the data collected with these observatories were extremely useful in acquiring information on individual sources, a major disadvantage resulted from the fact that only a very limited fraction of the sky was viewed (less than 10 percent of the sky was covered by pointed *Einstein* IPC observations). Consequently, the construction of large **unbiased** samples of objects was difficult if not impossible. The *Einstein* Medium Sensitivity Survey (EMSS) may be used as an example to demonstrate this: Many (but not all) of the nearby RS CVn systems were the target of pointed observations and hence – by definition – excluded from the EMSS, the same applies to the nearby stellar population. Thus the EMSS has a deficiency of nearby objects which makes a construction of the low-luminosity part of the X-ray luminosity distribution function impossible. Therefore, a major goal of the ROSAT mission was to carry out the first all-sky survey with an imaging X-ray telescope which - in particular - allows the construction of such large and unbiased samples of various classes of X-ray sources. This first major goal of the ROSAT mission has been accomplished: The ROSAT all-sky survey has been completed and almost the entire sky has been scanned down to a limiting flux of approximately 2×10^{-13} erg s^{-1} cm^{-2} in the pass band $0.1 - 2.0$ keV; in regions of deeper exposures near the poles of the ecliptic considerably fainter flux limits have been achieved.

2. The PSPC Background

A full understanding of the results of the ROSAT all-sky survey requires an understanding of the PSPC background (cf. Schmitt 1991). Point sources such as stars must be detected on a background typically composed of three components: Particle-induced background, scattered solar X-rays, diffuse cosmic X-ray background. Particle-induced background is produced by cosmic rays penetrating the PSPC without being rejected by the counter's veto electronics. Extensive measurements of this background have been carried out under a variety of orbital conditions (cf. Snowden *et al.* 1992); typical particle-induced background rates are $\sim 1 - 2 \times 10^{-5}$ cts s^{-1} arcmin^{-2}. The pulse-height spectrum of this component is essentially flat with a noticable increase towards the very lowest energies. The

J.F. Linsky and S. Serio (eds.), Physics of Solar and Stellar Coronae, 327–336.

second background component are solar X-rays scattered off ionospheric atoms into the beam of the XRT. The magnitude of this background depends very sensitively on the viewing geometry. It is significant only at large zenith angles during day-side viewing when count rates of up to 10^{-2} cts s^{-1} $arcmin^{-2}$ are encountered, while typical rates are between zero (night-side viewing) and a few times 10^{-4} cts s^{-1} $arcmin^{-2}$. Obviously, this background component essentially reflects the incoming solar X-ray spectrum and hence is very soft, with the majority of the events usually appearing at energies below 0.28 keV. However, quite frequently a strong oxygen K_α line is also observed in the scattered solar X-ray spectrum. Finally, sources must be detected against the general cosmic diffuse X-ray background; this backgound varies over the sky with rates of $\sim 5 \times 10^{-4}$ cts s^{-1} $arcmin^{-2}$ encountered at low galactic latitudes up to $\sim 20 \times 10^{-4}$ cts s^{-1} $arcmin^{-2}$ found at higher galactic latitudes. The spectrum of the diffuse X-ray background is also rather soft, with approximately 70 percent of its intensity appearing in the C-band below 0.28 keV. I would like to emphasise that the ROSAT PSPC's sensitivity is – at least for normal viewing conditions – determined by the diffuse X-ray background which represents a major advantage over all previously flown imaging X-ray instruments.

3. Sensitivity of the all Sky Survey

During the all sky survey the XRT scanned the sky once per orbit along great circles containing the north and south ecliptic poles. The scanned longitudes slowly moved with the apparent motion of the Sun along the ecliptic, i.e. at a rate of $\sim$ 1 degree/day. Since the field of view of the XRT/PSPC combination is 2 degrees, sources at low ecliptic latitudes could be viewed for 2 days; for sources at higher latitudes the viewing time increased roughly as $1/cos(\beta)$, with β being the ecliptic latitude. An observation of a source consists of a sequence of individual scans, each lasting between 20 to 30 seconds. During each scan the XRT/PSPC point response function varies rapidly. On-axis 50 percent of the photons are contained within a circle with 15 - 25 arcsec (dependent on energy) while at 50 arcmin off-axis an energy-independent circle with 160 arcsec radius is required. In the survey mode one finds about 80 percent of the photons in a 150 arcsec circle. Within such a detect cell the background rate (assuming "nominal" conditions of 10^{-3} cts s^{-1} $arcmin^{-2}$) 2×10^{-2} cts s^{-1} (detect cell)$^{-1}$. The minimum detectable count rates are typically of the same order as this mean background rate. Since the numbers of photons are so small, the Poissonian nature of the photon arrival times must be explicitly considered when dealing with source detection in the ROSAT all-sky survey. Highly significant sources may be detected with only a few counts and correspondingly large count rate errors.

As an example of a weak detection I would like to present the case of ϵ Tau. Obviously, as is the case in most surveys, the majority of sources is found close to threshold, and therefore will be similar in appearance to the case of ϵ Tau. The physical significance of the all-sky survey detection of ϵ Tau lies in the fact that it is the only one of the four Hyades giants not previously studied in X-rays; the four Hyades giants, i.e. δ^1 Tau, ϵ Tau, θ^1 Tau, and γ Tau are all very similar in their optical properties, but differ very much in their ultraviolet and – as shown

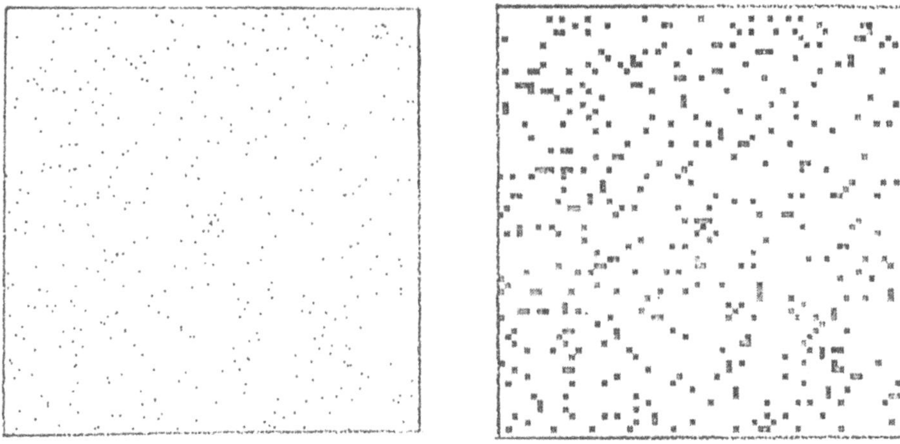

Fig. 1. Image to the left: Photon distribution in a 34 arcmin x 34 arcmin region as obtained in the ROSAT all-sky survey around the position of ϵ Tau; note the weak source present in the center of the image. Image to the right: The same data as in left image but now binned into 32 arcsec spatial bins; the presence of a source in the image center becomes obvious.

now with the all-sky survey data – in their X-ray properties; for a more detailed discussion and the physical implications of this detection see Stern *et al.* (1992).

In Figure 1a I show a photon plot of a 34 arcmin x 34 arcmin region centered on ϵ Tau; an experienced eye (such as mine) does (clearly) see a source; this fact becomes obvious when the same data as shown in Figure 1a is now plotted in a binned representation. Figure 1b shows the same data as Figure 1a, but now the data have been spatially binned into 32 arcsec x 32 arcsec pixels; the presence of a source in the center, i.e. at the position of ϵ Tau, is now evident. A detailed mathematical treatment of the photon distribution shown in Figure 1a shows that the significance of the source existence (in terms of existence likelihood) is 17 (with 10 being the threshold normally used). The all-sky survey count rate of ϵ Tau is 2.5×10^{-2} cts s^{-1}, which corresponds to an X-ray luminosity of $\sim 3.7 \times 10^{28}$ erg s^{-1}.

4. Coronal dividing lines

A so-called coronal dividing line was discovered by Ayres *et al.* (1982) and Haisch and Simon (1982) as a line separating X-ray emitting yellow giants and supergiants from non X-ray emitting red giants and supergiants using X-ray data obtained with the *Einstein Observatory*; the discovery of this X-ray dividing line was preceded by the discovery of a so-called wind dividing line with the IUE satellite (Linsky and Haisch 1979), i.e. a line separating giants with evidence for transition region temperature material in their ultraviolet spectra from stars showing evidence for massive cool winds (but no warm or hot material). The data collected in the ROSAT all-sky survey offers the possibility to test for the presence of a dividing line using large data samples.

Haisch, Schmitt and Rosso (1991) constructed a sample of non-binary giant stars for which X-ray data was available and which could be considered as single stars. Figure 2 shows an HR-diagram for 868 ROSAT non-detections, for 96 stars which were detected in the ROSAT all-sky survey, as well stars with measurabale cool winds. As is clear from Figure 2, almost all the detections are of spectral type K3 or earlier, only one star, HR 4289, is of spectral type K4 or later. Because of the large number of sample objects available, the statistical significance of this X-ray dividing line is extremely high. It is finally worthwhile noting that the star HR 4289, the only detection later than K3, appears to be somewhat unusual. A follow-up PSPC pointing on HR 4289 yields an unusual X-ray spectrum, which optical obserations indicate an incorrect classification (Reimers 1992); further studies of HR 4289 are definitely called for.

5. Hybrid stars

A particularly interesting class of stars are hybrid stars; these stars are located in the vicinity of the dividing line, yet their ultraviolet spectra show evidence both for cool wind material as found in the red giants to the right of the dividing line as well as warm material at tranition region temperatures. The X-ray properties of hybrid stars are unclear. Haisch (1987) reviewed the *Einstein Observatory* data on hybrid stars and concluded that no convincing detections of hybrid stars were obtained at X-ray wavelengths. Brown *et al.* (1990) present EXOSAT observations of the hybrid star αTrA, demonstrating that (at least some) hybrid stars also sustain coronae. Haisch, Schmitt and Rosso (1991) checked the ROSAT all-sky survey data for detections of hybrid stars; the detection of α TrA by Brown *et al.* (1990) could be confirmed, another hybrid star, δ And was detected, while all other hybrids listed by Haisch (1987) were not detected in the ROSAT all-sky survey. Unfortunately, δ And is a rather complex binary system and the observed X-ray emission cannot be unambiguously attributed to the giant star; however, Haisch, Schmitt and Rosso (1991) present arguments that the bulk of the X-ray emission does indeed come from the hybrid star. Finally, both the detections for α TrA and δ And were obtained in both the soft and hard energy band of the ROSAT PSPC, possibly indicating X-ray temperatures of $\sim 10^7$ K; however, the detections are quite weak, and the spectral results definitely require confirmation through pointing data (cf. Rosner 1992).

Reimers and Schmitt (1992) present detections of three more hybrid stars, i.e. β Ind, μUMa, and γ Aql, obtained in the ROSAT pointing program, while for ι Aur only an upper limit could be obtained. The deficiency of hybrid star detections in the ROSAT all-sky survey therefore appears to be mainly due its limiting sensitivity, while deeper pointings do yield detections. Thus hybrid stars as a class seem to be

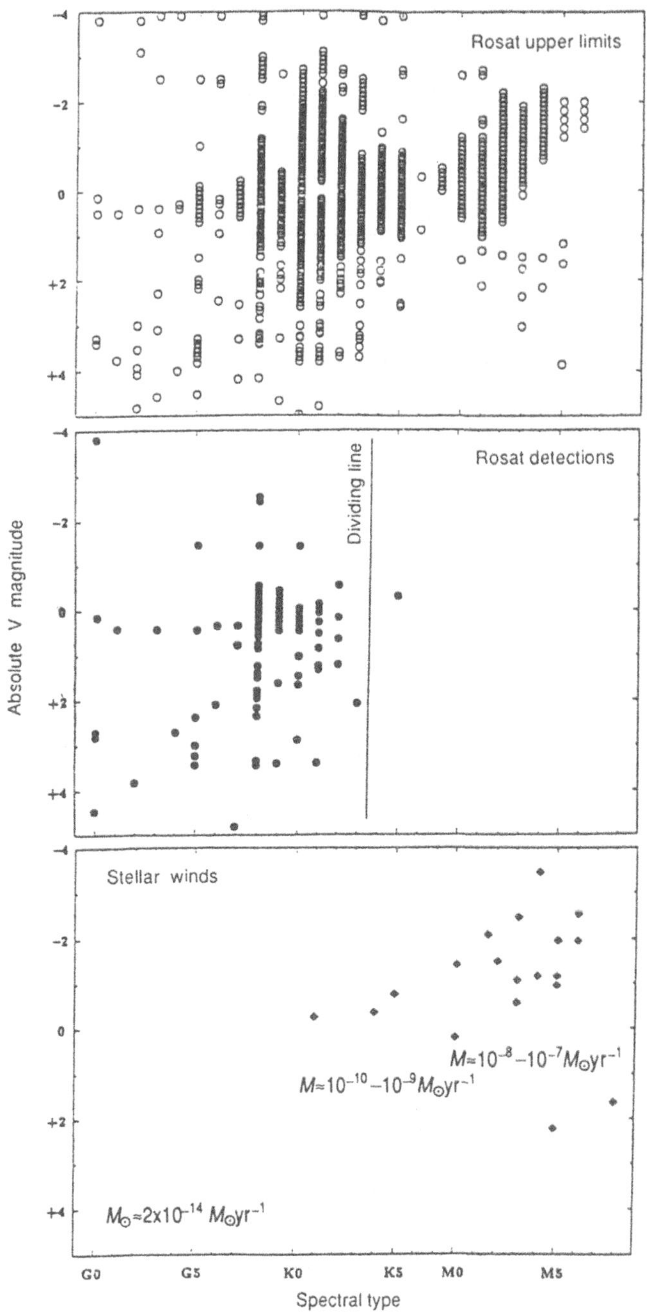

Fig. 2. HR diagrams for evolved stars showing the ROSAT all-sky survey non-detections (upper panel), the ROSAT all-sly survey detections (middle panel), and stars with measurable cool winds (bottom panel) vs. spectral type (taken from Haisch, Schmitt and Fabian 1992); note the mutually exclusive appearance of stellar coronae and cool winds.

X-ray sources, and hence it appears that whenever material at transition region temperatures is present, i.e. CIV emission is observed, a corona is also formed.

6. Stellar microflares

Most stellar X-ray detections so far discussed refer to quiescent X-ray emission, where I use the term "quiescent" purely operationally in the sense that the observed X-ray emission appeared to be steady during the individual observations. Flaring stellar X-ray emission has of course also been observed (see the review by Haisch, Strong and Rodonò 1991 and references therein) in quite a number of late-type stars. Specifically, the ROSAT all-sky survey observations allows an **unbiased** study of the flaring properties of late-type stars, and indeed X-ray flares have been found on all types of of late-type stars. Thus, the long known dichotomy between quiescent and flaring solar X-ray emission can be fully extended to the case of late-type stars in general.

The distinction between quiescent and flaring X-ray emission has so far been made only from the observational point of view; a far more interesting question is of course whether the physical processes leading to quiescent and flaring X-ray emission are the same or not. Spatially resolved X-ray observations of the solar corona at lower energies (typically in the range 0.2 - 5 keV) show - at least outside flares - all coronal structures to be steady on the time scales of the radiative cooling time or the Alfvén travel time (Vaiana and Rosner 1978); hence all changes in coronal structure can be interpreted in the context of the evolution of quasistatic models which require some form of stationary energy input ("heating") to balance the radiative energy losses. On the other hand, spatially unresolved observations of solar hard X-rays above 20 keV (cf. Lin *et al.* 1984) have revealed a wealth of variability on time scales of a few seconds; these hard X-rays are thought to be the thermal bremsstrahlung from non-thermal (i.e. somehow accelerated) electrons. Weaker hard X-ray flares ("microflares") are found to occur far more frequently than stronger flares; in fact, the cumulative distribution of the number of events ($N > F$) in excess of some peak flux (F) varies inversely proportional to the peak flux without any apparent cut-off at threshold. Therefore one may see only the "tip of the iceberg" and a lot of power can still be hidden in unresolved events of lower flux. As pointed out by Lin *et al.* (1984), a straightforward extrapolation of the observed log N - log F relationship can easily account for the total "quiescent" energy output of the solar corona. These observations together with theoretical considerations of the stability configuration of solar coronal magnetic fields have lead to an alternative view that quiescent and flaring X-ray emission may actually be manifestations of the same physical heating processes.

It is natural to extend these ideas also to the case of stellar X-ray emission. This has been done by Butler *et al.* (1986), who present evidence for stellar microflares from X-ray observations obtained with the EXOSAT satellite and simultaneously obtained H_γ spectra, by pointing out a high degree of correlation between the simultaneously recorded variations in the X-ray and H_γ flux. Since some of the observed variations appeared to occur on time scales as short as 30 seconds, they were coined as "microflares" by Butler *et al.* (1986) in analogy to the hard X-ray microflares, and they further proposed that a signifcant reduction in sensitivity would yield far more low level flares and that stellar X-ray emission might be viewed simply as a "succession of microflares."

The microflare interpretation presented by Butler *et al.* (1986) has been challenged by Collura, Schmitt and Pasquini (1988) and by Pallavicini, Stella and Tagliaferri (1990), who confirm the wide-spread X-ray variability found in flare stars, but argue that the statistically significant X-ray variability time scales are of the order of a few hundred seconds and hence much more comparable to the typical time scales of compact loop flares observed on the Sun. At any rate, with a pointed ROSAT observation on UV Ceti it is straightforward to check whether microflares as observed by Butler *et al.* (1986) can be found or not. The total energy fluence in those events was $\sim 7 - 10 \times 10^{-10}$ erg cm^{-2}, distributed over a 30 second interval. With the ROSAT PSPC, which is an order of magnitude more sensitive than the EXOSAT LE, this corresponds to $\sim$ 100 counts, again distributed over 30 seconds.

We carried out a 25 ksec pointed ROSAT observation on UV Ceti. In contrast to the EXOSAT observations, the ROSAT observations were not contiguous, but rather consist of about two dozen shorter contiguous observations of about 2000 seconds each; the observations were spread out over a period of two weeks between December 30, 1991 and January 15, 1992. In Figure 3 I show all of the data obtained in this observation as the number of counts recorded in a 30 second interval vs. a sequential bin number (in order to avoid having to show all the data gaps); the jumps in the light curve are thus produced not by flares, but rather by jumps in time and are indicated by vertical lines. Microflares such as those observed by Butler *et al.* (1986) should manifest themselves as bins containing $\sim$ 100 counts, surrounded by bins containing far fewer counts. As is clear from an inspection of Figure 3, such events cannot be found in this observation. 30 second bins with 100 or more counts can of course be found, however, such bins do not occur in isolation, but rather during the progress of large flares. The conclusion therefore is that microflares such as observed by Butler *et al.* (1986) in their EOXSAT observation were not present during the ROSAT observations. However, during part of the ROSAT observations we have simultaneous high-speed optical photometry during the occurrence of small flares; the morphological similarity between X-ray and optical flares is very high and is discussed in detail by Schmitt, Haisch and Barwig (1993).

In Figures 4 and 5 the basic result of these measurements are shown. In order to obtain the optical light curve, the simultaneously measured sky background was removed and the source signal was divided by the also simultaneously measured constant comparison star; further, a running mean of 5 seconds was applied to the data (which were originally recorded at 1 second time resolution) in order to reduce measuring noise. In Figure 4 we show the U-band light curve (vs. heliocentric Julian date - 2448844.0) obtained in the manner described above; the data are shown with the full time resolution (but a 5 sec running mean has been applied). Two obvious flares are visible: one event at HJD = 0.2895, and another event at HJD = 0.2951; we note in passing that both flares are also clearly visible in the simultaneously recorded U-band light curve. The first event is impulsive only, with rise and decay phase totalling less than 10 seconds; the second event shows an impulsive flare followed by a gradual phase which lasts at least five minutes. Both flares have

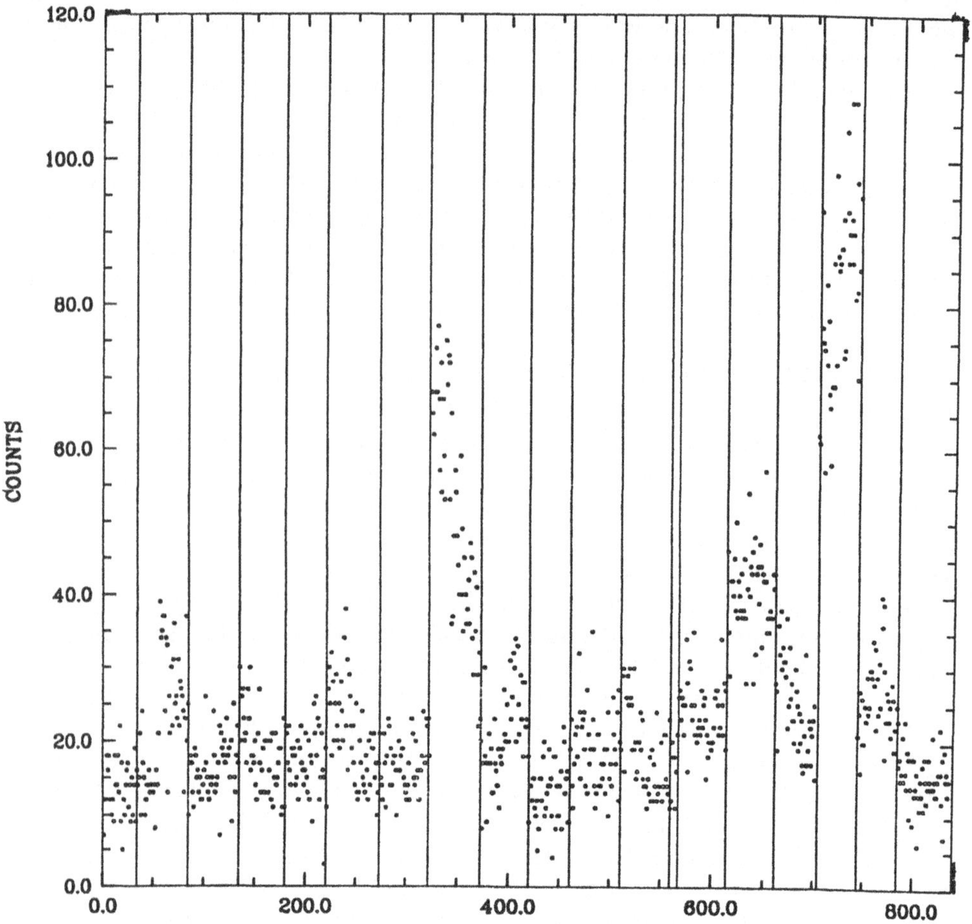

Fig. 3. Raw number of counts (recorded in 30 second integrations) vs. sequential bin number for a 25 ksec ROSAT pointed observation of UV Ceti; no data corrections have been applied, the vertical lines indicate data gaps between ROSAT observations. Note that bins containing 50 or more counts do not occur in temporal isolation, rather they only occur in the progress of longer lasting flare events; for details see text.

similar recorded peak magnitudes with $\Delta U \approx 1.2$ mag and $\Delta B \approx 0.4$ mag, but have obviously a rather different light curve morphology, indicating different intrinsic properties. It is of course possible that the first impulsive flare also had a gradual phase below our detection threshold; if this is the case, its gradual phase emission must have been surpressed relative to the impulsive phase.

In Figure 5 I show the simultaneously recorded X-ray data binned into 15 sec bins; the time axis is expressed in heliocentric Julian date as in figure 1. For the X-ray observations, UV Ceti was actually placed 40 arcmin off the optical axis of the X-ray telescope in order to avoid count rate fluctuations caused by the various

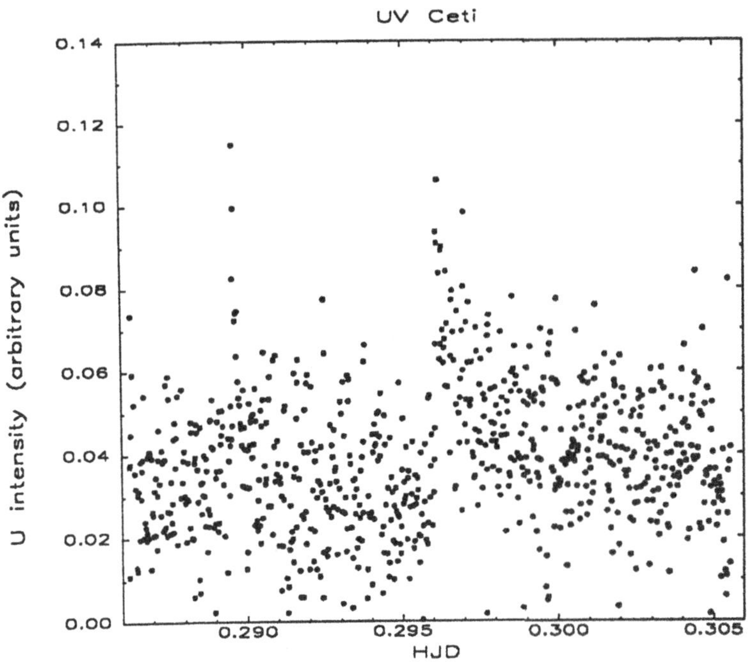

Fig. 4. U intensity (in arbitrary units) vs. heliocentric Julian Date (in units of HJD - 2448844.0) during the time period for which simultaneous X-ray data was taken; note the two impulsive flare events are HJD = 0.2895 and HJD = 0.2951.

wire grid systems inside the PSPC. The X-ray light curve shown in Figure 5 has been corrected for the count rate loss due to vignetting, but has not been background-subtracted.

The X-ray light curve appears - on first inspection - more or less constant in the first half of the observation until HJD = 0.296; after HJD = 0.298 we find an extremely significant enhancement in soft X-ray emission, i.e. $\sim$ 3 minutes after the optical onset of the second flare. Such delays between flare onset and soft X-ray enhancements have been previously observed in simultaneous soft X-ray observations (with the compared to ROSAT less sensitive EXOSAT LE telescope) and optical photometry of the flare stars UV Ceti and BY Dra (de Jager *et al.* 1986, 1989).

Alerted by the optical spikes (cf. Figure 4), we carefully investigated the X-ray light curve during and immediately after the times of optical flare onset. We found clear evidence of two short-lived soft X-ray bursts which are well above the noise level and are of high statistical significance (a statistical assessment of the data will be given below). Two data points (in Figure 5) prior to the soft X-ray enhancement at HJD = 0.298 clearly stick out; they both occur within 30 seconds after the optical flare onset. The statistical significance of this flare detection is approximately $\sim 10^{-3}$, we are thus confident to have found – for the first time – true impulsive microflares on another star.

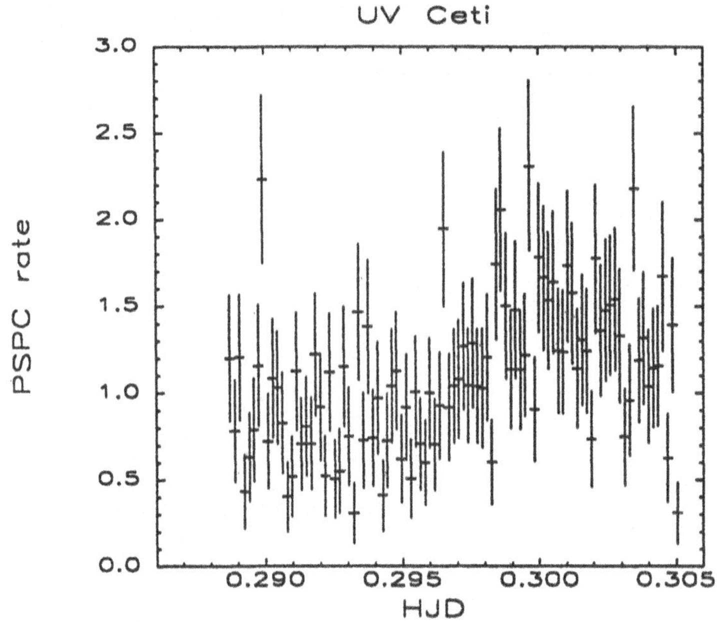

Fig. 5. PSPC X-ray light curve for the same time interval as in Figure 4; the X-ray data is integrated over 15 seconds and corrected for vignetting.

References

Ayres, T., 1981, *et al.*, *Ap. J.*, **250**, 293.

Collura, A., Pasquini, L. and Schmitt, J.H.M.M., 1988, *Astron. Ap.*, **205**, 197.

Brown, A., Drake, S.A., van Steenberg, M.E., and Linsky, J.L., 1991, *Ap. J.*, **373**, 614.

Butler, J.L., Rodonò, M., Foing, B.H., and Haisch, B.M., 1986, *Nature*, **321**, 679.

deJager, C. *et al.* (19 other authors) 1986, *Astron. Ap.*, **156**, 95.

deJager, C. *et al.* (18 other authors) 1989, *Astron. Ap.*, **211**, 157.

Haisch, B.M., 1987, in *Lecture Notes in Physics*, **291**, 69.

Haisch, B.M., *et al.*, 1981, **245**, 1009.

Haisch, B.M. and Simon, T., 1982, *Ap. J.*, **263**, 253.

Haisch, B.M., Schmitt, J.H.M.M. and Rosso, C., 1991, *Ap. J. Lett.*, **383**, L15.

Haisch, B.M., Schmitt, J.H.H.M. and Rosso, C., 1992, *Ap. J. Lett.*, **388**, L61.

Haisch, B.M., Schmitt, J.H.M.M., and Fabian, A.C., 1992, *Nature*, **360**, 239.

Haisch, B., Strong, K. T. and Rodonò, M. 1991, *Ann. Rev. Astron. Astrophys.*, **29**, 275.

Lin, R.P., Schwartz, R.A, Kane, S.R., Pelling, R.M., and Hurley, K. C., 1984, *Ap. J.*, **283**, 421.

Linsky, J.L. and Haisch, B.M., 1979, *Ap. J. Lett.*, **229**, L27.

Pallavicini, R., Tagliaferri, G., and Stella, L., 1990, *Astron. Ap.*, **228**, 403.

Reimers. D., 1992, private communication.

Reimers, D., and Schmitt, J.H.M.M., 1992, *Ap. J. Lett.*, **392**, 55.

Rosner, R., 1992, this volume.

Schmitt, J.H.M.M., 1991, *Adv. Space Res.*, Vo. 11, 11, 115.

Schmitt, J.H.M.M., Haisch, B.M.H., and Barwig, H., 1993, *Science*, submitted.

Snowden, S.L., Plucinsky, P.P., Briel, U., Hasinger G. and Pfeffermann, E. 1992 *Ap. J.*, **393**, 18.

Stern, R.A., Schmitt, J.H.M.M., Rosso, C., Pye, J.P., Hodgkin, S.T., and Stauffer, J.R., 1992, *Ap. J. Lett.*, **399**, 159.

Vaiana, G.S., *et al.*, 1981, *Ap. J.*, **244**, 163.

Vaiana, G.S. and Rosner, R., 1978, *Ann. Rev. Astron. Astrophys.*, **16**, 393.

X-RAY EMISSION FROM THE PLEIADES

F.R. HARNDEN JR.
Harvard-Smithsonian Center for Astrophysics

Abstract. With the advent of ROSAT observations, studies of the X-ray emission of the Pleiades open star cluster now span more than a decade. Combined with previous *Einstein* data, the pointed ROSAT observations obtained by Rosner, by Caillault, and by Stauffer, and the ROSAT All-Sky Survey (RASS) data analyzed by Schmitt *et al.*, can provide information on temporal variability for time scales up to the order of the solar cycle. Through a preliminary look at X-ray luminosity functions, we can address questions of cluster membership and completeness of the combined X-ray surveys, and the ROSAT data can provide new insights into the X-ray temperatures of some of the brighter Pleiades members, including one seen to undergo a "super-flare" during its RASS-visibility interval.

Key words: open star clusters – coronae

1. Introduction

X-ray investigators, like optical astronomers before them, are employing open star clusters as laboratories for isolating the effects of various parameters on observed stellar properties. With the addition of the ROSAT All-sky survey (RASS) and pointed observations, our X-ray experiment on the Pleiades has now moved into its second decade.

1.1. Observations

The first X-ray studies of the Pleiades were conducted in the early 1980's using the *Einstein* Observatory IPC (Caillault and Helfand 1985, Micela *et al.* 1985, 1990). Now in the 1990's, several teams of investigators are continuing the work with the ROSAT PSPC, as shown in Table I. (Note that this preliminary report is based upon just part of the total pointed data to be acquired: the Rosner and Caillault proposals are expected ultimately to yield the times shown in parentheses.) One can also hope that ROSAT HRI observations will eventually be conducted.

TABLE I
Journal of ROSAT Observations of the Pleiades

Investigator	R.A.	dec	date	exposure(ks)	detections
RASS	~3h 46m	~24d	~Aug 90	~0.4	24
Rosner	3h 46m 48s	23d 54m	Feb, Mar 91	15.8 (43)	>120
Caillault	3h 46m 59s	24d 09m 15s		7.6 (36)	
Stauffer NE	3h 50m 8.5s	24d 21m 40s		25.0	~200
Stauffer NW	3h 44m 24.5s	24d 47m		25.0	

The composite catalogue of over 400 Pleiades members compiled by Micela *et al.* (1990) for their *Einstein* investigations has now been expanded to include over 600 stars. The spatial coverage afforded by the Rosner ROSAT observation

J.F. Linsky and S. Serio (eds.), Physics of Solar and Stellar Coronae, 337–343.

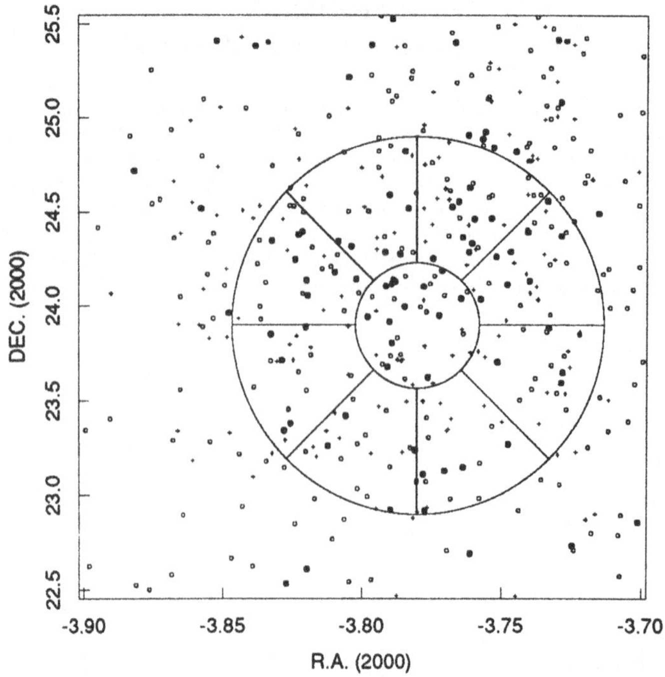

Fig. 1. Spatial distribution of stars from the adopted optical catalog: filled symbols, crosses, and empty circles represent stars with $V < 10$, $10 \leq V < 14$ and $14 \leq V$, respectively. Note that the window support structure is depicted only schematically, since the data were acquired at multiple roll angles.

is indicated in Fig. 1 for this catalogue of Pleiades members. Note that brighter stars are more concentrated toward the cluster center than are fainter ones. Given our present knowledge, it is unclear to what extent the difference in the spatial distributions of faint and bright members is due to a true mass segregation effect, as opposed to incompleteness (particularly in spatial coverage) of the available optical observations.

The distributions of apparent magnitudes from this catalogue are shown in Fig. 2. Note that the sampling of the PSPC changes the shape of the magnitude distribution, due to the different spatial distributions of catalogued stars of different spectral type. The central region of the PSPC field provides superior X-ray observations, not only because of its greater effective area, but also because the angular resolution of the ROSAT mirror is much better closer to the telescope axis.

Fig. 3 presents an H-R diagram for all stars surveyed with the Rosner PSPC observation (cf. Fig. 2).

2. Variability

One of our principal goals is to continue the study of Pleiades variability. Together with previous *Einstein* data, the ROSAT observations provide a temporally com-

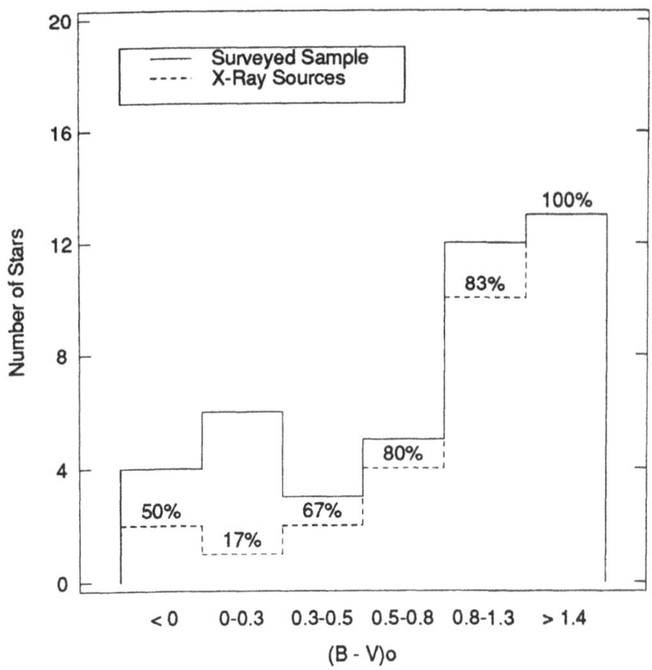

Fig. 2. B-V color index distributions for Pleiades stars falling in the central region of the PSPC field (where sensitivity is highest): solid line - all surveyed Pleiades members, dashed line - Pleiades X-ray detections. Percentages give fractions of subsample detected. Note that essentially all M stars (in this region of highest sensitivity) are detected.

plete data set for the central Pleiades cluster which permits studies of all three important time scales of stellar-activity: the stellar rotation period, the convective turn-over time ($\sim$ active-region life time), and the dynamo period.

2.1. *Einstein*-ROSAT COMPARISONS

Schmitt *et al.* (1992) have compared the RASS data for the Pleiades with the *Einstein* data of Micela *et al.* (1990) by compiling the ranked lists shown in Table II. Inspection of the rank lists reveals clear evidence for variability. For example, the brightest *Einstein* source went undetected in the RASS observations, while the brightest RASS source was eleventh in *Einstein* intensity. Overlap of the "top ten" of each with the other is only 50%.

2.2. ROSAT EVIDENCE

Further evidence for variability is available from the ROSAT data alone. A search for variations in quiescent emission observed with *Einstein* proved somewhat frustrating (cf. Micela *et al.* 1990), and this approach may only be viable in a statistical sense, even with the higher sensitivity of the ROSAT data. Nevertheless, as can be

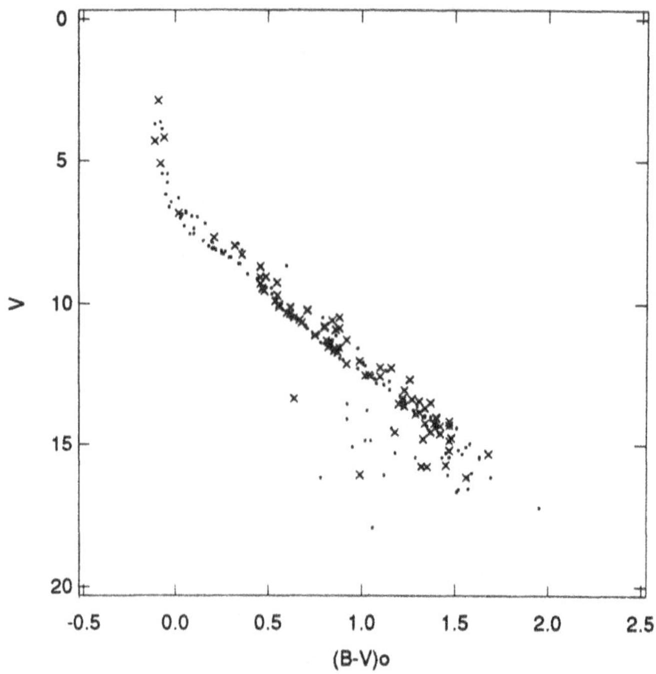

Fig. 3. H-R diagram of the Pleiades stars in the overall field of view; crosses indicate X-ray detections and small dots indicate undetected stars.

TABLE II
Pleiades Rankings (from Schmitt et al. 1992)

Einstein	ROSAT	ROSAT	*Einstein*
1	<24	1	11
2	8	2	3
3	2	3	48
4	5	4	6
5	24	5	4
6	4	6	7
7	6	7	62
8	11	8	2
9	19	9	15
10	15	10	12

TABLE III
Pleiades Flares

Survey	# flares	objects	flare factor
Einstein IPC (Caillault & Helfand - *Einstein*)	2	Hz 1136	
		Hz 1733	
RASS (Schmitt - ROSAT)	1	Hz 2034	
Pointed PSPC (Rosner - ROSAT data)	3	HCG 144	32 ± 4
		HCG 307	$\gtrsim 16$
		unknown	$\gtrsim 7$

seen from Table III, several flares have now been observed, leaving no doubt that at least the more active Pleiades members are highly variable.

3. Luminosity Functions

Another main goal of the Rosner ROSAT observations is to test our understanding of stellar X-ray luminosity functions, taking into account the "broadening" of these functions by the variations in coronal emission levels during the course of a stellar cycle and thereby sharpening the correlation of stellar X-ray emission levels with stellar rotation. This step is essential if we are to refine our understanding of the temporal evolution of stellar activity for stars of different mass, and improve the modeling of the (dynamo) processes which lead to stellar activity. Of particular interest is whether there is a discernible variation in the long-term variability levels as a function of stellar mass, as might be expected from the classic notion that stars of very low mass tend to be more highly variable (viz. flare stars).

In computing the preliminary luminosity functions of Fig. 4 from the Rosner pointed observations (Vinay Kashyap, private communication), a counts-to-ergs conversion factor of about 9 x 10^{-12} erg cm^2 ct^{-1} was used. This factor corresponds to luminosities of 2 x 10^{31} erg ct^{-1}, at the Pleiades' distance and column density, and was derived from a two-temperature spectral analysis of six stars in the central region. A large uncertainty in the mean for the luminosity functions of the last segment (dashed line) is likely due to the presence of flaring in a few of the stars.

A comparison of the mean luminosities given in Table IV with those previously published, reveals them to be much higher than those obtained for field stars.

4. Conclusions

Clearly, this is only a status report on work in progress: many challenges remain. Understanding the background behavior encountered by ROSAT in its high-inclination orbit is a task that will be important to many investigations. For the present report, we have simply excluded data from several intervals with unusual background, but an understanding of the various orbital effects could permit use of the entire data set.

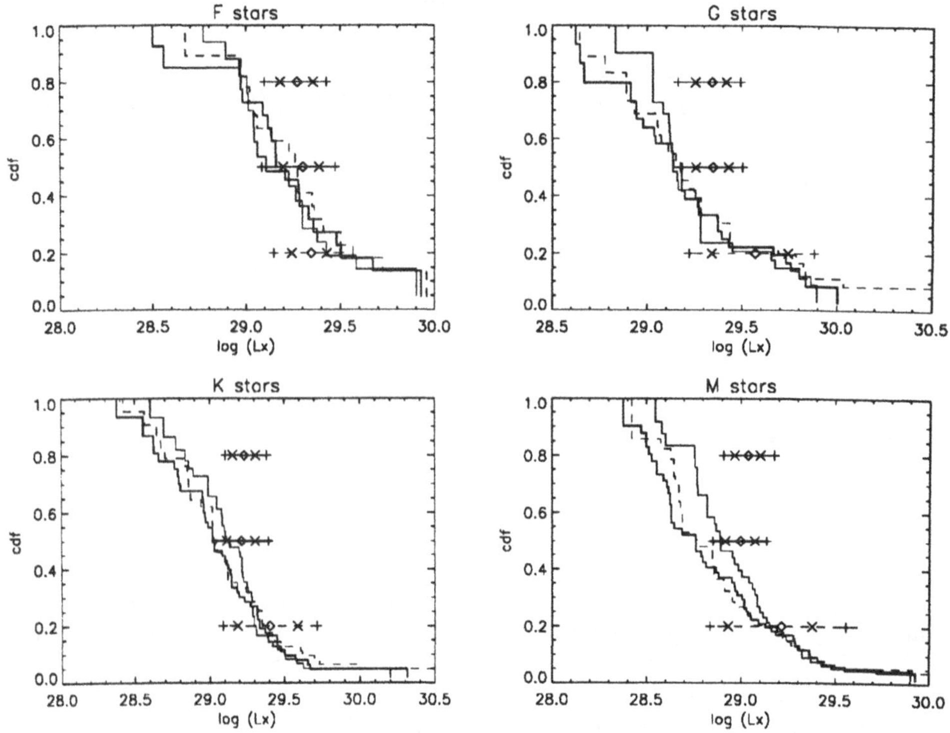

Fig. 4. X-ray luminosity functions versus spectral type for Pleiades stars detected with ROSAT; heavy solid line - entire observation, thin solid line - first segment only, dashed line - last segment only. Also plotted are the means (diamonds) and the 68% (X) and 95% (+) confidence limits on the means for each luminosity function (see Table IV).

Source detection is another difficulty with the analysis of regions like the Pleiades, where source confusion sets in, if not at the field center, then at most only a few arcmin from the telescope axis (as the spatial resolution degrades with off-axis angle). None of the presently available automated detection techniques is suited for such crowded fields.

In addition, more optical work will obviously be required for such tasks as identifying additional Pleiades members and addressing the question of X-ray emission from A-type stars. Given the "detection" of eight A-stars (of 22 surveyed as Pleiades members), the question of whether unseen companions are responsible for A-star X-ray emission demands resolution.

Acknowledgements

Much of the work reported here was performed by his collaborators, to whom the author is extremely grateful: Prof. R. Rosner, Department of Astronomy and Astrophysics, University of Chicago, the Principal Investigator of these ROSAT

TABLE IV
Pleiades mean luminosities (log L_X)

Spectral type	both segments	first segment only	last segment only
F	29.30	29.27	29.34
G	29.35	29.34	29.57
K	29.21	29.23	29.40
M	29.01	29.05	29.22

investigations, and his student Vinay Kashyap, and G. Micela and S. Sciortino, Istituto ed Osservatorio Astronomico, Palermo. The author and his collaborators acknowledge partial support from NASA, under NASA Contract NAS5-30934 and NASA Grant NAG5-1794, and from Agenzia Spaziale Italiana and Ministero della Università e della Ricerca Scientifica e Tecnologica.

References

Caillault, J.-P. and Helfand, D.J.: 1985, *Ap.J.* **289**, 279

Caillault, J.-P., Vilhu, O., and Linsky, J.L.: 1991, *Ap.J.* **383**, 594

Micela, G., Sciortino, S., Serio S., Vaiana, G.S., Bookbinder, J., Golub, L., Harnden, F.R.Jr., and Rosner, R.: 1985, *Ap.J.* **292**, 172

Micela, G., Sciortino, S., Vaiana, G.S., Harnden, F.R.Jr., Rosner, R., and Schmitt, J.H.M.M.: 1990, *Ap.J.* **348**, 557

Schmitt, J.H.M.M., Kahabka, P., Stauffer, J., and Peiters, A.: 1992, *Astron. Astrophys.* submitted,

Sciortino, S. and Micela, G.: 1992, *Ap.J.* **388**, 595

Stauffer, J.R., and Hartmann, L.W.: 1987, *Ap.J.* **318**, 337

Stauffer, J.R., Hartmann, L.W., Soderblom, D.R., and Burnham, N.: 1984, *Ap.J.* **280**, 202

Stauffer, J.R., Schmitt, J.H.M.M., Caillault, J.-P., Gagne, M., and Prosser, C.: 1992, *B.A.A.S.* ?,

ROSAT OBSERVATIONS OF THE HYADES

J.P. PYE and S.T. HODGKIN

X-ray Astronomy Group, Department of Physics and Astronomy, Leicester University, Leicester, LE1 7RH, UK

R.A. STERN

Lockheed Palo Alto Research Laboratory, O/91-30, Bldg. 252, 3251 Hanover St., Palo Alto, CA 94304, USA

and

J.H.M.M. SCHMITT and C.ROSSO

Max Planck Institut für Extraterrestriche Physik, D-8046 Garching bei München, Germany

Abstract. We present preliminary results from ROSAT all-sky survey and pointed observations of the Hyades star cluster. At typical distances of ~ 45 pc, and with low interstellar column densities, relatively low luminosity sources in the Hyades can be detected in both the soft X-ray and EUV. The ROSAT all-sky survey has provided an unbiased mapping of the whole cluster (diameter ~ 40°); a preliminary analysis of the PSPC data shows X-ray emission from 108 Hyads out of about 480 probable or possible members, nearly doubling the number reported from earlier *Einstein* Observatory measurements. Seven Hyads were also detected in the ROSAT EUV sky survey. A programme of long duration (~ 20 – 40 ks), pointed observations with the ROSAT PSPC is in progress; so far, 7 fields have been completed, yielding, for selected regions (total ~ 2.5 deg^2), a sensitivity improvement over the all-sky survey of more than a factor 5, and providing high quality spectra and time series for the brighter sources. In a preliminary analysis, 51 Hyads out of 79 catalogued have been detected, including 24 out of 36 Hyades dM-stars, allowing us to construct their X-ray luminosity function down to $L_X \sim 6 \times 10^{27}$ erg s^{-1}. We detect 4 out of 5 dM's with $m_V > 16$.

Key words: Stars – Open clusters – Hyades – X-rays – EUV

1. Introduction

Apart from the very extended U Ma cluster, the Hyades is the nearest open star cluster to the Sun, at a distance of about 45 pc. Its age is $\sim 7 \times 10^8$ years and it contains > 400 known members extended over ~ 40° diameter on the sky. The proximity of the Hyades to the Sun allows the detection of relatively low luminosity sources and with little absorption of low energy X-rays and EUV radiation. The first X-ray observations were made with the *Einstein* Observatory over ten years ago, mainly within a few degrees of the cluster centre. *Einstein* detected 66 members (out of 121 in the *Einstein* IPC fields) (Stern *et al.* 1981; Micela *et al.* 1988), allowing a good determination of the F- and G-type stellar X-ray luminosity functions. Recently, the ROSAT all-sky survey has provided an unbiased mapping of the *whole* cluster (Stern *et al.* 1992a,b). In addition, long duration, pointed ROSAT observations are needed: (i) to study the X-ray faint members, especially at spectral types K–M; (ii) to obtain 'high quality' ($\gtrsim$ 1000 counts) spectra for brighter members in order to perform coronal modelling; (iii) for variability studies.

In this paper we describe the all-sky survey observations and several pointed observations that we have obtained to date. As an illustration of the high quality

J.F. Linsky and S. Serio (eds.), Physics of Solar and Stellar Coronae, 345–351.

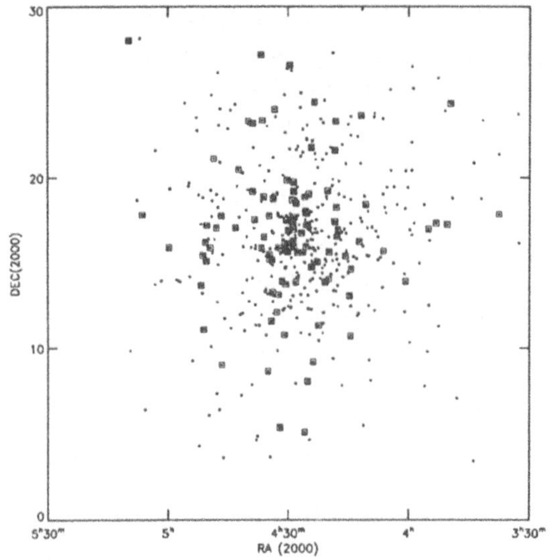

Fig. 1. Sky map (equatorial coordinates, equinox J2000) for $30° \times 30°$ of the Hyades region, showing the input catalogue members (dots) and all-sky survey X-ray detections (squares) (Stern *et al.* 1992b).

of these data, we present a Hyades dM-star X-ray luminosity function derived from the deep pointings.

2. All-sky Survey Observations

A preliminary analysis has detected 108 members in X-rays (with the PSPC) and 7 in the EUV (with the WFC S1a filter, $\lambda_{mean} \sim 100$Å) (Stern *et al.* 1992b). The corresponding sensitivity thresholds are $L_X \sim 3 \times 10^{28}$ erg s^{-1} (0.1–2.4 keV) and $L_{EUV} \sim 3 \times 10^{29}$ erg s^{-1} (90–190 eV). Fig. 1 shows the locations of the catalogued Hyades members, with the ROSAT X-ray detections indicated, clearly demonstrating the importance of the all-sky survey data in obtaining complete, X-ray flux limited coverage for the whole cluster. (It should be noted that the detection rate is biased by the absence, from this preliminary analysis, of 2 days of PSPC scan data which pass close to the cluster centre.) Fig. 2 shows the detected X-ray luminosities as a function of stellar B-V colour.

A number of short period, chromospherically active binary systems and the giants θ^1 Tau = VB 71 and γ Tau = VB 28 are among the most X-ray luminous objects in the cluster. The X-ray brightest cluster member is the well known red dwarf/white dwarf short period binary V471 Tau (Barstow *et al.* 1992). The second brightest X-ray source, HR 1394 = 71 Tau = VB 141, is a long-period lunar occultation binary. Six of the 13 brightest X-ray emitters are spectroscopic binaries with periods < 6 days, and of the 10 such systems known (from either Batten *et al.* 1989 or Griffin *et al.* 1985), only one has not yet been detected in the ROSAT X-ray

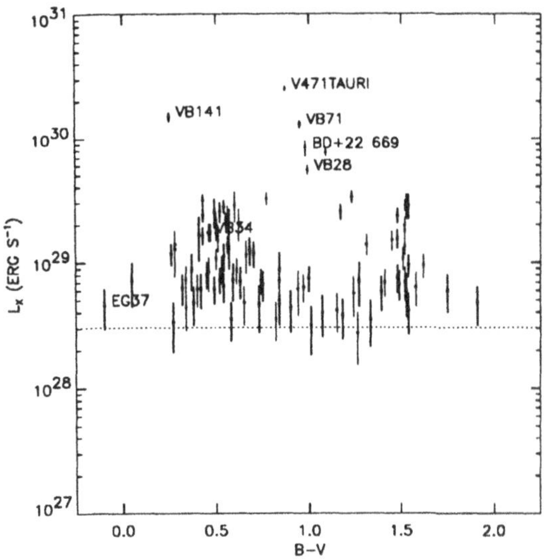

Fig. 2. All-sky survey: X-ray luminosity vs. B-V colour for detected X-ray sources. The WFC EUV detections are labelled by their catalogue designations. $\pm 1\sigma$ errors in L_X are shown. The *approximate* survey limit is shown as a horizontal dashed line. (Stern *et al.* 1992b).

survey (the Am system 60 Tau, a marginal X-ray detection by *Einstein*, Stern *et al.* 1981, and implying a flux close to the ROSAT survey limit).

Seven cluster members were also seen in the WFC EUV all-sky survey. Among the stars detected in both X-rays and EUV is the Hyades white dwarf EG 37(=VR16), confirming an earlier serendipitous EXOSAT detection (Koester *et al.* 1990); it is one of the two coolest DA white dwarfs so far detected at soft X-ray wavelengths ($T_{eff} \approx 24000$ K, Kidder *et al.* 1992).

Also reported is the first X-ray detection of the Hyades K0 giant, ϵ Tau, at roughly the survey limit. This new result establishes all four Hyades giants as X-ray emitters, although with $\approx$ 50:1 range in L_X. A comparison of *Einstein* and ROSAT data for three of the giants provides evidence for the presence of decade or longer X-ray variability, possibly due to activity cycles. Long-term X-ray variability might also account, to some degree, for the wide dispersion in L_X, though some, as-yet undiscovered differences in the giants' evolutionary states might also produce differing activity levels.

3. Pointed Observations

The following observations have been performed to date (i.e. in AO-1 and AO-2) by us, in the central region of the Hyades: 1 field at 40ks exposure (PI: R.Stern, USA) (Stern *et al.* 1992c); 6 fields at $\sim$ 20ks each (PI: J.Pye, UK). From an initial analysis of these 7 fields, we have so far detected (above 4.5σ significance) 51 (15)

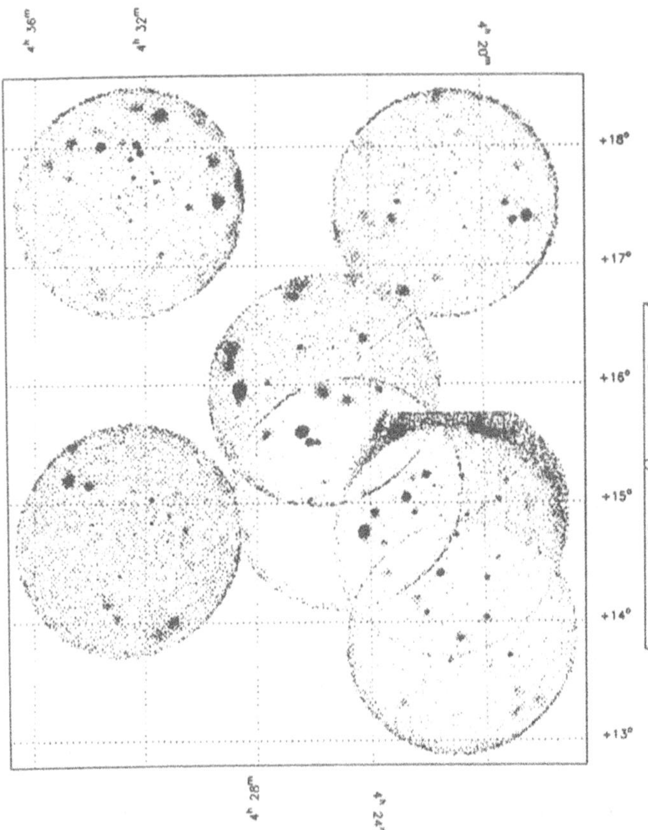

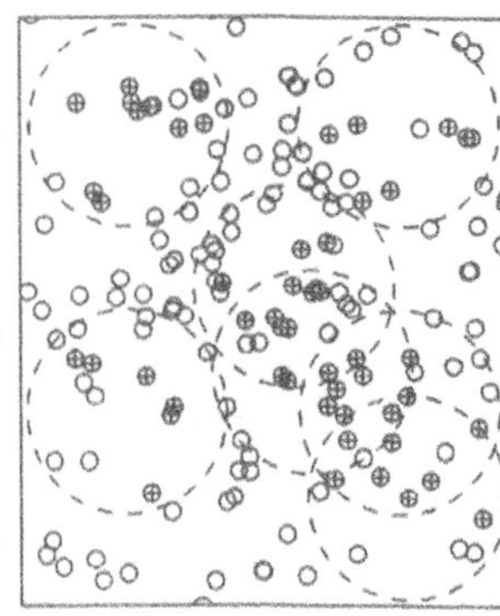

Fig. 3. **Left.** Mosaic image of the 7 deep AO-1/2 PSPC fields, in equatorial J2000 coordinates, for the 'total' PSPC energy band (0.1–2.4 keV). The grey-scale is linear in counts per second per unit solid angle, with the brightest sources being saturated (i.e. black). The 40ks exposure field (PI: Stern, US) is centred at $\alpha \approx 04^h25^m$, $\delta \approx +15°$. The exposure times of the other 6 fields (PI: Pye, UK) range from 15–23ks. Each field has been divided by an appropriate 'exposure map' to provide normalisation between the fields; this exposure map includes the effects of the telescope vignetting, however there are large uncertainties in the corrections at the field edges (i.e. > 50 arcminutes off-axis). **Right.** Map of the same area, showing positions of the Hyades input catalogue stars, indicated by circles (radius 5 arcminutes); the X-ray detected Hyads are marked with crosses (+). Only detections within 50 arcminutes of each field centre are shown. The PSPC field edges are indicated by dashed circles of radius 60 arcminutes.

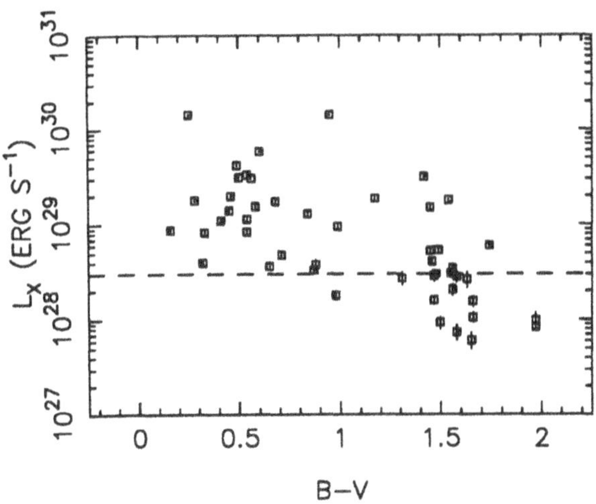

Fig. 4. The pointed observations: X-ray luminosity vs. B-V colour for detected X-ray sources. $\pm 1\sigma$ errors in L_X are shown. The *approximate* survey limit is shown as a horizontal dashed line.

members with the PSPC, out of 79 (18) within 50 (20) arcminutes of the centre of the PSPC field of view (full radius 60 arcminutes). (Detection sensitivity falls off towards the edge of the PSPC fields due to vignetting and increased point spread function.) We have also detected several Hyades stars (including VB 141 and VB 71) in the EUV, with the WFC S2b filter, $\lambda_{mean} \sim 140$Å. Fig. 3 is a 'mosaic' image of the 7 PSPC fields, while Fig. 4 shows the detected X-ray luminosities as a function of B-V (c.f. Fig. 2). The sensitivity of these pointed observations now allows us to investigate in some detail the activity levels of the Hyades dM-stars.

4. The M-dwarf X-ray Luminosity Function

In the 7 AO-1/2 fields we have so far detected (above a 4.5σ threshold) 24 out of 36 catalogued Hyades dM-stars (c.f. *Einstein* 14/46, Micela *et al.* 1988). The minimum detected X-ray luminosity is $\approx 6 \times 10^{27}$ erg s^{-1} (for H232 = VA 203, $m_V = 16.7$), and the upper limits (at 99% confidence) range from $\sim 2 \times 10^{27}$ to 2×10^{29} erg s^{-1}. We have constructed an X-ray luminosity function (Hodgkin *et al.* 1992) using 'survival analysis' methods to take account of the upper limits (Feigelson and Nelson 1985). The resulting cumulative luminosity distribution function is shown in Fig. 5. The mean and median values of L_X are both $\approx 1.6 \times 10^{28}$ erg s^{-1}. When tested over similar L_X ranges, the ROSAT and *Einstein* distributions are in good agreement. We note that binaries may be strongly influencing the derived X-ray luminosity function (c.f. Stauffer 1982).

Of particular interest for models of the convection zone and magnetic dynamo are indicators of coronal activity for the coolest known Hyades members with very deep (or even fully) convective interiors (e.g. Liebert and Probst 1987; Cox *et al.*

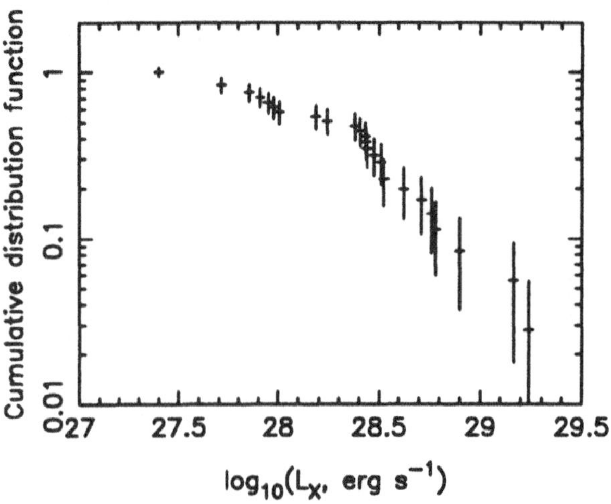

Fig. 5. The cumulative X-ray luminosity distribution function for the Hyades dM-stars, as derived from the 7 pointed observations.

1981; Dorman *et al.* 1989), i.e. mass $\lesssim 0.3M_{\odot}$, roughly corresponding to spectral type $\geq$ dM4, $M_V \gtrsim 12$, or at the Hyades distance, $m_V \gtrsim 15.3$. Table 1 shows our detection rates to date. The V-band faintest, X-ray detected Hyads so far are H232 = VA 203 and H280 = VA 260 both at $m_V = 16.7$ (Stauffer 1982).

TABLE I
ROSAT X-ray detections of 'late' dM-stars

m_v	Total known in Hyades	No. in 7 AO-1/2 fields	No. detected
15.0–16.0	≈ 65	7	4
16.0–	≈ 30	5	4
Totals:	≈ 95	12	8

5. Future ROSAT Pointed Observations

ROSAT Hyades observations approved to date (i.e. after the AO-3 selection) at high priority (A/B) and yet to be performed, include:

- 2 fields at 3ks each on the very active dK's seen in the all-sky survey, BD $+22^{\circ}669$ and BD $+23^{\circ}635$ (PI: R.Stern, US);
- 5ks on the K-giant γTau (PI: J.Pye, UK);
- 3 fields at 20ks each, to survey other parts of the central Hyades region and in particular to obtain high quality spectra of F and G stars (PI: J.Pye, UK);

– 3 fields at 30ks each for the study of faint dM stars (PI: N.Reid, US). (A 10ks observation was already performed in AO-2.)

Acknowledgements

We are glad to acknowledge use of the SIMBAD database, CDS, Strasbourg. The luminosity function calculation made use of the ASURV (Rev. 1.2) package of T.Isobe, M.LaValley and E.Feigelson (Penn. State University). The source detection in the pointed data was performed with the PSS program developed by David Allan and Trevor Ponman (Birmingham University). The ROSAT data analysis was performed at MPE, LPARL and the Leicester University *Starlink* node. STH and JPP acknowledge financial support of the UK SERC. RAS acknowledges support from NASA Contracts NAS5-31212 and NAS5-31828 and the Lockheed Independent Research Program.

References

Barstow, M.A., Schmitt, J.H.M.M., Clemens, J.C., Pye, J.P., Denby, M., Harris, A.W. and Pankiewicz, G.S.: 1992, *MNRAS* **255**, 369

Batten, A.H., Fletcher, J.M. and McCarthy, D.G.: 1989, *Publ.Dom.Astr.Obs.* **17**, 1

Cox, A.N., Shaviv, G. and Hodson, S.W.: 1981, *ApJ* **245**, L37

Dorman, B., Nelson, L.A. and Chau, W.Y.: 1989, *ApJ* **342**, 1003

Feigelson, E.D. and Nelson, P.I.: 1985, *ApJ* **293**, 192

Griffin, R.F., Gunn, J.E., Zimmerman, B.A. and Griffin, R.E.M.: 1985, *AJ* **90**, 609

Hodgkin, S.T., Pye, J.P., Stern, R.A. and Stauffer, J.R.: 1992, , in preparation

Kidder, K.M., Holberg, J.B., Barstow, M.A., Tweedy, R.W. and Wesemael, F.: 1992, *ApJ* **394**, 288

Koester, D., Beuermann, K., Thomas, H.C., Graser, U., Giommi, P. and Tagliaferri, G.: 1990, *A&A* **239**, 260

Liebert, J. and Probst, R.G.: 1987, *Ann.Rev.Astr.Ap.* **25**, 473

Micela, G., Sciortino, S., Vaiana, G.S., Schmitt, J.H.M.M., Stern, R.A., Harnden, F.R.,Jr. and Rosner, R.: 1988, *ApJ* **325**, 798

Stauffer, J: 1982, *AJ* **87**, 899

Stern, R.A., Schmitt, J.H.M.M., Rosso, C., Pye, J.P., Hodgkin, S.T. and Stauffer, J.R.: 1992a, in Proceedings of the 7th Cambridge Workshop on Cool Stars, Stellar Systems and the Sun, ed(s)., *M.Giampapa & J.Bookbinder*, Publ.A.S.P. Conference Series, 100

Stern, R.A., Schmitt, J.H.M.M., Rosso, C., Pye, J.P., Hodgkin, S.T. and Stauffer, J.R.: 1992b, *ApJ* , in press

Stern, R.A., Schmitt, J.H.M.M., Pye, J.P., Stauffer, J.R., and Simon, T.: 1992c, in Proceedings of the 7th Cambridge Workshop on Cool Stars, Stellar Systems and the Sun, ed(s)., *M.Giampapa & J.Bookbinder*, Publ.A.S.P. Conference Series, 103

Stern, R.A., Zolcinski, M-C., Antiochos, S.K. and Underwood, J.H.: 1981, *ApJ* **249**, 647

ROSAT HRI OBSERVATIONS OF THE ORION NEBULA

JEAN-PIERRE CAILLAULT and MARC GAGNÉ
University of Georgia, Athens, GA 30602 USA
and
JOHN STAUFFER
Center for Astrophysics, Cambridge, MA 02138 USA

Abstract. We report on three ROSAT HRI pointings of the Orion Nebula region. The observations range in exposure time from 2.4 to 29 kiloseconds. Over 200 sources are detected, ranging from the hot, massive O stars to the coolest PMS stars. We discuss these data in the context of the distribution of L_x/L_{bol} vs. M_{bol}, activity-rotation relations and the problem of the late B, early A stars.

Key words: Orion Nebula – X-ray emission – ROSAT

1. Introduction

The Orion Nebula is the best-studied site of recent star formation in our galaxy. A rich history of optical, infrared, and radio work provides an invaluable background for observations in the X-ray regime, while its age and proximity mark it as an essential target in studies of the evolution of stellar properties. The exciting center of the Nebula, the Trapezium Cluster, is the densest young cluster in the Galaxy and, thus, represents an important aspect of early stellar evolution that has only recently been recognized and has only recently begun to be examined in detail. In this paper we report on ROSAT HRI observations of this region.

2. Observations

The ROSAT HRI field-of-view is about 40' in diameter; its spatial resolution (close to the center of the field) is about 2". The HRI is sensitive to soft X-rays in the energy range 0.1-2.0 keV, but has no energy resolution. Three HRI observations of the Orion region were made in 1991. In Table I we list the relevant characteristics for these observations. Additional observations of fields 1 and 3 were performed in September of 1992. In Figure 1 we show the central portion of the Trapezium HRI image; the brightest star, θ^1 Ori C, is the brightest source in the image.

3. Results

The total number of X-ray sources detected within the ROSAT images (accounting for overlap between the fields) is about 220 (>5σ); the threshold sensitivity of each image is listed in Table I. The areal density of X-ray sources in Orion is >0.1/square arcmin, with the density increasing substantially toward the Trapezium. Unfortunately, the optical data are sparse: although about 200 of the 220 sources have corresponding optical candidates (within 5"), only 60 of the 200 have known spectral information. We have detected 10 O6-B5 stars, 1 B6-A5 star (see below), 1 A6-F5, 12 F6-G5, and 36 stars later than G6 (of which 33 are proper motion members of Orion).

J.F. Linsky and S. Serio (eds.), Physics of Solar and Stellar Coronae, 353–356.

TABLE I
ROSAT HRI Images of Orion

Field	Date	Position (2000)	Exposure (seconds)	# of sources (> 5σ)	Threshold log L_x (ergs/s)
1	03/1991	5 35 14 -4 38 24	2380	11	30.9
2	03/1991	5 35 24 -5 04 12	11285	65	30.4
3	10/1991	5 35 17 -5 23 24	28725	177	30.1

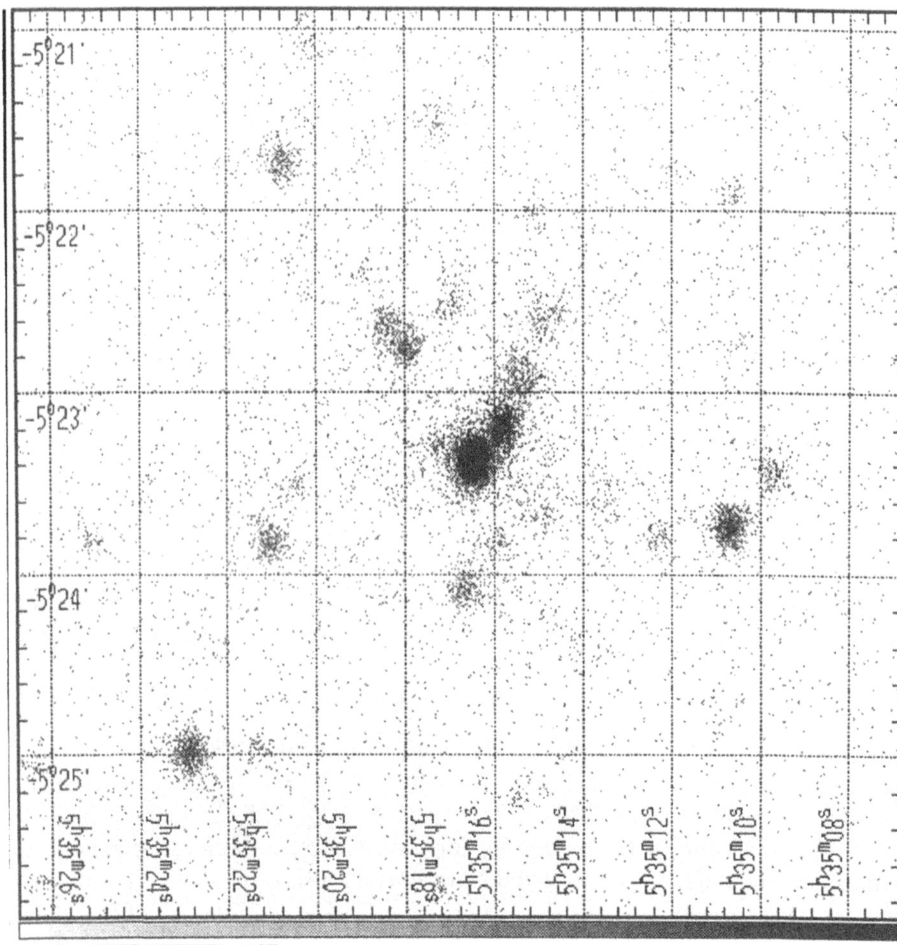

Fig. 1. ROSAT HRI image of the 5' x 5' region centered on the Trapezium.

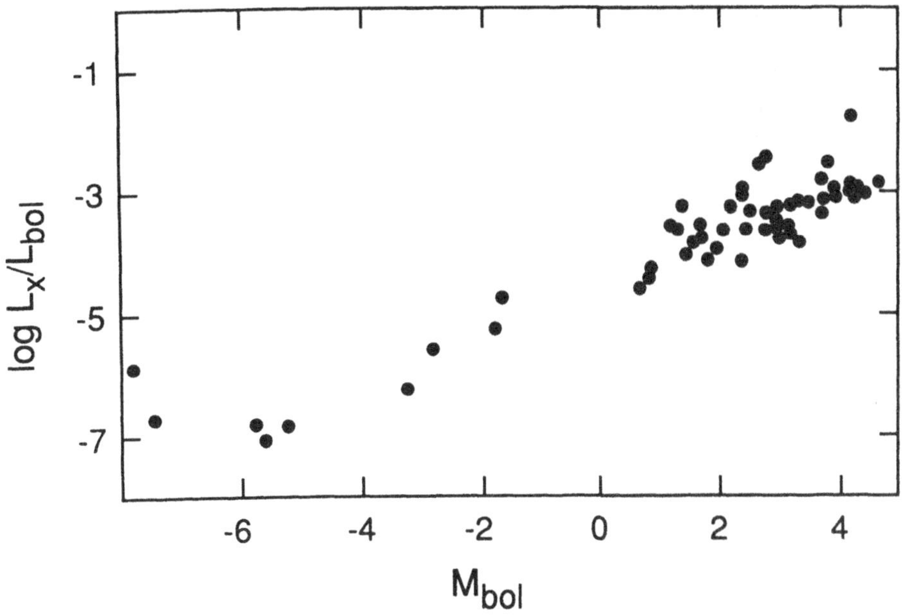

Fig. 2. L_x/L_{bol} vs. M_{bol} for stars with known spectral types.

4. Discussion

4.1. The L_x/L_{bol} Distribution

We have used the sub-sample of optical candidates with known spectral types to plot the L_x/L_{bol} values versus the bolometric magnitudes in Figure 2. In calculating the M_{bol} values, we assumed that the spectral types were indicative of the bolometric corrections, the distance to the Orion Nebula was 460pc (Walker 1969), and the average extinction was $A_v = 1$ mag (Herbig & Terndrup 1986). These PMS stars seem to have a distribution similar to the older cluster stars (see, e.g., the ROSAT study of the Pleiades by Stauffer et al. 1992), with only a slight deviance at the faint end where a few stars have L_x/L_{bol} values >-3. This value is thought to be the "saturation" limit for X-ray emission from stars, possibly attributable to the magnetic field filling factor being close to 1 (see, e.g., Vilhu & Walter 1987).

4.2. Activity and Rotation

Attridge & Herbst (1992) and Duncan (1992) have provided measurements of rotational periods and *V sin i* values, respectively, for a number of the late-type stars (later than F6) in Orion that have been detected by ROSAT. Plotting either L_x or f_x/f_v or L_x/L_{bol} versus either *V sin i* or P_{rot} results in scatter plots (correlation coefficients of $\rho = 0$ +/- 0.2 for 10-20 data points), implying again, perhaps, that these stars all rotate rapidly enough to be in the "saturation" regime.

4.3. Late B, Early A stars

Caillault & Zoonematkermani (1989) claimed to have detected about a dozen late B, early A main sequence stars in Orion with the EINSTEIN observatory. Walter et al. (1988) and Strom et al. (1990) have made similar claims for about 8 such stars in other regions of star formation. The emission from these stars is thought to be in conflict with current models for both hotter stars (see, e.g., Lucy 1982) and cooler stars (see, e.g., Rosner et al. 1985). In the former case, radiation-driven shocks in stellar winds are thought to be responsible for the x-ray emission, while in cooler stars the magnetic dynamo is thought to power the emission via coronal heating. The hot star model is a problem for late B, early A stars because their stellar winds are too weak to set up the requisite shocks, while the cool star model fails for these stars because they are thought to lack outer convection zones of sufficient depth for the dynamo to operate. Hence, one of our main goals for these ROSAT HRI observations was to confirm (or deny) the existence of these stars as candidates for the x-ray emission.

In the original Caillault & Zoonematkermani paper (1989), 9 out of 33 O6-B5 stars were detected, while 12 out of 51 B6-A5 stars were. However, the ROSAT HRI accuracy of about 5" is a vast improvement over that of EINSTEIN (about 60"), allowing us to (partially) refute the earlier claim. We have detected 10 out of 15 O6-B5 stars (a larger fraction than with EINSTEIN, attributable to ROSAT's greater sensitivity), but, surprisingly, only 1 out of 19 B6-A5 stars. About 10 of the original 21 EINSTEIN detections lay within the ROSAT fields: all 6 of the hotter group were redetected, while none of the 4 B6-A5 stars were. We conclude from this that the previous claim for the B6-A5 stars was in error (attributable mostly to too conservative a cutoff in f_x/f_v for other possible candidates) and that these stars are unlikely to be the emitters of detectable X-ray emission. If B6-A5 stars are not intrinsic X-ray sources, then our one detection, Parenago 1712, would best be explained by assuming that it has a lower mass binary companion, and that the companion is the true X-ray source.

5. Acknowledgements

This work was supported by NASA grants NAG5-1608 and NAG5-1610 to the University of Georgia.

References

Attridge, J. M. & Herbst, W.: 1992, *ApJ submitted* ,
Caillault, J.-P. & Zoonematkermani, S.: 1989, *ApJ* **338**, L57
Duncan, D.: 1992, *ApJ submitted* ,
Herbig, G. H. & Terndrup, D. M.: 1986, *ApJ* **307**, 609
Lucy, L. B.: 1982, *ApJ* **255**, 286
Rosner, R., Golub, L., & Vaiana, G. S.: 1985, *ARA&A* **23**, 413
Stauffer, J. et al.: 1992, *ApJ submitted* ,
Strom, K. et al.: 1990, *ApJ* **362**, 168
Vilhu, O. & Walter, F.: 1987, *ApJ* **321**, 958
Walker, M. F.: 1969, *ApJ* **155**, 447
Walter, F. et al.: 1988, *AJ* **96**, 297

ROSAT X-ray Study of the Stellar Population of the Chamaeleon I Dark Cloud

S. CASANOVA*, E. FEIGELSON**, T. MONTMERLE* and J. GUIBERT***

Abstract. Two soft X-ray images of the Chamaeleon I cloud obtained with the *ROSAT* PSPC are presented. Seventy reliable, and perhaps 19 additional X-ray sources are found. Up to 90% of these sources are certainly or probably identified with T Tauri stars formed in the cloud. Twenty to 35 are probably new "weak" T Tauri (WTT) stars. We summarize a statistical study of the relationships between X-ray emission and other physical properties for a number of Chamaeleon stars. We see no difference in spatial distributions, detection rate and luminosity functions between the WTTS and CTTS.

Key words: Stars: Pre-Main Sequence – X-rays

1. Introduction

X-ray surveys of nearby molecular clouds by the *Einstein Observatory* have detected "classical" T Tauri Stars (CTTS) at strong X-ray luminosities. The CTTS are low mass pre-main sequence stars, characterized by strong UV and optical emission lines, and a large IR excess widely interpreted in terms of a massive circumstellar accretion disk. X-ray surveys also revealed the "weak-line" T Tauri Stars (WTTS), which show little or no circumstellar material (Bertout 1989). The X-ray characteristics of T Tauri stars, including strong variability, point to a solar-like magnetic activity, enhanced by factors reaching up to 10^4 times the solar level or more.

The Chamaeleon I Cloud is a nearby (d = 140 pc) star forming region. Its dense dark material covers 1 x 2 degrees, allowing it to be completely surveyed by only two overlapping *ROSAT* images. Feigelson and Kriss (1989, here after FK) detected 21 X-ray emitting stars with the *Einstein* IPC in this region.

The *ROSAT* PSPC has an improved sensitivity over *Einstein* (by a factor of 10) and an improved angular resolution, so the data reported here (based on Feigelson et al. 1992) represent a substantial improvement over those presented by FK.

2. X-ray Observations of Chamaeleon PMS Stars

We summarize here our analysis of two overlapping images obtained with the *ROSAT* PSPC, with exposures $\sim$ 6,000 sec. The detection of the X-ray sources is based on *ROSAT* Standard Analysis Software System (*SASS*) lists, which we improved after visual inspection of the images: five sources close to a nearby stronger source (and two weak sources) were missed by the automatic detection process. Counts and signal-to-noise ratios were calculated in the "hard" band (0.4 - 2.5 keV). The final list comprises 89 sources: 70 strong sources and 19 fainter ones with a small signal-to-noise ratio (between 2.4 and 3.5), a fraction of which may be spurious.

* Service d'Astrophysique, Centre d'Etude de Saclay, 91191 Gif sur Yvette Cedex, France
** Department of Astronomy and Astrophysics, Penn State University, University Park PA 16802, USA
*** Observatoire de Paris, 61 av. de l'Observatoire, 75014 Paris, France

J.F. Linsky and S. Serio (eds.), Physics of Solar and Stellar Coronae, 357–360.

The search for stellar counterparts of the X-ray sources was conducted using published compilations of cloud members (Gauvin and Strom, 1992, hereafter GS; Schwartz, 1992), the Guide Star Catalog for all the stars brighter than 15 mag, and the inspection of digitized Schmidt plates from the ESO/SERC Sky Survey red plates. The digitization procedure (with astrometric and photometric analysis) of this plate was done with the MAMA automatic microdensitometer at the Observatoire de Paris. When the X-ray sources with the most accurate SASS positions were compared to the digitized optical star field, small systematic offsets of the X-ray source from likely stellar identifications were clearly seen. After correction of these small offsets probably due to boresight errors of the *ROSAT* satellite, X-ray and optical positions agree to within 3" (1 σ) for the strong centered sources. In the outer regions (> 20' from the axis) the sources are smeared by the *ROSAT* mirrors, and X-ray positions can be offset by as much as 20" - 45".

In 50 cases the identifications are certain or very probable (one single optical counterpart is found in the *ROSAT* error circle). In 25 cases, two or three optical counterparts are found in the *ROSAT* error circle. These X-ray sources have a probable stellar identification, but the exact counterpart is ambiguous. A few extragalactic sources were eliminated from our sample. X-rays are detected from 18 CTTS out of 30 known, 16 WTTS out of 30 known, and from 18 - 37 "New" stars, likely WTTS. Two of the new X-ray sources are coincident with *IRAS* Faint Source Catalog sources, and one or two are coincident with ground-based IR sources listed by Schwartz (1992). The Herbig-Haro objects and the "Infrared Nebula" of the Chamaeleon I Cloud are not detected. We have started an optical program to characterize the new stars at Mt Stromlo in collaboration with Dr. W. Lawson.

3. The X-ray Population

The spatial distribution of X-ray emitting stars is shown in Figure 1, along with *IRAS* contours of the cloud. Several points can be noted. First, the spatial distributions of the optically known young stars and of the X-ray sources are very similar: the X-rays are not selecting a particular type of star. Second, the distribution of the stars in the Northern and the Southern parts of the cloud are quite similar, despite the shape of the cloud (narrow North and larger South). This is consistent with the fact that only the dense cores of material (which are very similar in South and North) are relevant for star formation. Third, a spatial segregation is clearly seen when one considers the ages of known stars: we find that stars within the *IRAS* cores have a mean age $< t_* >= 7 \pm 2 \times 10^6$ yr, while stars outside the cores have $< t_* >= 15 \pm 2 \times 10^6$ yr. Last, the spatial distributions of CTTS, WTTS and proposed New stars do not differ. This means that at least in the vicinity of the cloud the two classes have the same age distribution.

The X-ray luminosities are approximately calculated from the hard band count rates, using the A_V values given in GS92. The analysis of our data on different orbits shows that the error due to source variability is about a factor of 2. The X-ray luminosity distribution of the T Tauri stars is a power law, over 3 orders of magnitude ($3 \times 10^{28} - 1.3 \times 10^{31}$ erg s^{-1}). No "minimum X-ray luminosity" is apparent. Figure 2 shows the resulting maximum likelihood estimators (taking

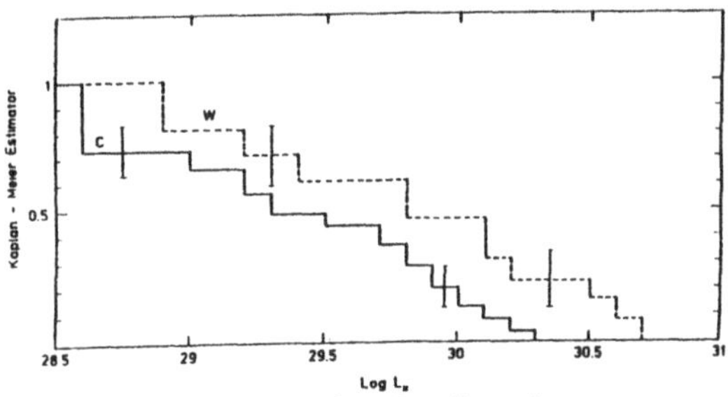

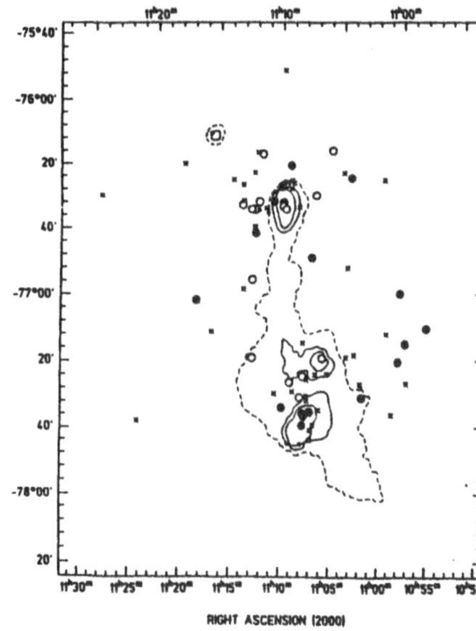

Left: **Figure2**. Luminosity function of the Cha I CTTS and WTTS
Right: **Figure1**.
Spatial distribution of the X-ray sources:
The filled circles are WTTS,
the empty circles are the CTTS,
the crosses are the "new" T Tauri.

into account nondetections) for the X-ray luminosity functions of the WTTS and CTTS subsamples. The mean luminosities for the subsamples are $6.1 \pm 1.5 \times 10^{29}$ erg s^{-1} and $2.3 \pm 1.3 \times 10^{29}$ erg s^{-1}, respectively. Nonparametric two-sample tests implemented in the ASURV statistical package (LaValley et al. 1992) indicate the difference is only marginally significant, with a probability $\sim$ 9-13% of belonging to the same population (1.5-1.7 σ effect).

4. Relations with other Physical Properties

The *Einstein* results showed that the X-ray emission of T Tauri stars is (like the Sun) of magnetic origin. The study of the new sample of *ROSAT* sources can give a more precise idea on this mechanism. The stellar magnetic fields are presumably the result of the dynamo effect. One of the key factors is rotation, but only a few data are available for the Chamaeleon I stars. Instead, we use the age of the star, which is a proxy indicator of rotation because of angular momentum loss. Figure 3 shows a significant anticorrelation between X-ray luminosity and age. Evolution can also be seen on a longer time scale, if we compare the mean *ROSAT* X-ray emission from the Chamaeleon TTS with the much lower value obtained in the Hyades, computed from the latest *ROSAT* data from Stern *et al.* (1992).

We next study the X-ray flux and luminosity in relation to other physical parameters determined by GS. We note a strong correlation of L_X with L_*, M_*, R_*. We find that the first relation can be expressed by the simple proportionality $L_X/L_* = 1.6 \times 10^{-4}$ within a factor of 2.3. Figure 4 shows the relation between L_X and R_*, with the line $L_X \propto R_*^2$ of Fleming et al. (1989) representing a possible "saturation" level observed in field and Orion stars. All of the Chamaeleon I stars lie below this line, and L_X shows a steeper relation scaling more like R_*^3 (or $F_X \propto R_*$), and shows no evidence for saturation. Looking at the relation between F_X and the H_α flux, we find no evidence of "smothering" of the X-rays by hot cir-

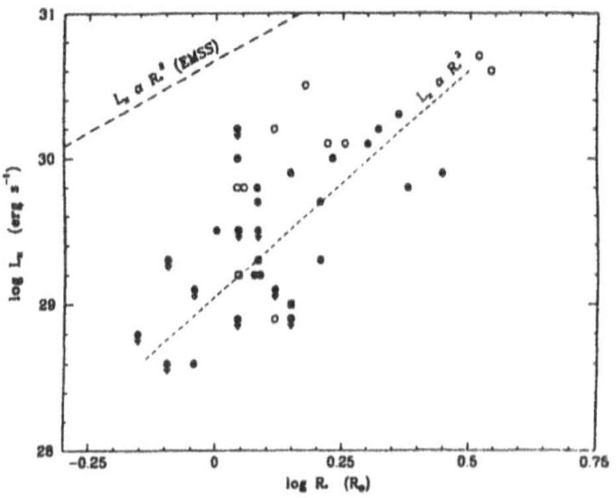

Figure 4: Log(L_X) vs. Log(R_*)

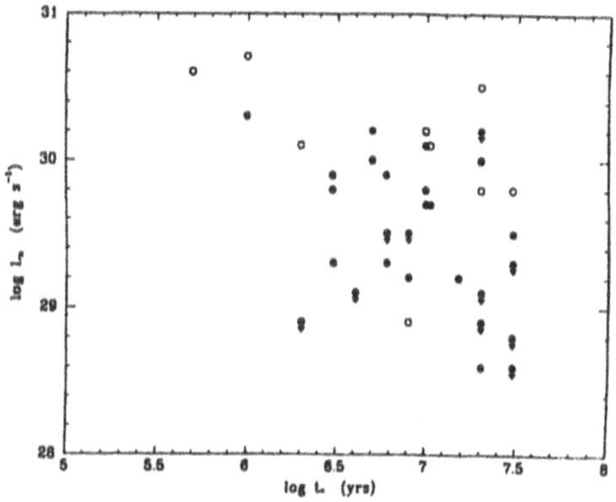

Figure 3: Log(L_X) vs. Log(t_*)

cumstellar material (Walter and Kuhi, 1981). A marginally significant correlation is found between L_X and the disk infrared luminosity, but it could be a byproduct of stellar relationships and have no physical relation to the disk.

5. Conclusion: Comparison with *Einstein Observatory* Results

All of the 21 *Einstein* X-ray sources reported by FK are recovered in the *ROSAT* images. The majority of these sources show quite similar fluxes (within 50%) between the two instruments. In addition, the improved sensitivity of *ROSAT* over *Einstein* has allowed the discovery of many new young stars on a very large range of X-ray luminosities. This wider range of L_X brings an important dispersion in all the correlations that the *Einstein* results did not have.

Another important difference between the two observations is the effect of the improved resolution of the *ROSAT* telescope. This allows many very precise identifications, and in some cases shows that *Einstein* sources must be broken down into two to five sources. This fact, added to the better sensitivity of *ROSAT*, drastically lowers the mean luminosity of the X-ray sources compared to that detected by *Einstein*, altering some previous conclusions.

References

Bertout, C.: 1989, *Ann. Rev. Astr. Ap* **27**, 351.

Feigelson, E.D., Casanova, S., Montmerle, T., and Guibert, J. : 1992, *ApJ* , submitted.

Feigelson, E.D., and Kriss, G.A.: 1989, *ApJ* **338**, 262 (FK).

Fleming, T.A., Gioia, I.M., and Maccacaro, T.: 1989, *ApJ* **340**, 1011.

Gauvin, L.S., and Strom, K.M.: 1992, *ApJ* **385**, 217 (GS).

LaValley, M., Isobe, T., and Feigelson, E.D.: 1992, *Bull. AAS* , in press.

Schwartz, R.D.: 1992, *Low Mass Star Formation in Southern Molecular Clouds*, B. Reipurth (ed.), Garching:ESO, p.93.

Stern, R. A., Schmitt, J. H., Rosso, C., Pye, J. P., Hodgkin, S. T., Stauffer J. R.: 1992, *ApJL* , in press.

Walter, F.M., and Kuhi, L.V.: 1981, *ApJ* **250**, 254.

The ROSAT All-Sky Survey of Active Binary Coronae: The RS CVn Systems

ROBERT C. DEMPSEY AND JEFFREY L. LINSKY*
Joint Institute for Laboratory Astrophysics Boulder, CO 80309-0440
and
THOMAS A. FLEMING, J. H. M. M. SCHMITT AND MARTIN KÜRSTER
Max-Planck-Institut für Extraterrestriche Physik
D-8046 Garching bei München, Germany

Abstract. We present observations of 136 RS CVn active binary systems obtained with the Position Sensitive Proportional Counter (PSPC) during the *ROSAT* All-Sky Survey (RASS) phase of the mission.

Key words: stars: Coronae – stars: X-rays – stars: Late-type – stars: Binaries

1. INTRODUCTION

During the RASS, more than 300 active binaries - RS CVn, BY Dra and Algol-types - were observed over intervals of several days with the PSPC. From these data we have calculated X-ray luminosities in the 0.1–2.4 keV bandpass, performed spectral modelling of the low resolution PSPC pulse height spectra for one third of the objects and studied variability on timescales of hours to days for all of these systems. Here we present results for the 136 systems classified as RS CVn binaries in the Strassmeier et al. (1988) catalogue of active binaries. Of this sample, 112 targets were detected in exposures of roughly 600 seconds or less. *This represents the largest, X-ray sample of RS CVn binaries observed to date.* With observational intervals of 2–11 days we have detected over 2 dozen flares with a wide range of energies and lifetimes in addition to variability due to rotational modulation and eclipses. Spectra for 44 RS CVn sources were fitted with Raymond-Smith thermal plasma models (Raymond & Smith 1977) allowing us to estimate coronal temperatures and emission measures. One- and two-temperature models were fit to the spectra as well as a "continuous" distribution where the emission is given by $(T/T_{\max})^{\alpha}$ as used in Schmitt et al. (1990).

2. RESULTS

One purpose of this research is to determine which, if any, parameters best determine the activity level of the active binaries. We therefore calculated X-ray luminosities (L_X) and surface fluxes (F_X) for the systems and then searched for correlations among stellar parameters such as $P_{\rm rot}$, R_*, $v_{\rm rot}$, and *(B-V)* as well as orbital properties such as $P_{\rm orb}$, separation, and $\Gamma = R_*/R_{\rm RocheLobe}$. As found in previous studies, we see in Figure 1 a decrease in $F_{\rm X}$ with increasing $P_{\rm rot}$ but a leveling off of $F_{\rm X}$ for $P_{\rm rot} \leq 3^d$. For the entire sample we find $F_{\rm X} \sim P_{\rm rot}^{-0.84\pm0.12}$ (solid line in Figure 1) compared to Walter's & Bowyer (1981) value of $L_{\rm X}/L_{\rm bol} \sim P^{-1.1\pm0.1}$ based on a smaller sample of RS CVn's . However, from Figure 1 it is apparent

* Staff Member, Quantum Physics Division, National Institute of Standards and Technology

J.F. Linsky and S. Serio (eds.), Physics of Solar and Stellar Coronae, 361–364.

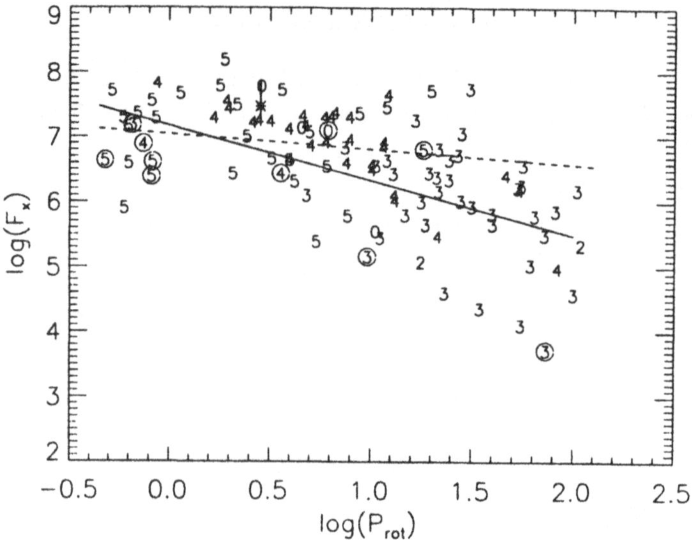

Fig. 1. Surface flux – rotational period. Numbers indicate the luminosity class of the active star, i.e. 3, 4, and 5 indicate giants, subgiants, and dwarfs, respectively. Zeros indicate systems with unknown luminosity class. Circled numbers are upper limits. The dashed line is a best fit to all the data. The solid line is the best fit to all the data while the dashed line is the fit for the dwarfs only. We indicate V711 Tau with an asterisk and include the likely range (vertical bar) in F_X assuming $F_X \sim F_{CIV}^{1.4}$ with C IV fluxes from Dorren & Guinan (1990).

that F_X is *independent* of period for *dwarf* systems (dashed line in Fig. 1) and that the negative slope results solely from the giants and subgiants. This has not previously been observed among the active binaries and is markedly different from the behavior of single dwarfs. Giants and subgiants alone yield $F_X \sim P^{-1}$. We do not see any correlation of F_X or L_X with v_{rot} or *(B-V)*.

How fundamental is the activity–period relation in determining the surface flux? As with F_X stellar radius is a strong function of rotational period showing the same relationship with period, including the leveling off for Log(P_{rot}) $\leq$ 0.5, as does F_X. Walter & Bowyer (1981) argue that the criteria used to select RS CVn systems introduce the additional bias of a period–radius relationship. This results primarily from the fact that the RS CVn systems are synchronous binaries covering a narrow range in effective temperature. However, the flat distribution of dwarfs out to periods of 30 days argues against a flux–period relation, unless all of the dwarf systems have surface fluxes near the saturation limit. This is not observed here with many dwarfs having $F_X/F_{bol} \sim 10^{-4} - 10^{-5}$, well below the saturation limit $\sim 10^{-3}$ (Vilhu & Walter 1987; Dempsey et al. 1992a). Furthermore, the observed flux–period relation is very different than that observed for *single* main-sequence stars (Montesinos & Jordan 1988). Therefore, we confirm the conclusion of Rengarajan & Verma (1983), based on a smaller sample, that the period–activity relation follows from the dependance of radius on period.

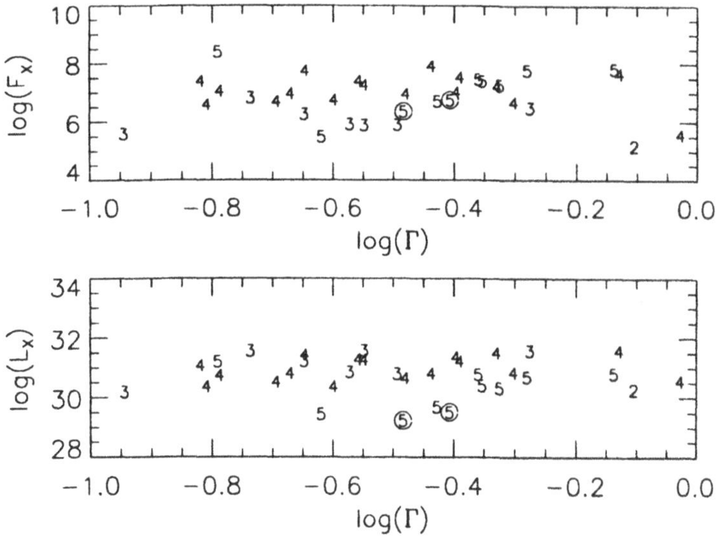

Fig. 2. X-ray surface flux (top) and luminosity (bottom) versus Γ.

In Figure 2 we present L_X and F_X as a function of Roche lobe filling fraction, Γ, for 37 systems. Remarkably, no correlation is evident. This indicates that the activity level is independent of tidal forces. We conclude, furthermore, that most important role played by the hotter star in the X-ray activity level of the system is to provide a mechanism to produce, and maintain, rapid rotation. Therefore, the more active cooler stars in binaries appear to have identical X-ray properties to their single counterparts. The absence of any correlation of F_X or L_X with Γ is remarkable, because the coronal heating mechanism for these active stars must be magnetic in character and the magnetic field depends on the interaction between convection and differential rotation inside the star. One would anticipate that tidal forces should play a major role in controlling circulation patterns and the dependance of rotation rate on radius and latitude by distorting the shape of the star as Γ approaches unity, but the data presented in Figure 2 indicate otherwise.

In addition to studying the quiescent properties of the active binaries, we have searched for variability. Figure 3 shows the folded light curve for V815 Herculis (G5V + M1-2V, $P_{\rm rot} = 1^d.8$, $v\sin i = 27$ km s^{-1}). Since this is a noneclipsing system, the periodic variations likely result from active regions rotating into and out of the line of sight. A brief flare is also seen near phase 0.36. At least 10 systems exhibit similar variations, indicating that inhomogeneous coronae are common among the RS CVn systems. Flares such as the one seen in Figure 3 are quite common in the survey light curves with many showing multiple flare events. For example, the V772 Herculis system flared at least 3 times in 2 days with $L_X(Peak)$ = 6.3–9.7 $\times 10^{30}$ ergs s^{-1}. Total flare energies in the range of 5–600 $\times 10^{30}$ ergs are observed.

Two-temperature models provide better fits to the *ROSAT* data than either single or continuous temperature models. The failure of the continuous emission measure model can be easily understood as due to the slowly changing spectrum

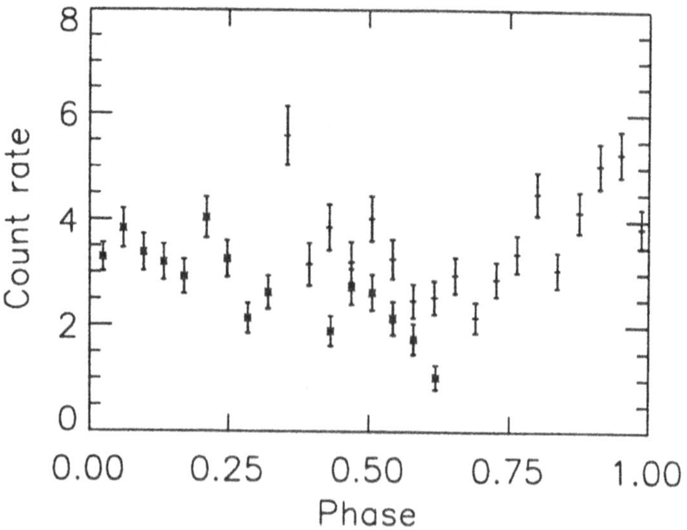

Fig. 3. Folded light curve of V815 Her.

over the PSPC bandpass for temperatures above $\sim 4 \times 10^7$ K. Typically, $T_{\rm high}$ is 1.6×10^7 K and $T_{\rm low}$ is $\approx 2.0 \times 10^6$ K. Two-temperatures were found in previous investigations, but typically the 2 values depended strongly on the detector and telescope used. However, we have reanalyzed the SSS data using the recent Raymond-Smith thermal plasma models and derive $< T_{\rm high} > \sim 2 \times 10^7$K, consistent with both the PSPC and IPC results (Dempsey et al. 1992b) [1]. The remaining differences in derived temperatures result from the different energy bands of the detectors used. Similar results have been obtained for 12 BY Dra systems and one Algol binary. We found evidence in the SSS data for a *third* temperature component near 50 MK for UX Ari and V711 Tau.

References

Dempsey, R. C, Linksy, J. L., Schmitt, J. H. M. M. & Fleming, T. A. 1992a, ApJS, submitted
Dempsey, R. C, Linksy, J. L., Schmitt, J. H. M. M. & Fleming, T. A. 1992b, in prep
Dorren, J. D. & Guinan, E. F. 1990, ApJ, 348, 703
Gehrels, N. & Williams, E. D. 1992, ApJL, submitted
Montesinos, B. & Jordan, C. 1988, in A Decade of UV Astronomy with the IUE Satellite, 1, 238
Raymond, J. C. & Smith, B. W. 1977, ApJS, 35, 419
Rengarajan, T. N. & Verma, R. P. 1983, MNRAS, 203, 1035
Schmitt, J. H. M. M., Collura, A., Sciortino, S., Vaiana, G. S., Harnden, F. R., Jr. & Rosner, R. 1990, ApJ, 365, 704
Strassmeier, K. G., Hall, D. S., Zeilik, M., Nelson, E., Eker, Z. & Fekel, F. C. 1988 A&AS, 72, 291
Vilhu, O. & Walter, F., M. 1987, ApJ, 321, 958
Walter, F. M. & Bowyer, S. 1981, ApJ, 245, 671

[1] See Gehrels & Williams (1992) for a counter discussion on the SSS data and for an explanation of the two-temperature structure of stellar coronae.

SOFT X-RAY EMISSION AND SPECTRA OF EVOLVED STARS FROM ROSAT POINTED OBSERVATIONS

A. MAGGIO and S. SCIORTINO
Istituto e Osservatorio Astronomico di Palermo
Palazzo dei Normanni, 90134 Palermo, Italy

and

F.R. HARNDEN, JR.
Harvard-Smithsonian Center for Astrophysics
60 Garnden Str., Cambridge, MA 02138

Abstract. We present preliminary results of ROSAT pointed observations of two $\sim$ 1.5 $M_{\odot}$ stars in different evolutionary phases beyond the main-sequence: η Sco (F3III-IV), and HD 74772 (G5III). The X-ray spectra of both stars can be fitted with Raymond thermal models with temperatures around 2 $\times$ 10^6 K. Evidence of hot plasma (T $\sim$ 10^7 K) has been also found for HD 74772 only. The X-ray luminosities of both sources are of the order of 10^{28} $erg\ s^{-1}$, below the known median values for evolved stars of similar spectral types.

Key words: Stars: coronae – Stars: evolution – Stars: X-rays

1. Data Analysis and Results

The aim of these observations was to monitor the variation of the magnetic activity level in late type stars during stellar evolution, to study the response of the outer atmosphere to the change of the internal structure, and its effects on stellar dynamo. In the present paper, we will focus the attention on the X-ray spectral characteristics of the observed stars. In particular, we will address the issue of what plasma temperatures can be found in the the coronae of giant stars.

We have performed spatial and spectral analyses of the ROSAT X-ray images (see Table 1 for details), using the IRAF/PROS (V2.0) software package, and adopting the most recent post-launch calibration data. We stress that the analysis of the image of HD 74772 resulted substantially more complex, since the target lies at the edge of the Vela SNR, which yields an intense X-ray background. In fact, the standard processing failed to detect the source, but its presence can be easily recognized in the X-ray image, and its significance is at more than 8σ, according to the "local" detection algorithm (DETECT) available in PROS, when the image is scanned with a $30''$ $\times$ $30''$ sliding cell.

Coronal luminosities and temperatures have been derived by fitting the soft X-ray spectra (0.1 - 2.4 keV) with Raymond thermal models (Fig. 1). In order to minimize the influence of the Vela SNR on the spectral analysis of HD 74772, we have collected source and background counts in a semi-circular region oriented approximately in the direction of the intensity gradient at the source position. Both spectra can be adequately fitted by single temperature Raymond-Smith thermal models, with log(T) = 6.34 $\pm$ 0.04 for η Sco, and log(T) = 6.38 $\pm$ 0.07 for HD 74772. In both cases the fitted hydrogen column density is less than 10^{19} cm^{-2}, in agreement with the expected values (5 - 6 $\times$ 10^{18} cm^{-2}) computed assuming an average hydrogen number density of 0.1 cm^{-3}, typical of the solar neighborhood in the direction of the galactic plane.

J.F. Linsky and S. Serio (eds.), Physics of Solar and Stellar Coronae, 365–368.

In the case of HD 74772, although the single temperature best fit model yields an acceptable reduced χ^2 (at the 80% confidence level), the distribution of the residuals indicates that the number of counts above $\sim$ 1 keV is underestimated (Fig. 1b). For this reason, we have also attempted to fit the data with a two-temperature model: in addition to the above 10^6 K component, we have found that a second component with $T_2 > 10^7$ K is allowed by the data, and increases the reliability of the fit. The emission measure of this high temperature component turns out to be about 40% that of the low temperature component.

We have also attempted a two-temperature fit to the spectrum of η Sco, but we were unable to find any evidence of an high temperature component: in fact, it turns out that the emission measure of any such component must be less than 10% the emission measure of the first one.

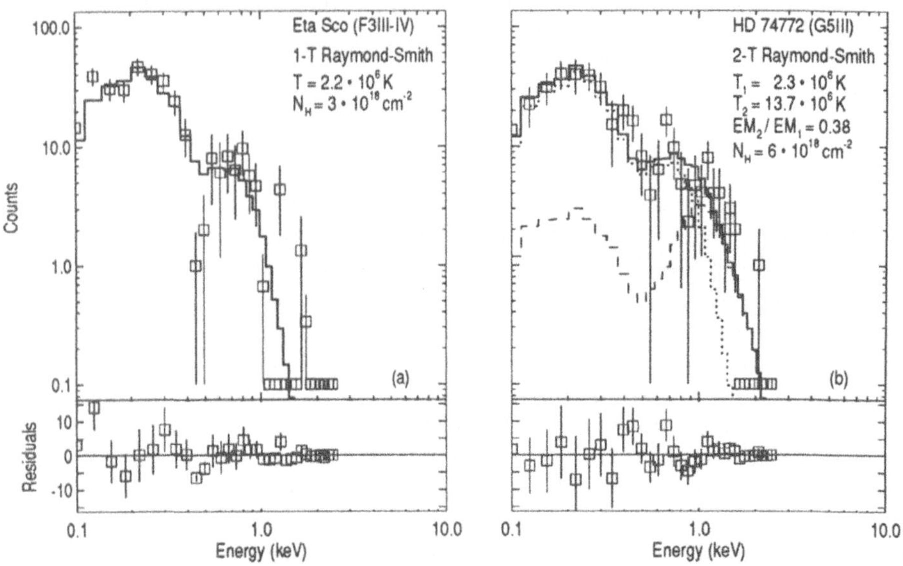

Fig. 1. Observed and best-fit thermal spectra (upper plots), and fit residuals (lower plots). In the case of HD 74772 (panel b), the observed spectrum has been fitted with a two-temperature model: the separate contributions of the low and high temperature components are also shown (dotted and dashed lines, respectively). Note the observed 1-2 keV counts in excess with respect to the prediction of the low temperature component.

2. Discussion

In a previous study, based on observations with the *Einstein* Observatory, Schmitt *et al.* (1990) suggest that main-sequence stars with spectral type F and G have low temperature coronae (T $\sim 3 \times 10^6$ K), while for the majority of the studied giant stars (all of which have X-ray luminosities above the median for this class; Maggio et al. 1990) there is evidence of hot coronae with T $> 10^7$ K (Fig. 2). The few giants which constitute exceptions to the above trend, are slightly evolved

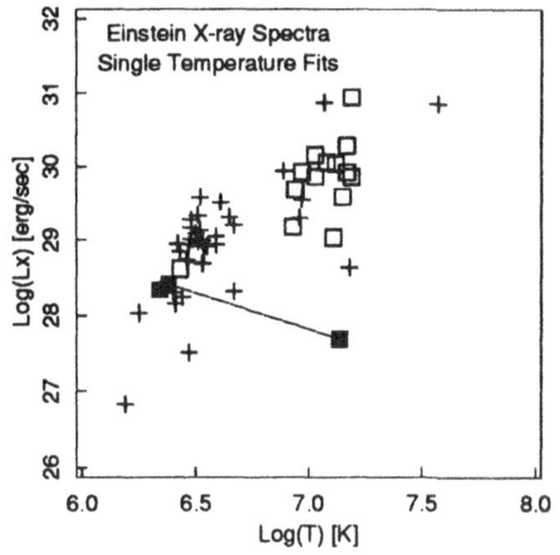

Fig. 2. X-ray luminosity vs. temperature for late-type stars observed with the *Einstein Observatory* IPC (whose X-ray spectra have been successfully fitted with isothermal models; adapted from Schmitt *et al.*, 1990), and for the two new ROSAT targets. Plus symbols denote F and G main-sequence stars, squares denote giants. The results of the new ROSAT observations are indicated by the filled symbols. A dotted line connects two points relative to the low and high temperature components of the best fit model of HD 74772.

F-type giants on the blue side of the Hertzsprung gap, such as our target η Sco. This is not the case of HD 74772, located near the base of the red giant branch, yet suggesting the presence of a low temperature corona (with perhaps some trace of higher temperature plasma).

We note that both targets constitutes interesting cases because of their relatively low X-ray luminosity ($L_X = 2 - 3 \times 10^{28}$ $erg\ s^{-1}$). In fact, the sensitivity of the *Einstein* IPC was usually not sufficient to collect enough counts to perform a reliable spectral analysis on giants with L_X below $\sim 10^{29}$ $erg\ s^{-1}$. Hence, the new ROSAT observations allow to explore in better detail the soft X-ray spectral characteristics of evolved stars with a low X-ray emission level. The present data suggest that such low X-ray luminosity sources do not share the property of having 10^7 K coronae with the higher X-ray luminosity stars of the same class.

The correlation between X-ray luminosity (0.16-4 keV) and coronal temperature – originally reported by Vaiana (1983) – seems to be confirmed by the present work, although further data are needed to firmly exclude observational biases. This behavior suggests that an increase in the level of magnetic activity is determined not simply by a larger filling factor, but also by the presence of higher density, higher temperature plasma.

Because of the importance of HD 74772 in the context of our study, we intend to further investigate whether the presence of the cool component may be partially due to residual contamination by the SNR soft X-ray background, despite the source mask adopted for the spectral analysis.

TABLE I
Summary of Observations and Spectral Fitting Results [1]

Target Name	η Sco	HD 74772
Spectral Type	F3III-IV	G5III
Distance (pc)	20.0	16.0
Observation Date	28-FEB-91	25-MAR-91
Net Observing Time (sec)	2343	3944
Spatial Analysis		
Source Region	circle: $r = 3'$	semi-circle: $r = 2'$
Source+Bkg Counts	459	1385
Bkg Region	annulus: $r_{1,2} = 3', 6'$	semi-annulus: $r_{1,2} = 2', 3'.5$
Bkg Counts	407	337
Source Net Counts	327 ± 22	380 ± 64
Spectral Fitting with Single Temperature Raymond Thermal Model		
log N_H (cm^{-2})	$15.3^{+4.0}_{-indef}$	$14.5^{+3.6}_{-indef}$
T (10^6 K)	$2.2^{+0.2}_{-0.2}$	$2.4^{+0.5}_{-0.3}$
EM (10^{50} cm^{-5})	7.2	8.6
χ^2 / dof	1.2 / 26	1.2 / 23
Spectral Fitting with Two-Temperature Raymond Thermal Model		
log N_H (cm^{-2})	18.5 (fixed)	18.8 (fixed)
T_1 (10^6 K)	2.2 (fixed)	$2.3^{+0.2}_{-0.4}$
EM_1 (10^{50} cm^{-5})	7.3	7.2
T_2 (10^6 K)	unconstrained	$13.7^{+indef}_{-4.6}$
EM_2 (10^{50} cm^{-5})	< 0.7	2.7
χ^2 / dof	1.2 / 26	1.0 / 22
X-ray Flux and Luminosity (0.16 - 4 keV)		
Emitted Flux (10^{-13} $erg\ cm^{-2}\ s^{-1}$)	7.1	5.6
Luminosity (10^{28} $erg\ s^{-1}$)	2.2	2.7

[1] Uncertainties on temperatures and hydrogen column densities are 68% joint confidence limits.

References

Maggio, A., Vaiana, G.S., Haisch, B.M., Stern, R.A., Bookbinder, J., Harnden, F.R., Jr., and Rosner, R. 1990, *Ap. J.*, **348**, 253.

Schmitt, J.H.M.M., Collura, A., Sciortino, S., Vaiana, G.S., Harnden, F.R., Jr., and Rosner, R. 1990, *Ap. J.*, **365**, 704.

Vaiana, G.S. 1983, in *Solar and Stellar Magnetic Fields: Origins and Coronal Effects*, ed. J.O. Stenflo (Dordrecht: Reidel), p. 165.

A ROSAT EUV SURVEY OF CORONAL ACTIVITY IN NEARBY STARS AND COSMIC RAY INJECTION INTO THE INTERSTELLAR MEDIUM.

L. MATTHEWS, J.J. QUENBY and T.J. SUMNER
Astrophysics Group, Imperial College, London SW7 2BZ, United Kingdom.

Abstract. It has been argued by Shapiro (Proc 21st ICRC Moscow 2, 260, 1987 and references therein), that the dominant population of red dwarf stars provides the seed particles for a subsequent, interstellar medium based cosmic ray acceleration process. Computation of the energy budget for the injection process has depended to date on sample estimates of the flaring activity of dMe and dKe stars. The ROSAT Wide Field Camera is providing the first systematic survey of general coronal activity of the nearby sky. Using results of the all sky survey, we are able to estimate the value for the x-ray input into the galaxy from coronally active stars to be 10^{36} ergs s^{-1} kpc^{-3}. Results of the analysis are related to the possible ISM proton input rate through the rough equality of particle and x-ray energy release demonstrated for the solar case.

Key words: Stars:red dwarf – flaring – cosmic ray acceleration

1. Introduction.

While supernova remnant shocks remain the chief candidates as sites for the bulk of the energisation of cosmic rays, it has been suggested that late type stellar coronal emission is an indicator of an attractive location for the injection of 'seed' particles into the interstellar medium (Shapiro 1990 and references therein). Coronal acceleration would avoid the problem of direct acceleration from thermal plasma and favour nuclei of atoms with relatively low first ionization potentials. Although the relationship between coronal activity and energetic particle injection can be obtained for the solar case by direct observation, there has, until now, been no systematic survey of stellar x-ray and EUV activity over the whole sky. The ROSAT satellite carried out a 6 month all-sky survey which is now able to supply this overall data. Here we will concentrate on the EUV emission from nearby stars taken with the WFC on ROSAT. From this data it is relatively easy to pick these out in the 85-180eV flux band and to explore the quantity of energy going into frequencies below the Einstein satellite cut off.

2. EUV Survey Results.

The ROSAT Wide Field Camera (WFC) consists of a set of 3 nested Wolter-Schwarzchild Type-1 mirrors, gold coated with a geometric collecting area of $456cm^2$ and a 2.5° radius field of view. Photons are reflected at a typical grazing incidence angle of 7.5° and recorded by a micro-channel plate pair of 45mm active diameter with CsI photocathode on the front surface. Two survey filters, S1 and S2, have passbands of 100-180eV, FWHM with peak response at 134eV, and 85-110eV, FWHM with peak response at 90eV respectively. A magnetic diverter reduces the contamination from $\leq$50keV electrons by a factor $\geq$670 but Van Allen belt particles, scattered solar x-rays and spacecraft ram direction pickup of solar induced local plasma all must be taken into account in extracting the cosmic photon countrate.

J.F. Linsky and S. Serio (eds.), Physics of Solar and Stellar Coronae, 369–372.

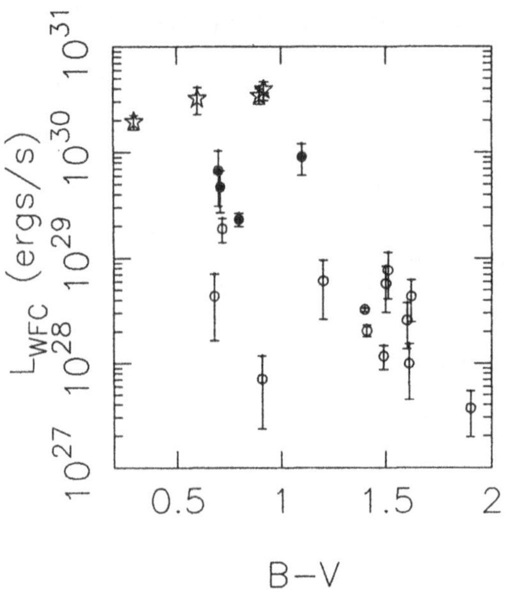

Fig. 1. The Relationship between WFC luminosity and B-V value. The sample includes RSCVn binary systems(⋆), late type giants(•) and dwarf stars(o)

During the survey, the spacecraft operated in a circular orbit, passing through the ecliptic poles, with a period of 96 minutes. It scanned at a rate of 1° of ecliptic longitude per day and with the WFC field of view of 2.5° radius, each detected source had an exposure of approximately 80 seconds per orbit. To obtain countrates each orbit was analysed separately obtaining, after correcting for the instrument vignetting function and dead time losses, a number of detected events above background level and exposure time. Data from all the orbits was then combined using poissonian statistics to give a mean countrate value.

A sample of 21 WFC sources, corresponding to coronally active nearby stars, is presented here. Source identification was obtained with the aid of the SIMBAD database. Because the interstellar medium EUV photon absorption cross section is in the order of 10^{-20}cm^2, the WFC results are sensitive to the line of sight column density and this in turn is not known accurately for many objects. To obtain the EUV flux from a source, we need both the column density and the coronal temperature of the emitting plasma, which is assumed to be thin and thermal. Distances to the sources were obtained through parallax catalogued measurements. Columns were then calculated from Paresce(1984). By folding a thin thermal plasma spectrum given by Landini and Fossi(1990) through the instrument response temperatures for each coronal source were then found from measuring th S1/S2 countrate ratio.

Fig. 1 plots the calculated EUV luminosities of the sources as a function of their B-V values. The empty circles represent dwarf stars, the solid circles are evolved stars, whilst the empty stars represent RSCVn like binary systems. There seems to be no variation in luminosity with spectral type. The evolved stars are clearly more luminous than the dwarf stars. The most luminous group appears to be those of the RSCVn class of synchronous binary. The luminosity values shown are comparable to those presented from an Einstein stellar survey by Vaiana *et al.*(1981) at higher energies.

From Paresce(1984) it can be seen that the hydrogen column remains small in all directions out to 10pc, and it is reasonable to assume that all active late type stars out to this distance would be seen by the WFC. Searching the complete catalogue of sources from the WFC all sky survey, 18 sources, identified as late type stars, were found with distances less than 10pc. Distances were again found from parallax measurements. These 18 included 1 RSCVn and 1 evolved star the rest were dwarf stars. From this and using the typical luminosities for each class of star shown in Fig. 1, an energy output density into the galaxy from coronal emission can be estimated to be 10^{36} ergs s^{-1} kpc^{-3}.

3. The Relationship between X-ray/EUV emission and Energetic Solar Proton Injection

For the remainder of this paper, we will assume that the average stellar coronal emission is created by a series of micro-flaring events with the same relationship between x-ray flux and energetic particle flux as occurs in the solar case.

In a solar flare, it is believed that an ion or neutral particle beam in the tens to hundreds of keV energy range is first accelerated and then becomes the prime carrier of the excess energies (Eg. Simnett and Haines 1990). The x-ray emission arises from the dissipation of the beam while shock or stochastic acceleration is likely to provide extra energisation of particles above 1MeV. Thus both the x-ray emission and the particle injections are secondary processes. However the ratio of proton to electromagnetic emission energies released in high energy solar events leading to a impulsive injections has been demonstrated to be between 0.1 and 1, (Priest 1981). Empirically this can also be shown. Cane et al(1985) observed 3×10^{33} particles $\geq$30MeV in space from a typical long duration flare corresponding to an energy release of 1.4×10^{29}ergs. Starr et al(1988) find energies in the thermal plasma NkT_{eff} by fitting a soft thermal and a hard non-thermal x-ray spectrum to flare data of about 10^{29}ergs. If all this latter energy goes into x-ray emission, the rough equality of particle and x-ray release is demonstrated.

4. Energy Budget for Cosmic Ray Injection

Shapiro(1990) estimated a power density, w_i for seed ions from late type flaring stars of magnitude

$$w_i = 8 \times 10^{25} \text{ ergs s}^{-1} \text{ pc}^{-3}$$

Using the results of section 2 and 3 we have a value

$$w_i \equiv \text{ EUV Output per pc}^3 \text{ or } w_i = 10^{27} \text{ergs s}^{-1} \text{ pc}^{-3}$$

The difference in these two results is arises because our work adopts the observed average coronal emission as a measure of cosmic ray input. Shapiro(1990) finds w_{cr} =1.2x10^{29}ergs s^{-1} pc^{-3} as the power required to maintain the galactic intensity against loss by escape, using an escape time of 1.3x10^{7}yr, while based on Ginsburg's halo model (Ginsburg 1987), this figure is reduced to 3 to 10x10^{27}ergs s^{-1} pc^{-3}. Hence SNR shock waves need to boost the energy flux of the stellar injected particles by at the most two orders of magnitude to obtain the Galactic flux.

References

Priest, E.R.: 1981, *Solar Flare Magnetohydrodynamics*, Gordon and Breach: New York, 1
Cane, H.V., McGuire, R.E. and Von Rosenvinge, T.T.: 1985, *Proc. 19th ICRC* **4**, 66
Ginsburg, V.L.: 1987, *Proc. 20th ICRC* **17**, 7
Landini, B.C. and Fossi, R.: 1990, *Astrophys.J* **82**, 22
Paresce, F.: 1990, *Astron.J* **89(7)**, 1022
Shapiro, M.M.: 1990, *Proc. 21st ICRC* **4**, 8
Simnett, G.M. and Haines, M.G.: 1990, *Solar Phys.* **130**, 253
Starr, R., Heindl, W.A., Crannell, C.J., Thomas, R.J., Batchelor, D.A., and Magun, A.: 1988, *Astrophys. J* **329**, 967
Vaiana, G.S., Cassinelli, J.P., Fabbriano, G., Giacconi, R., Golub, R., Gorenstein, P., Haisch, B.M, Harnden, F.R., Johnson, H.M., Linsky, J.L., Maxson, C.W., Mewe, R., Rosner, R., Seward, F., Topka, K., and Zwaan, C.: 1981, *Astrophys.J* **244**, 163

ROSAT EUV LUMINOSITY FUNCTIONS OF NEARBY LATE TYPE STARS

S.T.HODGKIN and J.P.PYE
Department of Physics and Astronomy, Leicester University
Leicester, LE1 7RH, UK

Abstract. The ROSAT EUV all-sky survey was conducted with the UK Wide Field Camera over the six month period July 1990 to January 1991, in two 'colours' using broadband filters to define wavebands centred at about 100 and 140 Å. We present EUV luminosity functions for nearby (distance $\lesssim$ 25 pc) late type (F-M) stars.

Key words: EUV – luminosity functions – nearby stars

1. The Rosat All-Sky EUV Survey

The ROSAT EUV all-sky survey was conducted with the UK Wide Field Camera (WFC; Sims *et al.*, 1990) over the six month period July 1990 to January 1991, in two 'colours' using broadband filters to define wavebands centred at about 100 Å(120 eV; S1A) and 140 Å(90 eV; S2A). The survey was fully imaging, with effective spatial resolution of $\sim$ 3 arcminutes and point source location accuracy of typically $\lesssim$ 1 arcminute. The survey achieved $\sim$ 95% coverage of the sky with the WFC. A catalogue of 384 relatively bright sources (the WFC-BSC) has recently been published (Pounds *et al.*, 1992).

2. The Third Catalogue of Nearby Stars

We have used a preliminary version of the Third Catalogue of Nearby Stars (CNS3; Gliese and Jahreiss, 1991) containing all known stars within 25 parsecs of the Sun (3803 stars). We have considered all stars in the spectral type range F–M that were observed by the WFC during the survey. As yet we have made no distinction between luminosity classes ($\simeq$ 85% of the WFC-BSC F–M star identifications are luminosity class V, the remainder are III–IV), nor have we examined the effects of binarity. We have investigated the space density of late-type stars in the catalogue as a function of distance; on this basis we have selected distance limits of 25 and 16 pc for the F–K and M stars respectively.

3. Measurements of Count Rates and Upper Limits

For each object in the CNS3 we have measured the count rates and upper limits directly from the $2° \times 2°$ images created during the standard processing of the WFC data (Pounds *et al.*, 1992). The threshold for detection has been set at the 99.9% Poisson confidence level. However, the same detection algorithm tested empirically on a set of random coordinates gives "detections" 10% of the time. For the non-detections we have applied Bayesan statistics to determine the 99% upper limits on count rates, using the techniques described by Kraft *et al.* (1991). The results of the detection procedure are listed in Table I. We have converted the WFC count rates to EUV luminosities using constant conversion factors appropriate for a cosmic

J.F. Linsky and S. Serio (eds.), Physics of Solar and Stellar Coronae, 373–376.

abundance, optically thin, hot plasma spectrum with T $\sim 10^6$ K, and a line-of-sight photoelectric absorption $n_H \lesssim 10^{18.5}$cm^{-2}. Where there are multiple stars for a single WFC detection we have removed the duplicate(s) using information from the WFC optical identification programme or from catalogues to estimate the most likely source of the EUV emission (c.f. Pounds *et al.*, 1992).

TABLE I
WFC Detections

Type	D_{LIM} (pc)	N_{CNS3}	NDET S1a	NDET S2a	$NDET_{RS}$ S1a	$NDET_{RS}$ S2a	$NDET_{BY}$ S1a	$NDET_{BY}$ S2a
F	25	220	20	28	0	0	0	0
G	25	399	22	22	8	9	0	0
K	25	684	32	22	1	1	6	6
M	16	734	40	32	0	0	4	3

D_{LIM} is the limiting distance for stars in our sample.
N_{CNS3} is the number of stars in CNS3 out to D_{LIM}.
NDET is the number of detections in each filter.
$NDET_{RS}$, $NDET_{BY}$ are the number of RS CVn and BY Dra systems (as listed in Strassmeier *et al.*, 1988) detected in each filter.

4. The EUV Luminosity Functions

We have constructed cumulative luminosity distribution functions for the F, G, K and M stars in both WFC filters using 'survival analysis' techniques to take account of the upper limits (Feigelson and Nelson 1985). The functions obtained for the S2a filter are shown here in Fig.1 (the derived luminosity functions for the S1a filter are similar) and the results are presented in Table II. We note that the functions are truncated when the lowest data point is an upper limit, thus the mean could be lower than quoted.

The F star luminosity function is 'steeper' than that of the later-type stars. The G star function is flattest of all, with a high luminosity tail. However, all but one of the RS CVn systems within our CNS3 study are classified as G stars, which will contribute heavily to this tail. We are unable to distinguish between the slopes of the luminosity functions in the two filters. All the functions are rather 'flat', implying that the source counts (logN-logS) will be dominated by the higher luminosity objects.

From the luminosity functions we have predicted the WFC source counts, and compared the latter with the measured (coverage corrected) source counts from the WFC-BSC. As an example, in Fig 2 we show this comparison for M stars in the S1a filter. The measured source counts yield the following coverage corrected values down to a count rate level of 0.015 counts.s^{-1} in the S1a filter: 21 F, 70 G, 70 K and 50 M stars. (These numbers have been updated slightly from those quoted in Pounds *et al.*, 1992, to take account of additional identifications.) Our luminosity functions predict 13 F, 73 G, 31 K and 60 M stars. In general the two sets are in

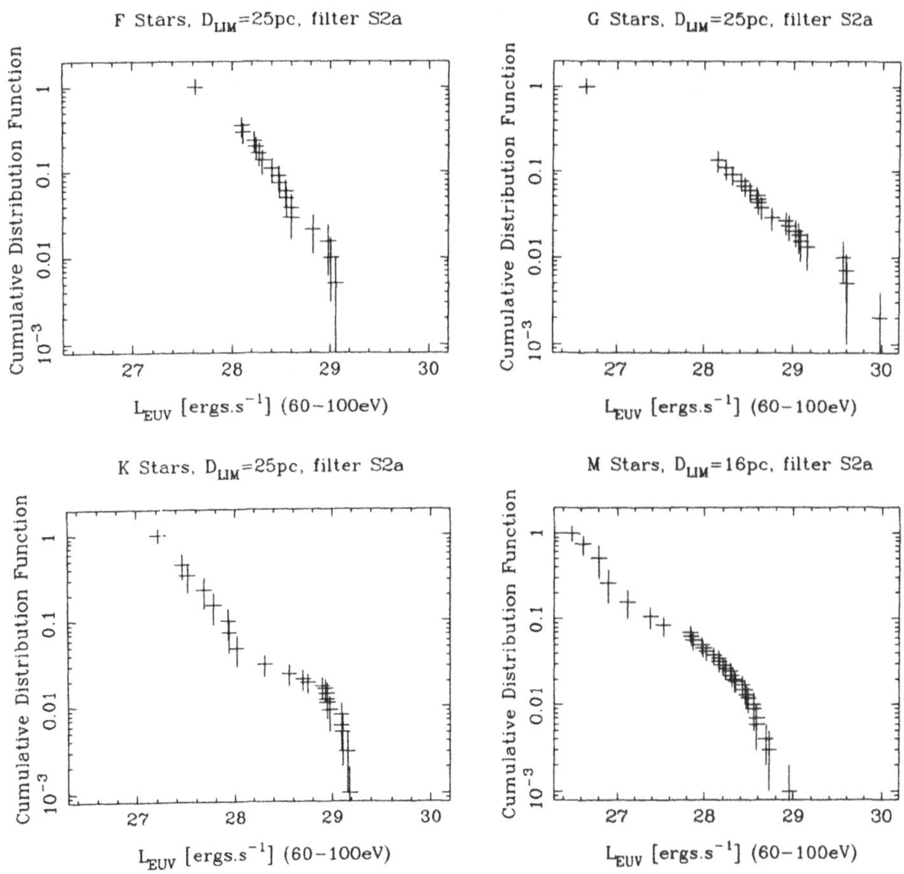

Fig. 1. Integral EUV luminosity functions for the F, G, K and M stars in the WFC S2a waveband. The error bars were computed according to the method of Feigelson and Nelson (1985).

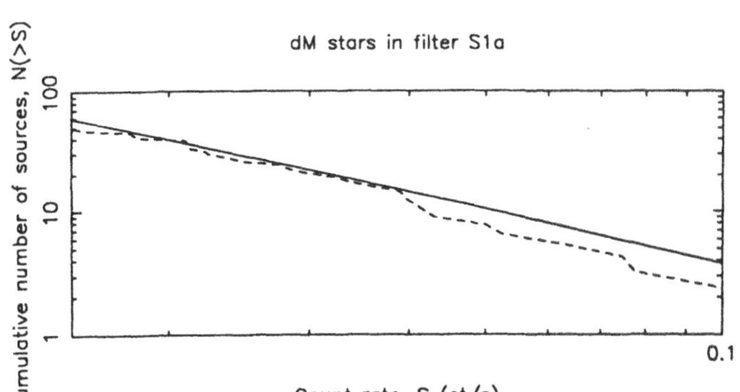

Fig. 2. dM star LogN-LogS (solid line) predicted from the WFC/CNS3 luminosity function for the WFC S1a filter compared with the measured (coverage corrected) LogN-LogS (dashed line) from the WFC-BSC (after Pounds *et al.*, 1992).

TABLE II
Results of Survival Analysis

Type	log (Mean L_{EUV})				log (Median L_{EUV})		α	
	S1a	(err)	S2a	(err)	S1a	S2a	S1a	S2a
F	27.78	0.05	27.86	0.06	27.5	27.7	-1.53	-1.60
G	27.12	0.05	27.92	0.06	26.9	26.7	-0.87	-0.82
K	27.26	0.08	27.44	0.06	27.2	27.2	-1.09	-1.11
M	26.77	0.05	26.83	0.07	26.5	26.3	-1.03	-0.97

α is the estimated slope of the cumulative luminosity function measured in log-log space.

good agreement, except for the K stars. In the WFC-BSC, many of the K stars are RS CVn systems outside the distance limit of the CNS3 and thus will have a large effect on the source counts, while not contributing to the CNS3 luminosity functions.

Pounds *et al.* (1992) noted the apparent lack of EUV-detected dM stars compared with pre-launch predictions. The present work supports that view. The prelaunch estimates (Pye and McHardy, 1988) were based on scaling from the *Einstein* X-ray luminosity functions (see e.g. Rosner *et al.*, 1985; Vaiana *et al.*, 1992 and references therein). The most striking difference between the latter and the EUV luminosity functions presented here, is the high luminosity 'tail' in the *Einstein*-derived dM distribution, which is lacking in the corresponding WFC function. It appears that the WFC source counts are not dominated by the dM stars; rather, that there are roughly equal contributions from all four spectral types.

Acknowledgements

The WFC project is supported by the UK Science and Engineering Research Council (SERC). The WFC instrument and Ground System developed by a consortium of five UK institutes: Leicester University, SERC's Rutherford Appleton Laboratory, Mullard Space Science Laboratory – University College London, Birmingham University, and Imperial College of Science, Technology and Medicine – London. The data analysis has made use of the STARLINK computing facilities and the ASURV package (Rev 1.2). The authors acknowledge the financial support of the SERC.

References

Feigelson, E.D., and Nelson, P.I.: 1985, *Astrophys.J.* **293**, 192

Gliese, W., and Jahreiss, H., 1991, "Third Catalogue of Nearby Stars", in preparation. Astron.Rechen-Institut, Heidelberg.

Kraft, R.P., Burrows, D.N., and Nousek, J.A.: 1991, *Astrophys.J.* **374**, 344

Pounds, K.A. *et al.*, 1992, *Mon. Not. R. astr. Soc.*, in press.

Pye, J.P., and McHardy, I.M.: 1988, *Astrophys. & Space Sci. Library* **143**, 241

Rosner, R., Golub.L., and Vaiana, G.S.: 1985, *Ann. Rev. Astron. Astrophys.* **23**, 413

Sims, M.R. *et al.*: 1990, *Optical Engineering* **29(6)**, 649

Strassmeier, K.G. *et al.*: 1988, *Astron. Astrophys. Suppl.* **72**, 291

Vaiana, G.S., Maggio, A., Micela, G., and Sciortino, S.: 1992, *Mem. Soc. Astr. It.* , in press. Palermo Observatory preprint 8/91

X-RAY/OPTICAL SURVEY OF LATE TYPE STARS

A.J.M. PITERS

Astronomical Institute "Anton Pannekoek", Amsterdam, The Netherlands
Center for High Energy Astrophysics, Amsterdam, The Netherlands

J.H.M.M. SCHMITT

Max Planck Institut für Extraterrestische Physik, Garching-bei-München, Germany

C.J. SCHRIJVER

Astronomical Institute, Utrecht, The Netherlands

S. BALIUNAS

Center for Astrophysics, Harvard University, Cambridge, USA

C. ZWAAN

Astronomical Institute, Utrecht, The Netherlands

and

J. VAN PARADIJS

Astronomical Institute "Anton Pannekoek", Amsterdam, The Netherlands
Center for High Energy Astrophysics, Amsterdam, The Netherlands

Abstract. We have obtained Ca II H&K line-core emission fluxes for ~ 200 F-, G- and K-type stars within at most a few days of the ROSAT all-sky survey observations. A tight relationship between the excess Ca II H+K line-core and soft X-ray surface flux densities has been derived earlier using observations of the EINSTEIN and EXOSAT satellites with much larger time intervals between the chromospheric and coronal flux measurements, but considerable scatter hampered a detailed analysis of this relationship. Part of this scatter is caused by stellar variability which is greatly reduced in the near-simultaneous measurements presented here. The stars in our sample cover a wide range of stellar activity, spectral type and luminosity class. With this sample we hope to unravel the connection between chromospheric and coronal heating in more detail. Here we show the first results: a comparison with EINSTEIN fluxes, and the correlation of the optical and X-ray observations.

Key words: Ca II H and K emission – late type stars – correlation of X-ray and Ca II emission

1. Introduction

Comparison of radiative flux densities from the outer atmospheres of cool stars frequently show power–law relationships, which appear independent of colour or luminosity class after subtraction of a so–called basal component that appears to be determined by the stellar effective temperature (e.g. Rutten et al. 1991). Stellar variability and intervals between the measurement of, e.g., chromospheric and coronal fluxes ranging up to several years result in a substantial scatter about the mean relationship that hampers its detailed description. In this study we present data with time intervals typically of the order of a few days or less.

2. The data

We have obtained Ca II H&K photometry of 238 stars at the Mt. Wilson Observatory typically within a few days of the observations of these stars in the ROSAT all–sky survey. The selected stars are late–F, G- and K-type stars of luminosity classes V, IV and III. We preferentially selected stars for which some information

J.F. Linsky and S. Serio (eds.), Physics of Solar and Stellar Coronae, 377–380.

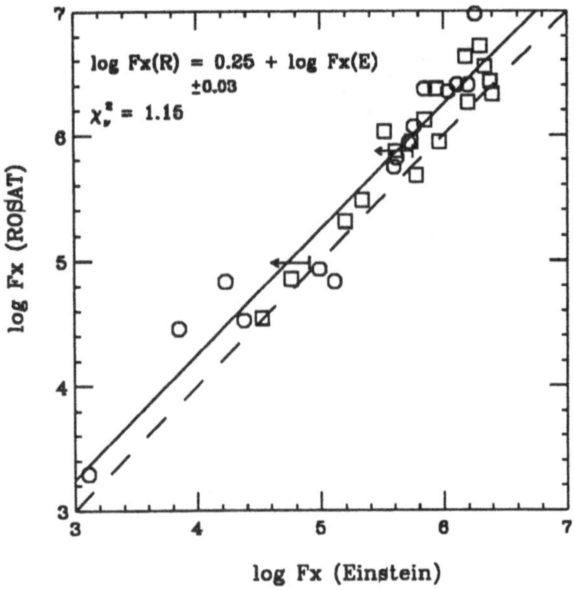

Fig. 1. ROSAT vs. EINSTEIN surface fluxes. Giants and dwarfs are denoted by circles and squares, respectively. The solid line represents a linear fit to the data, the dashed line shows where ROSAT and EINSTEIN fluxes are equal

on the mean activity level and rotation rate is available, while attempting to cover a large range of activity levels evenly. The ROSAT survey data for these stars are presently under investigation. Until now, 115 sample stars have been identified with X-ray sources. For each detected star the X-ray flux density has been derived from its countrate, using its hydrogen column-density (derived from Paresce 1984) and a measure for the mean coronal temperature (estimated from the spectral hardness, Piters et al. 1992). We used the apparent magnitude and the bolometric flux derived from effective temperatures to derive the flux densities on the stellar surface (cf. Rutten et al. 1991).

The surface flux density in the cores of the CaII H&K lines is derived from the Mt. Wilson data following Rutten (1984). By subtracting a basal flux component (in part line–wing emission and in part probably of acoustic origin; Schrijver 1987) we derive the so-called excess flux density.

3. Discussion

In Fig. 1 we show the X-ray surface fluxes for 34 stars for which EINSTEIN IPC and ROSAT PSPC survey observations are available. EINSTEIN surface fluxes are from Rutten et al. (1991), derived using a constant countrate-to-flux conversion factor. This is appropriate because the conversion factor in the EINSTEIN IPC energy-range (0.15–4.0 keV) is not very sensitive to differences in hydrogen column-densities and coronal temperatures (e.g. Giacconi et al. 1979). The solid line in Fig. 1

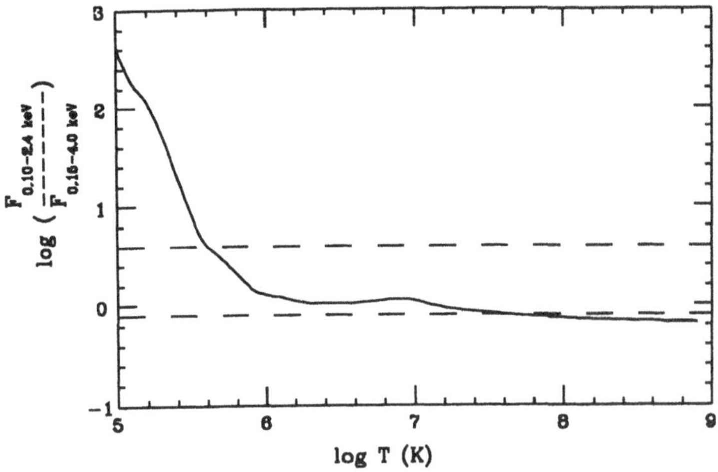

Fig. 2. Expected ratio of ROSAT and Einstein surface fluxes as a function of coronal temperature. The dashed lines show the range containing 90% of the stars from Fig. 1

is a linear fit to the data. The offset from the line F_X(ROSAT)= F_X(EINSTEIN) (the dashed line) is a factor 1.78 ± 0.12. There is no significant difference between giants and dwarfs. The overall uncertainties in the EINSTEIN- and ROSAT-fluxes are 20 to 30% and 10%, respectively, including uncertainties in countrate-to-flux conversion factors. Hence it appears that any systematic difference in the calibration can acount for at most 30% of the systematic difference between EINSTEIN- and ROSAT-fluxes. Since it is unlikely that all stars have increased their activity level since the EINSTEIN observations, at least part of the offset in Fig. 1 is caused by differences in the energy passbands of the IPC (0.15–4.0 keV) and the PSPC (0.1–2.4 keV).

The theoretically expected ratio of ROSAT vs. EINSTEIN fluxes for an optically thin isothermal plasma is shown in Fig. 2. At temperatures well below 1 MK the ROSAT passband is much more sensitive than the EINSTEIN passband, because much of the energy is radiated between 0.1 and 0.15 keV. The enhancement of ROSAT fluxes with respect to the EINSTEIN fluxes suggests that in addition to the component at several MK inferred from EINSTEIN and EXOSAT data there may be a cool component in many late–type stars (previously suggested by Schrijver et al. 1992), with a temperature below about $7 \cdot 10^5$ K and a strength of more than 50% of the total emission.

In Fig. 3 we show the ROSAT X-ray surface-fluxes versus the Ca II H&K line core excess fluxes for 41 stars in our sample. We selected these stars because they are all single (i.e. not classed as optical or spectroscopic binaries), have a well defined spectral type and $B - V > 0.45$. We determined a best power–law fit by minimizing χ^2, which is defined as the sum of the squared distances (in σ_i) of the individual error-ellipses to the fitted line.

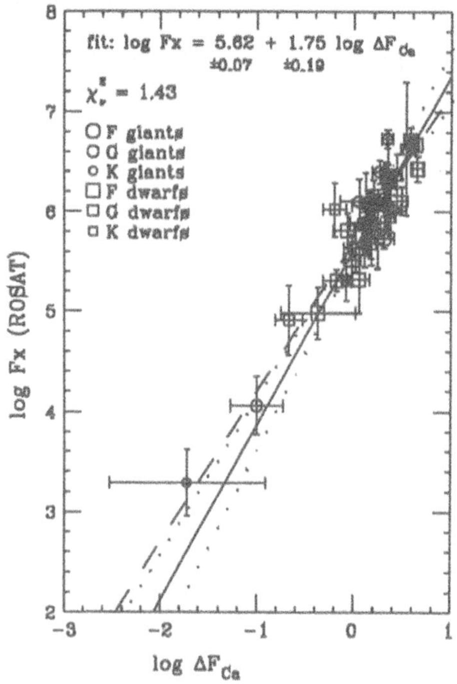

Fig. 3. ROSAT surface flux versus nearly-simultaneous Ca II H+K excess flux. The solid line is a power–law fit (dots are 1σ boundaries). The dashed line is the relation derived from EINSTEIN and EXOSAT data (Schrijver, 1992). We scaled the EXOSAT surface fluxes for this relation to ROSAT values, assuming F_X(EINSTEIN)= $0.4F_X$(EXOSAT) (Schrijver et al. 1992) and F_X(ROSAT)= $1.78F_X$(EINSTEIN) (cf. Fig. 1)

4. Conclusions

X-ray surface flux densities derived from the ROSAT all-sky survey are generally a factor 1.78 ± 0.12 higher than those derived from the EINSTEIN IPC observations. An additional soft component in the X-ray spectra of coronae, with a temperature less than $7 \cdot 10^5$ K and a strength of more than 50% of the total emission could explain this difference. The relationship between the X-ray surface flux density and the excess Ca II H&K line core excess flux density, as derived from the present data, is in good agreement with the relation that was previously found from EXOSAT and EINSTEIN observations.

References

Giacconi R., et al.: 1979, *ApJ* **230**, 540
Paresce: 1984, *Astron.J.* **89**, 1022
Piters A.J.M., Schrijver C.J., Schmitt J.H.M.M., 1992, in preparation
Rutten R.G.M.: 1984, *A&A* **130**, 353
Rutten R.G.M., Schrijver C.J., Lemmens A.F.P., Zwaan C.: 1991, *A&A* **252**, 203
Schrijver C.J.: 1987, *A&A* **172**, 111
Schrijver C.J., Dobson A.K., Radick R.R.: 1992, *A&A* **258**, 432

CORONAL ACTIVITY IN RELATION TO CHROMOSPHERIC CYCLES AND STELLAR ROTATION

A. HEMPELMANN, G. RÜDIGER, G. HILDEBRANDT
Astrophysikalisches Institut Potsdam, Germany

J.H.M.M. SCHMITT
Max-Planck-Institut für Extraterrestrische Physik Garching, Germany

A preliminary X-ray statistics is presented for a sample of late-type stars with known rotation rates usually taken from their short-term Ca H+K modulation. Also the Wilson stars which have been monitored over decades to find their chromospheric long-term activity are included in the sample. Our X-ray data base is formed by the ROSAT all-sky survey.

Possible correlations between the cycle characteristics and the X-ray emission are of main interest. The sample is thus divided into 3 groups, i.e. the regular or cyclic emitters, the chaotic emitters and those without temporal fluctuations. Significant differences of the X-ray emission levels are found between the various classes.

Similarly to earlier studies we have also attempted to correlate the main X-ray intensity characteristics (luminosity L_X, surface flux F_X and the ratio $L_X/L_{\rm opt}$) with the rotational velocities of the stars – here in form of their rotation rates Ω. The resulting plots show much less scatter than in previous studies made with the spectroscopic data $v \sin i$. We stress that there is a significant increase of all the fluxes with the rotation rate but the effect proves to be smaller than previously suggested.

The latter finding is in good agreement with results of the recent calculations of non-diffusive transport effects in rapidly rotating convection zones. Both the α-effect of the dynamo theory as well as that part of the Reynolds stress, which maintains the differential rotation, exhibit only a very slight increase with the angular velocity, provided the inverse Rossby number exceeds the value of (say) 1. There is, therefore, a distinct range of Rossby numbers with only a quite weak dependence of the dynamo activity on the related rotational period.

Key words: Rotation – Stellar cycles

J.F. Linsky and S. Serio (eds.), Physics of Solar and Stellar Coronae, 381.

CORRELATION BETWEEN RADIO AND X-RAY LUMINOSITIES AMONG LATE-TYPE STARS: A ROSAT-VLA SURVEY OF M DWARFS

MANUEL GÜDEL
Joint Institute for Laboratory Astrophysics
University of Colorado, Boulder, CO 80309-0440, USA

JAY A. BOOKBINDER
Harvard-Smithsonian Center for Astrophysics
Cambridge, MA 02138, USA

and

JÜRGEN H.M.M. SCHMITT and THOMAS A. FLEMING
Max Planck-Institut für Extraterrestrische Physik
8046 Garching, FRG

Abstract. We present results of one of our ongoing projects to study the relation between radio and X-ray emission from coronal stars, viz. a survey of dM/dMe stars obtained simultaneously with ROSAT and the VLA. We find that the X-ray and radio luminosities are tightly correlated over three orders of magnitude, irrespective of spectral type. The dependence is consistent with proportionality ($L_R \propto L_X$). This relation is the low-energy portion of a general relation between L_R and L_X that holds for many active coronal stars over five orders of magnitude.

Key words: stars: X-ray radiation – stars: radio radiation – stars: coronae

1. Introduction

For the lower main sequence, *radio emission* is widely considered as indicative of high energy, *non-thermal* processes. Accelerated electrons emit gyrosynchrotron radiation with $T_B \gtrsim 10^8$ K while spiraling in mostly closed magnetic coronal structures (loops). On the other hand, like on the Sun, *X-rays* are assumed to originate from hot *thermal* ($T > 10^6$ K) coronal plasma confined in these magnetic loops. Consequently, the radio and X-ray luminosities, L_R and L_X, are not expected to be directly correlated, since they are controlled by different electron populations.

For M dwarfs, an early study of a very limited sample indicated, however, a strong correlation (Bookbinder 1987). Also, White et al. (1989) concluded that for a given main-sequence spectral type the radio sources detected at the VLA tend to be those with the largest X-ray fluxes. In contrast, Caillault et al. (1988), based on non-contemporaneous observations in the radio and the X-ray regime, found that the quiescent L_X and L_R are essentially uncorrelated. Comparing late-type main-sequence stars with RS CVn binaries, Drake et al. (1991) concluded that the latter are, if scaled with the X-ray luminosities, significantly more radio luminous than main-sequence stars. RS CVn binaries indeed follow a well-defined correlation between L_R and L_X, reported by Drake et al. (1989).

The objective of our study was to obtain **simultaneous** and **highly-sensitive** diagnostics of both thermal and nonthermal plasma components of late-type stars. Here, we report a portion of our results on M dwarfs, focusing on the overall relation between L_X and L_R.

J.F. Linsky and S. Serio (eds.), Physics of Solar and Stellar Coronae, 383–386.

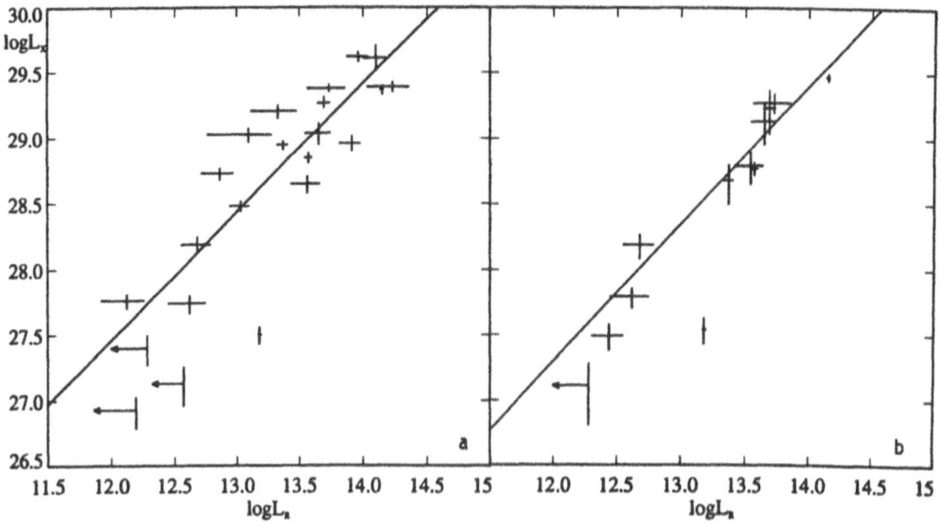

Fig. 1. L_X vs. L_R (a) for all M stars considered, (b) restricted to strictly simultaneous measurements. The solid lines are least-squares fits (see text).

2. Observations of L_R and L_X

We selected 22 dM/dMe targets on the basis of (1) an X-ray flux sufficiently high to be detected by ROSAT ($\gtrsim 10^{-13}$ erg cm^{-2} s^{-1} as previously observed with *Einstein* or EXOSAT); or (2) a previous microwave detection; additionally, (3) the stars are all relatively nearby, with a mean distance of about 8 pc. We note that the distribution of L_X of the targets is essentially identical to that of the parent X-ray sample for both the *old disk* and *young disk* stars, implying that our selection criteria have *not* biased this sample.

From the ROSAT database, we extracted X-ray count rates for 19 of our targets that were covered by the All-Sky Survey, obtained between August 1990 and February 1991; all targets were clearly detected. The VLA was used to obtain mostly (near-)simultaneous flux measurements at 6 cm (4.9 GHz) and 3.6 cm (8.5 GHz) for a total of 20 targets, with seven targets constituting visual binary systems. The detection rate was surprisingly high, i.e., 84% at both 6 cm and 3.6 cm. Eleven targets were observed strictly simultaneously with ROSAT and the VLA. To make the sample complete (missing ROSAT or VLA observations), we supplemented our X-ray and radio observations with three published L_X and two L_R values; a complete account of all observations will be given elsewhere.

In the complete sample, L_X varies between $10^{26.9}$ and $10^{29.6}$ erg s^{-1}, as compared to $L_X \approx 10^{27}$ erg s^{-1} for the quiet Sun. L_6 and $L_{3.6} \approx 10^{12.2} - 10^{14.3}$ erg s^{-1} Hz^{-1} for the detections at 6 cm and 3.6 cm, respectively, with little essential difference between the two wavelengths; most spectra were typically very flat, with many spectral indices around −0.5 between 6 cm and 3.6 cm.

In Fig. 1a we present a logarithmic plot of L_X versus L_R, L_R being defined

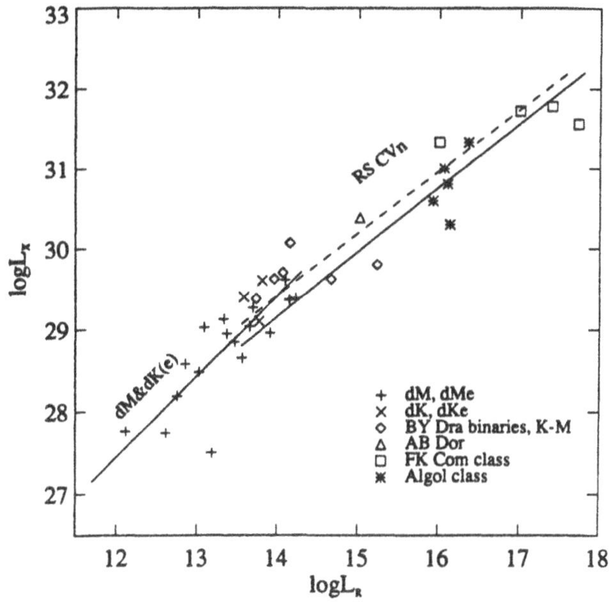

Fig. 2. A general correlation L_X vs. L_R valid for many classes of active stars. Approximate fits are for M and K dwarfs, and for RS CVn binaries (Drake et al. 1989). Symbols are used for individual detections in a given class, except RS CVn's (after Güdel & Benz 1992).

here as the average of L_6 and $L_{3.6}$. Our sample clearly shows that **the quiescent X–ray and radio luminosities are strongly correlated over three orders of magnitude close to proportionality $L_R \propto L_X$:** $\log L_X \approx (0.97 \pm 0.13) \log L_R + (15.9 \pm 1.76)$ [correlation coefficient $R = 0.87$ for 19 targets detected both in radio and X-ray regimes, excluding the microwave-strong Gl 65B]. From Fig. 1a and an estimated integration between 1 and 15 GHz, we find the approximate ratio between the total luminosities in X-rays and radio: $L_X/L_{R,\mathrm{tot}} \approx 10^{5.5}$, with $L_{R,\mathrm{tot}} = 10^{21.5}$–$10^{24.5}$ erg s^{-1}.

If we include only the 11 strictly *simultaneous* observations, the correlation appears to become tighter (Fig. 1b). For L_X versus L_R we again find proportionality, with $\log L_X \approx (1.05 \pm 0.12) \log L_R + (14.7 \pm 1.55)$ [$R = 0.96$ for 9 detections in both regimes, excluding Gl 65AB and the flaring Gl 406].

3. The general picture

We are currently studying the relation between L_R and L_X for several classes of active stars. Figure 2 is a compilation of $\log L_X$ versus $\log L_R$ (here, $L_R = L_6$) based in part on published values and in part on our ongoing campaigns. A comprehensive description of this study is given in Güdel & Benz (1992).

Evidently, L_X is positively correlated with L_R over five orders of magnitude: $\log L_X \approx 0.82 * \log L_R + 17.8$. The regression curve valid for M stars is **the low energy continuation of a general trend valid for most stellar classes that**

exhibit coronal activity. The *break* at $\log L_R \approx 14.5$ indicates that the more luminous stars possess slightly radio-overluminous (or X-ray deficient) coronal sources (as suggested by Drake et al. 1991).

The coherent trend of a positive correlation between L_X and L_R (close to proportionality) over five orders of magnitude is surprising; it is valid for a variety of stellar classes independent of *age* (young main-sequence stars to evolved giants), *spectral type* (G, K, and M), *binarity* (single stars or close binaries), *rotation* (periods between 12 hours and 100 days), or *photospheric/chromospheric activity* (spotted stars or unspotted stars, emission-line or non-emission-line stars).

4. Conclusions

We have found a proportionality between L_R and L_X of M dwarfs over three orders of magnitude. It seems to become tighter if the observational data are obtained closer in time. This clearly suggests that slow variations of the baseline flux level of L_R and L_X are correlated in time. Additionally, we find that the relation valid for M dwarfs is the low-energy continuation of a relation valid for many classes of active coronal stars.

The strong correlation points to an intimate physical relation between the two emissions. Standard models propose that flaring regions accelerate relativistic particles while at the same time heating the surrounding plasma. It is, however, not obvious why the fraction of the energy going into fast particles and into heat should be constant on many stars. Alternatively, it could be that fast particles are generated in a highly efficient accelerator. The lower energy portion of this population thermalizes and heats the plasma. The validity of the relation between the two coronal emissions over a wide range of stellar classes holds if the acceleration process and its high efficiency are similar in different stellar coronae (Güdel & Benz 1992).

Acknowledgements

The VLA is a facility of the National Radio Astronomy Observatory, which is operated by Associated Universities, Inc., under cooperative agreement with the National Science Foundation. This work was financially supported by the Swiss National Science Foundation, NASA grant NAG5-1887 through the University of Colorado, W17772 through NIST, and NASA grant NAGW-112 to the Smithsonian Astrophysical Observatory.

References

Bookbinder, J.A. 1987, in *Proceedings of the Midnight Sun Conference*, Tromsø, Norway, June 1987

Caillault, J.-P., Drake, S., & Florkowski, D. 1988, AJ, 95, 887

Drake, S.A, Simon, T., & Linsky, J.L. 1989, ApJ, 71, 905

Drake, S.A., Linsky, J.L., Judge, P.G., & Elitzur, M. 1991, AJ, 101, 230

Güdel, M., & Benz, A.O. 1992, ApJ, submitted

White, S.M., Jackson, P.D., & Kundu, M.R. 1989, A&AS, 71, 895

DISCUSSION FOLLOWING PAPERS ON ROSAT OBSERVATIONS

Discussion following paper by J. Schmitt

G. Tagliaferri: The number of X-ray selected stars in the EXOSAT High Galactic Latitude Survey is 37% (Giommi et al., Ap.J. **378**, 77 (1991)), which is still higher than the 25% of the EMSS and probably due to the softer energy band. The number that you quote (45%) for ROSAT probably includes also the B stars, whose detections are due to UV contamination.
F. Reale: Is it possible to have spectroscopic/temperature analysis in various phases of observed flares? (Are there enough photons?)
J. Schmitt: For most flares observed in the ROSAT All-Sky Survey it is not possible to do a detailed X-ray temperature analysis as a function of flare phase. However, in some cases, for example the flare on Speedy Mic, this has been done. With the All Sky Survey data one can model light curves for a large variety of flares.
B. Foing: The fast microflare seems reminiscent of solar hard X-ray spikes. Do you measure a similar time lag (30 sec?) versus the B light curve peak and hardness for the fast flare on UV Ceti compared to the more normal following flare?
J. Schmitt: The two B-band optical flares (for which we have simultaneous X-ray data) are followed by short X-ray "spikes" with a time delay of $\sim$ 30 sec. The second optical flare is, in addition, followed by an X-ray enhancement $\sim$ 5 min later. So the X-ray and optical light curves are very similar. The signal-to-noise of the X-ray data is insufficient to detect changes in hardness or temperature.
R. Mewe: Which spectroscopic model did you use in fitting the data?
J. Schmitt: We used the Raymond (1988) code for our spectral fits. We noted significant differences in our fit results when earlier versions of the same code were used, and we want to consider other X-ray emissivity codes such as yours.
R. Pallavicini: Have you detected any variability in early-type (O-B) stars?
J. Schmitt: So far I have not found any really convincing cases for variability in O-type stars, but we are still looking.
E. Tagliaferri: You have shown a flare coming from a B5+B5 star system. Could there be a previously unknown late-type star in the system?
J. Schmitt: It is always possible that the X-ray emission comes from an unseen later-type companion. This is precisely the problem in identifying unambiguous cases for X-ray emission from stars which "should" not be X-ray sources (such as A-type stars).

Discussion following paper by F. Harnden

J.-P. Caillault: The high percentage of M dwarf detections and the large number of faint optical candidates with unknown membership indicate that the ROSAT data may, in fact, supersede the optical database for finding cluster members.
J. Schmitt: Are there any Pleiades members detected with *Einstein* that are not seen in the 50 ksec ROSAT pointing?

J.F. Linsky and S. Serio (eds.), Physics of Solar and Stellar Coronae, 387–389.

R. Harnden: We will have better answers to this question when the data of all four pointed PSPC observations (cf. Table I) are combined, but I believe that this will be the case on the basis of the (partial) Rosner data.

Discussion following paper by J. Pye

J. Schmitt: John, a detection threshold of 4.5σ is very conservative. How does your detection rate increase by loosening this threshold to more realistic values?
J. Pye: We are currently analyzing the data at a detection threshold of $\sim 3\sigma$.

Discussion following paper by J.-P. Caillault

S. Drake: Can you explain the much higher *Einstein* IPC detection rate of late B- and A-type stars in Orion than was obtained by your ROSAT HRI observations?
J. Caillault: In our original paper, we excluded optical candidates fainter than $V \sim 13.5$, since, if they were the true X-ray sources, their F_x/f_V values would have been much higher than thought possible at that time. In retrospect, that was a mistake, i.e., those faint stars are likely candidates for the X-ray emission.
F. Walter: Of the 11 reported late-B star detections by Caillault and Zoonematkermani (1989), how many were reobserved and not detected in your HRI observations?
J. Caillault: Three (3).

Discussion following paper by S. Casanova

F. Walter: When you compared the luminosity distributions of the CTTS and the WTTS, you said these are stars with (the CTTS) and without (the WTTS) disks. Do the 18 WTTS not detected lack disks? Do they have IR excesses? A large number of PTTS stars lacking both disks and coronae would indeed be interesting. Note that WTTS selected solely by their Hα emission equivalent width do not necessarily lack large circumstellar disks and IR excesses (e.g., SU Aur and HDE 183447).
S. Casanova: I meant that the WTTS have no disk interacting with the star and thus no boundary layer. So they are selected on the criterion of Hα equivalent width which is available for most of the T-Tauri stars. About the importance of the disk, L_x is correlated with the infrared disk luminosity, but when using $F_x = L_x/4\pi R_*^2$, the surface flux, this correlation disappears.
J. Schmitt: Are there any differences in the PSPC spectra of the detected WTTS and the CTTS?
S. Casanova: Until now we have seen absolutely no difference in the spectra between WTTS and CTTS. They all have a best-fit temperature for the Raymond-Smith model between 1 and 3 keV, except for the superflare.

Discussion following paper by R. Dempsey

F. Walter: I take exception to your argument that there is no rotation-activity relation in RS CVn systems. By selecting RS CVn systems, you introduce a serious selection effect. In an RS CVn system the secondary, which produces most of L_x, must fill a significant fraction of its Roche radius (see Morgan and Eggleston's paper ~ 1980). Therefore, if RS CVn's have rotation activity relations anything like the single stars — and there is no significant difference between the activity levels of RS CVn's and single stars of comparable periods (see Basri, Laurent, & Walter, Ap.J. **298**, 761 (1985)) — then L_x is independent of P and $L_{\rm bol} \sim P$!
R. Dempsey: Yes, that is something I did not address and you may be correct. This will be clarified when I finish the analysis of the BY Dra binary systems.
A. Collura: *Einstein* IPC data showed that RS CVn spectra can be described by a continuous emission measure model with a very high $T_{\rm max}$ and a very steep slope, indicating that most of the emission measure is concentrated at high temperatures. If this holds also for the ROSAT sample, there is probably not enough leverage in the ROSAT band to constrain properly the model. How can you therefore conclude that the two temperatures determined by the fit are not an instrumental artifact?
R. Dempsey: Continuous models do fit the PSPC data, but not as well as the two-temperature models. This is because of the low sensitivity of the PSPC to plasma hotter than 4×10^7 K. Clearly, the detection process is affecting the measured temperatures. My point is that all of the detectors — the IPC, SSS, TGS & PSPC — yield essentially the same temperatures. I do not think this is solely a detector effect.

PART V

OTHER OBSERVATIONS

RADIO EMISSION FROM CORONAL-TYPE STARS

S. A. DRAKE
HEASARC, Code 668, NASA/GSFC, Greenbelt, MD 20771, USA

Abstract. In this short review I have limited the discussion to radio continuum emission from stars that have coronae or magnetospheres, i.e., having magnetically structured, X-ray emitting plasma in their outer atmospheres. I first summarize some of the recent improvements and enhancements that have occurred in radio instrumentation and observing capabilities, and then discuss the physical mechanisms and flux density levels of cm radio emission that may be expected to be important for coronal stars using the Sun as an example. I next briefly describe and categorize the various types of radio-emitting coronal stars that have been identified to date. I discuss the factors that appear to be correlated with enhanced levels of radio emission in coronal stars, and explore in detail the intimate linkage between (thermal) X-ray emission and (nonthermal) radio emission that several classes of active star (RS CVn stars, FK Comae stars, dMe stars, etc) exhibit. I finally suggest some potentially fruitful directions in which future research on radio emission in coronal stars might usefully be directed.

Key words: radio continuum: stars – stars: coronae

1. Introduction

I have narrowed the scope of this review to include only broad-band radio continuum emission at cm wavelengths from stars having magnetically structured, high-temperature ($T_e \sim 10^6 - 10^8$ K) outer atmospheres, e.g., coronae and/or magnetospheres. Such stars I will hereafter loosely refer to as coronal stars and include dwarf stars of spectral types F through M, F - K2 giant stars, and F - mid-G supergiants. Stars in this part of the H-R Diagram with enhanced levels of radio emission relative to normal stars include the RS CVn binary stars, FK Comae giant stars, AB Dor-like very young disk dwarfs (e.g., Pleaides members), BY Dra/flare stars (both binary and single stars), Algol binary stars having a cool, evolved component, W UMa contact binary stars, and low-mass pre-main sequence stars (T Tau stars and 'naked' or 'weak-lined' T Tau stars). A rather hotter group of active stars comprise the Magnetic Bp stars of the Helium-peculiar and Silicon-strong subclasses, and moderate-mass pre-main sequence stars such as the Herbig Ae/Be stars, FU Ori stars, and even an embedded B star ρ Oph S1 (André et al. 1988): these classes might be alternatively characterized as magnetospheric rather than coronal stars, inasmuch as X-ray emission has not been shown to be always present in such stars. Another possible group of coronal stars are the magnetic white dwarfs (Arnaud et al. 1992).

The restriction of this review to coronal stars implies the exclusion of stars that have no high-temperature plasma in their outer atmospheres, such as red giant and supergiant stars. Even though luminous OB stars and Wolf-Rayet stars do appear to have high-T_e plasma embeddded in their predominantly cooler massive winds, the conventional wisdom is that this hot plasma is in strong shocks propagating through their winds, and thus the structures are not magnetically controlled, and hence outside the scope of this review. (But see Usov and Melrose 1992 for an alternate view of the nature of the X-ray emission from hot stars). Finally, I will not consider radio emission from exotic classes of stars such as X-ray binaries, symbiotic binaries, or cataclysmic variables.

J.F. Linsky and S. Serio (eds.), Physics of Solar and Stellar Coronae, 393–400.

The field of stellar radio emission is now so wide that there have been few recent attempts to summarize its condition. A valuable resource that is continually being updated is Heinrich Wendker's catalog of radio-continuum emitting stars (e.g., Wendker 1987). As of 1992 July 16, this catalog contained radio data (flux densities and upper limits) on 2684 stellar objects, compared to 607 stars in its original (Wendker 1978) version. I estimate that about 25% of the stars in the 1992 update have at least one detection reported, which is about twice the 'success rate' in the 1978 version. This means that the number of definite, probable, and possible radio stars has increased by an order of magnitude in the last 14 years, and suggests why there have been few broad radio star reviews! The proceedings of the Boulder Radio Stars Workshop (Hjellming and Gibson 1985) summarize the state of the field as of 8 years ago. Other important general reviews since then include those of Dulk (1985), Hjellming (1988), and Bookbinder (1988); specialized reviews include those of Kuijpers (1989) on the properties of stellar radio flares, Stewart (1992) on the radio emission of late-type active stars, and of Bastian (1990) and Lang (1990) on dMe flare star radio emission.

One final general point to stress in a meeting like the present one that is primarily devoted to X-ray properties of stars is that stars at radio wavelengths are very weak (typically mJy flux densities) compared to the extragalactic background of strong (Jansky level) radio emitters. Few stars are discovered serendipitously in radio fields, unlike the situation in, say, the *ROSAT* All-Sky Survey where stars are expected to comprise $\geq$ 35% of the 5.4×10^4 X-ray sources that were discovered. For comparison, a recent 20-cm survey of the galactic plane (Zoonematkermani et al. 1990) with a threshold of $\sim$ 25 mJy produced a catalog of $\sim$ 2000 small-diameter radio sources of which only 4 to 8 are probably stars. This implies that stars comprise $\ll$ 1% of compact radio sources in the galactic plane, i.e., in that portion of the sky in which their density is highest. At high galactic latitudes, the fraction of stellar radio sources found in a serendipitous survey is probably no more than 0.01% of the total(e.g., see Drake et al. 1989).

2. Recent Improvements and Enhancements in Radio Instrumentation and Techniques

The field of VLBI (very long baseline interferometry) is about to enter its 'golden age' with the pending completion of NRAO's Very Long Baseline Array (VLBA). Even before this has happened, improvements in VLBI techniques such as phase-referencing observations of weak sources by frequent interleaving of observations of a nearby strong source have meant that it is possible to obtain information on the milli-arcsecond (mas) scale of stellar radio sources of quite low flux density ($\geq$ 1 mJy): this has enabled the nonthermal nature of several classes of radio stars to be directly confirmed by their having brightness temperatures T_B in excess of 10^8 K (e.g., Lestrade et al. 1990; Phillips et al. 1991; André et al. 1992). It is also now possible to obtain VLBI parallaxes of radio stars of superlative precision: e.g., Lestrade et al. (1992) derived a VLBI parallax for the RS CVn star σ^2 CrB of 44.23 mas that has a formal accuracy of $\pm$0.17 mas which is 20 times superior to the uncertainty in the optical parallax measured for this star.

Significant enhancements have also been made to the operational capabilities of existing arrays such as the VLA, MERLIN, and the Australia Telescope Compact Array (ATCA). The VLA now has all 27 antennas equipped with receivers for 1.3, 2.0, 3.6, 6.2, 20, and 90 cm, and 8 antennas equipped with an additional 400 cm receiver. Its most sensitive observing wavelength is X-band (3.6 cm): a 10-hour integration on source at this wavelength gives a $3\sigma \sim 20\mu$Jy detection threshold, implying that one could detect a flaring RS CVn star at a distance of 6 kpc. The MERLIN array has also recently been upgraded, so that, for example, one can now observe weak ($\sim$ 0.1 mJy) radio sources with its high spatial resolution capability (e.g., 50 mas beam at 6 cm). The completion of the ATCA means that it is now possible to observe weak southern hemisphere radio sources interferometrically at cm wavelengths on a routine basis: for example, an 8-hour integration can yield a 200μJy detection limit at 3.6 cm and 6 cm due to its ability to observe at two wavelengths simultaneously.

3. Possible Radio Continuum Emission Mechanisms in Coronal-Type Stars

There are three emission and/or opacity mechanisms that are anticipated to be important in these stars in the cm range of wavelengths:

Optically Thin or Thick Bremsstrahlung. At some height in a stellar atmosphere it will become opaque to radio waves due to free-free opacity, and the radio brightness temperature T_b (see Dulk 1985 for definition) will thus be equal to the local electron temperature at that location. For example, in the quiet solar atmosphere at 3.6 cm this occurs in the upper chromosphere, and the measured T_b is 1.3×10^4 K. In addition, a stellar corona will be a source of optically thin free-free emission: for a corona with $T_e \sim 10^7$ K, it can be shown (e.g., Drake and Linsky 1986) that the predicted radio flux density S_ν in mJy is given by

$$S_\nu = 1 - 2 \times 10^9 f_x \tag{1}$$

where f_x is the observed X-ray flux in the Einstein IPC bandpass in cgs units.

Gyroresonance or Cyclotron absorption. Although, strictly speaking, this is a narrow-band process being centered on the local gyrofrequency and its harmonics, because of the wide range of magnetic field strengths that are expected to be present in any realistic magnetic geometry (implying a wide range of gyrofrequencies) this can manifest itself as a quasi-continuous process. Above regions of a star having intense localized magnetic fields, gyroresonance absorption will provide an enhancement in the radio opacity; since coronal stars have temperature inversions above their photospheres, this can increase the radio brightness temperature considerably, perhaps even up to the coronal electron temperature of $10^6 - 10^7$ K. In the case of the Sun, at Solar Maximum the disk-averaged brighness temperature at 3.6 cm can be $\geq 3 \times 10^4$ K, i.e., 2 or 3 times the Solar Minimum value, due to the enhanced brightness temperatures associated with solar Active Regions that are covering a few per cent of the solar disk.

Gyrosynchrotron Emission. This is the semi-relativistic extension of the cyclotron emission process for electrons with temperatures of $5 \times 10^7 - 5 \times 10^9$ K, or nonthermal energies of $10-100$ keV. Gyrosynchrotron emission from high-energy (either super-hot thermal or power-law, nonthermal) particles is only confirmed to be present on the Sun during solar flares. In the most active stars, gyrosynchrotron emission appears to be a near-continual process rather than the highly sporadic process it is in the Sun, implying that near-continual acceleration of nonthermal electrons and injection into the coronal regions is taking place in such stars.

4. The Observed Radio Emission from Coronal Stars

I will use the monochromatic radio luminosity L_ν at a frequency ν as my basic unit for the measurement of the strength of a radio source. L_ν in erg s^{-1} Hz^{-1} is related to the observed flux density S_ν in mJy of a source at a distance of D parsecs by $L_\nu = 1.2 \times 10^{12} S_\nu D^2$. As a yardstick for comparison with other stars, I note that at X-band the Quiet Sun has an observed $\log L_\nu = 10.7$, the Active Sun of $\log L_\nu = 11.1$, and the biggest solar flares have $\log L_\nu = 12.4$. Using the 20μJy detection threshold at 3.6 cm that the VLA attains in 10 hours of observation on a source, this implies detection horizons (maximum observable distances) of 1.4, 2.3, and 10. pc for stars that have radio luminosities similar to the Quiet Sun, Active Sun, and the largest solar-level flares, respectively.

Given the very nearby detection horizons for stars emitting at levels similar to the non-flaring Sun, it is not surprising that few free-free radio emitters have been detected; in fact, the only 2 cases (among coronal stars) that I know of are Capella (G6 III + F9 III) with $\log L_\nu = 13.7$ at 6 cm (Drake and Linsky 1986) and Procyon (F5 IV-V) with $\log L_\nu = 11.7$ at 3.6 cm (Drake, Simon and Brown 1992). In the case of Capella, the 6-cm flux density is probably predominantly due to coronal emission, since $f_x = 10^{-10}$ erg s^{-1} cm^{-2} and thus the expected optically thin coronal emission (from equation 1) is $50 - 100\%$ of the observed value of 0.20 mJy. In the case of Procyon, the 3.6 cm emission is probably mostly optically thick chromospheric emission: correcting for the expected coronal contribution, we derive a 3.6-cm brightness temperature of 1.7×10^4 K, assuming that the angular diameter of the chromosphere is equal to the photospheric angular diameter.

The vast majority of radio-detected coronal stars appear to be nonthermal, gyrosynchrotron emitters, and detailed models of RS CVn and dMe stars' quiescent radio emission have been constructed based on this mechanism (cf. Morris et al. 1990 and White et al. 1989). The alternative explanation that they are gyroresonance emitters has occasionally been advanced (e.g., Gary and Linsky 1981) but appears to imply implausibly large sizes for the radio emission region. Furthermore, direct VLBI measurements of radio-emitting stars have yielded values and/or lower limits for their brightness temperatures that are much larger in most cases than the coronal temperatures (e.g., 2×10^9 K for the quiescent emission from the dMe star YZ CMi was measured by Benz and Alef 1991). Thermal gyrosynchrotron emission, although presumably present at some level, appears to be unimportant in most of the stars observed to date, with the possible exception of RS CVn stars at their lowest emission levels (Drake, Simon, and Linsky 1989 and 1992). In Table 1 I summarize the properties of the various classes of nonthermal radio emitters that are presently known. In the next section I discuss the common factors that characterize

TABLE I
The Known Classes of Nonthermal Gyrosynchrotron Emitters

Class Name	Spectral Type	Status	Period in days	$\log L_\nu$ erg/s/Hz	$\log L_x$ erg/s	Known Sources
Giant RSCVn	GK III	Binary	5 – 100	15.0 – 18.5	30.0 – 32.0	~ 35
Subgiant RSCVn	GK IV	Binary	1 – 18	14.5 – 17.5	30.0 – 32.0	~ 35
Dwarf RSCVn	FG V	Binary	0.5 – 3	14.5 – 16.3	29.5 – 30.5	~ 20
FK Com	GK III	Single	3 – 80	15.0 – 19.0	29.5 – 32.0	~ 10
AB Dor	GK V	Single	0.5 – 5	14.0 – 16.5	28.5 – 30.5	~ 5
W UMa	FGK V	Binary	0.2 – 1	13.5 – 15.5	29.5 – 30.5	~ 2
BY Dra/Flare	KM V	Both	0.5 – 5	12.5 – 15.0	28.5 – 30.0	~ 50
Pre-ms	B-M	Both	?	15.5 – 18.0	29.5 – 31.5	~ 50
Mag. Bp	Bp HeSi	Single	0.5 – 10	14.5 – 18.0	< 29 – 30.5	~ 25
Algol	GK III,IV	Binary	1 – 20	15.0 – 18.5	29.8 – 31.3	~ 25

and may indeed determine the radio emission of these stars.

5. The Factors that Correlate with Detectable (Nonthermal) Radio Emission

1. X-ray emission. Radio emission is stongly correlated with X-ray emission in all of the classes of strong radio emitters just described. (Since X-ray emission is known to be strongly correlated with TR and chromospheric emission in coronal stars, this implies a similar correlation must exist between these latter emissions and their **nonthermal** radio emission). For RS CVn stars, I find $L_\nu \propto L_x^{1.07\pm0.14}$ (using mean radio luminosity values) or $L_\nu \propto L_x^{1.20\pm0.17}$ (using maximum radio luminosity values), based on 83 systems including 24 with only radio upper limits, implying in either case $\log L_\nu \sim 15.0$ when $\log L_x = 30$. Similarly, Drake and Caillault (1992) find for dMe stars $L_\nu \propto L_x^{0.73\pm0.10}$ (using mean radio luminosity values) or $L_\nu \propto L_x^{0.91\pm0.13}$ (using maximum radio luminosity values), based on 66 stars including 38 with only radio upper limits, implying $\log L_\nu \sim 14.4$ [using mean radio values] and 14.9 [using peak radio values] when $\log L_x = 30$. The scatter of points representing individual stars about the RS CVn and dMe relations is ~ 0.7 dex. Notice that using peak rather than average radio values for dMe stars steepens the slope to a value close to unity and makes the X-ray versus radio relation for dMe stars basically identical to the RS CVn relation. There is much less information for the other classes of active stars, but it is the present author's impression that the radio versus X-ray properties of Naked T Tau, FK Com, and AB Dor stars are broadly consistent with the RS CVn/dMe relations discussed above. Contact binaries appear to be somewhat radio-weak for a given X-ray luminosity compared to the other types of active stars.

However, all coronal X-ray and radio emitters cannot be fit by a universal relation. For example, the free-free radio emitters like the non-flaring Sun, Capella, and Procyon lie $\sim 1.0 - 1.5$ dex below the RS CVn and dMe relations and their extensions to lower X-ray luminosities. Similarly, nearby moderately active giant stars like ι Cap and β Cet with $\log L_x \sim 30$ and active dwarf stars of the Hyades and U Ma cluster ages of $\sim 5 \times 10^8$ years with $\log L_x \sim 29.0 - 29.4$ appear to be radio-deficient compared to the radio versus X-ray relations obeyed by RS CVn and dMe stars.

2. The Presence of intense magnetic fields. Strong radio emission appears to be strongly correlated with the presence of strong magnetic fields, as either directly measured [magnetic Bp stars, some dMe stars], or inferred indirectly through the presence of large starspots. Notice that the presence of a convective envelope does not seem to be a necessary condition (since the He-pec/Si Bp stars and other stars lacking convective envelopes are radio sources: in these hot stars it is the interaction between the radiatively driven wind and the strong magnetic field that generates the nonthermal radio emission).

3. Rapid rotation. Rapid rotation appears to be a **necessary** but **not sufficient** condition, at least for the classic coronal stars (i.e., those having convective envelopes). All of these types of strong nonthermal radio emitters are more rapidly rotating than stars of similar spectral types that are not detected radio sources. This confirms the widely held belief that activity in such stars results from a convective-rotational dynamo mechanism. However, rapidly F main-sequence stars and F and early G post-Hertzsprung Gap giant stars are not strong radio sources, and hence rapid rotation *per se* is not a sufficient property: perhaps the dynamo mechanism is inhibited or altered in stars with thin convective envelopes. For the early-type nonthermal emitters such as the magnetic Bp stars where the magnetic fields are presumably primordial and not dynamo-related, there is no evidence for any correlation of radio emission with rotation (in fact, magnetic Bp and Ap stars rotate much more slowly on the average than similar non-magnetic stars).

4. Youth. It is difficult in many cases to separate the dependence on age from that of the rotation rate. Surveys of pre-main sequence associations have detected nonthermal radio (as well as X-ray emission) from types of stars such as late B stars which do not appear to have any counterparts among the field stars that have similar properties. In such cases, youth *per se* may be the important factor (e.g., perhaps in allowing the formation of a transient superficial convective zone). However, as a general rule (e.g., comparing young, single dMe stars with old, close binary dMe stars) it appears that the level of radio emission is determined by the (rapid) rotation rate and not by the mechanism (youth versus tidal synchronism) that is the origin of the rapid rotation.

5. Binarity. Again, the effect of a binary environment on stellar radio emission appears to be merely that of spinning one or both components up to a rapid rotation rate. The early emphasis on binarity as essential for strong radio emission in coronal-type stars might have been somewhat misleading; however, the final resolution of this problem will require careful comparisons of stars of near-similar properties that are single and in binary systems, and it is not possible at this time to rule out subtle differences related to binarity.

6. Flaring. A common feature of all of the active star types in the previous section is the property of flaring at optical, UV, radio, and/or X-ray ranges. Flaring presumably provides the reservoir of power-law electrons needed for the gyrosynchrotron emission process. The apparent existence of long-lasting (sometimes called quiescent or slowly varying) nonthermal radio emission in most if not all of the classes of active stars implies that there must be essentially continuous, low-level acceleration and injection of high-energy particles going on in such cases. Whether this process is actually low-amplitude flaring (sometimes called microflaring) is still controversial.

6. Potentially Fruitful Directions for Future Research

Now that nonthermal radio emission has been established as an important general property of the most active stars, I hope that any future discussion of stellar coronae will contain a much greater proportion of papers devoted to this area of coronal research. The importance of such radio emission is that it can ultimately be used to obtain information on the strengths and geometries of coronal magnetic fields that is not obtainable by any other technique (and has been done successfully for the Sun, e.g., Nitta et al. 1991). To adequately constrain the coronal models, however, it is important that simultaneous radio observations (at as wide range of wavelengths from the mm to the meter range as is possible) and X-ray observations (again of wide enough energy coverage and/or high enough spectral resolution) be obtained of the brightest (radio and X-ray) emitters.

It is also still painfully apparent that we do not know very much about the radio emission from coronal stars with average and below-average activity levels. Deep radio observations of the closest examples of 'normal stars' may yield surprises: examples of this phenomenon include the failure to detect 'quiescent' emission in a long ATCA exposure of the nearest dMe star Proxima Cen and its implications on the radio emission mechanism operating in such stars (Lim et al. 1992), and the detection (in 1 observation out of a total of 5) of a possible radio flare from the F star Procyon whose quiescent radio emission has been previously mentioned (Drake, Simon and Brown 1992). It is also surprising that, with the exception of the active binaries V471 Tau and HD 27130, no Hyades cluster or moving group stars have been detected as radio sources: their youth and more rapid rotation relative to solar-age stars would suggest that their radio emission should be enhanced. Does nonthermal radio activity 'turn off' abruptly when a star reaches some critical rotation rate and/or age, or does it decline gently? To answer these questions, deeper radio observations of Hyades or UMa stars are clearly required.

Finally, it is interesting to note that models of the radio emission from some of the fairly disparate classes of coronal stars seem to be converging, and that the concept of a unified magnetospheric model appears to be developing. In most of the classes of active stars which are believed to be nonthermal gyrosynchrotron sources, model calculations predict that the emission must be generated at a distance of say, a few to 10 stellar radii, at which location the magnetic geometry may well be predominantly determined by their global dipole fields and not by the more complex field structures that characterize the surface fields of solar-type stars (and which decay faster with radius than the global dipole). Also remember that the present evidence is that binarity is only an incidental property of many of these radio-strong stars, and not a necessary property. Thus, Morris et al. (1990) have developed a magnetosperic model for RS CVn binary stars, Linsky et al. (1992) one for the magnetic Bp stars, and André et al. (1992) one for pre-main sequence stars such as the naked T Tau stars and the radio-detected B and A stars like ρ Oph S1. However, White et al. (1989) have argued against magnetospheric models for dMe stars' radio emission, and prefer a model in which it comes from several small active regions. More observational data (fluxes, polarizations, photospheric magnetic field measurements) are clearly needed to test the validity of the magnetospheric model for the various classes of coronal stars, as well as more detailed quantitative modeling of both the radio and the X-ray properties of these stars.

Acknowledgements

I thank Jeffrey Linsky, Stephen White, and David Florkowski for their comments on a preliminary version of this review.

References

André et al.: 1988, *ApJ* **335**, 940
André et al.: 1992, preprint.
Arnaud, K.A., Zheleznyakov, V.V. and Trimble, V.L.: 1992, *PASP* **104**, 239
Bastian, T.S.: 1990, *Solar Phys.* **130**, 265
Benz, A.O. and Alef W.: 1991, *A&A* **252**, L19
Bookbinder, J.A.: 1988, in O. Havnes et al., ed(s)., *Activity in Cool Star Envelopes*, Kluwer: Dordrecht, 257
Drake, S.A. and Caillault, J.-P.: 1992, in preparation.
Drake, S.A. and Linsky, J.L.: 1986, *AJ* **91**, 602
Drake, S.A., Simon, T. and Brown A.: 1992, *ApJ*, in press.
Drake, S.A., Simon, T. and Linsky, J.L.: 1989, *ApJS* **71**, 905
Drake, S.A., Simon, T. and Linsky, J.L.: 1992, *ApJS* **82**, 311
Dulk, G.A.: 1985, *ARAA* **23**, 169
Gary, D.E. and Linsky, J.L.: 1981, *ApJ* **250**, 284
Hjellming, R.M.: 1988, in G.L. Verschuur and K.I. Kellerman, ed(s)., *Galactic and Extragalactic Radio Astronomy*, Springer-Verlag: Berlin, 381
Hjellming, R.M. and Gibson, D.M.: 1985, *Radio Stars*, Reidel: Dordrecht
Kuijpers, J.: 1989, *Solar Phys.* **121**, 163
Lang, K.R.: 1990, in L.V. Mirzoyan et al., ed(s)., *Flare Stars in Star Clusters, Associations and the Solar Vicinity*, Kluwer: Dordrecht, 125
Lestrade, J.-F. et al.: 1990, *AJ* **99**, 1663
Lestrade, J.-F. et al.: 1992, *A&A* **259**, 112
Lim, J. et al.: 1992, preprint.
Linsky, J.L., Drake, S.A. and Bastian, T.S.: 1992, *ApJ* **393**, 341
Morris, D.H., Mutel, R.L. and Su B.: 1990, *ApJ* **362**, 299
Nitta, N. et al.: 1991, *ApJ* **374**, 374
Phillips, R.B., Lonsdale, C.J. and Feigelson, E.D.: 1991, *ApJ* **382**, 261
Stewart, R.T.: 1992, preprint.
Usov, V.V. and Melrose, D.B.: 1992, *ApJ* **395**, 575
Wendker, H.J.: 1978, *Abh. Hamburger Sternw.* **10**, 1
Wendker, H.J.: 1987, *A&AS* **69**, 87
White, S.M.., Kundu, M.R. and Jackson, P.D.: 1989, *A&A* **225**, 112
Zoonematkermani, S. et al.: 1990, *ApJS* **74**, 181

MASS LOSS FROM COOL DWARFS: LIMITS ON DETECTABILITY

J.G. DOYLE
Armagh Observatory
and
D.J. MULLAN
Bartol Research Institute

Abstract. Recent spectroscopic evidence seem to suggest that certain cool dwarfs may have stellar winds with $\dot{M}$ values several orders of magnitude larger than the solar rate. A key test of the wind spectrum would be provided if the stars could be detected at $\lambda \approx 1$ mm, as an optically thick wind predicts a power-law at these frequencies.

Key words: millimeter emission – mass loss – optically thick wind

1. Introduction

The presence of hot coronal plasma favors the creation of a wind. The temperature of the plasma is a key parameter in determining the properties of the outflow. In fact, for a star where thermal pressure gradients are the driving mechanism, the mass loss rate in the wind is exponentially sensitive to coronal temperature (Parker 1963). Also, discrete magnetic events (CME's) may contribute significantly to the coronal dynamics of active dwarfs (Mullan et al. 1989a). In any case, the coronal evidence alone suggests that M dwarfs may be losing mass more efficiently than the Sun. The first report of direct detection of mass loss from a cool dwarf referred to the K2V star in the detached binary V471 Tauri (Mullan et al. 1989a).

2. Optically Thick Wind

The surest evidence of a wind is provided if the flux density S_ν is found to increase as a power law of frequency, ν^α, where α is positive and has a value of about 2/3. Such a spectrum indicates free-free emission from an optically thick wind (Wright & Barlow 1975). Assuming that the wind is optically thick, when a star at distance d (pc) emits a wind with mean molecular weight μ and rms ionic charge Z, and the wind is isothermal, spherically symmetric, flowing out steadily at constant velocity v_∞ ($km\ sec^{-1}$), the flux (mJy) detected at Earth at frequency ν Hz is

$$S_\nu = 2.32 \times 10^{10}(\dot{M}Z)^{4/3}(\gamma g_\nu \nu)^{2/3}(v_\infty \mu)^{-4/3}d^{-2} \tag{1}$$

Here, $\dot{M}$ is the mass loss rate ($M_\odot\ yr^{-1}$), γ is the number of electrons per ion, and g_ν is the free-free Gaunt factor. Values of μ, Z, and γ have been discussed for a sample of stars by Leitherer & Robert (1991): these stars were hot and some were Wolf-Rayet stars, where the composition departs significantly from cosmic abundances. In cool dwarfs, we have little evidence to guide us in the choice of these parameters. For simplicity, we assume $\mu = Z = \gamma = 1$. With wind temperatures of 10^{4-5} K (Mullan et al. 1989a), g_ν at a wavelength of 1 mm ranges from 2.81 to 4.72: we

J.F. Linsky and S. Serio (eds.), Physics of Solar and Stellar Coronae, 401–403.

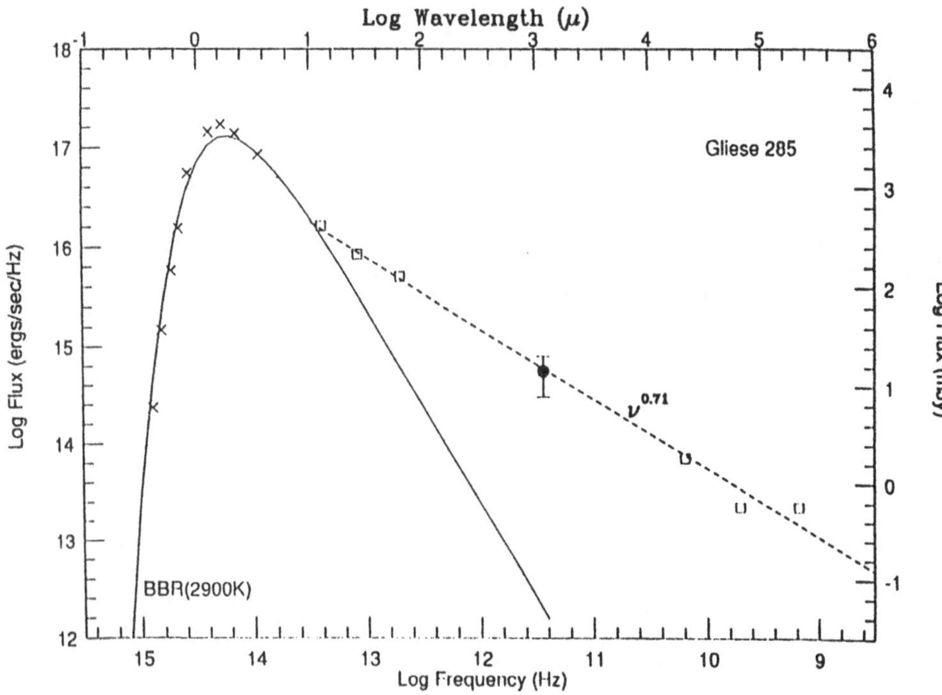

Fig. 1. Observational data for Gliese 285 (=YZ CMi)

adopt the value $g_\nu = 4$ here. The wind speed in V471 Tauri is observed to be 200-500 $km\ sec^{-1}$ (Mullan et al. 1989a), i.e. comparable to the solar wind, hence for present purposes we set $v_\infty = 300$. These values lead to the following prediction of the flux at a wavelength of 1 mm:

$$S_\nu(1mm) = 61(\dot{M}_{-10})^{4/3}/(d^2)\ mJy \tag{2}$$

where $\dot{M}_{-10}$ is the mass loss rate in units of $10^{-10}\ M_\odot\ yr^{-1}$.

3. Millimeter fluxes for M dwarfs

Using the JCMT, we observed for Gl 285 (a dM2.5e star) at the 2.2σ level a mean flux of 13 mJy at $\lambda = 1.1$ mm. If the only information available to us were our JCMT data, then we would probably not be in a position to draw any significant conclusions (as we are pushing the current bolometer detector on the JCMT to its limit.). However, when we place our JCMT data in the context of data obtained over a broader sample of wavelengths, the JCMT data become more valuable. In Fig. 1 we collate flux measurements for YZ CMi over a range of wavelengths. Optical and near infra-red data obtained from ground-based measurements (denoted by crosses) refer to UBVRIJHKL channels, IRAS data at 12, 25 & 60 μm are plotted as open squares and VLA data at $\lambda = 2$, 6 & 20 cm are shown by open squares. Finally, our measurement at 1.1 mm is shown by a filled circle. A distance of 6 pc has been used in constructing Fig. 1. Superimposed on the plot is black-body radiation with $T = 2900$ K. This photospheric emission fits the data shortward of a few microns quite well.

To characterize these long-wavelength excesses, we have used the six data points shown by open squares (3 IRAS plus 3 VLA) and fitted them by least squares. (Note that we do *not* include the JCMT point in the least squares fitting.) The result is the dashed line: this is a power law with index +0.71. We find that our JCMT data point falls very close to this line. If it were not for the JCMT point, it might be argued that our least squares fit is simply an artifact of attempting to connect two widely separated clusters of points, one at $\lambda = 10$-60μm, the other at $\lambda = 2$-20 cm. If the wind interpretation is correct, we can use Eq. (1) above to convert our JCMT flux (13 mJy) to a mass loss rate: we find $\dot{M} \approx 5 \times 10^{-10}$ $M_\odot$ yr^{-1}. The possibility that M dwarfs are losing mass efficiently is of interest in the context of coronal heating physics in stars other than the Sun. There is also a non-stellar reason for interest in this possibility: if even a few percent of all M dwarfs have mass loss rates as large as the above estimates, the mass balance of the interstellar medium will be dominated by M dwarf winds. More observations of M dwarfs at millimeter and sub-millimeter wavelengths are required in order to test whether or not these stars are really losing mass at high rates.

Acknowledgements

The JCMT is operated by the Royal Observatory Edinburgh on behalf of the Science and Engineering Research Council of the UK, the Netherlands Organisation for Scientific Research and the National Research Council of Canada. We also acknowledge the support provided in terms of both software and hardware by the STARLINK Project which is funded by the UK SERC. Armagh Observatory is granted-aided by the Dept. of Education for N. Ireland. The work of D.J.M. has been supported in part by NASA Grant NAGW-2456.

References

Leitherer, C. & Robert, C. 1991, ApJ 377,629
Mullan, D.J., Sion, E.M., Bruhweiler, F.C., Carpenter, K.G. 1989a, ApJ L 339,L33
Mullan, D. J., Stencel, R. E., & Backman, D. E. 1989b, ApJ 343,400
Parker, E. N., 1963, *Interplanetary Dynamical Processes*, J. Wiley, New York, pp. 74, 255
Wright, A.E. &Barlow, M.J. 1975, MNRAS, 170,41

RADIO AND X-RAY LUMINOSITY OF RS CVn BINARY SYSTEMS

F. CHIUDERI DRAGO and E. FRANCIOSINI
Department of Astronomy and Space Science, University of Florence, Italy

Abstract. Two different interpretations of the quiescent radio emission of UX Ari are considered and compared with available observations. It is assumed that the radiation mechanism is synchrotron and that the emitting electrons are distributed according to:

a) a Maxwellian distribution at the same temperature deduced from X-ray observations (suggested by the good correlation found between X-ray and radio luminosity);

b) a power-law distribution (suggested by the observed flatness of the radio spectrum).

The time evolution of a population of electrons following an initial power-law distribution is also studied. It is shown that the spectrum of the emitted radiation evolves, in a time scale of a few days, from the typical flare spectrum to the flat one observed in quiescent periods.

Key words: Stars–Binary stars–Radio emission of

1. The thermal synchrotron emission

Considering a large sample of RS CVn binary systems observed by the Very Large Array at $\lambda = 6$ cm, Drake et al. (1989) found an interesting correlation between the radio luminosity, L_6, and the X-ray luminosity, L_x, which was explained assuming that the same population of thermal electrons is responsible for both emissions. Although attractive, this model fails to reproduce the radio spectrum of the quiescent component (QC) of the few sources for which it has been measured (Pallavicini et al. 1985; Mutel et al. 1987; Massi and Chiuderi Drago 1992). The observed spectral index is in fact $\alpha \simeq -0.2$, whereas that for the thermal gyrosynchrotron emission in a constant magnetic field is $\alpha = -8$ (Dulk 1985).

In this paper we have computed the radio spectrum of UX Ari in a variable magnetic field, using the same parameters assumed by Drake et al. (1989), namely $T = 5 \times 10^7$ K, $N_e = 2 \times 10^8$ cm^{-3}. The source dimension along the line of sight has been assumed $L = 3 \times 10^{12}$ cm.

The magnetic field has been assumed to vary, in the plane perpendicular to the line of sight, according to $B(r) \propto 1/r^n$, with n=1, 2, 3 and $B(\phi)$ has been adjusted to satisfy the divergence-free condition. The maximum value of B at the solar surface has been assumed to be $B_{max} = 1000$ G (Giampapa et al. 1983; Gondoin et al. 1985). The magnetic lines of force have then been shifted along the line of sight according to the arcade approximation described by Klein and Chiuderi Drago (1987). The results of this calculation are shown in Fig. 1a): we see that a negative spectral index $\alpha \simeq -0.15$, in good agreement with the observations, is obtained for $n = 1$. The observed flux values can be easily reproduced by assuming a slight increase of T and/or B_{max}. Therefore the thermal interpretation can be in principle maintained, provided that the magnetic field decreases as r^{-1}. In spite of its ability to reproduce the observed flux spectrum, the thermal interpretation of the QC still strongly disagrees with the brightness temperature measurements inferred from VLBI observations by Mutel et al. (1985) that indicate values larger by at least one order of magnitude. Therefore, although the possibility of a thermal

J.F. Linsky and S. Serio (eds.), Physics of Solar and Stellar Coronae, 405–408.

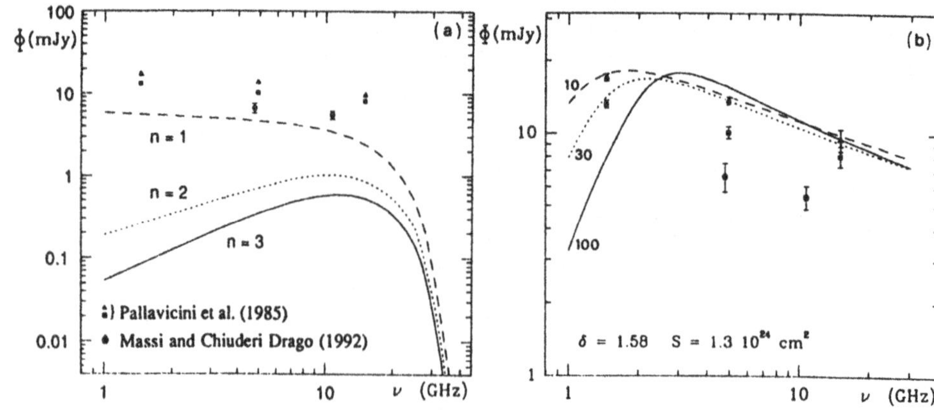

Fig. 1. Comparison of the observed radio fluxes with the computed spectrum using a thermal (a) and a non-thermal population of electrons (b)

origin of the QC cannot be ruled out, we will consider a different interpretation in the next sections.

2. Non-thermal synchrotron emission

The flaring radio emission observed in RS CVn systems has been interpreted by many authors as due to synchrotron emission by mildly relativistic electrons following a power-law energy distribution with an exponent $\delta = 2 - 3$.

In this section we will assume that the QC is also due to an electron population similar to that responsible for the flaring emission, which is assumed to be isotropic and uniform within the source.

The calculations show, however, that it is impossible to fit the QC radio spectrum using the same dipole field configuration used by Klein and Chiuderi Drago (1987) to fit the flare spectrum. A good agreement with the observations can be instead found assuming a uniform magnetic field within the source.

The results, obtained using magnetic field values of 10, 30 and 100 G, are shown in Fig. 1b): the exponent δ of the electron population has been derived from the empirical relations between α and δ given by Dulk (1985), the surface S from VLBI observations and the column density $(N_{\circ}L)$ has been adjusted, for each value of the magnetic field, in order to reproduce the observed spectrum with the largest flux.

Fig. 1b) shows that the spectral index decreases with decreasing B and that a perfect agreement is found for $B = 10$ G and a column density $N_{\circ}L = 7 \times 10^{14}\,\mathrm{cm}^{-2}$ (about three orders of magnitude lower than for the flare). Assuming a lower value of $N_{\circ}L$ we find that both the flux density and the peak frequency decrease, making it possible to reproduce also the other two observed spectra.

It appears, therefore, that a population of electrons following a power-law distribution is able to reproduce both the flaring and the quiescent component spectrum, although with different values of δ and of the column density N_oL. Also the magnetic field configuration derived in the two cases is very different. These disagreements however disappear in the framework of an evolving model.

An alternative interpretation of the L_6/L_x correlation, in the case that the population of emitting electrons is not the thermal X-ray emitting plasma can still be obtained, assuming with Golub et al. (1980) that the heating of a coronal loop is supplied by the release of magnetic energy accumulated by random motions with velocity v_ϕ of the lines of force at the photospheric level. Assuming moreover that all this energy is radiated away in the form of X-rays, which is the case at a temperature $T \gtrsim 2 \times 10^6$ K, we have:

$$L_x \propto B^2 \frac{dv_\phi}{dz} V ,$$

where V is the loop's volume.

On the other hand, the radio luminosity L_6 of an optically thin loop of volume V when the radiating electrons are distributed according to a power-law with exponent δ depends on the magnetic field according to (Dulk 1985):

$$L_6 \propto B^{-0.22+0.9\delta} V .$$

If we assume $dv_\phi/dz \propto v_\phi$, we obtain:

$$L_x \propto L_6^{0.45\delta-0.11} V^{1.11-0.45\delta}$$

since ~ 90 % of the considered objects belong to almost the same spectral type and, therefore, according to Belvedere et al. (1981), they have about the same velocity v_ϕ.

We see that L_x depends very weakly on the emitting volume for reasonable values of δ, thus introducing a modest scatter in the L_6/L_x correlation. For $\delta = 3$, and neglecting the volume correction, we find $L_x \propto L_6^{1.24}$, not far from $L_x \propto L_6^{1.41}$ found by Drake et al. (1989).

3. Time evolution of the spectrum

The time evolution of a population of electrons injected by the source with a power law distribution at a certain time $t_o = 0$, which is identified with the beginning of the flare, is studied, assuming that all electrons are trapped in the source, and that the energy losses are due to collisions and radiation. The computed spectra at different times after the injection are shown in Fig. 2a) and 2b) for an original power law distribution with $\delta = 2$.

An interesting scenario appears when one considers the time evolution of the brightness of the loop at a fixed frequency (not shown here for lack of space). It appears that, in the first hours after the injection, the dominant emission comes from regions where $B \gtrsim 30 - 50$ G, with dimension $R \lesssim 2R_*$, while the contribution

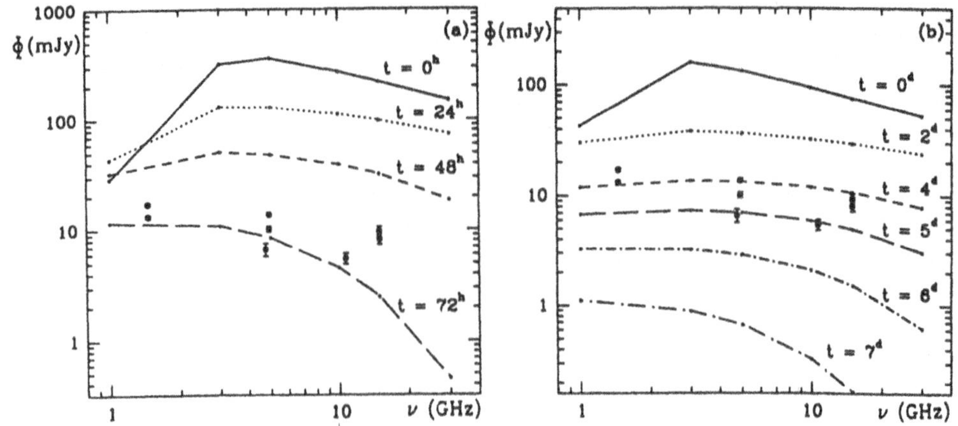

Fig. 2. Time evolution of the computed radio spectrum using an initial power-law distribution of electrons: the density of the thermal plasma is $N_e = 2 \times 10^8$ cm^{-3} and the magnetic field is 10 G (a) and 5 G (b).

of the regions where $B \lesssim 10$ G ($R \simeq 5\ R_*$) is about one order of magnitude lower. After one or two days the former disappears and only the low level emission remains.

In the light of these results, we may now give the following quantitative interpretation of the flaring and QC emission in UX Ari. At the beginning of a flare a large number of electrons are accelerated in a magnetic loop, with a power-law energy distribution and the flaring emission is observed. After a few hours the total number density of electrons strongly decreases and the QC is observed, which may present the "core-halo" structure observed by Mutel et al. (1985) in the first few days after the flare, or, later on, only the "halo" emission observed by Massi et al. (1988).

References

Belvedere, G., Chiuderi, C., Paternò, L.: 1981, *A&A* **96**, 369
Drake, S.A., Simon, T., Linsky, J.L.: 1989, *ApJS* **71**, 905
Dulk, G.A.: 1985, *ARA&A* **23**, 169
Giampapa, M.S., Golub, L., Worden, S.P.: 1983, *ApJ* **268**, L121
Golub, L., Maxson, C., Rosner, R., Serio, S., Vaiana, G.S.: 1980, *ApJ* **238**, 343
Gondoin, P., Giampapa, M.S., Bookbinder, J.A.: 1985, *ApJ* **297**, 710
Klein, K.-L., Chiuderi Drago, F.: 1987, *A&A* **175**, 179
Massi, M., Chiuderi Drago, F.: 1992, *A&A* **253**, 403
Massi, M., Felli, M., Pallavicini, R., Tofani, G., Palagi, F., Catarzi, M.: 1988, *A&A* **197**, 200
Mutel, R.L., Lestrade, J.-F., Preston, R.A., Phillips, R.B.: 1985, *ApJ* **289**, 262
Mutel, R.L., Morris, D.H., Doiron, D.J., Lestrade, J.-F.: 1987, *AJ* **93**, 1220
Pallavicini, R., Willson, R.F., Lang, K.R.: 1985, *A&A* **149**, 95

A REMARKABLE CIRCULARLY POLARIZED RADIO FLARE ON AR LAC

R.L. MUTEL
Dept. of Physics and Astronomy, University of Iowa, Iowa City IA 52242
J. E. NEFF
Dept. of Astronomy, Pennsylvania State University, University Park PA 16802
J. A. BOOKBINDER
Center for Astrophysics, Cambridge MA 02138
and
I. PAGANO
Institute of Astronomy, Catania University, 95125 Catania, Italy

Abstract. An intense radio flare was detected from the close binary system AR Lac during a 4 day multi-wavelength observing campaign in 1991 December. The flare lasted more than 6 hours and was preceded by a strong CIV flare one day earlier. The peak circular polarization was 70%, 38%, and 39% RCP at 1.4, 4.9, and 8.4 GHz respectively, with $\sim$ 15% LCP at 15 and 22 GHz. The high degree of circular polarization over such a large time scale and frequency range is highly unusual compared with previously observed radio flares from RS CVn binaries. Given these unusual characteristics, it is difficult to interpret the radiation mechanism either as a result of gyrosynchrotron emission or a coherent process such as an electron cyclotron maser.

Key words: Radio emission – Binary Stars

1. Introduction

The RS CVn system AR Lac consists of a GIV+KIV eclipsing binary in a 1.98 day orbital period. In several previous radio observations (Drake *et al.* 1990 and references) the flux density at centimeter wavelengths has varied from 2-16 mJy, corresponding to a mean radio luminosity at 50 pc of 2×10^{16} ergs s^{-1} Hz^{-1}. Doiron and Mutel (1984) observed the radio emission during a non-flare state (5-15 mJy) using the VLA during two secondary eclipses and found no reduction in the radio flux. They inferred a lower limit to the size of the emitting region of 5×10^{11} cm ($\sim$ 5 primary radii). In contrast, Kürster *et al.* (1992) report X-ray primary eclipses in ROSAT light curves, indicating that the X-ray emission arises entirely from the G star and has a spatial extent much less than the quiescent radio corona. Neff *et al.* (1989) used IUE observations to show that the chromospheric MgII emission has a complex spatial distribution with three emission centers on the G star.

In an effort to understand the physical connection between activity indicators at various levels of the stellar atmosphere, a simultaneous multi-wavelength campaign to observe AR Lac was undertaken over 4 days in late December 1991. The observations covered a wide range of wavelengths including radio (VLA), optical (HST, McMath, Penn State, OHP), ultraviolet (IUE), and X-ray (ROSAT). In this paper, we report preliminary VLA and IUE results involving a remarkable flare whose properties are not easily understood using emission mechanisms previously applied to radio emission from close binaries. A full description of the multi-wavelength results will appear later.

J.F. Linsky and S. Serio (eds.), Physics of Solar and Stellar Coronae, 409–412.

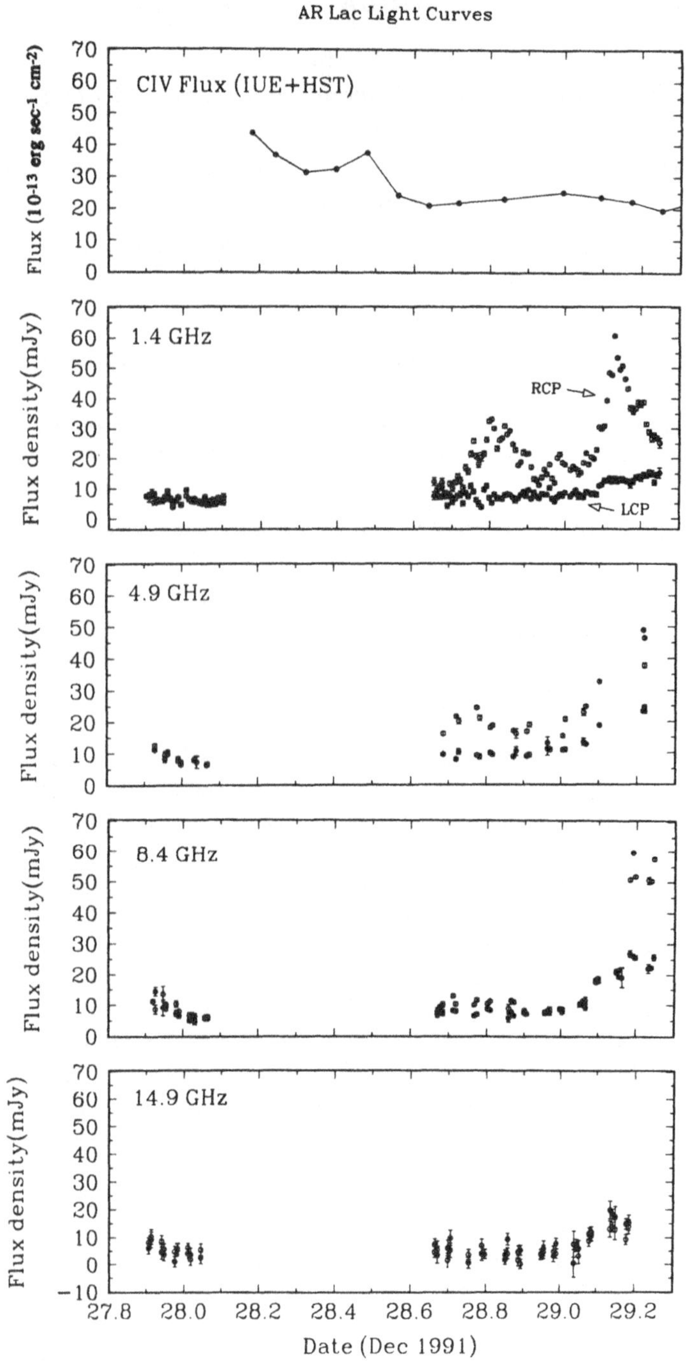

Fig. 1. Light curves for AR Lac during 27.8-29.2 December 1991 (orbital phase 0.10-0.81). The top panel is CIV flux density measured with IUE, while the lower four panels are radio fluxes measured at the VLA at 1.4, 4.9, 8.4, and 14.9 GHz. Open squares and filled circles represent right and left circular polarization respectively. The error bars represent $\pm 1\sigma$ uncertainties.

2. Observations

2.1. IUE Observations

The IUE satellite observed both CIV and MgII chromospheric emission lines from the AR Lac system for a total of 104 hours (2.17 orbits) during the period 28 Dec 1991- 01 Jan 1992. The CIV flux for December 27.9-29.2 is shown in the top panel of figure 1. There was a large flare beginning sometime before the first observing period, with a decay lasting until day $\sim$28.6 Dec. The flux density after the flare decay remained at a quiescent level near 20 ergs cm^{-2} s^{-1} until the end of the observing period. The MkII data also suggest a flare, probably on the G star.

2.2. VLA Observations

The VLA was divided into two subarrays, with the first subarray (13 antennas) observing at 1.4 GHz, and the second (14 antennas) sequenced among the 4.9, 8.4, 14.9, and 22.4 GHz bands with a $\sim$25 min cycling time. Both right and left circular polarization were recorded and each 25 MHz IF channel was observing in spectral line mode with eight channels per IF. We have not yet completely analyzed the spectral line data and will only report on the broad-band observations here.

The radio light curves for the first two days are shown in the lower four panels of figure 1. The large flare evident on 28 December lasted for more than 6 hours and had a spectral index α ($S \propto \nu^{\alpha}$) close to zero between 1.4 and 8.4 GHz, decreasing to α $\sim$-0.2 above 8.4 GHz. The flare was highly right-circularly polarized (up to 70% at 1.4 GHz) over the entire frequency range from 1.4 to 8.4 GHz, with a modest reversal to LCP at 14.9 GHz (the 22.4 GHz data also show a reversal). During the next two days, the emission returned to the level first observed (27.9-28.1 December), except for a smaller polarized flare on December 29.8 lasting $\sim$ 3 hr.

3. Emission Mechanisms

The flare characteristics described above are highly unusual for radio flares on RS CVn systems. More commonly the spectral index increases to α $\sim$+1 at least between 1.4 and 5 GHz (e.g. Mutel *et al.* 1987) and the circular polarization approaches zero. This is consistent with a gyrosynchrotron process in which the fractional circular polarization $\pi_c \sim 1/\gamma$, where γ is the Lorentz factor of the emitting electrons. As a flare develops, the average electron energy increases, leading to an increase in the peak frequency of emission and a decreased π_c. Detailed calculations based on gyrosynchrotron radiation in inhomogeneous plasma confirm this model and have agreed very well with previous observations of stellar flares (e.g. Klein and Chiuderi-Drago 1987; Morris, Mutel, and Su 1990).

A high degree of circular polarization over a large bandwidth is possible for a gyrosynchrotron process with a power-law electron energy distribution if the plasma is optically thin and at a small viewing angle to the magnetic field (Dulk and Marsh 1982). However, the polarization reversal at 15 and 22 GHz is *not* expected for an optically thin plasma; rather it suggests that the emission below 10 GHz is optically thick and that helicity reversal occurs at $\tau = 1$ between 8 and 15 GHz.

The extremely high circular polarization suggests a coherent emission mechanism such as an electron-cyclotron maser (e.g. Winglee and Dulk 1986; Dulk and Winglee 1987) or plasma radiation (Dulk 1985). Coherent radiation has been detected in similar systems: short timescale polarized centimeter bursts reported from dMe stars (Bastian and Bookbinder 1987; Güdel *et al.* 1989) probably result from a coherent process. Cyclotron maser or plasma emission should dominate if the ratio of electron plasma frequency to gyrofrequency ω_p/Ω_B is either $\gtrsim 3$ or $\lesssim 3$ respectively.

For a cyclotron maser, the observed radiation would be at the second to fourth harmonic of the gyrofrequency, implying a magnetic field strength of at least 800 Gauss for the 8.4 GHz emission. This is similar to surface magnetic fields on RS CVn systems as measured by e.g. Doppler imaging of the photosphere. However, it seems unlikely that the radio emission arises near the photosphere for two reasons. First, the plasma density is likely to be too high in the lower atmosphere to allow propagation at 1.4 GHz ($n_e < 2.4 \times 10^{10}$ cm^{-3}). Second, the CIV flare preceded the radio flare by $\sim$ 1 day, implying either that the flares were unrelated or that the flare propagated out to the outer corona, perhaps at the Alfven speed.

The very large bandwidth ($\Delta\nu/\bar{\nu} > 3$) of the polarized flare is also inconsistent with cyclotron maser emission, at least in a homogeneous medium (Bastian *et al.* 1990). It may be that a highly stratified inhomogeneous model could produce both the high polarization and large bandwidth, but no detailed models are yet available which demonstrate this.

References

Bastian, T. and Bookbinder, J.: 1987, *Nature* **326**, 6114

Bastian, T., Bookbinder, J., Dulk, G., and Davis, M.: 1990, *ApJ* **353**, 265

Doiron, D.J. and Mutel, R.L.: 1984, *AJ* **89**, 430

Drake, S., Simon, T., and Linsky, J.: 1989, *ApJS* **71**, 905

Dulk, G.A.: 1985, *ARAA* **23**, 169

Dulk, G.A. and Marsh, K.A.: 1982, *ApJ* **259**, 350

Dulk, G.A. and Winglee, R.M.: 1987, *Solar Physics* **113**, 187

Güdel, M. *et al.* : 1989, *A&A* **220**, 5

Klein, K.-L. and Chiudero-Drago, F.: 1987, *A&A* **175**, 179

Kürster, M., Schmitt, J.H.M.M., and Fleming, T.: 1992, in M. Giampapa and J. Bookbinder, ed(s)., *Cool Stars, Stellar Systems, and the Sun*, Astron. Soc. Pacific: San Francisco, 109

Mutel, R.L., Morris, D., Doiron, D., and Lestrade, J-F.: 1987, *AJ* **93**, 1200

Morris, D., Mutel, R., and Su. B.: 1990, *ApJ* **362**, 299

Neff, J.E., Walter, F., Rodonò, M., and Linsky, J.L.: 1989, *A&A* **215**, 79

Winglee, R.M. and Dulk, G.A.: 1986, *ApJ* **307**, 808

SINGLE-DISH MONITORING OF RS CVn BINARY SYSTEMS

C. TRIGILIO and G. UMANA
Istituto di Radioastronomia C.N.R. Bologna
VLBI Station, P.O. Box 169 Noto, Italy
S. CATALANO
Osservatorio Astrofisico Catania
Viale A. Doria 6, 95125 Catania, Italy
and
M. RODONÒ and A. FRASCA
Istituto di Astronomia Università di Catania
Viale A. Doria 6, 95125 Catania, Italy

Abstract. Radio monitoring of RS CVn systems is being carried out with the Noto VLBI antenna. Preliminary results on flare activity on HR 1099 are here presented. During one of these radio bursts we made VLBI observations which provide information about the source angular dimension and brightness temperature.

Key words: Radio continuum: stars – Stars: coronae – Stars: activity – Stars: flare

1. Introduction

RS CVn systems are stellar radio sources with 6 cm radio luminosities in the range between 10^{15} and 10^{18} erg s^{-1} (Morris and Mutel 1988). The radio flux is highly variable and it usually shows two different regimes: active periods characterized by a continuous strong flaring which can last for several days (i.e. Feldman et al. 1978), and quiescent periods. The radio flux can vary from a few mJy, measured during quiescent periods, up to almost 1 Jy, reached during very active phases. The non-thermal emission is generally taken to arise from the interaction between the magnetic field of one or both component stars and mildly relativistic electrons (gyrosynchrotron emission) (Mutel et al. 1987). However it is still not clear whether the most energetic activity events are localized in very compact regions of the corona of a star or whether the entire binary system is involved.

Since observations of flare time scales and VLBI measurements are extremely important to determine the characteristic dimensions of the radio source, we started in June 1990 single-dish monitoring of radio emission from active binary systems using the 32 m radiotelescope at the Noto VLBI Station of the Istituto di Radioastronomia of C.N.R..

2. The RS CVn program

The aims of systematic **total power** observations of RS CVn type systems are:
(1) to study the frequency of flares and the possible correlations of the radio flares occurrence with activity signatures in other wavelengths, in particular with Hα; and
(2) to catch the onset of a flaring activity and thus to organize VLBI observations to

J.F. Linsky and S. Serio (eds.), Physics of Solar and Stellar Coronae, 413–416.

measure the size of the radio source using the Noto-Medicina baseline and, possibly, additional stations of the European VLBI network.

Here we report on the preliminary results of the first two years of monitoring. In Table I, the list of the systems monitored for flares at Noto Station is given. Three systems have actually been detected: Algol, HR 5110, and HR 1099.

3. Observations

Single dish observations were obtained during the periods June–July 1990 and June–July 1991, with the on-off technique, using a cooled 5 GHz receiver (T_{sys} = 30 K, at the zenith) mounted at the secondary focus of the Noto VLBI 32 m radio telescope. We used a 4 MHz bandwidth and the basic observations consisted of 20 seconds on-source integration time preceded and followed by 20 seconds off-source intergration, with an offset of 18 arcmin. The basic cycle repeated several times permits us to reach a flux density limit of about 30 mJy in a total integration time of about one hour.

The temperature scale is fixed using a noise calibration source (T_{cal} = 17 K) every five on-source acquisitions. Pointing corrections and gain curves are checked by observing strong point sources, and the flux scale is determinated relative to standard calibrators; the source 3C123 (16.20 Jy) was adopted for the measurements of HR 1099 here reported.

TABLE I

Source	Name	Sp. type	S_6 (mJy)
0304+407	Algol	B8V/K2IV	20.
0323+285	UX Ari	G5V/K0IV	10.1–195.
0334+004	HR 1099	G5IV/K1IV	5.0–21.9
	EI Eri	G5IV	4.3
0734+310	YY Gem	dM1e/dM1e	0.5–2.
1308+362	RS CVn	F5IV/K0IV	2.96–7.0
	FK Com	G8III	2.1–6.1
1332+373	HR 5110	F2IV/K1IV	5.2–152.
1612+339	σ CrB	G0V/G0V	8.4–19.6
2159+436	RT Lac	G9IV/K1IV	0.76–2.8
2202+469	HK Lac	F IV/K0III	4.57
2206+456	AR Lac	G2IV/K0IV	2.39–16.5
	SZ Psc	F8V/K1IV	6.5–37.
	II Peg	K2-3V-IV	2.1–26.7

4. Results on HR 1099 monitoring

One of the most interesting results of this program came from the monitoring at 6 cm of HR 1099 (V711 Tau).

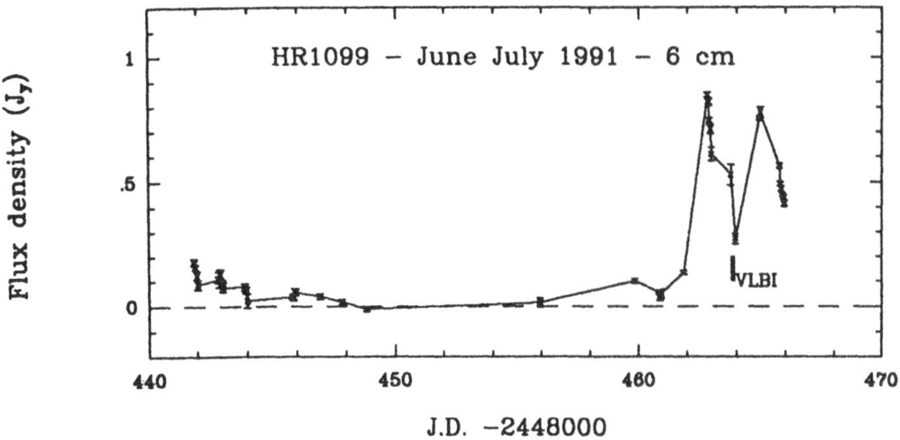

Fig. 1. Observations of HR 1099 at 6 cm obtained with the 32 m VLBI radio telescope of Noto. Data points are one hour averages, while error bars indicate the standard deviation.

During the 1990 observations the system was found almost all the time at the quiescent level. Only a small flare with a maximum flux of 250 mJy was detected.

In July 1991 HR 1099 experienced a series of strong radio flares, one of which reached a flux density of ≈ 800 mJy (Fig. 1). Just after the main flare event, which occurred on July 25, an *ad hoc* VLBI observation run was carried out.

The VLA observations were made on July 26, at 6 cm in standard VLBI MarkII mode, with a 2 MHz bandwidth. The Noto VLBI telescope N (32 m), and two of Jodrell Bank's telescopes, the MarkII J (25 m) and the Cambridge telescope C (32 m), were involved. The shortest baseline J-C is 199 km long while the two longest baselines N-J and N-C are about 2000 km.

From the single dish measurements performed just before and after the VLBI run, it was clear that we were following the flare decay phase. The flux went from ≈ 0.66 Jy measured at 6:30 U.T. to ≈ 0.36 Jy at 11:30 U.T. The angular size of the radio source was measured and found to increase during the flare decay. It reached about 4 m.a.s. of angular dimension which corresponds to a linear size comparable to that of the entire binary system. Figure 2 shows the increase of angular dimension with time and the contemporaneous decrease of the brightness temperature.

5. Conclusions

The first results of single dish observations here presented are quite encouraging. Although we plan to improve the single dish acquisition system further, there are limits that cannot be exceeded, i. e. the antenna size, which strongly influences the achievable sensitivity, and the limited time available for systematic radio flux monitoring over long periods because Noto is a dedicated VLBI antenna.

On the other hand, as shown here, such monitoring allows us to detect the beginning of flaring periods in active binaries and then to organize *ad hoc* VLBI sessions, using the Noto-Medicina baseline together with (possibly) other EVN

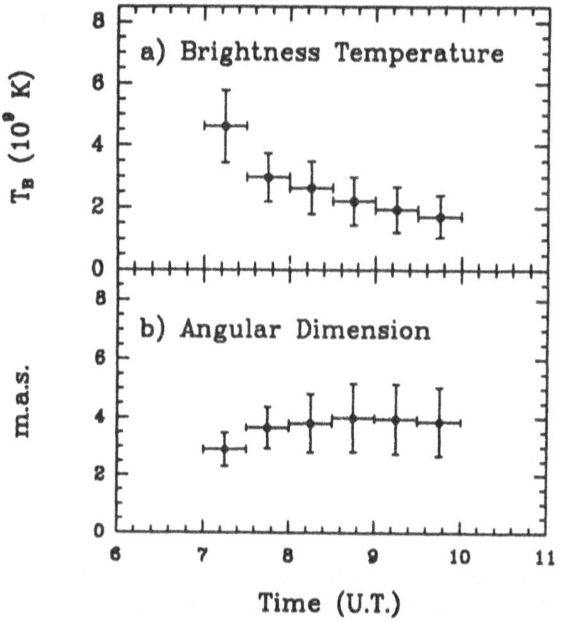

Fig. 2. Time evolution of the angular size and brightness temperature of HR 1099 for the flare decay measured during the VLBI run of July 26 1991.

stations, and to coordinate photometric and spectrographic (Hα) observations at the Catania Observatory.

Monitoring the radio flux for a sufficiently long time will give important information about the flare frequency, time scales for onset and decays of flares, and the energy budget.

Acknowledgements

This work was supported by the Catania Astrophysical Observatory, the Italian *Ministero dell' Universitá e della Ricerca Scientifica e Tecnologica,* the *Gruppo Nazionale di Astronomia* of the *Consiglio Nazionale delle Ricerche* and by the *Regione Sicilia*

References

Feldman P.A., Taylor A.R., Gregory P.C., Seaquist E.R., Balonek J.J., Cohen N.L.: 1978, *Astron. J.* **83**, 1471

Morris D.H., Mutel R.L.: 1988, *Astron. J.* **95**, 204

Mutel R.L., Morris D.H., Doiron D.J., Lestrade J.F.: 1987, *Astron. J.* **93**, 1220

SPECTRA OF CORONAL RADIO EMISSION FROM ALGOL BINARIES

G. UMANA AND C. TRIGILIO
Istituto di Radioastronomia C.N.R. Bologna VLB Station, P.O. Box 169 Noto, Italy
and
S. CATALANO AND M. RODONÒ
Osservatorio Astrofisico Catania
Viale A. Doria 6, 95125 Catania, Italy

Abstract. Radio continuum spectra of Algol-type binaries obtained with the VLA at four frequencies (1.49, 5, 8.4, and 14.9 GHz) are presented. The spectra of Algol, RZ Cas, HR 5110, TW Dra, 505 Sgr, and RT Lac can be fitted quite well by gyrosynchrotron emission from a compact core and a larger tenuous halo. The characteristic size of the core and the halo, and average values of the magnetic field are estimated.

Key words: Radio continuum: stars – Stars: activity – Stars: binaries – Stars: individual: Algol binaries

1. Introduction

Recent observations of radio, X-ray emission and flare-like events from Algol have raised the possibility of magnetic activity in the secondary component of these systems (Hall 1989). Even though Algol itself was one of the first stars detected at radio wavelengths (Wade and Hjellming 1972), very few studies of the radio emission from Algol-type systems have been done. From a limited survey of Algol-type systems carried out using the VLA at 5 GHz, we showed (Umana et al. 1991) that their radio properties compare quite well with those of the more extensively studied RS CVn systems. In this paper we present radio spectra of a small sample of Algol-type systems, and propose a two component model composed of a compact *core* and a larger *halo*.

2. Observations

The radio spectrum observations were carried out with the VLA mainly on February 18 and March 6 1989. We observed six Algol type systems (Algol, RZ Cas, HR 5110, TW Dra, V505 Sgr, and RT Lac) at four different frequencies 1.49, 4.9, 8.4, and 15 GHz. The flux density scale was determined by daily observations of 3C286 and 3C48. Appropriate phase calibrators were used for each system. The results on the observed systems are presented in Figure 1.

3. Gyrosynchrotron model analysis

The total power of synchrotron radiation $P_\nu(B,\theta)$ emitted by an ensemble of electrons is obtained from the emission coefficient $j_n(\gamma, B, \phi, \theta)$ by integration over all possible pitch angles, harmonics and energies. In the case of mildly relativistic electrons (gyrosynchrotron emission) it is not possible to derive a general expression for $P_\nu(B,\theta)$ and numerical calculations are required. Starting from the expressions

J.F. Linsky and S. Serio (eds.), Physics of Solar and Stellar Coronae, 417–420.

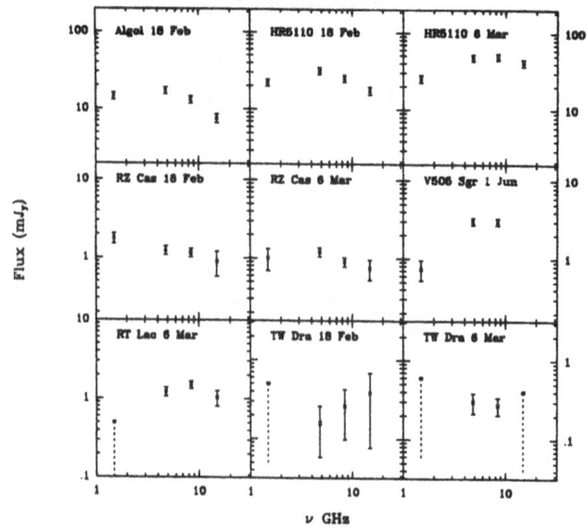

Fig. 1. Radio spectra of Algol binaries observed with the VLA.

given by Chiuderi Drago and Melozzi (1984), we performed a numerical integration of the coefficients over the energy assuming the following electron energy spectrum: $u(\gamma) = k \cdot (\gamma - 1)^{-\delta}$, where k is a normalization constant so that the integral of $n(\gamma)$ between the lowest energy value γ_1 and the highest value γ_2 gives the total number of electrons with energy in the range between γ_1 and γ_2, per unit volume. The spectrum slope is strongly dependent on the high energy cutoff (γ_2), and on the exponent of the energy spectrum δ, but is insensitive to the low energy cutoff. For our calculations we fixed γ_1 at 10 keV.

3.1. Homogeneous source model

To integrate over the entire emitting region, the first step is to assume a geometry for the source. We start with a homogeneous source model with the following basic assumptions: (i) the spatial distribution of energetic electrons inside the source is homogeneous and isotropic; (ii) the energy distribution can be represented by a power law; (iii) the magnetic field is uniform and forms an angle θ with the line of sight; (iv) the emission region is a spherically symmetric corona of thickness H, surrounding the active stellar component. We compute gyrosynchrotron spectra for different electron densities and magnetic field strengths. The effect of increasing the number density of the energetic electrons and of the magnetic field is to increase the radio flux and shift the emission peak to higher frequencies, but the shape of the spectrum always remains the same. A change in δ will strongly affect the shape of the spectrum, mainly at higher frequencies, but for any physically sensible choice of δ we cannot reproduce the flat spectra observed for our Algol systems (in Fig. 1). We conclude that models of homogeneous sources with constant magnetic field and uniform density of energetic electrons cannot reproduce the observed spectra for any values of B, $N_{\rm rel}$ and electron energy distribution.

3.2. Core–Halo Model

As seen in the previous section, while the value of magnetic field and energetic particle density cannot influence the shape of the radio spectrum, the spectral slope can probably be modified by introducing gradients of these two quantities into the model for the radio source. We have no *a priori* information on the actual distribution and geometry of the magnetic fields in these systems, so, as a first approximation to the probably very complex source structure, we consider two distinct regions for the radio emission souces; one which contributes to the spectrum mainly at low frequencies and another which dominates at higher frequencies. The existence of these extended coronal structures has been suggested by the X-ray observations of RS CVn-type systems and VLBI observations of Algol and UX Ari (Mutel *et al.* 1985; Lestrade *et al.* 1988).

Starting from the homogeneous model described above, we can easily obtain a two-component model corresponding to the physical situation just described, i.e.: (i) the *core* is described by a compact homogeneous source, with constant magnetic field and density number N_e; (ii) the *halo* is assumed to be a corona of thickness H, with constant magnetic field and the following radial dependence of the particle distribution function: $N_{\rm rel} = N_o(R_o/R)^\alpha$, where N_o is the number density at the stellar radius R_0, and the index α, which depends on the spatial dilution, is assumed to be $\alpha = 2$. We integrate the emission and absorption coefficient between $\gamma_1 = 1.02$ (E = 10 keV) and $\gamma_2 = 10$ (E = 5 MeV) and use $\delta = 2$ for the energy distribution function for both the halo and the core. Examples of the core-halo model are displayed in Figure 2, while the linear sizes and other relevant physical parameters, deduced from the fits, are summarized in Table 1.

TABLE I
Physical parameters for the observed systems

System		Mag. field (Gauss)	Dimension (cm)	Density (cm^{-3})
Algol	Core	60	1.1×10^{11}	$7{\times}10^5$
	Halo	20	7.4×10^{11}	1.4×10^4
V 505 Sgr	Core	40	1.8×10^{11}	2×10^6
	Halo	10	7×10^{11}	5×10^5
HR 5110	Core	80	1.2×10^{11}	2.7×10^6
	Halo	20	5.6×10^{11}	6.4×10^5
RZ Cas	Core	80	7.7×10^{10}	7×10^5
	Halo	10	6.4×10^{11}	2×10^5
RT Lac	Core	60	1.9×10^{11}	1×10^6
TW Dra	Core	80	4.8×10^{10}	2×10^7
	Halo	10	3.2×10^{11}	4×10^5

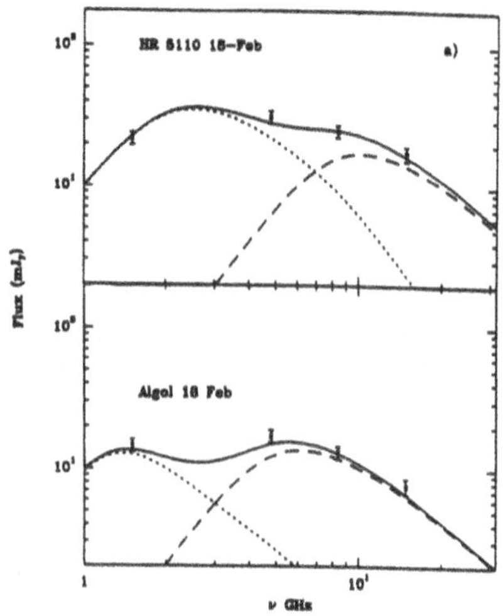

Fig. 2. Examples of comparisons of observed and computed spectra with a core-halo model. The dashed lines represent the contributions of the two components to the total spectrum (solid line).

4. Conclusions

We have here reported for the first time on a significant set of spectra of coronal radio emission from Algol–type binaries. Important conclusion can be drawn in defining the magnetic activity of these systems: (i) The nonthermal nature of the radio emission from Algol–type binaries, already suggested in our previous paper (Umana *et al.* 1991) is clearly demonstrated. (ii) The core-halo model we have developed, although a crude simplification, fits the observed spectra quite well.

Acknowledgements

This work was supported by the Catania Astrophysical Observatory, the Italian *Ministero dell' Universitá e della Ricerca Scientifica e Tecnologica,* the *Gruppo Nazionale di Astronomia* of the *Consiglio Nazionale delle Ricerche* and by the *Regione Sicilia.*

References

Chiuderi Drago, F. and Melozzi, M.: 1984, *Astron. & Astrophys.* **131**, 103.

Hall, D. S.: 1989, in *Algols, IAU Coll. 107,* A. Batten (ed.), Kluwer Academic Publisher Dordrecht, 219.

Lestrade, J. F., Mutel, R. L., Preston, R. A., Phillips, R. B.: 1988, *Astrophys. J.*, **328**, 232.

Mutel, R. L., Lestrade J. F., Preston, R. A., Phillips R. B., 1985, *Astrophys. J.*, **289**, 262.

Umana, G., Catalano, S., Rodonò, M., 1991 *Astron. & Astrophys.*, **249**, 217.

Wade, C. M. and Hjellming, R. M.: 1972, *Nature*, **235**, 270.

STELLAR CORONAL IMAGING

BERNARD H. FOING

Institut d'Astrophysique Spatiale (IAS), CNRS/Université Paris Sud, Bât. 121, F-91405 ORSAY Cedex, FRANCE

Abstract. We discuss the objectives of imaging stellar surfaces and coronae and possible methods and recent results that allow one to derive images of active structures and phenomena using the techniques of rotational modulation, tomography, eclipse mapping, Doppler imaging and interferometry with compact or long baselines. We also discuss the prospects for using these methods in the future to measure the spatial extent of stellar coronae at different temperatures, study circumstellar structures, trace flares and coronal energetic phenomena, and study extended coronae and mass transfer in close active binaries.

Key words: Stellar coronae - stellar activity - close binaries - Doppler imaging - tomography - interferometry

1. Why Image Stellar Surfaces and Coronae?

As demonstrated on the Sun, stellar observations at high resolution permit the study of hydrodynamic and plasma phenomena. The coupling of convection and rotation controls the emergence of magnetic fields at stellar surfaces, the interaction of magnetic fields with fluids, nonthermal plasma processes, and other active phenomena. Such observations allow one to observe the conversion of magnetic energy to thermal energy and the generation of magnetic fields in turbulent fields in different contexts. Stellar activity phenomena occur when strong magnetic fields perturb the structure and energy balance. Non-thermal energy deposition occurs in the chromosphere, transition region and corona. Active magnetic regions appear with excess emission in the core of strong lines such as Ca II, Mg II and Hα formed in the chromosphere, or Lα and C IV formed in the transition region (Foing et al. 1986), and in coronal lines. Observations from the Einstein X-ray satellite showed the presence of coronae in most of the convective stars. Also, ultraviolet spectra obtained with the IUE satellite permit one to diagnose the properties of solar-like chromospheres and transition region in stars.

The dynamo effect (Parker 1955), which involves the coupling of convective motions and differential rotation, has been applied to solar-like stars to explain the regeneration of magnetic fields. Also, the rotational modulation of the stellar flux and of some chromospheric activity indicators provides evidence for both photospheric dark spots and bright chromospheric active regions. The appearance of spots is attributed to intense sub-photospheric magnetic fields which inhibit the convective energy transport toward the surface. Also, long term periodic variations indicate solar-like activity cycles (Wilson 1978) on late-type stars. Empirical relations have been established (e.g. Mangeney & Praderie 1984; Noyes et al. 1984) between the coronal or chromospheric activity and the rotational velocity through the Rossby number, suggesting that the heating is related to the production of magnetic energy through a dynamo mechanism. In close binary systems, it is likely that the synchronization of orbital and rotational periods in short period binaries strengthens the dynamo effect and thus the activity.

J.F. Linsky and S. Serio (eds.), Physics of Solar and Stellar Coronae, 421–430.

Sunspots, plages, flares, and the non-radiatively heated chromosphere and corona all owe their existence to the presence of magnetic fields. Similar phenomena in late-type stars and in close binaries indicate that a substantial amount of magnetic flux exists in their photospheres (see Linsky 1985). The detailed physics behind the fundamental role of magnetic fields in these phenomena, however, remains unknown.

The distribution of magnetic fields and activity phenomena across these stellar surfaces is poorly determined, especially, their extension into coronal magnetic fields. Recent global photospheric magnetic field measurements (Saar et al. 1987) have permitted us to start deriving mean magnetic intensities and surface filling factors for active solar-type dwarfs. Although magnetic fields have been detected in only a few dMe stars (Saar & Linsky 1985; Saar, Linsky & Giampapa 1987) and RS CVn stars (Giampapa, Golub & Worden 1983; Donati et al. 1990), these stars hold the keys to understanding the extremes of stellar magnetic phenomena due to their high activity. Magnetic field strengths (B) scale with photospheric pressure, a finding which is consistent with the modeling of solar-like flux-tubes (e.g. Foing et al. 1986) and surface filling factors seem to be proportional to stellar angular velocity and to saturate at small periods (Saar et al. 1987). Measurements of B and f on rapidly rotating solar-like stars (e.g. BY Dra and RS CVn stars) will help verify these relations. A crucial issue is the spatial distribution of phenomena associated with these magnetic fields on different spatial and time scales: This information is needed to test the suitability of magnetic flux tube models, chromospheric thermal structures models (Vernazza, Avrett & Loeser 1981), coronal loops models (e.g. Rosner, Tucker & Vaiana 1978), dynamo models, empirical activity theories (e.g. Vilhu 1984), heating theories for stellar outer atmospheres, and rotational evolutional theories.

Flares are the most complex and violent phenomena occurring on solar-type, dMe, T-Tauri and active close binary stars. Simultaneous photometry and spectroscopy of flares have shown that emission line enhancements of different species occur as the response of the stellar atmosphere to the flare energy release. Fundamental issues for stellar flare research concern the energy transport mechanism (particle beam vs heat conduction, the atmospheric response to flares); mass motions (ejected components and momentum balance); microflaring, flaring and the heating of coronae; and statistics and recurrence of flares (see e.g. Foing 1989). Flares can often dominate the energy balance in stellar outer layers, and strongly enhance the observed X-ray and ultraviolet emission line fluxes that are used as diagnostics of stellar coronae and chromospheres.

2. Rotational Modulation by Active Magnetic Structures

We must know the geometric distribution of activity phenomena on stars to understand the differences compared with the Sun, to model the active and quiescent atmospheric regions, to study the correlation between the structures observed at different heights, and monitor the changes associated with active region behaviour, instabilities, cyclic activity, dynamo phenomena and differential rotation. The observation of solar-like activity phenomena suggests the existence of solar-like activity cycles on stars. The observed variability can be attributed to long term variations of the spot filling factor as is seen during the solar cycle. Also, systematic monitoring can allow one to observe the evolution of activity complexes, and the long-term arrangement of their geometry. This can be done by observing the same target over a period of a few months.

Periodic or quasi-periodic low-amplitude continuum and emission line fluxes are observed in late-type active stars. The flux variations appear as distortion waves which can vary in shape and amplitude over a few months. These variations are attributed to unevenly distributed surface inhomogeneities, whose visibility is modulated by stellar rotation. From analytical modeling of these light curves with surface distribution of starspots, typical results give 10–40% spotted area coverage, and spot temperatures 400–1500 K cooler than the immaculate photosphere (Rodonò et al. 1987). These parameters could arise from observational selection, as smaller or hotter spots would escape detection, and because rotational modulation methods based on integrated fluxes are biased toward large structures. The geometrical resolution can be improved and smaller structures detected by using interferometric information in addition to Doppler imaging techniques. Similar improvements are also needed to study stellar coronal structures, for which instantaneous imaging is often desirable.

Indirect imaging techniques are needed to complement interferometric imaging, which has limited angular resolution with the available baselines. Rotational modulation of broadband or line fluxes gives a one-dimensional projection in longitude of the stellar surface structures. This can be applied easily to broadband visible or UV bands and chromospheric lines, which do not show large rapid variations outside large flares. However, coronal fluxes in the XUV and X-ray broadbands, in high temperature ultraviolet lines, and in the radio range are intrinsically extremely variable. Yohkoh images show that the Sun is extremely dynamic, with the soft X-ray flux dominated by events due to the reconfiguration of the magnetic field between active complexes. Thus rotational modulation studies using only broadband or line fluxes are of limited use for deriving reliable images of stellar coronae.

3. Eclipse, Doppler and Spectral Tomographic Imaging

Information on quasistationary surface structures is available through the observation and interpretation of temporal photometric or spectroscopic variability as a function of stellar rotational phase and eclipse phase for a number of binary systems. Eclipse observations have been used to constrain the distribution of X-ray emission in the RS CVn system AR Lac (Walter et al. 1987) and the dMe binary YY Gem (Butler et al. 1993). A complete eclipse tomography method is feasible for locating active structures across and above the stellar disk from a detailed analysis of ingress and egress light curves, when the time scale for intrinsic variability is longer than the eclipse duration. However this method requires very high sensitivity and continuous observing that will be feasible only with the next generation of X-ray instruments. Fast time-resolved UV spectroscopy during the eclipses of YY Gem was attempted by Butler et al. (1993) at the limit of IUE performance in terms of S/N and the operational time scale of obtaining multiple exposures and alternating between the SWP and LWR cameras with about 10 min time resolution. This could be done far better with the HST/GHRS although the viewing constraints of the HST satellite will limit the amount of continuous coverage. Future UV instruments on FUSE or PRISMA will be particularly suited in sensitivity and observing efficiency for fast time-resolved spectroscopy of many fainter eclipsing systems.

The "Doppler imaging" technique allows one to obtain images of stellar surface structures with spatial resolution both in longitude and latitude from the variations in line profiles at high spectral resolution and with adequate phase coverage (e.g. Vogt & Penrod 1983; Jankov & Foing 1987, 1992). This technique has been applied to the analysis of ground based photospheric spectra to reconstruct images of spotted photospheres using both line profiles and photometric rotational modulation. This technique requires a rigorous treatment of the local profile to avoid systematic spurious effects at line center that translate into stable symmetric polar spots. Jankov & Foing (1992, 1993) have developed a method that makes use of reference profiles and does not neglect the continuum contribution of spots to reconstruct photospheric images of rapidly rotating stars. From a series of ESO CES spectra of HR1099, they obtained an image of the active primary star that shows a high latitude asymmetric spot that is significantly smaller and different in shape from previous images which showed a large symmetric polar spot (Vogt 1988). Jankov & Foing also found two spots located close to the equator that transit at phases 0.35 and 0.60. They also discovered bright emission surrounding the high latitude spot in the form of facular stripes with 200 K excess which would dominate the polarization signal in visible Zeeman measurements. We expect that joint Doppler and magnetic images (Donati et al. 1990) will give the photospheric magnetic boundary conditions for extrapolating coronal magnetic fields as is done in the solar case.

The "spectral imaging" technique has been applied to IUE high resolution Mg II spectra (Walter et al. 1987; Neff et al. 1989) to reconstruct the chromospheric surface of the RS CVn system AR Lac. These images cannot provide the full spatial resolution of photospheric Doppler images due to the intrinsic Wilson-Bappu width of Mg II lines and the lower spectral resolution and S/N possible with IUE and the chromospheric velocity fields. Char & Foing (1989, 1992) have developed a spectral imaging method for the Ca II H and K lines, which they applied to slow rotators to measure the rotational period, plage coverage, and chromospheric velocity of Alpha Cen B (Char et al. 1993). Spectral imaging of RS CVn systems in the Ca II lines requires an accurate subtraction of the photospheric contribution to the line profile from the two stars, but this subtraction is feasible using reference profiles. This technique provides residual chromospheric spectra which in spectral resolution and effective S/N are comparable to or better than the best available IUE Mg II spectra for chromospheric spectral imaging. Applying their spectral imaging method to Ca II spectra of HR1099, Char, Foing & Jankov (1993) found that the positions of plages are contiguous to the photospheric spots. For transition region or coronal spectral imaging, however, no spectral line is accessible from the ground and UV and X-ray observations are required. Spectral imaging techniques are less model dependent than flux rotational modulation and can distinguish intrinsic variations in resolved active regions, but they neglect possible large intrinsic velocities and they should be verified by an independent method.

One should combine tomographic, Doppler, and interferometric imaging methods using data obtained in different wavelength ranges. The three-dimensional configuration of the magnetic active structures on these stars (Jankov, Char & Foing 1992) can be obtained by combining observations of the rotational modulation of fluxes, spectroscopic profiles, interferometric images and the application of reconstruction techniques. This will permit one to model the chromospheric, transition region, and coronal emission structures associated with the magnetic fields, to calculate extrapolated magnetic fields from constraints at different atmospheric levels, and to derive the vertical stratification and energy balance of these magnetic structures.

4. Interferometric Imaging of Extended Stellar Coronae

Interferometry, even with modest baselines, can permit one to detect both dark spot-like and bright structures on active stars (Foing 1992). These results can be used in conjunction with the observed photometric rotational modulation to constrain the distribution of spots, even in the case of slowly rotating stars for which the Doppler imaging technique cannot be applied. In the case of fast rotators, the angular interferometer can give the centroid of intensity (along a line parallel to the stellar rotation projection) for each velocity stripe defined by a given position in the spectral line (corresponding to a given distance from the projected rotation axis). This can yield simultaneously the longitude and latitude information: the observation at a few phases will allow the complete reconstruction of surface structures using this generalized Doppler-interferometric tomography. The comparison of an interferometric image with a spectral image can also provide evidence for coronal velocity fields associated with dynamic and energetic phenomena.

The accuracy in measuring the FWHM of a stellar image with an interferometer determines the limit in measuring the angular extent. Below the Rayleigh resolution, one can still obtain information about diameters, provided the excess FWHM of the stellar image compared to the FWHM of the point spread function can be determined. The accuracy in determining the FWHM increases as FWHM/2σ, where σ is the square root of the number of counts. The accuracy will also be limited by all systematic sources of broadening of the stellar image such as stigmatic properties, pointing stability after correction, and proper calibration of the point spread function. This can be the nearby UV continuum around a coronal line, or a line formed in an extended atmosphere. Also the equivalent angular diameter can be measured as a function of wavelength in order to study structures with different spatial extents in different transition region and coronal lines.

For rapidly rotating stars, one can measure the angular diameter with high accuracy by measuring the shift of centroid of the image obtained in the blue and the red wing of a given spectral line by a technique analogous to differential speckle interferometry. With a spectral resolution of 0.1 Å at 1500 Å, the contribution due to redshifted and blueshifted stellar limbs can be separated for vsini greater than 10 km s^{-1}. One can make a differential measurement of stellar image centroid positions in the blue and red line wings, which is less affected by image broadening and other sources of systematic errors, since noise varies as FWHM/2σ for this difference. In that case it is possible to measure the sky projection of the rotation axis, and the angular diameter with an accuracy of a fraction of milliarcsecond.

5. Coronal Condensations and Circumstellar Phenomena

Extended coronal loops and condensations are also present at chromospheric and coronal heights. Circumstellar condensations can be observed in absorption due to ejected clouds of cool material transiting in front of a stellar disk, as described by Collier-Cameron & Robinson (1989). The distribution of absorbing corotating clouds can also be derived from an inverse "skewing" mapping technique. The structure of the clouds can be deduced from continuous repeated observations at high resolution in Hα, and a more recent campaign also showed absorption transients in lines of Ca II, Mg II and Na I and even through Hα photometry (Collier-Cameron et al. 1990). It is expected that such structures extending over a few stellar radii and between binary stellar surfaces can be resolved in the UV lines emitted at coronal and transition region temperatures.

For close active binaries, even with a modest baseline interferometer or 2m class telescope in the UV (Foing 1992) one can study the centroid positions in the continuum and in strong emission lines formed over a range of temperatures including Lα, O I, C II, C IV, He II, C I, Si II and Mg II, and from the ground the Ca II H and K and infrared triplet and hydrogen Balmer lines. The centroid position can be determined with an accuracy = resolution/2σ. For the strongest lines mentioned above, this leads to a significant increase in positioning accuracy. An interesting aspect of close binaries with orbital velocity separation up to 20 km s^{-1}, is that they can be separated with 0.1 Å resolution. Images can thus be obtained at different wavelengths corresponding to the orbital shifts of each stellar component, or to the Doppler shifts of intermediate material flowing at different velocities on the surfaces or residing above the stellar surfaces.

6. Coronae in RS CVn and Close Binary Stars

RS CVn systems are detached binary systems with periods typically between 1 and 14 days, which are synchronized by tidal effects and generally consist of a subgiant primary (with spectral type near K0 IV) and a dwarf secondary (near spetral type G5 V). The most important photometric characteristic of these systems (Hall 1980) is the quasisinusoidal distortion of the lightcurve attributed to the rotational modulation of a nonuniform distribution of photospheric spots. Rotational modulation of chromospheric emission in the Ca II H and K lines was shown for several RS CVn systems in the Mt. Wilson H and K variability survey (Vaughan et al. 1981). Results from joint IUE and ground based observations of II Peg have indicated that chromospheric plages cover, in a first approximation, an area correlated to the photospheric spots, but this is not yet completely clear from other observations.

From the adjustment of Hα emission profiles on the HR1099 system, Foing et al. (1990a,b) have distinguished two components which have radial velocity curves with amplitudes larger than the vsini limit corresponding to sources on the stellar equator. One component is narrow (FWHM = 40 km s^{-1}) and stationary near the gamma velocity, and the another is broad (FWHM 95 km s^{-1}) with a 240 km

s^{-1} velocity excursion. These components (Foing et al 1993a,b) are consistent with rigid corotation of circumstellar structures, one placed between the two stars and another more extended on the other side of the cool evolved primary. From velocity tomography at different phases, one can derive a height above the surface of 3.1 and 7.3 solar radii for these circumstellar structures. Such extended structures can be stable only at very high temperatures. In particular, one finds that a thermal scale height equivalent to the outer structure height and average gravity corresponds to a temperature of 4 million K. These hot extended structures may correspond to those derived from radio models of HR1099 (Chiuderi Drago & Klein 1992) or radio VLBI observations of the RS CVn system UX Ari. Thus one expects to see a spectral signature of these circumstellar components by observing global line shifts in coronal lines with a moderate spectral resolution of 1000 km s^{-1} or by resolving components and applying Doppler tomography with better than 100 km s^{-1} resolution. Such observations should be possible in UV coronal lines with HST/GHRS and in the future with FUSE-Lyman, the XMM grating experiment, and the AXAF spectrometer.

7. Flares, Coronal Mass Ejections and Energetic Phenomena

Flares can be observed from X-rays to the radio range. Flares occur sporadically in solar-like stars with spectral types F to K, and very frequently on young stars (T Tauri or post-T Tauri), dMe stars, and close active binaries (RS CVn, Algol, W UMa systems) with a significant chance of observing one large event in one day of monitoring and many small events in the UV emission lines. The scientific objectives of these flare programmes (Foing 1989) include (a) the determination of the energy budget for a typical sample of flares; (b) the respective roles of radiation from the corona (as shown by the X rays), conductive losses through the transition region (EUV) and expansion (as indicated from velocity fields measurements); and (c) measurement of the temperatures, densities and volumes of hot flaring plasma.

Coronal flares have been reported on RS CVn systems based on radio, UV or X-ray observations. A multi-site MUSICOS 89 continuous spectroscopy campaign involved telescopes in China, Crimea, Italy, France, Chile, USA, Hawaii (Catala & Foing 1990; Catala et al. 1993). This observing campaign aimed at studying surface structures and flares on the RS CVn system HR1099 (Foing et al. 1990a, 1993a,b). During this campaign, a remarkable episode with gigantic flares was observed that affected not only the upper coronal layers but also the photosphere, giving rise to white light flares with peak radiative losses up to 7 solar constants and integrated radiative losses up to 10^{38} erg (Foing et al 1993a,b). The spectral signature of the flare was observed in Hα simultaneous with the photometry. One white-light flare was occulted by the limb abruptly, while the Hα emission arising from coronal heights was nearly constant. Also velocity curves of the Hα flare components allowed the location of the flares and, in one case, provided evidence for a large matter ejection associated with the flare.

Flares can be studied spectroscopically and located on the star with an accuracy as small as a fraction of the stellar radius. In addition, a modest interferometer can separate those flares arising well above the photosphere at coronal heights from those occurring near the photosphere. Simultaneous monitoring of UV line flux variations, Balmer line fluxes and XUV fluxes allows one to study small compact flares and microflares, also in correlation with X-ray measurements (Butler et al. 1986). Large flare events are often associated with coronal mass ejections (Houdebine, Foing & Rodonò 1990; Foing et al. 1990a, 1993a,b) at high velocities, which can be studied by their spectral signatures and eventually followed in time when the flares enter the circumstellar environment.

An interferometer allows one to study departures from spherical symmetry for close binaries in which the evolved star fills its Roche lobe and for contact binaries. Even with the modest baseline of HST or a compact instrument, it should be possible to image directly the nearest prototype systems such as Algol, and to determine the effect of the surface distortion on the variation of the equivalent separation of the stars and the centroid positions with orbital phase. Also mass transfer phenomena can be studied, even below the 10 milliarcsec resolution limit of a 2.5m telescope or an interferometer operating at 1200 Å, by identifying the plasma flow at different velocity ranges and formation temperatures and by measuring the corresponding emission centroid in relation to the individual stars. This allows one to reconstruct the structure and dynamics of accretion, boundary layers and winds.

The previously mentioned programmes could be conducted *a fortiori* with the HST after its repair. Although HST is not diffraction limited in the ultraviolet, one can in principle calibrate its point spread function and deconvolve the images to restore the resolution limit of a 2.5m telescope by analyzing speckles in the point spread function that occur due to the medium-scale microripples on the primary mirror. After its repair the high spectral resolution and sensitivity of HST would be useful for Doppler tomography. For the future one may look forward to a 20m interferometer in space giving multi-wavelength and especially UV images, which could provide previews of the rich harvest of astrophysical discoveries expected with longer baseline instruments in space or on the Moon.

8. Conclusion

Stellar magnetic activity phenomena in the chromosphere, transition region and corona can be studied best using spectral lines in the UV and XUV. X-ray and XUV flux variability studies, due to rotational modulation (or to eclipses) in active stars, provide the spatial positions of discrete hot regions, but the large X-ray variability (due to active transients and flares) prevents any complete and reliable imaging. Using time resolved high resolution spectroscopy, Doppler tomography allows one to make images that are less sensitive to intrinsic variations. We have described some results obtained in UV and visible lines, indicating hot extended emission in active stars. Absorption spectroscopy allows one also to probe both the hot and cool circumstellar material.

Angular interferometry from the ground and with 2–20m baselines from space (especially in the UV) has the capability of resolving at the milliarcsec level the surface and coronal environment of stars in several temperature ranges instantaneously. Stellar coronal angular extents can be measured, even below the Rayleigh resolution of the interferometer, by proper calibration of the point spread function and by differential interferometric measurements for several lines formed at different coronal temperatures.

Interferometers in space will provide interesting spatial resolution for several stellar programmes in preparation for long baseline interferometers on the Moon. Space interferometers will also be able to study stellar flares, extended coronal structures, and circumstellar and mass transfer phenomena. Rotational modulation, Doppler tomography and differential interferometric imaging can be combined to map active stars in different lines formed at different temperatures and heights, giving a three-dimensional view of the global activity phenomena and coronae on nearby stars and binary systems.

9. Acknowledgements

I thank Dr S. Serio and the organizing committee of the G. Vaiana Memorial Palermo Colloquium for their invitation to present this paper, and for an enlightening programme. I also thank my former PhD students Drs E. Houdebine, S. Jankov and S. Char and my collaborators from the "Cool Star Consortium" for related discussions. Finally, I thank Dr J.L. Linsky for correcting the final version of the paper.

References

Butler, C.J., Rodonò, M., Foing, B.H., Haisch, B.: 1986, *Nature* **321**, 679

Butler, C.J. et al.: 1993, in preparation

Catala, C., Foing, B.H., eds.: 1988, *Proceedings of 1st Workshop on Multi-Site Continuous Spectroscopy*, Presses de l'Observatoire de Paris, Paris

Catala, C., Foing, B.H., eds.: 1990, *Proceedings of 2nd Workshop on Multi-Site Continuous Spectroscopy*, Presses de l'Observatoire de Paris, Paris

Catala, C., Foing, B.H., Baudrand, J. et al.: 1993, *Astron. Astrophys.*, in press

Char, S.: 1992, *PhD Thesis*, Institut d'Astrophysique Spatiale, Univ. Paris VI

Char, S., Foing, B.H.: 1989, in P. Delache, S. Laloe, C. Magnan, J. Tran Thanh Van, ed(s)., *Modeling the Stellar Environment*, Editions Frontieres, Gif-sur-Yvette, 211

Char S., Foing B.H.: 1992, *Astron. Astrophys.*, in press

Char S., Foing B.H., Beckman, J.E. et al.: 1993, *Astron. Astrophys.*, in press

Char S., Foing B.H., Jankov, S.: 1993, *Astron. Astrophys.*, in press

Chiuderi Drago, F., Klein, K.L.: 1992, *Astron.Astrophys.*, in press

Collier-Cameron, A., Robinson, R.D.: 1989, *MNRAS* **236**, 57

Collier-Cameron, A., Duncan, D.K., Ehrenfreund, P., Foing, B.H. et al.: 1990, *MNRAS* **247**, 415

Donati, J.F. et al.: 1990, *Astron. Astrophys.* **232**, L1

Foing, B.H., Bonnet, R.M., Bruner, M.E. : 1986, *Astron. Astrophys.* **162**, 292

Foing, B.H.: 1989, *Solar Phys.* **121**, 117

Foing, B.H., Char, S., Jankov, S., Houdebine E.: 1990a, in F. Sanchez & M. Vazquez, ed(s)., *New Windows on the Universe*, Cambridge University Press, 213

Foing, B.H. et al.: 1990b, in C. Catala & B.H. Foing, ed(s)., *Proceedings of 2nd Workshop on Multi-Site Continuous Spectroscopy*, Presses de l'Observatoire de Paris, Paris, 117

Foing, B.H.: 1992, *ESA SP* **344**, 123

Foing, B.H., Char, S., Ayres, T et al.: 1993a, *Astron. Astrophys.*, in press

Foing, B.H., Neff, J.E., Cutispoto, G. et al.: 1993b, *Astron. Astrophys.*, in preparation

Giampapa, M., Golub, L., Worden, S.P.: 1983, *Astroph.J.* **268**, L121

Hall, D.S.: 1980, in R.M. Bonnet & A.K. Dupree, ed(s)., *Solar Phenomena in Stars and Stellar Systems*, Reidel, Dordrecht, 253

Houdebine, E.: 1990, *PhD Thesis*, Institut d'Astrophysique Spatiale, Univ. Paris XI

Houdebine, E.R., Foing, B.H., Rodonò, M.: 1990, *Astron. Astrophys.* **238**, 249

Jankov, S.: 1992, *PhD Thesis*, Institut d'Astrophysique Spatiale, Univ. Paris VI
Jankov, S., Char, S., Foing, B.H.: 1992, *ESA SP* **344**, 113
Jankov S., Foing B.H.: 1987, in J.L. Linsky & R. Stencel, ed(s)., *Cool Stars, Stellar Systems and the Sun*, Springer-Verlag, Berlin-Heidelberg, 528
Jankov S., Foing B.H.: 1992, *Astron. Astrophys.* **256**, 533
Jankov S., Foing B.H.: 1993, *Astron. Astrophys.*, in press
Linsky, J.L.: 1985, *Solar Phys.* **100**, 363
Mangeney, A., Praderie, F.: 1984, *Astron. Astrophys.* **130**, 143
Neff, J.E. et al.: 1989, *Astron. Astrophys.* **215**, 79
Noyes, R.W. et al.: 1984, *Astrophys. J.* **279**, 763
Parker, E.N.: 1955, *Astrophys. J.* **122**, 293
Rodonò, M. et al.: 1987, *Astron. Astrophys.* **176**, 267
Rosner, R., Tucker, W.H., Vaiana, G.S.: 1978, *Astrophys. J.* **220**, 643
Saar, S., Linsky, J.L.: 1985, *Astrophys. J.* **299**, L47
Saar, S., Linsky, J.L., Beckers, J.M.: 1987, *Astrophys. J.* **302**, 777
Saar, S., Linsky, J.L., Giampapa, M.S.: 1987, in L. Delbouille, A. Monfils, ed(s)., *Observational Astrophysics with High Precision Data*, Institute of Physics, Liege, 103
Vaughan, A.H. et al.: 1981, *Astrophys. J.* **250**, 276
Vernazza, J.E., Avrett, E.H., Loeser, R.: 1981, *Astrophys. J.Suppl.* **45**, 635
Vilhu, O.: 1984, *Astron. Astrophys.* **133**, 117
Vogt, S.S.: 1988, in G. Cayrel & M. Spite, ed(s)., *Proc. IAU Symp. 132*, Kluwer, Dordrecht, 253
Vogt S., Penrod, H.: 1983, *P.A.S.P.* **95**, 565
Walter, F. et al.: 1987, *Astron. Astrophys.* **186**, 241
Wilson, O.C.: 1978, *Astrophys.J.* **226**, 379

A BRIGHT, BLUE SPOT ON THE dMe STAR, AU Mic

P.B. BYRNE

Armagh Observatory, Armagh BT61 9DG, N. Ireland

Abstract. We present light and colour curves of the flare/spotted star AU Mic taken on the Walraven photometric system. They show evidence for a highly variable blue feature on the mean spot light curve which persisted for at least a month.

Key words: Starspots – flaring – AU Mic

1. Introduction

AU Mic has a spectral type dM2.5e and lies at a distance of 8.85pc. Harding (1970) discovered it to be a flare star and Kunkel (1973) studied its flare activity extensively, finding that it was among the most active flare stars. Byrne et al. (1980) and Butler et al. (1981, 1987) examined its chromospheric and transition region emission lines and concluded that it is among the most active dMe stars known.

Torres, Ferraz-Mello and Quast (1972) found it to vary with a period ≈ 4.86 days and an amplitude of $\Delta V \approx 0.3$. Torres and Ferraz-Mello (1973) later modelled this behaviour as a rotating star with a large solar-like dark spot on one hemisphere, occupying $\approx 10\%$ of the total surface area. Torres, Busko and Quast (1983) later analysed observations taken over five consecutive seasons (1971-5) and concluded that the best fitting period was 4.8540 day.

In this paper we present spot photometry obtained in 1981, which gives interesting insight into the spot distribution on the star.

2. Observations

The major dataset was obtained between 1981 October 13 and November 11 using the Leiden 90 cm flux collector at ESO equipped with the Walraven photometer (Walraven and Walraven, 1960), which defines the Walraven VBLUW system (Lub et al. 1979). The Walraven bands are indicated in what follows by the subscript W and the Johnson bands by the subscript J, wherever there is likely to be confusion. AU Mic was observed on 28 nights, usually twice per night.

A second dataset was obtained using the 75 cm reflector of the South African Astronomical Observatory (SAAO) equipped with a Johnson UBV photometer and also an instrumental red passband, which we shall designate r.

To compare the datasets V_W has been transformed to V_J using the transformation of Pel (1976). $(V\text{-}B)_W$ and $(B\text{-}U)_W$ were also transformed to Johnson but these will not be as good a match as V_W. In Fig. 1 we plot the resulting light and colour curves, with $\varphi = (JD - 2440000.0)/4.8540$ (Torres, Busko & Quast 1983). There is a well-defined V_W modulation with an amplitude $\Delta V \approx 0.23$, maximum $V \approx 8.63$ at $\varphi \approx 0.2$ and minimum $V \approx 8.86$ at $\varphi \approx 0.5$. The curve is asymmetric with a more rapid fall to minimum than the subsequent rise to maximum. Furthermore the rising branch has a "standstill" between $\varphi \approx 0.7 - 0.8$.

$(B\text{-}V)_W$ varies in antiphase with V_W between $\varphi \approx 0.8 - 0.45$ (the star is bluest when at maximum ($\varphi \approx 0.2$) and reddest at minimum). The $(B\text{-}V)_W$ colour be-

J.F. Linsky and S. Serio (eds.), Physics of Solar and Stellar Coronae, 431–434.

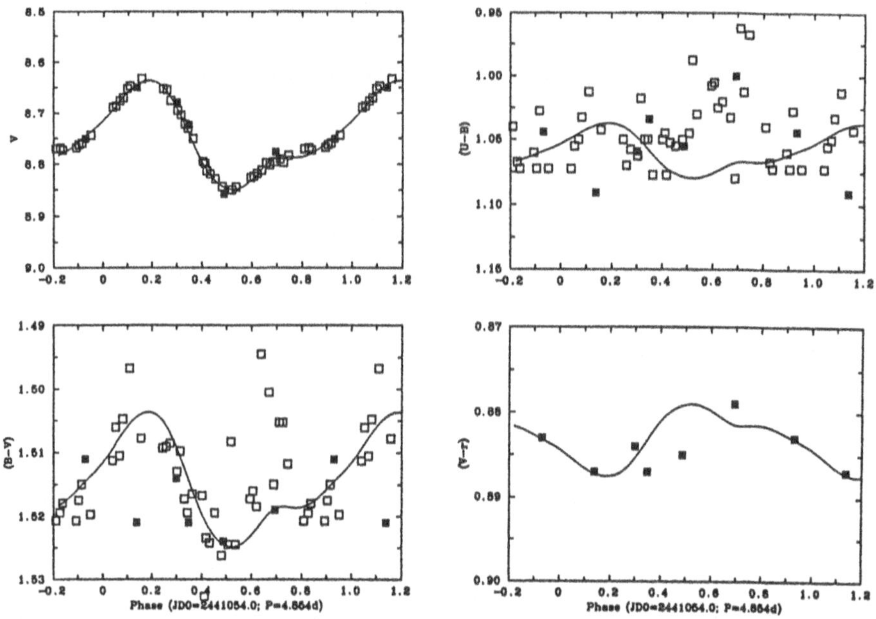

Fig. 1. The V (*upper left*), (B-V) (*lower left*), (U-B) (*upper right*) and (V-*r*) (*lower left*) curves for AU Mic. Open symbols are Walraven values transformed to Johnson, filled are Johnson data from SAAO. The continuous curve is the spot model discussed in the text.

haviour between light minimum and maximum, however, is not consistent with that between maximum and minimum in two ways. As V_W recovers the scatter becomes greater than before. All the Walraven colours show the same blueing near the V_W "standstill" at $\varphi \approx 0.8$ and very much increased scatter.

3. Discussion

Fig. 1 clearly shows that, at maximum light, AU Mic was fainter by ≈ 0.03 in V than the maximum of Torres et al. (1972) during 1971. This suggests that AU Mic is spotted at all phases and at no time do we see an unspotted hemisphere.

3.1. Model for the spot distribution

We have made a spot model using the computer program SPOTPIC which is based on the formulation of the spot problem by Friedemann and Gurtler (1975). A temperature difference between spot and unspotted photosphere of 1000°K has been used. We will justify this choice when discussing the $(B\text{-}V)_W$ colour variations below. We have also adopted 0° for the inclination of AU Mic's rotational axis. This seems reasonable in view of the large amplitude of spot modulation recorded by Torres et al. (1972) in 1971.

Generally two spots are sufficient to model a dMe V light curve but this is not so here. Fitting the steep slopes of the rise to and fall from maximum requires that the spots move rapidly off and then onto the disk. This is impossible with a two spot model if the spots are at a high latitude ($\geq$ 30°). Yet the prolonged minimum requires a combination of a spot at high latitude and an inclination of the rotational axis by $\geq$60° if we are confined to two spots. Therefore we require at least three circular spots if the inclination is to be $\leq$ 30°. Even with a three-circular-spot model it is impossible to fit the light curve exactly but, in view of the behaviour discussed below we have adopted the best-fitting three spot model for further discussion.

3.2. The colour behaviour

We have already remarked on the unusual behaviour of the various colours of the Walraven system. The minimum of the $(B\text{-}V)_W$ colour coincident with V_W light maximum (minimum spottedness) can be understood as a hotter average temperature when the coverage by spots is less. Unfortunately, neither $(B\text{-}V)_W$ nor $(U\text{-}B)_W$ are very sensitive to temperature at the spectral type of AU Mic. Thus the observed amplitude in $(B\text{-}V)_W$, while confirming a change in mean temperature between the hemispheres of AU Mic, does not constrain the temperature very effectively. For this reason we have adopted the canonical value of 1000°K for the temperature difference between spot and unspotted photosphere.

The remainder of the colour variations ($\approx$ 0.5 $\leq$ φ $\leq$0.8) are not related to the V_W light variation in so straightforward a manner. In Fig. 1 the inverse of the V_W model, scaled to the $(B\text{-}V)_W$ variation is plotted over the $(B\text{-}V)_W$ itself. It will be seen that the colour variation follows the light variation reasonably well between $\varphi \approx 0.2-0.5$. After $\varphi \approx 0.5$, however, the colour becomes much bluer than expected and displays about three times the scatter seen elsewhere.

The (B-L) and $(U\text{-}B)_W$ colour variations show the same behaviour, but with a weaker indication of temperature contrast between $\varphi \approx 0.2-0.5$, as one might expect. The blueing and increased scatter is again seen between $\varphi \approx 0.5-0.8$. $(U\text{-}W)_W$ is too scattered to reach definite conclusions.

Qualitatively this colour behaviour is similar to that observed by Torres and Ferraz-Mello (1973), also on AU Mic. Their observations, made on the Johnson UBV system, demonstrated a blueing and increased scatter in both $(B\text{-}V)_J$ and $(U\text{-}B)_J$ near minimum light in V_J. They concluded that this was due to low-level flares. The frequency of occurrence of optical flares on dMe stars increases with decreasing flare amplitude and they show greatest contrast in the bluer spectral regions (Byrne 1983).

Byrne et al. (1987) observed flares on the dMe star, FK Aqr, in the Walraven system. They demonstrated that flares hardly affected the V_W band while they showed increasing contrast in the B_W, L, and U_W bands, in keeping with results obtained here. On this view we would interpret the increased scatter between V_W minimum light ($\varphi \approx 0.5$; maximum spottedness) and $\varphi \approx 0.8$ as due to flares or flare-like events. This behaviour is similar to that reported by Butler et al. (1987) for AU Mic in the UV. They noted that the fluxes in UV transition lines, such as

CII(λ1335/6Å) and CIV(λ1548/51Å), were continually variable on a timescale of ≤ 1 hr. This they attributed to low-level flaring such as we are discussing here.

This does not, however, account for the "standstill" in the V_W light curve which has a distinctly blue colour. One possibility is that there was a solar-like active region near $\varphi \approx 0.7$ which emitted strong and highly variable Balmer emission.

4. Conclusions

The existence of a blue bump on the light curve of AU Mic in 1981 leads one to caution against the belief that only dark, red spots contribute to the light curves of active dMe stars.

Acknowledgements

Research at Armagh Observatory is supported by a grant-in-aid from the Department of Education of Northern Ireland. We wish to record our thanks to the Director and staff of the South African Astronomical Observatory for their help and support while carrying out part of the observations described herein.

References

Byrne, P.B., Butler, C.J., Andrews, A.D.: 1980, *Irish A.J.* **14**, 219
Byrne, P.B.: 1983, *Activity in Red-Dwarf Stars, eds. P.B. Byrne and M. Rodono*, D.Reidel; Dordrecht, 157
Byrne, P.B., Black, E., The, P.S.: 1987, *A&A* **186**, 261
Butler, C.J., Byrne, P.B., Andrews, A.D., Doyle, J.G.: 1981, *MNRAS* **197**, 815
Butler, C.J., Doyle, J.G., Andrews, A.D., Byrne, P.B., Linsky, J.L., Bornmann, P.L., Rodono, M., Pazzani, V., Simon, T.: 1987, *A&A* **174**, 139
Friedemann, C., Gurtler, J.: 1975, *Astron. Nachrichten* **296**, 125
Harding, G.A.: 1970, *Mon. Not. Sth. African Astron. Soc.* **XXIX**, 130
Kunkel, W.E.: 1973, *ApJS* **25**, 1
Linsky, J.L., Bornmann, P.L., Carpenter, K.G., Wing, R.F., Giampapa, M.S., Worden, S.P., Hege, E.K.: 1982, *ApJ* **260**, 670
Lub, J., Pel, J.W.: 1977, *A&A* **54**, 137
Pel, J.W.: 1976, *A&AS* **24**, 413
Torres, C.A.O., Ferraz-Mello, S., Quast, G.R.: 1972, *Ap Lett* **11**, 13
Torres, C.A.O., Ferraz-Mello, S.: 1973, *A&A* **27**, 231
Torres, C.A.O., Busko, I.C., Quast, G.R.: 1983, *Activity in Red Dwarf Stars", Ed. P.B. Byrne and M. Rodono*, D. Reidel: Dordrecht, 175
Walraven, T., Walraven, J.H.: 1960, *Bull. Astron. Inst. Netherlands* **15**, 67

IMAGING THE PHOTOSPHERE & CHROMOSPHERE OF THE RAPIDLY ROTATING dMe STAR, HK Aqr

P.B. BYRNE and M. MATHIOUDAKIS
Armagh Observatory, Armagh BT61 9DG, N. Ireland

Abstract. We present simultaneous starspot photometry and high resolution Hα spectroscopy of HK Aqr which provide information on the location of the starspots and the projected velocity of a flare.

Key words: starspots – Hα emission – chromospheres – flares – HK Aqr

1. Introduction

Young et al. (1984) showed that HK Aqr is the most rapidly rotating solar neighbourhood flare star (P = 10.35 hr, $v\sin i$ = 70 km sec^{-1}). Young et al. (1990) saw a shift of the Hα emission centroid, at low resolution, over a narrow range of phase, first to the red and then to the blue, with an accompanying drop in line flux. They suggested that near-polar, Hα-emitting plages at diametrically opposite longitudes were distorting the line profile as they were carried across the stellar disk.

Byrne and McKay (1990) found drops in MgII h&k flux at the same phase. Since rotational broadening dominates over radiation broadening in HK Aqr, they pointed out that obscuration by a transiting object would also account for the phenomenon. Doyle and Cameron (1990) suggested a link with AB Dor, where cool clouds of hydrogen condense at the apexes of loops extending to the corotation radius. Byrne et al. (1992) presented high resolution Hα profiles showing that the deviation of the centroid was due to absorption in the wings of Hα.

This paper presents broadband photometry and high resolution Hα spectroscopy and shows that the phenomenon is more complex than discussed hitherto.

2. Observations

HK Aqr was measured in UBV(RI)$_{KC}$ between 20 – 26 August 1991 using the 1 m telescope at the South African Astronomical Observatory. HK Aqr and its comparisons were measured 10-15 times on each of 5 clear nights. Typically, these data cover a phase interval, $\Delta\varphi \approx 0.6$ The star was also monitored for flares but none were recorded with $\Delta U \geq 0.1$ during ≈ 25 hr. The resulting V light curve, phased as in Byrne and McKay (1990), is given in Fig. 1.

Hα spectra were also obtained on 24 August 1991 using the University College London Echelle Spectrograph (UCLES) on the 3.9 m Anglo Australian Telescope (AAT). The resolution was $\Delta\lambda/\lambda \approx 50\,000$ and exposure times were 5 min. A continuous phase coverage corresponding to $\Delta\varphi \approx 0.86$ was achieved.

3. Spot distribution

Fig. 1 shows a spot modulation with amplitude $\Delta V \approx 0.09$, mean $V \approx 10.88$ and V_{min} (greatest spot coverage) at phase $\varphi \approx 0.5$. The variation is asymmetric. This V amplitude and mean are similar to those of previous observers. V_{mean} is ≈ 0.03

J.F. Linsky and S. Serio (eds.), Physics of Solar and Stellar Coronae, 435–438.

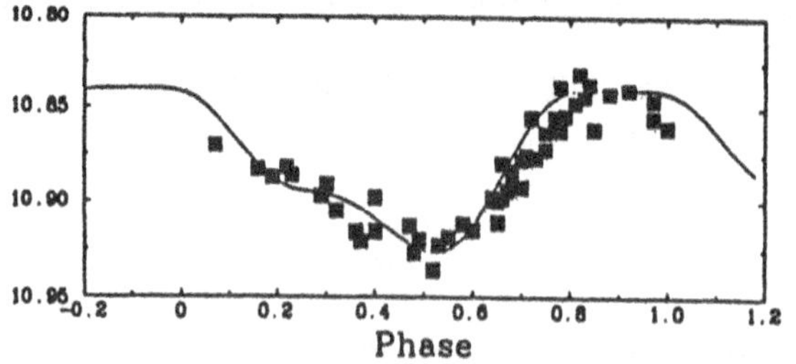

Fig. 1. V light curve for HK Aqr during the interval 20-25 August 1991. The continuous line is the model fit discussed in the text.

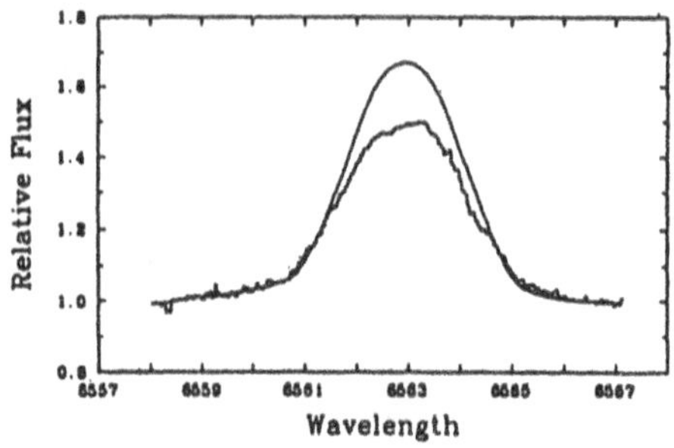

Fig. 2. The mean Hα emission line profile for HK Aqr for the 24 August 1991 (*histogram*) compared with the Hα profile for the more slowly rotating dMe star, AU Mic broadened to $v\sin i = 70\,\mathrm{km\,sec^{-1}}$ (*continuous*).

fainter than that given in Byrne and McKay (1990) but the latter curve was poorly sampled and so the difference is probably not significant. Pettersen et al. (1987) and Bopp et al. (1988), however, presented V curves for HK Aqr in 1983 which, when combined imply an amplitude of ≈ 0.16 with a $V_{max} \approx 10.72$ (using Bopp et al. to calibrate the differential observations of Pettersen et al.). Thus HK Aqr has been brighter than our maximum by $V \geq 0.1$.

A model of the spot distribution made using the program SPOTPIC, based on the formulation of Friedemann and Gurtler (1975), and an inclination of the rotational axis to the plane of the sky of 0°. To minimize spot sizes their temperatures were set to 0°K, they were required to cross disk centre and they were assumed circular in outline (see Byrne (1992) for a justification). Two spot groups are present, one centered at $\varphi \approx 0.5$ and occupying $\geq 2.3\%$ of the surface, the other near $\varphi \approx 0.25$ occupying $\geq 1.7\%$.

Photometric imaging measures only spot distributions which modulate light (Byrne 1992). It is insensitive to any distribution which is quasi-uniform in longitude or is at high latitude. HK Aqr had $V_{max} \geq 10.72$ in 1983 (Pettersen et al. 1987; Bopp et al. 1988). If unresolved spots have depressed V_{max} from 10.72 to the 1991 value of 10.88, we require a further area coverage of $\approx 20\%$.

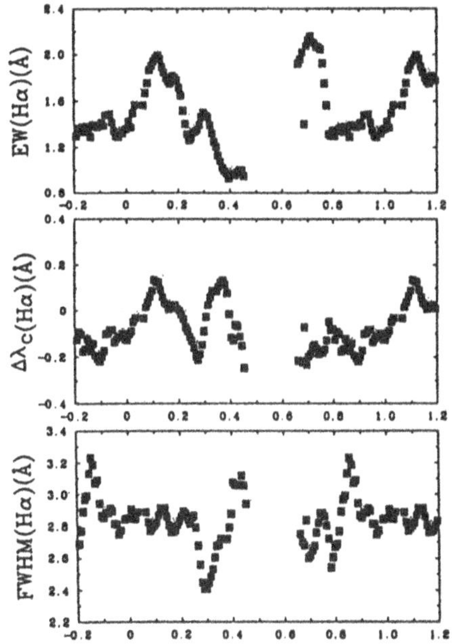

Fig. 3. Plots of the equivalent width (EW) (*upper panel*), centroid wavelength (λ_c) (*middle panel*) and full width at half maximum (FWHM) (*lower panel*) of Hα on HK Aqr with rotational phase.

4. The Hα line

Fig. 2 presents the mean Hα line for the entire night. We have also observed the Hα line of the more slowly rotating dMe star, AU Mic, with the same instrumental setup. To test whether HK Aqr's Hα is dominated by doppler, as opposed to radiation, broadening, we have broadened AU Mic's Hα with $v\sin i = 70\,\mathrm{km\,sec^{-1}}$ and plotted it over that of HK Aqr in Fig. 2. It will be seen that the FWHM's are the same.

Having established that doppler broadening dominates, we have tried to see if chromospheric features can be seen crossing the disk of HK Aqr. In Fig. 3 we plot the equivalent width (EW), full width at half maximum (FWHM) and the central wavelength (λ_c) of Hα as a function of phase.

All three quantities vary. Maximum EW is at $\varphi \approx 0.12$, when the lesser spot group was already on the disk (≈ 0.13 before meridian), and at $\varphi \approx 0.7$, when the major spot group has almost disappeared. Minimum EW occurs when the major spot group is close to meridian. Thus there is no evidence for an association of Hα-emitting plage and spots. We qualify this by repeating that there is probably a substantial spot distribution undetected by the photometry.

The correlation between EW and $\Delta\lambda_c$ is of interest in view of the earlier results described in the introduction. The largest excursion in EW occurred between $\varphi \approx 0 - 0.4$ where $\Delta\lambda_c$ also showed the greatest change. From $\varphi \approx 0 - 0.1$ Hα increased by $\approx 50\%$ while λ_c shifted ≈ 0.2Å ($10\,\mathrm{km\,sec^{-1}}$) to the red. Between $\varphi \approx 0.11 - 0.25$ Hα

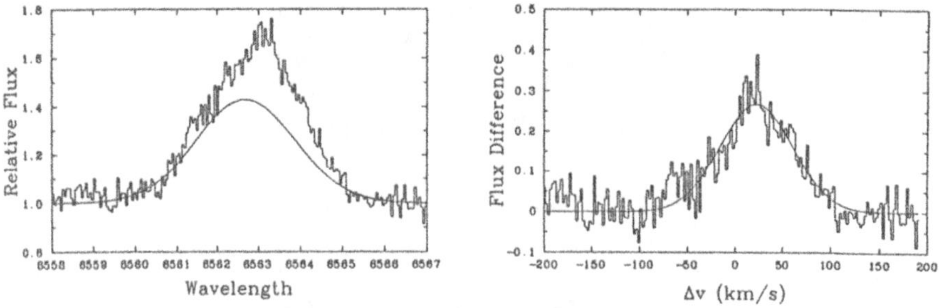

Fig. 4. (*a*) Plot of the Hα emission line profile at $\varphi \approx 0.11$ (maximum EW) (*histogram*) compared with a gaussian fit to the mean profile $\varphi \approx 0.8$-0.9 (*continuous curve*) and (*b*) the difference between these curves plotted on a velocity scale.

decreased again as did λ_c. A small "standstill" in EW at $\varphi \approx 0.18$ is also present in λ_c. Between $\varphi \approx 0.25 - 0.3$ Hα EW recovers and λ_c shifts to the red again. Between $\varphi \approx 0.3 - 0.4$ Hα continues to decrease but λ_c continues redward for a further ≈ 0.07 in phase before moving blueward.

Individual exposures at $\varphi \approx 0.1$ indicate an enhancement, strongly skewed to the red, and reaching a maximum in ≈ 20 min. Fig. 4a compares this maximum profile to the mean gaussian fit to the pre-outburst emission. There is an obvious excess flux to the red of line peak. Fig. 4b shows the difference between the pre-outburst profile and that at maximum EW plotted in velocity with respect to the Hα rest velocity. A gaussian fit indicates $v \approx +22.5\,\mathrm{km\,sec^{-1}}$. This is almost certainly a flare situated at a longitude such that its projected rotational velocity is $+22.5\,\mathrm{km\,sec^{-1}}$.

Acknowledgements

Research at Armagh Observatory is supported by a Grant-in-Aid from the Department of Education of Northern Ireland.

References

Bopp, B.W., Dempsey, R.C., Africano, J.L., Goodrich, B.D.: 1988, *PASP* **100**, 579

Byrne, P.B.: 1992, in P.B. Byrne & D.J. Mullan, ed(s)., *Surface Inhomogeneities in Late-Type Stars*, Springer Verlag: Heidelberg, 3

Byrne, P.B., McKay, D.: 1990, *A&A* **227**, 490

Byrne, P.B., Doyle, J.G., Mathioudakis, M.: 1992, in M.S. Giampapa & J. Bookbinder, ed(s)., *Proc. 7th Cool Star Workshop*, ASP Conf. Series, in press

Doyle, J.G., Cameron, A.C.: 1990, *MNRAS* **244**, 291

Friedemann, C., Gurtler, J.: 1975, *Astron. Nachr.* **296**, 125

Pettersen, B.R., Lambert, D.L., Tomkin, J., Sandmann, W.H., Lin, H.: 1987, *A&A* **183**, 66

Young, A., Skumanich, A., Harlan, E.: 1984, *ApJ* **282**, 683

Young, A., Skumanich, A., McGregor, K.B., Temple, S.: 1990, *ApJ* **349**, 608

IMAGING THE PHOTOSPHERE AND CHROMOSPHERE OF THE RS CVn STAR, II Peg

P.B. BYRNE, A.C. LANZAFAME and P.M. PANAGI
Armagh Observatory, Armagh BT61 9DG, N. Ireland

D.W. KILKENNY, F. MARANG, G. ROBERTS and F. VAN WYK
South African Astronomical Observatory, Observatory 7935, Sth. Africa

and

S. AVGOLOUPIS, L.N. MAVRIDIS and J.H. SEIRADAKIS
University of Thessaloniki, GR-54006 Thessaloniki, Greece

Abstract. We present simultaneous starspot photometry and high resolution Hα spectroscopy of II Peg which gives evidence of differential motions in the chromosphere.

Key words: starspots – chromospheres – II Peg

1. Introduction

II Peg is a single-line RS CVn binary of spectral type K4IV and period 6.72422d (Vogt, 1981). Light variations with the same period are caused by starspots whose growth and decay change the light curve with time (Byrne 1992).

UV studies show that, while chromospheric MgII h&k emission varies in antiphase with visible light, suggesting an association of chromospheric heating with spots, transition region emission lines vary more randomly (Rodonò et al. 1987, Byrne et al. 1987). Thus the localization of heating associated with spots changes between the temperature of formation of MgII h&k (≈ 8 000 K) and CII (≈ 30 000 K).

Hα emission, generally described as chromospheric, is formed in a broad region of the atmosphere extending up to ≈ 15 000 K (Houdebine and Panagi, 1990). Bopp and Noah (1980), Vogt (1981) and others have shown that Hα is highly variable. Byrne et al. (1989) described variations of Hα, CaII H&K and CIV(λ1548/51Å) over half a rotation, along with optical photometry to determine the spot distribution. They showed that H&K varied in antiphase with the optical (agreeing with the MgII results) but Hα was more erratic.

Here we describe a high spectral resolution study of the Hα, Hβ and HeI D3 lines on 6 consecutive nights along with contemporaneous optical photometry. We derive the spot distribution and discuss the appearance and variation of the emission lines.

2. Photometric spot distribution

II Peg was photometered on 25 nights at the South African Astronomical Observatory (SAAO) in UBV$(RI)_{KC}$ and on 6 nights in UBV at Stephanion Observatory, Greece. The V data are presented in Fig. 1.

A model spot distribution was made using the program SPOTPIC (based on Friedemann and Gurtler (1975)) with an inclination of the rotational axis to the plane of the sky of 30°. To minimize spot sizes their temperatures were set to 0 K, they were required to cross disk centre and were assumed circular (see Byrne, 1992).

J.F. Linsky and S. Serio (eds.), Physics of Solar and Stellar Coronae, 439–442.

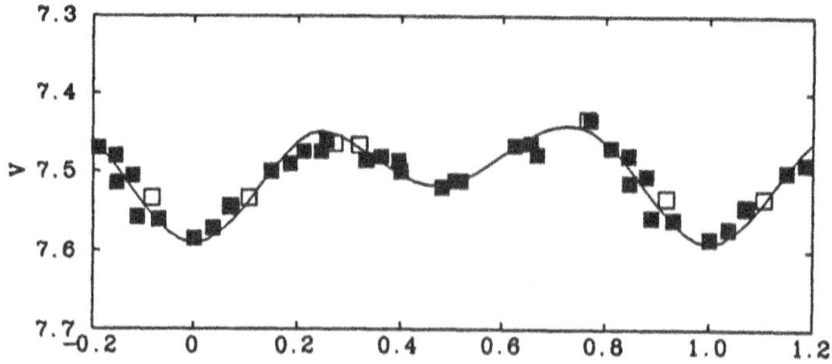

Fig. 1. V light curve for II Peg based on data taken at SAAO (*filled symbols*) and Stephanion Observatory (*open symbols*) phased according to Vogt's (1981) ephemeris. The solid line is the spot model described in Section 2.

Two major spot groups are present, one at $\varphi \approx 0$, covering $\geq 2.4\%$ of the surface, the other at $\varphi \approx 0.5$, occupying $\geq 1.3\%$.

Photometric imaging measures only spots which modulate visible light and not any distribution which is quasi-uniform in longitude or at high latitude (Byrne, 1992). II Peg had $V_{max} \approx 7.30$ in 1986 and 1989 (Byrne and Marang, 1987) and ≈ 7.18 in 1974 (Chugainov, 1976). Undetected spots could be responsible for depressing V_{max} from 7.18 to the 1991 value of 7.43 requiring a further, unresolved area coverage of at least 20% at the time of our observations.

3. The Hα line

Spectroscopy was obtained 24 – 30 August 1991 using the 1 m JKT on La Palma equipped with the QUB Echelle Spectroscope (QUBES) (Byrne et al. 1991) at a spectral resolution $\approx 50\,000$. 4 – 6 individual spectra of each line, each of exposure 20 – 30 min, were taken per night. Terrestrial water lines were removed by reference to B star spectra and the resultant spectra, summed to give nightly mean profiles, will be found in Fig. 2.

In Fig 3 we show a six-night mean profile which has a gaussian core, of FWHM $\approx$ 2Å and EW $\approx$ 0.9Å, with an asymmetric central reversal and excess flux in the wings, the blue wing being more prominent. Superimposed is a two-component gaussian fit to accentuate the asymmetric. In Fig 2 two-component gaussians, widths fixed to that of the mean profile, are superimposed on the data. A red asymmetry of the central reversal is present at all phases and is highly variable. The wings show a blue excess which also varies with phase.

It is difficult to interpret the Hα profiles with a stationary atmosphere. Rotational doppler broadening in II Peg ($v\sin i = 21\,\mathrm{km\,sec^{-1}}$, FWHM $\approx$ 0.8Å) is a small contributor to the observed profile and will be symmetric. Accordingly we are forced to invoke velocity fields in the Hα-forming region. The central reversal is formed near the top of the Hα-forming region, while the wings are formed deeper. Thus pending detailed calculations, these profiles suggest a downflow near 15 000 K

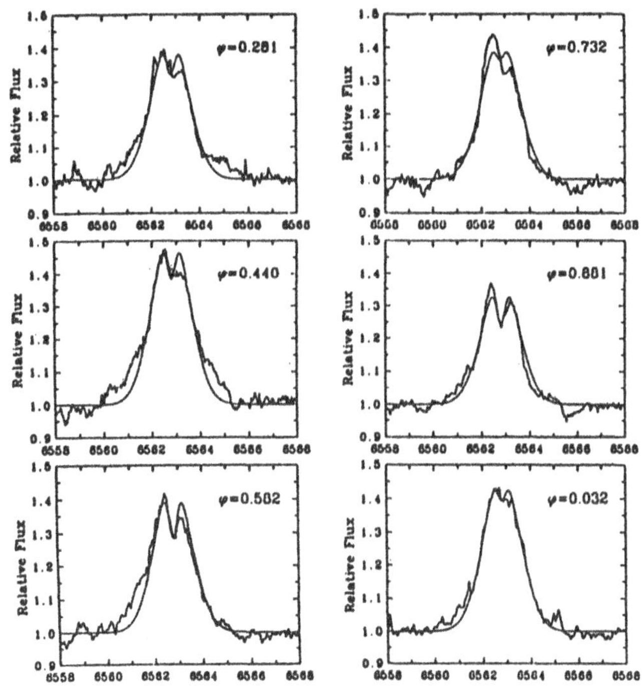

Fig. 2. The nightly mean Hα emission line profiles (*histograms*) with two-gaussian fits superimposed as described in the text (*smooth line*).

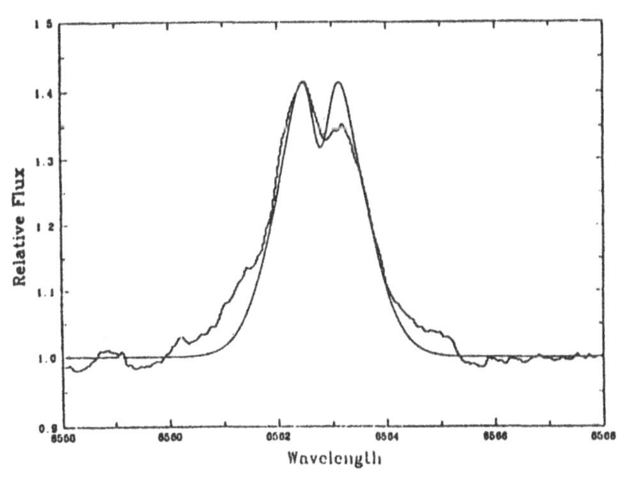

Fig. 3. The mean profile of the Hα emission line of II Peg in August 1991, corrected for the star's orbital velocity and then averaged over 6 nights of observation (*histogram*), compared to a symmetric two-gaussian fit centered on Hα's rest velocity (*smooth line*).

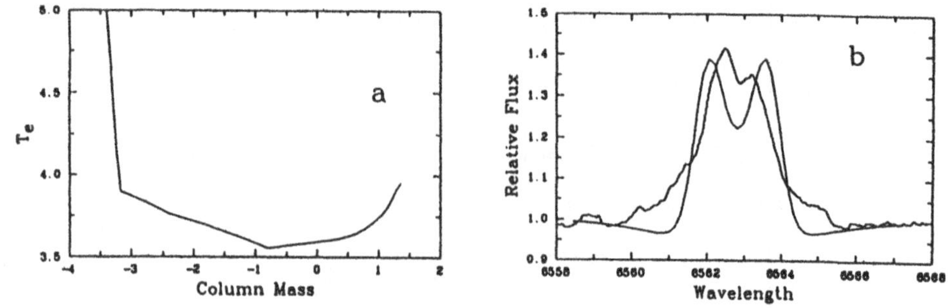

Fig. 4. (*a*) Atmospheric model for II Peg's chromosphere and lower transition region and (*b*) a comparison of the calculated and observed mean Hα profiles.

($v_{down} \approx 10\,\mathrm{km\,sec^{-1}}$) and upflow near 10 000 K ($v_{up} \approx 50\,\mathrm{km\,sec^{-1}}$) both variable in magnitude.

A model has been made of II Peg's atmosphere the NLTE code, MULTI (Carlsson, 1986), adopting the mean Hα profile and the non-detectibility of Hβ as constraints. The resultant atmospheric structure and Hα profile are shown in Fig. 4a, b respectively. The intensity and FWHM are reproduced but the asymmetries are not. Velocity fields are currently being included to improve these aspects of the fit.

The transition region electron density in this model ($N_e \approx 5\,10^8\,\mathrm{cm^{-3}}$ agrees well with that derived for II Peg by Byrne et al. (1987) from IUE observations.

Acknowledgements

Research at Armagh Observatory is supported by a Grant-in-Aid from the Department of Education of Northern Ireland.

References

Bopp, B.W., Noah, P.V.: 1980, *PASP* **92**, 333

Byrne, P.B., Coxhead, G., Doyle, J.G., Jordan, B.D., McCrory, M.: 1991, *QUBES User Manual* ,

Byrne, P.B.: 1992, in P.B. Byrne & D.J. Mullan, ed(s)., *Surface Inhomogeneities on Late-Type Stars*, Springer Verlag: Heidelberg, 3

Byrne, P.B., Marang, F.: 1987, *Irish AJ* **18**, 84

Byrne, P.B., Doyle, J.G., Brown, A., Linsky, J.L., Rodonò, M.: 1987, *A&A* **180**, 172

Byrne, P.B., Panagi, P.M., Doyle, J.G., Englebrecht, C.A., McMahan, R., Marang, F., Wegner, G.: 1989, *A&A* **214**, 227

Carlsson, M.: 1986, *Uppsala Astron. Obs. Publ.* , No. 33

Chugainov, P.F.: 1976, *Izv. Krimsk. Astron. Obs.* **54**, 89

Friedemann, C., Gurtler, J.: 1975, *Astron. Nachr.* **296**, 125

Houdebine, E.R., Panagi, P.M.: 1990, *A&A* **231**, 459

Huenemoerder, D.P., Ramsey, L.W.: 1987, *ApJ* **319**, 392

Rodonò, M., Byrne, P.B., Neff, J.E., Linsky, J.L., Simon, T., Butler, C.J., Catalano, S., Cutispoto, G., Doyle, J.G., Andrews, A.D., Gibson, D.M.: 1987, *A&A* **176**, 267

Vogt, S.S.: 1981, *ApJ* **247**, 975

SPATIAL CORRELATIONS BETWEEN CORONAL AND PHOTOSPHERIC ACTIVE REGIONS ON LATE-TYPE STARS

M. KÜRSTER and K. DENNERL

Max-Planck-Institut für extraterrestrische Physik, D-8046 Garching bei München, F.R.G.

Abstract. We report first results from a search for spatial correlations between coronal and photospheric active regions on late-type stars. This search is based on a comparison of ROSAT PSPC X-ray light curves with photospheric Doppler images. The observational data fits in the general picture that X-ray emission originates from loops connecting star spots or spot complexes.

Key words: Coronal X-ray emission – photospheric Doppler imaging

1. Introduction

On the Sun, the only star spatially resolvable by direct observation, the distribution of coronal X–ray emission is very patchy. Indications for similar inhomogeneities to exist on other active stars come from observations of variability due to changing visibility of active regions with stellar rotation (rotational modulation, RM).

From X–ray light curves representing the spatially unresolved flux of a star, only very crude models of the distribution of coronal regions are possible (*Ottmann et al.* 1992, *White et al.* 1990). We expect better constraints for coronal models to come from the comparison with observations performed at other wavelengths; however, the success of such multi–wavelength studies depends on the open question, if spatial correlations exist between activity regions on different atmospheric levels on stars other than the Sun. A comparison with the stellar photosphere is the 'natural' choice, since it can be mapped in detail by the Doppler imaging technique (e.g. *Vogt et al.* 1987, *Kürster* 1991, *Kürster and Schmitt* 1992).

We report here on our search for correlations of X–ray emission, as seen in ROSAT PSPC light curves, with photospheric spot distributions in near–simultaneous Doppler images. All Doppler imaging utilizes CaI λ6439 profiles obtained with the ESO CAT+CES within 6 weeks of the X–ray observations, a time span in which no substantial reconfigurations of the photospheric spot patterns should occur.

2. ROSAT X–ray Light Curves and Photospheric Doppler Images

2.1. AB Dor = HD 36705

AB Dor is a rapidly rotating single post–T Tau star of spectral type K0V. We present here (Fig. 1a) the X–ray light curve obtained from a 46 *ksec* pointed observation with the ROSAT PSPC (Oct 28–Nov 03, 1991). It has been folded with the rotation period (ephemeris HJD 244 4296.575 + 0.51479 *day*; *Innis et al.* 1988).

RM of the X–ray flux is quite evident. Contrary to our ROSAT all–sky survey light curve of AB Dor (*Kürster et al.* 1992a), which showed, at best, marginal evidence for RM, this new data clearly reveals phase–dependent variability with a maximum near phase $\Phi = 0.4$ (0.5 for the lower boundary of the light curve) and a minimum around $\Phi = 0.2$. Substantial erratic variation is superimposed. Scatter is large near the peak suggesting that much of it is caused by flaring events.

J.F. Linsky and S. Serio (eds.), Physics of Solar and Stellar Coronae, 443–446.

a)

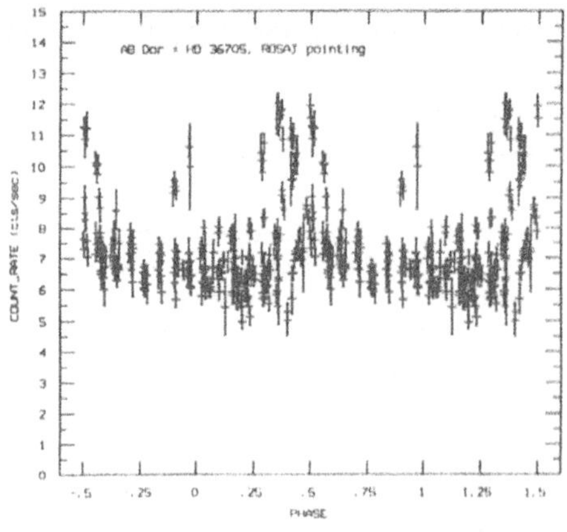

b)

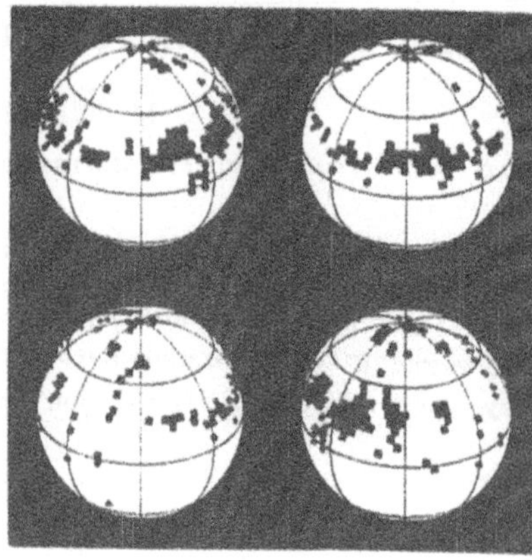

Fig. 1. AB Dor: a) A 46 *ksec* pointed ROSAT PSPC observation. X–ray light curve folded with rotation period and plotted vs. phases −0.5...1.5 (every data point is plotted twice). b) Doppler image of Nov 1991 (CaI λ6439). 4 different phases are shown: 0.00 (top left), 0.25 (top right), 0.50 (bottom left), and 0.75 (bottom right).

Fig. 1b shows a photospheric Doppler image of AB Dor produced by applying the CLEAN–like algorithm (*Kürster and Schmitt* 1992) to data from Nov 19–28, 1991. This new image (our 4th) confirms our earlier finding of an active latitude at $\approx +25°$ which we have observed since Feb 1989 (e.g. *Kürster et al.* 1992b).

Phase of the X–ray maximum coincides roughly with minimum spot visibility in the Doppler image. We suggest the following model for this behaviour. Assuming that the many spot features are connected by X–ray emitting loops, no strong RM of the observed X–ray flux should be produced by that section of the low–latitude belt that is almost gap–free. The gap itself can be bridged by quite extended loops which (since pressure scale height on AB Dor is ≈ 1 stellar radius for $T = 10^7\,K$) could be largely filled with X–ray emitting plasma. This feature would then produce the bulk of the observed RM with a maximum at $\Phi = 0.5$.

2.2. CF Tuc = HD 5303

CF Tuc is a partially eclipsing RS CVn–type binary with an inclination suitable for Doppler imaging (we find $i = 64°$). Spectral types are G0IV+K4IV. Fig. 2a shows the PSPC X–ray light curve extracted from the ROSAT all–sky survey and folded with the ephemeris HJD 244 4555.009 + 2.797672 *day* (period from *Budding* 1985; phase of conjunction refined with our radial velocity data). The observation is from Oct 19–24, 1990. Data points from 2 consecutive cycles exist for $\Phi = 0.0 - 0.73$.

Also shown in Fig. 2a is the light curve calculated from the best–fitting coronal two–spot model. The modelling assumes the X–ray emission to originate from the presumably more active K4IV–star only. Although the X–ray flux is low during eclipse phases the observed variation is produced by RM rather than actual eclipses which affect only a small fraction of the K–star surface.

Fig. 2b shows the first Doppler image ever presented for CF Tuc (K–star, obser-

a) 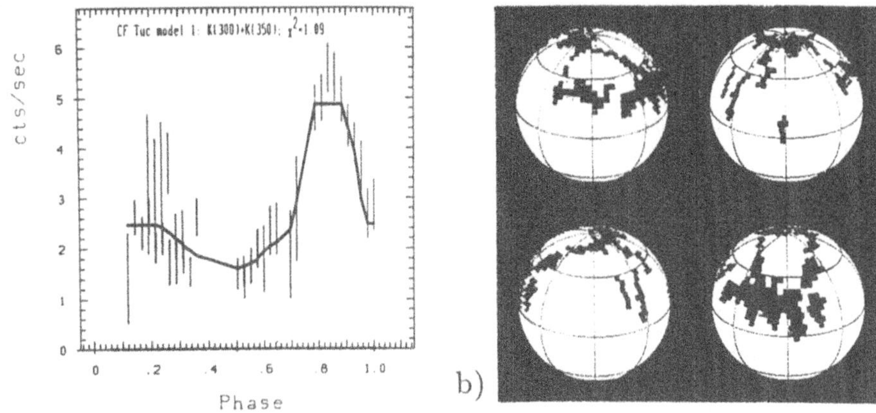

Fig. 2. a) ROSAT PSPC X-ray light curve of CF Tuc from the all-sky survey folded with the (synchronized) orbital/rotational period. Data points from two consecutive cycles overlap during $\Phi = 0.0 - 0.73$. The solid line is calculated from the best-fitting two-spot model. b) Doppler image of CF Tuc (Sep 1990, CaI λ6439). Phases are ordered as in Fig. 1.

vations from Sep 07–15, 1990). Most prominent is a large low-latitude spot complex, best visible at $\Phi = 0.8$, i.e. at the exact phase of X-ray maximum.

A comparison between the coronal model and the Doppler image reveals spatial proximity of the dominant features. In our modelling both coronal features are circular and have zero height. Feature 1 (2) is located at longitude 300° (350°) or phase 0.833 (0.972) and latitude −59° (+70°), its radius is 5° (25°), and it contributes 2.40 $cts\,sec^{-1}$ (2.45 $cts\,sec^{-1}$) to the total count rate.

This places feature 2 well between the close-to-polar spotted (photospheric) region and the large low-latitude spot complex. Feature 1 is centered exactly on the phase of maximum visibility of the large spot-complex, but at a lower latitude. This can be explained in view of the latitude uncertainty inherent in Doppler imaging of higher inclination stars ('mirroring effect'); it is possible that part of the large photospheric spot complex is actually near latitude −25° rather than +25°; hence, it would be closer to the coronal feature of same longitude which may consist of loops stretching towards the region near the invisible pole.

2.3. YY Men = HD 32918

YY Men is a single FK Com-type giant. Fig. 3a shows the ROSAT PSPC all-sky survey light curve of Sep 30–Oct 12, 1990 plotted vs. time in units of stellar rotation phase (ephemeris HJD 244 4155.64 + 9.5476 *day*; *Collier* 1982). Phase-dependent variability occurs with a maximum at $\Phi = 0.3 - 0.5$ and a minimum near $\Phi = 0.8$.

Fig. 3b shows the Doppler image based on data from Sep 09–18, 1990 (already presented by *Kürster et al.* 1992b). It contains a low-latitude activity belt and a polar spot with a long appendage extending almost all the way down to the active belt. Regions in the image where the belt seems to form a continuous band are likely to be somewhat affected from effects of incomplete phase coverage in the data. The

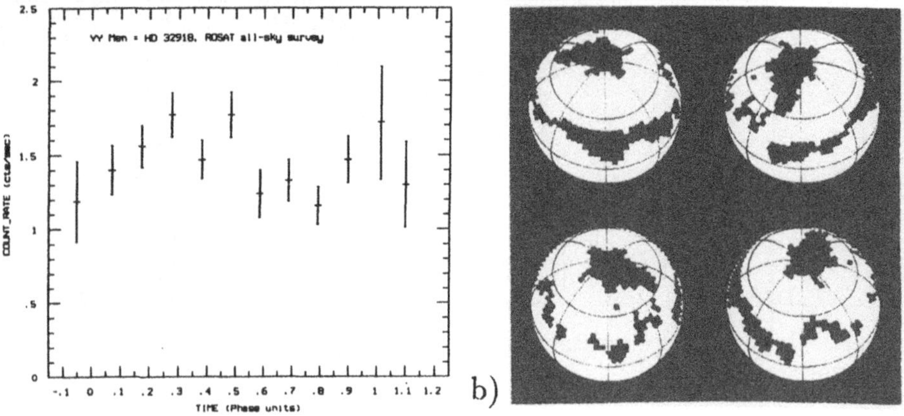

Fig. 3. a) ROSAT PSPC X-ray light curve of YY Men from the all-sky survey plotted vs. time in units of the stellar rotation period. b) Doppler image of YY Men (Sep 1990, CaI λ6439). Phases are ordered as in Fig. 1.

belt is probably dissolved into a multitude of spots as seen in the view of $\Phi = 0.5$.

We suggest the following scenario for YY Men: X-ray minimum coincides exactly with the phase at which the polar spot appendage points away from the observer (0.8). At this phase, an emission region, located between the end of the appendage and the active belt disappears on the back side of the star. The belt itself, possibly accompanied by coronal regions, does not contribute much RM.

3. Conclusion

Our data supports the concept of spatial correlations to exist between coronal and photospheric active regions, i.e. X-ray bright loops connecting star spot complexes. The fact that X-ray light curve and photospheric spot visibility are 'in phase' for CF Tuc and YY Men while they are 'in anti-phase' for AB Dor is the result of a differing spot distribution pattern and can be explained within the same model.

References

Budding, E.: 1985, *I.B.V.S. No.* 2729.

Collier A.C.: 1982, *M.N.R.A.S.*, **200**, 489.

Innis, J.L., Thompson, K., Coates, D.W., and Lloyd Evans, T.: 1988, *M.N.R.A.S.*, **235**, 1411.

Kürster, M.: 1991, in *Proc. ESO Workshop on Rapid Variability of OB-stars: Nature and Diagnostic Value*, Garching, 15–17 Oct 1990, D. Baade (ed.).

Kürster, M., and Schmitt, J.H.M.M.: 1992, in *Surface Inhomogeneities on Late-type Stars, Armagh Observatory, Northern Ireland* 1990, P.B. Byrne and D.J. Mullan (eds.), *Lecture Notes in Physics Series* 397, Springer, Berlin, Heidelberg, New York.

Kürster, M., Schmitt, J.H.M.M., and Fleming, T.A.: 1992a, in *Proc. Seventh Cambridge Workshop on Cool Stars, Stellar Systems, and the Sun*, Tucson, AZ, 9–12 Oct 1991.

Kürster, M., Hatzes, A.P., Pallavicini, R., and Randich, S.: 1992b, in *Proc. Seventh Cambridge Workshop on Cool Stars, Stellar Systems, and the Sun*, Tucson, AZ, 9–12 Oct 1991.

Ottmann, R., Schmitt, J.H.M.M., and Kürster, M.: 1992, *Ap. J.*, in press.

Vogt, S.S., Penrod, G.D., and Hatzes, A.P.: 1987, *Ap. J.*, **321**, 496.

White, N.E., Shafer, R.A., Horne, K., Parmar, A.N., and Culhane, J.L.: 1990, *Ap. J.*, **350**, 776.

MAGNETIC ACTIVITY OF LATE TYPE STARS

S. CATALANO

Osservatorio Astrofisico Catania
Viale A. Doria 6, 95125 Catania, Italy

Abstract. Several diagnostics are currently used to investigate the physical characteristics of magnetic activity in late type stars and their functional dependence on global stellar parameters. Chromospheric radiative losses have been shown to be one of the most conspicuous signatures of the surface magnetic flux, which in turn is determined by the dynamo efficiency. The empirical emission-rotation relation demonstrates how the dynamo efficiency depends on the rotation rate. The magnetic braking that causes the time dependence of the rotation rate of evolving stars actually determines the distribution of magnetic activity of single stars across the H-R diagram. The presence of boundary lines delineating regions in the H–R diagram with different activity and rotation behavior is discussed with major emphasis on the boundary for the onset of convection. It is argued that acoustic energy dissipation may be the dominant heating mechanism for A-F stars.

Key words: Stars: late type – Stars: activity

1. Introduction

The existence of solar–type activity in stars has been inferred on the basis of observed Ca II H and K emission in the spectra of late type stars. This simple analogy with the solar spectrum has stimulated in the last two decades a variety of studies, leading to an interesting interplay of solar and stellar investigations. The combination of solar and stellar observational material has deepened our understanding of the phenomenon now known as "magnetic activity". The association of solar bright chromospheric and transition region plages with local magnetic field and the brighter X-ray emission with closed magnetic loops suggests that these emission enhancements are due to large nonthermal heating rates that involve magnetohydrodynamic processes. The presence of chromospheric, transition region and coronal emission in the spectra of late type stars is considered as evidence for similar processes occurring in the atmospheres of stars other than the Sun. Moreover, activity signatures observed on the Sun, including dark photospheric spots and bright chromospheric and coronal plages, are probably the cause of variations in broad-band light and spectral line profiles that vary with rotational phase for solar type stars. Evidence has also been found for stellar activity cycles.

The discovery of tight empirical correlations between rotation and activity indicators (e.g. chromospheric and transition region emission line flux and coronal X-ray flux) has stimulated the developement of dynamo theory for the generation of magnetic fields in stellar interiors. Although, by analogy with the Sun, it has been assumed that stellar activity is also magnetic in nature, only in recent years have magnetic field strengths and surface filling factors been measured in solar type stars. This review will highlight some of the observational evidence for stellar dynamos and will discuss the limitations and possibilities for quantitative developments that could extend the present qualitative scenario.

J.F. Linsky and S. Serio (eds.), Physics of Solar and Stellar Coronae, 447–456.

2. Is Stellar Activity Really Magnetic?

Is the activity in stars really magnetic in character? To what extent is it similar to solar activity? What is the observational evidence for stellar activity? The answers to these questions are given by the magnetic field measurements now available for several cool stars (Robinson 1980; Marcy 1984; and the summary by Saar 1992).

The average magnetic field strengths show a significant range in fB, where f is the filling factor, for dwarfs of spectral type G–M, but no fields have yet been detected in F-type and non-emission M-type dwarfs or in giants and supergiants (Saar 1992). This may not mean that fields are absent, because they could be below the present detection threshold. In any case, reduced magnetic activity should be considered for these stars as we shall see later. There is an apparent anti-correlation between B and T_{eff}, which is interpreted as a consequence of the equipartition condition where the gas pressure (P_{gas}) confines the magnetic field in the photosphere (i.e. $B{=}B_{eq} = (8\pi P_{gas})^{0.5}$) (Linsky and Saar 1987; Saar 1991).

The existence of a relationship between the mean absolute magnetic flux density and the atmospheric radiative losses has been found to hold also in the stellar case (Marcy 1984; Linsky and Saar 1987). The best fit power law relations given by Saar (1992) are $\Delta F_{\rm CaII} \propto (f_p B)^{0.56}$ and $F_{\rm X} \propto (f_p B)^{0.92}$. This Ca II correlation is consistent with the $\Delta F_{\rm CaII} \propto (f_p B)^{0.60}$ found by Schrijver *et al.* (1989) for solar active regions. Knowing the relationship between atmospheric radiative losses and the magnetic flux density one may use any atmospheric diagnostic (e.g. luminosity, surface flux, or normalized flux to bolometric flux) as a measure of the magnetic activity to make quantitative comparisons and to infer scaling parameters.

An important step toward determining that solar and stellar magnetic activity are the same phenomenon is provided by observations of the structured distribution of stellar surface magnetic fields. Analysis of the profile modulations of lines sensitive to temperature and magnetic field has been succesful in mapping the temperature and the magnetic field in the photospheres of rapidly rotating stars. The greatest concentrations of magnetic flux fB are found to be near, but somewhat displaced from, the centers of cool regions; for example, on HD 82558 (Saar *et al.* 1992). The connection between the photospheric magnetic field and the chromospheric activity tracers (Ca II, Mg II, C IV, C II) was seen in the active dwarfs ϵ Eri and ξ Boo A. The emission lines and magnetic flux appear to be modulated in phase with the rotation, confirming a solar-type structure in which bright plages overlie photospheric magnetic regions.

3. Activity Diagnostics and Dynamo Models

Convection and rotation appear to be the most significant parameters that control dynamo action in stars and the associated magnetic phenomena. This statement suggests that a dynamo process operates whenever a rotating star has an outer convection zone, i.e. it implies a dependence of dynamos on the internal structure and age of stars. Dynamo models can be characterized by dimensionless parameters: the dynamo number N_D (Parker 1979), given by the squared ratio of the diffusion time (T_D) to the amplification time (T_A), and the Rossby number (R_o), defined as the ratio of the rotation time of the convective zone (P_{conv}) to the convection turnover time (τ_c). These two parameters are simply related by $N_D \approx R_o^{-2}$ (Durney and Latour 1978). Magnetic activity is expect to be related somehow to these parameters.

Several activity indicators can be used to study dynamos, to isolate the effect of a given parameter, and to identify functional dependencies. Let me now analyze the various pieces of evidence for the dynamo parameters and check the consistency of results from different magnetic activity indicators including the magnetic field itself, the associated excess flux, spots, and flares.

3.1. Magnetic Fields

The magnetic field intensity B does not appear to show any clear dependence on the stellar angular velocity, Ω, or on the inverse Rossby parameter, $\tau_c\Omega$ (Linsky and Saar 1987). On the other hand, the magnetic flux density, fB, correlates well with both Ω and $\tau_c\Omega$ (Saar 1987, 1991) with power law dependences. Actually the correlation is with the magnetic filling factor f.

3.2. Excess Emission Flux

The simple proportionality between the Ca II emission and the angular rotation Ω suggested by Skumanich (1972) has been studied further using more accurate data. Noyes *et al.* (1984) showed that the Ca II HK line flux normalized to the surface flux R'_{HK} measured with the Mt. Wilson photometer and corrected for the photospheric contribution outside H_1 and K_1) is a smooth function of the Rossby number. The correlation appears to be independent of the stellar mass, and it applies to main sequence stars. A power law relation would represent equally well the corrected Ca II chromospheric flux F_{HK} as a function of the rotation period P_{rot} (Noyes *et al.* 1984), but with a slightly larger scatter. A similar relation was found by Hartmann *et al.* (1984) using Mg II h and k line fluxes normalized to the bolometric flux.

Controversial correlations have been presented for coronal diagnostics. Power law dependences of X-ray luminosity on vsini, with different exponents (between 2 and 3) have been proposed (Ayres and Linsky 1980; Pallavicini *et al.* 1981; Walter 1982). An alternative description of the data involves an exponential decay of R_X with Ω (Walter 1982; Caillault and Helfand 1985).

Marilli and Catalano (1984) showed that when an exponential relation between emission luminosity and rotation period is adopted a unique functional dependence holds for all chromospheric, TR, and coronal diagnostics with a slope that is consistent with the flux–flux correlations. Recently Barry (1991) improved this unified view by proposing correlations with the Rossby number.

An important advance in our understanding of the role played by the Rossby number was made by Marilli, Catalano and Trigilio (1986) who showed that the H and K line luminosity can be represented by exponential relations of the type:

$$\log L_{HK} = -aP_{rot} + b,$$

where both the a and b coefficients are color dependent. The inverse of coefficient a, which has the dimension of time, has been related to the convective turnover time. Excellent agreement was found with the value τ_c computed by Gilman (1980) for a ratio of the mixing length to the pressure scale height $\alpha = 2$. The high degree of correlation found for each of the B-V bins suggests that an exponential relation could be the most suitable representation of the magnetic activity dependence on rotation. This result could be improved to determine an empirical convective turnover time.

3.3. Star spots

Large–scale dark spots are detected from light variability in several types of stars, including RS CVn, BY Dra, and K and M dwarfs in young open clusters. Taking the maximum amplitude of light variation as a spot coverage parameter, it is possible to infer spot areas and locations and their dependence on rotation.

The largest amplitude $\Delta V > 0.2$ mag is found in the rapidly rotating ($P \leq 5^d$) stars found in the field and in young open clusters (Stauffer *et al.* 1986; Radick *et al.* 1987). Field G – K single stars are characterized by marginal variability with amplitudes never exceeding a few hundredths of magnitudes (Giglas *et al.* 1962; Blanco, Catalano and Marilli 1979; Lockwood *et al.* 1984). On the average these stars are relatively slow rotators with $5 < P < 20$ days.

Hall's (1991) extensive analysis of the variability in dwarf, subgiant and giant stars showed that spottedness occurs when the Rossby number is less than about 2/3. For his analysis he adopted the convective turnover time computed by Gilliland (1985) as a function of B–V for dwarfs, but he had to scale the corresponding values by a factor of 2 and 4 for subgiants and giants.

3.4. Flares

Flaring activity has been reported in many types of stars across the HR diagram at different wavelengths (see Pettersen 1989). Flares are seen at early phases in the evolution of stars and flaring is a common phenomenon in young stars. The largest numbers of flare stars have been found among stars of spectral type dMe (i.e. the classical UV Ceti-type flare stars). Statistical analyses of flare activity (flare frequency or maximum flare luminosity) as a function of stellar rotation are not yet available. However, the high frequency of flare stars among the dMe stars, the almost total absence of flaring activity among the nonemission and more slowly rotating dM stars, and the decrease in flaring activity with age all clearly indicate that rotation is an important parameter for flare activity. Stellar convection parameters appear also to affect the flaring activity. Pettersen (1989) found a well defined empirical relationship between flare luminosity and the volume of the convection zone for a number of solar neighbourhood stars.

Although a number of empirical correlations between activity indicators and rotation have been proposed, the situation is complicated by the mutual dependence of global stellar parameters upon each other. This makes it very difficult to disentangle the effects of individual crucial parameters. For example, strong evidence for the Rossby number as the key parameter for magnetic activity has been shown by Noyes *et al.* (1984), Mangeney and Praderie (1984), Marilli, Catalano and Trigilio (1986) and Basri (1987). However, the convective turnover time that defines the Rossby number is strongly dependent on the adopted characteristic length $L = \alpha H$, where α is the ratio of mixing length to pressure scale height. The present data on activity-rotation correlation, even with some uncertainty seem to support the value $\alpha = 2$ (Noyes *et al.* 1984; Marilli, Catalano and Trigilio 1986; Catalano 1988).

The relevance of a Rossby number dependence becomes uncertain, however, when giants are included. Although Rutten (1987) claims that dwarfs and single giants fall on the same curve of rotation vs B–V color, no convincing correlation of chromospheric and coronal flux with rotation or the Rossby number has been found for giants (Basri 1987).

4. Magnetic Braking

The stellar wind flowing in the magnetosphere along field lines that open to the interstellar space drains angular momentum and thus brakes stellar rotation. Low–mass stars with convective envelopes (i.e. $M \leq 1.3\ M_{\odot}$) lose a substantial fraction of their angular momentum during their main–sequence life time. More massive stars experience angular momentum loss only after they cross the *granulation boundary* when they develop a solar–like dynamo. The decrease in rotation rate for these stars is affected by evolutionary changes in addition to the angular momentum loss. The relative effects and the interpretation of the observational scenario have been recently summarized by Gray (1991).

Since no significant structural changes take place during the main–sequence in low–mass stars, their rotational decay rate is controlled by the dynamo action with interesting consequences. The rotational decay with time of low–mass main–sequence stars, displayed in Fig. 1 can be summarized as follows (Catalano *et al.* 1988):

- the decay of rotation proportional to $t^{-1/2}$, as proposed by Skumanich (1972) for solar mass stars, holds for low–mass stars at least in the range 1.1–0.5 $M_{\odot}$;
- the rate of spin–down increases as the mass decreases;
- the rotation period for a given star is tightly determined by its mass and age.

Theoretical rotation evolution models based on a simple wind model, with axisymmetric radial structure and dynamo generated surface magnetic fields, explain quite well the observed decay with $t^{-1/2}$ (Weber and Davis 1967). Models that include internal angular momentum transport and mixing give reasonable fits to the rotational evolution of old solar–type stars, but do not explain satisfactorily the angular momentum evolution of young rapidly rotating stars (Pinsonneault *et al.* 1989; Collier–Cameron *et al.* 1991; Sofia *et al.* 1991). Critical questions that models must address are the magnetic field dependence on rotation, the field topology as derived from dynamo models, and the internal angular momentum transport at different rotation regimes.

The main physical quantities entering the rotational braking model equations,

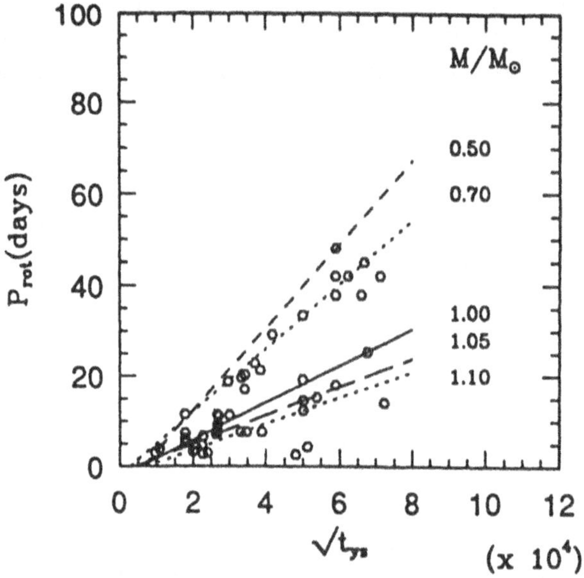

Fig. 1. Observed rotation period for a subset of stars with known age. Lines are fits for stars of constant mass (Trigilio *et al.* 1991).

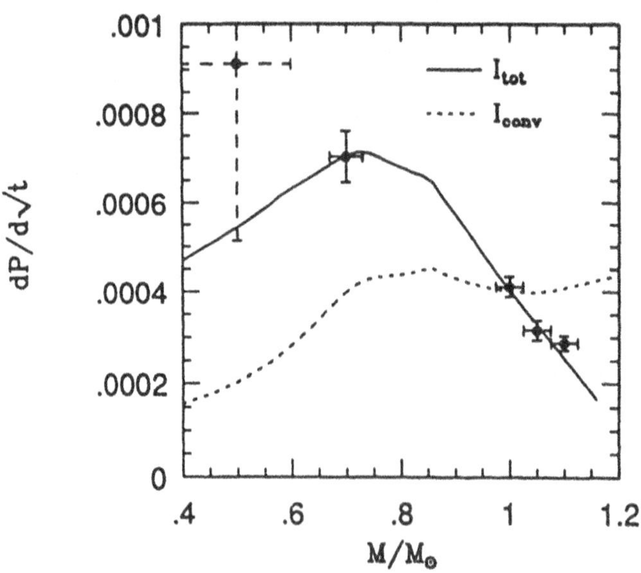

Fig. 2. Empirical coefficients of the rotation decay for different mass groups and computed coefficients for braking confined to the convection zone (dotted line) and braking of the whole star (solid line).

like the magnetic field strength, the covering factor, and convective turnover time, are now available from observations; therefore, it is possible to check these models for consistency with the empirical dependencies deduced from independent observations. Trigilio *et al.* (1991) made simple calculations to compute the rotation decay rate for different stellar masses using the observed magnetic field strengths $fB_{\rm eq}$ and the empirical correlation between the filling factor f and rotation from Saar (1990). The results based on a radial field model show that (Fig. 2) the observed rotation decay rate is fairly well reproduced for stars up to 0.7 $M_\odot$ under the assumption that the whole star is slowed down during the main-sequence life time. The assumption that only the convection zone is braked leads to results that are completely discordant with respect to the observed decay rates. The result that the whole star is slowing down during the main-sequence is indicative of a time scale for the internal angular momentum transport that is much shorter than the main-sequence life time. A core-envelope synchronization time of 10 Myr has been estimated by MacGregor and Brenner (1991) from the evolution time scale for rotation in the young α Per, Pleiades and Hyades clusters, but longer time scales may characterize slowly rotating stars (Sofia *et al.* 1987).

5. Dividing Lines and the Onset of Activity

Linsky and Haisch (1979) first discovered the absence of high excitation temperature UV lines in the spectra of late K giant and supergiants. They proposed that the disappearance of these line occurs along a well-defined boundary called the *coronal dividing line.* Several other *dividing lines* have been proposed to identify boundaries in the H–R diagram where various magnetic activity indicators first appear or disappear. Gray (1991) introduced the *rotation dividing line* that divides typically fast rotators (to the left) from slow rotators (to the right). He also proposed the so-called *rotostat* mechanism in which magnetic activity disappears once the rotational velocity in the convection zone drops below a minimum value, but can reappear if the transfer of angular momentum from the core raises the convection zone rotation above that limit. More likely, this complex behavior is the star's response to the combined effects of magnetic braking and evolutionary changes rather than to any fundamental changes in the dynamo.

An important dividing line is the *granulation boundary*, which should mark the onset of convection and dynamo. However there is some evidence for the existence of a weak dynamo immediately blueward of the *granulation boundary.* The discussion relies on the reality of the proposed dependence of activity indicators on rotation for early spectral types. The activity–rotation correlation has been suggested to start at middle F main–sequence stars (Walter 1983; Wolff *et al.* 1986; Simon and Landsman 1991). However, Walter and Schrijver (1987) present direct evidence for a relationship between rotation and C II emission for B–V as small as $\approx$0.40 (spectral type F4), after an appropriate correction for the basal flux. Moreover Schmitt *et al.* (1985) found that coronal X-ray emission may occur in stars as early as spectral type A5, and they show that a fairly good correlation with the Rossby number may hold also for these stars. Since it is well established that X–ray emission originates in magnetic coronal loops of trapped plasma, the detection of x–rays should be

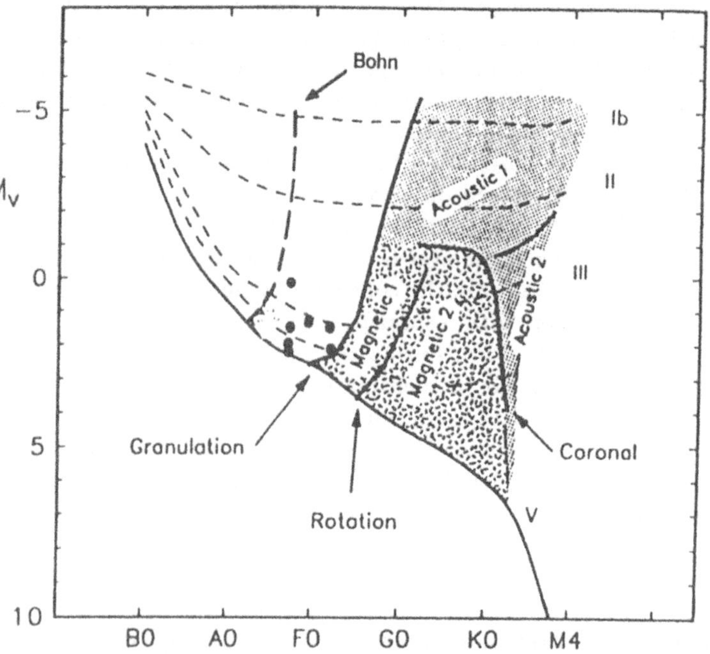

Fig. 3. Magnetic activity on the H–R diagram, after Gray (1991). The dashed lines have been added to indicate the limits for acoustic energy generation computed by Bohn (1984). Filled circles are stars with detected Ly–α emission

considered as evidence that magnetic fields are present in stars as early as spectral type A5.

The *granulation boundary* should be considered as a boundary for the onset of deep convection zones, not for the convection itself. Recent IUE observations of late A-type stars (Catalano *et al.* 1991; Marilli *et al.* 1991; Freire Ferrero *et al.* this meeting) show that Ly–α emission is present in stars as early as spectral type A7, far away from the granulation line (Fig. 3). They also found a slow decrease in the Ly-α flux for B–V < 0.5, consistent with the more pronounced decrease observed in the X–ray flux (Micela *et al.* 1988) and indicating a change in the relative behavior of chromospheres and coronae. This different behavior of chromospheres and coronae at early spectral types is better described by the ratio of the EUV luminosity to the Ly–α luminosity (cf. Fig. 2 in the paper by Freire Ferrero *et al.* in this volume). Note that the emission decrease starts at about the same spectral type as where the emission–rotation correlation seems to cease. This may indicate a real decrease of the dynamo efficiency in bluer stars due to the decrease of the convection zone depth. The acoustic energy release would compete with the magnetic heating and become the principal nonthermal heating source in early-type stars.

Chromospheres and coronae seem to vanish as one proceeds from the early F to the A-type stars and probably completely disappear around spectral type A6 or

A5, where the *convection boundary* may in fact be located, as indicated by Bohn's (1984) calculations.

Acknowledgements

This work was supported by the Catania Astrophysical Observatory, the Italian *Ministero dell' Università e della Ricerca Scientifica e Tecnologica,* the *Gruppo Nazionale di Astronomia* of the *Consiglio Nazionale delle Ricerche*, and by the *Regione Sicilia*

References

Ayres, T.R., Linsky, J.L.: 1980, *Astrophys. J.* **241**, 279.

Barry, D.J.: 1991, preprint.

Basri G.S.: 1987, *Astrophys. J.* **316**, 377.

Blanco, C., Catalano, S., Marilli E.: 1979, *Astron. Astrophys. Suppl. Series* **36**, 297.

Bohn, H.U.: 1984, *Astron. Astrophys.* **136**, 338.

Caillault, J.P., Helfand, D.J.: 1985, *Astrophys. J.* **289**, 279.

Catalano, S.: 1988, *Irish Astron. J.* **18**, 265.

Catalano, S., Marilli, E., Trigilio, C.: 1988, in *NATO ASI Conf. Proc., Formation and Evolution of Low-mass Stars*, A.K. Dupree, M.T.V.T. Lago, eds., Kluwer, Dordrecht, p. 377.

Catalano, S., Gouttebroze, P., Marilli, E., Freire Ferrero, R.: 1991, in *The Sun and Cool Stars: Activity, Magnetism, Dynamos*, I. Tuominen, D. Moss, G. Rüdiger, eds., Springer-Verlag, Berlin, p. 466

Collier-Cameron, A., Jianke, L., Mestel, L.: 1991, in *Angular Momentum Evolution of Young Stars*, S. Catalano, J.R. Stauffer, eds, Kluwer, Dordrecht, p. 297.

Durney, B.R., Latour, J.: 1978, *Geophys. Astrophys. Fluid Dyn.* **9**, 241.

Giglas, H., Burnham, R., Thomas, N.: 1962, *Lowell Obs. Bull.* **5**, 257.

Gilliland, R.:1985, *Astrophys. J.* **299**, 286.

Gilman, P.: 1980, in *Stellar Turbulence, IAU Coll. N. 51*, D.F. Gray, J.L. Linsky, eds, Springer-Verlag, New York, p.19.

Gray D.F.: 1991, in *Angular Momentum Evolution of Young Stars*, S. Catalano, J.R. Stauffer, eds, Kluwer, Dordrecht, p. 183.

Hall, D.S.: 1991, in *The Sun and Cool Stars: Activity, Magnetism, Dynamos*, I. Tuominen, D. Moss, G. Rüdiger, eds., Springer-Verlag, Berlin, p. 353.

Hartmann, L., Baliunas, S.L., Duncan, D.K., Noyes, R.W.: 1984, *Astrophys. J.* **279**,778.

Linsky, J.L.,Saar, S.H.: 1987, in *Cool Stars, Stellar Systems and the Sun*, J.L. Linsky, R.E. Stencel, eds., Springer-Verlag Berlin, p. 44.

Linsky, J.L., Haisch, B.M.: 1979, *Astrophys. J.* **229**, L27.

Lockwood, G.W., Thompson, D.T., Radick, R.R., et al.: 1987, *Publ. Astron. Soc. Pacific* **96**, 714.

MacGregor, K.B., Brenner, M.: 1991, *Astrophys. J.* **376**, 204

Mangeney, A., Praderie, F.: 1984, *Astron. Astrophys.* **130**, 143.

Marcy, G.W.: 1984, *Astrophys. J.* **276**, 286.

Marilli, E., Catalano S.: 1984, *Astron. Astrophys.* **133**, 57.

Marilli, E., Catalano, S., Trigilio, C.: 1986, *Astron. Astrophys.* **167**, 297.

Marilli, E., Catalano, S., Freire Ferrero, R., Gouttebroze, P., Bruhweiler, F., Talavera, A.: 1991, in *Cool Stars, Stellar Systems and the Sun*, M.S. Giampapa, J.A. Bookbinder, eds., ASP Conf. Ser., San Francisco, p. 178.

Micela, G. et al.: 1988, *Astrophys J* **325**, 798.

Noyes R.W., Hartmann L.W., Baliunas S.L., Duncan D.K., Vaughan A.H.: 1984, *Astrophys. J.* **279**, 763.

Pallavicini R., Golub L., Rosner R., Vaiana G.S., Ayres T.R., Linsky J.L.: 1981, *Astrophys. J.* **248**, 279.

Parker E.N.: 1979, in *Cosmic Magnetic Fields: their origin and their activity*, Clarendon Press, Oxford.

Pettersen, B.R.: 1989, in *Solar and Stellar Flares, IAU Coll 104*, B.M. Haisch, M. Rodonò, eds., p. 299.

Pinsonneault, M. H., Kawaler, D.S., Sofia, S., Demarque,P.: 1989, *Astrophys. J.* **338**, 424.

Radick, R.R., Thompson, D.T., Loockwood, G.W., Duncan, D.K., Bagget, W.E.: 1987, *Astrophys. J.* **321**, 459

Robinson, R.D.: 1980, *Astrophys. J.* **239**, 961.

Rutten, R.G.M.: 1987, *Astron. Astrophys.* **177**, 131.

Saar, S.H.: 1987, in *Cool Stars, Stellar Systems and the Sun*, J.L. Linsky, R.E. Stencel, eds., Springer-Verlag, Berlin, p. 10.

Saar, S.H.: 1990, in *Solar Photosphere: Structure, Convection and Magnetic Field, IAU Coll. 138*, J.O. Stenflo, ed. Kluwer, Dordrecht, p. 427.

Saar, S.H.: 1991, in *The Sun and Cool Stars: Activity, Magnetism, Dynamos*, I. Tuominen, D. Moss, G. Rüdiger, eds., Springer–Verlag, Berlin, p. 389.

Saar, S.H.: 1992, *Mem. Soc. Astron. It.*, in press.

Saar, S.H., Piskunov, N.E., Touminen, I.: 1992, in *Cool Stars, Stellar Systems and the Sun*, M.S. Giampapa, J.A. Bookbinder, eds., ASP Conf. Ser., San Francisco, p. 255.

Schmitt, J.H.M.M. et al.: 1985, *Astrophys J* **290**, 307.

Schrijver, C.J., Coté, J., Zwaan, S.H., Saar, S.H.: 1989, *Astrophys. J.* **337**, 964.

Simon,T., Landsman, W.: 1991, *Astrophys. J.* **380**, 200.

Skumanich A.: 1972, *Astrophys. J.* **171**, 565.

Sofia, S., Pinsonneault, M., Kawaler, S.D., Demarque, P.:1987 in *Cool Stars, Stellar Systems and the Sun*, J.L. Linsky, R.E. Stencel, eds., Springer-Verlag, Berlin, p. 192.

Sofia, S., Pinsonneault, M., Deliyannis, C.P.: 1991, in *Angular Momentum Evolution of Young Stars*, S. Catalano, J.R. Stauffer, eds, Kluwer, Dordrecht, p. 333

Stauffer, J., Dorren, J., Africano, J.: 1986, *Astron. J.* **91**, 1443.

Trigilio, C., Umana, G., Catalano, S., Marilli, E.: 1991, in *Angular Momentum Evolution of Young Stars*, S. Catalano, J.R. Stauffer, eds, Kluwer, Dordrecht, p. 183.

Walter, F.M.: 1982, *Astrophys. J.* **253**, 745.

Walter, F.M.: 1983, *Astrophys. J.* **274**, 794.

Walter, F.M., Schrijver, C.J.: 1987, in *Cool Stars, Stellar Systems and the Sun*, J.L. Linsky, R.E. Stencel, eds., Springer-Verlag, Berlin, p. 262.

Weber, E.J., Davis, L.: 1967, *Astrophys. J.* **148**, 217.

Wolff, S.C., Boesgaard,A.M., Simon,T.:1986, *Astrophys. J.* **310**, 360.

RIASS OBSERVATIONS OF AB DORADUS (HD 36705)

I. PAGANO, M. RODONÒ and G. CUTISPOTO

Institute of Astronomy of Catania University and Catania Astrophysical Observatory, v.le A.Doria 6, I-95125 Catania, Italy

A. COLLIER CAMERON

Astronomy Center, University of Sussex, Falmer, Brighton BN19QH, United Kingdom

M. KÜRSTER

Max-Planck-Institut für extraterrestrische Physik, D-8046 Garching bei Munchen, Germany

B.J. KELLET and G.E. BROMAGE

Rutherford Appleton Laboratory, Didcot, Oxfordshire OX11 OQX, United Kingdom

R. JEFFRIES

University of Birmingham, Edgbaston, Birmingham B15 2TT, United Kingdom

B. FOING

Institut d'Astrophysique Spatiale IAS, Bat. 121, F-91405 Orsay Cedex, France

and

P. EHRENFREUND

Service d'Aeronomie du CNRS, BP 3, 91371 Verrieres le Buisson Cedex, France

Abstract. We present the analysis of the data acquired in the fall of 1990 during coordinated multiwavelength observations of the K2 IVp rapidly rotating, very active, single star AB Doradus (HD 36705). These observations did involve, within the RIASS program, the IUE and ROSAT satellites, and various ground based Observatories. Table I summarizes the observation dates, the telescopes and instruments we used.

Our aim is to study the physical and geometrical parameters of discrete structures at various heights in the atmosphere of this pre-main sequence active star. In pursuing this objective we have investigated the rotational modulation of continuum and line fluxes over a wide range of wavelengths covering the temperature interval from 10^3 to 10^7, i.e. from photospheric up to coronal levels.

Remarkable variability versus rotational phase has been found at all wavelengths, and the characteristics of spots, plages and coronal features are discussed.

Key words: Stellar Activity – Multiwavelength Observations

TABLE I

The Observations

Telescope	Measurement Type	Dates
IUE	UV spectr./FES phot.	Dec,30 1990
ROSAT	PSPC (0.1-2.4 keV)	Dec,14 1990-Jan,17 1991
ROSAT	WFC (0.08-0.18 keV)	Dec,28 1990-Jan,26 1991
CASPEC-ESO	optical spectroscopy	Dec,28 1990-Jan,2 1991
ESO (0.5m)	$UBVR_cI_c$ and H_α phot.	Dec,23 1990-Jan,2 1991
Mt. Stromlo (1.9m)	H_α spectroscopy	Dec,23-25-27-30 1990
SAAO (0.5m)	$UBVR_cI_c$ photometry	Dec,28-29 1990

J.F. Linsky and S. Serio (eds.), Physics of Solar and Stellar Coronae, 457–462.

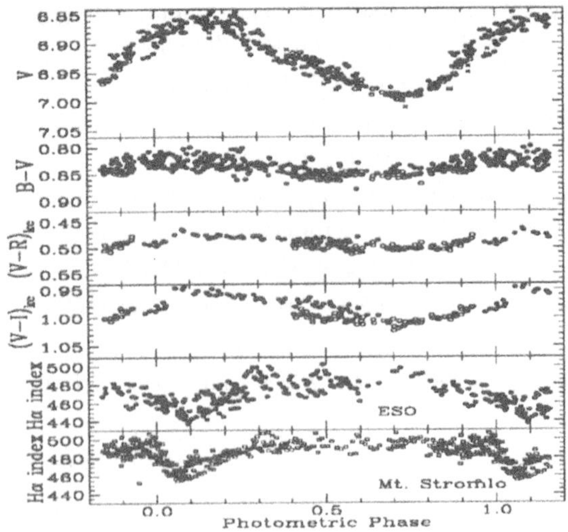

Fig. 1. UBVR$_c$I$_c$ and H$_\alpha$ photometry of AB Dor from SAAO (open circles), ESO (closed circles) and IUE Fine Error Sensor (FES) (crosses) observations.

1. Optical photometry and H$_\alpha$ spectra

The observed V light curve and color indices versus rotation phase are shown in Figure 1. The photometric phases were calculated according to the ephemeris: HJD=2444296.575 + 0.51423 E (Pakull, 1981). The V band curve has an amplitude of about 0.13 mag. The V-I curve shows that the star is cooler at light minimum, as expected when the V-band rotational variability is attributed to the transit of cool solar-like spots on the visible hemisphere. The V light curve was modelled (see Rodonò *et al.*, 1986) by assuming the inclination of the rotation axis to the line of sight i=53 degrees (Collier Cameron *et al.*, 1992), the quiescent photosphere and spot temperatures T_{ph}=5200 K and T_{sp}=4300 K, respectively (Kürster and Schmitt, 1992), and 0.68 for the limb darkening coefficient. The model solution is shown in Figure 2. Only the light modulation due to these two spots has been considered, without taking into account the overall depression of the light curve with respect to the unspotted level. Since the magnitude of the light maximum (V=6.85) in 1990 was fainter then the absolute light maximum (V=6.76) reported in the literature (see Cutispoto and Rodonò, 1992), the spot covering factor of 0.06 resulting from our model should be increased by at least an additional 5 per cent.

The ESO H$_\alpha$ indices in Figure 1 were obtained photometrically and are defined as 10000 [H$_\alpha$(narrow)/H$_\alpha$(wide)]. The Mt. Stromlo H$_\alpha$ indices were derived from dynamical spectra. Both data sets clearly show variability versus phase, suggesting the presence of bright high-lying prominences possibly related to photospheric spots.

The H$_\alpha$ dynamical spectra of AB Dor obtained at Mt. Stromlo indicate a major nest of prominence activity at phase 0.1, which corresponds to the phase where the

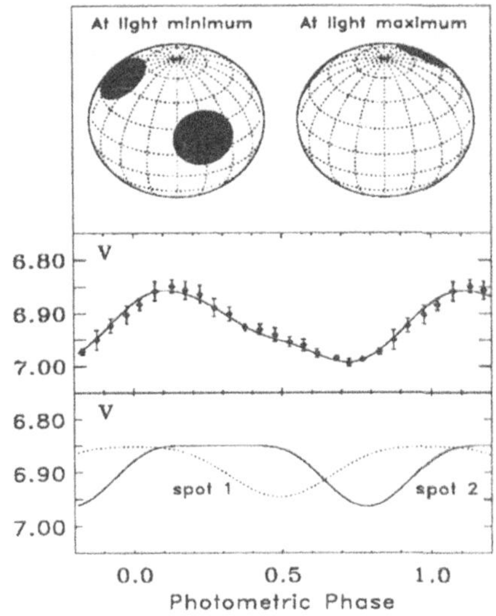

Fig. 2. Results of the spot model: the two spots have radii 19 and 21 degrees. Rotation is clockwise. *Upper panel*: spot configuration at light maximum and minimum. *Middle panel*: comparison between observations (diamonds) and computed light curve (continuous line). *Bottom panel*: the contribution of the individual spots to the computed light curve.

deepest minimum of the H_α index occurs. The continuous monitoring from Dec 24 to Dec 31 has shown that this feature had persisted for at least a week. Weaker prominence features are seen at phases 0.2, 0.3, 0.45, 0.67 and 0.95. Note that the major prominence complex at phase 0.1 is best visible at the phase of minimum spot visibility. However, as discussed above, we note that the present light maximum depression of about 0.1 mag requires about 50% of spots to be always on the visible hemisphere.

2. The UV data

We observed HD 36705 by both the short (SWP) and long wavelength (LWP) cameras of the IUE satellite, which cover the wavelength ranges 1150-2000 Å and 2000-3200 Å, respectively. The NASA-ESA observations covered about 1.5 star rotations. Most of our data were taken in the low dispersion mode (resolution ~ 6 Å). The spectra were extracted and calibrated with the IUE standard software in the RDAF libraries in use at Goddard Space Flight Center. The line analysis was performed with the ICUR-IDL package (Walter, 1989).

The V-band photometric data from the IUE Fine Error Sensor (FES) counts which were acquired before and after each spectroscopic image, are included in Figure 3.

The UV line fluxes were obtained by integrating the emission features with respect to a local background continuum defined on each side of the lines. The

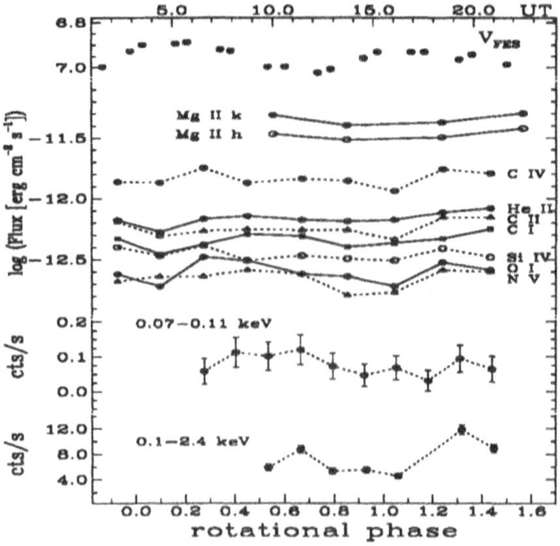

Fig. 3. *From the top*: the IUE FES photometric light curve; the line fluxes in the Mg II *h* and *k* lines (2795.5 Å, and 2803.0 Å); the integrated fluxes at Earth in the strongest lines of the SWP IUE spectra; the counts rates in the S2A filter of the WFC, and the PSPC count rates. Only simultaneous data from the available data sets are included. Both UT times (*upper scale*) and rotation phases (*bottom scale*) are given. The initial UT time corresponds to MJD 48255.0.

integrated fluxes for the most prominent chromospheric and transition region lines are plotted in Figure 3 versus time (*upper scale*) to show the temporal behaviour of UV lines. The rotation phase is shown in the lower scale. Most of the emission line fluxes show a maximum close to phase 0.25, both during the first and second orbit.

3. The EUV data

During the ROSAT all-sky survey AB Dor was detected by the WFC in the two bands: 0.085-0.180 keV (S1A filter) and 0.070-0.110 keV (S2A filter). The whole data set in both filters is given in Figure 4A, while Figure 4B (b',c',d') shows the WFC S1 and S2 data folded by the Pakull (1981) rotational period and grouped in three distinct time intervals. The large scatter in the data is partially due to the closeness of the point in direction of which AB Dor was seen by ROSAT to the southern auroral particles zones of the Earth. However, the scatter can be attributed, at least in part, to intrinsic activity on AB Dor. Line flux modulation versus rotation phase is apparent, expecially for the data plotted in Figure 4B(b' and c').

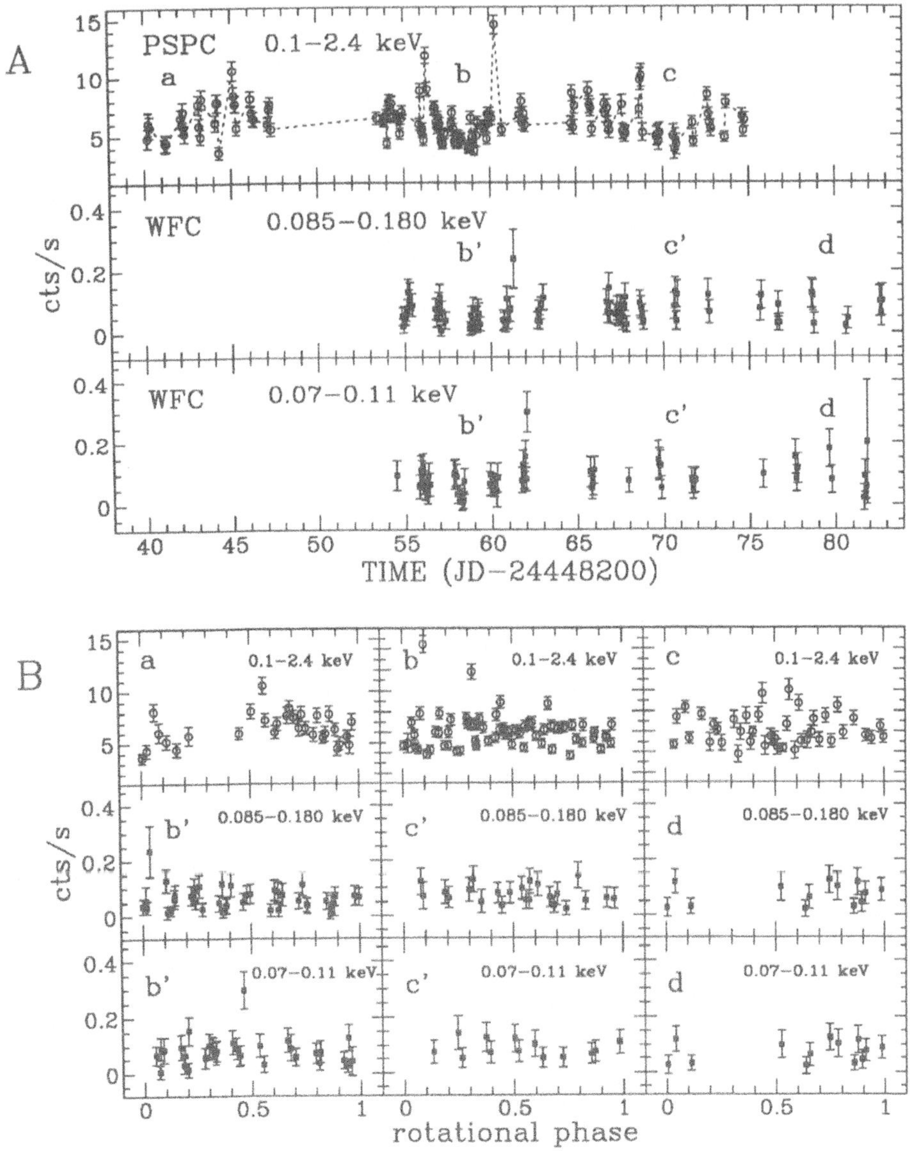

Fig. 4. *Panel A (top)*: the PSPC count rates in the 0.1-2.4 keV band; *(middle and bottom)*: the EUV count rates detected by ROSAT through the S1A and S2A filters of the WFC, respectively. *Panel B (top)*: the (a), (b) and (c) data sets in panel A are plotted versus rotational phase; *(middle and bottom)*: the (b'), (c') and (d) data sets in panel A are plotted versus rotation phase, respectively.

4. The X-ray data

The X-ray data in the 0.1-2.4 keV band were obtained, as part of the RIASS program, during the ROSAT PSPC all-sky survey. The count rate is extremely variable with frequent flare-like events (Figure 4A). A Fourier analysis of the whole data set, as well as of the data points below 8 cts/s (i.e., outside of flares), did not show any conclusive evidence of periodicity. However, by dividing the data sets in three groups (Figure 4B) it appears that some fraction of the X-ray emission is rotationally modulated. This modulation is perhaps best seen in Fig.4B(a), were most of the data points are grouped around a sinusoidal band with maximum near phase 0.65. A similar wave appears also in Fig.4B(b). However, many other points seem to group around another sinusoidal band, almost in antiphase with respect the former one (maximum near phase 0.25). Also the subsequent data set in Figure 4B(c) seems to show the same effect.

5. Discussion

In addition to the UV line fluxes, Figure 3 shows the FES V curve, the PSPC count rates and the S2A WFC data acquired simultaneously with the IUE observations. The two X-ray flares detected between MJD 48255 and 48256 fall at phases ~0.66 and ~1.30, respectively. Both enhancements are also clearly visible in the 0.07-0.11 keV band. However, no significant enhancement of the UV lines during the first of these two flares is apparent, while the second flare shows a 30 percent enhancement only in some of the UV lines, including C IV. The Mg II chromospheric lines do not show any flare increase. However, short-lived faint flares can not be excluded because of the long exposure times of the LWP-HI spectra.

One possible explanation of this behaviour is that the soft X-ray emission comes from a loop structure or structures extending over several stellar radii, as suggested by Collier Cameron *et al.* (1988) from a two-hour flare rise time seen in the EXOSAT data. The loop footpoints could be located on the far side of the star during the first flare, and on the limb during the second flare, so that only the optically thin high-lying TR emission was visible. Another possibility is that the observed flares did not occur on AB Dor, but on Rst137B, a dMe common proper motion companion, 10 arcsec apart. Actually, from the fast photometry collected at SAAO, there is some evidence that two small U band flares might have occured on Rst137B.

References

Collier Cameron *et al.*: 1988, *MNRAS*, **231**, 131.

Collier Cameron *et al.*: 1992, *in preparation.*

Cutispoto, G., Rodonò, M.: 1992, In *Cool Stars, Stellar Systems and the Sun; Seventh Cambridge Workshop*, M.Giampapa and J.Bookbinder (eds.), A.S.P. Conf. Ser., in press.

Kürster, M., Schmitt, J.H.M.M.: 1992, In *Surface Inhomogeneities in Late-Type Stars*, P.B. Byrne and J.D. Mullan (eds.), Spinger-Verlag, 69.

Pakull, M.W.: 1981, *A&A*, **104**, 33.

Rodonò, M., Byrne, P.B., Neff, J.E., Linsky, J.L., Simon, T., *et al.*: 1987, *A&A*, **176**, 267

Walter, F.M.: 1989, private comunication.

OBSERVATIONS OF LOW LEVEL OPTICAL VARIABILITY IN ACTIVE STARS

R. VENTURA, I. PAGANO, G. PERES AND M. RODONÒ
Catania Astrophysical Observatory and Institute of Astronomy, Catania University, v.le A.Doria,6 I-95125, Catania, Italy

Abstract. As part of a broad program dedicated to the study of low level stellar activity, in particular microflares and quasi-periodic variations, we have started an exploratory program to detect low level ($\sim$ 0.005 m) variability over time scales of the order of seconds in the optical emission of active stars by the double channel photometer URSULA fed by the 91 cm telescope of the Catania Astrophysical Observatory. Among the several possible targets, we have considered some active flare stars (UV Ceti and BY Dra-type). We describe the first observations and the preliminary results.

Key words: active stars – microvariability

1. Instrumental apparatus and observation method

The observations of active star low level variability and microflares in the optical band require very high photometric sensitivity and a signal-to-noise ratio sufficient to detect variations sometimes of the order of milli-magnitudes. We have carried out a feasibility study on detection of low level events in stellar optical emission using an improved version of the twin-beam photon counting photometer *URSULA* (De Biase *et al.* 1988) fed by the 91-cm Cassegrain telescope at the Mount Etna station of Catania Astrophysical Observatory. Such a system allows us to achieve a very high photometric precision by reducing several sources of inaccuracy which typically affect ground based photoelectric photometry.

The *URSULA* photometer operates with two virtually identical optical channels, allowing the simultaneous observation both of the target star and a nearby reference star. This operation mode offers two great advantages:

- the ratio of photon count rates, after background light subtraction, is largely independent of the atmospheric transmittance variations, thus eliminating one of the most serious sources of inaccuracy for photoelectric photometry;
- uninterrupted observations of both stars, at variance with the sequential procedure of single-beam photometry, allow a continuous time coverage, which is very important for detecting elusive rapid events.

The photometric accuracy and the instrumental limitations of the observational system have been carefully and extensively checked in very different conditions of atmospheric transmittance variation and for a wide sample of stellar magnitudes (De Biase *et al.* 1988). The whole system was found to be significantly better than standard photometric single-beam techniques. Typical accuracies are found to be better than a few milli-magnitudes for stars of about 9th magnitude, provided that an appropriate choice of the integration time is made and good photometric weather conditions occur.

Positive indications of the reliability of the *URSULA* photometer in detecting low level variability in active stars come from Rodonò *et al.* (1979). The authors were able to present for the first time conclusive evidence for strong pre-flare enhanced

J.F. Linsky and S. Serio (eds.), Physics of Solar and Stellar Coronae, 463–466.

activity, short-lived spike-type flares and pre-flare negative dips in the flare star YZ CMi.

2. Observations

Our test observations of the flare star FF And (BD +34 106, m_v=9.4) were carried out during two nights in October 1991. On the first night the photometric quality was not good and observations were dedicated to setting up the experimental apparatus and to providing a noise template for a "bad" observing case.

In order to obtain a high photon count rate, to minimize the statistically intrinsic photon count error which, as is well known, decreases with the square root of the integration time, and to maintain good time resolution, we chose to observe without a filter. The spectral response was therefore mainly determined by the blue sensitive photomultiplier tubes EMI 9789 QA with a spectral response similar to Bialk/B peaked at 3600 Å. The resulting photometer response was similar to the B-colour of the standard system. An integration time of 3 sec was chosen as a compromise between the requirements of high temporal resolution and low noise level.

During the night of October 18, 1991 almost one hour of good photometric data on FF And was collected. Thereafter clouds prevented us from continuing the observations.

In Figure 1a we show separately the light curves of the variable star and its comparison star. The differential light curve of the variable star against the comparison star (BD +34 103, m_v=9.5) is presented in Figure 1b.

A large flare was detected after a nonflaring period. The differential light curve can be divided in four parts each with different characteristics, thanks to the sensitivity of the apparatus:

- part A : quiet phase (lasting about 9^m)
- part B : microactivity (lasting about 15^m)
- part C : rising trend (lasting about 13^m)
- part D : precursor and a large flare (lasting about 5^m).

Since colour variations were not measured, temperature variations during the flare cannot be studied. We focused our attention on the principal subject of our investigation, an analysis of the light curve outside of the flare.

First of all we derived the noise level for each segment of the light curve and in order to detect any possible periodic microvariability, each part of the light curve was separately Fourier-analyzed.

The derived rms variations of data for the first three data sets are, at least, of the order of 8×10^{-3} (part A in Figure 1). The power spectra concerning the A and C parts present no significant peaks other than noise. The B part (see Figure 1) exibits a power spectrum rather flat, except two peaks at about 4^m and 2^m (Figure 2), the meaning of which should be investigated further. In the light of analogous findings by Andrews (1990 and references therein) and Mullan *et al.* (1992) microvariability in part B appears to be significant.

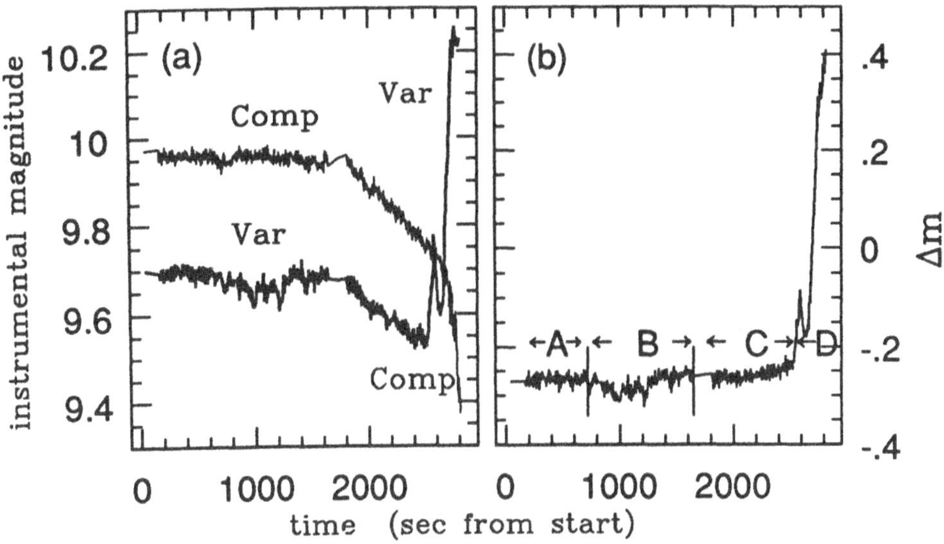

Fig. 1. a) Light curves of FF And (Var) and its comparison star (Comp); b) Differential light curve of FF And against the comparison star (BD +34 103)

3. Discussion

This feasibility study on the detection of low level variations in optical emission of active stars, by using the *URSULA* twin-beam photometer, has pointed out the capability of the instrument to attain sufficient photometric accuracy and high time resolution to permit the detection of this kind of phenomenon. However, even if it is evident that the present technique can offer new perspectives in stellar microvariability research, many open questions still remain.

In particular: it is not clear whether the variability shown by FF And in Part B of the light curve could be interpreted in terms of the microvariability searched for; namely small flares. On the other hand, it should be related to the following precursor and flare part and, therefore, considered as some kind of pre-flare activity analogous to pre-flare dips. The present data do not allow us to distinguish between the two cases.

The uncertainty largely stems from the lack of an adequately large database of observations of this quality which, for instance, could provide typical amplitude and temporal scales of stellar microvariability in different active stars.

On the other hand, the observational approach used for the present work is rather different from the standard photometric measurements which are usually employed. Therefore, other photometric observations are of little - if any - help on this issue.

In conclusion, there is a definite need to provide an extended basis of observa-

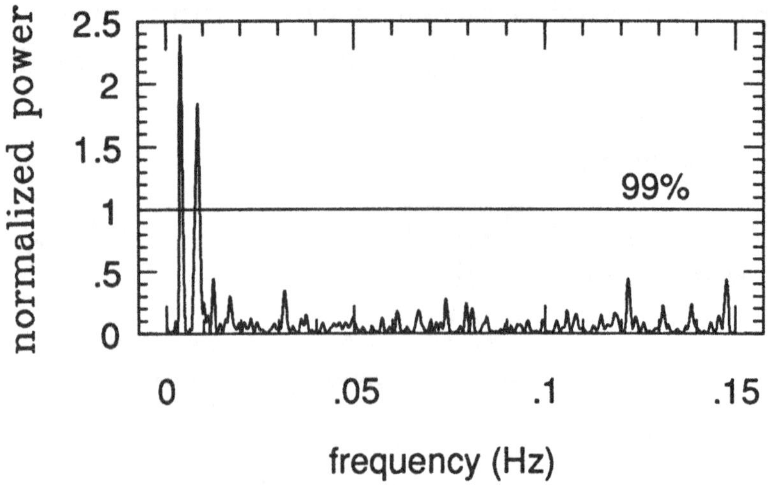

Fig. 2. Power spectrum of the FF And differential light curve (part B), normalized to the value corresponding to the 99% confidence level. The two prominent peaks of 4^m and 2^m are both well above the 99% confidence level.

tions so as to fill the gap of our knowledge on time scales and amplitudes of optical variability of active stars. Theoretical studies, on the other hand, could help in many respects, for instance in predicting the low energy tail in the distribution of stellar flares and to provide the expected number of small optical flares of given amplitude.

References

Andrews, A.D.: 1990, *Astron. Astrophys.* **229**, 504

De Biase, G.A., Paternó, L., Fedel, B., Santagati, G., Ventura, R.: 1988, in "Second Workshop on Improvements to Photometry," NASA Conference Publication *10015*, ed. Borucki, p. 17 - 33

Mullan, D.J., Herr, R.B., Bhattacharyya, S.: 1992, *Astrophys. J.* **391**, 265

Rodonò, M., Pucillo, M.,Sedmak, G., De Biase, G.A.: 1979, *Astron. Astrophys.* **76**, 242

EXTENDING THE PRESENCE OF CHROMOSPHERES AND CORONAE TO MIDDLE A-TYPE STARS

R. FREIRE FERRERO

Strasbourg Observatory, 11, Rue de l'Université, F-67000 Strasbourg, France

S. CATALANO and E. MARILLI

Catania Astrophysical Observatory, Catania, Italy

D. WONNACOTT

Rutherford Appleton Laboratory, Chilton, Didcot, UK

P. GOUTTEBROZE

Insitute d'Astrophysique Spatiale (IAS), Orsay, France

F. BRUHWEILER

Physics Department, The Catholic University of America, Washington, USA

and

A. TALAVERA

ESA IUE Observatory, VILSPA, Madrid, Spain

Abstract. Recently, we presented new evidence for the presence of high temperature layers (chromospheres, transition zones, and coronae) above the photosphere of late A-type stars. Here we summarize some of these results and in particular we present our detections of the chromosperic Lyman-α emission in a set of four late A-type stars and three early F-stars observed at high resolution with IUE. Alternative chromospheric heating scenarios are proposed from our comparison of X-ray (EINSTEIN satellite), EUV (ROSAT satellite) and Lyman-α (IUE, COPERNICUS satellites) emission luminosities.

Key words: Lyα emission – EUV emission – Chromospheres – Coronae – A-type stars

1. Introduction

The precise location in the H-R diagram where chromospheric emission begins is of great importance for stellar structure theory, because it should indicate the boundary for the onset of convection. With this objective we have searched for Ly-α chromospheric emission from IUE high resolution spectra. The observed set of four late A-type stars and three early F-type stars are the brightest and closest in this spectral type range. Among them are included α Aql (A7 IV-V), sometimes considered as a standard star, and the δ Scuti star γ Boo (A7 III). Our detections of Ly-α chromospheric emission unambiguously move the onset of chromospheres and convection zones to as early as spectral type A7 (B-V = 0.19) (Freire Ferrero et al. 1990, 1992a; Catalano et al. 1991a, b; Marilli et al. 1991, 1992).

The chromospheric character of the emission is proved by semiempirical NLTE atmospheric models. Such models include chromospheres with a steep temperature rise from the temperature minimum between 5000 and 6000 K and a plateau between 10000 and 20000 K (Freire Ferrero et al. 1992b). Our stellar sample does not show any evidence for a shell either in the visible or the UV that might explain the origin of the observed Ly-α emission.

J.F. Linsky and S. Serio (eds.), Physics of Solar and Stellar Coronae, 467–470.

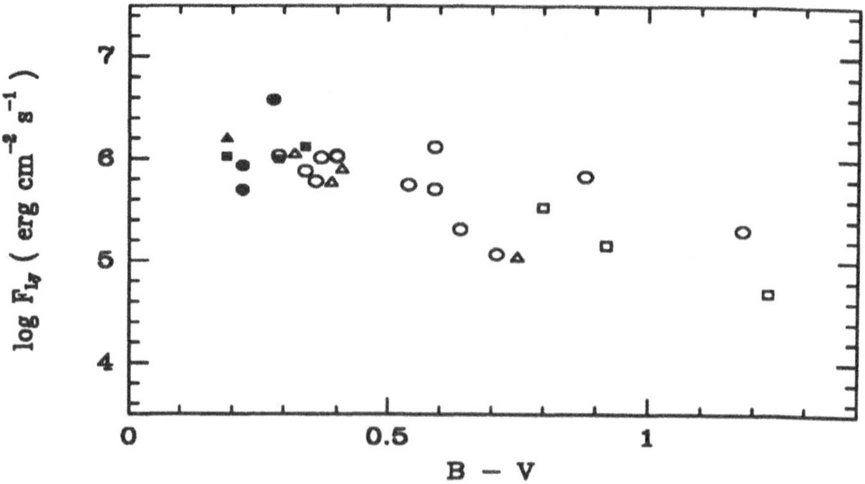

Fig. 1. Lyα integrated flux as a function of B-V. Filled symbols are present data, open symbols are from Bruno et al. (1992) (low dispersion IUE data) and other literature (high dispersion IUE data). Circles, triangles and squares refer respectively to: main sequence, subgiant, and giant stars.

2. Results

We first assume that the Ly-α emission profile is a Gaussian curve and the interstellar HI absorption is a simple Lorentzian profile to estimate both the stellar Ly-α emission and the HI column density. From the Ly-α emission line profiles, we determine the emission luminosity and the emission flux, starting from the flux at the Earth and the stellar parameters and parallaxes.

To explore the behaviour of Ly-α, we plot in Figure 1 the Ly-α flux vs B-V colour using data from both high resolution and low resolution (Bruno et al. 1992) IUE spectra. We can see that: *i*) there is no clear difference between main sequence and giant stars; *ii*) the average Ly-α surface flux has a curved trend with a maximum at B-V equal to 0.3–0.4; *iii*) this trend is reminiscent of the decrease in X-ray emission for B-V$\leq$0.5 (Schmitt et al. 1985; Micela et al. 1988); *iv*) the average flux drops towards larger values of B-V.

Plotting the X-ray luminosity taken from the EINSTEIN satellite data against the Ly-α luminosity, we have shown (Catalano et al. 1991b) that the A stars Altair and α Cep lie below the correlation line, indicating a clear deviation from the flux-flux correlation. This may result from a lower coronal heating efficiency at this spectral type, probably due to a change in the atmospheric structure or in the heating mechanism.

Better insight into this question can be obtained by comparing the Ly-α luminosity with EUV data from the British Wide Field camera on board the ROSAT satellite (Wonnacott 1992). Figure 2 shows the ratio of the EUV luminosity in the S2a band (111$\pm$200 Å) and the Ly-α luminosity as a function of B-V. Although for

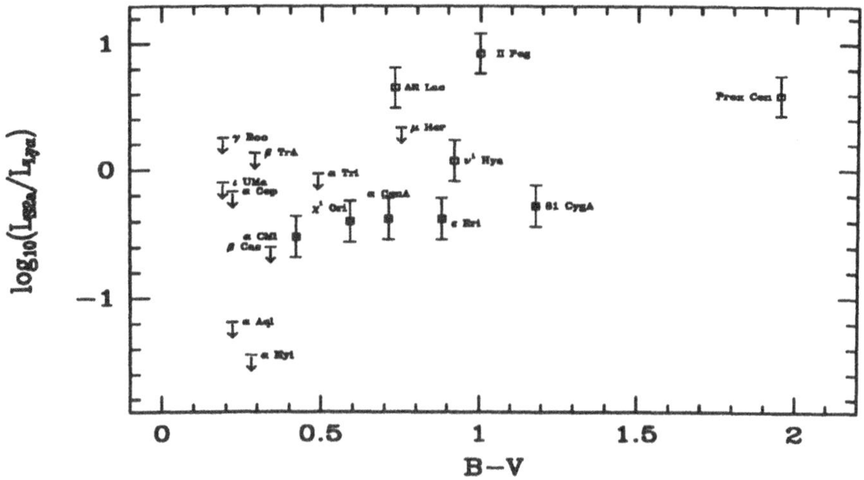

Fig. 2. Ratio of ROSAT WCF S2a luminosity to the Ly α luminosity as a function of B-V. Filled squares refer to solar type main sequence stars.

many stars only EUV upper limits are available, we can draw the following conclusions: *i*) very active stars, including RS CVn systems and some giants, occupy the upper region of the plot; *ii*) the main sequence (solar-type) stars have EUV emission somewhat smaller than Ly-α, with a clear decreasing trend toward low B-V values; *iii*) the luminosity ratio rapidly decreases for B-V$\leq$0.5.

The 111–200 Å EUV band is composed mainly of emission lines formed below and above 5 $\times 10^6$ K. The spectrum of very active stars is dominated by emission from high excitation temperature species, from Fe XIX to Fe XXIII, in the region 100–140 Å with very little emission in the band 140–200 Å, as shown by Lemen et al. (1989) for the RS CVn binaries Capella and σ CrB. The EUV spectra of low activity stars like the quiet Sun (Kastner et al. 1974) and α CMi (Lemen et al. 1989) are dominated in the 150–200 Å region by emission from lower temperature species like Fe IX and Ni XII to Ni XIV.

Stars with high S2a/Ly-α luminosity ratios should have coronae dominated by the high temperature component, probably confined in gigantic loop structures. The EUV emission of low activity solar type main sequence stars may be due to low excitation temperature species in the 150–200 Å region, and should originate mainly in lower temperature coronal holes.

The drastic drop in the luminosity ratio for B-V smaller than 0.5 can be interpreted in terms of a decay in the scale height of the corona and/or a rapid temperature decrease to earlier spectral types as shown by Schrijver et al. (1984).

3. Conclusions

From the Ly-α line core emission we have established unambiguously that the onset of chromospheres occurs at least as early as spectral type A7. This observational result may be biased by instrumental limitations and the visibility of very weak chromospheric and coronal indicators, and we can argue that the true onset of chromospheres could occur at an earlier spectral type. Coronae and chromospheres appear to vanish rapidly between early-F and A-type stars and probably completely disappear at spectral type A6 or A5.

The existence of chromospheres and coronae in late type stars is usually ascribed to magnetic heating. At the boundary with early-A type stars, chromospheres and coronae become weak, indicating a lower nonthermal energy release in the external atmospheric layers. This trend is consistent with the decrease of the acoustic flux calculation by Bohn (1984), suggesting that the acoustic energy release rather than magnetic processes may be the principal mechanism for heating at these early spectral types.

Whatever the heating mechanism of chromospheres and coronae may be, magnetic or acoustic, convective energy transport through the photospheric layers is generally believed to be a necessary part of the heating process. Our results provide clear evidence for the existence of convective layers in the A7 spectral types and suggest that its emergence limit should be placed in the middle A-type stars.

Acknowledgements

This work was supported by the DRCI-CNRS (France) and the CNR (Italy). The Catania Astrophysical Observatory and ASTRONET facilities are also aknowledged.

References

Bohn, H.U.: 1984, *Astron Astrophys* **136**, 338.

Bruno, M., Catalano, S., Marilli, E.: 1992, *in preparation.*

Catalano, S., Marilli, E., Freire Ferrero, R., Gouttebroze P.: 1991a, *Astron Astrophys* **250**, 573.

Catalano, S., Gouttebroze, P., Marilli, E., Freire Ferrero, R.: 1991b, *Proc IAU Coll 130: The Sun and cool stars: activity, magnetism, dynamos*, Tuominen I. et al. (eds.), Springer-Verlag, p. 466-470.

Freire Ferrero, R., Catalano, S., Gouttebroze, P., Marilli, E.: 1990, *Proc Int Symp: Evolution in Astrophysics*, Rolfe E.J. (ed.), ESA SP-310, p. 315-318.

Freire Ferrero, R., Catalano, S., Marilli, E., Gouttebroze, P.: 1992a, *CCP7/IoA Workshop: Physics of Chromospheres, Coronae and Winds*, Institute of Astronomy, Cambridge, UK, 25-27 March 1992, in press.

Freire Ferrero, R., Gouttebroze, P., Catalano, S., Marilli, E., Bruhweiler, F., Kondo, Y., Van der Hucht, K.A.: 1992b, submitted to *Astrophys J.*

Kastner, S.O., Neupert, W.M., Swartz, M.: 1974, *Astrophys J* **191**, 261.

Lemen, J.R. et al.: 1989, *Astrophys J* **341**, 474.

Marilli, E., Catalano, S., Freire Ferrero, R., Gouttebroze, P.: 1992, *Astron Astrophys*, in press.

Marilli, E., Catalano, S., Freire Ferrero, R., Gouttebroze, P., Bruhweiler, F., Talavera, A.: 1991, *Proc 7th Cambridge Workshop on Cool Stars, Stellar Systems and the Sun*, Giampapa M.K. and Bookbinder J.A. (eds.), ASP Conf. Ser., San Francisco, p. 178.

Micela, G. et al.: 1988, *Astrophys J* **325**, 798.

Schmitt, J.H.M.M. et al.: 1985, *Astrophys J* **290**, 307.

Schrijver, C.J., Mewe, R., Walter, F.M.: 1984, *Astron Astrophys* **138**, 258.

Wonnacott, D.: 1992, *CCP7/IoA Workshop: Physics of Chromospheres, Coronae and Winds*, Institute of astronomy, Cambridge, UK, 25-27 March 1992, in press.

DISCUSSION FOLLOWING PAPERS ON OTHER OBSERVATIONS

Discussion following paper by S. Drake

G. Field: Why not go to wavelengths longer than 20 cm to take advantage of the fact that self-absorption ($\sim \lambda^2$) will give $\tau = 1/2$ high in the corona where $T = 10^6 - 10^7$ K, rather than the lower brightness temperatures you mentioned?
S. Drake: Clearly, observations at wavelengths $\gg$ 21 cm that sample the outer coronal regions can yield important constraints on the plasma parameters of the emission region for nonthermal radio sources. For optically thick thermal sources, the λ^{-2} dependence of the observed flux density cancels out the increase due to the higher brightness temperatures at long wavelengths. For the brightest RS CVn and dMe stars, Arecibo observations can be useful, but for most stars source confusion will be a big problem. The VLA P band (327 MHz) is probably the best long-λ system for stellar observations because of its good sensitivity and spatial resolution.
R. Rosner: You said that the Ap stars with unusually high rare earth surface abundances have not been seen as radio sources. Do you (or Jeff Linsky, or anyone else here) know whether any of these stars have been detected as X-ray sources? The reason I ask is that the presence of the rare earths places strong bounds on surface (i.e. photospheric) velocities. Thus, it may be that magnetic field perturbations due to surface motions could after all be relevant also to the A stars.
S. Drake: We only detected one such classic Ap star (a known binary) in our ROSAT All-Sky Survey sample, and we believe that the secondary component (rather than the Ap star candidate) is the more plausible X-ray source. In the Cash and Snow (Ap.J. **263**, L59 (1982)) sample of Ap stars observed by *Einstein* there were several detections of relatively weak ($L_x \sim 10^{28} - 10^{29}$ergs sec^{-1}) X-ray sources. Again, low-mass coronal-type companions may be the actual X-ray emitters, but the alternate explanation cannot be ruled out. A deep ROSAT pointing at one or more of those Ap stars that are not known to be binaries might help.

Discussion following paper by F. Chiuderi Drago

S. Drake: Did you calculate the circular polarization predicted by your thermal and nonthermal gyrosynchrotron models and compare it with observed values?
F. Chiuderi Drago: No, we did not. Since the radio transfer equation was solved separately for the ordinary and extraordinary modes, the circular polarization can be easily calculated for any given magnetic field configuration. We plan to do it.

Discussion following paper by P. B. Byrne

G. Peres: You mentioned dips in the K band during flares, but there are also preflare dips. Are they different? Have you observed any K band preflare dips?
P. Byrne: No preflare K-band dips have been observed to date — only dips during

J.F. Linsky and S. Serio (eds.), Physics of Solar and Stellar Coronae, 509–511.

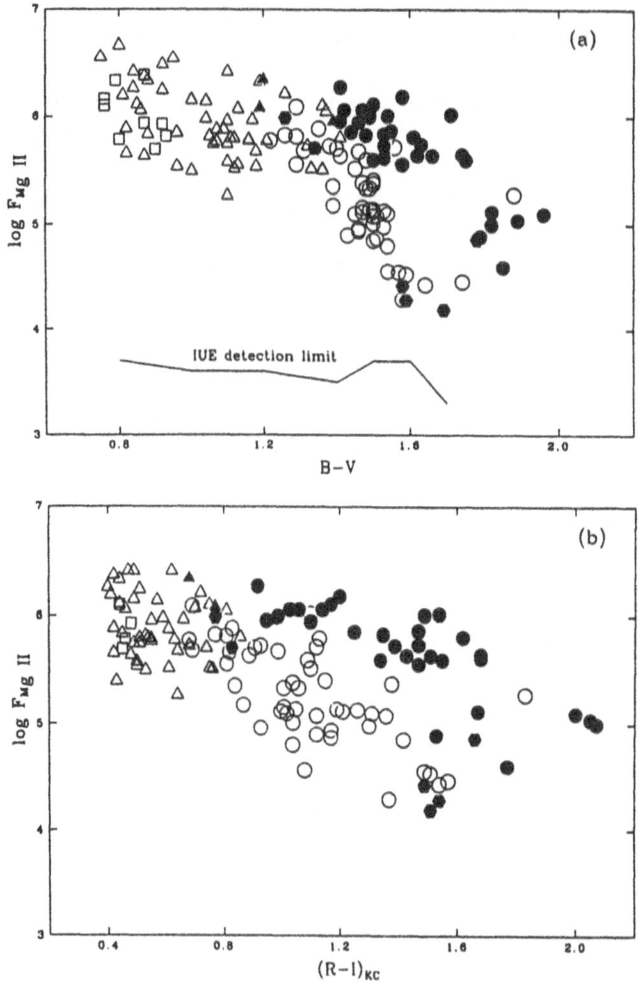

Fig. 1. (a) Log F_{MgII} versus (B-V) and (b) versus (R-I) in late-type dwarf K & M stars, ● [dMe], ○ [dM], ▲ [dKe], △ [dK], ◆ [dM(e)], □ [dK] Upper limits

3. Acoustic heating of the chromosphere

The lower Mg II flux limit in Figs. 1a,b suggests that there is a **basic** non-radiative heating mechanism in the atmosphere of late-type dwarfs. The fact that the lower limit strongly depends on effective temperature (which is an intrinsic stellar parameter), suggests that it could measure the contribution of acoustic heating in the chromosphere of cool dwarfs. The most recent calculations on the generation of acoustic energy from the convection zone of late-type stars are those of Bohn (1984). The improvements in the theory were accomplished by using new opacity tables. The empirical Mg II minimum flux density is given in Fig. 2 (dotted line) together with the mechanical flux generated by the model of Bohn (full line). The

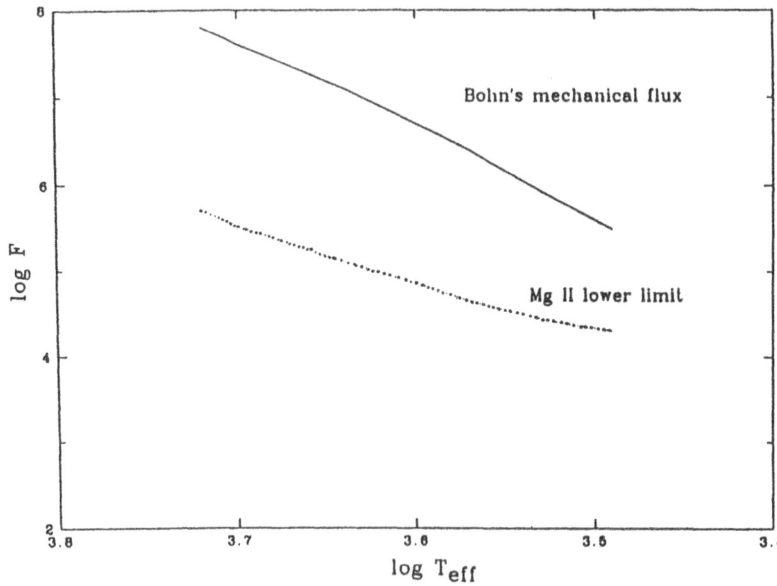

Fig. 2. Log $F_{acoustic}$ versus T_{eff} based on the model of Bohn [—], plus the observed minimum Mg II h&k flux from the late-type stars considered here [....]

flux generated by Bohn's models lie comfortably higher than that required, being two orders of magnitude higher for the early spectral types and one order of magnitude for the late spectral types. An interesting point with the new calculations is the general trend of a decrease in the acoustic flux with decreasing temperature, this being in reasonable agreement with the decrease of the Mg II minimum flux. The fact that the calculations of the mechanical fluxes are considerably larger than the minimum flux does not distract from the suggestion that the minimum flux is due to acoustic heating. There are two main reasons for this; (i) the Mg II lines, although they constitute an important source of the chromospheric radiative losses, are not the only radiative loss source and (ii) the fluxes calculated by Bohn (1984) is the mechanical energy generated in the convection zone, without treating the behaviour of acoustic waves in stellar atmospheres. Only a fraction of this energy will heat the outer atmosphere, in fact as much as 90% of the wave energy could be removed in the photosphere (Stein & Leibacher 1980, Ulmschneider 1991).

4. The nature of dM(e) stars

Model calculations (e.g. Cram & Mullan 1979, Cram & Giampapa 1987), predict that the Hα absorption equivalent width of an M dwarf in pure radiative equilibrium, varies from –0.08Å to –0.24Å for effective temperatures between 3500–4000 K. By introducing a chromosphere and gradually increasing its strength will make the absorption first deeper then filling in makes the absorption shallower until the line finally goes into emission. The theory therefore suggests that no detectable Hα in a late M dwarf, say later than M3, could be attributed either to a very weak/zero chromosphere or to a chromosphere of moderate strength (i.e. the line is filled-in). However for the early M dwarfs, no detectable Hα would simply mean that the line is filled-in. Therefore the activity levels of these stars can not be determined in terms of Hα only. Mg II line fluxes for 6 dM(e) stars are reported here. These are plotted in Fig. 1 (filled hexagons). The conclusion which can be drawn from these observations is that the activity levels of these stars can be either of intermediate type, between the active dMe and the less active dM, this would be the case for Gl 447, Gl 900, Gl 907.1, or of extremely low activity, Gl 105B, Gl 643, Gl 628. The position of the dM(e) stars in Fig. 1 suggest that the predictions of the Cram & Mullan theory may very well be correct. In fact the very low activity dM(e) stars are at the minimum flux level which implies that their temperature structure could be very close to radiative equilibrium.

If we are to combine the above conclusions with our discussion on the two component atmosphere, it would imply that the magnetically heated atmospheres dominate the emission in Gl 447, Gl 900, Gl 907.1, while in the remaining dM(e) stars it vanishes, perhaps due to a ceasing 'shell' dynamo and there the acoustic component can be detected. Since the dynamo generated magnetic activity depends on rotational velocity, the more active dM(e) stars are expected to be relatively fast rotators, while the in-active stars are the slowest rotators. Since most of the period determinations in late-type dwarfs rely on the rotational modulation of white light due to spots covering the stellar surface, it is difficult to determine periods of low activity stars. A possibility however exists through the relation between Mg II flux and rotational period presented by Doyle (1987). Rotational periods determined from this relation were tested for one star (Gl 735) by Duquennoy & Mayor (1988), and good agreement was obtained with the period determined through observation. The rotational periods determined this way are around 8 $\pm$ 2 days for Gl 900 and Gl 907.1, 50 $\pm$ 12 days for Gl 447 and 70 $\pm$ 15 days for Gl 105B, Gl 628 and Gl 643. This is consistent with the previously described scenario.

Acknowledgements

Research at Armagh Observatory is grant-aided by the Department of Education for N. Ireland. We acknowledge the support provided in terms of both hardware and software by the STARLINK Project which is funded by the UK SERC. MM would like to thank Armagh Observatory for a research studentship.

References

Bohn, H.U., 1984, A&A 136,338
Cram, L.E., & Mullan, D.J., 1979, ApJ 234,579
Cram, L.E., & Giampapa, M.S., 1987, ApJ 323,316
Doyle, J.G., 1987, MNRAS 224,1p
Duquennoy, A. & Mayor, M., 1988, A&A 200,135
Rutten, R.G.M., Schrijver, C.J., Lemmens, A.F.P, & Zwaan, C., 1991, A&A 252,203
Schrijver, C.J., 1987, A&A 172,111
Stein, R.F., Leibacher, J.W., 1980, *IAU Colloquium 51 on Stellar Turbulence*, eds. D.F. Gray & J.L. Linsky, p.225
Ulmschneider, P., 1991, *Mechanisms of Chromospheric and Coronal Heating*, Springer-Verlag, eds. P. Ulmschneider, E.R. Priest & R. Rosner, p.328

ANALYSIS OF CHROMOSPHERIC ACTIVITY INDICATORS IN MM Her AND AR Psc

D. MONTES, E. DE CASTRO, M. CORNIDE, AND M.J. FERNÁNDEZ-FIGUEROA
Dpto. Astrofísica. Facultad de Físicas.
Universidad Complutense. E-28040 Madrid (Spain)

Abstract. High resolution observations of Hα and CaII H and K have been carried out at Calar Alto and Roque de Los Muchachos Observatories. The H and K observations at two different phases of MM Her has allowed us to derived that the two components are chromospherically active being the activity level of the cool component much higher. On the contrary AR Psc shows H and K emission from only one component at all the orbital phases, this star presents also an important Hϵ emission. The Hα observations confirm the different chromospheric behaviour above mentioned.

Key words: RS CVn binary systems – Chromospheric activity

1. Introduction

MM Her and AR Psc are two RS CVn systems, with spot wave, and with coronal X-ray emission, (Strassmeier et al., 1988, hereafter referred to as CABS). In this work we study the behaviour of the Hα and CaII H and K lines as chromospheric activity indicators in these systems. In Table I we show the adopted stellar parameters, (CABS).

MM Her (HD 341475) is a double-lined spectroscopic binary with partial eclipses. The components have spectral types G2IV and G8IV (Popper, 1980). The orbital and photometric periods, are similar and the orbit is practically circular (e=0.04), (CABS). Popper (1976) found that CaII H and K emissions belong to the cool component. The Balmer Hα line behaviour has not been studied.

AR Psc (HD 8357) is a double-lined spectroscopic and non eclipsing binary. The spectral type of the cool and more massive component is G8IV. This is an unusual chromospherically active system, it has a large eccentricity (0.19) (CABS) and a short orbital period (14.30 days). Due to the orbital eccentricity it rotates pseudosynchronously, (Hall, 1986).

Bopp (1984) observed strong CaII H and K lines and also noted Hϵ in emission. Fekel et al. (1986) showed that the Hα line of the more massive star is in emission, while in the less massive star is a very weak absorption feature or at times may also be in emission.

TABLE I
Stellar Parameters

Name	S	T_{sp}	$R(R_\odot)$	d(pc)	V	V-R	P_{orb}(d)	P_{rot}(d)
AR Psc	15	/G8IV	≥1.5	70	7.28	0.69	14.300	12.245
MM Her	118	G2IV/G8IV	1.58/2.83	190	9.51	0.84	7.960322	7.936

S: Catalogue number from CABS, $\mathbf{T_{sp}}$: Spectral type and luminosity class of the Hot/Cool component,

J.F. Linsky and S. Serio (eds.), Physics of Solar and Stellar Coronae, 475–478.

2. Observations and Data Reduction

High resolution observations of Hα and CaII H and K lines, of this two systems, and several inactive stars of similar spectral type and luminosity class have been carried out on the 2.2 m Telescope at the German Spanish Astronomical Observatory in Calar Alto (Almería, Spain), using the Coudé spectrograph and a CCD as detector, (November 1986, July 1989), and on the Isaac Newton Telescope (INT) at the Observatorio del Roque de los Muchachos (La Palma, Spain), using the IDS and an IPCS as detector, (July 1988).

The reciprocal dispersion achieved is 0.2 Å/pixel. The spectra have been extracted using MIDAS package. The spectra in the region of CaII H and K have been flux calibrated using standards stars and the emission fluxes are obtained by reconstruction of the absorption line profile below the emission peak, following the method given by Blanco et al. (1974) and used also by Fernández-Figueroa et al. (1986).

In the region of Hα line the spectra have been normalized to the flux of the continuum at 6540 Å and the equivalent width (EW) have been measured following the method of Bopp et al. (1988).

3. Results and Discussion

3.1. MM Her

Two spectra were taken in the region of the CaII H and K lines, one in July 1988 at orbital phase 0.27, and other in July 1989 at phase 0.98. The orbital phases were computed with the ephemeris: MinI = (JD. Hel.) 2444500.6665 + 7.960322E (CABS)

At phase 0.27 (outside of eclipse) are visible two emissions peaks coming from both components (Fig. 1a): the strongest red-shifted emission belongs to the cool component and the other belongs to the hot component. Table II shows the CaII H and K observed fluxes of the two components and the surface fluxes.

At phase 0.98 (Fig. 1b), a fraction of 0.30 of the hot component is hidden, and it is not possible deblend the emissions of the two components so we have measure the total flux in H and K emissions (See Table II). It is normally reported (CABS) that the emission is only from the cool star, but our observations show that both components are active.

Hα line of the two components is visible in absorption in the July 1989 spectrum at phase 0.24. According to the orbital phase, the strongest absorption belongs to the cool component, (See Fig. 1c). The very different residual intensity of these two absorptions is due to the different contributions to the continuum of the hot and cool components (f_H/f_C=0.43).

3.2. AR Psc

Three spectra in the region of CaII H and K lines have been obtained in November 1986 (See Fig. 1d, 1e, 1f), at orbital phases: 0.33, 0.39, and 0.67, computed using the ephemeris: MinI = (JD. Hel.) 2446078.642 + 14.300E (CABS). In these three

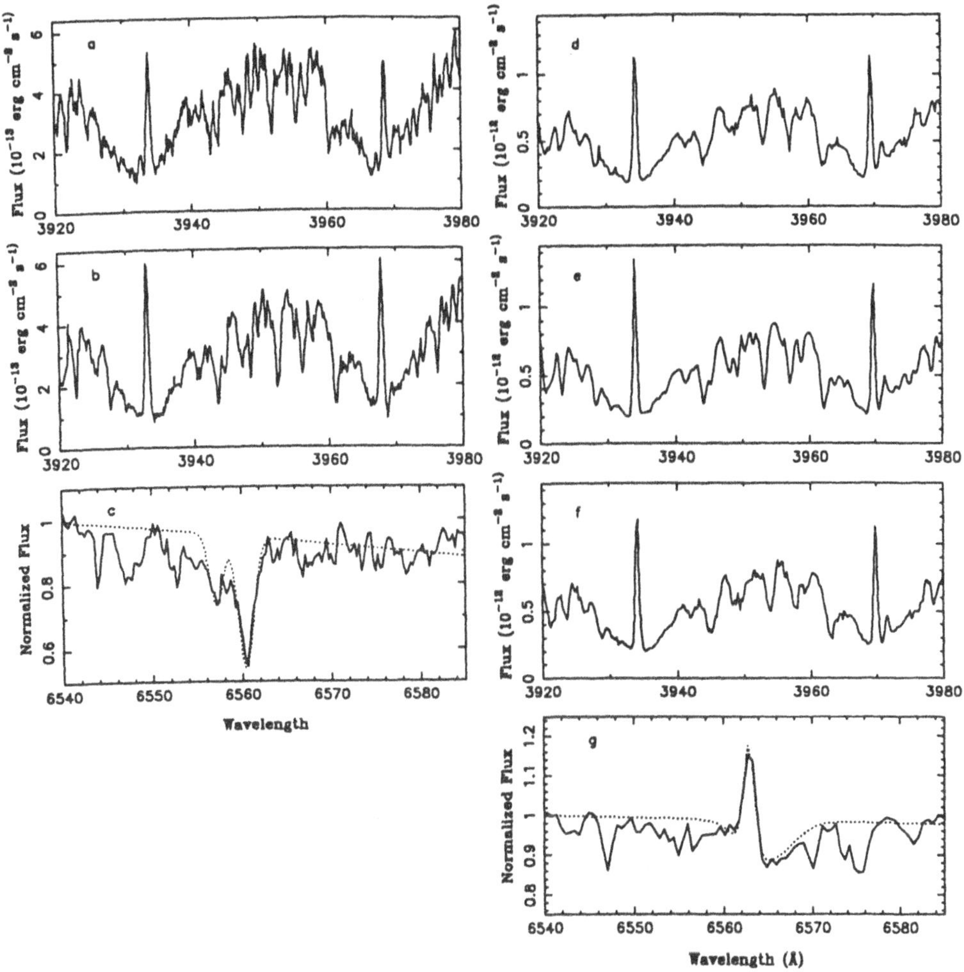

Fig. 1. CaII H and K, and Hα in MM Her (a, b, c) and AR Psc (d, e, f, g)

spectra strong CaII H and K emissions can be seen, which remain unshifted during the orbital period. It is remarkable the strong Balmer Hϵ emission. This behaviour points out that both the emission and continuum belong to the cool component (G8IV). In Table II we present the CaII H and K, and Hϵ measures. Since the radius of the components are unknow,the surface flux has been calculated using the relation from Pasquini et al. (1988).

In the region of the Balmer Hα line we have taken only one spectrum in November 1986 at phase 0.61. As it can be seen in Figure 1g the Hα line appears in emission above the continuum and overlapped to the absorption of the other component, we have performed a two-gaussian fit (absorption and emission) in order to obtain the equivalent width. The emission equivalent width was computed by substracting the absorption equivalent width corresponding to a star of the same spectral type and luminosity class. Both values are listed in Table III.

TABLE II
CaII H and K, and Hε Fluxes

Name	S	φ		F_{obs}(K) (E-13)	F_{obs} (H) (E-13)	F_S (K) (E+6)	F_S (H) (E+6)	F_{obs}(Hε) (E-13)	F_S(Hε (E+6
AR Psc	15	0.33	C	7.253	5.958	3.171	2.605	1.375	0.60
		0.39	C	7.054	6.468	3.084	2.828	1.415	0.61
		0.67	C	7.347	6.109	3.212	2.671	1.240	0.54
MM Her	118	0.27	H	0.509	0.285	1.447	0.810		
			C	3.009	2.237	2.666	1.982		
		0.98	T	3.453	3.338	-	-		

φ: Orbital phase, F_{obs}(H,K,Hε): Observed fluxes (CaII H and K, and Hε) in units of 10^{-13} erg cm^{-2} s^{-1}, F_S(H,K,Hε): Surface fluxes (CaII H and K, and Hε) in units of 10^{-6} erg cm^{-2} s^{-1}, **H,C**: Refers to Hot and Cool component respectively, **T** : Total flux has been measured.

TABLE III
Hα equivalent widths

Name	S		φ	FWHM (Å)	EW_α (Å)	R_c	EW_e (Å)
AR Psc	15	H	0.61	6.349	0.416	0.898	....
		C		1.405	-0.688	-0.727	-1.807
MM Her	118	H	0.24	1.919	0.464	0.782	....
		C		2.311	1.025	0.606	....

FWHM: Full width half maximum, $\mathbf{EW_\alpha}$: Measured Hα equivalent width, $\mathbf{R_c}$: Hα Residual Intensity, $\mathbf{EW_e}$: Emission Hα equivalent width.

These two RS CVn systems present strong CaII H and K emission, however the behaviour of the Hα and Hε lines is very different since AR Psc presents both lines in emission, whereas Hα and Hε appear in absorption in MM Her.

Acknowledgements. This work has been partially supported by the Universidad Complutense de Madrid.

References

Blanco, C., Catalano, S., Marilli, E., Rodono, M.: 1974, *Astron.Ap.* **33**, 257
Bopp, W.B.: 1984, *Ap.J.Supp.* **54**, 387
Bopp, W.B., Dempsey, R.C., Maniak, S.: 1988, *Ap.J.Supp.S.* **68**, 803
Fekel, F.C., Moffett, T.J., and Henry, G.W.: 1986, *Ap.J.Supp.S.* **60**, 551
Fernández-Figueroa, M.J., Montesinos, B., De Castro, E., Rego, M., Giménez, A., Reglero, V.: 1986, *Astron.Ap.* **169**, 219
Hall, D.S.: 1986, *Ap.J.Lett.* **309**, L83
Popper, D.M.: 1980, *Ann.Rev.Astron.Ap.* **18**, 115
Strassmeier, K.G., Hall, D.S., Zeilik, M., Nelson, E., Eker, Z., Fekel, F.C.: 1988, *Astron.Ap.Supp.S.* **72**, 291 (CABS)

ULTRAVIOLET STUDY OF 71 TAU, THE BRIGHTEST X SOURCE OF THE HYADES CLUSTER*

L. PASTORI and E. PORETTI
Osservatorio Astronomico di Brera, Via E.Bianchi 46, 22055 Merate (Como), Italy

and

L.E. PASINETTI FRACASSINI and F. DE NILE
Dipartimento di Fisica, Universitá degli Studi di Milano, Via Celoria 16, 20133 Milano, Italy

Abstract.
In a survey of δ Scuti variables with the IUE satellite, we observed 71 Tau, the X-ray source in the Hyades cluster according to the results of the Einstein Observatory. Our survey was planned to study the possible correlations between chromospheric activity and dynamics of pulsation, convection, rotation and to search for evidence of a possible mass loss hypothesized by the theory for these variables. The MgII k,h lines of 71 Tau show an anomalous absorption profile. Variations during the pulsation phases are questionable. A spectrum of the IUE archive, taken some years ago, shows an interesting feature in the Mg lines whose interpretation is also discussed.

Key words: δ Scuti Stars – X-ray sources – Stars: 71 Tau.

1. Introduction

IUE observations of δ Scuti variables were planned to study the main characteristics of the stars, the possible correlations between chromospheric activity and dynamics of pulsations, convection, rotation (Fracassini and Pasinetti 1982) and to search for evidence of possible mass loss hypothesized by Willson *et al.* (1987) in these stars. The behaviour of the chromospheric activity was found to be correlated with the pulsation phase, but with different characteristics in each star: an increase in the MgII emission during part of the cycle occuring near minimum light was observed in β Cas (Teays *et al.* 1989), in contrast to what observed in ρ Pup where, as in the classical Cepheids, a peak near maximum light was found (Fracassini *et al.* 1983). Moreover, in spite of its spectral type, no chromospheric activity was observed in the multiperiodic variable o^1 Eri (F2II-III). The observed results could raise the question about the influence of the variability mode and its connection with the convection on the chromospheric activity (Pasinetti Fracassini *et al.* 1990a). No clear evidence of emission was detected in the MgII *h,k* doublet nor in the highly ionized transition-region lines of 69 Tau (Pastori *et al.* 1990). However, variations of the *h* line profile during the pulsation cycle were observed, in particular asymmetries and the appearance of violet and red components near the line core. It is not possible to establish whether these peculiarities are due to absorption or emission, mass loss or pulsation. Other results and discussions on our survey may be found in Castelli

* Based on observations collected at the IUE European Satellite Tracking Station, Villafranca, Spain

J.F. Linsky and S. Serio (eds.), Physics of Solar and Stellar Coronae, 479–482.

et al. (1982), Pasinetti Fracassini *et al.* (1990b), Fracassini *et al.* (1991), Pasinetti Fracassini and Pastori (1991).

2. Discussion

We included in the survey two stars of the Hyades cluster, 71 Tau and 69 Tau, which exhibit interesting characteristics. 71 Tau is the most intense X-ray emitter in the cluster (Stern *et al.* 1981). The outstanding character of its X-emission was stressed by Micela *et al.* (1988) and discussed exhaustively by Antonello (1990). The log $\frac{L_x}{L_{bol}}$ value is -4.8, which is higher than the value found in normal early F stars and close to that of late F stars. 71 Tau is a wide spectroscopic binary; however, the dwarf G4 type companion cannot explain the strong X-ray emission, as remarked by Micela *et al.* . So far, there is no explanation of such emission unless we ascribe the X-ray emission of 71 Tau to the interaction between pulsation and rapid rotation (Antonello 1990).

69 Tau, which is very similar to 71 Tau as regards rotation velocity and other physical properties, was not observed with the Einstein satellite. No clear evidence of chromospheric emission was detected by our UV observations in the MgII *h,k* doublet nor in the highly ionized transition-region lines. However, we observed variations of the *h* line profile (less affected by noise) during the pulsation cycle, in particular asymmetries and the appearance of violet and red components near the line core. It is not possible to establish whether these peculiarities are due to absorption or emission, mass loss or pulsation (Pastori *et al.* 1990).

Our IUE observations of 71 Tau (F0V, B-V=0.25, V*sin i*=192 km s^{-1}), cover 1.5 cycles of the pulsation period. The MgII doublet exhibits variations spanning for about 0.5 Å, during the pulsation phases. Such variations are also confirmed by the statistical analysis which gives deviations larger than 3σ in some spectra. These features could be due to the presence of chromospheric emissions, which are modulated by the pulsation, as found in β Cas by Teays *et al.* (1989). This hypothesis questionable, is supported by the detection of CII emission made by Walter *et al.* (1988). On the contrary, Zolcinski *et al.* (1982) did not find chromospheric and transition-region line emission and suggested that the apparent absence of emission could be due to a smearing effect by the high rotational velocity. The presence of chromospheric emission in 71 Tau is, therefore, an open problem.

An IUE archive spectrum taken in 1980 shows a narrow absorption core in the MgII *k* line, which we ascribe to the physical characteristics of the star rather than to interstellar absorption, as suggested by Zolcinski *et al.* (1982) and Antonello (1990); indeed, this hypothesis is not supported by our observations of 69 Tau belonging to the same cluster.

Finally, if we compare the spectra of 71 Tau and 69 Tau of similar spectral type and rotation velocity, we find differences in the profiles and behaviour of the MgII doublet, which are so far not explained.

The figure reports our spectra of 71 Tau.

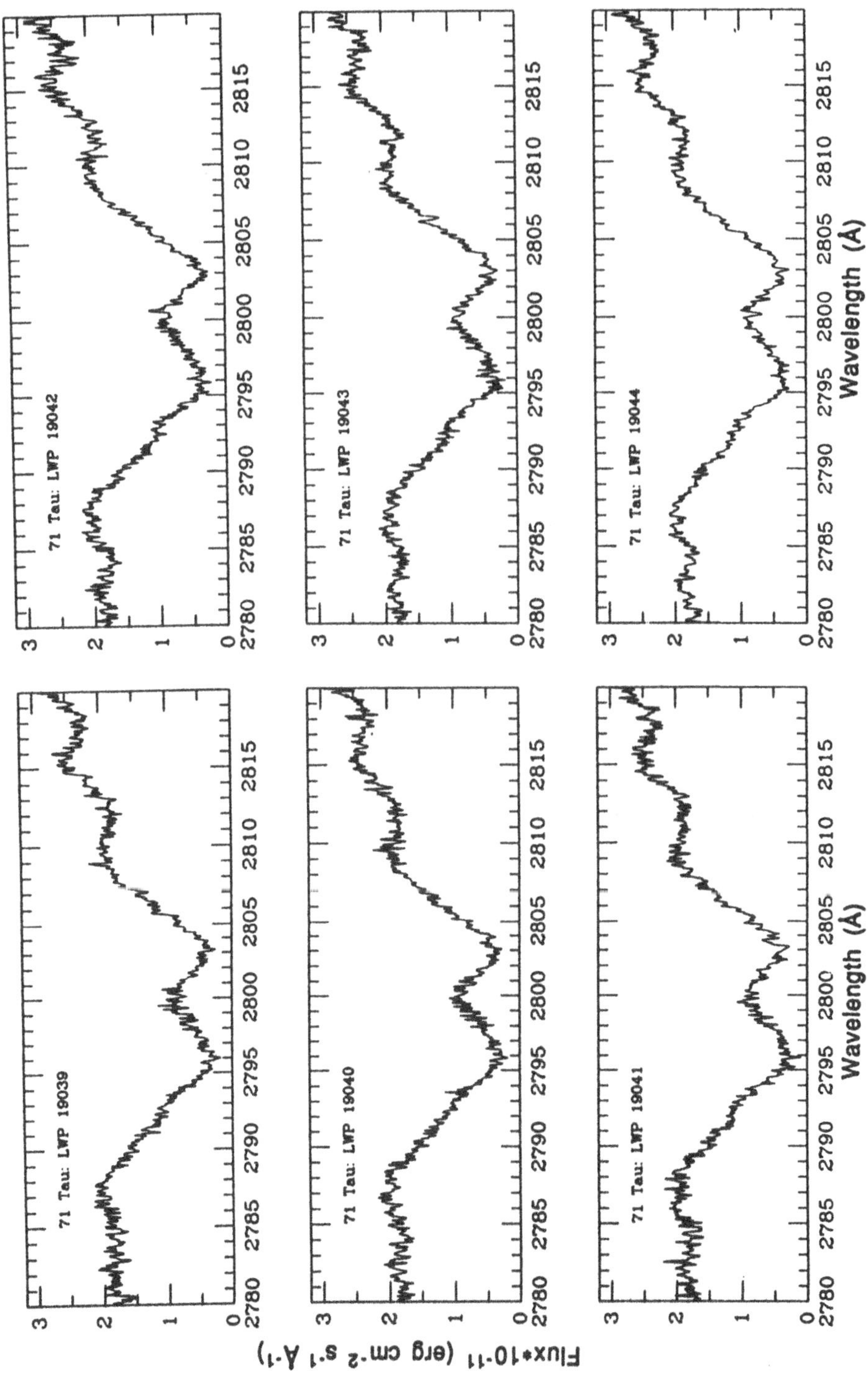
71 Tau: LWP 19039
71 Tau: LWP 19040
71 Tau: LWP 19041
71 Tau: LWP 19042
71 Tau: LWP 19043
71 Tau: LWP 19044
Flux*10^-11 (erg cm^-2 s^-1 Å^-1)
Wavelength (Å)

References

Antonello,E.: 1990, in "Angular Momentum and Mass Loss for Hot Stars", eds. L.A.Willson and R.Stalio, Kluwer Academic Publishers, p.97.

Castelli,F., Fracassini,M., and Pasinetti,L.E.: 1982, *Proc. Third European IUE* Conf., Madrid, ESA SP-176, p.135.

Fracassini,M., and Pasinetti,L.E.: 1982, *Astron.Astrophys.* **107**,326.

Fracassini,M., Pasinetti,L.E., Castelli,F., Antonello,E., and Pastori,L.: 1983, *Astrophys. Space Sci.* **97**,323.

Fracassini,M., Pasinetti Fracassini,L.E., Pastori,L., Teays,T.J., and Mariani,A.: 1991, *Astronom.Astrophys.* **243**, 458.

Micela,G., Sciortino,S., Vaiana,G.S., Schmitt,J.H.M.M., Stern,R.A., Harnden,F.R. Jr., and Rosner,R.: 1988, *Astrophys.J.* **325**,798.

Pasinetti Fracassini,L.E., Pastori,L., Teays,T.J., and Schmidt,E.G.: 1990a, in "Confrontation between stellar pulsation and evolution", eds. C.Cacciari, G.Clementini, Bologna May 28-31 1990, p.340.

Pasinetti Fracassini,L.E., Pastori,L., Schmidt,E.G., and Teays,T.J.: 1990b, in "Confrontation between stellar pulsation and evolution", eds. C.Cacciari, G.Clementini, Bologna May 28-31 1990, p.230.

Pasinetti Fracassini,L.E., and Pastori,L.: 1991, δ Scuti Star Newsletter **4**,12.

Pastori,L., Pasinetti Fracassini,L.E., and Bestetti,M.: 1990, in "Confrontation between stellar pulsation and evolution", eds. C.Cacciari, G.Clementini, Bologna May 28-31 1990,p.344.

Stern,R.A., Zolcinski,M., Antiochos,S.K., Underwood,J.H.: 1981, *Astrophys.J.* **249**,647.

Teays,T.J., Schmidt,E.G., Fracassini,M., and Pasinetti Fracassini, L.E.: 1989, *Astrophys.J.* **343**,916.

Walter,F.M.,Schrijver,C.J., and Boyd,W.: 1988, in "A Decade of UV Astronomy" with IUE, *ESA SP-281*, Vol.1, p.323.

Willson,L.A., Bowen,G.H., and Struck-Marcell,C.: 1987, *Comm.Astrophys.* **12**,17.

Zolcinski,M.C.S., Anthiocos,S.K., Stern,R.A., and Walker,A.B.C.: 1982, *Astrophys.J.* **258**, 177.

NEW EFFECTS IN THE IRAS DATA AND THEIR CONSEQUENCES FOR CORONAE OF ACTIVE LATE-TYPE DWARFS

MARIA KATSOVA

Sternberg State Astronomical Institute, Moscow State University, 119899 Moscow V-234, RUSSIA

VASSILIKI TSIKOUDI

Physics Department, University of Ioannina, 45110 Ioannina, GREECE

and

MOISSEI LIVSHITS

Institute of Terrestrial Magnetism, Ionosphere, and Radio Wave Propagation, Academy of Sciences, Troitsk, 142092 Moscow Region, RUSSIA

Abstract. The observed excess far-IR and mm-wave radiation from active K and M dwarfs could be either thermal or nonthermal in origin. We argue that the most likely explanation is thermal emission from cool winds.

Key words: infrared emission – coronae

1. Detection of the IR-excess from K and M dwarfs

Preliminary analysis of the IRAS data provided by Tsikoudi (1988, 1989, 1990) and Mullan et al. (1989) for late-type dwarfs indicates differences between the behavior of IR-radiation for active and non-active K and M dwarfs. There are small differences in the mean ratios of the observed to the calculated fluxes $F_{\rm obs}/F_{\rm calc}$ for these stars, which are comparable to the measurement errors. This means that some stars with surface activity show excess far-IR radiation compared to the blackbody radiation of their photospheres.

We have carried out a more detailed analysis of the IRAS data at 12μ and 25μ for several tens of the K and M dwarfs. We introduce the ratio E of observed IR-luminosity per unit frequency, $L_{\rm obs} - 4\pi d^2 F_\nu$, where F_ν is the observed flux at a given frequency in ergs cm^{-2} s^{-1} Hz^{-1}, to the luminosity calculated from the blackbody radiation of a stellar photosphere with the known effective temperature, $L_{\rm theor} = 4\pi R^2 \pi B(T_{\rm eff})$. If $L_{\rm bol}$ is included to avoid a strong dependence on $T_{\rm eff}$, we obtain the ratio E in the following form:

$$E(12) = 4\pi d^2 F(12) T_{\rm eff}/1.05 \cdot 10^{-11} L_{\rm bol}(F_\nu/F_{\nu_{\rm max}}).$$

Some stars shown in Fig. 1 show enhanced far-IR radiation compared to that predicted from blackbodies. Values of E(12) range from 1 to 3, and the mean ratios are E(12)=1.69 ± 0.68 for 41 stars and E(25)=2.15 ± 1.02 for 25 stars, respectively.

2. Relation Between the IR-excess and Activity

The observed IR-excesses are weakly dependent on activity as will be demonstrated below. There are at present 22 late-type stars for which measurements of magnetic

J.F. Linsky and S. Serio (eds.), Physics of Solar and Stellar Coronae, 483–487.

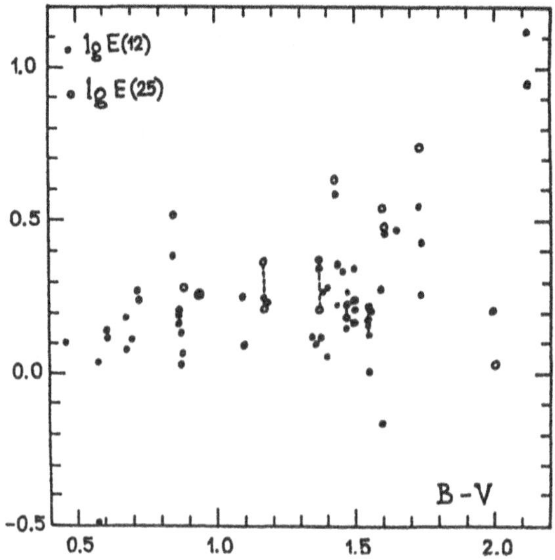

Fig. 1. The ratios E(12) and E(25) for the K and M dwarfs versus their color index.

fields as well as far-IR observations are available. We compare the values of logE(12) and logE(25) with stellar magnetic fluxes $f \times B$, where f is the filling factor and B the magnetic field strength (Saar 1989; Bruning et al. 1986) as shown on Fig. 2.

Further, we have analyzed the X-ray luminosities for the IRAS-detected dwarfs. Part of the X-ray data, mainly for the G and K stars, is analyzed here for the first time on the basis of the *Einstein* IPC archive data (Katsova & Tsikoudi 1992). The X-ray luminosities of the M dwarfs are taken from Agrawal et al. (1986) and Pallavicini et al. (1990). All available X-ray data for the stars considered are presented in Fig. 3. It is worthwhile to note that some stars with larger values of E(12) are components of binary systems.

3. Discussion

Departures of the solar radiation from a blackbody occur at $\lambda > 1$ cm. For the active late-type stars we have observed these departures from $\lambda \geq 10\mu$. If we consider the 1 mm fluxes measured by Mullan et al. (1992) as real, then additional emission in the $10\mu - 1$ cm region is described by the law $F \sim \nu^{0.7}$ for the stars YZ CMi and EV Lac and, with a somewhat steeper slope, for V1054 Oph and Gl 725 (an inactive dM4 star without emission lines in its optical spectra) (see Fig. 4). Thermal radio emission of optically thick gas in a stellar wind with the density distribution $n \sim r^{-2}$ is described by the law $F \sim \nu^{2/3}$ (Zheleznyakov 1983).

For an emission source of size on the order of R_*, the brightness temperatures of the additional radiation varies from several thousands K at $\lambda = 10\mu$ to several millions K at $\lambda = 1$ cm. This indicates that this radiation is most likely thermal.

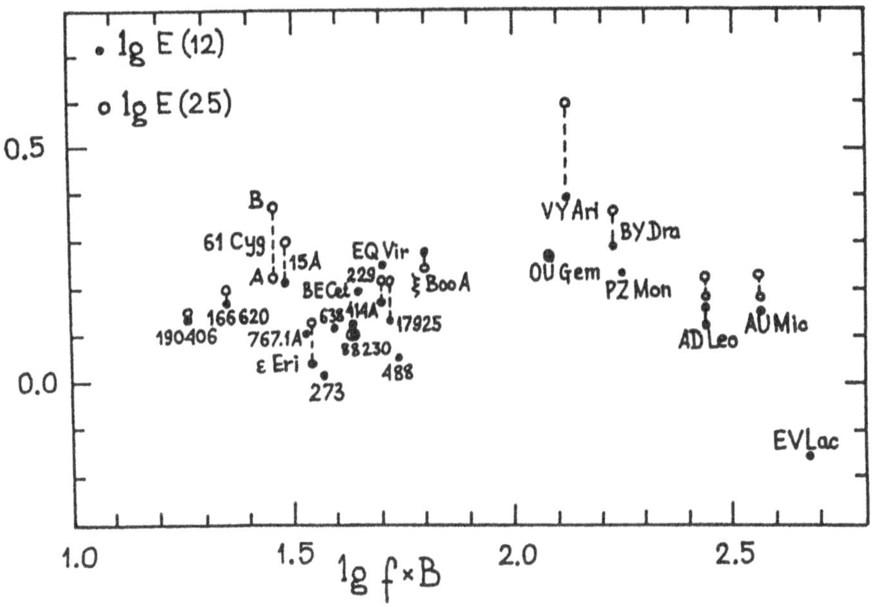

Fig. 2. Infrared flux ratios are plotted vs. the magnetic fluxes for K and M dwarfs.

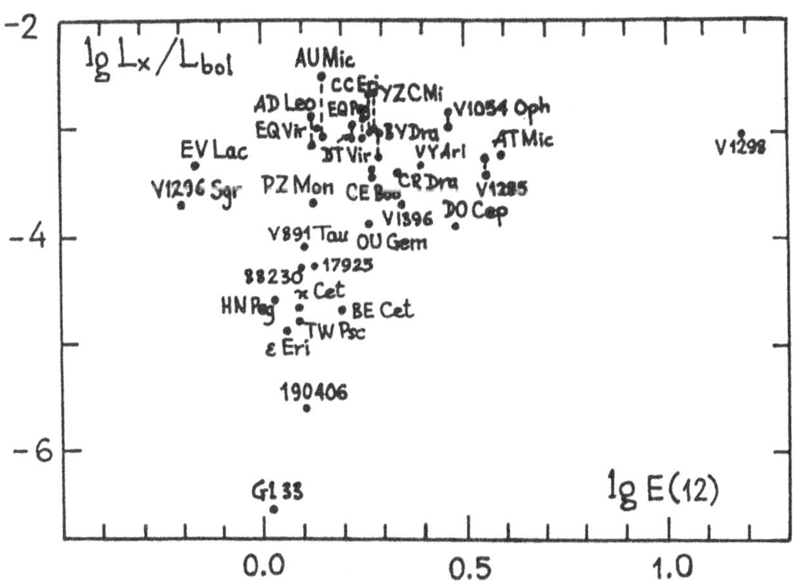

Fig. 3. Comparison of X-ray and infrared flux ratios.

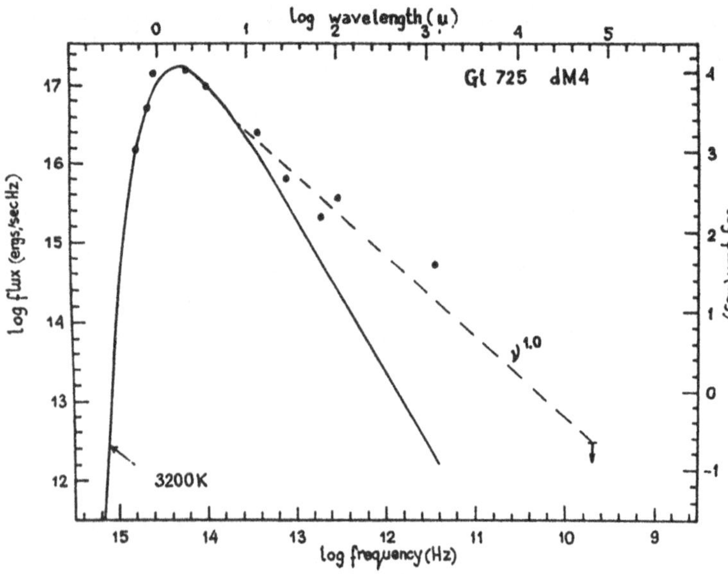

Fig. 4. Comparison of observed and blackbody fluxes for the star Gl 725

The intensity of the mm-radiation from an optically thin plasma turns out to be lower than that observed, when it is restricted to the volume that is typical for circumstellar space. Since hot coronal winds are in general optically thin (even if one adopts the large mass loss rate of $10^{-11} M_{\odot}$ yr^{-1}, estimated by Badalyan & Livshits 1992), they cannot be a source of this additional radiation.

For an optically thick cool plasma with $T = 10^4$ K and $h\nu << kT$, we estimate the size of the YZ CMi source to be $9.9 R_*$ from the observed 1.1 mm-flux of 13 mJy (Mullan et al. 1992). For the 60μ-flux of 120 mJy, the size of the source is $1.65 R_*$ if $T = 10^4$ K. Such source sizes and spectral shapes can be explained by a model in which the electron density distribution is proportional to r^{-2} as in the case of an outflow of constant speed. Densities of such a wind in the region of its formation must reach 10^{11} cm^{-3} if the wind is formed close to the stellar surface, or several times lower if the wind arises at greater heights. Mullan et al. (1992) assumed that the wind is cool and obtained the size of the 1.1 mm source as 2–3 R_*. However, the brightness temperature in this case must be 10^5 K, which contradicts the hypothesis of cool winds and the UV observations.

One may explain excess emission in the $10\mu - 1$ cm spectral range as radiation by accelerated particles. Güdel & Benz (1989) explain the 1–3 cm radiation of active red dwarfs by cyclotron radiation of hot plasma. In principle, this radiation can be extended up to $\lambda = 1$ mm, but this requires the presence of very strong magnetic fields of 10 kG and large numbers of accelerated electrons emitting at the 9–10th harmonic of the gyrofrequency.

To explain radiation in the range of tens of microns one can, in principle, invoke the synchrotron radiation of electrons with energies greater than 100 MeV. But it is difficult to understand the spectrum of the excess radiation in the far-IR, and there is an absence of observed effects at the radio frequencies. Even strong emission of several lines of neutral elements, for example MgI, which are observed at $\lambda > 8\mu$, cannot explain the observed spectrum.

4. Conclusion

Deviations of the far-IR and mm radiation from that predicted on the basis of the blackbody law have been found, together with a weak dependence of the IR excess on magnetic fluxes and, perhaps, also on the activity level. When stars are components of binary systems these effects are enhanced.

The most likely explanation for the excess radiation in the $10\mu - 1$ cm range is the existence of cool winds, extending to tens of stellar radii. It is not clear how this phenomenon can exist concurrently with hot coronae and hot winds.

An alternative interpretation is to invoke radiation of accelerated particles in magnetic fields. However, for the far-IR range, there is a problem with the acceleration of particles up to relativistic energies and, for the mm-range, the presence of large numbers of subrelativistic particles.

Spectral energy distributions in the $10\mu - 1$ cm region can be used to distinguish between thermal and nonthermal radiation mechanisms. In the first case the shape of the spectrum should be smooth and the flux approximately proportional to $\nu^{2/3}$. In the second case, the spectrum should contain structure, as would be predicted by the different emission mechanisms discussed in this paper. The detection of polarization at these wavelengths would be helpful in identifying the important emission mechanisms.

References

Agrawal, P.C., Rao, A.R., Sreekantan, B.V.: 1986, *MNRAS* **219**, 225

Bruning, D.H., Chenoweth, R.E., Jr, Marcy, G.W.: 1986, in M. Zeilik and D.M. Gibson, ed(s)., *Cool Stars, Stellar Systems, and the Sun*, Springer-Verlag: Berlin, 36

Güdel, M., Benz, A.O.: 1989, *Astron & Astrophys* **211**, L5

Katsova, M., Tsikoudi, V.: 1992, *Soviet Astron* **36(4)**,

Mullan, D.J., Stencel, R.E., Beckman, D.E.: 1989, *Ap.J* **343**, 400

Mullan, D.J., Doyle, J.G., Redman, R.O., Mathiodakis, M.: 1992, *Armagh Obs. Prepr. Ser.*, No. 127

Pallavicini, R., Tagliafferri, G., Stella, L.: 1990, *Astron & Astrophys* **228**, 403

Saar, S.: 1989, *Center for Astrophysics Preprint* No. 2970

Tsikoudi, V.: 1988, *AJ* **95**, 1797

Tsikoudi, V.: 1989, *AJ* **98**, 290

Tsikoudi, V.: 1990, *Ap. Sp. Sci* **170**, 690

Zheleznyakov, V. V.: 1983, *Izv. vuzov, Radiofizika* **6**, 647

STELLAR FLARES

P.B. BYRNE

Armagh Observatory, Armagh BT61 9DG, N. Ireland

Abstract. We review recent observational advances in the field of stellar flares in the optical and ultraviolet wavebands. In particular we make an estimate of the kinetic energy budget of stellar flares and point out that, as in the solar case, a very significant fraction of a flare's total energy may be in the form of kinetic energy of bulk mass motions.

Key words: Stellar flares

1. Introduction

Several other speakers at this colloquium will be discussing X-ray observations of flares, especially those made recently by ROSAT. So, in this review, I intend to concentrate on results obtained from other wavelength regimes, especially those from optical and ultraviolet wavelengths. Several reviews on stellar flares have appeared in recent years and the reader is referred to these for information on topics not covered here (Byrne, 1989, Haisch 1989).

First, I feel I should declare my prejudices and state what I mean by "flare" stars. The Sun is obviously a flaring star. Yet, if the Sun were placed at typical flare star distances (5–50 pc), its flares would not be detectable by present techniques. Those stars which we will classify as flare stars, however, exhibit such large flares that they have on occasion been detected by eye (Andrews, 1969). Thus, for the purposes of this review, stellar flares must be sufficiently energetic to be detectable against the global photospheric background and we shall use this as a working definition of a flare star.

This definition includes a range of mid- to late-type rapidly rotating stars. Our comments here will largely be confined to two groups of objects, the dMe or "Classical" flare stars and the sub-giant RS CVn binaries. This restriction has been adopted basically because of observational selection, in that they have been the most intensively studied of these objects.

2. Optical Photometry

Broad band optical photometry has provided a huge database on stellar flares. As measured in the Johnson/Cape $UBV(RI)_{KC}$ system stellar flare light is markedly blue in colour, reflecting a higher temperature in the flaring region than in the background photosphere, as well as a prominent Balmer line contribution to flare light (see below). Lacy et al. (1976) analysed UBV light curves of 223 flares on 8 dMe flare stars of different spectral types and concluded that *on average* the time-integrated energies in the three bands were related by $E_U = 1.2\,E_B = 1.8\,E_V$ but with a large scatter in individual flares. Hydrogen bound-free radiation contributes significantly to some flares (Doyle et al. 1989a) but a substantial number of flare continua cannot be explained in this way. Houdebine (1992) and Eason et al. (1992) conclude that free-free emission can account for the observed continuum distributions.

J.F. Linsky and S. Serio (eds.), Physics of Solar and Stellar Coronae, 489–494.

Although the study of optical flares on dMe stars is a long established observational activity, it is only recently that it was realised that continuum optical flares occur also on RS CVn stars (Doyle et al. 1991, Zhang et al. 1991). Furthermore, recent studies by Mathioudakis et al. (1992) have shown that the RS CVn star, II Peg, flares in broad band optical once every ≈ 6 hr with a U band energy $E_U \geq 10^{33}$erg. In a typical dMe star, e.g. YY Gem, such a flare would be recorded at most every $\approx 60-100$ hr (Lacy et al. 1976).

2.1. FLARE ENERGIES

Lacy et al. developed a relationship between the time-averaged energy released in U band flaring, L'_U, and the quiescent U band flux, q_U. This took the form

$$\log L'_U = 0.60 \log q_U + 10.3$$

Although the above relationship was developed for dMe flare stars, it is interesting to enquire if it might apply to optical U band flares in RS CVn stars. Mathioudakis et al. (1992) have observed a sufficient number of optical flares on the RS CVn star, II Peg, to be able to attempt to answer that question. Combining II Peg's mean non-flare U magnitude (= 9.0, Cutispoto et al. 1987) with the calibration of the U band in Allen (1973) and a distance to the star of 30 pc, we derive a quiescent U band luminosity of $\log q_U = 31.89$ erg sec^{-1}. Inserting this into Lacy et al's dMe flare star equation above yields an expected time averaged U band flare "luminosity" of $\log L'_U = 29.3$.

Mathioudakis et al. have followed Lacy et al's methods and summed the individual energies of their observed flares to derive a mean, time averaged, flare U-band luminosity of $\log L'_U = 30.3$ erg sec^{-1}, one order of magnitude larger than Lacy et al.'s prediction. Examination of their Fig. 2, however, reveals that their derived L'_U is heavily influenced by two unusually large flares, a topic addressed previously in Byrne et al. (1984). Eliminating these from the mean flare luminosity determination and summing between the observed energy yields $\log L'_U = 29.1$ erg sec^{-1} in much better agreement with the results of Lacy et al. This result would suggest that the underlying conversion of photospheric energy to flare energy is similar in these two different groups of stars.

2.2. FLARE MORPHOLOGY

The forms of late-type stellar flare light curves differ widely from one to another. At the one extreme impulsive flares exhibit two distinct parts to their light curves. They have a very rapid rise to maximum, often as short as 1 sec, followed by an initially equally rapid decay but which then slows to a more gradual phase which evolves on a time scale of 10 min or longer. Slow flares have rise times of anything up to 10 - 15 min and may take hours to decay fully. Furthermore, the two stages of the light curve decay are not seen in the latter type. The former are commonly likened to solar compact flares and the latter to two-ribbon flares on the basis of their time histories. In general, the slow flares are the more energetic and are more likely to be found on earlier type stars.

Kunkel (1973) made a comprehensive study of the light curve morphology of optical broad band flares in dMe stars. He concluded, among other things, that the durations of flares were shorter in later type stars. Specifically, he found that

$$\overline{T_{0.5}} \approx -0.15M_V + 1.6$$

where $\overline{T_{0.5}}$ is the mean time spent by the flare above half its maximum intensity in the U band. Lacy et al. (1976), on the other hand, showed that less luminous stars produce relatively more less energetic flares than their earlier type relatives. They demonstrated that flare rates, in terms of U-band energy, E_U, could be expressed as a power law and that the power law index was correlated with the stellar luminosity. Lacy et al. further showed that there is a high energy limit to flare energy which is itself dependent on spectral type.

These observations are consistent with the proposal stated above that flare stars of later spectral types produce less two-ribbon type flares and more compact, impulsive flares than do the more massive stars. On the Sun compact flares, as their name suggests, usually occur in relatively simple, compact loop structures. Two-ribbon flares, on the other hand arise in larger-scale, more complex regions involving many individual loops, the two ribbons referring to the Hα appearance of the flare.

Why the tendency for the latter to be favoured in the earlier type stars? One of the obvious differences between the two groups of stars is their gravities and, therefore, the scale heights of the non-magnetic portions of their coronae. Since magnetic loops are maintained in part by external pressure, larger loops, extending to greater heights, may disrupt in the later type stars where scale heights are smaller, favouring compact loops.

3. Optical spectroscopy

All flare stars are characterized by strong Balmer emission, even in their quiescent states. In flares these lines are usually greatly enhanced and broadened. Higher excitation lines of HeI (λ4026Å) and even HeII (λ4686Å) are sometimes seen during the rise phase. Other chromospheric lines are also enhanced, especially CaII H&K, but they are not broadened as much as the Balmer lines, as might be expected from Stark broadening. Using the solar analogue, a combination of macroscopic turbulence and Stark broadening are invoked to explain the excess broadening of Balmer lines (Doyle et al. 1988b). Balmer line broadening during the impulsive phase of flares, however, is not always symmetric, especially in the wing and explanations involving bulk motions of the flaring plasma must be evoked (see below).

4. Ultraviolet spectroscopy

The advent of the International Ultraviolet Explorer (IUE) satellite revolutionised the study of cool star activity in the ultraviolet. Many studies of the flaring behaviour of both dMe and RS CVn stars have been made (Butler et al. 1981, Butler et al. 1987, Doyle et al. 1989c, Haisch et al. 1989a,b, Linsky et al. 1989, Mathioudakis et al. 1991) and the general characteristics of UV flare radiation are now well known. During flares, the chromospheric and transition region emissions, already present in quiescence, are seen to increase by factors of 2-10. It should be borne in mind, however, that, because of IUE's limited aperture (it is after all only 45 cm) and the relative faintness of flare stars as UV sources, exposure times were long (typically 30 – 60 min) compared with impulsive flare timescales. Early results from the Hubble Space Telescope have underlined this caution with the observation of a CIV flare on AD Leo whose rise time was ≤ 25 sec and whose total duration was probably ≤ 10 min with the bulk of the variation taking place during the first 3 min (Bookbinder et al. 1992).

IUE has also revealed the presence of strong ultraviolet continua in dMe flares (Butler et al. 1981, Byrne & McKay, 1989, Mathioudakis et al. 1991) whose energy content is a very sizeable proportion of the combined chromospheric and transition region radiative losses and may even equal or exceed these. Phillips et al. (1992) have successfully interpreted the observed continua in the 1200 – 1900Å range in some flares as due to SiI(λ1521Å) and CI(λ1444Å) continua excited by backradiation from EUV line radiation. They quote the correlation of UV continuum energy and that in CIV(λ1548/51Å) (Byrne & McKay 1990) in support of their hypothesis. The relatively flat, or even blue, continua observed over the same spectral range in some dMe flares (see e.g. Byrne and McKay, 1989, 1990) are unlikely to be explained by this hypothesis alone.

IUE observations first suggested the unexpectedly high frequency of RS CVn flares as short-lived enhancements in transition region emission lines (Doyle et al. 1989b). Some of these flares are hugely energetic and long-lived by dMe standards and may even represent the release of energy stored by a different mechanism related to the orbital motion of the binary (Doyle et al. 1989c).

5. Mass motions

Although Stark broadening may account successfully for the broadened cores of the higher Balmer lines in flares, asymmetric wings extending out to as much as 700 km sec^{-1} to both blue and red have been observed (Bromage et al. 1986, Doyle et al. 1988b), especially during the impulsive rise and early decay. Furthermore, strong red wing enhancements are ubiquitous in Hα during flares (Doyle et al. 1988a) and frequently found in MgII *h&k* emission (Doyle et al. 1989, Linsky et al. 1989). These cannot be accounted for as single component Stark or turbulent broadening. Systematic motions in the flaring atmosphere have been invoked, similar to solar flare surges and sprays. This evidence in favour of mass motions generally occurs during the rise phase of the flare and, perhaps, during maximum. When observing relatively faint dMe stars it is often difficult to get adequate signal-to-noise spectra in short enough exposure times to study these events.

Recent observations with the Hubble Space Telescope of red wings extending to $\approx$2000 km sec^{-1} in the ultraviolet CIV resonance doublet emission in an AD Leo flare (Bookbinder et al. 1992) have dramatically demonstrated the reality of these effects. They recorded strong downflow velocities for ≈ 20 sec during the rise phase and flare maximum.

These observations have raised the question as to whether the very large energies *radiated* during flares represent the bulk of the energy of the process. IUE observations of time averaged flare spectra suggest typical electron densities of $N_e \geq 1\,10^{11}\,cm^3$. From Bookbinder et al.'s published light curve, we have estimated the emission measure in CIV at the peak of the HST flare as $N_e^2 dV \approx 4.9\,10^{51} cm^{-3}$. Using the arguments of Byrne et al. (1987) yields a CIV emitting volume of $\leq 5\,10^{29}\,cm^{-3}$. As in Byrne and McKay (1989) we adopt a model of the flare involving N_{loop} loops, each of footpoint separation D and uniform cross section 0.1 D. Then the total flaring volume may be stated to be $V_{loops} = 0.1\pi^2\,N_{loop}(D/2)^3$. Equating this to the derived volume above yields a footpoint separation, $D \approx 3.4\,10^{10}/N_{loop}^{1/3}$ cm and a height, $H = D/2$.

Since red-shifted CIV was observed for 25 sec with a maximum velocity of $1800\,km\,sec^{-1}$ the CIV surge would have travelled $\leq 4.5\,10^9$cm in this time. This is of the same order as H for $1 \leq N_{loop} \leq 55$, assuming, of course, that the downflowing CIV stays within the loop.

We may further investigate the consequences of this flare based on the above assumptions. The emission measure of the red-shifted component is about half that of the unshifted component and its half-intensity velocity is $650\,km\,sec^{-1}$. Therefore its volume is a third that above or $1.6\,10^{29} cm^3$. In the transition region where CIV is formed, hydrogen is completely ionized so $N_e \approx N_p$ and so a mass density in flares of at least $\rho \geq 1.7\,10^{-13} gr\,cm^{-3}$ is suggested. We would estimate the mass involved in the above flows to be greater than $2.7\,10^{16}$gr and this, at a mean velocity of $650\,km\,sec^{-1}$, represents $5.7\,10^{31}$erg of kinetic energy. This should be compared to $\approx 2\,10^{30}$erg of energy radiated in CIV.

Bruner and McWhirter (1988) showed that, for solar flaring plasmas, the total radiative losses over the entire range of temperatures $4 \leq \log T_e \leq 8$ could be calculated from the observed CIV flux. Doyle et al. (1989c, 1992) argued that this relationship also held for late-type stellar flares. We apply this here to calculate the total radiated losses from the AD Leo flare and derive a figure of 10^{33}erg. Thus the kinetic energy of the flare material at CIV temperatures (10^5K) alone is 6% of the *total* energy radiated by the flare and 25 times that radiated by CIV itself.

Mass flows have been observed also in chromospheric Balmer lines (Bromage et al. 1986, Doyle et al. 1988b). An analysis similar to that carried out for the optically thin CIV lines above cannot be so easily repeated for the optically thick Balmer lines. Nevertheless velocities $300 - 700\,km\,sec^{-1}$ are reported. Chromospheric densities are expected to be higher than those in the hotter transition region and so the contributions to kinetic energy content of the flare are likely to be at least as large as those above.

We conclude that mass flows in stellar flares are likely to contain significant amounts of kinetic energy compared to those radiated.

Acknowledgements

Research at Armagh Observatory is supported by a Grant-in-Aid from the Department of Education of Northern Ireland. The author is grateful to the organizers of the conference for financial support to attend.

References

Allen, C.W.: 1973, *Astrophysical Quantities*, Athlone Press, London

Andrews, A.D.: 1969, *IBVS* , No. 326

Bookbinder, J., Walter, F., Brown, A.: 1992, *Cool Stars, Stellar Systems and the Sun*, J. Bookbinder & M. Giampapa, ASP Conf. Series, p. 27

Bromage, G.E., Phillips, K.J.H., Dufton, P.L., Kingston, A.E.: 1986, *MNRAS* **220**, 1201

Bruner, M.E., McWhirter, R.W.P.: 1988, *ApJ* **326**, 1002

Butler, C.J., Byrne, P.B., Andrews, A.D., Doyle, J.G.: 1981, *MNRAS* **197**, 815

Butler, C.J., Doyle, J.G., Andrews, A.D., Byrne, P.B., Linsky, J.L., Bornmann, P.L., Rodonó, M., Pazzani, V., Simon, T.: 1987, *A&Ap* **174**, 139

Byrne, P.B.: 1989, *Solar Phys.* **121**, 61

Byrne, P.B., Doyle, J.G., Butler, C.J.: 1984, *MNRAS* **206**, 907

Byrne, P.B., Doyle, J.G., Brown, A., Linsky, J.L., Rodonó, M.: 1987, *A&Ap* **180**, 172

Byrne, P.B., McKay, D.: 1989, *A&Ap* **223**, 241

Byrne, P.B., McKay, D.: 1990, *A&Ap* **227**, 490

Cutispoto, G., Leto, G., Pagano, I., Santagati, G., Ventura, R.: 1987, *IBVS* **No. 3034**,

Doyle, J.G., Butler, C.J., Callanan, P.J., Tagliaferri, G., de la Reza, R., White, N.E., Torres, C.A.O., Quast, G.: 1988a, *A&Ap* **191**, 79

Doyle, J.G., Butler, C.J., Byrne, P.B., van den Oord, G.H.J.: 1988b, *A&Ap* **193**, 229

Doyle, J.G., van den Oord, G.H.J., Butler, C.J.: 1989a, *A&Ap* **208**, 208

Doyle, J.G., Butler, C.J., Byrne, P.B., Rodonó, M., Swank, J., Fowles, W.: 1989b, *A&Ap* **223**, 219

Doyle, J.G., Byrne, P.B., van den Oord, G.H.J.: 1989c, *A&Ap* **224**, 153

Doyle, J.G., Kellett, B.J., Byrne, P.B., Avgoloupis, S., Mavridis, L.N., Seiradakis, J.H., Bromage, G.E., Tsuru, T., Makashima, K., McHardy, I.M.: 1991, *MNRAS* **248**, 503

Doyle, J.G., Byrne, P.B., Avgoloupis, S.,Seiradakis, J., Bromage, G.E., Kellett, B.J.: 1992, *A&Ap* , in press

Eason, E.L.E., Giampapa, M.S., Radick, R.R., Worden, S.P., Hege, E.K.: 1992, *AJ* **104**, 1161

Haisch, B.M.: 1989, *Solar Phys.* **121**, 3

Haisch, B.M., Schmitt, J.H.M.M., Rodonó, M., Gibson, D.M.: 1989a, *A&Ap* **230**, 419

Haisch, B.M., Butler, C.J., Foing, B., Rodonó, Giampapa, M.S.: 1989b, *A&Ap* **232**, 387

Houdebine, E.R.: 1992, *Irish AJ* , in press

Kunkel, W.E.: 1973, *ApJS* **25**, 1

Lacy, C.H., Moffett, T.J., Evans, D.S.: 1976, *ApJS* **30**, 85

Linsky, J.L., Neff, J.E., Brown, A., Gross, B.D., Simon, T., Andrews, A.D., Rodonó, M., Feldman, P.A.: 1989, *A&Ap* **211**, 173

Mathioudakis, M., Doyle, J.G., Rodonó, M., Gibson, D.M., Byrne, P.B., Avgoloupis, S., Linsky, J.L., Gary, D., Mavridis, L.N., Varvoglis, P.: 1991, *A&Ap* **244**, 155

Mathioudakis, M., Doyle, J.G., Avgoloupis, S., Mavridis, L.N., Seiradakis, J.H.: 1992, *A&Ap* **255**, 48

Phillips, K.J.H., Bromage, G.E., Doyle, J.G.: 1992, *ApJ* **385**, 731

Zhang, R-X., Zhai, D-S., Zhang, J.T., Li, Q-S.: 1991, *IBVS* **No. 3456**,

AN EXTENDED CORRELATION BETWEEN THE BALMER AND SOFT X-RAY EMISSION FROM SOLAR AND STELLAR FLARES *

C.J. BUTLER

Armagh Observatory, College Hill, Armagh, BT61 9DG, N. Ireland

Abstract. The integrated soft X-ray (8-12 A) fluxes for solar flares, given by Thomas and Teske (1971), have been scaled to the equivalent EXOSAT (0.04-2.0 KeV) fluxes using spectra obtained from a variety of rocket-based experiments. A model by Mewe for a 1.5 KeV plasma gives essentially similar results. The data show good agreement with the soft X-ray - $H\gamma$ correlation established earlier by Butler, Rodonò and Foing (1988) for flares on dMe stars. Simultaneous GINGA and optical spectroscopy of a flare on the RS CVn star II Peg extends the correlation a further two orders of magnitude and confirms that a common, linear, relationship applies to all stellar and solar flares for which data are currently available (see figure 1).

Key words: Flares: solar and stellar, Balmer emission, X-rays, dMe stars, RS CVn stars

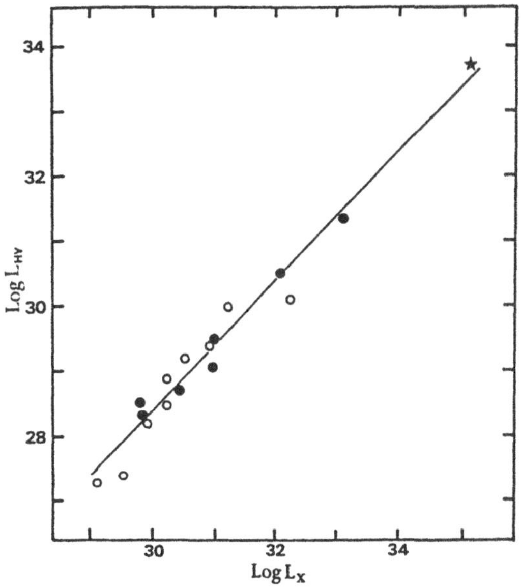

Figure 1. The integrated flux in $H\gamma$ against the integrated flux in soft X-rays (0.04-2.0 KeV) for flares on: dMe stars (*filled circles*), the Sun (*open circles*) and the RS CVn star II Peg (*star*). The line has slope unity and corresponds to the relation $L_X = 40L_{H\gamma}$. The solar flare X-ray fluxes from OSO III have been converted to the equivalent EXOSAT(CMA) fluxes using the relation: $F_{CMA} = 15F_{OSO}$.

References

Butler, C.J., Rodonò, M. and Foing, B.F.: 1988, Astron. and Astrophys. 206, L1

Thomas, R.J. and Teske, R.G.: 1971, Solar Phys. 16, 431

* The full text of this paper is to be published by Astronomy and Astrophysics

J.F. Linsky and S. Serio (eds.), Physics of Solar and Stellar Coronae, 495.

ENERGETICS AND DISTRIBUTION OF GAMMA RAY BURSTS AND FLARES ON STARS

A. R. RAO, M. N. VAHIA and R. K. SINGH
Tata Institute of Fundamental Research
Homi Bhabha Road, Bombay 400 005, India

Abstract. We had suggested (Vahia and Rao,1988) that the Gamma Ray Bursts (GRBs) are energetic flares occurring on binary systems with Sun like magnetic activity, i.e. the Magnetically Active Stellar Systems (MASS) consisting of Flare Stars, RS CVn binaries and Cataclysmic Variables. We reinforce this suggestion by discussing the possible gamma ray emission from MASS by extrapolating from the X-ray emission from these objects. We also show that the new results by the Burst and Transient Source Experiment (BATSE) on the GRO are consistent with known distribution and properties of the MASS systems. Using this result we further predict the anisotropy that should become visible in the GRO data as its data base increases and specific object identification should become possible as the localization improves.

Key words: Gamma Ray Bursts – Flares on stars – Cataclysmic Variables – RS CVn Binaries – Flare Stars

1. Introduction

The main characteristics of the Gamma Ray Bursts (GRB) is their large variation in structure, duration as well as spectra (Mazets et al, 1981). Attempts to identify specific objects in other wave bands from the regions of the GRB error boxes have met with little success. GINGA (Murakami, 1991) has reported a few Gamma Ray Bursts whose spectra have a distinct absorption feature.

The new results by the Burst and Transient Source Experiment (BATSE) on the Gamma Ray Observatory suggest that the distribution of the γ-ray burst sources is isotropic about the Earth's position but are radially non uniform. It was suggested that these constraints can be reconciled if there are at least two classes of bursts (Lingenfelter and Higdon, 1992). More importantly, the GRO/BATSE results have shown that the GRB distribution is approximately isotropic with a peculiar log(N) - log (P) distribution (Meegan et al, 1992) that is inconsistent with either isotropic distribution or Galactic Disc distribution (Fenimore et al, 1992).

We discuss a suggestion made by us some time ago (Vahia and Rao, 1988, hereafter Paper I) that the GRBs are flares on magnetically active stellar systems, namely Flare Stars, RS CVn binaries and Cataclysmic Variables which have been known to show sun like activity of inter binary scales. The positional association between the two as well as the large variety of time structures were discussed in the earlier paper (Paper I). We show that this suggestion is consistent with all the observations of GRB and make predictions for some future observations based on this suggestion.

2. Suggestion

Magnetic activity on the sun shows a large variety of phenomena from sun spots to prominences and flares. Studies during the last couple of decades have shown that the sun is not unique in showing such activity and that there are several classes of

J.F. Linsky and S. Serio (eds.), Physics of Solar and Stellar Coronae, 497–500.

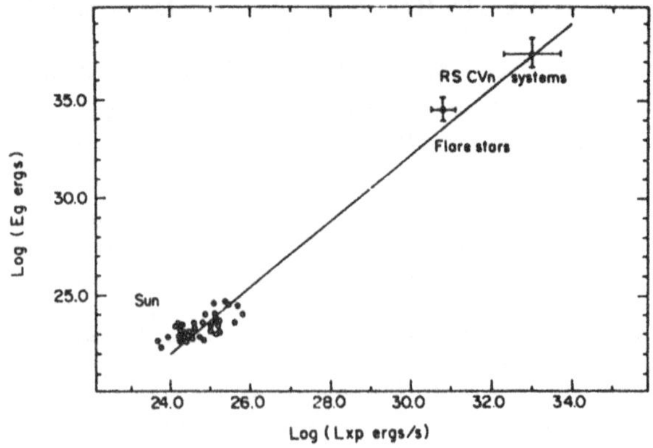

Fig. 1. *Correlation between (E_γ) and the peak X-ray luminosity of the L_{X_p} for Solar flares (Starr et al, 1988) extended to the observed L_{X_p} in Active stars (Rao and Vahia, 1987) to estimate the (E_γ) of the gamma ray transients on Active stars*

stars that show sun like activity. In particular Flare stars and RS CVn binaries have been known to have spots as well as flaring activity. These, largely binary systems, consist of late type stars with active dynamo and strong magnetic fields. Apart from these the magnetic Cataclysmic Variables are also known to show magnetic fields and activity of the order of 10^5 Gauss. Since most of these objects are in binaries with objects having active coronae and dynamo, it has been suspected that these objects may have flares of inter binary scale size. This suggestion was confirmed from direct VLA and VLBI observations as well as from soft X-ray flares observed on these objects (Kundu et al, 1988). It is therefore essential that some fraction of the energy from these objects should be emitted in gamma rays as transient gamma ray events or gamma ray bursts. Such events, would have a large variety of time structures as well as a large variation in spectra and duration depending on the line of site of the observer with respect to the burst event.

Rao and Vahia (1987) have analyzed the observations of the Ariel V satellite and shown that the Fast Transient X-ray events recorded by the satellite are from flares on RS CVn binaries and Flare Stars (Active stars). Starr et al (1988) have shown that in Solar flares the energy emitted in gamma rays (E_γ) and the peak X-ray luminosity of the L_{X_p} of the flares are correlated. Using that correlation and extrapolating to the peak X-ray luminosity of the Active stars we can estimate the expected Gamma ray energy from such flares. This is shown in figure 1. As can be seen from the figure, the (E_γ) of the flares on Active stars is comparable to the GRB energies if they occur in the Solar neighborhood. For the third type of star in our grouping of MASS, namely the Cataclysmic Variables are more energetic due to an accretion disc and magnetic dragging and the energy release of such systems is considerably higher (cf. References in Paper I and Vahia and Rao 1991).

In the present study we take the distribution, densities and luminosities of the MASS systems from the literature and we show that the Log(N) - Log(P) graph follows naturally from the suggestion that GRB are flares on MASS. Using this result we further predict the anisotropy that should become visible in GRO as its data base increases.

3. Observational Features of GRB

The recent GRO/BATSE results have shown that the GRB population is inconsistent with the isotropic or entirely Galactic Disc Population (Meegan et al, 1992; Fenimore et al, 1992, Lingenfelter et al, 1992).

Results from SIGNE, APEX and LILAS (Atteia et al, 1991) have shown that a) deviation from a uniform distribution indicating a decrease in number density with distance and b) Concentration of the weak sources in the Galactic Disc c)Strong Bursts have harder spectra.

During a brief period when the previous generation of interplanetary network of Gamma Ray satellites were active, the Ariel V satellite was also active in the 0.1 - 5 keV band and observed about 27 fast transient X-ray (FTX) events. Of these, one FTX event arose from the same region of sky and at the same time as a GRB and another was temporally coincident though the GRB was not localized (Paper I). The FTX as a class are overwhelmingly associated with Flare Stars and RS CVn binaries (Rao and Vahia, 1987). Hence, even though the specific objects associated with these two FTX event have not been identified, this could very well be due to the catalogue incompleteness of such objects(Rao and Vahia, 1987; Paper I).

In our earlier study we had attempted a positional coincidence between the GRB positions (for GRBs localized to less than 100 arc min^2 area) and the MASS. There is a statistically significant (confidence level better than 10^{-5}) number of MASS in the neighborhood of the GRBs.

4. Fitting the GRO/BATSE results with MASS/GRB distribution

One of the highlights of the GRO/BATSE observations of GRB has been the fact that while the number of GRB decreases significantly with distance, their distribution continues to be isotropic(Fenimore et al, 1992).

Using the parameters of the distribution of these objects from Harwit et al (1988) and Paper I we predict the Log(N) - log(P) distribution as well as anisotropy expected for GRB if they are flares on stars. In figure 2 we have plotted the contribution from various objects to GRB Log(N) - Log(P) as well as the total expected log(N) - Log(P) distribution. In the lower panel we have plotted the anisotropy in GRB expected from these suggestions. As can be seen from the figure, the predictions are in good agreement with the current observations but predicts a distinct anisotropy at higher intensity GRB. Preliminary results from a recent Franco Russian observation of GRB shows a clear indication that the higher energy GRB are concentrated towards the galactic plane (Atteia et al, 1991) which is in good agreement with the present suggestion.

5. Discussion

In Paper I we have discussed various models for the GRB and have concluded that flares on stars is the most promising scenario for GRB. Subsequent studies in Solar Physics, MASS systems as well as GRB have brought out further evidence that the

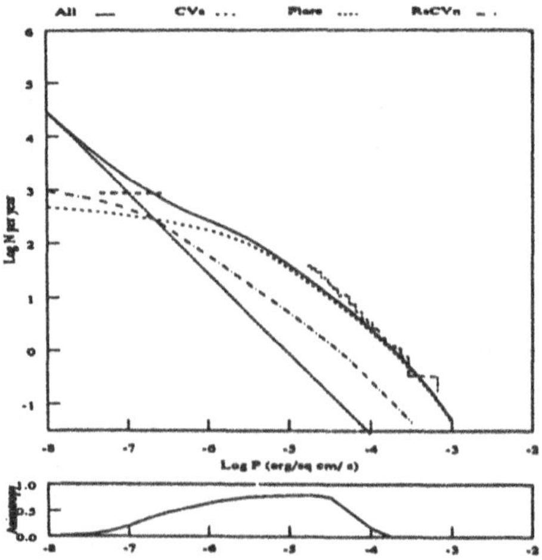

Fig. 2. *Observed and predicted Log(N) - Log(P) distribution of GRBs (Fenimore et al, 1992) if they arise from flares on MASS. The lower panel shows the expected anisotropy.*

GRBs are flares on stars. Such objects can account for the following features: a) the large variety in the time structures of GRB, b) energetics of the GRB c) observed positions of GRB d) Log(N)- Log(P) distribution as well as e) the observed isotropy.

One of the most conspicuous features of this model fitted to the GRB log(N) - Log(P) data is the high degree of anisotropy with a clear enhancement in the disc for the intermediate luminosity. The GRB of this luminosity are less frequent and the GRO/BATSE should start to observe statistically significant number of them in the next few years. However, the indications from the SIGNE, APEX and LILAS (Atteia et al, 1991) indicate that there is an increase in the anisotropy in the lower intensity bursts of the order of 10^{-5} to 10^{-6} ergs/s. Also, with the expected improvement in the localization the GRB sources, the positional association (currently at a confidence level of 10^{-5} (Paper I), partly due to positional localization of GRB and partly due to catalogue incompleteness of MASS, will also improve.

References

Atteia, J. L., et al., 1991, Int. Cosmic Ray Conference, Dublin 1, 93
Fenimore, E. E. et al. 1992, *Nature*, **357**, 140
Harwit, M. et al., 1988, in *Astrophysical Concepts*, **195**, 159
Kundu, M. R. et al., 1988, *A&A*, **195**, 159
Lingenfelter and Higdon, 1992, *Nature*, **356**, 132
Mazets, E. P., et al., 1981, *Astrophys. Space Sci.*, **80**, 3
Meegan, C. A. et al., 1992 *Nature*, **355**, 143
Murakami, T., 1991, *Adv. Space Res.*, **11**, 119
Rao, A. R. and Vahia, M. N. 1987, *A&A*, **188**, 109
Starr, R., et al., 1988 *ApJ*, **329**, 967
Vahia, M. N. and Rao, A. R., (Paper I) 1988, *A&A*, **207**, 55
Vahia, M. N., Rao, A. R., and Singh, R. K., 1991, *A&A*, **250**, 424

OBSERVATIONAL EVIDENCE FOR GALACTIC ORIGIN OF COSMIC RAYS ABOVE 10^{19} eV AND RECONNECTION ON GALACTIC SCALES

M. N. VAHIA, R. K. SINGH and A. R. RAO
Tata Institute of Fundamental Research
Homi Bhabha Road, Bombay 400 005, India

Abstract. The space distribution of 10^{19} eV cosmic rays obtained from the four major observatories at Haverah Park, SUGAR, Volcano Ranch and Yakutsk are compared with the expected distribution by taking the zenith angle distribution of the cosmic rays in each station individually and predicting the distribution in other coordinate systems. The observations are consistent with uniform distribution of cosmic rays along RA and Dec on the angular size greater than 10°. We then use the same procedure to predict the distribution that is expected to be seen along the galactic coordinates (l_{II} & β_{II}). While there is no marked difference between the observed and expected values in the β_{II} plane, there is a clear anisotropy when plotted along the l_{II} plane with a chance probability of 5 10^{-6}. We then plot the l_{II} distribution along the galactic mass distribution as well as the magnetic field distribution. This plot shows that there is a preference for these cosmic rays to arrive from the Galactic arm region (l_{II} around 120 and 300).

Key words: Cosmic Rays – Galactic Dynamics

1. Introduction

Several recent studies (cf. Vahia, Rao and Singh, 1991, Chi et al 1992) have concentrated on the possible origin of the highest energy cosmic rays, a total of 943 of which have been observed from Haverah Park SUGAR, Volcano Ranch, and Yakutsk and are published by the World data center C2 on Cosmic Rays. Vahia, Rao and Singh (1991) have shown that while the composition and other studies of the low energy cosmic rays are consistent with stellar origin, there is also a clustering of highest energy cosmic rays indicating that they arise from sources within the Galaxy. Chi et al (1992) in a series of papers have shown that 1) The highest energy cosmic rays have a clear excess in the galactic plane in the anti center direction, 2) The observations indicate a clear clustering of the cosmic rays with 22 clearly identified sources, 3) The composition of the highest energy cosmic rays is possibly iron rich.

In the present study we concentrate on the distribution of the cosmic rays in the galactic plane. We first confirm that the cosmic rays show statistically significant deviation from the uniform distribution *only* when plotted on the galactic plane. We then superpose the l_{II} distribution of the cosmic rays on the galactic mass distribution and the magnetic field distribution and discuss the inferences from this observation.

2. Procedure

The problem of converting the cosmic rays catalogued from one observatory to another is largely a problem of being able to correct for the exposure characteristics of each station such as exposure time, field of view, geometrical factor, and detector sensitivity with respect to zenith angle. For the present study we follow a procedure

J.F. Linsky and S. Serio (eds.), Physics of Solar and Stellar Coronae, 501–504.

similar to that of Chi et al (1992). In order to extract the detector response of various detectors, we use the zenith angle distribution of the observed cosmic rays in each station. Using the latitude and longitude of each station, we convert the zenith angle distribution to the expected flux in the RA-DEC plane assuming the isotropic distribution of cosmic rays. Since the detectors cover a strip along a declination band covering all RA with each rotation, we would expect the flux would be uniformly distributed in RA. The observed and expected distribution are fitted and the χ^2 of the fit are given in table 1. As can be seen from the table, the fit is quite acceptable.

TABLE I

Difference between the observed and expected values of cosmic rays in various coordinates:

Coordinate	χ^2	Degrees of freedom	Confidence level
Right Ascension	34.4	35	0.54
Declination	24.7	35	0.9
l_{II}	87.8	35	5.2 10^{-6}
β_{II}	47.3	35	0.1

An accurate test of the fitting procedure however is that it fits the observed declination distribution since each detector covers a limited declination range. The declination fit for the complete data was normalized for each station by the total flux observed by the station. This procedure ensures that the fit corrects itself for different operational time scales for various detectors. As can be seen from table 1 the fit is very satisfactory indicating that the assumption and procedure of data fitting is satisfactory.

Finally, the data is converted to the galactic coordinate system and the data and the expected distribution are superposed on each other. As can be seen from table 1, the data fits well with the expected β_{II} distribution. However the results from l_{II} distribution are significantly different and we discuss them in detail in the next section.

Results

The results from the l_{II} are plotted in figures 1.The first point is that the expected observation itself shows a double peaked structure. This is because of the fact that these l_{II} regions are covered by Yakutsk and SUGAR which have observed much larger fluxes than the other two stations. The smooth line is the distribution expected from isotropic distribution of cosmic rays while the histogram shows the observed distribution. The lines on the top mark identification of one or more (indicating by stacking of the line) cosmic ray sources identified by Chi et al (1991).

Another feature that is equally prominent is that the observed and the expected values do not fit with a chance scatter of 5 10^{-6} indicating that there is a clear deviation from uniform distribution in the data. The data continues to show significant deviation even when the cosmic rays from high galactic latitude are ignored (cf. Chi et al 1992). In order to extract the physical significance of it, we have plotted the

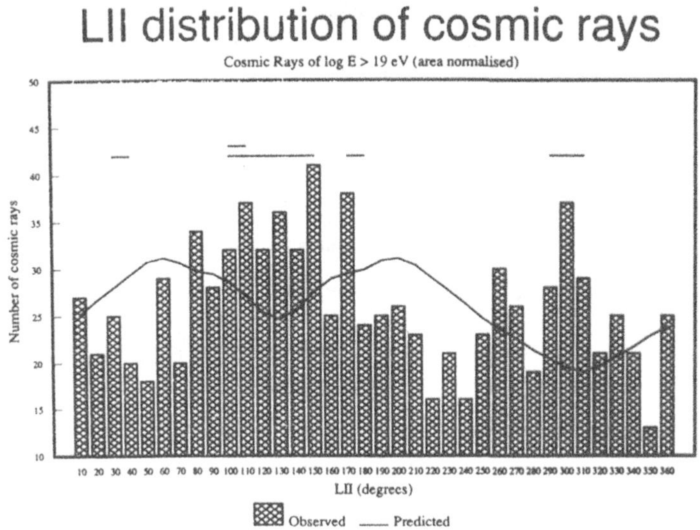

Fig. 1. *Cosmic ray distribution along l_{II} (see text).*

l_{II} distribution on the Galactic mass distribution (Puget, 1983) and magnetic field distribution (Rand and Kulkarni, 1989) in figure 2. In the figure, the galactic mass distribution is shown by a solid line while the magnetic field distribution (which shows a cylindrical symmetry) is indicated by dotted lines and the field direction in each region shown by an arrow.

As can be seen from the figure, the maxima from the cosmic ray distribution is concentrated along those regions where the galactic magnetic field neutral line crosses the galactic mass distribution in both the arms that is in regions with l_{II} 130° and 300°. We discuss the significance of this below.

3. Discussion

There are several features in the studies discussed above which strongly indicate that the cosmic rays of highest energy are from galactic sources. We list each of these below.

a) The distribution of the cosmic rays of energy $> 10^{19}$ eV clearly shows a markedly different distribution from that expected from isotropic arrival direction only when plotted in galactic coordinates.

b) The excess shows up from regions where magnetic activity of galactic scale is likely.

In view of this we feel that there is a strong case for considering the possibility that the cosmic rays of highest energy also arise from galactic sources. The fact that the cosmic ray fluxes are seen from regions where the galactic field inversion takes place i.e. where the galactic magnetic field gradients are large, indicates some significant physical processes. Even relatively weak stellar sources emitting cosmic

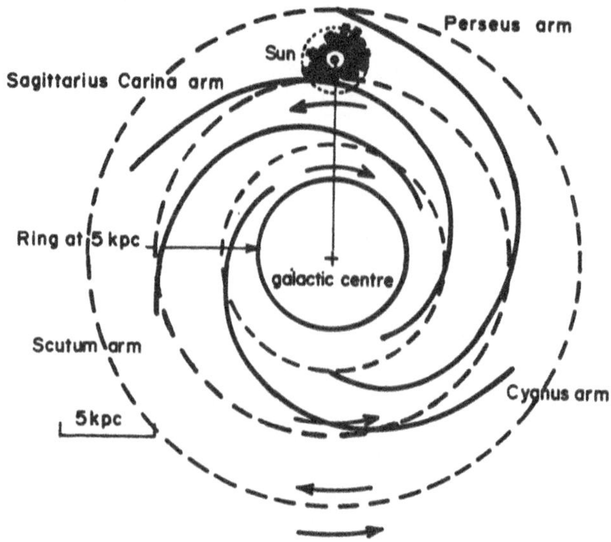

Fig. 2. l_{II} *distribution of cosmic rays plotted along with the mass distribution of the galaxy (Puget, 1983) as well as the magnetic field distribution (Rand and Kulkarni, 1989) of the galaxy.*

rays in these regions could act as injectors to this region of steep field gradients allowing particles to be accelerated to the highest energies.

However the problems associated with the acceleration of particles to these energies has remained difficult to track. Vahia et al (1991) had suggested that these could arise from inter binary flares in magnetically active stellar systems (MASS). Another possibility is that the cosmic rays are accelerated in compact regions (a few pc across) around the neutral line of the Galactic magnetic field reversal that occurs in the cylindrically symmetric magnetic field where relatively small regions would show high gradients and a stellar systems existing in such a region would be excellent injectors of high energy cosmic rays.

References

Chi X *et al.*, 1992, J. Phys. (G), **18**, 539, 553, 567

Puget J L, 1983, in *Birth and Infancy of Stars*, ed. R. Lucas, A. Omont and R. Stora, Les Houches Session XLI, page 77

Rand R. J. and Kulkarni S R, 1989, *Astrophys. J.* **343**, 760

Vahia, M. N. Rao, A. R. and Singh, R. K. 1991, *Astron. Astrophys.*, **250**, 424.

PRISMA: PROBING STARS FROM CORE TO CORONA *

T. APPOURCHAUX

ESA ESTEC/SSD, P.O. Box 299, NL-2200 AG Noordwijk, The Netherlands

C. CATALA

DESPA Observatoire de Paris-Meudon, F-92100 Meudon, France

S. CATALANO

Osservatorio Astrofisico, Cittá Universitaria, I-95125 Catania, Italy

S. FRANDSEN

Astronomisk Institut, Univ. Langelandsgade DK-8000 Aarhus, Denmark

C. FRÖHLICH

PMOD/WRC, POB 173, Davos Dorf, Switzerland

D.O. GOUGH

Institute of Astronomy, Madingley Road, Cambridge CB3 OHA, UK

P. HOYNG

SRON, Sorbonnelaan 2, 3584 CA Utrecht, The Netherlands

A. JONES

Astronomisk Institut, Univ. Langelandsgade DK-8000 Aarhus, Denmark

P. LEMAIRE

IAS, BP 10, F-91371 Verrières-le-Buisson Cedex, France

G. TONDELLO

Dip. Elettronica ed Informatica, Via Gradenigo 6A, 35100 Padova, Italy

and

W. WEISS

Institut für Astronomie, Türkenschanzstrasse 17, A-1180 Wien, Austria

Abstract. We describe here the space mission PRISMA (**P**robing **R**otation and **I**nterior of **S**tars: **M**icrovariability and **A**ctivity), which represents a global approach to the study of stars from core to corona. The scientific goals of PRISMA are: stellar structure and evolution, stellar internal dynamics and rotation, stellar dynamo, and surface magnetic field structure. Asteroseismology and activity monitoring are the two complementary tools of PRISMA designed to meet the scientific objectives. The scientific objectives, the mission concept, the main characteristics of the model payload and observation strategies are briefly described. PRISMA is an ESA mission, presently in phase A study.

Key words: – Stellar physics – Asteroseismology – Stellar activity

1. Introduction

Stellar evolution is currently described by models which reproduce fairly well the different phases in the life of a star on the H–R diagram, but a careful examination reveals a large number of problematic cases and the many simplifications and approximations that the models are based on. A significant role is played by uncertainties in the equation of state, opacities and the theory of convection.

* Presented by S. Catalano

J.F. Linsky and S. Serio (eds.), Physics of Solar and Stellar Coronae, 505–508.

Standard stellar models assume spherical symmetry, steady state structure and surface homogeneity, while real stars show surface inhomogeneities and variability on several time scales. Stellar interior dynamics, in particular the internal rotation and the relevant hydrodynamics, must be properly treated to have a more realistic description of stars. Here are some of the problems for which PRISMA can effectively contribute solutions.

1. The interior rotation pattern of the Sun determined from the observed splitting of p–modes was found to differ from the one predicted for a convective shell and by the presently assumed chemically mixed core. There is no fast spinning core in the Sun. The rotation rate in solar–type stars is known to decrease with age because of the angular momentum loss through the magnetic wind, but surface and internal rotation rates for different types of stars are needed to provide a solid observational basis for models of angular momentum transport at different levels of magnetic braking and to include core–envelope coupling that can justify the internal rotation pattern of the present day Sun.
2. Clear evidence has been found for the stellar equivalent of sunspots and for stellar activity cycles. Despite vigorous efforts, we do not yet understand how a solar–type dynamo works, although differential rotation and helical convection are known to be important ingredients.
3. By analogy with the Sun, the magnetic field generated by the dynamo action is thought to channel energy from the subphotospheric layers to the outer atmosphere. Little is known, however, about the mechanisms heating the chromospheres and coronae of late–type stars, although many theories exist. The details of the distribution of magnetic flux over stellar surfaces are also unknown: the properties of stellar butterfly diagrams and the possible existence of active longitudes or polar spots or plages remain a mystery.

The acquisition of high–quality data, which cannot be obtained from either ground based observatories or other current satellites, makes PRISMA a unique facility suitable for significant advances in the studies of the evolution and internal structure of stars, as well as surface activity and coronal plasma.

2. Measurements and observation strategy

PRISMA is designed to provide the necessary observational data by means of two complementary sets of measurements: white light measurements to detect oscillation, modes and activity tracers monitoring.

The numerous efforts to observe oscillations in solar–type stars from the ground have only recently given a positive result for the bright star α Cen (Pottash et al. 1991), whose power spectrum of the Doppler velocity clearly displays evenly spaced peaks. Observations of brightness fluctuations from space have proven to be a very powerful tool to sound the solar interior (see Fröhlich and Toutain 1990).

High precision photometry will be used to observe the frequencies, amplitudes and lifetime of eigenmodes of stellar oscillations with PRISMA. This method appears to be the most promising to perform an exstensive survey of stars in the H–R diagram, because the use of a wide wavelength band provides the large photon flux needed to reach faint stars, including those in some nearby open clusters.

Pulsation eigenmodes carry a wealth of information on the state of the interior of stars. By measuring the width and the amplitude of p–modes, one can study the excitation of the modes and the conditions in the upper part of the star. Frequency splitting and broadening carry information on the interior rotation rate as a function of the radius.

Rotation and magnetic activity mutually interact in stars with convective envelopes. The measurements of activity tracers should permit the derivation of the surface rotation and the surface structure in various parts of the external atmosphere. Analysis of the rotation modulation of the photospheric, chromospheric, transition–region, and coronal signals should provide a crude image of the magnetic field structure in three dimensions.

Since one of the primary goals of the PRISMA mission is to understand the structure of the Sun, F–G–K stars are primary targets, but a reasonable sampling of the H–R diagram is essential to put the Sun in the more general context of stellar evolution. The observation of more than 100 targets in a two–year mission, with exposure times of at least 1 month each imposes a strategy in which several stars must be observed simultaneously. To optimize the results on stellar structure models, it is essential that one or more of the basic stellar parameters be known; therefore the high priority targets are:

- members of open clusters for which a good estimate of the age is available;
- members of some binary systems for which the masses, and possibly the radii, are well known;
- stars observed by HIPPARCOS, for which absolute luminosities are well determined;
- stars observed by IUE and ROSAT.

Classical pulsating stars as well as non–radial pulsators are also of high interest for the mission.

3. The proposed payload

The model payload proposed to achieve these scientific objectives, includes four instruments:

1. Two photometers dedicated mainly to asteroseismology will measure the intensity of the visible spectrum within a band 300 nm wide. The Large Photometer (**LP**), with a 40–cm diameter effective aperture and a 1.5° x 1.5° field of view, is dedicated to stars brighter than eighth magnitude. The star image blurred to about $1'$ will be spread over about 100 pixels of the CCD detector, allowing some 10^7 photoelectrons to be collected for each exposure. The **LP** drives the pointing axis of the UV Spectrometer and XUV Telescope imager near the anti–Sun direction. A Small Photometer (**SP**) with a 15–cm diameter effective aperture has a 3° x 3° field of view. It is designed to observe stars brighter than sixth magnitude using a custom–made diode array as a detector. A movable mirror allows a wide selection of field offsets from the line of sight of the other instruments. A precision of 1 ppm per unit frequency–band is the objective of PRISMA measurements.

2. An Ultra Violet Spectrometer (**UVS**), with a 60 x 45 cm aperture observes objects in a 1.5° x 1.5° field of view. The wavelength range 120–280 nm is covered by a cross–dispersed echelle system yielding a resolving power of about 29000. Although the whole spectrum is recorded, only the more important chromospheric and transition region lines will be monitored: Mg II k (λ = 279.6 nm), He II (λ = 164.0 nm), C IV (λ = 155.0 nm) and Ly α (λ = 121.6 nm). The photometric precision of 10% along the line emission profile can be reached with integration times much shorter than the stellar rotation period, allowing a good resolution of the rotation modulation.
3. An eXtreme Ultra Violet Telescope (**XUVT**) images the field of the LP at the high coronal temperature (10^6 K) plasma sampled by the Fe VIII–XI lines included within a bandpass of 3 nm near 17.0 nm. The opportunity to have a second band including the Fe XXII–XXIII lines near 13.0 nm formed at T = 10^7 K is under consideration. A set of four coaligned concave gratings will image the flux selected by the multilayer coating.

4. Conclusions

PRISMA appears to be a space experiment well suited for probing the interior of stars by measuring microvariability and activity. Long time series of continuous observations (typically one month) will provide the resolution needed to detect eigenmodes of stellar oscillations, to completely cover the phase of stellar rotation for the activity measurements, and to define time scales for the lifetimes of spots, plages and flares.

To remain in the field of this meeting let us summarize some important contributions that PRISMA would give to coronal physics: sampling the high and low coronal temperature plasma with the XUV bands during the stellar rotation will allow us to estimate the relative distribution of the quiet and active coronal regions; the scale height of the emission from coronal loops can hopefully be derived; relative fluxes from the chromosphere, transition region and corona will provide scaling laws for energy deposition in different atmospheric environments.

The phase **A** study now in progress will further define the scientific impact of PRISMA, already summarized in the Assessment Study (Appourchaux et al. 1991).

References

Appourchaux,T., Catala,C., Catalano, S., Frandsen, S., Jones,A., Lemaire,P., Pace,O., Volonté,S., and Weiss,W.: 1991, 'Probing Rotation and Interior of Stars: Microvariability and Activity – Report on the Assessment Study', ESA, SCI(91)5,1991.

Fröhlich, C., Toutain, T.: 1990, ' P–mode analysis of IPHIR data', in *Progress of Seismology of the Sun and the Stars*, Y. Osaki, H. Shibahashi, Springer–Verlag, Berlin, p. 215.

Pottash,E:M:, Butcher, H.R. and van Hoesel,F.H.J: 1992, *Astron. Astrophys.*, in press.

DISCUSSION FOLLOWING PAPERS ON OTHER OBSERVATIONS

Discussion following paper by S. Drake

G. Field: Why not go to wavelengths longer than 20 cm to take advantage of the fact that self-absorption ($\sim \lambda^2$) will give $\tau = 1/2$ high in the corona where $T = 10^6 - 10^7$ K, rather than the lower brightness temperatures you mentioned?
S. Drake: Clearly, observations at wavelengths $\gg$ 21 cm that sample the outer coronal regions can yield important constraints on the plasma parameters of the emission region for <u>nonthermal</u> radio sources. For optically thick thermal sources, the λ^{-2} dependence of the observed flux density cancels out the increase due to the higher brightness temperatures at long wavelengths. For the brightest RS CVn and dMe stars, Arecibo observations can be useful, but for most stars source confusion will be a big problem. The VLA P band (327 MHz) is probably the best long-λ system for stellar observations because of its good sensitivity <u>and</u> spatial resolution.
R. Rosner: You said that the Ap stars with unusually high rare earth surface abundances have not been seen as radio sources. Do you (or Jeff Linsky, or anyone else here) know whether any of these stars have been detected as X-ray sources? The reason I ask is that the presence of the rare earths places strong bounds on surface (i.e. photospheric) velocities. Thus, it may be that magnetic field perturbations due to surface motions could after all be relevant also to the A stars.
S. Drake: We only detected one such classic Ap star (a known binary) in our ROSAT All-Sky Survey sample, and we believe that the secondary component (rather than the Ap star candidate) is the more plausible X-ray source. In the Cash and Snow (Ap.J. **263**, L59 (1982)) sample of Ap stars observed by *Einstein* there were several detections of relatively weak ($L_x \sim 10^{28} - 10^{29}$ergs sec^{-1}) X-ray sources. Again, low-mass coronal-type companions may be the actual X-ray emitters, but the alternate explanation cannot be ruled out. A deep ROSAT pointing at one or more of those Ap stars that are not known to be binaries might help.

Discussion following paper by F. Chiuderi Drago

S. Drake: Did you calculate the circular polarization predicted by your thermal and nonthermal gyrosynchrotron models and compare it with observed values?
F. Chiuderi Drago: No, we did not. Since the radio transfer equation was solved separately for the ordinary and extraordinary modes, the circular polarization can be easily calculated for any given magnetic field configuration. We plan to do it.

Discussion following paper by P. B. Byrne

G. Peres: You mentioned dips in the K band during flares, but there are also preflare dips. Are they different? Have you observed any K band preflare dips?
P. Byrne: No <u>pre</u>flare K-band dips have been observed to date — only dips during

J.F. Linsky and S. Serio (eds.), Physics of Solar and Stellar Coronae, 509–511.

the flare. However, very few observations of stellar flares in the K-band have been made. The Catania group (see papers in this Proceedings) have observed U-band preflare dips in as much as 40% of flares. Since these preflare and flare dips occur in very different wavebands, I think that they are unlikely to be related.

J. Linsky: Are the U-band and UV continuum observed during the impulsive phases of flares thermal or nonthermal (e.g. gyrosynchrotron emission) in character?

P. Byrne: Analyses of stellar flare continuum distributions in terms of nonthermal emission have been made (Doyle et al. 1991, A.&A.) and the conclusion is that the required particle numbers are unacceptably high. We prefer neutral atom ionization edge radiation (H I, Si I, etc.) as the possible source (Phillips et al. 1991).

P. Ulmschneider: Are there studies connecting the frequency and time-behavior of flares with the level of the quiescent chromospheric activity?

P. Byrne: Yes. Integrated flare energy budgets are, in a general sense, correlated with the mean global chromospheric activity. This is true in the sense that low chromospheric flux stars have very weak flaring activity (Byrne, Doyle, & Merzies, MNRAS (1985); Byrne, MNRAS (1983)). However a more detailed correlation has not been made for the more active stars.

F. Reale: What is the relative timing of flares in X-rays with respect to UV?

P. Byrne: Unfortunately, typical IUE flare exposure times are $\approx$ 10-30 min. Since this is much longer than typical stellar flare timescales, no conclusions can be drawn on relative timings. With HST, however, we have the possibility to make such timings with an accuracy $\sim$ 1 sec, but I do not know the results.

Discussion following paper by M. Kürster

P. Byrne: How did you locate the X-ray emitting region on CF Tuc near the pole? Since such a polar component is not modulated in the light curve and the plasma is optically thin, you cannot derive its location from an X-ray light curve!

M. Kürster: It is a model that fits the data. The polar feature accounts for the minimum level, but since it is very decentered it also produces some modulation.

R. Mutel: (1) Did you decompose the X-ray light curve for CF Tuc as a function of energy? (2) How confident are you that the light curve is not at least partially caused by activity on the G star?

M. Kürster: (1) I subdivided the total light curve into three energy bands yielding "three-color photometry." No differences between the individual light curves are evident perhaps due to low S/N. (2) This is the usual ambiguity for binaries. I can produce similar spot models by putting the spot(s) on the G star. What appears to be the primary eclipse (phase 0.0) could never be modeled by an eclipse effect of the G star alone; some rotational modulation must be included.

Discussion following paper by S. Catalano

N. Weiss: Comment: I was glad to see agreement between your spindown models, assuming a uniform angular velocity, and the observations. If there is any significant magnetic field in the radiative zone, then we would expect a star to rotate

approximately as a solid body, as suggested by recent helioseismic measurements. Radial variations in angular velocity should only persist on timescales short compared with the star's main-sequence lifetime. Question: Are there any stars in which the spatial structure of the magnetic field and the location of starspots have both been measured? If so, how are they related?

S. Catalano: I agree with you. We mainly made the computations to check the consistency of two different, but related, observed quantities. The magnetic structures and star spots for this star and one other appear to be cospatial.

S. Drake: How accurately known are the Rossby numbers for evolved stars that you showed in one of your figures? What is the uncertainty in R_o? I am asking this because (P/τ_c) is a mixture of an observed (P) and a theoretical quantity (τ_c), and thus is known much less precisely than P.

S. Catalano: As far as I know the Rossby number, or better the τ_c, is reliable for main-sequence stars but is very uncertain for evolved stars. Thus correlations for evolved stars do not work. I did not show it during my talk, but we found that the inverse of the slope in the log L_{HK} - $\mathrm{P}_{\mathrm{rot}}$ relations agrees perfectly with τ_c as computed by Gilman and Rucinski and Van den Berg.

A. Maggio: I would like to comment further on Drake's question. There are few theoretical computations of convective turnover times for evolved stars. The most recent ones are from Gilliland (1985) and from Rucinski and Van den Berg (1986). We have used this last work to compute Rossby numbers for a selected sample of stars in the evolutionary phases between the main sequence and the base of the red giant branch. We find a statistical correlation between the X-ray luminosity and the Rossby number for most of these stars, whose quality is reasonably good given the many uncertainties in the computation of this activity parameter. (See the poster paper by Maggio et al. on this subject.)

Discussion following paper by M. Katsova

R. Pallavicini: An infrared excess has been observed also in several RS CVn binaries. Could this be related to what you found for dMe flare stars? Or are the two phenomena completely unrelated (e.g. binarity versus activity)?

M. Katsova: Since the components of the RS CVn binaries demonstrate activity phenomena, the IR excess could be related to activity. However, binarity may independently produce IR excesses.

Discussion following paper by T. Appourchoux

R. Mewe: Do you expect that current technology for metallic multi-layer coatings will be able to resolve the iron VIII-XI lines in the 170 Å band of the XUV telescope to provide temperature diagnostics in the corona?

S. Catalano: The proposed dispersion will not allow us to resolve the Fe lines. We proposed grating mirrors to avoid the use of a filter for eliminating the light reflected at longer wavelengths by the multi-layer mirror.

PART VI

THEORETICAL ASPECTS OF CORONAL PHYSICS

CORONAL HEATING MECHANISMS

ERIC R. PRIEST

Mathematical & Computational Sciences Dept
The University, St Andrews
KY16 9SS, Scotland, UK

Abstract. Pippo Vaiana was a pioneer in showing us the complexity and beauty of the solar corona and recent rocket and satellite views have done nothing to dampen our sense of wonder. The solar corona has a three-fold nature of bright-point, coronal hole and coronal loop structure. A brief review is given of some of the mechanisms which have been proposed to explain the heating of these structures. It is suggested that bright points are heated by reconnection submergence, phase mixing of Alfven waves and coronal loops by ohmic heating in many small current sheets. Whether a wave or a current-sheet mechanism is at work, a highly turbulent state is set up and this can be described by a new self-consistent MHD turbulence theory for coronal heating.

Keywords: Coronal heating – reconnection – Alfven waves – turbulence

1. Introduction

For many of us Pippo Vaiana's pioneering pictures from rockets in 1969 and later from Skylab revealed a new world. No longer could we see the solar corona indirectly via Thomson scattering at eclipses, but at last we were able to view it in all its glory directly in soft X-rays. This transformed our view of the corona and gave us a sense of wonder at its incredible beauty and complexity. Indeed, Pippo challenges us to this day to try and understand the subtle and complex ways in which the magnetic field interacts with and creates this fascinating plasma environment. In this review I shall try and summarise some of the different ways that have been discovered so far in which the corona may perhaps be heated.

Coronal loops have a variation in their properties. Active region loops have, according to old Skylab measurements, typical lengths (L) of 10 - 200 Mm, temperatures (T) $2.2-2.8\times10^6\,K$, densities (n) of $5\times10^{14}-5\times10^{15}\,m^{-3}$, while quiet region loops have $L\sim 20-700\,Mm$, $n\sim 2\times10^{14}-10^{15}\,m^{-3}$ and interconnecting loops have $L\leq 700\,Mm$, $n\sim 2\times10^{6}\,K$, $n\sim 7\times10^{14}\,m^{-3}$. However, we now know from SMM and from YOHKOII (Tsuneta, private communication) that normal active region loops, for instance, can be very much hotter with temperatures up to $4-6\times10^6\,K$. In coronal holes the plasma is cooler with $T\sim 1.4-1.8\times10^6\,K$ and rarer with densities about one-third of quiet coronal values (n_c). Bright points are between 4 and 22 Mm in scale and last for typically 8 hours with temperatures $1.3-1.7\times10^6\,K$ and densities between 2 and 4 times greater than n_c.

Many coronal structures are in equilibrium under a balance

$$\mathbf{j}\times\mathbf{B}-\nabla p+\rho\mathbf{g}=0 \tag{1.1}$$

between magnetic, pressure gradient and gravitational forces, where $\mathbf{j}=\nabla\times\mathbf{B}/\mu$. The ratio of the second to the first term in (1.1) is the *plasma beta*

$$\beta\equiv\frac{2\mu p}{B^2}=3.5\times10^{-21}\,\frac{nT}{B^2} \tag{1.2}$$

with n in m^{-3}, T in degrees K and the magnetic field B in Gauss. Adopting typical values of $n=10^{15}$ and B = 100 we find $\beta\simeq 7\times10^{-4}$. The ratio of the third to the second term is Λ/H where H is the height of a structure and $\Lambda=p/(\rho g)\simeq 50\,T$ metres is the scale-height, which is about 100 Mm for a temperature of $2\times10^6\,K$.

J.F. Linsky and S. Serio (eds.), Physics of Solar and Stellar Coronae, 515–532.

In the corona usually $\beta << 1$ and $\Lambda \simeq H$, so the force balance (1.1) may be approximated by

$$(\nabla \times \mathbf{B}) \times \mathbf{B} = \mathbf{0} \tag{1.3}$$

and we have a force-free field. Thus, since $(\nabla \times \mathbf{B})$ is parallel to **B**, we may write

$$(\nabla \times \mathbf{B}) = \alpha \mathbf{B} \tag{1.4}$$

where α is a function of position. However, the divergence of (1.4) together with the condition $\nabla.\mathbf{B} = 0$ gives $\mathbf{B}.\nabla\alpha = 0$ so that α is constant along a field line. (1.4) may be solved to give models of coronal loops or arcades. The simplest have $\alpha \equiv 0$ so that $\nabla^2\mathbf{B} = 0$ and the field is potential. The next simplest have α uniform, but very little is known about the more general case. Exceptions to (1.3) are: in weak-field regions, where $\beta \sim 1$ and the pressure gradient becomes important; and in large structures where $H \sim \Lambda$ and gravity plays a role.

Coronal magnetic field lines are anchored in the dense photosphere where $\beta \gtrsim 1$, and their feet move about due to a variety of motions. Granulation has typical horizontal velocities of $0.25-2$ kms^{-1} and lifetimes for individual granules of about 8 min., while supegranules involve horizontal motions of about 0.3 km s^{-1} for typically a day, and also there are motions associated with general active region evolution.

The way the corona responds to such motions on a timescale τ depends on the value of τ relative to the time $(10L/v_A)$ it takes an Alfven wave to propagate about ten times at the Alfven speed $\left(v_A = B/(\mu\rho)^{1/2}\right)$ along the length (L) of a coronal structure. If $\tau < 10L/v_A$, which is typically 2 min for $B = 100\ G$, $n = 10^{15} m^{-3}$, $L = 100\ Mm$, then the effect is to produce wave motions. Whereas if $\tau > 10L/v_A$ there is a slow evolution of the coronal field through a series of equilibria.

Now, the energy is injected upwards from the photosphere as a Poynting Flux

$$\mathbf{S} = \frac{\mathbf{E}\times\mathbf{B}}{\mu} = \frac{-(\mathbf{v}\times\mathbf{B})\times\mathbf{B}}{\mu} \tag{1.5}$$

and Priest and Forbes, (1990a) have recently considered the consequences of such injection for a coronal arcade. In particular, they find that slow photospheric motions can easily produce much more rapid motions in the corona.

But what happens to the energy? The answer is provided by Poynting's theorem

$$\int \mathbf{E}\times\mathbf{H}.d\mathbf{S} = \frac{\partial}{\partial t}\int \frac{B^2}{2\mu}\, dV + \int \frac{j^2}{\sigma}\, dV + \int \mathbf{v}.\mathbf{j}\times\mathbf{B}\, dV \tag{1.6}$$

which shows that the inflow of electromagnetic energy (the term on the left) produces several effects (on the right), namely, in turn a rise of magnetic energy, ohmic heating and work done by the magnetic force. Thus some energy is stored, and may be eventually released as for example an erupting prominence or a solar flare, while some energy is continuously dissipated and heats the corona, and a final part of the energy accelerates the plasma in a wide variety of ways. The latter may either escape (up or down) or eventually be dissipated by viscous or resistive damping.

In the pre-Vaiana era the corona was quiet, almost-uniform and heated by acoustic waves. Now it is a highly turbulent environment with a general unresolved "turbulent" velocity of 10-40 km s^{-1} in coronal lines. Mason (1991) has summarised our present knowledge of turbulent velocities (Figure 1). Transition region lines from Skylab show an increase from 16 to 24 km s^{-1} as T rises from 5×10^4 to $2\times10^5 K$. At $10^6 K$ UV coronal lines give about 25 km s^{-1} and at $3\times10^6 K$ recent SMM observations in Mg XI give 40 km s^{-1}. Sometimes one finds turbulent events at 250 km s^{-1} in regions smaller than 1.5 Mm (Dere et al, 1989). Also, intermittent heating events are observed in UV from low-lying loops (Porter et al, 1986) and Lin and Schwartz (1987) observed tiny transient hard X-ray events.

Heat is required to balance radiation, conduction and outflow. In a coronal hole this amounts to 600 Wm^{-2} and is required mainly for outflow. In a quiet region it is half this value and is needed to balance conduction and radiation, whereas in an active region it is 5000 Wm^{-2}, again for conduction and radiation. The mechanism is universally thought to be magnetic since the acoustic flux in the transition region is only 10 Wm^{-2} and more strongly magnetic regions are observed to be hotter. Also the Poynting Flux of

$$S \sim \frac{vB^2}{\mu} \tag{2.1}$$

is typically $10^4\, Wm^{-2}$ for a flow speed of, say, 0.1 $km\, s^{-1}$ and a field of $100G$ and so is more than adequate for coronal requirements. It is highly likely, however, that different mechanisms are acting in different parts of the corona, as follows

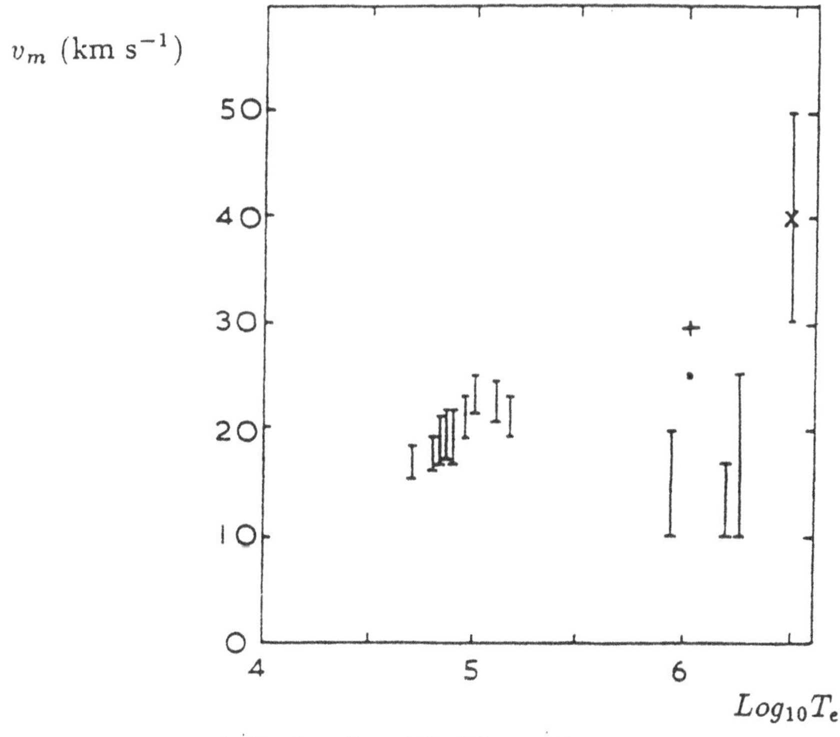

Figure 1. Turbulent velocities from line widths (Mason, 1991).

2. X-Ray Bright Points

Bright points (BP) are situated in the corona above pairs of opposite-polarity photospheric magnetic fragments. They were discovered by Vaiana et al (1970) from rocket X-ray images and their basic properties from Skylab images studied by Golub et al (1974, 1976, 1977). They form as diffuse clouds which grow at 1 $km\, s^{-1}$ and then a bright core of width 3 Mm forms and later fades, followed by the fading of diffuse cloud. At sunspot minimum they are uniformly distributed over the surface at 200 and are seen to be present at anyone time. The number as a function of their lifetime increases with decreasing lifetime.

Later *ephemeral active regions* (ER) were discovered (Harvey and Martin, 1973, 1975; Martin and Harvey, 1979). They are small emerging magnetic dipoles with a flux of

$2-3\times10^{11}\,Wb$ $(2-3\times10^{19}\,Mx)$ and with random orientations. Also it was discovered by Harvey et al (1975) and Golub et al (1989) that *Helium 10830 dark points* (DP) are proxies for large BP's (the 30% largest) The lifetimes are 2 - 48 hrs for BP's, 1 - 2 days for ER's and a few hours for DP's. The sizes are typically 20 *Mm* for BP, 30 *Mm* for ER's and 7 - 20 *Mm* for DP's, while the number born per day is estimated to be 1500 for BP's (Golub et al, 1974) and 100 for ER's.

It was natural to assume that BP's are regions of newly emerging flux, but this suggestion is not tenable because the number of BP's is observed to be out of phase with the solar cycle (Davis et al, 1977; Golub et al, 1979), while the number of ER's is in phase (Harvey et al, 1975; Martin and Harvey, 1979): it doubles from solar minimum (when they are uniformly distributed over the solar surface) to solar maximum (when they have peaks at latitudes $\pm\, 30^\circ$ and $\pm\, 50^\circ$. This major puzzle was solved in two key papers by Karen Harvey (1984, 1985), who showed that, although one-third of the DP's and (therefore presumably BP's) overlie emerging ER's, two-thirds of them lie above so-called *cancelling magnetic features* (CMF). These are opposite polarity magnetic fragments that are seen in the Big Bear video-magnetograms to be approaching on another and cancelling out (Martin et al, 1984, 1985; Martin, 1986, 1988, 1990a,b). Karen Harvey suggested therefore that most BP's are due to chance encounters of opposite polarity fragments in the network or of emerging flux with opposite polarity network. She pointed out that such cancellations should depend on the amount of mixed-polarity areas, which decreases by a factor of 6 from 95% at solar minimum (1976) to 14% at solar maximum (1979-80), while the number of BP's also decreases by a factor 6 from 90 to 14. This explains the anticorrelation of the BP's with the solar cycle.

Many time variations are observed in BP's. Golub et al (1974) report that 5 - 10% of them flare with their intensity increasing by a factor of 10. More recently a study of Strong et al (1992) from YohKoh has shown that BP's fluctuate in intensity by 30 - 200% over time-scales from a few minutes to hours. Also, they find that BP flares involve neighbouring loop brightenings at CMF's with the intensity increasing by a factor of 10 to 100. Furthermore, radio BP have been described by Habbal et al (1986) and Nitta et al (1992). It would be interesting to know whether there is a relation between BP's and the transition region brightenings in CIV from the UVSP instrument on SMM (Porter et al, 1987; Habbal and Withbroe, 1981) and the HRTS explosive events in CIV (Brueckner, 1980) which have time-scales of 20 sec., velocities usually of 10 - 20 *km* s^{-1}, sizes of 1 *Mm* and densities of$10^{16}\,m^{-3}$.

Thus, although some BP are caused by emerging flux most are due to CMF's, so what is happening during a cancellation? The natural explanation is at first sight that they represent regions of *simple submergence*, where a flux tube is descending through the photosphere, but I would like to propose that they are instead regions of *reconnection submergence* where field lines are first reconnecting and then submerging. This idea was first proposed in cartoon form by Priest (1987) and Zwaan (1987), but quantitative details have been worked out recently by Priest and Parnell (1992).

The difficulties with simple submergence are that it gives no explanation for the overlying coronal energy release (the X-ray brightening) and the brightening starts before the cancellation. Also, although emerging flux with separating magnetic fragments in the photosphere are observed, such fragments are never observed to reverse direction and cancel. The fragments which cancel are initially very widely separated and therefore unconnected magnetically. So we need to include the effect of the ambient magnetic field.

The keypoint about *reconnection submergence* is that initially the magnetic fragments are unconnected and so they need to reconnect before they can submerge. The top panel in Figure 2 shows the initial situation with, for simplicity two opposite-polarity fragments connected not to each other but to an overlying horizontal field. They then approach one another and, if no reconnection takes place, a vertical current sheet will be created (middle panel). The magnetic energy then exceeds that of a purely potential field by an amount that can be released by reconnection to give the configuration in the lower panel, and the lower newly connected field lines are then free to submerge. In practice reconnection will continuously take place as the fragments approach one another and the magnetic energy is converted to the heat of the X-ray bright point. At the same time the reconnection process will inject hot plasma into the reconnected field lines and

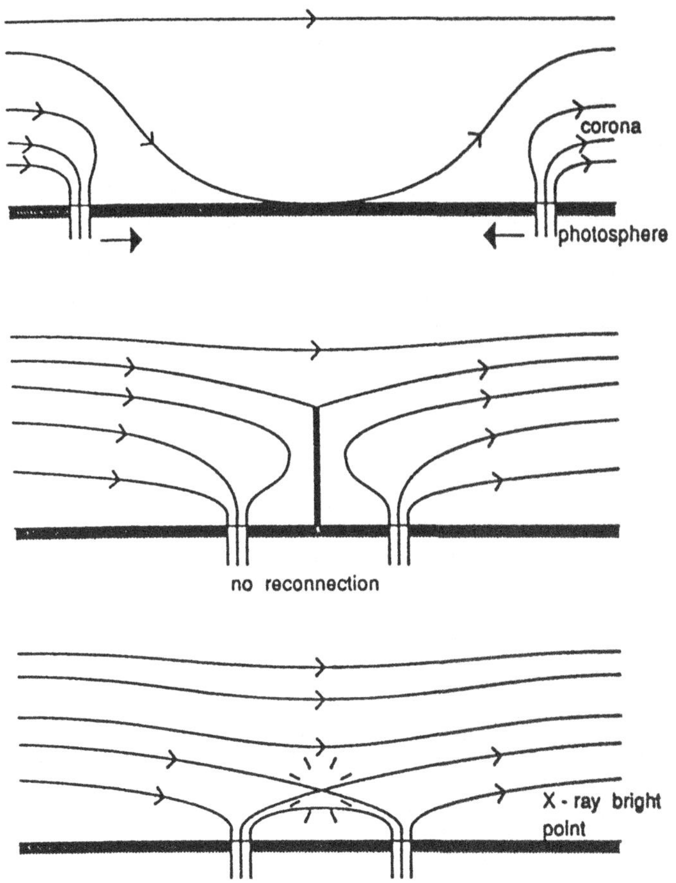

Figure 2. Reconnection Submergence.

create the hot loops which have been observed by the Soft X-ray Telescope on YohKoh in association with microflares (Uchida, 1992; Tsuneta, 1992) and flaring BP's (Strong et al, 1992).

3. Coronal Holes

Coronal holes are in my view probably heated by Alfven waves (Hollweg, 1983; Roberts, 1984; Goossens, 1991) and, in particular, seminal papers on their behaviour in a nonuniform medium have been written by Goedbloed (1975, 1983).

The main way in which Alfven waves propagating outwards in coronal holes are likely to dissipate is by *phase mixing* (e.g. Heyvaerts and Priest, 1983; Sakurai, 1985). The simplest way to see this is to consider a unidirectional field $B_0(x)\,\hat{\mathbf{z}}$ with a z directed vertically upwards and the Alfven speed increasing with x. Suppose the footpoints oscillate to and fro in the y-direction and so create a plasma velocity $v(x, z, t)\,\hat{\mathbf{y}}$ and an extra magnetic field $B_1(x, z, t)\,\hat{\mathbf{y}}$ in the corona. The ideal MHD equations then reduce to a simple wave equation

$$\frac{\partial^2 v}{\partial t^2} = v_A{}^2(x)\,\frac{\partial^2 v}{dz^2} \tag{3.1}$$

with solution of the form

$$v \sim \exp\left[i\left(\omega t - k_z(x)\right)z\right] \tag{3.2}$$

where (3.1) determines the wave number as

$$k_z(x) = \frac{\omega}{v_A(x)} . \tag{3.3}$$

From (3.3) we see that there is a *continuous spectrum* of wavelengths, with waves propagating along different lines with different wavelengths - short wavelengths where the field is weak and long where it is strong. The result is that the phases of the waves on neighbouring field lines become mixed in space. If all of the field lines are shaken in phase at their footpoints, they will become more and more out of phase as they propagate upwards and steep gradients in the x-direction will be generated. Indeed differentiating (3.2) gives

$$\frac{\partial v}{\partial x} \sim v \frac{dk_z}{dx} z, \tag{3.4}$$

so that the gradient increases linearly with height until eventually it is so strong that dissipation becomes important. Typically, the waves dissipate at a height of a few wavelengths.

Several points may be noted about this basic process of phase mixing.

(i) The dissipation may be enhanced by small-scale Kelvin-Helmholtz or tearing-mode instabilities (Browning and Priest, 1984).

(ii) The same process may take place for the other polarization, when the motions $(v\hat{\mathbf{x}})$ are in the direction of the inhomogeneity, but the equations are much more complicated (Goedbloed, 1970). Thus, if one assumes incompressibility and Fourier analyses in t and z, so that

$$v(x,z,t) = \tilde{v}(x)e^{i(\omega t - kz)} \tag{3.5}$$

then $\tilde{v}(x)$ satisfies

$$\frac{d}{dx}\left[\rho_o\left(\omega^2 - k^2 v_A{}^2\right)\frac{d\tilde{v}}{dx}\right] - k^2\rho_0\left(\omega^2 - k^2 v_A{}^2\right)\tilde{v} = 0, \tag{3.6}$$

where $\rho_0 = \rho_0(x)$, $v_A = v_A(x)$ and this contains a singularity where $\omega^2 - k^2 v_A{}^2 = 0$.

(iii) If the Alfven speed varies with height then standing waves may be set up by reflection at a height where the wavelength

$$\lambda = \frac{v_A}{\partial v_A / \partial z} \tag{3.7}$$

according to Musielak et al (1992).

(iv) Closed magnetic regions such as coronal loops and arcades may be heated by phase mixing of standing waves. For such waves one imposes k and replaces (3.3) by

$$\omega(x) = k_z v_A(x) \tag{3.8}$$

so that we now have a continuous spectrum of wave frequencies rather than wavelengths and the waves phase mix in time rather than space. The usual way of treating such phase mixing is to Fourier analyse in time and have great (and exceedingly complicated) fun with the Fourier inversion in the complex plane. However, Cally (1991) has recently undertaken an equivalent but much simpler treatment by Fourier analysing in space. He imposes an initial perturbation $\xi(x,0) = \sin x$ between $x = 0$ and $x = \pi$ and writes the subsequent plasma displacement as

$$\xi(x,t) = \sum_{1}^{\infty} a_n(t)\sin nx. \tag{3.9}$$

Numerical solution of the resulting equations for $a_n(t)$ show that after several periods the resulting displacement has many oscillations in space as the different harmonics are excited. The energy of the fundamental is converted into that of the first harmonic and then the second, the third and so on as the waves progressively phase mix.

Magnetically closed regions may, however, be heated by another but related mechanism, namely ***resonant absorption***. Acoustic or fast-mode waves may propagate upwards, say in the centre of a supergranulation cell or below a magnetic arcade and may then dissipate their energy at a critical surface where

$$v_A(x) = \frac{\omega}{k} \tag{3.10}$$

so that the local Alfven speed equals the phase speed of the wave (Tataronis and Grossman, 1973) and the singularity in (3.6) is reached. Several points may be noted about this process.

(i) The dissipation is essentially independent of η, but the width of the dissipation layer increases with η like $\left[\eta/(kv_A')\right]^{1/3}$.

(ii) There also exist so-called global (or quasi-, or collective, or surface) modes with a frequency

$$\omega_s = k\sqrt{v_{A1}^2 + v_{A2}^2} \tag{3.11}$$

if the density is uniform and the Alfven speed increases from v_{A1} to v_{A2} over a distance a, say. These global modes decay in time (Tataronis and Grossman, 1973; Goedbloed, 1983) at a rate

$$\gamma = -\frac{ka\pi k^2(v_{A2}^2 - v_{A1}^2)}{8\,\omega_s} \tag{3.12}$$

and as they do so they give their energy to Alfven modes in the continuous spectrum with small wavelengths.

(ii) For steady driving of a system at frequency ω_0, say, Poedts et al (1989) have shown how the magnetic perturbation is peaked near the critical surface. They find that the fraction of the energy absorped peaks at a value of unity when the driving frequency (ω_0) equals the surface mode frequency (ω_s). It takes about 100 driving periods to reach a steady state (Poedts et al, 1990).

(iii) Compressibility adds a continuous spectrum of slow magnetoacoustic modes with the dominant velocity component parallel to the magnetic field (Goedbloed, 1975).

(iv) In two-dimensional equilibria, the continuous spectrum continues to exist (Goedbloed, 1975; Goossens et al, 1984; Poedts and Goossens, 1987). The poloidal wave numbers are coupled together and the eigenfunctions are localised near the loop summits, which would tend to give preferential heating there.

(v) Recently, Halberstadt and Goedbloed (1992) have made the surprising discovery that the effect of line tying is to remove the Alfven continuum, which, at first sight, is worrying, since both phase mixing and resonant absorption depend on the existence of a continuum. However, thankfully there is then a new so-called "line-tied continuum", which is independent of the poloidal magnetic field and has a global ballooning character.

4. Coronal Loops and Arcades

In coronal loops and arcades, slow footpoint motions make the coronal magnetic field try to evolve through a series of equilibrium configurations. Often, however, the equilibria are not smooth but contain singularities (current sheets and filaments) where dissipation can occur. There are several ways of forming such sheets, namely driven reconnection and flux interaction, as we have described in §3, braiding, X-point collapse and shearing, so let me say a little about each as follows.

4.1 BRAIDING BY RANDOM FOOTPOINT MOTIONS

Parker (1979, 1990) has pointed out that, if a series of flux tubes are closely packed together and each twisted in the same direction, then current sheets will form at the boundaries of the tubes (Figure 3). If instead the footpoints are braided around one another a smooth solution exists when the footpoint motions are small, but he argues persuasively that finite-amplitude footpoint displacements lead to nonequilibrium and current sheet formation at the boundaries of the braids (see also Van Ballegooijen, 1985; Rosner and Knobloch, 1982; Zweibel and Li, 1987; Sakurai and Levine, 1981).

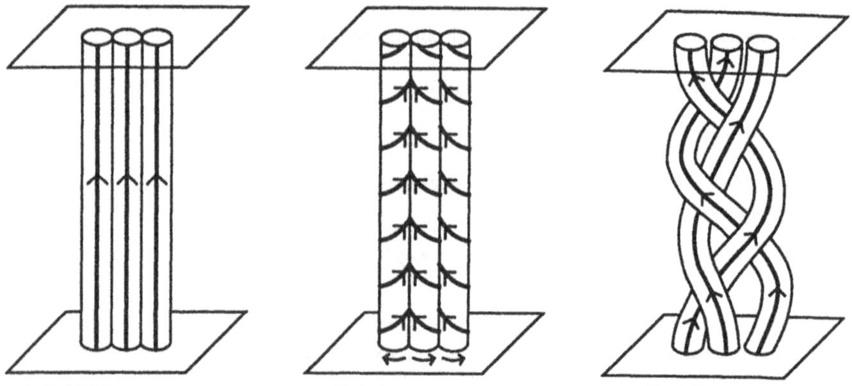

Figure 3. Twisting and braiding of flux tubes.

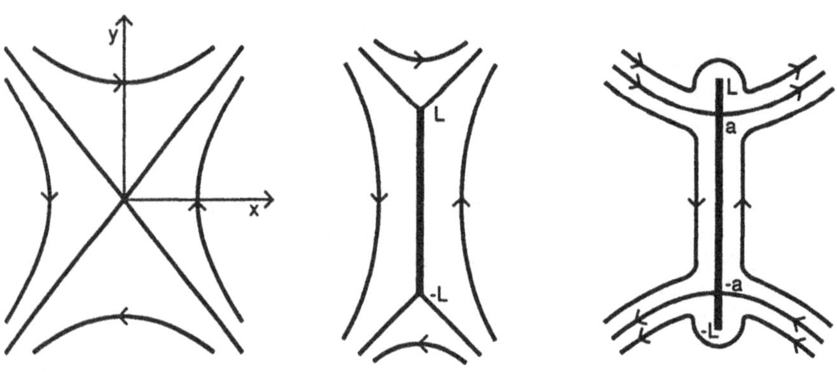

Figure 4. X-point Collapse

4.2 X-POINT COLLAPSE

Somov and Syrovatsky (1976), Aly and Amari (1992) have used complex variable theory to describe current sheet formation. For example, if one starts with an X-point field

$$B_y + iB_x = z \equiv x + iy\ , \tag{4.1}$$

it may evolve into a field

$$B_y + iB_x = \left(z^2 + L^2\right)^{1/2} \tag{4.2}$$

with a cut in the complex plane (a current sheet) stretching from $z = -iL$ to $z = -iL$ and Y-points at the ends of the sheet (Figure 4). It may also instead evolve to the field

$$B_y + iB_x = \frac{z^2 + a^2}{(z^2 + L^2)^{1/2}} \tag{4.3}$$

with reverse currents and singularities near the ends of the sheet.

The process of flux interaction and restructuring, such as has been observed with Yohkoh (Uchida, 1992; Tsuneta, 1992; Lemen, 1992), has been described by Priest and Raadu, (1975) and Tur and Priest (1976). For example, they consider a simple model of the interaction of two bipoles, which initially have a potential field

$$B_y + iB_x = \frac{iD}{(z+a)^2} + \frac{iD}{(z-a)^2} \ . \tag{4.4}$$

When the dipoles approach one another or more flux emerges, the field becomes

$$B_y + iB_x = \frac{E(z^2 + p^2)^{1/2}\,(z^2 + q^2)^{1/2}}{(z^2 - a^2)} , \tag{4.5}$$

where E, p and q follow from flux conservation. Reconnection then occurs and hot plasma is injected into the reconnected field lines in a way that is reminiscent of the loops that are energised in the restructuring events seen by Yohkoh. The analysis has also been undertaken in 3D, when the sheet becomes an annulus, and with unequal dipoles, when the sheet is curved.

A simple model for flux emergence, which may model the minority of bright points which appear to be caused by newly emerging flux, is to start with a dipole in a uniform field such that

$$B_y + iB_x = \frac{iD}{z^2} + i\,B_0. \tag{4.6}$$

Then, as more flux emerges, a current sheet is formed, curving from $z = z_1$ to $z = z_1^*$ (Tur and Priest, 1976) and having a field

$$B_y + iB_x = \frac{E\left[\left(z^2 - z_1^2\right)\left(z^2 - z_1^{*2}\right)\right]^{1/2}}{z^2} \ . \tag{4.7}$$

All of the above complex-variable solutions model a slow evolution through a series of equilibria. Recently, however, Titov and Priest (1992) have discovered some nonlinear, self-similar, compressible solutions for dynamic time-dependent formation of a current sheet from an X-point, building on previous work of Somov and Syrovatsky (1976). When the plasma velocity is much greater than the sound speed and much less than the Alfven speed the dimensionless equation of motion is of the form

$$\varepsilon\,\rho\,\frac{D\mathbf{v}}{Dt} = \mathbf{j}\times\mathbf{B} \ . \tag{4.8}$$

Expanding $\mathbf{v}$ and $\mathbf{B}$ in powers of ε we then have in two dimensions a series of states with

$$\mathbf{j} = 0 \tag{4.9}$$

around the sheet to lowest order. The plasma velocity ($\mathbf{v}_\perp$) perpendicular to the magnetic field is given from the motion of the field lines, i.e. from the induction equation

$$\frac{\partial \mathbf{B}}{\partial t} = \nabla\times(\mathbf{v}\times\mathbf{B}). \tag{4.10}$$

Also the velocity parallel to the magnetic field is determined by the condition that $D\mathbf{v}/Dt$ be perpendicular to the zeroth order magnetic field; this is a consequence of the first-order part of (4.8). What this means physically is that, as the magnetic field lines move, they rotate and the flow along them is given by a balance between the Coriolis and centrifugal forces associated with the rotation of the field lines. The potential magnetic field sequence may not, however, be imposed at will: it has to be determined in a self-consistent way from (4.9), (4.10) and

$$\frac{D\mathbf{v}}{Dt} \,.\, \mathbf{B} = 0 . \tag{4.11}$$

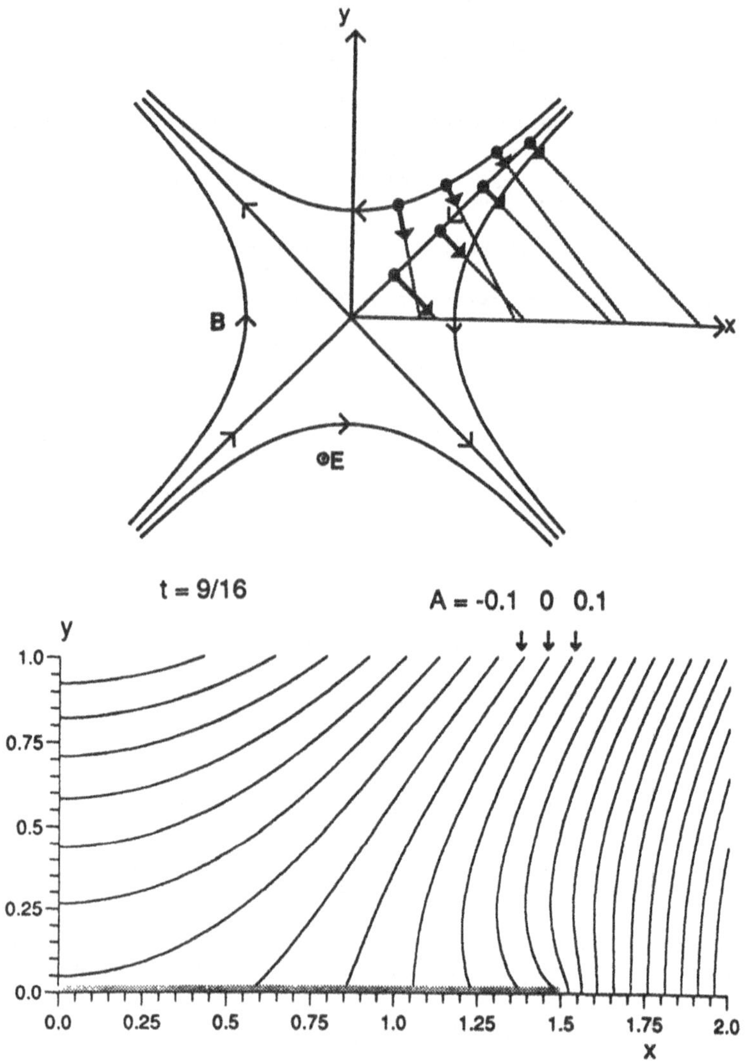

Figure 5. Magnetic field lines at $t = 0$ and 9/16 (Titov and Priest, 1992)

The simplest solution we have found of this type has magnetic and velocity components

$$B_y + iB_x = \frac{\left(\frac{1}{2}z + \sqrt{\frac{1}{4}z^2 - t}\right)^2}{2\left(\frac{1}{4}z^2 - t\right)} \tag{4.12}$$

$$v_x + iv_y = \frac{1}{\overline{\frac{1}{2}z + \sqrt{\frac{1}{4}z^2 - t}}}$$

where $z = x + iy$ is the usual complex variable. This is an elegant formalism, which to me is almost as beautiful as some of the soft X-ray pictures we have been gazing at from Yohkoh! Individual plasma elements converge on the x-axis along straight lines (Figure 5a) and they force a current sheet of length $4/\sqrt{t}$ to form (Figure 5b). As the magnetic field collapses, the current sheet grows in length and the magnetic dissipation increases. The ends of the sheet move with speed $1/\sqrt{t}$. They swallow up part of the magnetic flux and cause the remainder of it to pile up around a region of reversed current.

An interesting numerical experiment on X-point collapse was performed by Biskamp and Welter (1989), which shows how current sheets can themselves filament and produce much fine structure (Figure 6).

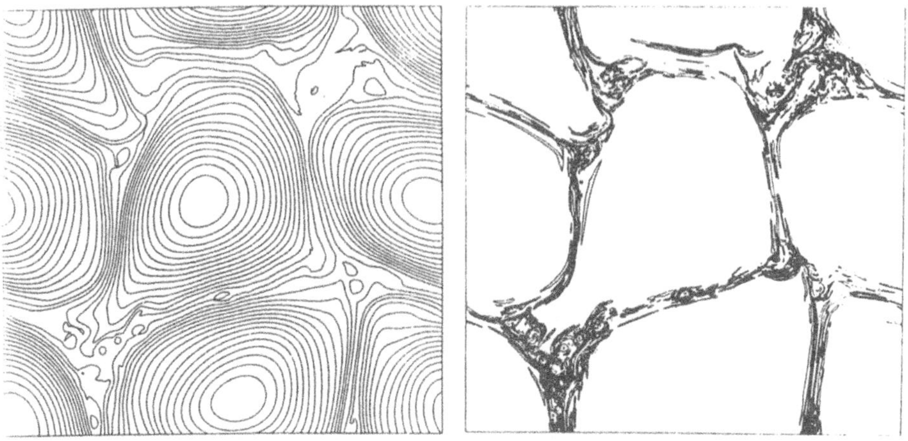

Figure 6. Numerical experiment on X-point collapse (Biskamp and Welter)

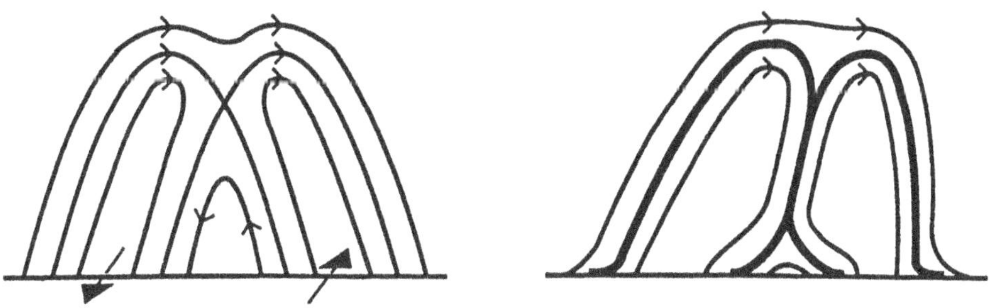

Figure 7. Current Sheets produced by shearing.

4.3 CURRENT SHEETS NEAR SEPARATRICES

An X-point is structurally unstable in the sense that it splits into a pair of Y-points joined by a current sheet under the action of converging motions. However, shearing motions can instead produce current sheets all along the separatrices, the field lines which link to the X-point (Low and Wolfson, 1988; Vekstein and Priest, 1990; Vainshtein 1990; Vekstein and Priest, 1991). In a 2.5 dimensional field its Cartesian components are in terms of the flux function (A)

$$\left(B_x, B_y, B_z\right) = \left(\frac{\partial A}{\partial y}, -\frac{\partial A}{\partial x}, B_z(A)\right) \tag{4.13}$$

and the force-free equation $(\mathbf{j}\times\mathbf{B}=\mathbf{0})$ reduces to

$$\nabla^2 A + B_z \frac{dB_z}{dA} = 0. \tag{4.14}$$

If the footpoint positions $\xi_z(X)$ at the photosphere are imposed, the toroidal field $(B_z(A))$ is given by

$$B_z = \frac{d(A)}{V(A)}, \tag{4.15}$$

where d is the difference in footpoint displacement between the ends of the field and $V(A)=\int ds/B_p$ is a property of the poloidal field. $B_z(A)$ is constant along a given field line but it may be very different on field lines just above and below the separatrix - thus the whole separatrix must become a current sheet.

Finn and Lau (1990) suggested that, in response to shearing, the X-point would just close up slightly, but in general this cannot remain in equilibrium since B_p tends to zero as one approaches the X-point from any direction and so the separatrix cannot support a jump in magnetic pressure across it associated with the jump in B_z. The answer is to use a cusp (Vekstein and Priest, 1990, 1992) because it has the property that B_p tends to zero from one side and to constants from the other two sides, so there is a jump in ${B_p}^2$ across the separatrix which can balance the jump in ${B_z}^2$.

Consider the simplest case where there is shearing present only in the region, (I) say, below the X-point so that in the regions, (II) and (III), to either side $B_z\equiv 0$ and the field is potential with $\nabla^2 A=0$.

In (I) near the cusp there is a self-similar solution

$$A = r^\alpha f(\xi), \tag{4.16}$$

where

$$\xi = \frac{\theta}{r^\beta}, \tag{4.17}$$

so the separatrix ($A=0$, say) is $\xi=1$, say; in other words, it is not a straight line but a curve $\theta=r^\beta$. Then the field components are

$$B_r = \frac{1}{r}\frac{\partial A}{\partial\theta} = r^{\alpha-1-\beta} f'(\xi) \tag{4.18}$$

$$B_\theta = -\frac{\partial A}{\partial r} = -\alpha r^{\alpha-1} f(\xi) + \beta r^{\alpha-1} f'(\xi)\xi . \tag{4.19}$$

The equilibrium equation

$$\nabla^2 A = -B_z \frac{dB_z}{dA} \tag{4.20}$$

must have the right-hand side of the form $-\varepsilon A^{-n}$, where substitution of (4.16) gives

$$n = \frac{2\beta+2-\alpha}{\alpha} \tag{4.21}$$

and to lowest order the function $f(\xi)$ is given by

$$f'' = -\varepsilon f^{-n} \quad . \tag{4.22}$$

In region II the field is potential and an appropriate form for A is

$$A = B_0 r \sin\theta + b r^p \sin p(\theta-\pi) \tag{4.23}$$

which has $A = 0$ on $\theta = \pi$, the vertical arm of the separatrix, as required. As far as region II is concerned the curved part of the separatrix is given by

$$\theta = \frac{b}{B_0} r^{p-1} \sin p\pi \tag{4.24}$$

and so, by comparing with the form $\theta = r^{\beta}$ in region I, we see that

$$p = 1+\beta\ . \tag{4.25}$$

Finally, magnetic pressure balance across the separatrix separating regions I and II gives

$$B_{z0}^2 + cr^{2(\alpha-1-\beta)} = B_0^2 + 2pb\, B_0\, r^{\beta} \cos p\pi \tag{4.26}$$

so that in order that the variations with r should agree we need

$$\alpha = 1 + \frac{3\beta}{2}\ . \tag{4.27}$$

The shapes that are produced when shearing is present in region I alone are shown in Figure 8a. For comparison, the effect of shearing in regions I and IV, regions II and III, and regions I and II are shown in Figures 8b, 8c and 8d, respectively.

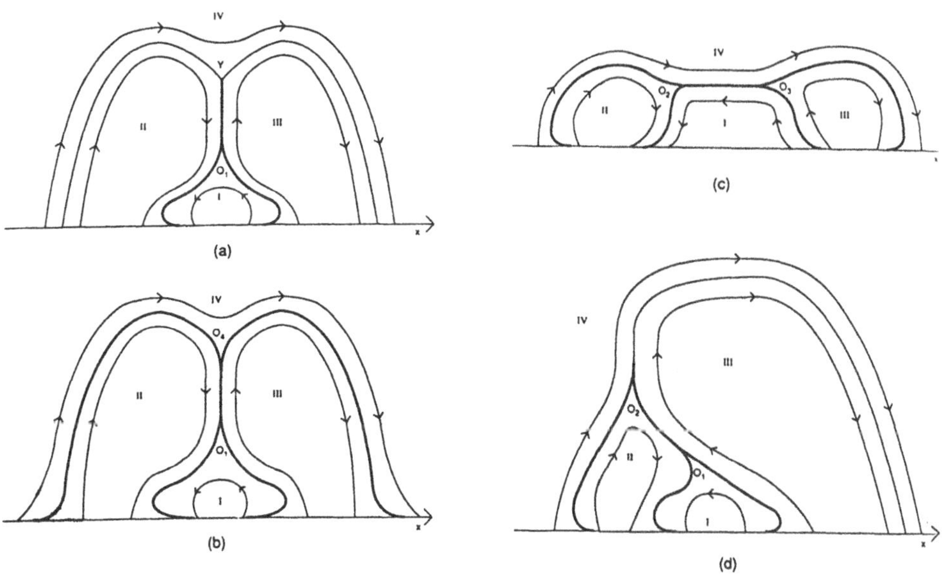

Figure 8. The shapes of the separatrix current sheets due to shearing in different regions (Vekstein and Priest).

5. Self-Consistent Model for Heating by MHD Turbulence

Many coronal heating mechanisms, such as braiding and current sheet formation or resistive instabilities or waves, all lead to a state of MHD tubulence, so how can we analyse it? Heyvaerts and Priest (1984) made a start by adapting Taylor's relaxation theory to the coronal environment, in which the field lines thread the boundary rather than being parallel to it.

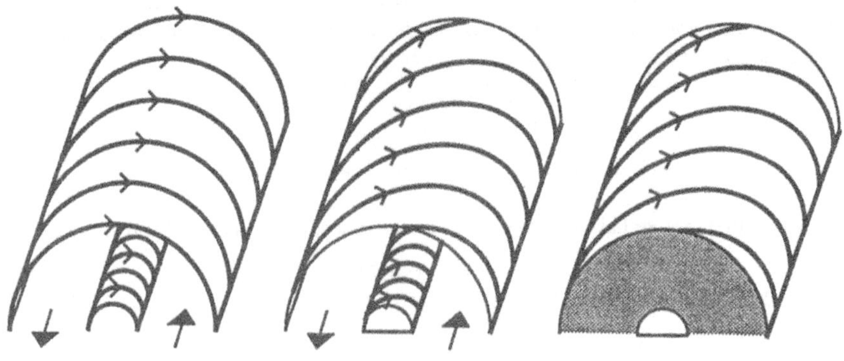

Figure 9. The scenario for Taylor-Heyvaerts-Priest relaxation.

In Taylor's model the global magnetic helicity is conserved, but in our model footpoint motions make the coronal field evolve through a series of linear force-free fields, satisfying

$$\nabla \times \mathbf{B} = \alpha_0 \ \mathbf{B}. \tag{5.1}$$

The footpoint connections are not preserved, but instead the constant α_0 is determined from the evolution of the *magnetic helicity*

$$K = \int \mathbf{A}.\mathbf{B} \ dV \tag{5.2}$$

where $\mathbf{B} = \nabla \times \mathbf{A}$, due to the injection by boundary motions according to

$$\frac{dK}{dt} = \int (\mathbf{A}.\mathbf{v}) \ \mathbf{B}.d\mathbf{S} \ . \tag{5.3}$$

The resulting heating flux is of the form

$$F_H = \frac{B^2 v}{\mu} \ \frac{\tau_d}{\tau_0} \tag{5.4}$$

where τ_d is the dissipation time and τ_0 the time-scale for footpoint motions.

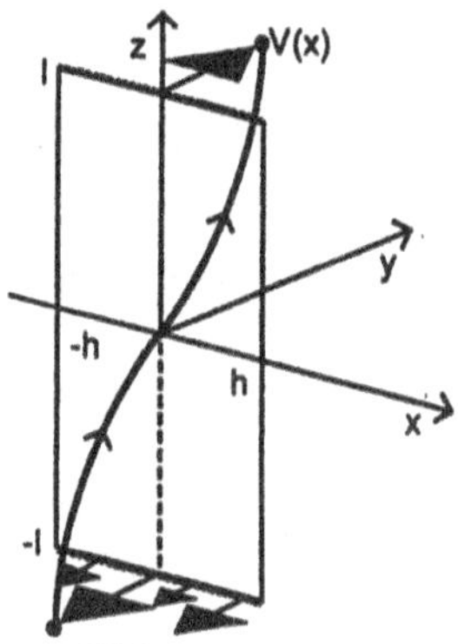

Figure 10. Nomenclature for example of turbulent heating.

Although many mechanisms produce a turbulent state, they are incomplete in the sense that there is a free parameter present, such as τ_d in (5.4) or a correlation time or a relaxation time. In other words, they don't determine the heating flux (F_H) in terms of photospheric motions alone. Heyvaerts and Priest (1992) have therefore begun a new approach in which we assume photospheric motions inject energy into the corona and maintain it in a turbulent state with a turbulent magnetic diffusivity (η^*) and viscosity (ν^*). There are two parts to their theory. First of all, they calculate the global MHD state driven by boundary motions, which gives F_H in terms of ν^*. Secondly, they invoke cascade theories of MHD turbulence to determine the ν^* and η^* that result from F_H. In other words the circle is completed and F_H is determined independently of ν^* and η^*. They apply their general philosophy to a simple example of one-dimensional random photospheric motions producing a two dimensional coronal magnetic field

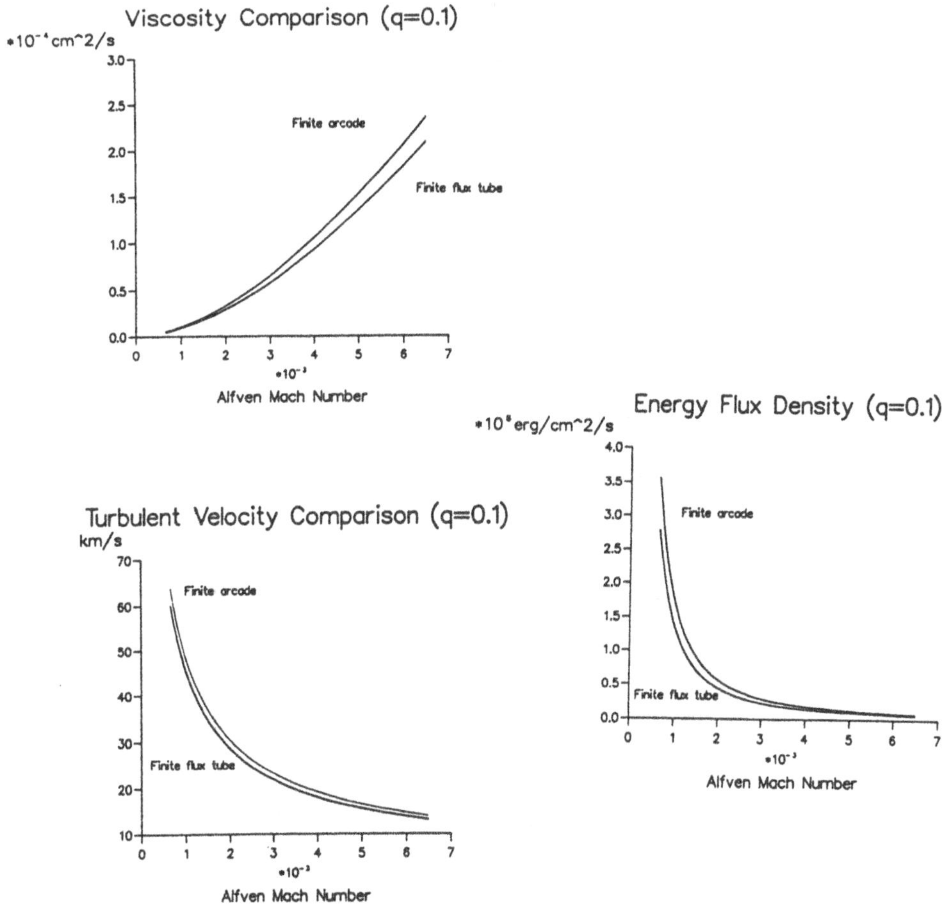

Figure 11. Turbulent viscosity, velocity and heat flux (Inverarity, Priest and Heyvaerts, 1992)

Suppose the boundary motions are $\pm V(x)\,\hat{\mathbf{y}}$ at $z = \pm Ú$ and produce motions $v(x,z)\,\hat{\mathbf{y}}$ and field $B_0\,\hat{\mathbf{z}} + \mathbf{B}_y\,(x,z)\,\hat{\mathbf{y}}$ within the volume between $z = -Ú$ and $z = Ú$. Then the steady MHD equations of motion and induction reduce simply to

$$0 = B_0\,\frac{\partial B_y}{\partial z} + v^*\,\nabla^2\,v\,, \tag{5.5}$$

$$0 = B_0\,\frac{\partial B_y}{\partial z} + \eta^*\,\nabla^2\,B_y. \tag{5.6}$$

The solutions may be found and the resulting Poynting energy flux through the boundary is

$$F_H = \frac{B_0^2\,v\,A0}{\mu}\sum_0^\infty \frac{V_n^2 H}{\eta}\left(1+2\lambda_n^2/\sqrt{1+4\lambda_n^2}\right)$$

$$\frac{\sinh\left(\sqrt{1+4\lambda_n^2}\,/\,H\right)+\sinh(1/H)}{\cosh\left(\sqrt{1+4\lambda_n^2}\,/\,H\right)-\cosh(1/H)} \tag{5.7}$$

where

$$H = \frac{\sqrt{\eta^*\,v^*}}{Ú\,v_{A0}}\,. \tag{5.8}$$

For the second step, invoking Pouquet theory gives $v^* = \eta^*$ and

$$F_H = \frac{27\,v_*^2\,\pi^3}{2h^3/Ú^2}\;\frac{B_0^2\,v_{A0}}{\mu\,Ú^2\,v_{A0}^2} \tag{5.9}$$

so that equating (5.7) and (5.9) gives a single expression for v^*. We find typically for a quiet-region loop that a density $2\times10^{16}\ m^{-3}$ and a magnetic field of 30 - 50G produces a heating $(F_H)$of $2.4-5.5\times10^2\ Wm^{-2}$ and a turbulent velocity of $24-33\ km\ s^{-1}$, whereas values of $5\times10^{16}\ m^{-3}$ and 100 G for an active-region loop give $2\times10^3\ Wm^{-2}$ for the heating and 40 $km\,s^{-1}$ for the turbulent velocities. Given the limitations of the model, we find these reasonable values very encouraging.

Inverarity et al (1992) have also applied the theory to the twisting of a flux tube by azimuthal motions $v_0(R)$ instead of $v_y(x)$. The equations for $v_\theta(R,z,t)$, $B_\theta(R,z,t)$ are similar to those for v_y and B_y and the solutions are Fourier -Bessel expansions. Again, the turbulent diffusion coefficients are related to the rate of energy transfer through the turbulent cascade, which in a steady state is equal to the rate of dissipation and the rate of injection from the photosphere. Also, the equation for v^* has been solved numerically as a function of the root-mean square photospheric velocity $(\bar{v})$, non dimensionalised in terms of the Alfven speed and shown in Figure 11a for the case when the inverse aspect ratio $(h\,/Ú)$ is 0.2. Thus, for example, for $\bar{v} = 1\,km\ s^{-1}$ the run of Alfven Mach number in the Figure corresponds to the magnetic field decreasing from 100G to 10G. Also shown in Figure 11 are the way the turbulent velocity in the corona and the heat flux increase with magnetic field strength for the case when the loop length $Ú = 10\,Mm$ and $n = 2\times10^{16}\ m^{-3}$. The comparison between the arcade and the flux tube is given as well.

6. Conclusion

Whereas at first, when Pippo Vaiana showed us the new corona we could only wonder, now theorists have been highly imaginative in proposing a variety of ways in which the corona may be heated. X-ray bright points are probably powered by reconnection submergence or by emerging flux. Coronal holes are likely to be heated by phase mixing of propagating Alfven waves. Coronal loops, on the other hand may be heated by resonant absorption of Alfven waves or by many small current sheets formed by X-point collapse or separatrix shearing. The resulting wave or current sheet state can be described by a new self-consistent model of MHD turbulence. However, in future, we hope to continue the voyage of discovery about the solar corona which Pippo inspired many of us to take and, in particular, to use YOHKOH and SOHO to try and determine just which mechanisms are at work in our Sun.

Acknowledgment

I would like to thank Salvatore Serio most warmly for organising this meeting so well. We shall long remember the breathtaking theatre at Segesta, the evening at his farmhouse and the truly memorable banquet in the botanic gardens. I am sure that Pippo would have been delighted at the great care with which Salvatore has looked after us.

References

Aly J J and Amari T (1992) Preprint.
Biskamp D and Welter (1989) *Phys. Fluids* **B1**, 1964.
Browning P K and Priest E R (1984) *Astron. Astrophys.* **131**, 283.
Cally P (1991) *J Plasma Phys*.
Davis J, Golub L and Krieger A (1977) *Astrophys. J.* **214**, L141
Dere K, Bartoe J and Brueckner G 91989) *Solar Phys.* **123**, 41.
Dungey J W (1958) *Cosmic Electrodynamics*, Camb. Univ. Press.
Finn J and Lau Y-T (1991) *Phys. Fluids* **B3**, 2675.
Goedbloed J P (1970) PhD thesis.
Goedbloed J P (1975) *Phys. Fluids* **15**, 1090.
Goedbloed J P (1983) Lecture Notes on Ideal MHD, Rijnhuizen Report 83-145.
Golub L, Krieger A, Silk J, Timothy A and Vaiana G (1974) *Astrophys. J.* **189**, L93.
Golub L, Krieger A, Harvey J and Vaiana G (1977) *Solar Phys.* **53**, 111.
Golub L, Davis J and Krieger A (1979) *Astrophys. J.* **229**, L145.
Golub L, Harvey K, Herant M and Webb D (1989) *Solar Phys.* **124**, 211.
Goossens M (1991) in *Advances in Solar System MHD* (ed E R Priest and A W Hood) Cambridge, p137.
Habbal S and Withbroe G (1981) *Solar Phys.* **69**, 77.
Habbal S, Ronon R, Withbroe G, Shevgaonkar R and Kundu M (1986) *Astrophys. J.* **306**, 740.
Halberstadt, G and Goedbloed J P (1992) these proceedings
Harvey K L (1984) *Proc. 4th European Meeting on Solar Phys.*, ESA SP 220, 235.
Harvey K L (1985) *Aust. J. Phys.* **38**, 875.
Harvey K L and Martin S F (1973) *Solar Phys.* **32**, 389.
Harvey K L, Harvey J W and Martin S F (1975) *Solar Phys.* **40**, 87.
Harvey K L, Harvey J W and Martin S F (1975) *Solar Phys.* **40**, 87.
Heyvaerts J and Priest E R (1983) *Astron. Astrophys.* **117**, 220.
Heyvaerts J and Priest E R (1984) *Astron. Astrophys.* **137**, 63.
Heyvaerts J and Priest E R (1992) *Astrophys. J.* in press.
Hollweg J V (1983) in *Solar Wind 5*, NASA CP 2280, 1.
Inverarity G and Priest E R (1992) in preparation.
Lemen J (1992) these proceedings.
Lin R and Schwartz R (1987) *Astrophys. J.* **312**, 462.

Low B C and Wolfson R (1988) *Astrophys. J.* **324**, 574-581.
Martin S F (1986) *Coronal and Prominence Plasmas* (ed A Poland) NASA CP 2442, p73.
Martin S F (1988) *Solar Phys.* **117**, 243.
Martin S F (1990a) *IAU Symp.* **138**, 130.
Martin S F (1990b) *Mem S A It.* **61**, 293.
Martin S F and Harvey K L (1979) *Solar Phys.* **64**, 93.
Martin S F et al (1984) *Adv. Space Res.* **4**, 61.
Martin S F, Livi S H B and Wang J (1985) *Aust. J. Phys.* **38**, 929.
Mason H (1991) *Proc 6th European Meeting on Solar Phys.* p232.
Nitta N, Bastian T S, Aschwanden M J, Harvey K L and Strong K T (1992) *Pub. Astron. Soc. Japan.* in press.
Musielak Z, Fontenla J and Moore R (1992) *Phys. Fl.* **B4**, 13.
Parker E N (1979) *Cosmical Magnetic Fields,* Oxford University Press.
Parker E N (1990) *Geophys. Astrophys. Fluid Dyn.* **52**, 183-210.
Poedts S, Kerner W and Goossens M (1989) *J. Plasma Phys.* **42**, 27.
Poedts S, Goossens M and Kerner W (1990) *Comp. Phys. Comm.* **59**, 75.
Poedts S and Goossens M (1987) *Solar Phys.* **109**, 265.
Porter J, Reichmann E, Moore R and Harvey K (1986) *Coronal and Prominence Plasmas* (ed A Poland) NASA CP2442, p383.
Porter J G, Moore R L, Reichman E J, Engvold O and Harvey K L (1987) *Astrophys. J.* **323**, 380.
Priest E R (1987) in *The Role of Fine-Scale Magnetic Fields on the Structure of the Solar Atmosphere* (ed E Schroter, M Vazquez, A Wyller) Camb. Univ. Press. p297.
Priest E R and Parnell C (1982) submitted.
Priest E R and Raadu M A (1975) *Solar Phys.* **43**, 177.
Priest E R, Hood A W and Anzer U (1989) *Astrophys. J.* **344**, 1010.
Roberts B (1984) in *Hydromagnetics of Sun,* ESA SP-220, 137.
Rosner R and Knobloch E (1982) *Astrophys. J* **262**, 369.
Sakurai T and Levine R (1981) *Astrophys. J.* **248**, 817.
Sakurai T (1985) in *Th. Probs in High Resoln S. Phys.* (ed H Schmidt), 263.
Somov B and Syrovatsky S I (1976) Proc. Lebedev Phys Inst **74**, 13.
Tataromis J and Grossman W (1973) *Zs. Physik* **261**, 203.
Titov V and Priest E R (1992) submitted.
Titov V, Demoulin P and Priest E R (1992) in preparation.
Tsuneta S (1992) these proceedings.
Tur T J and Priest E R (1976) *Solar Phys.* **48**, 89.
Uchida Y (1992) these proceedings.
Vaiana G S, Krieger A S, Van Speybroeck L P and Zehnpfennig T (1970) *Bull. Am. Phys. Soc.* **15**, 611.
Vainshtein S I (1990) *Astron. Astrophys.* **230**, 238.
Van Ballegooijen A (1985) *Astrophys. J.* **298**, 421.
Vekstein G, Priest E R and Amari T (1990) *Astron. Astrophys.* **243**, 492.
Vekstein G, Priest E R (1992) *Astrophys. J.* **384**, 333.
Zweibel E ansd Li H (1987) *Astrophys. J.* **312**, 423.

CHROMOSPHERIC HEATING

P. ULMSCHNEIDER

Institut für Theoretische Astrophysik, Universität Heidelberg, Im Neuenheimer Feld 561, D–6900 Heidelberg, Germany

Abstract. Chromospheres consist of magnetic and non-magnetic areas, which are heated by different mechanisms. For the non-magnetic areas a satisfactory acoustic wave heating picture emerges, where the wave generation calculations, the solar wave observations, the solar acoustic heating calculations and the solar chromospheric cooling observations are all roughly consistent. This picture applies for most of the solar surface and for slowly rotating late-type stars. The heating of magnetic areas is presently not well understood.

Key words: chromosphere – acoustic waves – heating

1. Introduction

Quite different from late-type stellar coronae which probably are always magnetic regions (Stepień and Ulmschneider 1989), late-type stellar chromospheres are essentially two-component regions, consisting of areas with intense vertically directed magnetic flux tubes and of essentially field-free areas. The larger the rotation rate of the star, the greater is the relative size of the magnetic areas as compared to the field-free areas. For the heating of the magnetic and non-magnetic areas different heating mechanisms apply.

UV observations show that for all stars, except for the very slowly rotating ones, the heating of the magnetic areas is dominant. This is also found when comparing network (magnetic) regions with interior (non-magnetic) regions of supergranulation cells on the sun. What heats the magnetic areas is not well known. A multitude of different (longitudinal=acoustic, transverse, torsional) magnetic tube-wave (AC = alternating current) mechanisms is likely to be involved in the heating, but also direct heating by magnetic field reconnection (DC = direct current) mechanisms seem to be at work. For recent reviews of chromospheric and coronal heating mechanisms see Narain and Ulmschneider (1990) as well as Ulmschneider, Priest and Rosner (1991).

In principle, chromospheres and coronae of stars without a mass transfering companion or without protostellar accretion, depend only on the internal structure of the underlying star: for given effective temperature T_{eff}, surface gravity g and rotation period P_{rot} the average chromospheric and coronal structures of a star must be completely determined. The logic of such a physical dependency is so far quite elusive for the magnetic areas and progress on the many magnetic heating mechanisms has been slow and not very conclusive. Because of this I primarily concentrate on the heating of field-free areas in this work. It is in the field of chromospheric heating of slowly or non-rotating stars and of the non-magnetic regions (within supergranulation cells) of the sun that now a fairly consistent acoustic heating picture has emerged in the last few years which allows for the first time to see the complicated logic which connects the physics of chromospheres with the underlying structure of the star.

For stars later than spectral type A, the acoustic heating picture works as follows: acoustic waves generated near the top of the surface convection zone run

J.F. Linsky and S. Serio (eds.), Physics of Solar and Stellar Coronae, 533–540.

down the steep density gradient of the outer stellar atmosphere, and, due to energy conservation, grow to large amplitude and form shocks. Shock dissipation heats the outer layers to high temperatures and thus produces a chromosphere. From this picture it is clear that acoustic heating must always be present in stars, because the turbulent motions of a convection zone will always produce acoustic energy and the rapid density decrease in the outer atmosphere will always lead to shock dissipation of the acoustic waves. The crucial question is, however, whether this acoustic heating is sufficient to balance the observed chromospheric emission. It is the important realization in the last few years that the acoustic heating is indeed sufficient and consistent with the observations and is able to explain the heating of the non-magnetic areas of the chromospheres.

It is important to note that acoustic heating also operates in O- and B-stars, where surface convection zones no longer exist. In these stars intense radiation fields amplify microscopic acoustic disturbances until strong acoustic shock waves develop. These shocks very likely occur in individual blobs propagating on top of the rapidly expanding winds of these stars, a situation which is very different from that in chromospheres of late-type stars. For earlier reviews of acoustic heating see Narain and Ulmschneider (1990) and Ulmschneider (1990, 1991).

2. Acoustic energy generation

In 1948 Biermann realized that the solar convection zone produces acoustic waves which in turn might heat the corona. Since the pioneering work of Lighthill (1952) the methods of computing acoustic energy generation from turbulent gas flow fields have been extensively developed and tested particularly in the aircraft industry. Fig. 2.13 of Goldstein (1976) shows an excellent agreement of Lighthill's u^8 quadrupole sound generation law with experimental values from noise of jet engines, u being the flow velocity. Theoretical acoustic energy generation rates, F_M, from stellar surface convection zones of late-type stars have been computed by Bohn (1984), using Stein's (1967) theory of sound generation which includes additional multipole (monopole and dipole) terms. These calculations depend on three stellar parameters: T_{eff}, g, and the mixing length parameter α.

In Fig. 6 of Bohn (1984) the acoustic fluxes F_M are shown to rise rapidly with T_{eff} up to a maximum value near $T_{eff} \approx 10^4$ K of A-stars where the disappearing convection zones abruptly decrease the acoustic flux. This rise is due to the fact that the acoustic energy generation depends on a high power of the convective velocity u. In fairly efficient convection zones one finds that the total flux is $\sigma T_{eff}^4 \approx \rho u^3$, where ρ is the density. Thus with increasing T_{eff} and decreasing ρ the acoustic energy generation increases from M- to F-stars. The acoustic energy generation also increases when going from dwarfs to giant stars, because for a given T_{eff} the density in the atmospheres of giants is much smaller than in dwarfs, resulting in larger convective velocities u. Bohn finds

$$F_M \approx 7.08 \cdot 10^{-27} T_{eff}^{9.75} g^{-0.5} \alpha^{2.8} \quad . \tag{1}$$

Bohn (1981, unpublished Ph. D. thesis, Univ. of Würzburg, Germany) has also computed the acoustic frequency spectra of main-sequence stars (see Fig. 4 of Ulm-

schneider 1991). These spectra extend roughly over a decade in the period range of $P_A/10 < P < P_A$, where $P_A = 4\pi c_S/(\gamma g)$ is the acoustic cut-off period. γ is the ratio of specific heats, c_S the sound speed. The maximum of the acoustic spectrum is found near $P_{max} \approx P_A/6$ and shifts longerwards for late-type dwarf stars. From theoretical considerations one thus finds that for the sun with $P_A \approx 220$ s acoustic waves with typical periods of $P \approx 35$ s (Bohn (1984), Stein (1968) find $P_{max} = 29, 38$ s for an EE turbulence spectrum, respectively) and fluxes of about $F_M = 1.8 \cdot 10^8$ erg cm^{-2} s^{-1} for $\alpha = 1.0$ should be present at the top of the convection zone. For $\alpha = 1.5$, the flux F_M would be three times higher.

3. Solar observations of acoustic waves

Are such theoretically inferred waves observed on the sun? Work by Endler and Deubner (1983) and recently by Deubner (1988) and Deubner et al. (1988) indeed shows that propagating acoustic waves, similar to the theoretically inferred waves, are present in the solar atmosphere. Cross correlating simultaneous observations of velocity fluctuations in different spectral lines allows to compute the cross power spectrum $CP(\omega)$ and the phase spectrum $\Delta\phi(\omega)$ (see also Ulmschneider 1990). It can be shown that for propagating waves the phase differences $\Delta\phi$ of the velocity fluctuations in two considered spectral lines depends linearly on frequency, while for standing waves $\Delta\phi$ is constant and jumps between the values zero and 180°. By removing the influence of seeing, Endler and Deubner (1983) showed on the basis of their corrected phase spectra that there are propagating acoustic waves in the solar atmosphere with periods as low as 40 s. They found that this period limit is a detection limit, caused by the solar atmosphere itself. To detect an acoustic wave as frequency fluctuation of the central absorption core of a spectral line the wavelength of the wave should be considerably larger than the width of the line contribution function which has typical values of about 300–400 km. With a sound speed of 7 km s^{-1} this condition is violated for waves with periods less than 40 s. In addition, detecting waves with periods $P < 40$ s by line broadening is difficult, because due to the limiting strength property these waves have velocity amplitudes which decrease proportional to P.

From the crosspower spectrum Deubner (1988) finds for the low and middle photosphere acoustic fluxes of $F_M = 2.0 \cdot 10^7$ erg cm^{-2} s^{-1}, for the height of NaI 5896Å, $F_M = 1.2 \cdot 10^6$ erg cm^{-2} s^{-1} and for the height of Ca II 8542Å roughly $F_M = 4.5 \cdot 10^5$ erg cm^{-2} s^{-1}. He notes that these values tend to be lower bounds because the acoustic flux of very short period waves is not included. Taking his estimates of the heights for these lines, 300 km, 800 km and 1500 km, respectively, the acoustic fluxes have been plotted in Fig. 1a. The flux F_M at 300 km agrees roughly with Bohn's theoretical solar value at the top of the convection zone, if one allows for radiation damping by a factor of 6 (Ulmschneider et al. 1978, Fig. 5).

4. Solar chromospheric emission

The magnitude and height dependence of the empirically determined solar chromospheric radiation loss rates are powerful tests for the validity of the chromospheric

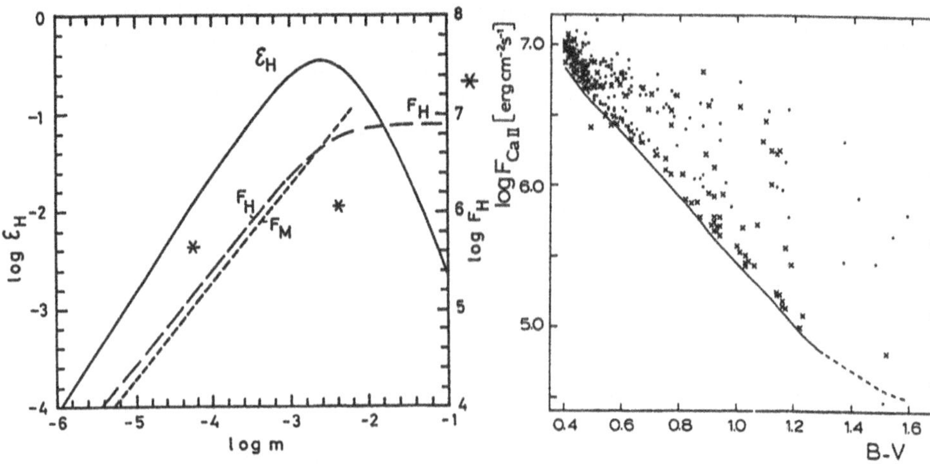

Fig. 1. a. Net heating rate ϵ_H (erg cm^{-3} s^{-1}) and heating flux F_H (erg cm^{-2} s^{-1}) versus mass column density m (g cm^{-2}) after Anderson and Athay (1989b), together with a theoretical limiting strength acoustic flux F_M, and directly observed acoustic wave fluxes by Deubner (1988), labeled by *. b. Chromospheric Ca II K-line emission core fluxes F_{CaII} of giants (x) and dwarfs (•) after Rutten (1987).

heating mechanisms and permit to clarify the relative importance of the heating in magnetic and non-magnetic areas. Anderson and Athay (1989a) argue for acoustic heating as the dominant heating mechanism of the solar chromosphere: The plage and magnetic network regions ".. do not dominate the average chromosphere and thus do not require that magnetic heating be dominant for the chromosphere as a whole". Assuming a height-dependent mechanical heating rate ϵ_H, Anderson and Athay (1989b) construct semi-empirical models which conserve the total (mechanical and radiative) energy flux. Adjusting ϵ_H such that the resulting temperature distribution is consistent with that of model C of Vernazza et al. (1981), which was derived by matching observed spectral features, Anderson and Athay empirically determine the required heating rate to balance the chromospheric losses $\epsilon_R = \epsilon_H$. Integrating this heating rate over height, Anderson and Athay (1989b) find a necessary heating flux F_H which a correct chromospheric heating mechanism must provide. Both ϵ_H and F_H are shown in our Fig. 1a. It is seen that F_H and the observed acoustic fluxes by Deubner are in reasonable agreement.

5. Stellar chromospheric emission

Solar observations indicate that the areas of largest chromospheric emission seen in the core of the Ca II H+K lines, are strongly correlated with the magnetic network regions at the supergranulation boundaries and with plage regions (Schrijver et al. 1989). Fig. 1b taken from Rutten (1987) shows observations of the stellar F_{CaII} line core emission flux from dwarfs (dots) and giants (crosses). It shows three important facts. First, Fig. 1b clearly demonstrates that all late-type stars have chromospheres, because the values of F_{CaII} show a definite lower limit, the so called *basal flux line*. Theoretical calculations of radiative equilibrium atmospheres without chromospheres show no core emission flux. The observed finite flux F_{CaII} is thus a clear indication for the existence of stellar chromospheres.

The second fact is the large variability of F_{CaII} for essentially identical stars which demonstrates the existence of at least two different heating mechanisms (for magnetic and non-magnetic areas). Dwarf stars of a given T_{eff}, which are indicated by dots in a vertical slice in Fig. 1b, all have similar gravity $log\ g \approx 4.5$ and thus, except for their rotation period P_{rot}, represent identical stars. As discussed above one gets identical values of F_M and $P = P_{max}$ from acoustic energy generation calculations for these stars. Acoustic wave calculations then lead to identical theoretical chromosphere models and consequently to unique Ca II line profiles. These emission core fluxes could be the observed basal flux values F_{CaII} which apply for the stars with largest rotation period P_{rot}.

The surface of these basal flux stars thus appears to consist mainly of non-magnetic areas which are heated acoustically. However, with decreasing P_{rot} the fraction of the magnetic areas increases and with it the emission due to the increasingly dominant magnetic heating. As shown by Vilhu (1987) this increase of the chromospheric emission flux reaches a maximum at a saturation boundary where the magnetic areas cover essentially the entire surface of the star. The emissions from T Tau-stars which lie above this saturation boundary is explained by external processes, like accretion.

Third, the range of variability of F_{CaII} caused by P_{rot} decreases markedly with increasing T_{eff} in Fig. 1b. This can be explained by the fact that the total emission at the earlier stars has a much larger acoustic contribution (due to the strong T_{eff}-dependence of the acoustic energy generation) relative to the emission contribution from the magnetic areas.

That the acoustic heating is a good explanation for the basal chromospheric emission has also been found by Mathioudakis and Doyle (1992) who recently showed that the basal Mg II fluxes of M-stars are between 1 and 2 orders of magnitude lower than Bohn's acoustic fluxes for these stars. Because Mg II is only one of the many chromospheric emitters and because acoustic waves are damped during transit from the top of the convection zone to the Mg II line emitting height, it is clear that Bohn's fluxes must be considerably larger.

An additional hint for the validity of the acoustic heating picture was given by Middelkoop (1982, Fig. 4a-c), who showed for giant stars that the chromospheric emission variability decreases very much toward late spectral types and there becomes a basal emission. As the rotation period of giants is expected to become very large due to the increasing radius in the course of their post-main-sequence evolution, the fraction of magnetic areas on these stars is supposed to decrease strongly and only the non-magnetic acoustic heating component appears to remain.

6. Acoustic wave calculations and the limiting shock strength

It has long been recognized (Ulmschneider 1970, 1991) that only by shock heating can realistic heating rates be achieved from acoustic waves and that with the typical fluxes F_M and periods P_{max} by Bohn (1981, 1984) realistic shock formation heights and shock dissipation rates are found. There are two effects which severely influence the behaviour of acoustic waves in the outer stellar atmospheres. First, the acoustic waves are strongly affected by *radiation damping*, when the wave propagates through the radiation damping zone, which in the sun extends to heights of about 200 km, but for other stars can be much more extended (Ulmschneider 1988).

Second, it is a persistent result of time-dependent monochromatic acoustic wave calculations that acoustic shock waves, once formed, tend to quickly reach *limiting shock strength*. This behaviour, where the wave amplitude becomes essentially constant with height, and independent of the initial amplitude, results from the balance of shock dissipation which decreases the wave amplitude, and amplitude growth, which is caused by the steep density decrease. In an isothermal atmosphere the limiting shock strength property can be derived analytically (Ulmschneider 1970, 1989). Time-dependent acoustic wave calculations show that this behaviour is well established for non-isothermal chromosphere models (Ulmschneider 1991, Fig. 6).

The important property of acoustic waves, which have reached limiting shock strength, is that their flux F_M and dissipation rate $\epsilon_M = -dF_M/dh$ become independent of the wave flux initially injected into the atmosphere and can be given analytically by

$$F_M = \frac{1}{12}\frac{\gamma^3 g^2}{(\gamma+1)^2 c_S}\, P^2\, p \quad , \tag{2}$$

where p is the gas pressure. For the sun both the gravity g and the wave period $P \approx P_{max}$ are given. Moreover plotting F_M versus the mass column density $m = p/g$, Eq. (2), does not permit much freedom in a comparison with the empirically determined heating fluxes F_H by Anderson and Athay (1989b). Fig. 1a shows that for $P \approx P_A/6 = 35$ s and $log\ m < 10^{-3}$ indeed a remarkable agreement exists of the magnitude and height dependence of F_M with that of F_H. The discrepancy at larger m is due to the fact that here limiting strength has not yet been reached. Please note that even if Bohn's computed theoretical fluxes F_M were wrong or if the measurements of the acoustic fluxes were inaccurate the existence of the limiting flux waves indicates that the wave has completely forgotten its origin and attains a flux which, at a given m, is only determined by the wave period. For the sun one has $F_M \approx 7.1 \cdot 10^4\, p$ for $P = 35$ s.

7. Problems of the acoustic heating picture

Despite the fact that there exists now a fairly consistent overall picture of the chromospheric heating, the details are still far from clear. The computation of acoustic sound generation depends on the mixing length theory, which has the greatest uncertainty just at the top of the convection zone, where the sound generation is at its maximum.

In addition one needs more accurate fluxes from the observation of solar acoustic waves. Here observations by Fleck and Deubner (1989) lead to a puzzle. Comparing the phase differences of velocity fluctuations in the Ca II infrared triplet (IRT) lines 8542Å and 8498Å these authors find large phase speeds, using a height difference

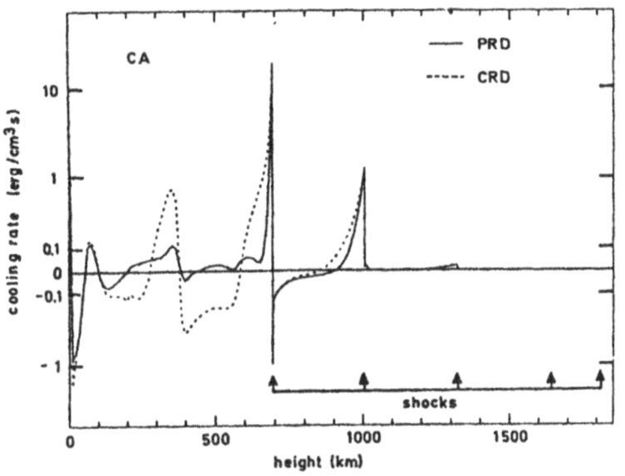

Fig. 2. Ca II K-line cooling rate (erg cm^{-3} s^{-1}) versus height for a solar acoustic wave calculation assuming CRD and PRD.

of 300 km between these lines which they infer from the Vernazza et al. (1981) model. They interpret their observation by postulating a "magic height" at 800 km below which there are running acoustic waves, but above which one has standing waves. It is difficult to picture how sawtooth shock waves should become standing waves and moreover how this standing wave hypothesis could be reconciled with the shock heating behaviour as seen in Fig. 1a. Presumably a sawtooth shock wave will have essentially lost all its energy before it reaches the reflecting layer which is essential for the standing wave picture. From our time-dependent wave calculation we suspect that, different to the VAL model, the emission of the IRT lines is located behind the shocks and thus might be spacially correlated. An essentially cospatial formation of the IRT lines would greatly reduce the inferred phase speed.

Another uncertainty is the use of monochromatic waves in our arguments. This usage is fine in order to get a rough overall picture and it is clear that for studying the period-dependent properties of acoustic waves it is essential to use monochromatic waves. But in reality the sound generation process will be intermittent and produce wave packets. Thus it would be more appropriate to use acoustic spectra. It is not clear whether the limiting strength behaviour of Eq. (2) will be retained on the average when acoustic spectra are used. In the propagation of acoustic spectra shock overtaking will occur which occasionally leads to strong shocks. This behaviour may necessitate very long time-dependent wave calculations to generate averaged results.

Moreover the semi-empirical analysis of Vernazza et al. (1981) and of Anderson and Athay (1989b) have to be taken as highly idealized. These authors use a rather smooth chromospheric temperature dependence to match the observed spectrum. But the waves which are postulated from their inferred heating rate have a very inhomogenous, spiked temperature distribution, with the emission occuring primarily behind the shocks. Fig. 2 shows the Ca II K-line cooling rate from a typical solar wave calculation. Note the extreme concentration of the emission behind the shocks, which is further accentuated when the computation is made using partial redistribution (PRD) instead of complete redistribution (CRD).

Finally even for purely non-magnetic areas there probably is an additional heating mechanism in red giant stars. Here shocks generated by nonradial oscillations, generated from the κ-mechanism seem to be operating (Ulmschneider 1991).

8. Conclusions

Chromospheric heating is different in magnetic and in non-magnetic areas. While the heating of magnetic areas is not well known, a consistent picture of the heating of non-magnetic areas on late-type stars has emerged in the last few years. This is valid for most of the solar surface and essentially for the entire surface of very slowly rotating stars. It is found that:

- Acoustic heating occurs in all stars except possibly the A- stars, it is the main heating mechanism for non-magnetic areas.
- Magnetic areas are heated by different magnetic mechanisms which are presently poorly understood. Are AC or DC mechanisms at work?
- For comparable size of magnetic and non-magnetic areas the magnetic heating is dominant.
- The acoustic wave generation calculations, the solar acoustic wave observations, the acoustic heating calculations and the chromospheric cooling observations are all roughly consistent with each other.
- There are many unanswered questions in the detail.

References

Anderson, L.S., Athay, R.G..: 1989a, *Astrophys. J.* **336**, 1089.
Anderson, L.S., Athay, R.G..: 1989b, *Astrophys. J.* **346**, 1010.
Bohn, H.U.: 1984, *Astron. Astrophys.* **136**, 338.
Deubner, F.-L.: 1988, in: *Pulsation and Mass Loss in Stars*, R. Stalio, L.A. Willson Eds., Kluwer: Dordrecht, p. 163.
Deubner, F.-L., Reichling, M., Langhanki, R.: 1988, in: *Advances in Helio- and Asteroseismology, IAU Symp.* **123**, J. Christensen-Dalsgaard, S. Frandsen Eds., p. 439.
Endler, F., Deubner, F.-L.: 1983, *Astron. Astrophys.* **121**, 291.
Fleck, B., Deubner, F.-L.: 1989, *Astron. Astrophys.* **224**, 245.
Goldstein, M. E.: 1976, *Aeroacoustics*, McGraw Hill: New York.
Lighthill, M.J.: 1952, *Proc. R. Soc. London* **A211**, 564.
Mathioudakis, M., Doyle, J.G.: 1992, *Astron. Astrophys.* in press.
Middelkoop, F.: 1982, *Astron. Astrophys.* **113**, 1.
Narain, U., Ulmschneider, P.: 1990, *Space Science Reviews* **54**, 377.
Rutten, R.G.M.: 1987, *Astron. Astrophys.* **177**, 131.
Schrijver, C.J.: 1987, in: *Cool Stars, Stellar Systems and the Sun*, J.L. Linsky, R.E. Stencel Eds., Lecture Notes in Physics **291**, Springer: Berlin, p. 135.
Schrijver, C.J., Coté, J., Zwaan, C., Saar, S.H.: 1989, *Astrophys. J.* **337**, 964.
Stein, R.F.: 1967, *Solar Phys.* **2**, 385.
Stein, R.F.: 1968, *Astrophys. J.* **154**, 297.
Stepień, K., Ulmschneider, P.: 1989, *Astron. Astrophys.* **216**, 139.
Ulmschneider, P.: 1970, *Solar Phys.* **12**, 403.
Ulmschneider, P.: 1988, *Astron. Astrophys.* **197**, 223.
Ulmschneider, P.: 1989, *Astron. Astrophys.* **222**, 171.
Ulmschneider, P.: 1990, in: *Cool Stars, Stellar Systems and the Sun*, Astr. Soc. Pacific Conf. Ser. **9**, G. Wallerstein Ed., 1990, p. 3.
Ulmschneider, P.: 1991, in: *Mechanisms of Chromospheric and Coronal Heating*, P. Ulmschneider, E. Priest, R. Rosner Eds., Springer: Berlin, p. 328.
Ulmschneider, P., Schmitz, F., Kalkofen, W., Bohn, H.U.: 1978, *Astron. Astrophys.* **70**, 487.
Ulmschneider, P., Priest, E., Rosner, R.: 1991, *Mechanisms of Chromospheric and Coronal Heating*, Springer: Berlin.
Vernazza, J.E., Avrett, E.H., Loeser, R.: 1981, *Astrophys. J. Suppl.* **45**, 635.
Vilhu, O.: 1987, in: *Cool Stars, Stellar Systems and the Sun*, J.L. Linsky, R.E. Stencel, Eds., Lecture Notes in Physics **291**, Springer: Berlin, p. 110.

SOLAR AND STELLAR DYNAMOS

N.O. WEISS

Department of Applied Mathematics and Theoretical Physics
University of Cambridge
Cambridge CB3 9EW, UK

Abstract. Hydromagnetic dynamos provide the only viable mechanism for producing magnetic activity in cool stars like the Sun. Helioseismic data strengthen the arguments for placing the solar dynamo at the base of the convective zone, where there is a strong rotational shear and flux expulsion can hold the toroidal magnetic fields down against the effects of magnetic buoyancy. Simple models show that magnetic action can appear initially as symmetric, periodic dynamo waves which develop more complex temporal and spatial structure as the dynamo number (which depends on the angular velocity of a star) is increased. Thus magnetic fields in more active, rapidly rotating stars should exhibit less regular behaviour than the Sun. In extremely active stars the magnetic field is strong enough to control the whole pattern of convection and any dynamo action is likely to be very different from what occurs in the Sun. In fully convective late M stars the pattern should again be different from that produced by a shell dynamo.

Key words: dynamos – dynamo number – magnetic cycles

1. Introduction

Fourteen years ago, I spent a happy year at the Center for Astrophysics. It was a year that changed the course of my research and, among other things, I had the good fortune to collaborate with Pippo Vaiana, Bob Rosner and Leon Golub in a study of the implications for the solar dynamo of X-ray observations of emerging magnetic flux (Golub *et al.* 1981). Working with Pippo was a remarkable experience. As Riccardo Giacconi has said, Pippo realised the unity between theory and experiments. His approach to theory was intuitive: deep ideas emerged in a rather incoherent form and were gradually moulded into shape. He always demanded that we should establish the observational consequences of any theoretical statement, and insisted that a theoretical result was useless unless it had observational consequences. This was an excellent discipline for theoreticians – though rather a severe constraint.

In this review I shall try to bring the ideas of Golub *et al.* up to date and to extend them to cover stellar coronal activity. So I shall start with the Sun, then discuss aspects of dynamos that are currently of interest, and finally go on to consider dynamos in stars.

2. The Solar Dynamo

The essential ingredients for a stellar dynamo are convection and rotation. It is not surprising, therefore, that late-type stars, with deep convection zones, are magnetically active. In such stars poloidal (meridional) field lines are drawn out by differential rotation to form toroidal (azimuthal) magnetic flux, while cyclonic eddies act on the toroidal field to produce a regenerated – or reversed – poloidal field (e.g. Parker 1979). The latter process typically requires a locally non-vanishing mean helicity and is represented by the parameter $\alpha = -\frac{1}{3}\tau_c\langle \mathbf{u}\cdot \mathrm{curl}\mathbf{u}\rangle$, where τ_c is the convective turnover time and $\mathbf{u}$ is the fluid velocity.

J.F. Linsky and S. Serio (eds.), Physics of Solar and Stellar Coronae, 541–547.

Dynamo theory provides the only satisfactory explanation of the systematic features of the solar cycle and a new consensus has recently emerged (Weiss 1989; Belvedere 1990; Stix 1991; De Luca and Gilman 1991; Rosner and Weiss 1992). In this picture the dynamo is located in a thin shell at the base of the convection zone. This is precisely the region where helioseismology has revealed a steep radial gradient in the angular velocity Ω (Dziembowski, Goode and Libbrecht 1989; Rhodes *et al.* 1990), which should lead to the formation of a strong toroidal field. Earlier discussions (e.g. Spiegel and Weiss 1980; van Ballegooijen 1982) had predicted that such a field would be excluded from the convection zone and concentrated in a layer of weak convective overshoot, where flux expulsion could counteract magnetic bouyancy. The toroidal field is currently estimated to occupy a layer about 10^4 km thick and to reach a magnitude of $10^4 - 10^5$G. This would require a poloidal field of about 200 G, generated by helical motion (the α-effect) either within the magnetic layer or just above the base of the convection zone. As the cycle proceeds, the toroidal field is built up through differential rotation and the magnetic layer expands to reclaim part of the convection zone. Eventually the layer becomes unstable to modes driven by magnetic buoyancy (Hughes 1992) and loops of flux (with fields approaching 10^5G) rise up through the convection zone, emerging through the surface to form active regions (Moreno-Insertis 1992). The toroidal field dwindles and the convection zone expands until a reversed field is generated and the cycle is repeated.

The solar cycle has a mean period that is well-defined but it is not strictly periodic. Furthermore, it is modulated irregularly by grand minima, such as the Maunder minimum of the seventeenth century, with a characteristic timescale of about 200 years. The evidence is consistent with the hypothesis that both the cycle and its modulation are chaotic, though the record is too short for this to be definitely established (Weiss 1990).

Large-scale fields, including active regions, on the Sun conform to this systematic pattern but on a small scale magnetic flux emerges chaotically. We should expect the combination of turbulent convection and rotation within the convection zone to act as a dynamo, generating an irregular fibril field as in the model of Nordlund *et al.* (1992). In the sun, this turbulent dynamo is coupled to the cyclic dynamo, which feeds flux into the convection zone. So the magnetic flux that emerges in ephemeral active regions and gives rise to X-ray bright points in the corona may be produced either by shredding larger flux tubes or locally by a fibril dynamo.

3. Dynamo Models

The development of the magnetic field $\mathbf{B}$ is governed by the induction equation

$$\frac{\partial \mathbf{B}}{\partial t} = \mathrm{curl}(\mathbf{u} \times \mathbf{B}) + \eta \nabla^2 \mathbf{B}.$$

where the magnetic diffusivity η is assumed constant. The linear (or kinematic) dynamo problem requires the construction of some velocity $\mathbf{u}$ that allows a seed field to grow exponentially – as a magnetic instability. It can be proved that any axisymmetric field must eventually decay (Cowling 1976; Hide and Palmer 1982) and the prescribed velocity must be sufficiently complicated; nevertheless, there are many examples of kinematic dynamos.

Growth of the field is limited by the nonlinear action of the Lorentz form, so it becomes necessary to solve an equation of motion of the form

$$\rho[\frac{\partial \mathbf{u}}{\partial t} + (\mathbf{u} \cdot \nabla)\mathbf{u}] = \rho \mathbf{g} - \nabla p + \mathbf{j} \times \mathbf{B} + \mu[\nabla^2 \mathbf{u} + \tfrac{1}{3}\nabla\nabla \cdot \mathbf{u}],$$

where ρ, p are the density and pressure, respectively, $\mathbf{g}$ is the gravitational acceleration, μ is the Newtonian viscosity and the electric current $\mathbf{j} = \mu_0^{-1}\nabla \times \mathbf{B}$. Gilman (1983) and Glatzmaier (1985) have computed self-consistent nonlinear dynamo models, which demonstrate that the process really works and can led to cyclic behaviour, although the details do not match the pattern of activity on the Sun. Since we do not yet understand the purely hydrodynamic processes that lead to the measured differential rotation of the Sun there is little point in constructing more elaborate self-consistent models at this time.

Instead, much effort has gone into investigating simplified models of solar and stellar dynamos. Mean field dynamo theory allows us to construct axisymmetric models in which the field is separated into a poloidal component $\mathbf{B}_{\mathrm{P}}$ and a toroidal component $\mathbf{B}_{\mathrm{T}}$ such that

$$\mathbf{B} = \mathbf{B}_{\mathrm{P}} + \mathbf{B}_{\mathrm{T}}, \quad \mathbf{B}_{\mathrm{P}} \cdot \mathbf{e}_\phi = 0, \quad \mathbf{B}_{\mathrm{T}} = B_\phi \mathbf{e}_\phi,$$

referred to spherical polar co-ordinates (r, θ, ϕ), with $\mathbf{e}_\phi$ a unit vector in the ϕ-direction. Since $\nabla \cdot \mathbf{B}_{\mathrm{P}} = 0$ we can introduce a vector potential $\mathbf{A} = A\mathbf{e}_\phi$ and set $\mathbf{B}_{\mathrm{P}} =$ curl $\mathbf{A}$. Then the induction equation can be reduced to two scalar equations for the poloidal and toroidal fields:

$$\frac{\partial A}{\partial t} = \alpha B_\phi + \tilde{\eta}(\nabla^2 - 1/r^2 \sin^2\theta)A,$$

$$\frac{\partial B_\phi}{\partial t} = r \sin\theta \; \mathbf{B}_{\mathrm{P}} \cdot \nabla\Omega + \tilde{\eta}(\nabla^2 - 1/r^2 \sin^2\theta)B_\phi.$$

Toroidal flux is generated from the poloidal field by differential rotation (the ω-effect), while the averaged effect of helicity (cyclonic eddies) is represented by the parameter α and $\tilde{\eta}$ is a turbulent diffusivity (Moffatt 1978; Parker 1979). This procedure can be rigorously justified if there is an appropriate separation of scales – but that is not the case in stars like the Sun. Nevertheless, the mean field dynamo equations contain the essential physics of a stellar dynamo.

For a given configuration, solutions of these equations depend on a single dimensionless parameter, the dynamo number $D = \alpha\omega d^4/\tilde{\eta}^2$, where ω is a measure of the gradient in angular velocity and d is a characteristic length scale. The dynamo equations possess a trivial solution ($\mathbf{B} = 0$) which can undergo an oscillatory (Hopf) bifurcation when $D = D_{\mathrm{crit}}$. For turbulent eddies with a lengthscale l and turnover time τ_{c} we estimate $\alpha \approx \Omega l, \omega \approx \Omega/l, \tilde{\eta} \approx l^2/\tau_{\mathrm{c}}$, so that $D \approx \Omega^2\tau_{\mathrm{c}}^2(d/l)^4$. If we assume that there is a unique lengthscale, then we can set $d = l$ and the dynamo number $D \approx \sigma^2$, where the inverse Rossby number $\sigma = \Omega\tau_{\mathrm{c}}$ (Durney and Latour 1978).

For $D > D_{\rm crit}$, growth of the magnetic field is limited by nonlinear effects which can, for instance, be represented by progressively quenching either the α-effect or differential rotation, or by inserting nonlinear dissipative terms, corresponding to losses by magnetic buoyancy. There are many examples of nonlinear mean-field dynamos, designed to reproduce the solar magnetic cycle or to illustrate other aspects of magnetic activity (e.g. Parker 1979; Stix 1989; Schmitt and Schüssler 1989; Moss, Tuominen and Brandenburg 1990; Schmalz and Stix 1991; Belvedere, Proctor and Lanzafame 1991).

All these models depend on arbitrary assumptions: an alternative approach is to consider highly simplified (toy) models that can be investigated in detail in order to illuminate generic features of the nonlinear dynamo process. Parker's (1955, 1979) dynamo waves provide a prototype for this approach. He reduced the mean field dynamo equations to the linear system

$$\frac{dA}{dt} = 2DB - A, \quad \frac{dB}{dt} = \mathrm{i}A - B,$$

which describe the amplitude of one-dimensional plane waves propagating in an infinite domain. (More recently, Parker (1992) has developed a two-dimensional model, which describes surface waves propagating in the neighbourhood of a discontinuity, representing the base of the convection zone.) Dynamo action sets in at $D_{\rm crit} = 1$, giving rise to exponentially growing waves with frequency $D^{\frac{1}{2}}$. The inclusion of nonlinear terms in these equations makes it possible to obtain finite amplitude nonlinear solutions. Moreover, it is possible to find regimes where the nonlinear solutions undergo sequences of bifurcations that lead to chaos. Furthermore, there exist solutions that are aperiodically modulated in a manner reminiscent of grand minima in solar activity (Weiss, Cattaneo and Jones 1984). Although this result cannot be applied directly to the Sun, it is significant that such behaviour can be found even in such a simple truncated model, where chaotic modulation arises from the breakdown of a torus in phase space. Complicated time-dependent behaviour seems likely to arise in any dynamo if D is sufficiently large.

This approach can be extended to cover spatial structure by imposing lateral boundaries on the system (Stix 1970). Then it is possible to explore the interactions between solutions with dipole and quadrupole symmetry (Jennings 1991; Jennings and Weiss 1991). The resulting bifurcation structure is fairly complicated: symmetry-breaking leads to mixed-mode solutions and it is possible for several stable nonlinear solutions to coexist. Moreover, the period of the oscillatory solutions decreases as the dynamo number is increased. These features all appear in spherical dynamo models too.

The combination of convection and rotation in a star or planet acts readily as a dynamo. However, dynamo theory raises many tricky problems, which have engaged the interest of mathematicians. One such issue concerns the existence of so-called fast dynamos: are there kinematic dynamos in which the growth rate of the solutions tends to a finite positive limit as the magnetic Reynolds number $R_{\rm m} = ul/\eta \to \infty$? This problem has so far resisted analytical treatments, though Galloway and Proctor (1992) have provided a convincing numerical example of a fast dynamo. That is reassuring, since stellar dynamo models all rely implicitly on the existence of fast dynamos. Turbulent dissipation and reconnection raise other difficulties that have not been fully dealt with yet (cf. Rosner and Weiss 1992). In particular, small-scale turbulent transport may be halted locally even by magnetic fields that are globally weak (Vainshtein and Rosner 1991; Vainshtein, Parker and Rosner 1992). Owing to these complications the interpretion of numerical simulations is a delicate matter. Whenever dynamo action is claimed it is important to establish whether the computation has been pursued for a sufficiently long time (Cattaneo, Hughes and Weiss 1991).

4. Stellar Dynamos

Observations show that magnetic activity in late-type stars depends principally on the inverse Rossby number σ (Noyes *et al.* 1984a; Baliunas and Vaughan 1985), while theory predicts that nonlinear behaviour is determined by the dynamo number $D \approx \sigma^2$. This agreement is comforting and suggests not only that dynamos are indeed responsible for stellar activity but also that $\alpha\omega$-dynamos do capture the essential physics. On the other hand, it is surprising that all the relevant structure of a star is contained in the convective turnover time τ_c, which is rather arbitrarily computed, and that, for instance, the ratio l/R, where R is the stellar radius, does not appear significant. Perhaps this is just a lucky coincidence.

Solar-type dynamos, with regular cycles, can be distinguished from the behaviour of more active, rapidly rotating stars. These slow rotators, with rotation periods greater than about 20 days, have cycle periods of around 10 years and the cycle frequency increases with increasing angular velocity (Noyes, Weiss and Vaughan 1984b; Saar and Baliunas 1992). This agrees with predictions from theoretical models and we should expect spatiotemporal behaviour to become more complicated as Ω is increased. In fact the angular velocity of a star decreases with age, owing to magnetic braking, but the rate of spindown ($|\dot{\Omega}|$) itself progressively decreases, so that the critical value of the dynamo number seems never to be attained (Baliunas and Jastrow 1990). Comparison of the solar record with the measured Ca^+ emission from similar stars suggests that grand minima, corresponding to aperiodic modulation of activity cycles, are a common feature, affecting all these stars for about 30% of the time (Baliunas and Jastrow 1990;, Saar and Baliunas 1992). The range of activity shown by these stars lies between extremes corresponding to the minimum and maximum emission from the Sun, which appears in all respects to be a typical member of this family (White *et al.* 1991).

Rapid rotators pose more difficulties. There is observational evidence indicating a different relationship between the cycle period and rotation period (Saar and Baliunas 1992). Theoretical arguments also indicate that their behaviour should differ from that of the Sun. First of all, there may be qualitative changes in the properties of the nonlinear dynamo as D is increased. Secondly, rapid rotation should lead to a different pattern of convection (Knobloch, Rosner and Weiss 1981): for $\sigma \gg 1$ the Coriolis force becomes dominant, leading to the appearance of banana cells and differential rotation with Ω constant on cylindrical surfaces. Finally, in very active stars the magnetic field must itself interfere with convection. In such circumstances we might expect the field to be segregated from the motion so as to produce the large flux tubes which emerge to form prominent starspots. A further possibility is that there may be a balance between the Lorentz and Coriolis forces, again leading to a different pattern of convective motion (Roberts 1991).

In Section 2 a distinction was drawn between two different dynamo processes that can operate in the Sun. The shell dynamo seems to be responsible for systematic cyclic behaviour, while a weak fibril dynamo generates chaotic fields in the convection zone. It is worth speculating on the relationship between these processes in more active stars. The shell dynamo may be a feature of slow rotators (with $\sigma \approx 1$) where there is a steep radial gradient in Ω at the base of the convection zone. These are the stars that exhibit clear magnetic cycles with periods around 10 yr. Rapid rotators (with $\sigma \gg 1$) have a different rotation profile and are much more active. In such stars the dynamo probably operates within the convection zone and has very different properties, leading to aperiodic behaviour with large isolated flux tubes for large values of σ. Thus the field structure becomes more ordered as Ω is increased, and the fibrils develop into starspots. I conjecture, therefore, that as a star evolves on the main sequence there is a transition from a dynamo within the convection zone, typical of very active stars, to a shell dynamo, typical of slow rotators like the Sun. In late M-stars, which are fully (or almost fully) convective, activity must depend on the fibril dynamo alone.

5. Conclusion

Dynamo theory provides a satisfactory explanation for the magnetic activity observed in cool stars, including X-ray emission from their coronae. After considerable effort we have attained a limited understanding of magnetic cycles in the Sun and in similar, slowly rotating stars. However, there are still no detailed models of the solar dynamo. In more rapidly rotating stars we expect enhanced activity and more complex spatiotemporal behaviour as Ω increases but it is dangerous to extrapolate from solar models.

In the long run, future progress must rely on detailed computation, involving idealized models as well as large-scale simulations. The first challenge is to explain the detailed structure of convection in a late-type star; then we have to describe the interaction between convection and rotation, and the resulting distribution of angular velocity; finally, we must add magnetic fields in order to describe a stellar dynamo. That is an ambitious programme for the next decade.

Acknowledgements

I thank Salvatore Serio for arranging the Symposium, and several participants for helpful comments which have improved this text. I am also grateful to SERC for financial support while I held a Senior Fellowship.

References

Baliunas, S.L. and Jastrow, R.: 1990, *Nature* **348**, 520

Baliunas, S.L. and Vaughan, A.H.: 1985, *Ann. Rev. Astr. Ap.* **23**, 379

Belvedere, G.: 1990, in G. Berthomieu and M. Cribier, ed(s)., *Inside the Sun*, Kluwer: Dordrecht, 371

Belvedere, G., Proctor, M.R.E. and Lanzafame, G.: 1991, *Nature* **350**, 491
Cattaneo, F., Hughes, D.W. and Weiss, N.O.: 1991, *MNRAS* **253**, 479
Cowling, T.G.: 1976, *Magnetohydrodynamics*, Adam Hilger: Bristol
DeLuca, E.E. and Gilman, P.A.: 1991, in A.N. Cox, W.C. Livingston and M.S. Matthews, ed(s)., *Solar Interior and Atmosphere*, U. of Arizona Press: Tucson, 275
Durney, B. and Latour, J.: 1978, *Geop. Ap. Fl. Dyn.* **9**, 241
Dziembowski, W.A., Goode, P.R. and Libbrecht, K.G.: 1989, *Ap. J.* **337**, L53
Galloway, D.J. and Proctor, M.R.E.: 1992, *Nature* **356**, 691
Gilman, P.A.: 1983, *Ap. J. Supp. Ser.* **53**, 243
Glatzmaier, G.A.: 1985, *Ap. J.* **291**, 300
Golub, L., Rosner, R., Vaiana, G.S. and Weiss, N.O.: 1981, *Ap. J.* **243**, 309
Hide, R. and Palmer, T.N.: 1982, *Geop. Ap. Fl. Dyn.* **19**, 301
Hughes, D.W.: 1992, in J.H. Thomas and N.O. Weiss, ed(s)., *Sunspots: Theory and Observations*, Kluwer: Dordrecht, 371
Jennings, R.L.: 1991, *Geop. Ap. Fl. Dyn* **57**, 147
Jennings, R.L. and Weiss, N.O.: 1991, *MNRAS* **252**, 249
Knobloch, E., Rosner, R. and Weiss, N.O.: 1981, *MNRAS* **197**, 45P
Moffatt, H.K.: 1978, *Magnetic field generation in electrically conducting fluids*, Cambridge U. Press: Cambridge
Moreno-Insertis, F.: 1992, in J.H. Thomas and N.O Weiss, ed(s)., *Sunspots:Theory and Observations*, Kluwer: Dordrecht, 385
Moss, D., Tuominen, I. and Brandenburg, A.: 1990, *Astr. Ap.* **228**, 284
Nordlund, Å., Brandenburg, A., Jennings, R.L., Rieutord, M., Ruokolainen, J., Stein, R.F. and Tuominen, I.: 1992, *Ap. J.* **392**, 647
Noyes, R.W., Hartmann, L.W., Baliunas, S.L., Duncan, D.K. and Vaughan, A.H.: 1984a, *Ap. J.* **279**, 769
Noyes, R.W., Weiss, N.O. and Vaughan, A.H.: 1984b, *Ap. J.* **287**, 769
Parker, E.N.: 1955, *Ap. J.* **122**, 293
Parker, E.N.: 1979, *Cosmical Magnetic Fields: their Origin and their Activity*, Clarendon Press: Oxford
Parker, E.N.: 1992, *Ap. J.* , in press
Rhodes, E.J., Cacciani, A., Korzennik, S., Tomczyk, S., Ulrich, R.K. and Woodward, M.F.: 1990, *Ap. J.* **351**, 687
Roberts, P.H.: 1991, in I. Tuominen, D. Moss and G. Rüdiger, ed(s)., *The Sun and Cool Stars: Activity, Magnetism, Dynamos*, Springer:Berlin, 37
Rosner, R. and Weiss, N.O.: 1992, in K.Harvey, ed(s)., *The Solar Cycle*, Astron. Soc. Pacific.: San Francisco, 511
Saar, S.H. and Baliunas, S.L.: 1992, in K.Harvey, ed(s)., *The Solar Cycle*, Astron. Soc. Pacific.: San Francisco, 150
Schmalz, S. and Stix, M.: 1991, *Astr. Ap.* **245**, 654
Schmitt, D. and Schüssler, M.: 1989, *Astr. Ap.* **223**, 343
Spiegel, E.A. and Weiss, N.O.: 1980, *Nature* **287**, 616
Stix, M.: 1970, *Astr. Ap.* **20**, 9
Stix, M.: 1989, *The Sun: an Introduction*, Springer: Berlin
Stix, M.: 1991, *Geop. Ap. Fl. Dyn.* **62**, 211
Vainshtein, S.I., Parker, E.N. and Rosner, R.: 1992, *Ap. J.* , in press
Vainshtein, S.I. and Rosner, R.: 1991, *Ap. J.* **376**, 199
Van Ballegooijen, A.A.: 1982, *Astr. Ap.* **113**, 99
Weiss, N.O.: 1989, in G. Belvedere, ed(s)., *Accretion Disks and Magnetic Fields in Astrophysics*, Kluwer: Dordrecht, 11
Weiss, N.O.: 1990, *Phil. Trans. Roy. Soc. Lond. A* **330**, 617
Weiss, N.O., Cattaneo, F. and Jones, C.A.: 1984, *Geop. Ap. Fl. Dyn.* **30**, 305
White, O.R., Skumanich, A., Lean, J., Livingston, W.C. and Keil, S.L.: 1992, *PASP* , in press

MASS LOSS AND X-RAY EMISSION FROM GIANTS AND SUPERGIANTS

R. ROSNER

Department of Astronomy and Astrophysics, University of Chicago

Abstract. Although the basic theory for mass loss from late-type stars has been understood for almost three decades, the detailed theory for why particular types of stars lose mass – and how such mass loss relates to other indicators of stellar activity such as (coronal) X-ray emission – has proved to be remarkably elusive. Evolved stars such as giants and supergiants are an especially challenging puzzle from this perspective: Roughly speaking, these stars seem to show an anti-correlation between significant mass loss and coronal and transition region emissions. An intriguing aspect of this puzzle is the existence of "hybrid stars," which show evidence for both significant mass loss and coronal/transition region activity. I review the status of the observations and theory, concluding that a completely satisfactory explanation for mass loss from giants and supergiants is not quite in hand. What is needed is a better understanding of the X-ray spectra of hybrid stars (in order to constrain the relative geometries of the coronae and winds of these transitional-type stars); and further work on the magnetic dynamo activity of evolved stars.

Key words: Late-type stars – Giant and Supergiants – Coronae – Winds

1. Introduction

Part of our fascination with the "activity" of late-type evolved stars undoubtably derives from its apparently paradoxical nature: These stars are obviously quite different from our Sun in the most basic of stellar attributes – those that place a star in the Hertzsprung-Russell diagram – yet in many ways the nature of stellar activity on these evolved stars seems to mirror (albeit at an extreme) phenomena that are familiar from solar observations. Thus, these stars are also of late spectral type, and therefore also have surface convection zones; however, they are significantly evolved beyond the main sequence phase, and therefore have bolometric luminosities and radii that are far in excess of the solar value. Whereas the Sun is an extremely weak source of outflowing matter, to the point that we are unable at this time to detect stellar winds from Sun-like stars with comparable mass loss rates (e.g., $\dot{M} \approx 10^{-14}$ $M_\odot$ yr^{-1}), the stars I will be discussing instead have mass loss rates far in excess of the solar rate – rates that can reach values of 10^{-8} $M_\odot$ yr^{-1}. These extreme values are, it must be made plain, largely the simple result of the far larger stellar radii: The energy and mass losses, when expressed as fluxes, are in fact very much solar-like. The key differences in activity characteristics between these stars and the Sun turn out to lie in the temperatures of the winds (cool in the former, and hot in the latter), and in the terminal speeds (slow in the former, and of order the escape speed in the latter). This review is in the nature of a progress report on this subject, which I have previously reviewed at Oslo (Rosner 1991), and from which the present paper has been adapted. My focus is decidedly and unapologetically on the theoretical end of things; and I therefore highly recommend the excellent discussion, from a somewhat different point-of-view, by Hammer (1990). It should also be said that my own interests in this subject stem directly from my interactions with Pippo Vaiana, who introduced me to the pleasures of stellar X-ray astronomy, and with whom I had the privilege of collaborating on stellar activity problems for

J.F. Linsky and S. Serio (eds.), Physics of Solar and Stellar Coronae, 549–561.

many years. It is for this reason, among many others, that I am very pleased to be able to participate in this Symposium honoring Pippo's memory, though saddened by the circumstances.

Why focus on the evolved stars? Quite aside from the intrinsic interest presented by the winds from these stars, these winds directly challenge our claimed understanding of the solar wind: Although other types of stars also show observational evidence for prodigious mass loss – such as the OB stars and the Mira variables – these other cases without exception represent winds which are with high probability nothing like the solar wind. In particular, winds from these stars are most likely driven by radiative pressure (acting either on the gas itself in the OB star case, or on dust grains lying in the outer atmosphere of the Miras), and thus the physics of these winds is quite distinct from what we believe takes place in the solar wind. In the case of the stars we shall be discussing, these more exotic processes do not function; and by what is essentially a process of elimination, it is today believed that the physical processes underlying the winds we shall be discussing are very likely to be similar to those underlying the solar wind – despite the observed fact that, as pointed out above, the parameters describing the winds from these stars are not at all solar-like. This means that if we believe we have a theory for the solar wind, it should enable us to make sense of the winds from these giants and supergiants; this confrontation of (solar) theory with (stellar) observations forms a powerful test of our understanding of either type of wind.

There are of course a number of other reasons for focussing on the winds from these stars. For one thing, the total contribution of the stars in question to the interstellar medium is roughly $0.3\ M_\odot\ yr^{-1}$, and therefore is a significant contributor to the chemical evolution of our galaxy. It is well-known from solar observations that the chemical composition of the solar wind is not identical to that of the solar surface, so that the wind acceleration process acts as a kind of abundance filter. Thus, in order to understand the chemical evolution of our galaxy, it will ultimately be necessary to understand the mechanisms which determine the chemical composition of stellar winds from giants and supergiants. Still other reasons for focussing on these stars have to do with the fact that evolved stars of this type are often found in binary systems containing a degenerate star (i.e., a white dwarf or a neutron star); and in such systems, the mass outflow from the giant or supergiant plays a central role in "feeding" the accretion process onto the compact object – clearly, it behooves us to understand the mass loss process in this case as well.

Returning to the problem at hand, the essential astrophysical problem has been nicely discussed by Haisch (Haisch 1987): Evolved stars segregate themselves in the H-R diagram so that a number of so-called "dividing lines" exist. These dividing lines roughly separate evolved late-type stars into distinct classes based on a number of well-defined observational criteria:

(a) The "coronal dividing line," separating stars with and without emission, such as X-rays, characteristic of high temperature material (Linsky & Haisch 1979; Simon, Linsky & Stencel 1982; Ayres *et al.* 1981; Maggio *et al.* 1990; Haisch *et al.* 1990).

(b) The "mass loss dividing line," separating stars with and without substantial mass loss, as indicated by (for example) circumstellar absorption lines (cf. Reimers

1977; and review by MacGregor 1983).

There are other, similar, dividing lines; but these seem to be largely tied to one or the other of the two dividing lines just defined. For example, the "transition region" dividing line, separating stars with and without emission characteristic of the solar transition region, seems to be very closely linked to the coronal dividing line (cf. Haisch, Schmitt, & Rosso 1992). For this reason, I shall in the following assume that there are only two distinct "dividing lines," namely those defined by the presence or absence of X-ray emission and large mass loss, respectively.

The central questions to be answered are then: First, can we understand the existence of these "dividing lines"; second, can we understand the rough co-incidence of these separatrices; and third, can we understand the detailed properties of the outflows, including the observed mass loss rates, wind temperatures, and terminal wind speeds. One of the "holy grails" of this subject is to find a unified explanation for these various phenomena; and my purpose in this talk is to discuss how one can try to reconcile the coincidence of these transitions in behavior – the coincidence of the relatively sudden onset of large mass loss rates with the equally sudden disappearance of emission associated with high temperature ($> 10^6$ K) material as one moves along the giant and supergiant branches in the H-R diagram – with the physical properties of the observed winds.

Perhaps the most puzzling of these problems are the physical characteristics of the observed winds (see MacGregor 1983 for an excellent review of this general subject). The observed winds from late-type evolved stars tend to have rather low terminal speeds – of the order of several tens of kilometers per second, as opposed to the solar value of 300-800 km s^{-1} (we do need to caution that the determination of terminal speeds for the evolved stars is by no means uncontroversial). Holzer, Flå, & Leer (1983) were the first to point out precisely what the issue is, and how it differs from the solar case: They noted that *any* theory for accelerating winds must satisfy two constraints:

1. The wind acceleration mechanism must deposit enough energy below the sonic point of the outflow in order to balance observed radiative losses.

2. The wind acceleration mechanism must deposit enough energy to lift the gas out of the star's gravitational potential well.

In addition, a theory for *these* winds must also satisfy the third constraint that

3. The wind acceleration mechanism cannot deposit significant momentum above the sonic point (because if it did, the outflow could not remain relatively slow, as is observed).

In contrast, in the solar case the key issue is to understand why the terminal outflow speeds in, for example, coronal hole-related wind streams can be far above the surface escape speed; here the problem is to understand how to reconcile massive outflows with very low terminal speeds. Finally, we note a fourth puzzle, namely

4. The winds from giants and supergiants tend to be cool, whereas the solar wind is relatively hot.

In the following, we shall examine each of these central problems in turn.

2. The Origin of "Dividing Lines"

As a first step, we consider the lack of high temperature matter when crossing the "coronal dividing line." The earliest suggestions for why this might occur arose right at the time of discovery of this phenomenon, and were based on a simple analogy with the solar corona. That is, Haisch & Linsky (1979) and others argued that this dividing line marks a change-over from a "closed" magnetic field configuration in the outer stellar atmosphere (in which the bulk of stellar magnetic field lines emerging at the photospheric level were re-entrant) to an "open" configuration (in which most emerging field lines were connected to the circumstellar medium). This suggestion thus drew an analogy to solar coronal holes, with the idea being that the change in field connectivity would lead to the disappearance of confined hot (and visible) coronal matter, and to the appearance of an outflow which itself had little emission measure at coronal temperatures. The difficulties with this appealing suggestion are several: First, it leaves the explanation of the change in field topology unanswered; second, it does not explain why the resulting wind is cool; and third, it is a remarkable fact that the upper bounds on the *surface X-ray flux* obtained from *Einstein* for a number of nearby giants and supergiants on the cool side of the coronal dividing line are far below the observed surface X-ray flux for solar coronal holes – that is, the upper bounds on outflowing matter which is at X-ray temperatures are well below the observed solar values (cf. Vaiana *et al.* 1981). (These observations suggest strongly that little, if any, of the outflowing gas in these winds can be hot.)

A rather different suggestion is due to Antiochos & Noci (1986) and Antiochos, Haisch, & Stern (1986): They argued that careful investigation of the possible equilibrium solutions for atmospheres in coronal magnetic structures shows the existence of both the (usual) hot "coronal" solution, and a cool "chromospheric" solution, which is preferred for small coronal structures (where by small one means substantially smaller than the local gravitational scale height). Antiochos and collaborators then argued that the dividing line in question may be regarded as a kind of marginal stability line for the "hot" solutions, so that to the low effective temperature side of the dividing line, only the chromospheric solution is allowed. This explanation however leaves the wind, and its properties, entirely unexplained (although it must be said that this model does not even attempt an explanation).

A third distinct explanation for the coronal dividing line suggests that it marks the onset of winds which are sufficiently dense so as to be able to obscure any coronal X-ray emission that might be present. Thus, in this picture, hot coronae do not really disappear at the dividing line, but rather simply become invisible. This line of argument is particularly appealing because it is the only one discussed so far that makes an attempt at connecting the two observed dividing line transitions. Unfortunately, there still no conclusive evidence to support this line of argument (cf. Maggio *et al.* 1990). Indeed, a few years ago, it was thought that this issue would be settled by ROSAT observations of the so-called "hybrid" stars (which show evidence for both circumstellar absorption by a massive wind, and evidence for emission from hot chromospheric or coronal matter; Hartmann, Dupree, & Raymond 1980 and, more recently, Brown, Drake, Van Steenberg, & Linsky 1991) by examining the spectrum of the coronal emission: Does one see the expected absorption of soft X-rays by the far cooler wind? At the time of my previous review, plans were already in place for a thorough all-sky survey for hybrid stars, and in addition two groups had planned pointed observations of a number of hybrid stars with the position-sensitive proportional counter (PSPC) on ROSAT. The results of these observations will be discussed further below; here let it suffice to say that the results in hand are unfortunately still inconclusive.

Finally, Castor (1981) has suggested that the dividing line may mark the transition between solar-type winds (for which the radiative cooling time is much longer than the typical dynamical crossing time) to winds for which these two time scales are comparable. This type of explanation, which focuses on the characteristics of the wind, regards the disappearance of the hot corona as an entirely separate problem (which it does not address).

To summarize: Of the models discussed above, only the absorption model explicitly recognizes the implausibility of an accidental coincidence between the coronal and wind dividing lines. The difficulty is that the observations can in principle readily falsify this model (and may well do so shortly!). Can one formulate a model which provides a physical basis for the observed coincidence between the dividing lines? I will argue below that it may be possible to construct such a model.

3. The Physics of Winds From Late-type Stars

It is possible to construct models based on wave heating and momentum deposition which satisfy the observational constraints imposed by winds from giants and supergiants. These models, which are based on the standard stellar wind equations, and incorporate momentum and energy deposition by Alfvén waves (with damping parametrized, rather than solved for from first principles), were first extensively explored by Hartmann & MacGregor (1980), who showed that if the wave damping length λ_D were of order the stellar radius R_*, then models with low terminal velocity and large mass loss rates could be constructed. This result was encouraging, if puzzling, since it was not obvious how the wind "knew" what the stellar radius was – for example, the gravitational stratification depends on the total stellar mass and the distance to the star's center, and not on the intrinsic radius of the star.

However, a detailed parameter study of this problem by Holzer, Flå, & Leer (1983; see also Leer, Holzer, & Flå 1982) showed that such a model suffers from a far more fatal defect. The key point recognized by these authors is that if one wants to obtain low terminal wind velocities, then one must insure that there is little if any direct momentum deposition beyond the sonic point (since any such driving can only lead to an increase in the mean velocity of wind particles, and not to an increase in the mass flux). It is therefore essential to any model for winds from these stars to make sure that the wind acceleration is largely confined to the region below the sonic point, a feature which Hartmann & MacGregor's model satisfied as long as $\lambda_D \approx R_*$. Now, the central question is, how do the mass loss rate and the terminal wind speed depend in detail on the ratio λ_D/R_*? The parameter study showed that both the mass loss rate and the terminal wind speed have a transition, from low values to large values, at roughly $\lambda_D/R_* \approx 1$ (this is why Hartmann &

MacGregor obtained the result they did), but that these transitions were very sharp – indeed, in order to obtain a solution with large mass loss rate but small terminal speed, the control parameter λ_D/R_* had to be tuned remarkably precisely. This fine-tuning is physically very unappealing, and suggests a fundamental flaw in the application of this model to the observations.

4. A Unified Model?

It is easy to state exactly what is needed to solve the puzzle posed by the "dividing lines": One requires a way of "turning off" hot coronae at the same time as one initiates a wind, and furthermore, the mechanism which then drives the wind ought to have the property that it deposits virtually all of the momentum in the wind below the sonic point. What I will now do is to sketch just such a model, based on the recent work of An *et al.* (1990) and Rosner *et al.* (1991, 1993), from which some of the following has been excerpted.

We begin with what we know for certain: It is absolutely incontrovertible that when coronal emission is seen from giants and supergiants, the emitting gas *must* be magnetically confined. The argument is as follows: Consider the escape temperature for a typical G0 giant, say with $M_* \approx 10^{0.4} M_\odot$ and $R_* \approx 10^{0.8} R_\odot$ (Allen 1973),

$$T_{escape} \approx 10^{-1.2} T_{escape,\odot} , \qquad (1)$$

where $T_{escape,\odot}$ ($\approx 10^7$K) is the escape temperature for the Sun; for a G5 giant, the corresponding escape temperature is approximately $10^{-1.5} T_{escape,\odot}$. Such stars *cannot* have a solar-like multi-million degree corona in the absence of some confining force other than gravity, i.e., stellar magnetic fields (cf. Rosner, Tucker, & Vaiana 1978, Holt *et al.* 1979). That is, such gas would be simply entirely unconfined in the absence of non-gravitational forces because the mean energy per particle would be larger than the gravitational binding energy per particle, i.e., the average coronal particle would not be gravitationally bound and hence the fluid would expand freely into space. This is an entirely different phenomenon than a thermally-driven wind, which exists for coronal temperatures well below the escape speed, and hence for mean coronal particle energies well below the gravitational binding energy per particle; viz., Parker 1963). It is inescapable that X-ray-emitting giants and supergiants *must* have the X-ray emitting gas confined by stellar magnetic fields.

What about the evolved stars on the other side of the coronal dividing line, those without observed X-ray emission, or rather, those for which the upper bounds on observed X-ray emission lie well below typical detection levels for the observed "coronal" stars? The crucial fact to recall, already mentioned, is that for a number of giants and supergiants, the observed upper bounds on the X-ray surface flux (in ergs s^{-1} cm^{-2}) from the *Einstein* Observatory are well below the observed X-ray surface flux in solar coronal holes. Thus, we already know observationally that the coronal hole analogy cannot strictly apply – there is simply far less X-ray-emitting gas in the giant/supergiant stellar outflows than in the solar coronal hole case. To put it another way: The prevailing evidence is that the outflow is relatively cold, a state of affairs supported by the observations of circumstellar absorption lines, which are characteristic of cool matter, and not hot coronal matter (cf. Reimers 1977).

As an aside, it is useful to note that the argument regarding the temperature of the unconfined plasma for the non-coronal evolved stars applies equally well to plasma attached to open field lines in the coronae of the X-ray emitting evolved stars. That is, whatever winds flow from the "coronal" giants and supergiants, it is very likely that the associated gas is cool, in contrast to the solar case (where the gas in, for example, coronal holes does reach temperatures of order 10^6 K; Withbroe 1988) – gas at solar coronal temperature would have a mean energy per particle far above the gravitational binding energy per particle. Thus, the explicit assumption made by many that winds from the "coronal" giants and supergiants have solar wind temperature (viz., Drake 1988, Hammer 1990) is not likely to be correct, and for this reason the upper bounds obtained on the wind mass loss in these stars are extremely weak.

To summarize: What one needs to explain – aside from the rough coincidence of the two dividing lines – is the existence of a change in the temperature of the dominant plasma component in the outer stellar atmosphere of these evolved stars as one traverses the dividing lines.

Let me then sketch a model that has the ingredients to answer these various questions, which I have adapted from the discussion in Rosner *et al.* (1991). To begin with, I recall that it has been long known that reflection of Alfvén waves in stellar atmospheres becomes significant when the wavelength of the propagating wave becomes of order of (or larger than) the local Alfvén speed scale height of the background medium (see earlier discussions by, for example, Ferraro & Plumpton 1958; Hollweg 1978; Heinemann & Olbert 1980). Indeed, as pointed out by Leer *et al.* (1982), the WKB approximation usually employed in problems involving wave propagation in stellar atmospheres breaks down precisely under such circumstances for waves with periods of order a minute or longer. For the simplest case, namely an exponential atmosphere which is also plane parallel, Ferraro & Plumpton (1958) showed that analytical solutions for Alfvén waves may be obtained, which depend on the gradient of the Alfvén speed

$$V_A{}' \equiv \frac{dV_A}{dR} \equiv \frac{V_A}{h} \tag{2}$$

(the last relation may be regarded as the definition of the Alfvén speed scale height h). These solutions have the property that in the short wavelength limit (in which $h/(V_A/\omega) >> 1$), there is essentially no reflection; this limit is obtained for atmospheres in which the typical scale height is large compared to the wavelength of the dominant energy-carrying waves, e.g., the wavelength of waves at the peak of the wave power spectrum; and conversely, there is strong reflection in the opposing limit $h/(V_A/\omega) << 1$.

Now, consider a giant or supergiant with a corona: The atmosphere for such a star can be modeled to lowest order by a two-layer description (Leer *et al.* 1982), consisting of two distinct adjacent layers: The lower layer, corresponding to the photospheric and transition region layers in an actual star, is characterized by a small Alfvén speed scale height ($h = h_1 << V_A/\omega$ for ω at the peak of the wave power spectrum) because of both the low temperature of the photospheric region, and the small spatial extent of the temperature transition region. (Since the transition region thickness is much smaller than the local pressure scale height, the density varies inversely with the temperature in this layer, and hence experiences a 100-fold decrease in the same region in which the temperature increases 100-fold). In contrast, the overlying layer, corresponding to the coronal region, is characterized by the Alfvén speed scale height $h = h_2$, with $h_2 >> V_A/\omega$ for ω as above. One can then show that all waves with wavelength $\lambda > h_1$ will be strongly reflected as long as h_1 obeys this inequality; but that the reflection efficiency quickly drops to zero as $h_1 \rightarrow h_2$. The strong inequality would seem to apply to the evolved stars with coronae, and so one would expect a significant wave pressure exerted by the reflected waves at the transition region level – but will all this really matter to the structure of these stars' coronae? Rosner *et al.* contend that the answer is "no": That is, since magnetic fields must be present to confine the hot gas against pure thermal expansion (i.e., for these stars, the presence of a transition region implies the existence of high-temperature matter, matter which because of its high temperature barely notices the gravitational well due to the underlying star), the wave pressure just discussed (exerted by the longer-period waves) very likely does nothing of consequence except to modify the radial stratification of the hot matter. The shorter-period waves which can penetrate into the coronal gas, however, may well be involved in the coronal heating process.

Consider now the case of giants and supergiants with no coronae. By the above argument, these are stars for which magnetic fields have proven to be ineffective in confining hot gas, had any been present. As above, we can now consider the effect of Alfvén waves on the resulting cool outer atmospheres (e.g., we still assume that the stars in question have surface magnetic fields, albeit fields which are too weak to confine hot coronal gas). This sort of case has been studied explicitly by means of numerical calculations by An *et al.* (1989, 1990), in which the characteristics of Alfvén waves in a hydrostatic, isothermal, and spherically-symmetric atmosphere with purely radial background magnetic fields were considered. In this atmosphere, the density falls off exponentially with distance from the stellar surface,

$$\rho = \rho_* \exp\left[-\alpha\left(1 - \frac{R_*}{R}\right)\right] , \tag{3}$$

where α is a constant, given by

$$\alpha = \frac{GM_*}{kT/m}\frac{1}{R_*} = \frac{R_*}{H}\left(\frac{R}{R_*}\right)^2 = \frac{R_*}{H_*} , \tag{4}$$

and where R is the radius, H ($\equiv kT/mg$) is the density scale height at R, and all other quantities have their customary meaning, e.g., T is the temperature, and so forth. Quantities evaluated at the base of the stellar atmosphere, i.e., at the stellar radius $R = R_*$, are indicated by a "$*$" subscript. The Alfvén speed then varies with radius as

$$V_A = V_{A*}\left(\frac{R_*}{R}\right)^2 \exp\left[\frac{\alpha}{2}\left(1 - \frac{R_*}{R}\right)\right] . \tag{5}$$

Thus, we see that in such an atmosphere, the Alfvén speed increases steeply with radius above the base for sufficiently large α (which, for given M_* and R_*, requires a sufficiently low atmospheric temperature); this suggests the possibility of reflection, as discussed above. Indeed, detailed calculations for such an atmosphere by An *et al.* (1990), based on numerical solution of the linearized wave propagation equations, show evidence both for the continuous reflection process in such an atmosphere (this distinguishes reflection in a cool, uniform temperature, stratified medium from reflection in an atmosphere with a transition zone, where the reflection is completely dominated by the sharp spatial gradient in density at the temperature transition point) and for the associated strong local body force, which results from the locally-enhanced wave pressure gradient in the reflection region for $H/(V_A/\omega) << 1$ (An *et al.* 1990).

Rosner *et al.* show by illustration that evaluation of α for various types of evolved stars leads to a segregation of the evolved stars into two classes, located on either side of the dividing line (namely stars with $\alpha > 20$ and $\alpha < 10$). This segregation largely reflects the difference in the peak temperature reached in the outer atmospheres of the stars on either side of the dividing line. The key point made by Rosner *et al.* is that wave reflection should be much stronger for values of $\alpha > 20$ than for $\alpha < 10$. Thus, it appears that although wind acceleration effects resulting from mode reflection near the base of the atmosphere occur for stars both to the right and to the left of the dividing line in the H-R diagram, only in the case of the "open" atmospheres to the right of the dividing line is it possible for this reflection to lead to an outflow.

This picture is seductively appealing, but I must alert the reader of one important caveat: A fully self-consistent calculation – of the type carried out for example by Cuntz (1987; see also Cuntz & Ulmschneider 1988) for purely acoustic waves – which treats Alfvén wave propagation in an outward-flowing wind from these stars without the imposition of *ad hoc* assumptions (such as fixed/arbitrary wave damping lengths, or parametrized damping models) is not yet in hand, and remains to be carried out.

5. The Case of the Hybrid Stars

For some time, it has been hoped that observations by ROSAT of what are perhaps the most critical stars – the hybrid stars which show aspects of both coronal and wind phenomena – would help us in exploring models for the outer atmospheres themselves. The reason is straightforward: the position-sensitive proportional counter (PSPC) on ROSAT has proved to be a remarkable sensitive imager, with excellent spectroscopic characteristics for this type of device (cf. Trümper *et al.* (1991) and Pfeffermann *et al.* (1990)), with sufficient spectral resolution and sensitivity to in principle allow one to study the possible absorption of X-ray-emitting material by the winds in these stars.

The first step was to address the question: Are these stars as a class X-ray sources? The EXOSAT observations of the hybrid star α TrA by Brown *et al.* (1991) clearly showed that this star was an X-ray source; but was this typical, or atypical? The more recent observations by Haisch, Schmitt, & Rosso (1992) made as part of the ROSAT All-Sky Survey now provide considerable support for the proposition that these stars are indeed a new class of stellar X-ray sources. Indeed, subsequent pointed ROSAT observations of such stars by Reimers & Schmitt (1992) and Kashyap *et al.* (1993) are consistent with this proposition.

The second question concerns the nature of the emitting atmosphere, the hybrid's corona: Is the solar analogy appropriate? By and large, the only observations sufficiently sensitive to have obtained a high signal-to-noise PSPC (or any other) spectrum are those of α TrA by Kashyap *et al.* (1993). Although the spectral analysis is still too preliminary for definitive quantitative results, the following facts are not likely to change as our understanding of the spectrum improves:

1. The X-ray spectrum can clearly not be modeled by a single-temperature corona; instead, as is the case for virtually all other active late-type stars, the PSPC spectrum contains evidence for at least two spectral components, one at "solar coronal" temperatures of a few million degrees K, and the other at a much higher temperature, on the order of 10^7 K.

2. There is considerable evidence for absorption by cool material *in excess* of the absorption expected on the basis of the known interstellar matter column density. Preliminary estimates place this absorption at a level of roughly twice the expected ISM value for α TrA.

These data are clearly extremely tantalizing, since they support the picture that hybrid stars indeed do represent a transitional class of objects. That is, it appears that these stars do not just happen to lie near the dividing line locus, but actually do show the transitional physical characteristics one would expect if they represented a class of stars making the physical transition between "corona-dominated" and "wind-dominated" behaviors. Unfortunately, we do not yet understand the X-ray spectra of these stars well enough to be able to test models for the geometric relation between the coronal material and the outflowing wind; there is however considerable hope that such an understanding will be reached shortly.

6. Summary and Conclusions

I have discussed the main constraints imposed by observations of coronal emission and wind outflows associated with giants and supergiants, and considered the various types of models proposed to explain the existence of "dividing lines" separating stars with significant coronal emission, but no evidence for mass loss, from stars with no coronal emission, but with abundant evidence for massive winds. This overview allows us to identify the principal observational points which any theory that seeks to explain the observations must confront:

(i) The coincidence of the coronal and wind "dividing lines."

(ii) The change in temperature of the dominant plasma component above the stellar surface as one traverses the dividing lines, from hot coronal matter to cool outflowing wind gas.

(iii) The solar-like mass outflow fluxes, coupled with the non-solar, very low, terminal wind speeds.

I then reviewed a recent proposal for a unified model to account for these observational constraints, which is based on the idea first broached by Linsky & Haisch

(1979) that changes in surface magnetic field configuration as one moves along the giant and supergiant tracks in the H-R diagram can play the central role in accounting for the coincidence of relatively sudden absence of stellar X-ray emission and the onset of large mass loss rates observed for stars located on the giant and supergiant branches in the H-R diagram. In this picture, the dividing line in mass loss rate is roughly coincident with the coronal dividing line, because the absence of high temperature matter implies both an absence of effective magnetic confinement and the presence of a cool outer atmosphere (an atmosphere which thus has a small Alfvén speed scale height, and hence experiences strong Alfvén wave reflection at its base). This suggestion has the additional attraction that the momentum deposition by reflected Alfvén waves naturally occurs below the sonic critical point; that the position of maximal momentum deposition (as well as the amplitude of maximal momentum deposition) is entirely decoupled from the natural damping length of these waves; and that the mode damping beyond the reflection point may well be very small (so that the further acceleration of the wind is minimal, as required by the observed low terminal velocities). It is important to note, however, that this model is at present more a suggestion than a quantitative prescription – for one thing, it remains to be shown quantitatively how Alfvén wave reflection occurs in a full wind model for these stars; for another, the model at present makes no pretense to have resolved the energetics problem.

This model furthermore leaves one major issue unresolved: It does not identify the reason why evolved stars to the right of the coronal dividing line ought to have "open" magnetospheres, and stars to the left ought to have "closed" magnetospheres. The answer to this question lies outside the domain of coronal and stellar wind physics, and instead lies in the domain of stellar magnetic dynamo theory. This key point has been recently re-investigated by Rosner *et al.* (1993), who have made the following point: The observations indicate that the "mechanical" energy input (measured in terms of the surface energy flux) above the stellar surface for giants and supergiants is comparable for stars with coronae and transition regions but no significant winds, and stars without coronae and transition regions, but with significant winds – a relation which also holds approximately for solar quiet sun regions and coronal holes (cf. Holzer 1988). Thus, if we believe that magnetically-related heating is at work in the former case, it is hard to believe that it is not in the latter case as well; this is indeed the presupposition of the models of An *et al.* (1990) and Rosner *et al.* (1991). The issue is thus why the topology of the stellar magnetic fields would change systematically when crossing a particular locus in the H-R diagram, without any significant change in the magnetically-related mechanical energy input. Rosner *et al.* (1993) suggest one possibility: As stars cross the dividing lines, it is possible for the dynamo processes responsible for the large-scale (mean) stellar fields to become linearly stable (i.e., for the dynamo number to fall below its critical value), but yet for small-scale dynamo action to continue (see also Vainshtein *et al.* 1993). In that case, the dominant spatial structure of surface magnetic fields changes from large-scale (i.e., in solar terms, scales comparable to active regions and larger) to the very small scale (again, in solar terms, scales comparable to granules or yet smaller) as one crosses the dividing lines; and because of the very rapid falloff of magnetic field strength with height for the case in which surface fields are very small scale, it then becomes possible for coronal gas to "open" field lines and to escape. This possibility is only preliminary, and has emerged only recently as new (numerical) advances in dynamo theory have been made; we are clearly at the beginnings of a new understanding of stellar dynamos, magnetic fields, and their relations to the manifestations of stellar activity.

Acknowledgements

Much of the work reported in this paper was supported by NASA through the *Einstein* and ROSAT Guest Observer Programs and the Space Physics Theory Program. This work would not have been possible had it not been for the Pippo Vaiana, who not only motivated my own interests in this subject matter, but through his efforts virtually created the field of stellar X-ray astronomy. I would also like to thank my many other collaborators in this work, including F.R. Harnden, Jr., V. Kashyap, A. Maggio, R. Moore, Z. Musielak, J.H.M.M. Schmitt, and S. Sciortino, for the many enjoyable discussions and arguments that have molded my views on this subject. I would also like to acknowledge the very useful comments of R. Hammer.

References

Allen, C.W.: 1973, *Astrophysical Quantities*, London: Athlone Press
An, C.H., Musielak, Z.E., Moore, R.L., & Suess, S.T.: 1989, *ApJ* **345**, 597
An, C.H., Suess, S.T., Moore, R.L., & Musielak, Z.E.: 1990, *ApJ* **350**, 309
Antiochos, S.K., Haisch, B.M., & Stern, R.A.: 1986, *ApJ (Letters)* **307**, L55
Antiochos, S.K., & Noci, G.: 1986, *ApJ* **301**, 440
Ayres, T.R., Linsky, J.L., Vaiana, G.S., Golub, L., & Rosner, R.: 1981, *ApJ* **250**, 293
Brown, A., Drake, S.A., Van Steenberg, M.E., & Linsky, J.L.: 1991, *ApJ* **373**, 614
Cuntz, M.: 1987, *AA (Letters)* **188**, L5
Cuntz, M., & Ulmschneider, P.: 1988, *AA* **193**, 119
Drake, S.A.: 1986, in Cool Stars, Stellar Systems, and the Sun, ed(s)., *M. Zeilik & D.M. Gibson*, Springer Verlag, 369
Drake, S.A.: 1988, in Proc. 6th International Solar Wind Conference, ed(s)., *M. Neugebauer*, NCAR TN-306, 129
Dupree, A.K.: 1983, in Solar Wind 5, ed(s)., *M. Neugebauer*, Washington, DC, NASA CP-2280,
Ferraro, V.C.A., & Plumpton, C.: 1958, *ApJ* **127**, 459
Haisch, B.M.: 1987, in Cool Stars, Stellar Systems, and the Sun, ed(s)., *J.L. Linsky & R.E. Stencel*, Berlin: Springer, 269
Haisch, B.M., *et al.*: 1990, *ApJ* **361**, 570
Haisch, B.M., Schmitt, J.H.M.M., & Rosso, C.: 1992, *ApJ (Letters)* **388**, L61
Hammer, R.: 1990, in New Windows to the Universe, ed(s)., *F. Sanchez & M. Vazquez*, Cambridge Univ. Press, 77
Hartmann, L., Dupree, A.K., & Raymond, J.C.: 1980, *ApJ (Letters)* **236**, L143
Hartmann, L., & MacGregor, K.B.: 1980, *ApJ* **242**, 260
Heinemann, M., & Olbert, S.: 1980, *JGR* **85**, 1311
Hollweg, J.V.: 1978, *Solar Phys.* **56**, 305
Holt, S.S., *et al.*: 1979, *ApJ (Letters)* **234**, L65
Holzer, T.E., Flå, T., & Leer, E.: 1983, *ApJ* **275**, 808
Kashyap, V., *et al.*: 1993, *in preparation* ,
Leer, E., Holzer, T.E., & Flå, T.: 1982, *Space Sci. Rev.* **30**, 161
Linsky, J.L., & Haisch, B.M.: 1979, *ApJ (Letters)* **229**, L27
Maggio, A., *et al.*: 1990, *ApJ* **348**, 253

MacGregor, K.B.: 1983, in Solar Wind 5, ed(s)., *M. Neugebauer*, Washington, DC, NASA CP-2280, 241

Parker, E.N.: 1963, *Interplanetary Processes*, New York: Interscience

Pfeffermann, E., *et al.*: 1990, *Proc. SPIE* **733**, 519

Reimers, D.: 1977, *AA* **57**, 395

Reimers, D., & Schmitt, J.H.M.M.: 1992, *ApJ (Letters)* **in press**,

Rosner, R.: 1991, in Oslo Workshop on Solar and Stellar Winds, ed(s)., *P. Maltby & E. Leer*, Univ. of Oslo, Norway,

Rosner, R., An, C.H., Musielak, Z.E., Moore, R.L., & Suess, S.T.: 1991, *ApJ (Letters)* **372**, L91

Rosner, R., Musielak, Z.E., Cattaneo, F., An, C.H., Moore, R.L., & Suess, S.T.: 1993, *in preparation* ,

Rosner, R., Low, B.C., & Holzer, T.E.: 1986, in Physics of the Sun, ed(s)., *P.A. Sturrock*, New York: Reidel, Ch. 11 (Vol. II, pp. 135-80)

Simon, T., Linsky, J.L., & Stencel, R.: 1982, *ApJ* **257**, 225

Trümper, J., *et al.*: 1991, *Nature* **349**, 579

Vainshtein, S.I., Cattaneo, F., Tao, L., & Rosner, R.: 1993, in Cambridge Workshop on Magnetic Dynamos, ed(s)., *in press*, ,

RADIATION DRIVEN WINDS AND X-RAY EMISSION FROM O STARS

R. P. KUDRITZKI and D. J. HILLIER

Institute of Astronomy and Astrophysics, Scheinerstrasse 1, W-8000 Munich 80, Fed. Rep. of Germany

Abstract. We briefly review our current knowledge concerning the winds of massive stars. New ROSAT X-ray observations of the O4 f star ζ Pup are presented and analyzed. The observed X-ray fluxes are consistent with a model in which the X-rays originate in shocks distributed throughout the stellar wind. Detailed fitting indicates a shock temperature of 3 to 4×10^6 K corresponding to shock velocities of 500 km s^{-1}. The *importance* of using a reliable wind model to take into account absorption of the X-rays by the cool wind is illustrated. Different assumptions regarding the wind opacity can alter conclusions regarding the X-ray formation process, and additionally can make a significant difference to the deduced shock temperatures.

Key words: ζ Pup, stars: early-type – stars: mass loss –Xrays: stars

1. Introduction

Through their large ionizing luminosity and their massive stellar winds, O stars, and their evolutionary descendents, play a key role in determining the energy budget, chemical enrichment, and dynamics of the interstellar medium (e.g., Leitherer (1992)). While our knowledge of O stars, their evolution, and their winds has increased dramatically over the last 2 decades (see reviews by Kudritzki and Hummer (1990), Conti and Underhill (1988)), many key aspects remain uncertain, and many significant observations remain unexplained.

Early observations by the X-ray satellite EINSTEIN showed that O stars were soft X-ray emitters (Seward et al.(1979), Harnden et al. (1979)), with a flux significantly in excess of that expected on the basis of the star's effective temperature. These, and subsequent observations, showed that the observed X-ray flux scales with the bolometric luminosity. The ratio of L_X to L_{BOL} is typically 10^{-7}, but shows a large scatter of $\pm$1dex (Chlebowski et al. (1989)). The observed X-rays, while representing only a small fraction of the star's energy balance, are important for 2 reasons. Firstly, the X-ray flux may be sufficiently large to influence the ionization of the stellar winds, and hence affect fundamental parameters derived from atmospheric models. Secondly, the X-rays provide diagnostics regarding processes occurring at the photosphere, or within the stellar winds. Such processes may be important for determining mass loss rates and the dynamics of the stellar wind.

With the advent of the X-ray satellite ROSAT, with its enhanced soft energy sensitivity and increased resolution with respect to EINSTEIN, we have a new opportunity to gain insights into X-ray production in O stars. We have therefore undertaken a large project in order to re-investigate the X-ray properties of a moderate sized sample of O stars. Importantly, we will use the theory of stellar winds to examine the influence of wind absorption on the emergent spectrum.

Below (Sect. 4) we discuss and analyse new ROSAT observations of the O4 f star ζ Pup. To lay the ground for this analysis, we present (Sect. 2) a brief summary of our current knowledge and ideas concerning the atmospheres and winds of massive

J.F. Linsky and S. Serio (eds.), Physics of Solar and Stellar Coronae, 563–571.

stars, while in Sect. 3 we discuss the various mechanisms that have been proposed to explain soft X-ray production in O stars. Variability is another important diagnostic which may place constraints on the X-ray production mechanism, and is examined in Sect. 5. Finally, in Sect. 6, we discuss future work and directions.

2. The Winds of Massive Stars

Mass loss plays an important role in the evolution and spectral appearance of O stars. While it had been realized for a long time that Wolf-Rayet and P Cygni stars were undergoing extensive mass loss, the first direct evidence for mass loss from O stars came from Ultraviolet observations. Observations by Morton (1967) showed P Cygni profiles on numerous resonance lines in several O stars, with the absorption component indicating velocities 2-3 times the escape velocity.

Since that time, further UV observations (most notably by Copernicus and IUE) have shown that all stars with initial mass greater than $15M_{\odot}$ show evidence for winds (Abbott (1982b)). Extensive mass loss also reveals its presence in other ways, notably emission at Hα (either a pure emission line or a weakened photospheric absorption profile), or by an infrared and radio excess.

Numerous observational studies to determine mass loss rates of O stars, crucial for evolutionary studies, have been made. The most reliable method is based on the radio flux, and has often been used to calibrate mass loss rates derived, for example, from UV observations. These observations show that the mass loss rate correlates strongly with luminosity. The extensive observational data set by Garmany and Conti (1984), for example, shows that

$$\log \dot{M} = -6.87 + 1.62 \log[L(L_{\odot})/10^5] \qquad (1)$$

although there is a large scatter ($\sigma = 0.5$ dex) about this trend. Other dependences in the data may be present, however the observed mass loss rates have large uncertainties — typically a factor of 2. Another important finding of these studies is that the terminal velocity of the wind (as measured from the blue edge of saturated P Cygni profiles) is strongly correlated with the escape velocity (Abbott (1982a)). To first order, $V_{\infty}=3V_{esc}$.

The use of a coronal model, by analogy with the sun, cannot explain the large observed mass loss rates for O stars — other mechanisms need to be invoked. One such mechanism, proposed by Lucy and Solomon (1970), is that a wind can be driven by radiation pressure action through resonance lines in the UV. Because Lucy and Solomon considered driving by only the strongest line, their predicted mass loss rates were much smaller than observed.

A major breakthrough in understanding the winds of O stars came from the work of Castor et al. ((1975); hereafter CAK). They considered an ensemble of lines, both optically thick and optically thin, and showed that the line force could be parameterized by a power law. The constants in this power law are related to the total number of lines, and the ratio of the number of thick to thin lines. With their simple force law they were able to solve the momentum equation, and hence solve for the expected mass loss rates and wind terminal velocities of O stars.

Their results provided strong support for radiation pressure acting on lines as the mechanism responsible for mass loss in O stars.

A detailed comparison of the CAK theory with observation was given by Abbott (1982a). The standard CAK theory was found to produce mass loss rates a factor of 2-3 too high, while the predicted terminal velocities were a factor of 2-3 too low. The discrepancies were removed by the calculations of Friend and Abbott (1986), and by Pauldrach et al. (1986), both of whom relaxed the assumption that the driving radiation came from a point source. This assumption had been introduced by CAK because it led to an enormous simplification of the hydrodynamical equations.

A considerable effort is currently being spent in order to improve the steady-state radiation driven wind theory. Improved line lists have been created, and further assumptions (none as important for the dynamics as the point source assumption mentioned previously) have been relaxed. Puls (1987), for example, treated the complex case of line overlap self-consistently, and showed that it must be taken into account for detailed quantitative analyses of individual stars.

The current Munich code, a joint project between Pauldrach, Puls, Kudritzki and Butler (Pauldrach et al. (1992)), currently solves the full non-LTE transfer problem for all relevant ions self-consistently with the radiation driven wind hydrodynamics. In all, 26 elements, 133 ionization stages, 4000 levels and 10000 bound-bound transitions are taken into account. The effects of multiple scattering, and the continuous radiation field are also included. More than 100000 lines contribute to the final non-LTE line force.

The steady-state line driven wind theory, however, has 2 important failings. It cannot explain

1. the observed variability, particularly of the UV P Cygni resonance lines, or
2. the observed X-ray emission.

3. X-Ray Production in O Stars

Two basic scenarios have been proposed to explain the soft X-ray emission from single O stars. The first proposes that the X-rays arise from a hot corona located at the base of the stellar wind, while in the second the X-rays are assumed to arise from shocks distributed throughout the wind.

The suggestion that a hot corona exists in O stars predates the detection of X-rays from O stars. It was invoked, for example, by Cassinelli and Olson (1979) to explain the anomalous ionization observed in O and B stars. In many O and B stars, lines from high ionization species (e.g., O VI) are seen to occur in the UV spectrum (Rogerson and Lamers (1975), Lamers and Snow (1978)). These high ionization lines cannot be produced by the normal photospheric radiation field. Auger ionization by X-rays, in which 2 electrons are ejected from the dominant ion, provides a viable explanation for their existence.

Work by Cassinelli et al. (1978) had shown from Hα observations that any hot coronal zone must be geometrically thin, and located at the base of the wind. Since the thin corona lies at the base of the wind, the emitted X-rays are strongly attenuated by the opacity of the overlying wind. In particular, the coronal model predicts strong K shell absorption — absorption which is not seen in any of the

EINSTEIN PSPC spectra (Cassinelli et al.(1981)) or in the higher resolution solid state observations (Cassinelli and Swank (1983)).

The coronal model has been elaborated and refined by Waldron (1984), but the lack of K shell absorption in the observed X-ray spectra probably rules it out. There are also other arguments against it (see Hillier et al. (1992)). The ROSAT spectrum of ζ Pup (see Sect. 4), also shows no evidence for strong K shell absorption, and hence confirms that the coronal model is inapplicable.

The most viable scenario for producing the X-ray emission from O stars is to assume that they arise from shocks in the stellar wind. These shocks are assumed to arise from instabilities in the wind — it has been known for a long time that radiation driven winds are inherently unstable (e.g., Lucy and Solomon (1970)), Owocki and Rybicki (1986)). A phenomenological model for X-ray production was developed by Lucy and White (1980), and further refined by Lucy (1982b). Detailed comparisons of this model with observation by Cassinelli and Swank (1983) showed that the preferred shock structure was unable to match the observations.

Since that time Owocki et al. ((1988), hereafter OCR) have run detailed hydrodynamical calculations to investigate instabilities in line-driven winds. Their model confirms the concept that radiation instabilities generate strong shocks within the stellar wind, however, their calculated shock structure differs from that of Lucy and White. In the OCR model, the strongest shocks are reverse shocks, with the high velocity gas having low density — the opposite of the Lucy and White model. Because the original OCR calculations assume an isothermal wind and shocks, they are currently unable to predict the X-ray spectrum. This problem is currently being addressed by both Owocki (Cooper and Owocki, (1992)), and by Feldmeier and Puls in Munich (Puls (1992)).

4. Model and Interpretation of Observations

On the basis that the wind shock model appears the most promising for explaining the origin of X-rays from O stars, we have adopted the following scenario. We assume that the X-rays arise from a uniform distribution of sources in the stellar wind above some minimum radius R_{min}. Two further parameters are required — the shock temperature which characterizes the X-ray emission, and a filling factor e_s .

The filling factor e_s is defined such that the (total) emission, from a volume increment dV, is given by

$$16 e_s^2 N_p N_e \Lambda_{RS} \, dV \qquad (2)$$

where Λ_{RS} is the volume emission coefficient of an X-ray plasma calculated using the Raymond-Smith code (1977), N_p is the ionized hydrogen density of the ambient wind, and N_e the electron density.

To compute the observed X-ray spectrum we must take into account absorption by the cool stellar wind, for which a detailed atmospheric model is required. Unfortunately, the X-ray opacity is sensitive to the ionization state of He in the wind, which is uncertain. It has been generally assumed, given the effective temperature

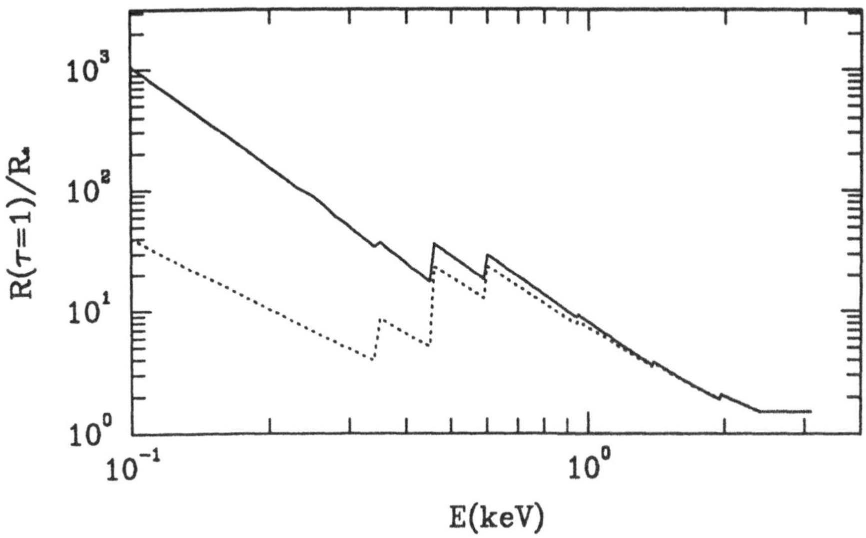

Fig. 1. Radius of optical depth unity as a function of X-ray energy. (solid line - Model 1; dotted line - Model 2). The observed X-ray flux scales approximately as the inverse of $R(\tau = 1)$.

of ζ Pup, that He would remain doubly ionized in the stellar wind. However, calculations by Pauldrach et al. (1992) suggest that line blanketing in the extreme UV can cause He to recombine. Therefore we have considered 2 models — one in which He recombines to He^+ beyond $0.8V_\infty$(M1), while in the second model He is always doubly ionized (M2).

The final parameters we adopted for ζ Pup are d= 440 pc, $R_* = 19.0\,R_\odot$, $\dot{M} = 5.0 \times 10^{-6}\,M_\odot\,\mathrm{yr}^{-1}$, $V_\infty = 2200\,\mathrm{km\,s}^{-1}$, $T_{eff} = 42500\,\mathrm{K}$, and N(He)/N(H)= 0.12. Further and more complete details about the X-ray opacities and modeling procedure are given in Hillier et al. (1992).

The very different X-ray absorption properties of the two models are illustrated in Fig. 1 where we have plotted the radius at which a (radial) optical depth of unity is reached in the wind. In the recombination model M1, the soft X-ray opacity is greatly enhanced.

In addition to wind absorption, we need to allow for absorption by interstellar material. Fortunately the interstellar column density towards ζ Pup is low, and well determined. From fitting of the wings of Ly α, Shull and van Steenberg (1985), and Kudritzki et al. (1989), have determined $\log N_H = 20.0$.

For each model (and assumed value of R_{min}) we adjusted the shock temperature and filling factor to obtain the best fit as determined by the minimum of the χ^2 value.

Two ROSAT PSPC observations of ζ Pup were obtained. Complete details of the observations and reduction procedure will be given by Hillier et al. (1992). Here we confine ourselves to interpretation of the 60 ksec observation for which

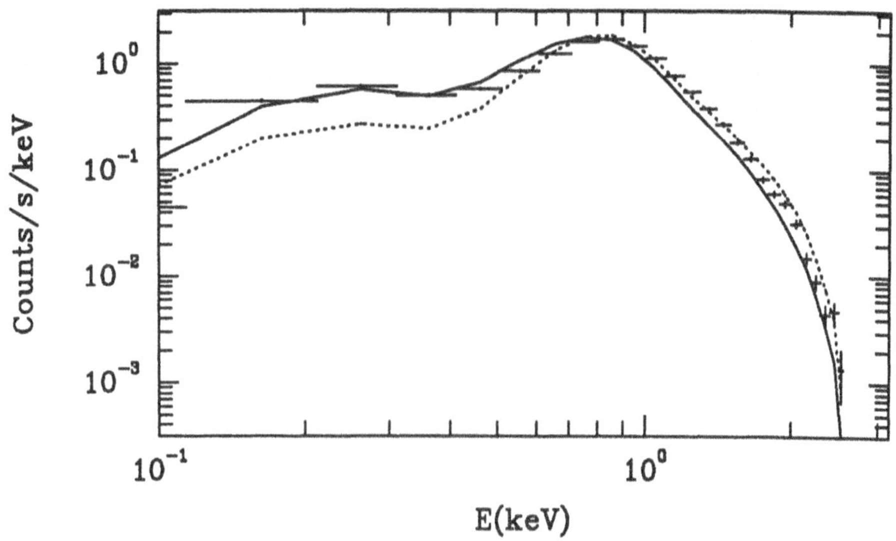

Fig. 2. Comparison of observations (horizontal intervals with error bars) with the predicted X-ray flux from M1 assuming R_{min} = 1.5 and $\log T$ = 6.5 (solid line) or $\log T$ = 6.6 (dotted line). The two models approximately bracket the observation, suggesting that the emission from ζ Pup can be characterized by temperatures of 3 to 4 $\times$ $10^6 K$, implying shock velocities of around 500 km s^{-1}.

no Boron filter was used. Fits to the X-ray fluxes obtained with the Boron filter, presented by Hillier et al. (1992), are qualitatively similar to those without the Boron filter. Following the philosophy of EXSAS, we compare the theoretical and observed spectra by convolving the model, corrected for the effects of interstellar absorption, with the instrumental response function. The best fits for M1, adopting $R_{min} = 1.5R_*$, are shown in Fig. 2.

As readily apparent, a shock temperature of 3 to 4 $\times$ $10^6 K$ provides a moderately good fit to the observations. The lower temperature underestimates the flux at energies above 1 keV, while the higher temperature model underestimates the flux at lower energies. Discrepancies are not unexpected given the overly simplistic assumptions of the model — notably, a single shock temperature and a constant filling factor. The inferred shock velocities are around 500 km s^{-1}.

As has been pointed out previously (e.g., Lucy and White (1980)), the observed X-ray fluxes represent only a small part of the total shock X-ray emission. For the above models, less than 5% of the X-ray flux (defined to have E< 0.1 keV) escapes to the observer — the rest is absorbed by the cool wind.

The same procedure was carried out for M2, and the best fit is shown in Fig. 3. The fit is qualitatively worse, with the theoretical spectrum showing enhanced absorption around 0.5 keV due to K shell absorption. There is no evidence in the observed spectrum for such an absorption feature. In addition, a much higher shock temperature of 6.3 $\times$ $10^6 K$ is inferred, corresponding to shock velocities of

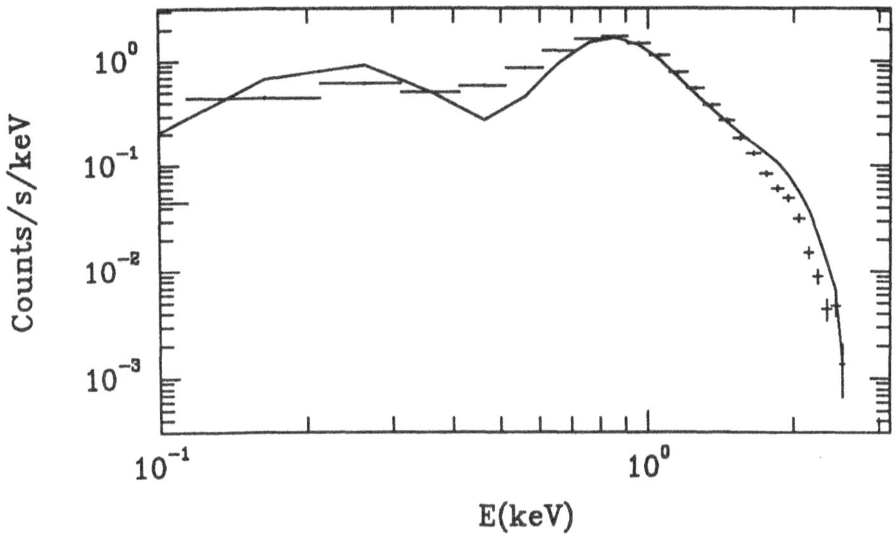

Fig. 3. Comparison of observations (horizontal intervals) with the predicted X-ray flux from M2 assuming $\log T = 6.8$ and $R_{min} = 1.5$. The most striking discrepancy between the model and observed fluxes is the presence of the strong absorption at 0.5 keV in the model fluxes. This discrepancy occurs in all fits, independent of R_{min} and suggests that the M1 models provide a more viable scenario for explaining the X-ray emission from ζ Pup.

650 km s^{-1}. Such high shock velocities (with a sufficiently large filling factor) may be difficult to reproduce in the standard instability models.

5. Variability

Variability of the X-ray emission is another key observation which may provide significant constraints on X-ray production in O stars. Collura et al. (1989) analyzed the X-ray emission (detected with the EINSTEIN PSPC) from 12 OB stars. Significant variations were detected in 3 O stars, including ζ Pup. For ζ Pup the amplitude of the variations was 40% in the soft energy band, with an associated timescale of days. The total band amplitude was 18%.

Our 60 ksec observation, spread over 12 days, was designed specifically to search for such variations of the X-ray flux with rotation period. A preliminary analysis finds no significant variations although a 20% amplitude variation would have been seen easily. In addition the X-ray flux, determined from the Boron observations taken 6 months earlier, appears to be consistent (at the 10% level) with the later observations.

We can also compare our current X-ray fluxes with those from EINSTEIN. Such an analysis is difficult due to different sensitivities and has not yet been performed. However, in our modeling procedure we used the EINSTEIN X-ray flux to provide a first estimate of the filling factor. This estimate was surprisingly accurate (within 10% for our best fit model), and suggests that there has been no long term variation in the X-ray flux from ζ Pup.

The lack of variability of the X-ray emission in ζ Pup is very surprising, especially in view of the observations of Collura et al. (1989). However these results may not be so inconsistent if we recall 2 important points that have emerged from analysis of the variability of UV resonance lines.

Firstly, the absorption part of the profile tends to be much more variable than the emission component. As has been noted by many authors, this is most likely a selection effect in the sense that the absorption samples only that part of the wind between us and the star, while the emission component samples a much larger volume. The X-ray emission, like the P Cygni emission component, will also tend to come from a large volume of the wind.

Secondly, variations in the UV and their nature are very sporadic. Some stars appear to show stable profiles over long periods of time, followed by marked activity (see Conti and Underhill (1988)). Thus the UV observations suggest that very long data sets, much larger than those currently available, are needed to address and understand the variability of X-ray emission from O stars.

A detailed analysis of the ROSAT variability of ζ Pup is being performed (Baade et al. (1992)), and will provide tighter and more reliable limits on the variability.

6. Discussion

ROSAT observations of the bright O4 f star have revealed new insights into X-ray emission from O stars. Simple model fits suggest that the shocks in the wind of ζ Pup can be characterized by temperatures of 3 to $4 \times 10^6\,K$, with corresponding shock velocities of $500\mathrm{km\,s^{-1}}$. The models indicate the *importance* of using a reliable wind model to take into account absorption of the X-rays by the cool wind. Different assumptions regarding the wind opacity can alter conclusions regarding the X-ray formation process, and additionally can make a significant difference to the deduced shock temperatures.

In an effort to place tighter constraints on X-ray emission from O stars the Munich group has undertaken a project to measure X-ray fluxes (and spectra when possible) from a well defined sample of 40 O stars. Initially, each of the spectra of these stars will be analyzed in a manner similar to that for ζ Pup. This will allow a search for correlations between X-ray properties and the fundamental parameters of the star (e.g., T_{eff}, g, R). Such correlations might explain the large dispersion in the ratio of X-ray to bolometric luminosity. While such correlations have not been seen previously in EINSTEIN data this may be due to a combination of observational errors, and the fact that the observed X-ray flux represents only a small portion of the total shock emission.

References

Abbott, D.C.: 1982, *Ap. J.*, **259**, 282
Abbott, D.C.: 1982, *Ap. J.*, **263**, 723
Baade, D., et al., 1992 (in preparation)

Cassinelli, J.P., Olson, G.L.: 1979, *Ap. J.*, **229**, 304

Cassinelli, J.P., Olson, G.L., Stalio, R.: 1978, *Ap. J.*, **220**, 573

Cassinelli, J.P., Swank, J.H.: 1983, *Ap. J.*, **271**, 681

Cassinelli, J.P., Waldron, W.L., Sanders, W.T., Harnden F.R., Rosner, R., Vaiana, G.S.: 1981, ApJ, 250, 677

Castor, J.I., Abbott, D.C., Klein, R.I.: 1975, *Ap. J.*, **195**, 157

Chlebowski, T., Harnden, F.R., Jr., Sciortino, S.: 1989, *Ap. J.*, **341**, 427

Collura, A., Sciortino S., Serio, S., Vaina, G.S., Harnden, F.R., Rosner, R.: 1989, *Ap. J.*, **338**, 296

Conti, P.S., Underhill, A.B.: 1988, 'O Stars and Wolf-Rayet Stars', NASA SP-497

Cooper, R.G., Owocki, S.P.: 1992, in 'Nonisotropic and Variable Outflows from Stars', eds. L. Drissen, C. Leither, A. Nota, A.S.P. Conf. Ser, Vol 22, p281

Friend, D.B Abbott, D.C.: 1986, *Ap. J.*, **311**, 701

Garmany, C.D., Conti, P.S.: 1984, *Ap. J.*, **284**, 705

Harnden, F.R., Branduardi, G., Elvis, M., Gorenstein, P., Grindlay, J., Pye, J.P., Rosner, R., Topka, K., Vaiana, G.S.: 1979, *Ap. J.*, **234**, L51

Hillier, D.J., Kudritzki, R.P., Pauldrach, A.W.A, Schmitt, J.H.M.M., Baade D., Puls, J.: 1992, in preparation for *Astr. Ap.*

Kudritzki, R.P., Puls, J., Gabler, R., Schmitt, J.H.M.M.: 1991, Proc. Berkeley Workshop on EUV Astronomy, eds. R.F. Malina, and S. Boyer, Pergamon Press, p 130

Kudritzki, R.P., Hummer, D.G.: 1990, *Ann. Rev. Astr. Ap.*, **28**, 303

Lamers, H.J.G.L.M., Snow, T.P.: 1978, *Ap. J.*, **219**, 504

Leitherer, C, in Star Forming Galaxies and Their Interstellar Medium, ESO/EIPC Workshop (in press)

Long, K.S., White, R.L.: 1980, *Ap. J.*, **239**, L65

Lucy, L.B.: 1982, *Ap. J.*, **255**, 286

Lucy, L.B., Solomon, P.M.: 1970, *Ap. J.*, **159**, 879

Morton, D.C.: 1967, *Ap. J.*, **147**, 1017

Owocki, S.P., Rybicki, G.B.: 1986 ,*Ap. J.*, **309**, 127

Owocki, S.P., Castor, J.I., Rybicki, G.B.: 1988, *Ap. J.*, **335**, 914

Pauldrach, A.W.A.: 1987, *Astr. Ap.*, **183**, 295

Pauldrach, A.W.A., Puls, J., Kudritzki, R.P.: 1986, *Astr. Ap.*, **164**, 86

Pauldrach, A.W.A., Puls, J., Butler, K., Kudritzki, R.P., Hunsiger, J.: 1992, in preparation for *Astr. Ap.*

Puls, J.: 1987, *Astr. Ap.*, **184**, 227

Puls, J.: 1992, in *Rev. in Mod. Astr.* (in press)

Raymond, J.C., Smith, B.W.: 1977 *Ap. J. Suppl.*, **35**, 419

Rogerson J.B., Lamers, H.J.G.L.M.: 1975, *Nat*, **256**, 190

Seward, F.D., Forman, W.R., Giacconi R., Griffiths, R.E., Harnden, F.R., Jones, C., Pye, J.P.: 1979 *Ap. J.*, **234**, L55

Shull, J.M., Van Steenberg, M.E.: 1985, *Ap. J.*, **294**, 599

Waldron, W.L.: 1984, *Ap. J.*, **282**, 256

MAGNETIC ACCRETION DISKS AND CORONAE IN ACTIVE GALACTIC NUCLEI

GEORGE FIELD*
Osservatorio Astrofisico di Arcetri
and
ROBERT ROGERS
Institute for Geophysics and Planetary Physics, Lawrence Livermore National Laboratory

Abstract. A model is presented for accretion disks around supermassive black holes in which the dynamics is dominated by magnetic fields. The work done by magnetic stresses in the disk is dissipated in a low-density corona by magnetic reconnection, which accelerates electrons to relativistic energies. The synchrotron and inverse-Compton emission by the electrons covers the range from the far infrared to gamma rays. The emitted spectrum qualitatively resembles that observed for moderate-luminosity active galactic nuclei (AGNs) and is consistent with the diffuse X-ray and gamma-ray background being composed of numerous faint AGNs. A key component of the model is the demonstration by Galeev, Rosner, and Vaiana (1979) that magnetic fields in accretion disks must form loops, and hence coronae.

Key words: accretion disks –black holes – magnetic reconnection

1. Introduction

It is believed that AGNs are powered by accretion onto supermassive ($M \sim 10^8 M_\odot$) black holes (Blandford and Rees 1992), and powerful arguments can be given that accretion occurs through a disk (Collin-Souffrin 1992). However, it is not certain how matter loses angular momentum, allowing it to accrete. Lynden-Bell (1969) pointed out that magnetic fields can transfer angular momentum efficiently, and also provide the means for accelerating electrons to the energies needed to explain the nonthermal emission of AGNs. Pursuing his suggestion, we have constructed a model for moderate-luminosity AGNs in which angular momentum is transferred by strong magnetic fields ($\sim 10^4$ G) in the disk, which buckle to form a magnetic corona (Galeev, Rosner, and Vaiana 1979) with loops of field in which $B \sim 10^2$ G. Reconnection of the fields in loops of opposite polarity is efficient, and results in acceleration of electrons to high energies. The emission by these electrons can account for the observed nonthermal spectrum of AGNs.

This model contrasts with the conventional "α disk" model, in which angular momentum transfer takes place by turbulent viscosity; in such disks, dissipation heats the disk itself, leading to thermal emission, but not to nonthermal emission.

2. Assumptions

We assume:

(i) that ideal MHD (flux freezing) applies throughout, except in current sheets associated with reconnection in the corona;

(ii) that magnetic fields are present in the accreting gas (*e.g.*, interstellar matter);

* On leave from the Harvard-Smithsonian Center for Astrophysics

J.F. Linsky and S. Serio (eds.), Physics of Solar and Stellar Coronae, 573–577.

(iii) that all of the energy released in magnetic reconnection goes into accelerating electrons, positrons, and ions;
(iv) that acceleration occurs near the tops of coronal loops; and
(v) that the differential energy spectrum of the accelerated electrons and/or positrons is a power law (γ^{-2}).

All of these assumptions are broadly consistent with what we know about solar flares, where magnetic energy is converted to the energy of fast particles (Kundu and Woodgate 1986; Dröge et al. 1989; Rieger 1989; Vlahos 1989). They are discussed in detail in forthcoming papers (Field and Rogers 1992, 1993). In particular, it is shown there that the γ^{-2} power law follows if the electrons are accelerated by MHD shock waves generated by reconnection.

In our model, the gravitational energy released by accreting matter does work on the magnetic field of the disk; in turn, the resulting magnetic energy propagates by field-line buckling (Parker instability) to the corona, where it dissipates by reconnection. Thus, all of the gravitational energy is converted to the energy of fast particles. Thermal radiation is generated indirectly by the particles when nonthermal energy interacts with the disk, but about 20% of the energy finally emerges as nonthermal radiation.

As in all steady accretion disks (Pringle 1981) the flux from one side of the disk at distance R from the black hole is:

$$F = \frac{3\,GM\dot{M}}{8\pi R^3}\left[1 - \left(\frac{R_i}{R}\right)^{1/2}\right]$$

where we take R_i, the radius of the inner edge of the disk, to be 6 GM/c^2. Half of the emitted radiation originates within $R_{1/2} = 4R_i = 24\ GM/c^2$, which equals 3.6×10^{14} cm for $M = 10^8 M_\odot$; there $F = 2.3\times10^{13}$ ergs cm^{-2} sec^{-1} for $L = \frac{1}{12}\dot{M}c^2$, which equals 10^{44} ergs sec^{-1} for $\dot{M} = 0.02 M_\odot$ yr^{-1}.

We have calculated the spectrum at $R_{1/2}$ as typical of that of the disk as a whole.

3. Emitted Spectrum

Accelerated ions do not radiate, but move down the field lines until they encounter sufficient material to stop them; we assume that the resulting energy is ultimately reemitted thermally by the dense gas of the disk, and arbitrarily assume that half of the energy goes into the ions. Electrons with $\gamma > \gamma_1 \sim 30$ lose energy rapidly by both synchrotron emission in the coronal field ($\sim$ 100 G) and by inverse-Compton scattering of the thermal photons from the disk. As a result, they do not thermalize, but lose all their energy to nonthermal radiation. In doing so, their steady-state energy distribution steepens to γ^{-3} above γ_1. We have introduced an arbitrary upper energy cutoff at $\gamma_2 = 500$.

The inverse-Compton emission is in the X-ray region, and because of relativistic beaming, the X rays are directed along the electron velocity. Because acceleration occurs at the tops of loops, the electrons are moving downward along magnetic field lines, and most of the X rays encounter the disk below, to be either absorbed or reflected with a probability that depends on energy. This "Compton reflection" process (Guilbert and Rees 1988; Lightman and White 1988) modifies what would otherwise be a power-law spectrum, flattening it in the 10 - 100 keV region in agreement with observations of individual AGN (Pounds *et al.* 1990) and of the diffuse X-ray background (Fabian *et al.* 1990; Rogers and Field 1991a).

Some synchrotron photons are also inverse-Compton scattered, as are some photons produced by Compton scattering. The calculated composite spectrum (Figure 1) therefore contains synchrotron radiation scattered zero (S), one (SC), or two (SC^2) times, as well as thermal radiation from the disk scattered zero (D), one (DC), or two (DC^2) times. Higher-order scatterings are negligible because the scattering cross section falls rapidly at very high energies.

The calculated spectrum depends on the coronal magnetic field strength B, for which the minimum value for confinement is calculated to be 160 G, and γ_1, the break in the electron energy distribution. We find that a sharp cutoff at $\gamma_1 \sim 30$ enables us to interpret the observed diffuse γ-ray background as the superposition of moderate-luminosity AGN (Rogers and Field 1991b). The same model (with the numbers of AGN depending on redshift z as $(1+z)^{2.8}$ out to $z_m = 4.6$) gives a good fit to the spectrum of the diffuse X-ray background.

We find that the number density of electrons with $\gamma > \gamma_1$ is 10^7 cm^{-3} if the height of the corona is $10^{-2}R = 3.6 \times 10^{12}$ cm. The corresponding Thomson depth and Compton y-parameter are 2×10^{-5} and 0.03, respectively.

Electron-positron pairs are created by $\gamma - \gamma$ collisions in the corona. We estimate that $2n_p = 10^9$ cm^{-3}, about 100 times the density of relativistic electrons; it is possible that the relativistic "electrons" in our model are actually pairs that have been accelerated. The total Thomson depth in pairs, including those outside of magnetic loops, is less than 0.06, so for a moderate-luminosity AGN with $L = 10^{44}$ ergs sec^{-1}, it is unlikely that the energy input is sufficient to generate an optically thick pair plasma at $T \sim 100$ keV that will Comptonize low-energy photons into X rays by repeated scattering. This mechanism, discussed by Lightman and Zdiarski (1987) and Zdiarski *et al.* (1990) among others, may well dominate for higher-luminosity objects.

As the model is based on reconnection in a finite number of coronal loops, we predict that both synchrotron and inverse-Compton emission should vary on time scales of hours, and with amplitudes of a few percent. All nonthermal components should be affected.

4. Discussion

Magnetic accretion disks offer a possible way to understand the nonthermal radiation of moderate-luminosity AGNs. The spectrum of Figure 1 exhibits five features that agree qualitatively with the observations of individual such objects and/or the diffuse high-energy background:

(i) a far infrared continuum that is cut off sharply at wavelengths greater than 100 μm by synchrotron self absorption; (ii) a pronounced "UV bump" carrying 80% of the flux, due to thermal radiation from the disk, heated from above; (iii) a ν^{-1} continuum below $\sim$ 1 keV, caused by Compton-scattered synchrotron radiation;

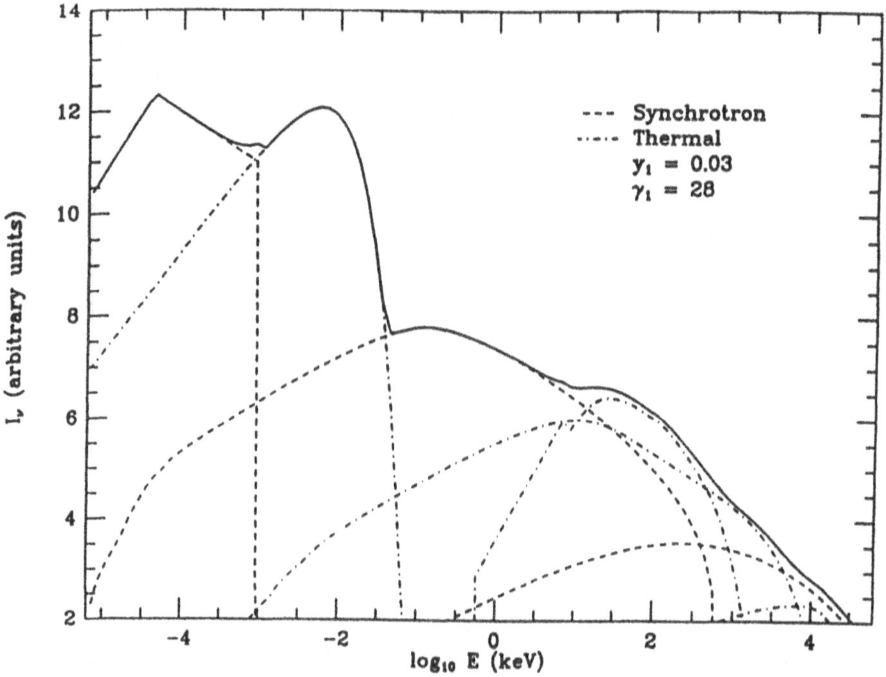

Fig. 1. **Calculated Spectrum of an AGN.** The calculated spectrum emitted by a point on an accretion disk 3.6×10^{14} cm from a $10^8 M_\odot$ black hole for a coronal magnetic field of equipartition strength (160 G), a minimum electron Lorentz factor $\gamma_1 = 28$, and a maximum value $\gamma_2 = 500$. Synchrotron emission (dashed line) is cut off at wavelengths greater than $\sim 100\mu$m by self absorption. Other dashed lines are synchrotron radiation that has been inverse-Compton-scattered once (peaking at 3×10^{-2} keV) or twice (peaking at 20 keV). The component peaking at 0.01 keV or 1200A (dash-dot line) is thermal emission from the disk at 24,000 K, due to heating of the disk from above by nonthermal radiation and energetic ions from the corona. Other dash-dot lines are disk radiation that has been inverse-Compton-scattered once (peaking at 3 keV) or twice (peaking at 2 MeV); also shown is a component Compton reflected from the disk (peaking at 20 keV). Features in the composite spectrum are consistent with observations of individual AGNs and of the diffuse high-energy background radiation.

(iv) an X-ray spectrum which is flat below ~ 30 keV, and steep above ~ 100 keV, due to inverse-Compton radiation that has been Compton-reflected from the disk; and (v) an "MeV tail" due to double inverse-Compton scattering.

Much work remains to be done. The calculations performed here for $R = R_{1/2}$ should be extended to other radii, and the resulting spectra averaged for detailed comparison with observation. The dependence of the model upon accretion rate (hence luminosity) should be analyzed, and in particular, the onset of an optically thick pair plasma should be considered. Many AGNs have jets, which are not accounted for by the model. Finally, possible sources of magnetic field should be studied.

Given that the theme of this meeting is solar and stellar coronae, and particularly the heating of these coronae by magnetic fields, it is appropriate to remark that the bulk of the X rays observed from the universe may be produced in magnetic coronae. However, these coronae belong to accretion disks in AGNs rather than to stars.

Acknowledgements

One of us (G.F.) thanks Prof. Franco Pacini for his hospitality at the Arcetri Observatory, where this paper was written, Dr. Franco Drago for helpful references, and NASA for support of this research under NAGW-931.

References

Blandford, R.D. and Rees, M.J. 1992, in *AIP Conference Proceedings* 254, *Testing the AGN Paradigm,* ed. S.S. Holt, S.G. Neff, and C.M. Urry (American Institute of Physics, New York), p. 3.

Collin-Souffrin, S. 1992, in *AIP Conference Proceedings 254, Testing the AGN Paradigm,* ed. S.S. Holt, S.G. Neff, and C.M. Urry (American Institute of Physics, New York), p. 119.

Dröge, W., Meyer, P., Evenson, P. and Moses D. 1989, in *Solar and Stellar Flares,* ed. B.M. Haisch and M. Rodonò (Dordrecht: Kluwer Academic Publishers) p. 95.

Fabian, A.C., George, I.M., Miyoshi, S., and Rees, M.J. 1990, MNRAS, 242, 14p.

Field, G.B. and Rogers, R.D. 1992, in preparation.

Field, G.B. and Rogers, R.D. 1993, Ap.J., January 1993.

Galeev, A.A., Rosner, R., and Vaiana, G.S. 1979, Ap.J. 229, 318.

Guilbert, P. and Rees, M.J. 1988, MNRAS, 233, 475.

Kundu, M. and Woodgate, B. 1986, *Energetic Phenomena on the Sun* (Washington, D.C.: NASA Scientific and Technical Branch).

Lightman, A.P. and White, T.R. 1988, Ap.J. 335, 57.

Lightman, A.P. and Zdiarski, A.A. 1987, Ap.J. 319, 643.

Lynden-Bell, D. 1969, Nature, 223, 690.

Pounds, K.A., Nandra, K., Stewart, G.C., George, I.M., and Fabian, A.C. 1990, Nature, 344, 132

Pringle, J.E. 1981, ARA and A, 22, 471.

Rieger, E. 1989, in *Solar and Stellar Flares,* ed. B.M. Haisch and M. Rodonò (Dordrecht: Kluwer Academic Publishers), p. 323.

Rogers, R.D. and Field, G.B., 1991a, Ap.J., 370, L57.

Rogers, R.D. and Field, G.B., 1991b, Ap.J., 378, L17.

Vlahos, L. 1989, in *Solar and Stellar Flares,* ed. B.M. Haisch and M. Rodonò (Dordrecht: Kluwer Academic Publishers), p. 431.

Zdiarski, A., Ghisellini, G., George, I.M., Svensson, R., Fabian, A.C., and Done, C., 1990, Ap.J., 363, L1.

MAGNETIC FIELDS AND CORONÆ IN LATE-TYPE STARS

B. MONTESINOS * and C. JORDAN
Department of Physics (Theoretical Physics)
University of Oxford
1 Keble Rd., Oxford OX1 3NP, UK

Abstract. In this paper we give an outline of a detailed study by Jordan and Montesinos (1991) and Montesinos & Jordan (1992) (JM and MJ hereafter respectively) on the relations between the surface magnetic fluxes $B_s f_s$, filling factors f_s and coronal fields B_c, and Rossby numbers for a sample of late-type main-sequence stars. Some other relations, involving the Ca II H and K fluxes and coronal parameters are also studied and analysed.

Key words: Late-type stars – Magnetic fields – Coronæ

1. The surface magnetic fields

Spatially averaged magnetic fields can be measured in the photospheres of cool main-sequence stars and the magnetic fields and filling factors can be found by modelling (see e.g. Saar 1990). In the last few years a fairly large number of reliable measurements of magnetic fields, B_s, and filling factors, f_s, have become available, and data for 19 stars are given in Tables 1 and 2 of MJ. The observed magnetic fields are thought to arise from dynamo action at the lower part of the sub-photospheric convection zone. We have investigated correlations between the magnetic fluxes, $B_s f_s$, filling factors, f_s, and the Rossby numbers.

Of the possible forms of correlation we find that $\log B_s f_s$ and $\log f_s$ are best fitted with a linear dependence on Ro, such that

$$\log B_s f_s = 3.27(\pm 0.08) - 0.97(\pm 0.08) Ro \tag{1}$$

and

$$\log f_s = -0.01(\pm 0.07) - 0.86(\pm 0.07) Ro \tag{2}$$

2. Ca II activity indicators

Correlations between several activity indicators based on the Ca II H and K fluxes and rotation periods and Rossby numbers have been investigated by several authors (e.g. Noyes *et al.* 1984, Rutten 1987, Stepień 1989). We find that the best correlation is between $\Delta R_{\rm HK} \equiv \Delta F_{\rm HK}/\sigma T_{\rm eff}^4$ and Ro in a log–linear fit

$$\log \Delta R_{\rm HK} = -3.94(\pm 0.03) - 0.50(\pm 0.03) Ro \tag{3}$$

where $\Delta F_{\rm HK} \equiv F_{\rm HK} - F_{\rm basal}$ is the so-called 'excess flux density' as defined by Rutten (1987).

From spatially resolved solar observations of active regions Schrijver *et al.* (1989) found that

* Present address: LAEFF, ESA-IUE Satellite Station, P.O. Box 50727, 28080 Madrid, SPAIN

J.F. Linsky and S. Serio (eds.), Physics of Solar and Stellar Coronae, 579–582.

$$\log \Delta F_{\rm HK} = 4.8 + 0.60(\pm 0.10) \log B_s f_A \qquad (4)$$

where f_A is the filling factor for the active region. With the solar value of $T_{\rm eff}$ this converts to

$$\log \Delta R_{\rm HK} = -6.0 + 0.60(\pm 0.10) \log B_s f_A \qquad (5)$$

The fit to the stellar data is

$$\log \Delta R_{\rm HK} = -5.89(\pm 0.17) + 0.61(\pm 0.06) \log B_s f_s \qquad (6)$$

The stellar correlation (6) is consistent with the solar one (5) to within the uncertainties.

3. Coronal parameters and scaling laws

Although the mechanisms for the coronal heating are unknown, it is widely accepted that the heating of stellar coronæ is controlled by the magnetic field. The coronal field must be controlled by the surface field and filling factor. It is of interest to write all the observable parameters as a function of the coronal magnetic field.

The observed emission measure $\mathrm{Em}(T_c) = \int N_e^2 dr$ can be written (see Jordan *et al.* 1987, JM, MJ) as

$$\mathrm{Em}(T_c) = \frac{P_c^2 H}{2k^2 T_c^2} \qquad (7)$$

where P_c is the coronal electron pressure, $H/2$ is the isothermal pressure-squared scale height, and T_c is the coronal temperature. Equation (7) refers to a spherically symmetric corona. In MJ we also discuss the effects of a filling factor.

The electron pressure can be expressed in terms of the plasma β and the coronal magnetic field, B_c, as $1.8P_c = \beta(B_c^2/8\pi)$, and equation (7) can be written as

$$B_c = 8.6 \times 10^{-10} \frac{(\mathrm{Em}(T_c) T_c g_*)^{1/4}}{\beta^{1/2}} \quad \mathrm{Gauss} \qquad (8)$$

JM postulated that for a fixed value of β, the coronal field might depend simply on *Ro*. We have used a sample of F, G and K main-sequence stars, for which a one-temperature fit to their coronal spectra is satisfactory (Schmitt *et al.* 1990). The revised values of their emission measures are given in Table 2 of MJ, updating those in Table 1 of JM which have been rederived (see MJ for details). The fit of $(\mathrm{Em}(T_c) T_c g_*)^{1/4}$ with *Ro* is

$$0.25 \log(\mathrm{Em}(T_c) T_c g_*) = 10.26(\pm 0.04) - 0.39(\pm 0.04) Ro \qquad (9)$$

and using expression (8), it gives

$$\log B_c = 1.20(\pm 0.04) - 0.39(\pm 0.04) Ro - 0.5 \log \beta \qquad (10)$$

Further scalings between $\mathrm{Em}(T_c)$, T_c, P_c and g_* can be obtained if some assumption is made concerning the energy balance in the corona. For example, Hearn

(1975, 1977) proposed that coronæ would tend to a minimum energy loss configuration. He assumed that below a certain temperature at the base of the corona there is no external heating and the radiation losses are balanced by thermal conduction from the corona. This provides a relation between P_c, T_c and g_* (see MJ for details). The precise powers in the derived scaling laws depend on the functional form adopted for the radiative power losses. For temperatures above 2×10^5 K one can write

$$P_{\rm rad} = AT_c^{-\delta} \tag{11}$$

where A is a constant that depends on the value of δ adopted; $\delta = 1/2$ or $\delta = 1$ have been used in the literature. For either value of δ the minimum energy loss approach plus the assumption that the energy density in the coronal gas is determined by the available energy density in the magnetic field (see JM) lead to a fixed value of $\beta = 1/6$ for an energy dissipation $< \delta B_c^2 > / B_c^2 = 1/4$. Using this, equation (10) can be written as

$$\log B_c = 1.59 - 0.40 Ro \tag{12}$$

With Hearn's method we also obtain the following scalings of $\mathrm{Em}(T_c)/g_*$ with T_c

$$\log(\mathrm{Em}(T_c)/g_*) = 1.32 + 3.5 \log T_c \quad \text{for} \quad \delta = 1 \tag{13}$$

and

$$\log(\mathrm{Em}(T_c)/g_*) = 4.52 + 3 \log T_c \quad \text{for} \quad \delta = 1/2 \tag{14}$$

these can be compared to the direct fit to the data for the dwarf stars listed in Table 2 of MJ

$$\log(\mathrm{Em}(T_c)/g_*) = 1.41(\pm 3.33) + 3.50(\pm 0.51) \log T_e \tag{15}$$

suggesting $\delta = 1$ but also allowing $\delta = 0.5$.

Two further scalings involving $\mathrm{Em}(T_c)$, T_c, g_* and Ro result. The predicted relations for $\delta = 1/2$ are $\mathrm{Em}(T_c) g_*^{1/2} \propto 10^{-3x Ro}$ and $T_c g_*^{1/2} \propto 10^{-x Ro}$, where x is the dependence on Ro in equation (12). The fits to the observations are

$$\log(\mathrm{Em}(T_c) g_*^{1/2}) = 32.02(\pm 0.10) - 1.26(\pm 0.09) Ro \tag{16}$$

and

$$\log(T_c g_*^{1/2}) = 9.09(\pm 0.07) - 0.38(\pm 0.07) Ro \tag{17}$$

Equation (17) allows a *prediction* of the coronal temperature T_c for a star provided its rotation period and spectral type are known, since these allow the surface gravity, g_*, and the Rossby number, Ro, to be found. For several stars with two temperature fits it predicts coronal temperatures in the 'gap' found by Schmitt *et al.* (1990).

4. Comparison between photospheric, chromospheric and coronal parameters

Because there are only 6 stars for which both $B_s f_s$ and B_c are known (see Table 1 of JM and Table 1 of MJ) it is not possible to find direct relations between them with any accuracy. However we can combine (12) with (1) to obtain

$$\log B_c = 0.28(\pm 0.29) + 0.40(\pm 0.07) \log B_s f_s \qquad (18)$$

and with (2) to give

$$\log B_c = 1.62(\pm 0.06) + 0.47(\pm 0.09) \log f_s \qquad (19)$$

Because of the different dependence on *Ro* of $B_s f_s$ and B_c it is clear that with a coronal filling factor $f_c = 1$, magnetic flux conservation from the photosphere to the corona *cannot* be satisfied. If f_e is defined as the fraction of the stellar surface which contains magnetic fields that extend to the corona, i.e. $f_e = B_c f_c / B_s$, it can be shown from our scalings that

$$\log \frac{f_e(\max)}{f_s} = -1.58 + 0.42Ro \qquad (20)$$

thus although both f_e and f_s *increase* with decreasing *Ro*, their ratio *decreases* with decreasing *Ro*. Also f_e tends to f_s at in the limit where Ro is 3.7.

5. Conclusions

A study of the relations between the observed magnetic fields and filling factors, the chromospheric activity, the coronal magnetic fields, other coronal parameters and the Rossby numbers has been carried out for a sample of late-type dwarfs. Although the dependence of chromospheric and coronal parameters on *Ro* is not yet understood, coronal scaling laws based on the hypothesis of minimum energy loss can account for the observed trends. Comparisons of the implied coronal field B_c and the surface magnetic flux $B_s f_s$ show that only part of the flux reaches the corona. We refer to the papers by JM and MJ for further details.

References

Hearn, A.G.: 1975, *A&A* **40**, 355
Hearn, A.G.: 1977, *Sol. Phys.* **51**, 159
Jordan, C. and Montesinos, B.: 1991, *MNRAS* **252**, 21p (JM)
Jordan, C. *et al.*: 1987, *MNRAS* **225**, 903
Montesinos, B. and Jordan, C.: 1992, submitted to *MNRAS* (MJ) ,
Noyes, R.W. *et al.*: 1984, *ApJ* **279**, 763
Rutten, R.G.M.: 1987, *A&A* **177**, 131
Saar, S.H.: 1990, *Proc. IAU Symp. 138, Solar Photosphere: Structure, Convection and Magnetic Fields*, ed: J.O. Stenflo, Kluwer: Dordrecht , 427
Schmitt, J.H.M.M. *et al.*: 1990, *ApJ* **365**, 704
Schrijver, C.J. *et al.*: 1989, *ApJ* **337**, 964
Stepień, K.: 1989, *A&A* **210**, 237

RESONANT HEATING OF LINE-TIED CORONAL LOOPS

G. HALBERSTADT and J.P. GOEDBLOED

FOM-Instituut voor Plasmafysica, P.O. Box 1207, 3430 BE Nieuwegein, The Netherlands

Abstract. One of the mechanisms that may contribute considerably to the heating of the solar and stellar coronae is resonant absorption of Alfvén waves. In this paper we discuss the incorporation of the photospheric boundary conditions in the theory of resonant heating of coronal loops. It is shown that the Alfvén continuum of coronal loops which are line-tied to the photosphere has entirely different properties than the conventional one. This new line-tied Alfvén continuum is found to be responsible for resonant heating of coronal loops which are bounded by the photosphere and excited at their foot points by photospheric motion.

Key words: hydromagnetics – wave motions – Sun: corona

1. Introduction

One of the viable coronal heating mechanisms that is investigated is heating of magnetically closed coronal regions by resonant absorption of Alfvén waves. Most work done on resonant heating of coronal loops has been restricted to unbounded (i.e., periodic) plasma cylinders (Davila 1987, Grossmann and Smith 1988, Poedts et al. 1989). However, as we show in a forthcoming paper (Goedbloed and Halberstadt 1992), the presence of the photosphere entirely alters the behaviour of MHD waves in coronal loops, thus revealing a wealth of new phenomena, which are likely to appear in resonant wave heating as well. We therefore consider a coronal loop that is line-tied to the solar photosphere and show that the conventional Alfvén continuum disappears and that it is replaced by a new one, the line-tied continuum, which has entirely different properties. We then pay attention to numerical simulations in which Alfvén continuum waves are excited at the foot points of the loop, thus modelling the excitation by photospheric motion. It turns out that the line-tied continuum is of fundamental importance for resonant heating of foot point excited coronal loops.

2. Resonant heating

Resonant heating hinges on the existence of a continuous Alfvén spectrum in linearized ideal MHD. The associated continuum modes are singular on magnetic surfaces where the excitation frequency matches the local Alfvén frequency, which is given by

$$\omega_A^2(r) = \frac{1}{\rho(r)}(\frac{m}{r}B_\theta(r) + \frac{2\pi}{L}nB_z(r))^2, \tag{1}$$

for the case of a periodic cylindric plasma. Here, B_θ and B_z are the magnetic field components, ρ is the mass density, L is the length of the cylinder and m and n are the mode numbers in the poloidal and longitudinal direction, respectively. It is well established, both theoretically and numerically, that the nearly singular modes that remain if resistivity is added, yield *finite* dissipation in the limit of *vanishing* resistivity. For this reason, excitation and absorption of these modes may lead to effective heating of the highly conductive coronal plasma.

J.F. Linsky and S. Serio (eds.), Physics of Solar and Stellar Coronae, 583–586.

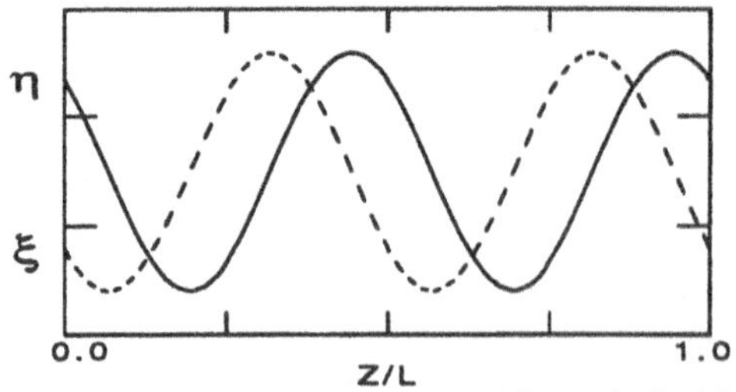

Fig. 1. Radial displacement component ξ (dashed line) and perpendicular component η (drawn line), in periodic cylinder; η is perpendicular to **B** in the magnetic surfaces.

3. Line-tied Alfvén continuum

As a result of the 'frozen-in' effect, the magnetic field lines of coronal loops are *line-tied* to the high-density photospheric plasma and forced to follow the photospheric fluid motions. This effect is usually modelled by requiring that both ξ and η vanish at the photosphere (i.e., at $z = 0$ and $z = L$). It is easily seen from Fig.1 that this condition is a non-trivial one: ξ and η are in general out of phase, so that the line-tying condition cannot be met by applying a simple phase shift. Hence, the singularities somehow have to vanish at the photosphere if line-tying is incorporated and the question arises whether this is possible at all. Does the Alfvén continuum disappear (and therewith the resonant heating mechanism) under the influence of line-tying?

In a more extensive paper (Goedbloed and Halberstadt 1992) we have shown that the eigenmodes of a homogeneous plasma slab, with magnetic field $(0, B_y, B_z)$, thickness L and which is subject to the line-tying constraint, behave entirely different from the solutions in an unconstrained slab. The behaviour of solutions with large k_x (the wave number in the x-direction), was considered explicitly in a related paper (Halberstadt and Goedbloed 1992). The infinitely rapid varying solutions in a homogeneous plasma ($k_x \to \infty$) to some extent represent the behaviour of the singular continuum modes that may exist in an inhomogeneous plasma. In the limit $k_x \to \infty$ the spectrum of the homogeneous line-tied slab accumulates at a fixed value ω_0^2. If *inhomogeneity* in the x-direction is added, the eigenfrequency of the individual field lines varies *continuously* in the x-direction, so that the accumulation point is spread out and turned into a genuine continuum. The similarity between a slab with magnetic field $(0, B_y, B_z)$ and a cylinder with magnetic field $(0, B_\theta, B_z)$ combined with the expression for ω_0^2 then reveals that a continuous Alfvén spectrum *does* exist in a line-tied cylindric plasma and that it is described by

$$\omega_A^2(r) = \frac{1}{\rho(r)}\left(\frac{n'\pi}{L}B_z(r)\right)^2. \tag{2}$$

The difference with the conventional continuum (1) is striking: the line-tied continuum (2) no longer depends on the poloidal mode number and the poloidal magnetic field (and is therefore independent of the equilibrium current J_z). The eigenmodes, however, *do* depend on m and B_θ. As shown in Fig. 2, they consist of an Alfvén contribution and a fast wave contribution which have to cooperate to satisfy the line-tying condition: the separate contributions do not satisfy the constraint, but the total solution does. The line-tied continuum modes are ballooning modes in the sense that the Alfvén contribution follows the inclination of the magnetic field:

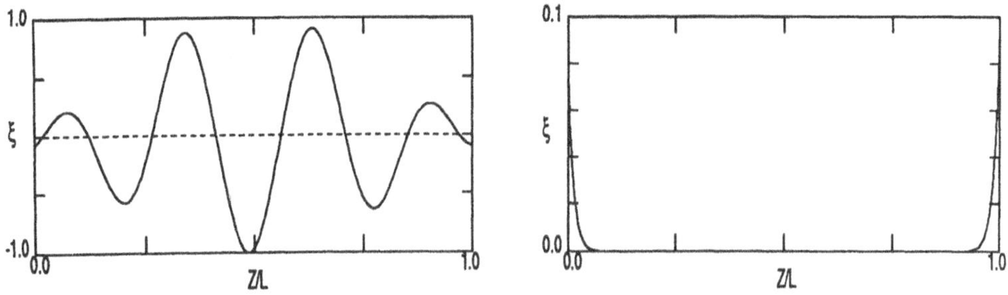

Fig. 2. Alfvén (left) and Fast (right) contribution for large k_x and $B_y \neq 0$.

the phase variation along magnetic field lines is quantized to $n'\pi$. This was first suggested by Strauss and Lawson (1989) on the basis of their numerical results. Furthermore, the fast wave contribution becomes increasingly localized towards the photospheric boundaries at $z = 0$ and $z = L$ as $k_x \to \infty$ and thereby gives rise to an ideal boundary layer in which the gradients in the z-direction become arbitrarily large. Both the ballooning feature and the boundary layer are absent if $B_\theta = 0$.

4. Numerical simulations

Numerical calculations on resonant heating of bounded coronal loops were done with the linearized resistive MHD code POLLUX. The POLLUX-code models a coronal loop as a cylindrical plasma with a general sheared magnetic equilibrium and it includes the photospheric end effects. In a first approach, Alfvén continuum waves were excited sideways by imposing a perturbation along the surface of the loop which varies harmonically in time, whereas the displacement at the foot points was constrained to zero. The results are in perfect agreement with the expression (2) for the line-tied continuum.

A more physical approach is excitation at the foot points of the loop. here one can identify the external excitation source with convective motions and MHD waves in the photospheric plasma. In Fig. 3, the results of a calculation in which the velocity fields at the foot points are constrained to arbitrary and finite values are shown. The resonant behaviour occurs at the radial position which corresponds to the line-tied continuum. Furthermore, the longitudinal behaviour clearly displays the ballooning character (if B_θ is increased the phase variation in the last frame is increased, whereas the resonance position does not change). This means that *excitation of coronal loops at the foot points*, such that the velocity field is constrained to a finite value, *generates line tied continuum waves in the loop*, which are resonantly absorbed and heat the loop. The calculation of the efficiency and time-dependence of this process is presently under investigation.

5. Conclusion

The conventional Alfvén continuum (1) is no longer present in photospherically bounded coronal loops. Instead, the new line-tied Alfvén continuum (2) arises. The

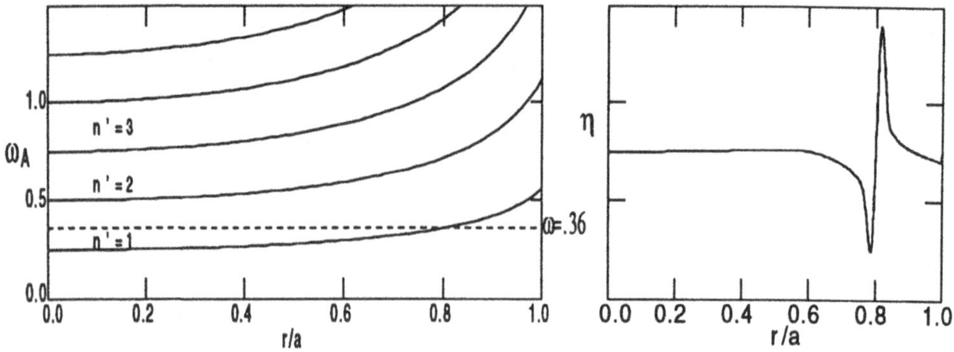

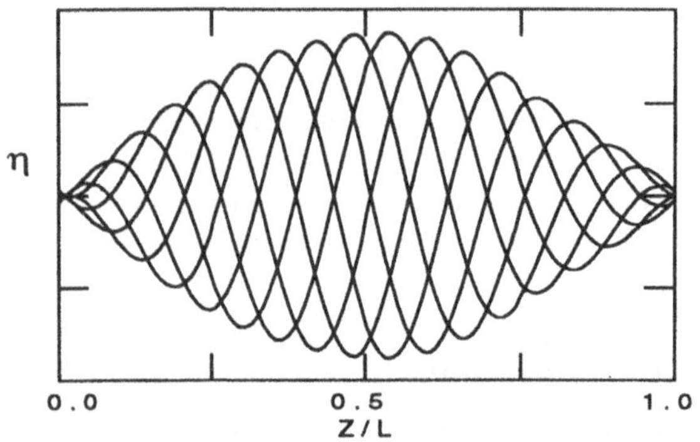

Fig. 3. Foot point excitation at $\omega = 0.36$ (dashed line). The first frame shows the *line-tied* continua (2). In the last frame, curves of constant θ are drawn. The number of nodes in the envelope corresponds to the value of the integer n' in the expression (2) for the line-tied continuum.

most important feature of this continuum is that it no longer depends on m and B_θ. In relation to this the associated continuum modes have a global ballooning character. Foot point excitation of coronal loops yields resonances which correspond with the line-tied spectrum. We therefore conclude that resonant heating of foot point excited coronal loops is due to the line-tied Alfvén continuum and the associated global ballooning modes.

Acknowledgements

The authors are indebted to Stefaan Poedts, Ronald van der Linden, Guido Huysmans, Marcel Goossens, Wolfgang Kerner, and Elisabeth Schwarz for fruitful and stimulating cooperation. This work was supported by the European Community Scientific Co-operation (grant No SC1 - 0255 - C), the 'Nederlandse Organisatie voor Wetenschappelijk Onderzoek' (NWO) and by the Euratom association agreement with the 'Stichting voor Fundamenteel Onderzoek der Materie' (FOM).

References

Davila, J. M.: 1987, Ap. J. **317**, 514.

Goedbloed, J. P., Halberstadt, G.: 1992, 'Magnetohydrodynamic waves in coronal flux tubes', submitted for publication in *Astron. Astrophys.*

Grossmann, W., Smith, R. A.: 1988, *Ap. J.* **332**, 476.

Halberstadt, G., Goedbloed, J.P.: 1992, 'The continuous Alfvén spectrum of line-tied coronal loops', submitted for publication in *Astron. Astrophys.*

Poedts, S. M., Kerner, W., and Goossens, M.: 1989, *J. Plasma Physics* **42**, 27 – 58.

Strauss, H. R., and Lawson, W. S.: 1989, *Ap. J.* **346**, 1035 – 1040.

ON THE HEATING MECHANISM OF CORONAL HOLES

R. HAMMER

Kiepenheuer-Institut für Sonnenphysik, Freiburg, FRG

R. L. MOORE

NASA-Marshall Space Flight Center, Huntsville, AL

Z. E. MUSIELAK

University of Alabama at Huntsville, AL

and

S. T. SUESS

NASA-Marshall Space Flight Center, Huntsville, AL

Abstract. We explain why the properties of coronal holes depend not only on the total amount, but also on the spatial distribution of coronal heating. Observations can thus be used to constrain the unknown heating mechanism. According to current semiempirical models, most of the heat input in solar coronal holes must be deposited over a characteristic length of about half a solar radius – which is too long for spicules to be the main heating mechanism, and too short for propagating monochromatic Alfvén waves. In the light of this empirical constraint, we examine the possibility that coronal heating is an inherently intermittent process, with quiet and active phases. In active phases, the inner, quasi-hydrostatic part of the corona expands as a result of magnetic field reconfigurations or of the wave pressure associated with MHD wave packets. During quiet phases, the plasma sinks back and releases its potential energy, which is ultimately converted into heat. We show that the required velocities are consistent with empirical findings. Therefore, this process could well be an important heating mechanism in the magnetically open corona.

Key words: Sun: corona – Sun: solar wind

1. Location of the Heat Input

The energy that is deposited in the open solar corona is conducted away from the site of coronal heating, both inwards and outwards. Inward conduction balances mainly radiation from the inner corona and transition region, and the enthalpy and part of the potential energy of the wind. Outward conduction balances the remainder of the potential energy and part of the kinetic energy of the wind.

The ratio of inward to outward conduction, and therefore also the properties of the corona, depend on the location where coronal heating takes place. This can be understood most easily in terms of a very simple analytical model by Unsöld (1960). He assumed spherical symmetry and classical heat conduction, neglected radiation and wind, and considered the special case where all heating takes place at radius $r = r_H$. Then the outward and inward conductive fluxes $\Phi_{C\pm}$, normalized to the solar surface (at radius R), are constant,

$$\Phi_{C\pm} = -\left(\frac{r}{R}\right)^2 \kappa_0 T^{5/2} \frac{dT}{dr} = \text{const.}$$

With the boundary conditions $T \to 0$ for $r \to R$ and for $r \to \infty$, this equation can be used to determine $T(r)$, Φ_{C+}, and Φ_{C-}. The maximum temperature T_m varies like $\Phi_{M0}^{2/7}$, where $\Phi_{M0} = \Phi_{C+} + \Phi_{C-}$ is the total heat input,

$$T_m^{7/2} = (\Phi_{C+} + \Phi_{C-}) \frac{7R^2}{2\kappa_0} \frac{r_H - R}{r_H^2}.$$

J.F. Linsky and S. Serio (eds.), Physics of Solar and Stellar Coronae, 587–590.

T_m has a broad maximum when the heating is deposited at 1 solar radius R above the surface. This result was qualitatively confirmed by much more sophisticated numerical models, where radiation and wind were taken into account and the heating was distributed over an extended range (Hammer 1982). These numerical models gave a smaller exponent for $T_m(\Phi_{M0})$, only 0.20 rather than $2/7 \approx 0.29$. Nevertheless, the general agreement shows that Unsöld's analytical model, despite its simplicity, includes the essential physics for understanding the dependence of the open corona on the location of the heat input.

The ratio of outward to inward conduction is

$$\left|\frac{\Phi_{C+}}{\Phi_{C-}}\right| = \frac{r_H - R}{R}.$$

When the heating occurs at heights below (above) 1 solar radius, most of the energy is conducted inward (outward). Since the inward and outward fluxes balance different types of energy losses, the properties of the corona depend indeed on the location of the heat input.

This makes it possible to derive information on the spatial distribution of heating in solar coronal holes by comparing theoretical models with various kinds of empirical information (e.g., density profiles, radiation from the inner corona and transition region, temperatures derived from frozen-in ionization states, and wind mass fluxes). Withbroe (1988) concludes from such an analysis that solar coronal holes are heated by an energy flux of about $4\,10^5\,\mathrm{erg\,cm^{-2}\,s^{-1}}$, which is absorbed in the corona over an e-folding damping length $L \approx 0.5\,R$.

2. Heating Mechanisms

Many previously suggested coronal heating mechanisms can be excluded since they do not operate on this empirically determined length scale. For example, the maximum heights of spicules are at least an order of magnitude too small. Parker (1991) reviewed damping mechanisms of monochromatic propagating Alfvén waves. He concluded that for reasonable wave periods ($\gtrsim 100\,\mathrm{sec}$) known damping mechanisms (e.g., linear viscous or Joule damping, nonlinear mode coupling, phase mixing and resonant absorption) are inefficient; they operate on length scales that are at least an order of magnitude too large.

On the other hand, there is empirical evidence for the presence of hydromagnetic waves in the solar corona. Hassler et al. (1990) measured the widths of Mg X lines above the solar limb. They found significant nonthermal broadening, which they interpreted as due to MHD waves. Moore et al. (1992) showed that this broadening is consistent with Alfvén waves of sufficient amplitude to supply the energy in Withbroe's (1988) coronal hole models.

A way out of this dilemma might lie in the fact that Parker's (1991) analytical results apply strictly only to waves that are *(i)* propagating and *(ii)* monochromatic. Both these simplifications are questionable in the real solar atmosphere.

It is well known that long-period Alfvén waves undergo partial reflection in the inner corona because the Alfvén speed and its derivative change with height over scales comparable to the wavelength. This breakdown of the WKB approximation causes wave reflection (e.g., Musielak et al. 1992). Moore et al. (1991) concluded that Alfvén waves of periods as low as a few minutes should be significantly reflected in solar coronal holes. Reflected waves dissipate more efficiently than propagating ones, for example by turbulent decay or by phase mixing. In addition, they impart their momentum to the atmosphere and perturb the hydrostatic equilibrium.

Moreover, MHD waves generated in the lower solar atmosphere are probably not monochromatic but rather are emitted as short wave packets. During and after such bursts, a large amount of wave energy is contained in the inner part of the corona along a given magnetic field line. Since the propagation speed is not constant as a function of height, and since wave reflection and refraction occur, wave packets exert a significant wave pressure, which leads to a temporary expansion of the inner, quasi-hydrostatic part of the corona ("magnetic levitation"). Similar upward forces occur during reconfiguration processes of the magnetic field. Time series obtained with the YOHKOH satellite show that upward acceleration of plasma, associated with a change of the magnetic field structure, is a ubiquitous phenomenon in the solar corona (Uchida, these proceedings).

During such active phases, the upward moving plasma gains potential energy. In quiet phases, when the wave pressure and other upward forces are reduced, it can sink back, thereby converting potential energy into kinetic energy and eventually into heat. Since the dissipation of the compressive energy occurs mainly via thermal conduction, its efficiency depends on the size and temperature of the compressed regions. Moore (1991) and Moore et al. (1992) have suggested that magnetic levitation due to Alfvén waves plays an important role in the heating of coronal holes.

3. Velocity Fluctuations

We now estimate the velocities that might be involved. Sinking plasma releases potential energy at a rate $\rho g v_{sink}$, where v_{sink} is the sinking speed. By equating this term with the heating requirements in Withbroe's (1988) coronal hole models,

$$\rho g v_{sink} = \nabla \mathbf{F}_M,$$

we can determine the required sinking speed under the extreme assumption that *all* coronal heating is due to this mechanism. The result is shown in Fig. 1.

We typically find sinking speeds of about $10\,\mathrm{km\,sec^{-1}}$ near the base of the models. This value increases to about $30\,\mathrm{km\,sec^{-1}}$ at $r \approx 1.2R$, i.e. within the first 3 density scale heights above the base, where essentially all of the plasma resides. These vertical speeds are of the same order as the horizontal velocity fluctuations inferred from observations above the limb. Within the first damping length L, where most of the heating takes place, the required speeds increase further to between 30 and $100\,\mathrm{km\,sec^{-1}}$, depending on the corona model.

The increase of v_{sink} in the inner corona is due to the fact that there the density ρ decreases outward more rapidly than the heating requirements $\nabla \mathbf{F}_M$, simply because the density scale height is by a factor of 5 smaller than the damping length. In the outer corona, ρ decreases more slowly than $\nabla \mathbf{F}_M$; and the curves decrease again. The outer parts of these curves are not realistic, however, for various

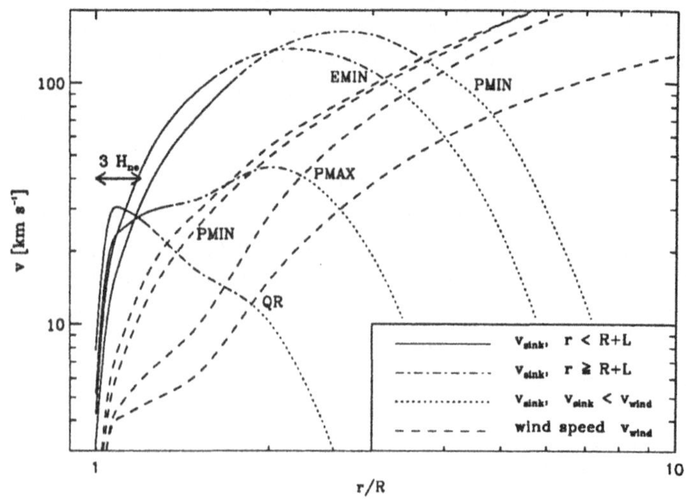

Fig. 1. The dashed curves show the wind speed as a function of radius for the four regions modeled by Withbroe (1988), *viz.*, polar coronal holes at activity minimum (PMIN) and maximum (PMAX), an equatorial coronal hole at activity minimum (EMIN), and a quiet region (AR). The other curves show the average sinking speed that would be required if the entire heat input in these models were supplied as potential energy of downflowing plasma. They are fully drawn within the first damping length L, where most of the heating occurs, and dash-dotted at larger radii. In the outer parts, some of the simplifying assumptions underlying the current estimates break down, in particular the assumption that the wind speed is small. We typically find $v_{wind} > v_{sink}$ for $r \gtrsim R + 3.5L$; these parts of the curves are shown dotted. The double arrow indicates the first three density scale heights above the base ($T = 5\,10^5$ K) of the models; virtually all of the observable emission comes from the lower part of this interval, where the sinking speeds are small *(from Hammer et al. 1992, in preparation)*.

reasons (e.g., wave reflection and other acceleration processes become unimportant; and the wind speed is no longer negligible).

To summarize, if coronal heating is intermittent, plasma is occasionally lifted up and then sinks back. The associated potential energy release could be an important energy source in the main heating region ($R \leq r \lesssim L$) of the open solar corona.

References

Hammer, R.: 1982, *Ap.J.* **259**, 779
Hassler, D.M., Rottman, G.J., Shoub, E.C., Holzer, T.E.: 1990, *Ap.J.Lett.* **348**, L77
Moore, R.L.: 1991, *BAAS* **23**, 1037
Moore, R.L., Musielak, Z.E., Suess, S.T., An, C.-H.: 1991, *Ap.J.* **378**, 347
Moore, R.L., Hammer, R., Musielak, Z.E., Suess, S.T., An, C.-H.: 1992, *Ap.J.Lett.* **397**, L55
Musielak, Z.E., Fontenla, J.M., Moore, R.L.: 1992, *Phys. Fluids B* **4**, 13
Parker, E.N.: 1991, *Ap.J.* **372**, 719
Unsöld, A.: 1960, *Z.Ap.* **50**, 57
Withbroe, G.L.: 1988, *Ap.J.* **325**, 442

THE CORONAL ENERGY BALANCE IN ACTIVE REGIONS: THEORETICAL ASPECTS

R. HAMMER

Kiepenheuer-Institut für Sonnenphysik, Freiburg, FRG

Abstract. The observed nonlinear correlation between X-ray and EUV emissions from solar active regions is discussed in terms of analytical models. It is shown that the nonlinearity cannot be attributed to a volume *vs.* area effect in shape-invariant coronal loops, as a result of global energy balance constraints. Nonlinear correlations between coronal and transition region emissions are inevitable whenever the length of field lines, the shape of coronal loops, or their base pressure change. The effects of nonconductive heating in the transition region, either directly or indirectly by enthalpy deposition of downflowing plasma, are briefly discussed.

Key words: Stars: activity – Sun: activity – Sun: corona – Sun: X-rays

1. Introduction

When the emission from solar active regions is integrated over the projected area, one finds a nonlinear relationship between the coronal X-ray emission and the emission in a given transition region (TR) line, $F_X \propto F_{TRL}^{\beta}$, with an exponent $\beta \approx 1.5$ (Schrijver et al. 1985, Table 7). This paper discusses theoretical aspects of the nonlinearity in terms of analytical coronal loop models. Sect. 2 gives an explicit account of the simplest possible model, which nevertheless provides important insight. Sect. 3 summarizes more complex situations.

2. Simple Case

Let us first assume the TR to be sufficiently thin that it can be approximated as plane-parallel and isobaric. We neglect flows and direct heating and restrict ourselves to temperatures where hydrogen is fully ionized. Then the energy equation describes the balance between conductive heating and radiative cooling,

$$\frac{d}{ds}F_C + \left(\frac{p}{2kT}\right)^2 P_{Rad}(T) = 0 \tag{1}$$

$$F_C = -\kappa_0 T^{5/2}\frac{dT}{ds}, \tag{2}$$

where F_C is the (classical) conductive energy flux and s the spatial coordinate along the loop. For analytical studies it is common to approximate the emissivity by a power law, $P_{Rad}(T) \propto T^{\lambda}$. Since $dF_C/ds = (dF_C/dT)(dT/ds)$, (1) and (2) can be combined into a first order differential equation for $F_C(T)$, with the solution

$$F_C \propto pT^{(2\lambda+3)/4}, \tag{3}$$

where the temperature and conductive flux at the base of the TR have been neglected. Equations (2) and (3) can now be used to calculate the emission F_{TRL} in a given optically thin line,

$$F_{TRL} = \int_{\Delta V} n_e^2 f(T)dV = A\int_{\Delta T}\left(\frac{p}{2kT}\right)^2 f(T)\frac{ds}{dT}dT \propto Ap, \tag{4}$$

J.F. Linsky and S. Serio (eds.), Physics of Solar and Stellar Coronae, 591–594.

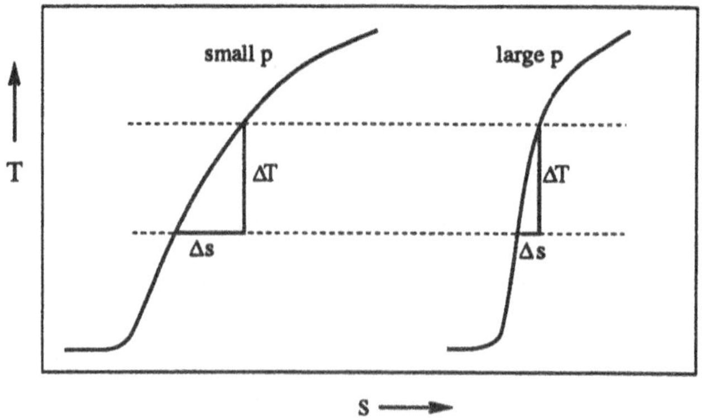

Fig. 1. The emission from the transition region increases with the base pressure p. In a conductively heated transition region, this increase of the emission necessitates a steepening of the temperature gradient in order that the energy losses can still be balanced by thermal conduction. As a result, the height interval Δs over which a given line is emitted, decreases ($\Delta s \propto p^{-1}$). Therefore, the emission is not proportional to p^2, as one would naïvely expect, but only linearly proportional to p. When the transition region is heated not only by conduction, but also by the enthalpy of downflowing hot matter, the pressure dependence of F_{TRL} is even reduced.

where $f(T)$ contains atomic data, A is the loop cross section and ΔT the temperature interval over which the line is formed. It is important to note that F_{TRL} is independent of the power law fit to the emissivity and depends only linearly on p in a conductively heated TR model (Fig. 1).

In order to estimate F_X, we approximate the coronal part of the loop as isothermal with temperature T_c, and we assume the height of the loop to be smaller than the pressure scale height at the base, so that $p \approx$ const within the entire loop. Then the total coronal emission is

$$F_X = \int_V \left(\frac{p}{2kT_c}\right)^2 P_{Rad}(T)dV \propto Vp^2T_c^{\lambda-2}, \tag{5}$$

where V is the coronal volume of the loop. Temperature and pressure of a coronal loop are related to its semilength L via a so-called scaling law, $T_c \propto (pL)^{1/3}$ (*e. g.* Rosner et al. 1978). Various authors have shown that this relationship is extremely robust – it depends only very weakly, if at all, on the geometry, the presence of flows, and the heating and cooling laws. It can be used to eliminate T_c from (5),

$$F_X \propto VL^{(\lambda-2)/3}p^{(4+\lambda)/3}. \tag{6}$$

With (4) and (6) we can now discuss how the relationship between coronal and TR emission depends on the properties of coronal loops. We do this first for homologous (shape-invariant) volume changes at constant pressure. Cram (1991) suggested that the observed nonlinearity, $\beta \approx 1.5$, can be understood if with increasing size of active regions the coronal loops become larger, but maintain their shape. Then the coronal volume (which determines the coronal emission) scales as $V \propto L^3$, while

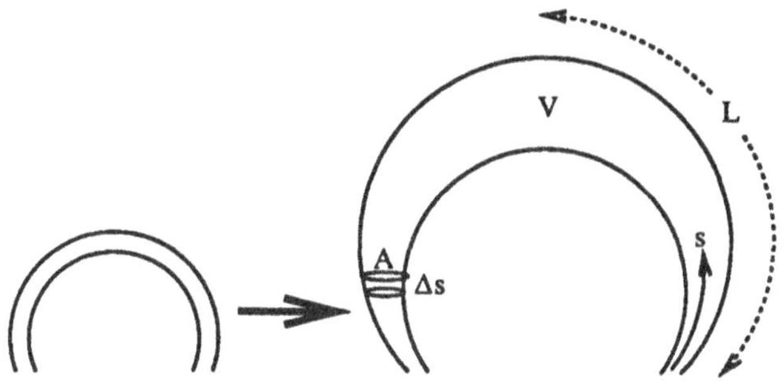

Fig. 2. Idealized geometry of coronal loops. Nonhomologous volume changes, in which the cross section increases more rapidly at the apex than at the footpoints as we go to larger active regions, could explain the observed nonlinear flux-flux relation.

the footpoint area (which determines the TR emission) scales as $A \propto L^2$, so that $V \propto A^{3/2}$ – *i. e.*, one might indeed expect $\beta \approx 1.5$. The current calculation shows that this is not the case. With $A \propto L^2$ and $V \propto L^3$ we obtain from (4) and (6) $F_X \propto F_{TRL}^{(7+\lambda)/6}$, so that $\beta = (7+\lambda)/6 \approx 1.08$ for typical coronal temperatures, for which $\lambda = -1/2$ provides a reasonable approximation to the emissivity. Most of the expected nonlinearity disappears because the TR emission depends not only on A, but also on p, which is coupled to the temperature and length of the loop via a scaling law.

If the *shape* of an average coronal loop changes with the size of active regions, the TR cross section A in (4) and the coronal volume V in (6) are no longer coupled, thus any kind of nonlinearity is possible (Fig. 2).

On the other hand, if the loop geometry remains constant, but the pressure increases with increasing activity, (4) and (6) yield $\beta = (4+\lambda)/3 \approx 1.17$ for $\lambda = -1/2$ (Hammer and Linsky 1984).

3. Generalizations

The above estimates can easily be generalized to loops that are no longer small compared to the scale height. Detailed calculations show that stratification tends to increase β.

The effects of rapidly varying geometry and nonconductive heating in the TR, where the latter could occur either directly or indirectly by enthalpy deposition of downflowing plasma, are more difficult to assess. For the sake of tractability we parametrize the loop cross section and the heat input as power laws in T, $A = A_0 T^\alpha$ and $H = H_0 T^\epsilon$. Then the energy equation can be written in the form,

$$\phi_C \frac{d\phi_C}{dT} + a\phi_C = bp^2 T^{2\alpha+\lambda+1/2} - cT^{2\alpha+\epsilon+5/2}, \tag{7}$$

where $\phi_C = -\kappa_0 T^{\alpha+5/2} dT/ds = F_C A/A_0$ is the normalized conductive flux, a is proportional to the (constant) mass flux, b is a constant depending on the emissivity and conductivity coefficients, and $c = \kappa_0 H_0$ depends on the heat input. In its general form, Equation (7) cannot be solved analytically. But a number of special cases can be treated, some of which have been discussed in the literature:

- $a = 0$ (no flows). Such models have been constructed at various levels of complexity, *e. g.* by Landini and Monsignori Fossi (1981), Kuin and Martens (1982), and Bray et al. (1991).
- For $a \neq 0$, an exact analytical solution is possible if the right-hand side of equation (7) becomes constant ($\alpha = 0$, $\lambda = -1/2$, and either $H = 0$ or $\epsilon = -5/2$). This solution was discussed by Endler et al. (1979). Even if these requirements are not met, *approximate* analytical solutions can still be found (Craig and McClymont 1986).

Examination of these more complex models shows that the nonlinearity factor β is independent of H_0 and weakly dependent on ϵ, but increases when downflows become important.

4. Discussion and Conclusions

A nonlinear correlation between coronal and transition region emissions is inevitable whenever the length L of the field lines, the shape of coronal loops, or their base pressure p change. The distribution of heating along the loop may affect the correlation; in particular enthalpy heating tends to increase the nonlinearity.

Neither homologous volume changes nor pressure changes alone suffice to explain the observed nonlinearity quantitatively. One probably needs a combination of several effects. If volume changes are responsible for the observed correlations, they must be substantially *non*homologous. Moreover, part of the TR emission has been suggested (*e. g.*, Feldman 1983, Rabin and Moore 1984) to come from cool loops that do not reach coronal temperatures, rather than from the feet of hot loops. This would affect the observed relationship between coronal and TR emission. This relationship could also be influenced by strongly time-dependent phenomena (cf. the reviews by Golub and Uchida, these proceedings); but their role needs to be studied in nonlinear time-dependent calculations, which are beyond the scope of the present work.

References

Bray, R.J., Cram, L.E., Durrant, C.J., Loughhead, R.E.: 1991, *Plasma Loops in the Solar Corona*, Cambridge Univ. Press, chapter 5.4

Craig, I.J.D., McClymont, A.N.: 1986, *Ap.J.* **307**, 367

Cram, L.E.: 1991, in *Mechanisms of Chromospheric and Coronal Heating*, eds. P. Ulmschneider, E.R. Priest, R. Rosner, Springer, p. 282.

Endler, F., Hammer, R., Ulmschneider, P.: 1979, *Astr.Ap.* **73**, 190

Feldman, U.: 1983, *Ap.J.* **275**, 367

Hammer, R., Linsky, J. L.: 1984, in *Proc. Fourth European IUE Conference*, ESA SP-218, p. 25

Kuin, N.P.M., Martens, P.C.H.: 1982, *Astr.Ap.* **108**, L1

Landini, M., Monsignori Fossi, B.C.: 1981, *Astr.Ap.* **102**, 391

Rabin, D., Moore, R.: 1984, *Ap.J.* **285**, 359

Rosner, R., Tucker, W. H., Vaiana, G. S.: 1978, *Ap.J.* **220**, 643

Schrijver, C. J., Zwaan, C., Maxson, C. W., Noyes, R. W.: 1985, *Astr.Ap.* **149**, 123

THERMAL TRANSPORT IN STATIC STELLAR CORONAL LOOPS

A. CIARAVELLA

Istituto ed Osservatorio Astronomico, Palermo

G. PERES

Osservatorio Astrofisico, Catania

S. SERIO

Istituto ed Osservatorio Astronomico, Palermo

and

Istituto per le Applicazioni Interdisciplinari della Fisica - CNR, Palermo

Abstract. We have studied the thermal structure of static stellar coronal loops with different surface gravity and a wide range of characteristic parameters of loop (base pressure and loop length), including cases in which the ratio of the electron mean free path to the temperature scale height (λ_0/L_T) can be so large to invalid classical plasma thermal conduction (Spitzer, 1962). We have used, whenever appropriate, the non local thermal conduction of Luciani, Mora and Virmont's (1983), and the limited free streaming formulation (Mannheimer, 1977), valid for progressively higher values of λ_0/L_T.

We show how plasma pressure at the base of the loop p, loop length L, and surface gravity, determine the different regimes of thermal conduction, and how the temperature stratification and the differential emission measure distribution depend on the mechanism of thermal conduction. In particular we have found that the range of applicability of the Luciani Mora and Virmont formulation is limited to a very small region between Spitzer and free-streaming conduction and that, when free-streaming conduction applies, the apex temperature decreases with increasing base pressure at variance from Spitzer conduction, thus defining a locus of minimum coronal temperature in the $p - L$ plane.

Key words: Late-type stars – Coronae

1. Introduction

Traditional loop models of stellar coronae (Giampapa *et. al.*, 1985; Schmitt *et al.*, 1985; Antiochos and Noci, 1986; Landini *et al.*, 1985) have used classical Spitzer conduction. The latter is valid when the electron mean free path λ_0 is much less than the temperature scale height ($L_T = T/\nabla T$), more specifically, when $\lambda_0/L_T < 2 \times 10^{-3}$. Since the thermal conduction plays a crucial role in the energy balance of coronal plasma and in the determinig the plasma stratification inside the loop, it is important to model coronal loop use a formulation of thermal conduction appropriate to plasma physical conditions. The loops we study here cover a large range of plasma physical conditions; in particular, in some cases the ratio λ_0/L_T can be so large as to invalid Spitzer theory. For this reason we have also used in our models, when appropriate, either the Luciani, Mora and Virmont's (1983; henceforth LMV) non-local formulation and the limited free-streaming formulation of thermal flux (Mannheimer, 1977) to complement Spitzer's formulation.

2. Hydrostatic Model

Our static model of stellar coronal loop used describes the plasma confined in a semi-circular loop of semilength L and mirror symmetric with respect to the apex.

J.F. Linsky and S. Serio (eds.), Physics of Solar and Stellar Coronae, 595–598.

The plasma is described as a fluid in hydrostatic equilibrium, with energy balance among heat input E_H constant along the loop, optically thin radiative losses $E_R(s)$, and thermal conduction. For further details cf. Serio *et al.*, (1981), Ciaravella, Peres and Serio (1991) and Ciaravella, Peres and Serio (1992). The equation of energy balance is:

$$\frac{d}{ds}F(s) = E_H + E_R(s) \tag{1}$$

where for $F(s)$, the thermal flux, we use the formulation appropriate to the value of λ_0/L_T. More specifically for $\lambda_0/L_T > 2 \times 10^{-3}$ we use the Spitzer formulation of thermal flux:

$$F(s) = F_{SH}(s) = -\kappa T^{5/2}\frac{dT}{ds} \tag{2}$$

where $\kappa = 9.2 \times 10^{-7} erg\ sec^{-1} cm^{-1} K^{-1}$, T is the temperature and s the field line coordinate.

For $2 \times 10^{-3} < \lambda_0/L_T < 6.67 \times 10^{-3}$ the appropriate formulation is according to LMV:

$$F(s) = F_{LMV}(s) = \int ds' F_{SH}(s') w(s, s') \tag{3}$$

where $F_{SH}(s)$ is the Spitzer-Harm's thermal flux, and the kernel $w(s, s')$ is:

$$w(s, s') = \frac{1}{2\lambda(s')} exp[-| \int_{s'}^{s} \frac{ds'' n(s'')}{\lambda(s') n(s')} |] \tag{4}$$

and $\lambda(s') = 32(Z + 1)^{1/2} \lambda_0(s')$ is the effective mean free path of heat carrying electrons.

For $\lambda_0/L_T > 6.67 \times 10^{-3}$ we use the limited free-streaming conduction:

$$F(s) = F_{LFS} = Min(F_{SH}(s), Q(s)) \tag{5}$$

with

$$Q(s) = f(2/\pi)^{1/2} n_e k_B T v_e \tag{6}$$

where n_e and $v_e = \sqrt{k_B T/m_e}$ are the electron number density and velocity, respectively. We use $f =$ 0.1, a value proposed by Mannheimer (1977) as appropriate to astrophysical conditions.

3. Results and Discussion

We have modeled static stellar coronal loops for three different values of stellar surface gravity 0.3, 3, 10 $g_\odot$ (the solar case was already studied in Ciaravella, Peres and Serio, 1991); for each star we have explored a wide set of models parameterized in terms of loop-semilength (ranging from 2×10^4 to $5 \times 10^{10} cm$) and base pressure (from 0.03 to 30 $dyne\ cm^{-2}$). For high gravity, a significant fraction of models needs the free-streaming treatment in part of loop; for low gravity, all models can

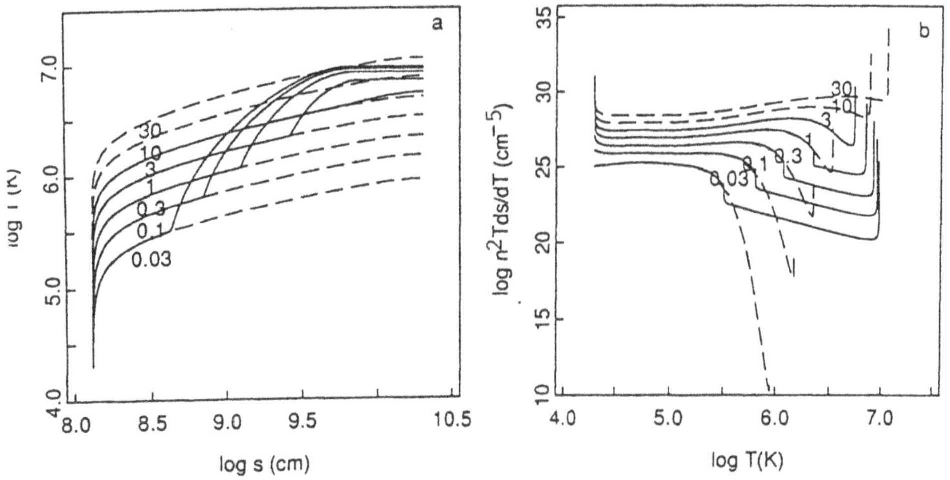

Fig. 1. Fig. 1. Distribution of temperature (panel a) and differential emission measure (panel b) for a set of loop models with loop semi-length $2 \times 10^{10} cm$ and gravity $g = 10 g_{\odot}$. The base pressure, in *dyne* cm^{-2}, is reported on each curve. All dashed curves are computed using Spitzer conductivity, the solid curve, corresponding to base pressure 3 *dyne* cm^{-2}, is computed with the LMV formulation of conduction. The remaining solid curves refer to models computed with limited free-streaming conduction.

be entirely described with Spitzer conduction, while LMV conduction applies in a very small range between Spitzer and limited free-streaming conduction.

Figure 1 shows, for $g = 10 g_{\odot}$, the profile of temperature (panel a) and differential emission measure (panel b) for loops of the same length but different base pressure.

Figure 2 shows the dependence of maximum temperature (located at the apex of the loop) on loop semilength and base pressure for all gravities explored. The valley present in the $T(p, L)$ surface marks the lower bound of maximum temepratiure for which limited free-streaming conditions occur. We found the following scaling law for this valley:

$$T(k) \simeq 10^{-9} Lg \quad (c.g.s.) \tag{7}$$

The maximum temperature predicted with limited free-streaming conduction is higher than that obtained with Spitzer (Fig. 2), thus the profiles of differential emission measure (solid line in Fig. 1b) are flatter and extend to higher temperature than those computed with Spitzer conduction (dashed lines in Fig. 1b). In conclusion, in such cases models based on Spitzer or LMV conduction give wrong predictions of coronal plasma conditions.

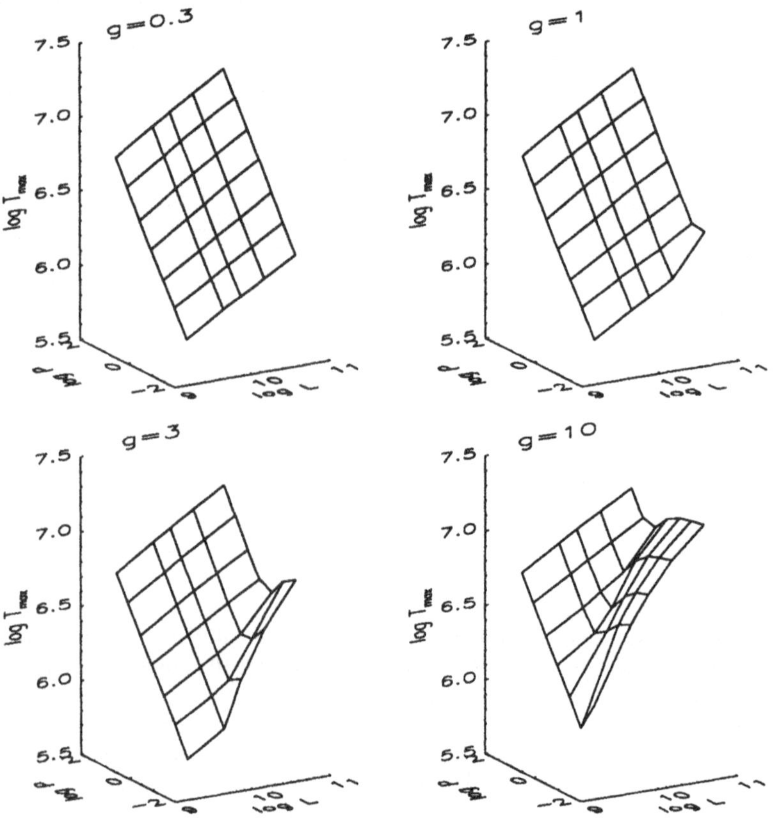

Fig. 2. Fig. 2. Maximum loop plasma temperature T_{max} as a function of loop semilength L and base pressure p for all models and gravities (in units of solar gravity) studied.

References

Antiochos, S. K., and Noci, G.: 1986, *Astrophys. J.* **301**, 440

Ciaravella, A., Peres, G., and Serio, S.: 1991, *Solar Phys.* **132**, 279

Ciaravella, A., Peres, G., and Serio, S.: 1992, *submitted to Solar Phys* ,

Giampapa, M.S., Golub, L., Peres, G., Serio, S., and Vaiana, G.S.: 1985, *Astrophys. J.* **289**, 203

Landini, M., Monsignori-Fossi, B. C., Paresce, F., and Stern, R. A.: 1985, *Astrophys. J.* **289**, 709

Luciani, J.F., Mora, P., and Virmont, J.: 1983, *Phys. Rev. Lett.* **51**, 1664

Mannheimer, W.,M.: 1977, *Phys. Fluids* **20**, 265

Schmitt, J. H., Harnden, F. R., Jr., Peres, G., Rosner, R., and Serio, S.: 1985, *Astrophys. J.* **288**, 751

Serio, S., Peres, G., Vaiana, G. S., Golub, L., and Rosner, R.: 1981, *Astrophys. J.* **243**, 288

Spitzer, L.: 1962, *Physics of Fully Ionized Gases* **p.143**,

SCALING LAWS FOR THE DECAY PHASE OF STELLAR LOOP FLARES

F. REALE and S. SERIO*

Istituto ed Osservatorio Astronomico, Palazzo dei Normanni, 90134 Palermo, Italy

and

G. PERES

Osservatorio Astrofisico, Città Universitaria, 95125 Catania, Italy

Abstract. We have generalized the analysis of the decay phase of X-ray flares inside coronal loops of constant cross section to flares on stars with non solar gravity, and investigated the implications on the diagnostics obtained with X-ray observations. We have performed hydrodynamic simulations of the free decay of compact stellar flares, for various values of stellar surface gravity, loop length and peak temperature. We have analysed the corresponding slope of the decay path on the density-temperature diagrams, in particular as it would be derived from observations with ROSAT PSPC.

Key words: Hydrodynamics - Stars: atmospheres - Stars: coronae - Stars: flare - X-rays: stars

1. Introduction

In a series of previous papers (Serio et al. 1991, Jakimiec et al. 1992 and Sylwester et al. 1992, hereafter Paper I, II and III respectively) we have studied the decay phase of X-ray flares in solar coronal loops, by means of a hydrodynamic model of coronal loop with constant cross-section.

We have shown that, during a relatively long phase early in the decay of solar flares, the entropy per particle $S = k \ln(T^{5/2}/p)$ computed at the top of the loop decays linearly (*linear phase*), and that the average temperature and density are related through a power law $T \propto n^{\zeta}$.

Here we extend these results to stellar coronal loops, and consider observational diagnostics in the ROSAT PSPC passband.

2. The Models

We have used the Palermo-Harvard 1-D hydrodynamic numerical code to compute the evolution of plasma flows inside rigid semicircular loops of constant cross-section (see Peres et al. 1982, Peres and Serio 1984 for details on the code) in a variety of gravity environments.

In most cases we have considered evolution starting from steady-state hydrostatic loop profiles computed according to Serio et al. (1981) and already at typical flare temperature and density; the loop plasma has then been let evolve by switching off the heating term abruptly (see Reale et al. 1992 for a similar analysis with heating gradually switched off).

We have considered models for two loop semilengths $L = 2 \times 10^9$ cm and $L = 10^{10}$ cm, and two initial temperatures $T_{top} = 2 \times 10^7$ K and $T_{top} = 10^7$ K, and

* also at IAIF, CNR, Via Archirafi 36, 90100 Palermo

J.F. Linsky and S. Serio (eds.), Physics of Solar and Stellar Coronae, 599–602.

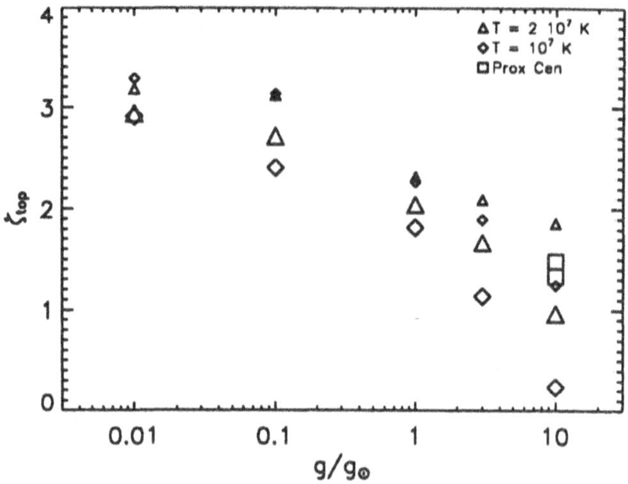

Fig. 1. Ratio ζ_{top} of the density to the temperature flare decay time, at the top of the loop, vs. the gravity (in solar units). Symbols identify different temperatures at the loop apex at the beginning of the decay, and the symbol size identifies the size of the model flaring loop. Results from the decay of dynamic models of a Proxima Centauri flare are also reported (see text for details).

five values of gravity in the range 0.01 to 10 $g_\odot$, therefore exploring conditions from very low gravity such as those of giant stars, to the high gravity for instance typical of M stars. We have also included the decay phase of two flare models specifically developed to describe a flare on the M star Proxima Centauri observed by the *Einstein* satellite (Models II and IV in Reale et. al. 1988), which evolve from hydrostatic active region loops with $L = 7 \times 10^9$ cm.

3. Results

We first note that the result given in Paper I for the entropy decay time can be generalized to loops comparable in length with the pressure scale height (h), by means of the scaling laws of Serio et. al. (1981), as follows:

$$\tau_{\rm th} = \frac{3.7\ 10^{-4}\ L}{\sqrt{T_0}}\ e^{0.5L/h} \tag{1}$$

where T_0 is the initial temperature at the top of the loop.

We have found that the entropy decay time (τ_S), as derived from numerical simulations, increases slightly with respect to $\tau_{\rm th}$ both with the gravity and as the ratio of the loop semilength to the pressure scale height (L/h), up to a value $\tau_S/\tau_{\rm th} \sim 1.4$.

As for the dependency of the temperature and density exponential decay times (τ_T and τ_n, respectively) during stellar flares, in the solar case we typically find

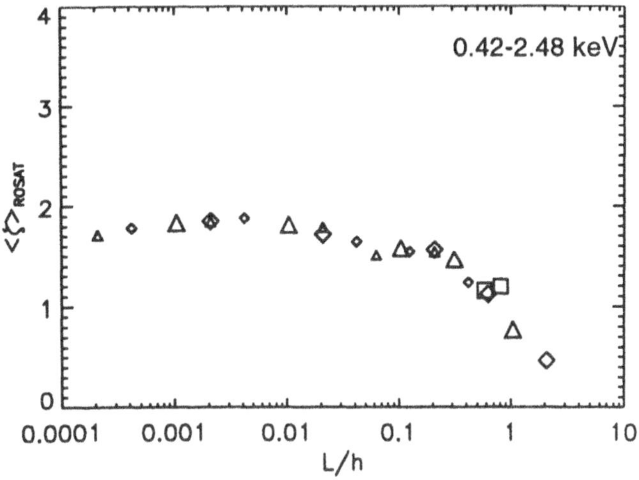

Fig. 2. Ratio $<\zeta>_{\rm ROSAT}$ of the density to the temperature decay time, as computed by folding model results with PSPC energy response in the band 0.42-2.48 keV, vs the ratio L/h of the loop semilength to the pressure scale height. Symbols as in Fig. 1.

$\zeta \equiv \tau_{\rm n}/\tau_{\rm T} \approx 2$ when density and temperature are sampled at the top of the loop ($\zeta_{\rm top}$). In Fig. 1 we report $\zeta_{\rm top}$ as a function of gravity. Although the data points show a significant scatter which increases with g, there is a clear decrease of $\zeta_{\rm top}$ with increasing gravity. We may identify in Fig. 1 different regions for different star surface gravities. We notice that for solar gravity the data points are concentrated around $\zeta = 2$,

In the perspective of comparing the model results with observations, we have folded the results obtained from the calculations with the instrument response of the ROSAT PSPC in the 0.42 – 2.48 keV band, deriving effective density ($< n >$) and temperature ($< T >$) as they would be inferred from observations in the instrument band-pass.

The resulting $<\zeta>_{\rm ROSAT} = \tau_{\langle n\rangle}/\tau_{\langle T\rangle}$, reported in Fig. 2, is particularly well represented as a function of L/h, rather than of gravity. In the comparison with Fig. 1 we notice that $<\zeta>_{\rm ROSAT}$ is typically smaller than $\zeta_{\rm top}$, and remains typically below ~ 2. This effect is expected for any kind of folding of the emission from the whole loop and has been already pointed out in Paper III, where a typical value $\zeta \approx 1.5$, significantly different from $\zeta_{\rm top} = 2$, is indicated when filtering typical solar flares in the Ca XVIII/XIX energy window.

4. Discussion and conclusions

The value of ζ at the top of the loop indeed depends on gravity, and the value $\zeta \approx 2$, invariably obtained from the solar studies, is typical of solar gravity and solar flare loop dimensions. This value of $\zeta_{\rm top}$ is intermediate between very low gravity environments, which yield the maximum possible value $\zeta \approx 3$, and very long and relatively cold loops on high gravity stars, which tend to evolve along the steady state cooling curve in the n-T diagram with $\zeta \sim 0.5$.

When temperature and density evolution are folded through the ROSAT/PSPC response in the energy band 0.42 - 2.48 keV, given the inability to resolve different *spatial* regions in the flaring stellar loop, the combination of the non-uniform distribution of density and temperature and the complicated instrument spectral response makes the results definitely non-trivial.

In general, we find that $< \zeta >$ monotonically decreases with increasing L/h, including the two cases of decay of the model of flare on Proxima Centauri. The value $< \zeta > \sim 2$ is therefore an upper limit for the possible values of $< \zeta >$ to be obtained from ROSAT observations according to our models. Values significantly lower than 2 may indicate the presence of large flaring loops, of size comparable to the local flare pressure scale height (but see Reale et al. 1992 for a discussion of the effect of a prolonged heating during the decay). This kind of information may be useful when combined with the one provided by the analysis of the flare light curve. The fitting of the light curve, for instance as performed by Reale et al. (1988), in fact yields a diagnostics complementary to that provided by the n-T diagram, since it contains time information. The two independent constraints are expected to give strong insight on the geometry of the flaring region, and indirectly on the general stellar environment (for instance on gravity conditions).

Therefore we propose a new approach to interpretation and diagnostics of stellar X-ray flares, which involves the combined analysis of the n-T diagrams and of the flare light curve. Stellar flare data collected by the ROSAT PSPC instrument may be very suitable to undertake such kind of investigation.

Acknowledgements

We acknowledge partial support from the Agenzia Spaziale Italiana and from the Italian Ministero dell'Università e della Ricerca Scientifica e Tecnologica. We also acknowledge support from Italian National Research Council (CNR) for calculations performed on the CRAY-YMP at CINECA (Bologna, Italy).

References

Jakimiec J., Sylwester B., Sylwester J., Serio S., Peres G., Reale F.: 1992, *A&A*, **253**, 269 (Paper II).

Peres G., Rosner R., Serio S., Vaiana G. S.: 1982, *Ap.J.*: **252**, 791.

Peres G., Serio S., 1984, *Mem. SAIt.*: **55**, 749.

Reale F., Peres G., Serio S., Rosner R., Schmitt J.H.M.M.: 1988, *Ap.J.*, **328**, 256.

Reale F., Serio S., Peres G: 1992, *A&A*, submitted.

Rosner R., Tucker W. H., Vaiana G. S.: 1978, *Ap.J.*, **220**, 643.

Serio S., Peres G., Vaiana G. S., Golub L., Rosner R.: 1981, *Ap.J.*, **243**, 288.

Serio S., Reale F., Jakimiec J., Sylwester B., Sylwester J.: 1991, *A&A*, **241**, 197 (Paper I).

Sylwester B., Sylwester J., Serio S., Reale F., Bentley R. D., Fludra A.: 1992, *A&A*, in press (Paper III).

MAGNETIC RECONNECTION AND PARTICLE ACCELERATION IN THE SOLAR CORONA

B.V. SOMOV and YU. E. LITVINENKO

Sternberg Astronomical Institute, Moscow State University, Moscow, Russia

Abstract. It was believed for a long time and now it is confirmed by Yohkoh observations that magnetic reconnection (i.e., an interaction of magnetic fluxes having pair of antiparallel components) plays a key role in the global dynamics of coronal plasma as well as in conversion of the so-called "free magnetic energy" to other forms: thermal and supra-thermal energy of coronal plasma, hard electromagnetic radiation, accelerated particles. Some new results concerning the theory of electron acceleration in solar flares are presented for the case of high-temperature turbulent current sheets in the solar corona.

Key words: Solar Flares – Magnetic Fields – Particle Acceleration

1. Introduction

Magnetic fields determine many phenomena in the solar corona: X-ray bright points, prominences and flares, transients and mass ejections into the interplanetary space. It was believed for a long time (e.g., Syrovatskii and Somov, 1980) that reconnection (i.e., interaction of magnetic fluxes having antiparallel components) plays a key role in the global dynamics of coronal plasma. Now, this concept seems to be confirmed by Yohkoh observations. The theory of magnetic reconnection in high-temperature current sheets (HTCS) explains the total amount of accumulated energy in solar flares, the power of energy release and some other parameters of solar flares (Somov, 1992). Future development of the theory should include the models which could explain the total number of accelerated particles, their maximum energy, the rate of particle acceleration (see Bai and Sturrock, 1989).

Litvinenko and Somov (1991) showed that acceleration by DC electric field and scattering of particles by ion-acoustic turbulence in a HTCS lead to appearance of about 10^{36} electrons with a power-law spectrum and with energies higher than tens of keV. In the present paper, we consider the question about the maximum energy which can be reached by particles accelerated in a HTCS.

The problem of particle motion in the magnetic field which changes the sign of its direction and in the electric field related to reconnection was considered several times. For example, Speiser (1965) found particle trajectories near the "neutral plane" where the magnetic field equals zero. Particle can spend an infinite time near such a plane and can take an infinite energy from the electric field. However, in the real conditions of the solar atmosphere, the probability of such a situation is rather small; usually, the magnetic field in the "reconnecting plane", i.e. the CS, has non-zero transverse and longitudinal components.

Speiser (1965) showed that even a very small transverse field changes the particle motion in such a way that after a finite time the particle leaves the CS having a finite energy. In what follows, we show that this energy is not sufficient in the

J.F. Linsky and S. Serio (eds.), Physics of Solar and Stellar Coronae, 603–606.

context of solar flares.

How to increase the time which the particles spend inside the CS? The influence of the longitudinal component has not been considered yet. We show that the longitudinal field increases the efficiency of particle acceleration to such an extent that allows us to interpret the first step of acceleration in solar flares. To show this we develop some iterative method which gives an approximate general solution of the problem for particle motion inside a non-neutral CS. It also gives the results obtained by Speiser as a partial solution.

2. Particle trajectories inside the current sheet

Let us consider the CS placed in the plane (xz). The electric field $\mathbf{E}$ and current density $\mathbf{j}$ are parallel to the axis z; so, the associated magnetic field components are parallel to the axis x and change their sign in the plane $y = 0$. Therefore, we prescribe the electric and magnetic fields inside the current sheet as follows:

$$\mathbf{E} = (0, 0, E), \qquad \mathbf{B} = B(-y/a, \xi_\perp, \xi_\parallel). \tag{1}$$

The non-relativistic equation of motion for the particle with mass m and charge q is

$$m \frac{\partial \mathbf{v}}{\partial t} = q\left(\mathbf{E} + \frac{1}{c}\mathbf{v} * \mathbf{B}\right) \tag{2}$$

Let us take the half-thickness a of the CS as the unit of length and the inverse gyro-frequency $\omega_B^{-1} = mc/qB$ as the unit of time. Then, equation (2) can be re-written in the dimensionless form:

$$\frac{\partial^2 x}{\partial t^2} = \xi_\parallel \frac{\partial y}{\partial t} - \xi_\perp \frac{\partial z}{\partial t}, \tag{3}$$

$$\frac{\partial^2 y}{\partial t^2} = -\xi_\parallel \frac{\partial x}{\partial t} - y \frac{\partial z}{\partial t}, \tag{4}$$

$$\frac{\partial^2 z}{\partial t^2} = \varepsilon + \xi_\perp \frac{\partial x}{\partial t} + y \frac{\partial y}{\partial t}, \tag{5}$$

where $\varepsilon = mc^2 E/(aqB^2)$ is the dimensionless electric field.

In equation (2) we have neglected the influence of plasma turbulence on particle motions. This can be done under the following condition:

$$\nu_{ef}\,(v/\sqrt{kT/m})^{-3} < \xi_\perp \omega_B. \tag{6}$$

For typical parameters of HTCS in solar flares (Somov, 1992), the effective frequency of ion-acoustic wave-particle collisions is $\nu_{ef} \simeq \xi_\perp \omega_B \simeq 10^6\, sec^{-1}$. Hence, the condition (6) can be well satisfied for supra-termal particles.

The character of the particle motion is determined by two parameters: $\xi_{\parallel}$ and $\xi_{\perp}$. Depending on them, two limiting cases are considered.

A. The case $\xi_{\parallel} = 0$. A particle can spend inside the CS only the time $\tau = \pi / \xi_{\perp}$ After this time, the particle moves quickly out of the CS (Speiser, 1965).

B. The case $\xi_{\parallel} \neq 0$, the CS with longitudinal magnetic field. Particle trajectories are stable under the condition:

$$\xi_{\parallel}^2 > \varepsilon / \xi_{\perp} . \tag{7}$$

If the inequality (7) is valid then the maximal velocity of particles $v_{\max} \sim \xi_{\parallel}$. Therefore, the longitudinal component of a magnetic field qualitatively changes the character of the particle motion in the CS.

3. Electron acceleration in a HTCS during solar flares

The model of a HTCS allows one to express the characteristics of a high-temperature turbulent current sheet through the external parameters of a magnetic reconnection region: concentration of plasma n_0 outside the HTCS, electric field $\mathbf{E}$, magnetic field gradient h_0, and the relative value $\xi_{\perp}$ of a transverse magnetic field (see Somov, 1992). By applying this model to the results obtained in the previouos section, one can find that in the case $\xi_{\parallel} = 0$, the maximum energy of electrons is:

$$\mathcal{E}_{\max} = 2 m c^2 (E / \xi_{\perp} B)^2 , \tag{8}$$

or, after using the HTCS model,

$$\mathcal{E}_{\max} (keV) = 5 \; 10^{-9} T (K) . \tag{9}$$

The formula (9) shows that electron acceleration in the CS without longitudinal magnetic field is not efficient: for the temperature of electrons inside the CS, $T \sim 10^8$ K the maximum energy becomes equal only to 0.5 keV.

Let us consider now the case of a non-zero longitudinal magnetic field. The stabilization condition (7) can be re-written in dimensional units as follows:

$$\left(\frac{B_{\parallel}}{B} \right)^2 > \frac{m c^2 E}{\xi_{\perp} a q B^2} . \tag{10}$$

In the frame of the HTCS model the last inequality becomes especially simple:

$$B_{\parallel} > 0.1 \, B . \tag{11}$$

Therefore, the longitudinal component can be one order of magnitude smaller than the reconnecting components related to the electric current in the CS.

The maximum energy (written also in dimensional units) of accelerated electrons in the CS is

$$\mathcal{E}_{\max} = \frac{1}{2m} \left(\frac{q a B}{c} \xi_{\parallel} \right)^2 , \tag{12}$$

or, in the HTCS model,

$$\mathcal{E}_{\max}(keV) = 10^{-5}\,\xi_{\parallel}^{2}\,T(K)\,. \tag{13}$$

If $T = 10^8\,K$ and $\xi_{\parallel}^2 = 0.1$, formula (13) gives $\mathcal{E}_{\max} = 100\,keV$. Therefore, the longitudinal magnetic field increases the acceleration efficiency to such an extent that it becomes possible to explain the first stage of electron acceleration in solar flares as the particle energization process in a non-neutral HTCS.

4. Concluding comments

From the physical point of view, the results obtained appear very clear. On the one hand, a transverse magnetic field turns a particle trajectory in the plane of the CS. At some point, where the projection of particle velocity v_z on the direction of electric field changes its sign, the Lorentz force component associated with the magnetic field component $B_x = (-y/a)B$ pushes a particle out of the CS. This process is described by equation (4) with $\xi_{\parallel} = 0$. On the other hand, a non-zero longitudinal magnetic field tries to turn a particle back to the CS. This effect is related to the first term in the right-hand side of equation (4); and, when it is not small, the stabilization of particle motion takes place. That is why the maximum velocity of a particle is proportional to the gyro-frequency in the longitudinal magnetic field.

In the real case of the solar atmosphere, zero lines of magnetic field, where all three components of the field equal zero, do not frequently occur (see Gorbachev et al., 1988). Reconnection usually takes place at the so-called separators where a longitudinal (parallel to the current associated with the reconnection process) magnetic field is present. This effect was already considered in the MHD approximation from the point of view of the energetics of a current sheet (see Somov and Titov, 1985). We have shown that the longitudinal field has a strong influence on the kinetics of supra-thermal particles: the non-neutral HTCS does efficiently work as an electron accelerator and, at the same time, as a trap for fast electrons in solar flares.

References

Bai, T. and Sturrock, P.A.: 1989, *Annual Rev. Astron. Astrophys.* **27**, 421

Gorbachev, V.S., Kel'ner, S.R., Somov, B.V. and Shwarz, A.S.: 1988, *Soviet Astron. AJ* **32**, 308

Litvinenko, Yu.E. and Somov, B.V.: 1991, *Soviet Astron. Letters* **17**, 835

Somov, B.V.: 1992, *Physical Processes in Solar Flares*, Kluwer Academic Publ.: Dordrecht, London, 280

Somov, B.V. and Titov, V.S.: 1985, *Solar Phys.* **102**, 79

Speiser, T.W.: 1965, *J. Geophys. Res.* **70**, 4219

Syrovatskii, S.I. and Somov, B.V.: 1980, in Solar and Interplanetary Dynamics, ed(s)., *Dryer, M. and Tandberg-Hanssen, E.*, D. Reidel Publ. Co.: Dordrecht, 425

DEPTH OF FORMATION AND EXCITATION BY ELECTRON IMPACT FOR He I LINES IN THE ATMOSPHERE OF SOLAR TYPE STARS

ALESSANDRO C. LANZAFAME

Armagh Observatory, College Hill, Armagh BT61 9DG, N. Ireland

Abstract. Using line contribution functions for the emergent radiation it is shown that current models for the atmosphere of solar-type stars predict electron temperatures for the region of formation of He I lines approximately between 6,000 and 80,000 K. The precise region of formation of a particular line depends on the transition involved and on the model atmosphere. At present, it is possible to compute accurate electron collision rates for all the transitions between levels with principal quantum number $n \leq 4$, but only for electron temperatures below 20,000 K. Here a method to extrapolate these values to the entire region of formation of He I lines in a physically sensible way is presented, together with the consequences for the formation of the resonance line at 584Å.

Key words: transition region – collision rate – radiative transfer

1. Introduction

Electron-helium scattering calculations (R-matrix, e.g. Berrington and Kingston, 1987, Sawey *et al.*, 1990) have provided a tool for an extensive and accurate description of collisional transitions for levels with principal quantum number $n \leq 4$. However, these calculations may be unreliable at energies above the ionization threshold (24.58 eV in helium). This implies that it is possible to compute accurate collision rates only for temperatures below 20,000 K. Furthermore, it is not possible to take properly into account the effects of the departure from the Maxwellian distribution of the electron energy (e.g. Shoub, 1983, Ljepojevic and Burgess, 1990) since the tail of the distribution lies well above the limit imposed by the R-matrix calculations.

The sensitivity of helium lines to coronal X-rays, particle diffusion effects and to the local electron energy distribution, has made an understanding of the formation of these lines difficult. On the other hand, these peculiarities can be used as a powerful activity indicator and in particular as diagnostics for the transition region between the chromosphere and the corona, provided we have a better insight into the physics of formation of these lines.

2. The Depth of Formation of He I Lines in the Atmospheres of Solar-type Stars

An analysis of the depth of formation has been carried out using line contribution functions of the emergent radiation C_F^l as defined by Achmad *et al.* (1991). The line contribution function for the He I(584Å), normalized to unity, is shown in Fig. 1 for the atmospheric model VAL3C (Vernazza, Avrett and Loeser, 1981). The computations are performed under the assumption of a radiation field in non-LTE but a plasma in LTE. The combined equations of radiative transfer and statistical equilibrium are solved using MULTI (Carlsson, 1986) for a 20-level He I model atom.

J.F. Linsky and S. Serio (eds.), Physics of Solar and Stellar Coronae, 607–610.

The region of formation of He I lines has an electron temperature approximately between 6,000 and 80,000 K. Three model atmospheres are considered as examples: VAL3C, the Thatcher *et al.* (1991) model for ϵ Eridani (dK2) and the Houdebine (1991) model for AU Mic (dM2.5e). In the present computations, the model VAL3C gives a flux for the resonance lines that is roughly 70% of that observed and a flux for He I(10830Å) that is roughly 25%, without including the over-ionization pro duced by coronal X-ray back-radiation. This behavior is qualitatively in agreement with the work of Milkey *et al.* (1973).

The model for ϵ Eridani produces fluxes in He I(10830Å) and He I(5876Å) that agree with the observed quantities within the observational uncertainties (Lanzafame, A.C., 1991). In this model, the atmospheric parameters are such that a significative fraction of the lines is formed in a relatively broad region in the higher chromosphere, producing a region of formation of He I lines broader than solar-type models.

The AU Mic model produces fluxes in He I(10830Å) and He I(5876Å) that are roughly 40% of those observed. This star has a high X-ray emission so it is possible that the effects of the over-ionization by coronal back-radiation are higher than in other stars. However, the very high temperature gradient of the transition region in the model causes a very thin region of formation for the He I lines.

Since there is no constraint imposed on the lower transition region temperature gradient for the ϵ Eri and AU Mic models, a modification that takes into account the energy balance in the lower transition region or a semi-empirical modelling that makes use of the observed hydrogen Ly_α or He lines is required.

3. Extrapolation of Collision Strengths for Electron Impact

Theoretical investigations for the calculation of the collision strengths, Ω, for electron energies near and above the ionization threshold are going on, but it will be a while before any results are available. The extrapolation method proposed here is based on a fit to the collision strengths, Ω, using the code OMEUPS (Burgess and Tully, 1992). The collision strengths are scaled in order to remove the main energy dependence in the high energy limiting behaviour. A five-point spline is then used to fit these scaled quantities. From this fit it is possible to reconstruct the original quantities and in particular to extrapolate the Ω values for electron energies near and above the ionization threshold. Since the fit is not suitable for collision strengths near the excitation threshold in neutral species (for which $\Omega = 0$ at the excitation threshold) and it smooths over the atomic resonances, only the extrapolation is taken from the fit and the original collision strengths from the R-matrix calculations are maintained where available.

The effective collision strengths, Υ, (i.e. Ω averaged over a Maxwellian), are then numerically computed using a trapezoidal rule to take properly into account the effects of the atomic resonances.

In Fig. 2 is shown the resulting Υ for the transition $1s^2\,{}^1S - 1s2p\,{}^1P$ corresponding to the line He I(584Å), compared with other available sets of data.

The extrapolation method presented here for Υ allows us to make an estimate based on physical considerations for the collision rates in the region above the

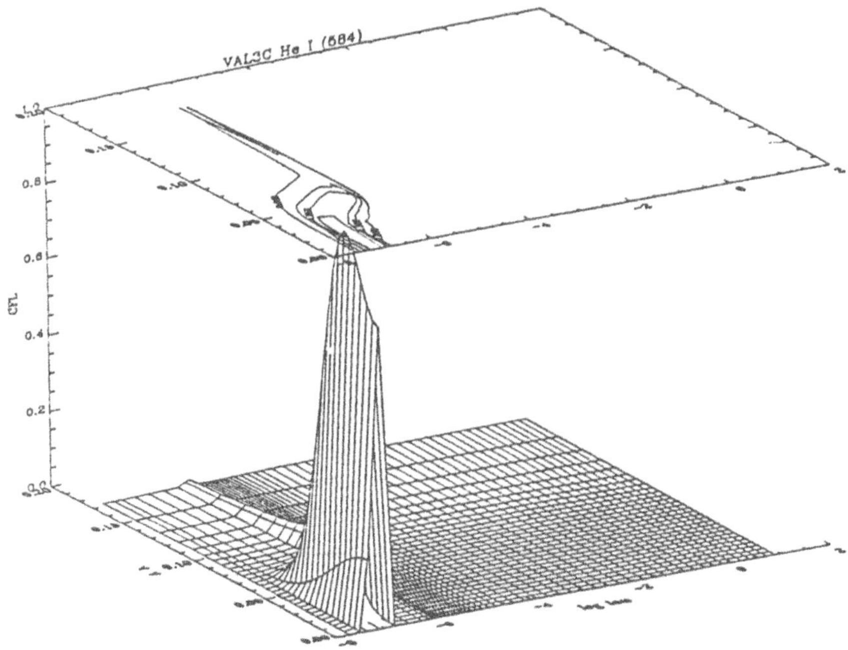

Fig. 1. Line contribution function to the emergent radiation for He I(584Å) in VAL3C model atmosphere. The region of formation, defined as the region where the contribution function is greater than 10% of its maximum, corresponds to electron temperatures between 10,000 and 50,000 K.

present limit of the theory. Furthermore this method allows us to take properly into account the effects of non-Maxwellian electron energy distribution on the collision rate, providing a more accurate estimate for the contribution to the rate from the tail of the distribution.

4. Results for the Resonance Line He I(584Å)

The effective collision strengths have been extrapolated with the method outlined above for all the electric dipole transitions included in the model atom. Work is in progress to extend the extrapolation to the other transition types.

A comparison has been made with two sets of collisional data. In the first set, the effective collision strengths from Berrington and Kingston (1987) have been considered for all the transitions between levels with $n \leq 3$, and from Sawey *et al.* (1990) (29-state computations) for the remainder. The Υ values for temperatures above 30,000 K have been set equal to the tabulated Υ(30,000 K). In the second set, all the Υ values for the electric dipole transitions have been recomputed with

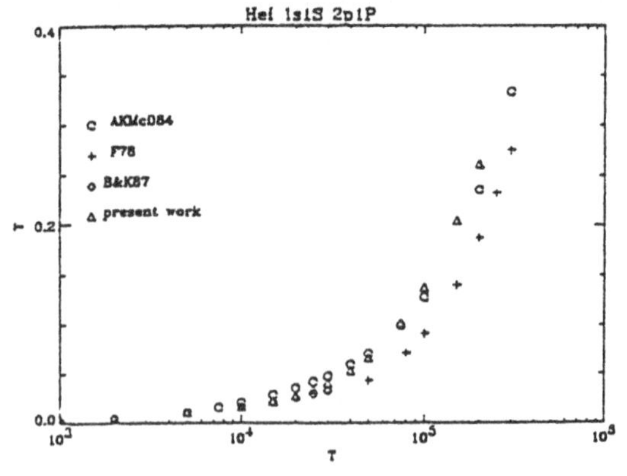

Fig. 2. Effective collision strengths for the transition $1s^2\,^1S - 1s2p\,^1P$ in He I.AKMcD84 are data from Aggarwal *et al.* (1984), F78 from Fujimoto (1978), B&K87 from Berrington and Kingston (1987).

the method outlined. The Υ values are essentially the same for T $\leq$ 20,000 K in the two sets of data. The center of the He I(584Å) line turns out to have a flux that is 27% higher using the latter set of effective collision strengths.

Acknowledgements

I would like to tank J. A. Tully, for his untiring help with the analysis of the collision strengths and the use of OMEUPS. I tank also A. Burgess for making available his code and P.B. Byrne, P.D. Dufton, K.A. Berrington and P.M.J. Sawey for useful discussions. Research at Armagh Observatory is grant-aided by the Department of Education for N. Ireland.

References

Achmad, L., de Jager, C., and Nieuwenhuijzen: 1991, ***Astron. Astrophys.* 250**, 445
Aggarwal, K.M., Kingston, A.E., and McDowel, M.R.C.: 1984, ***Astrophys. J.* 278**, 874
Berrington, K. A. and Kingston, A. E.: 1987, ***J. Phys. B* 20**, 6631
Burgess, A., and Tully, J.A.: 1992, ***Astron. Astrophys.* 254**, 436
Carlsson, M.: 1986, *Uppsala Astronomical Observatory report* No 33.
Fujimoto, T.: 1978, *Semi-Empirical Cross Sections and Rate Coefficients for Excitation and Ionization By Electron Impact and Photoionization of Helium*, Institute of Plasma Physics, Nagoya, Japan, Rep. IPPJ-AM-8.
Houdebine, E.R.: 1991, in Mechanisms of Chromospheric and Coronal Heating, ed(s)., *Ulmschneider, P., Priest, E.R., Rosner, R.*, Springer-Verlag: Berlin-Heidelberg, 182
Lanzafame, A.C.: 1992, in 7th Cambridge Workshop on Cool Stars, Stellar Systems and the Sun: Tucson 1991, ed(s)., M. Giampapa and J. Bookbinder, *in press*.
Ljepojevic, N.N., and Burgess, A.: 1990, ***Proc. R. Soc. Lond. A* 428**, 71
Milkey, R.W., Hesley, J.N., and Beebe, H.A.: 1973, ***Astrophys. J.* 186**, 1043
Shoub, E.C.: 1983, ***Astrophys. J.* 266**, 339
Sawey, P.M.J., Berrington, K.A., Burke, P.G., and Kingston, A.E.: 1990, ***J. Phys. B* 23**, 4321
Thatcher, J.D., Robinson, R.D., and Rees, D.E.: 1991, ***M.N.R.A.S.* 250**, 14
Vernazza, J.E., Avrett, E.H., and Loesner, R.: 1981, ***Astrophys. J. Suppl.* 45**, 635

THREE-DIMENSIONAL MAGNETIC RECONNECTION IN A CORONAL NEUTRAL SHEET

R. B. DAHLBURG and S. K. ANTIOCHOS
Naval Research Laboratory, Washington, D.C., USA
and
T. A. ZANG
NASA Langley Research Center, Hampton VA, USA

Abstract. We review our research on the transition to turbulence of a 3d neutral sheet. The existence of a 3d secondary instability which grows on an ideal time-scale is demonstrated numerically. When the secondary mode attains sufficient amplitude to become nonlinear, there is a breakdown of familiar laminar structures such as electric current filaments and vortex quadrupoles. Our results suggest that magnetic reconnection might be a flexible enough process to account for the three energy release phases observed in two-ribbon flares. Estimates of growth times compare well with impulsive phase data.

Key words: Solar flares – magnetic reconnection – turbulence

1. Turbulent Magnetic Reconnection

It is well known that the classical rates for magnetic reconnection are too slow to explain many solar phenomena, such as coronal heating and flares. In addition, it is difficult to see how magnetic reconnection could account for events, such as two-ribbon flares, in which there exist phases which occur on widely disparate timescales. Turbulence has often been invoked as a process which could boost up the rate of energy release by magnetic reconnection (*e.g.* Strauss, 1988; Biskamp and Welter, 1989). However, the physics of turbulent magnetic reconnection is still unclear. We here describe results of our recent investigation of a route by which magnetic reconnection can become turbulent (Dahlburg *et al.*, 1992), *viz.*, by means of a 3d secondary instability (*e.g.*, Bayly *et al.*, 1988).

There are three steps in the MHD secondary instability process: [1] the 1d neutral sheet primary quasi-equilibrium is first destabilized by a linear 2d resistive primary disturbance (Furth *et al.*, 1963); [2] the primary disurbance grows until it saturates to form a 2d secondary equilibrium state (Rutherford, 1973); and then [3] the secondary equilibrium state is destabilized by a 3d secondary instability which grows on an ideal, *i.e.*, nonresistive, timescale. When the secondary mode attains a large enough amplitude, there is a breakdown of familiar 2d reconnection structures such as electric current filaments and vortex quadrupoles, as well as an enhanced generation of small scale structure.

2. Methodology

Our investigation of the transition to turbulence in 3d neutral sheets combines analysis and numerical computations, which we describe briefly here. A more detailed account is found in Dahlburg *et al.* (1992). We consider a sheared magnetic field in a viscoresistive, incompressible 3d magnetofluid. The mean magnetic field is specified to be $B_0(z)\hat{e}_x = \tanh z\hat{e}_x$. Primary perturbations of this system are determined by

J.F. Linsky and S. Serio (eds.), Physics of Solar and Stellar Coronae, 611–614.

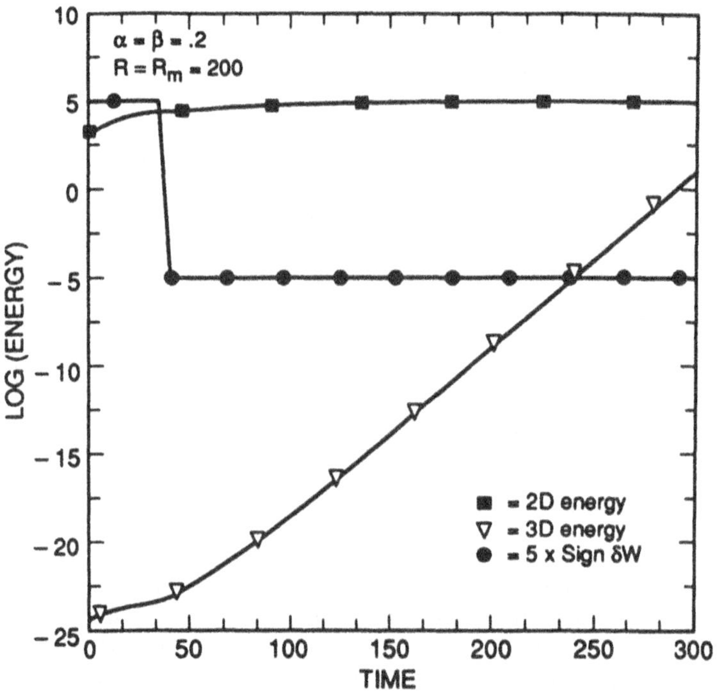

Fig. 1. 2d and 3d energies and δW as a function of time for a typical run.

an equation of the Orr-Sommerfeld type (Dahlburg *et al.*, 1983; 1986). An important linear result is that, for a given set of wavenumbers and Lundquist numbers, the 2d primary mode grows at a rate faster than the 3d primary modes, similar to Squire's result for Navier-Stokes systems (Squire, 1933).

To investigate the linear characteristics of the secondary mode, we first initialize out 3d calculations with a dose of the unstable 2d primary mode large enough that the system is close to a secondary equilibrium. Then we add a very small dose of a 3d primary mode to initiate activity in the third direction.

3. Results

Figure 1 shows the time evolution, for a typical run, of the 2d and 3d energies, *i.e.*, the energy of modes which vary only in x and z and in x, y and z, respectively. The figure shows that when the 2d mode saturates, the 3d mode starts to grow at a faster rate which turns out to be independent of Lundquist number. Calculation of the perturbed potential energy, δW, shows that the ideal instability is predictable by the classical energy principle (Bernstein *et al.*, 1958).

When the secondary mode becomes nonlinear there is a breakdown of the familiar structures of 2d reconnection, such as electric current filaments and vortex

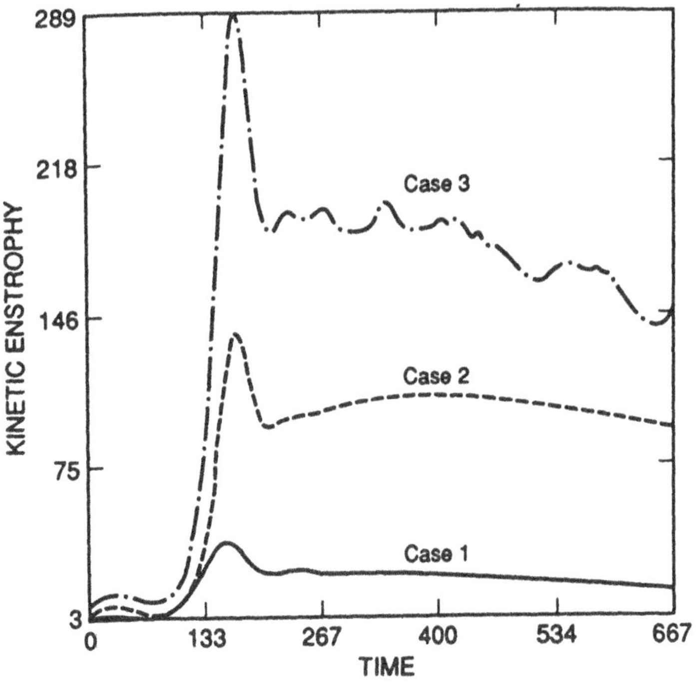

Fig. 2. Enstrophy as a function of time for highly nonlinear runs; Lundquist numbers: Case 1: 50, Case 2: 100, Case 3: 200.

quadrupoles. Furthermore, in the nonlinear phase of the secondary mode there is a significantly higher level of excitation of small scale spatial structure than exists in the nonlinear or saturated phases of the 2d primary mode. Figure 2 is a plot of the enstrophy (integrated squared vorticity) as a function of time for three fully nonlinear simulations at different Lundquist numbers. The plot shows that, following the 2d saturated state, there is an explosive growth of small-scale structure in the velocity field, followed by a plateau in the enstrophy value at a higher level than that attained before the onset of the secondary instability.

4. A Conjecture on the Two-Ribbon Flare

Two-ribbon flares are believed to occur when, following the eruption of a prominence, the resulting open magnetic field configuration closes down *via* magnetic reconnection. Three distinct energy release phases occur during the evolution of a two-ribbon flare, *viz.*, a startup phase, an impulsive phase, and a late phase (Sturrock, 1980). It has been argued that magnetic reconnection alone might account for all three of these phases, but it has been problematical to explain how one process can account for three kinds of activity with such widely disparate timescales.

On the basis of our present results, we conjecture that three-dimensional reconnection might provide an answer. Consider a magnetic field which is line tied at the photosphere and which extends straight out to infinity. Furthermore, let this magnetic field reverse polarity along some inversion line on the photospheric boundary. Let the neutral sheet be 100 km wide, the coronal Alfvén speed be 1000 km/sec, and the magnetic diffusivity be 1 m^2/sec. This gives a coronal Lundquist number of 1×10^{11}, and an Alfvén time of 1/10 sec.

When this neutral sheet configuration is perturbed it will begin to reconnect, a process which has been simulated in 2d (*e.g.*, Forbes and Malherbe, 1991). We speculate that the growth of the resistive 2d mode could correspond to the startup phase of the two-ribbon flare. We compute a resistive growth time of about 3 hours, and hence some decrease of this value must be found. This could involve forcing the reconnection, or enhancing the resistivity *via* kinetic effects. Next, the impulsive phase could correspond to the fast growth of a secondary 3d mode. The run shown in Figure 1 implies a secondary mode growth time of 2 seconds, which is reasonable for the impulsive phase. Finally, the late phase could correspond to the time in which the secondary mode has become nonlinear. It is difficult to make a time estimate for this phase, but Figure 2 indicates that the energy release for the later stages of 3d reconnection is relatively steady at a rate decreased from the initial burst. We are currently trying to bound these estimates and compare them more rigorously to observed timescales.

Acknowledgements

The numerical calculations reported here were performed on the NASA Ames Research Center Cray Y-MP under a grant from the NAS program. This work was supported by the NASA Space Physics Theory Program, the NASA SMM Guest Investigator Program, and by the Office of Naval Research.

References

Bayly, B. J., Orszag, S. A., and Herbert, Th.: 1988, *Ann. Rev. Fluid Mech.* **20**, 359

Bernstein, I. B., Frieman, E. A., Kruskal, M. D., and Kulsrud, R. M.: 1958, *Proc. Roy. Soc. (London) A* **244**, 17

Biskamp, D., and Welter, H.: 1989, *Phys. Fluids B* **1**, 1964

Dahlburg, R. B., Zang, T. A., Montgomery, D., and Hussaini, M. Y.: 1983, *Proc. Nat. Acad. Sci. USA* **80**, 5798

Dahlburg, R. B., Zang, T. A., and Montgomery, D.: 1986, *J. Fluid Mech.* **169**, 71

Dahlburg, R. B., Antiochos, S. K., and Zang, T. A.: 1992, *Phys. Fluids B* **4**, in press

Forbes, T. G., and Malherbe, J. M.: 1991, *Solar Phys.* **135**, 361

Furth, H., Killeen, J., and Rosenbluth, M. N.: 1963, *Phys. Fluids* **6**, 459

Rutherford, P. H.: 1973, *Phys. Fluids* **16**, 19003

Squire, H. B.: 1933, *Proc. Roy. Soc. A* **142**, 621

Strauss, H. R.: 1988, *Astrophys. J.* **326**, 412

Sturrock, P. A.: 1980, Flare Models, in *Solar Flares*, ed(s). P. A. Sturrock, Colorado Associated University Press: Boulder, 411

OUTFLOW FOCUSING IN ROTATING STELLAR MAGNETOSPHERES

K. TSINGANOS [1], E. TRUSSONI [2] and C. SAUTY [3]
[1] *Dept. of Physics, Un. of Crete, GR-71409, Heraklion, Crete, GREECE*
[2] *Osservatorio Astronomico di Torino, I-10025, Pino Torinese, ITALY*
[3] *Observatoire de Paris, DAEC, F-92195 Meudon Cedex, FRANCE*

Abstract. Loss of mass, angular momentum and magnetic flux from stars and stellar systems is occuring either in the form of a spherically symmetric wind, or, in the form of bipolar flows and jets. In this article we discuss a family of exact MHD solutions for such outflows from the central gravitational field of those stellar systems. The topology of the solutions is studied in detail and is controlled by several critical points. For the characteristic solution through these critical points it is shown that when the azimuthal components of the hydromagnetic field are small and the central object is a *Slow Magnetic Rotator* (SMR) the outflow is conical and asymptotically we obtain the classical picture of a radial wind. On the other hand, when the poloidal electric current and rotation are high and the central object is a *Fast Magnetic Rotator* (FMR) the outflow is collimated and the wind is focused toward the rotation and magnetic axis with wavy and knotty streamlines along this axis. Thus we conclude that when a young stellar object due to gravitational collapse condenses and therefore posseses high rotational rates due to angular momentum conservation, it is likely to have jet-like outflows. Bipolar flows from young stellar objects and protostars, T-Tauri stars, compact stellar objects like SS433, and also AGN may be understood in this manner. On the other hand, as the star ages and angular momentum is lost through its wind and associated magnetic field, the star slows down and its mass loss takes the form of a roughly radial classical solar/stellar wind.

Key words: Hydromagnetics – Stars: coronae, magnetic fields, winds – Galaxies: jets

1. Introduction

Spherically symmetric astrophysical outflows have been studied to a considerable extent in connection with their central role in several astrophysical phenomena, such as fast solar wind streams and their association with coronal holes, (Parker 1958, Zirker 1977, Wang, Sheeley and Nash 1990 and references therein), stellar winds and their association with stellar angular momentum loss, (Weber and Davis 1967, Sakurai 1990 and references therein), optically thick winds and their association to novae outbursts (Kato 1991, Orio *et al*, 1992 and references therein), etc. On the other hand, the nonspherically symmetric phenomenon of collimated astrophysical outflows and jets has been extensively studied observationally as well as theoretically in the last three decades, also in the content of many astrophysical systems, such as bipolar flows and their association with protostellar objects (Pudritz and Norman 1983, Mundt *et al.* 1990 and references therein), cosmic jets and their association with compact galactic and extragalactic objects (Blandford and Rees, 1974, Hughes 1991, and references therein), relativistic beams and their association with pulsar magnetospheres (Michel 1969, Camenzind 1989, and references therein), etc.

In the approach that we review in the following, by means of an *a priori* physically sound choice of the angular dependence of the field and streamline pattern, the MHD equations are explicitly solved without imposing any polytropic relation. Then, through the energy equation, the form of the heating/cooling distribution along the flow is calculated *a posteriori* and *selfconsistently* with the assumed angular distribution of the field and flow structure. The analytical nature of the study

J.F. Linsky and S. Serio (eds.), Physics of Solar and Stellar Coronae, 615–622.

enables us to have a detailed display of the various complicated topologies of the MHD solutions for magnetized winds, an experience useful in more sophisticated numerical studies. The interested reader may find a more detailed presentation in Low and Tsinganos (1986), Tsinganos and Low (1989), Tsinganos and Trussoni (1990, 1991), Tsinganos and Sauty (1992a,b).

2. An Interesting Analytical Class of Axisymmetric MHD Outflows

Our starting point is the following set of the MHD equations for the steady dynamical interaction of an inviscid but compressible fluid flow of high electrical conductivity with a magnetic field **B** in the spherically symmetric gravitational field surrounding a central gravitating object,

$$\nabla \cdot \mathbf{B} = 0, \quad \nabla \cdot (\rho \mathbf{V}) = 0, \quad \nabla \times (\mathbf{V} \times \mathbf{B}) = 0\,, \tag{1a}$$

$$\rho(\mathbf{V} \cdot \nabla)\mathbf{V} = -\nabla P + \frac{1}{4\pi}(\nabla \times \mathbf{B}) \times \mathbf{B} - \frac{\rho G M(1-\ell)}{r^2}\mathbf{e}_r\,, \tag{1b}$$

for the bulk flow speed $\mathbf{V}(r,\theta)$, the magnetic field $\mathbf{B}(r,\theta)$, the density $\rho(r,\theta)$, and the pressure $P(r,\theta)$ in spherical coordinates $(r,\ \theta,\ \phi)$, with M the mass of the central body, G the gravitational constant and $\ell = L_r/L_{Edd}$, the constant ratio of the diffusive luminosity L_r and Eddington luminosity L_{Edd} (Ruggles and Bath 1979).

Equations (1) may be closed with a polytropic relationship between pressure and density, or with an energy equation. Since specification *a priori* of a constant polytropic index γ may not correspond to simultaneously correct conditions at the base of the outflow and at large distances from the central object (Weber and Davis 1967), as well as to artificial heating/cooling distributions along the flow, we shall follow the approach of allowing for a variable index γ and calculate the net heating rate required to support the deduced flow pattern (Tsinganos, Trussoni and Sauty, 1992).

In the case of our Sun, it has long been established that the distribution of the plasma density and pressure in the solar atmosphere is clearly nonspherically symmetric and at least two-dimensional. For example, coronal holes from where the solar wind emanates are observed to be lower density regions than the surrounding streamers (Zirker 1977). In the case of other stars and in the presence of magnetic fields, it is also expected to have a meridionally anisotropic density distribution. For example, a latitudinal decrease of the density has also been inferred to exist in the chromospheres and cool envelopes of Be stars (Ringuelet and Iglesias 1991).

Theoretical arguments have long predicted that disks are a natural part of the stellar formation process and all current models of bipolar outflows require the existence of a pre-main sequence circumstellar disk (Königl 1982, Hartmann and MacGregor 1982). On the other hand, high angular resolution optical, infrared and radio images have accumulated observational evidence for the ubiquitous presence of axially symmetric circumstellar clouds around young stars (Cassen *et al.* 1986). With the above motivations in mind, it is inevitable to start our study by considering nonspherically symmetric density distributions. Thus, assume the following axisymmetric but meridionally anisotropic functional dependence of the density on R and θ

$$\rho(R,\theta) = \rho_o \frac{M_o^2}{M_a^2}[1 + \delta\, f(R)\sin^2\theta] \,, \tag{2}$$

where ρ_o and M_o are the density, and poloidal Alfvén number at the polar base $(R = 1, \theta = 0)$, and $M_a(R)$ the poloidal Alfven number at some distance R. The constant δ controls the meridional density asymmetry, such that for $\delta > 0$, the equatorial parts of the atmosphere are denser than the polar ones, at the same radial distance.

In dealing with the r- and θ-components of the force balance equation (1b), we may separate the variables R and θ, if we assume that the pressure $P(R, \theta)$ has the following functional dependence on the two variables R and θ:

$$P(R,\theta) = P_o(R) + P_1(R)\sin^2\theta = \frac{\rho V_o^2}{2}\left[Q_o(R) + Q_1(R)\sin^2\theta\right] \,. \tag{3}$$

Note that if $P_1(R) > 0$, the pressure has a polar deficit relative to its equatorial value at the same radial distance. This pressure distribution may be similar then to a *de Laval* nozzle type configuration where the pressure is lower along the polar direction of the steepest density gradient, than along the equatorial plane, as it has been assumed in models of bipolar outflows (Königl 1982).

We have suggested that a physically interesting choice of the magnetic and velocity field that satisfies the MHD equations (1) is,

$$V_r = V_o \frac{M_a^2}{M_o^2}\frac{f}{R^2}\frac{\cos\theta}{\sqrt{1+\delta\, f\sin^2\theta}} \,, \quad V_\theta = -V_o \frac{M_a^2}{M_o^2}\frac{1}{2R}\frac{\mathrm{d}f}{\mathrm{d}R}\frac{\sin\theta}{\sqrt{1+\delta\, f\sin^2\theta}} \,, \tag{4}$$

$$B_r = \frac{1}{r^2\sin\theta}\frac{\partial A}{\partial\theta} = \frac{B_o f(R)}{R^2}\cos\theta \,, \quad B_\theta = -\frac{1}{r\sin\theta}\frac{\partial A}{\partial r} = -\frac{B_o}{2R}\frac{\mathrm{d}f}{\mathrm{d}R}\sin\theta \,, \tag{5}$$

$$V_\varphi = \lambda V_o \frac{M_a^2}{M_o^2}\frac{f}{R}\frac{\sin\theta}{\sqrt{1+\delta\, f\sin^2\theta}}\left(\frac{1-(f_\star/R_\star^2)(R^2/f)(1/M_a^2)}{1-M_a^2}\right) \,, \tag{6}$$

$$B_\varphi = -\lambda B_o \frac{f}{R}\sin\theta\left(\frac{1-(f_\star/R_\star^2)(R^2/f)}{1-M_a^2}\right) \,. \tag{7}$$

In the above expressions, $R = r/r_o$ is the dimensionless radial distance in terms of some base radius r_o corresponding to the stellar base, and $f_\star = f(R_\star)$, where $R_\star$ is the radial distance of the poloidal Alfvén transition of the flow, *i.e.*, where $M_a = 1$. Also, $A(R,\theta) = f(R)\sin^2\theta$ specifies the magnetic and streamlines on the poloidal plane.

Note that the constant δ controls also the degree of velocity collimation around the axis of the flow $(\theta = 0)$. Thus, $\delta = 0$ corresponds to a sinusoidally symmetric flow pattern, while $\delta \to \infty$ corresponds to a flow along the polar axis only, $(\theta = 0)$. Take for example the highly–collimated bipolar flows and jets from young stars. Their majority have opening angles of about 3 to 10 deg (Mundt 1986), corresponding to large values of δ. Thus, for fixed R and $\delta \sim 100(1000)$, at $\theta = 10(3)$ deg the radial velocity in Eq. (4) drops to less than one half its value at $\theta = 0$.

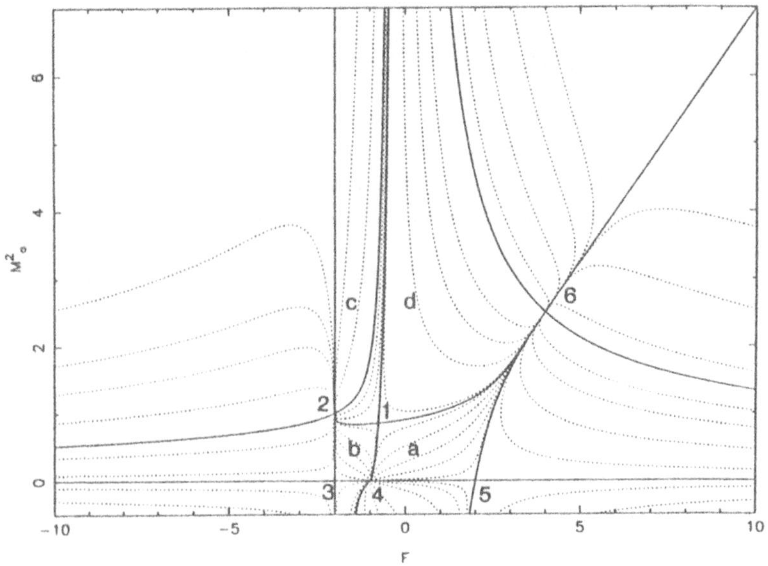

Fig. 1. Solution topology for $\delta = 0$ on the plane–(F, M_a^2), containing three saddle points, (1,3,5) and three nodal points (2,4,6). Critical solutions are indicated by solid lines and noncritical ones by dotted or dashed lines.

Finally, in the remaining r- and θ-components of the force balance equation (1b), the variables r and θ can be separated with the previous angular dependences yielding a set of three nonlinear differential equations for the four unknowns $f(R)$, $M_a(R)$, $Q_o(R)$, and $Q_1(R)$. We are then free to specify one of them. In the following we shall illustrate a possible set of solutions by relating the pressure-inhomogeneity function $Q_1(R)$ to $Q_o(R)$.

3. Asymptotically Radial Outflows in Slow Magnetic Rotators

A class of solutions can be generated by setting $\lambda = 0$ and $Q_1(R) = \kappa f(R) Q_o(R)$ and solving the system for $f(R)$, $M_a(R)$, and $Q_o(R)$. This assumption of relating $Q_1(R)$ and $Q_o(R)$ essentially amounts to saying that the pressure, or the temperature, have the same basic radial profile along each streamline. We are dealing then with a class of solutions controlled by the five parameters κ, δ, M_o, $s = \mathrm{d}f(R=1)/\mathrm{d}R$ and ν^2, where ν is the ratio of the escape speed and polar base speed, $\nu = (2GM(1-\ell)/r_o V_o^2)^{1/2}$. Thus, a single nonlinear second order differential equation is obtained for $f(R)$ whose solutions are controlled by several novel critical points that, for example, select a unique wind-type solution. In particular, the simpler subcase $\kappa = 0$ corresponding to nonrotating flows with a spherically symmetric pressure distribution, is interesting on its own and we shall illustrate it further in the following.

To study the topologies of the solution, we shall use an appropriate new variable,

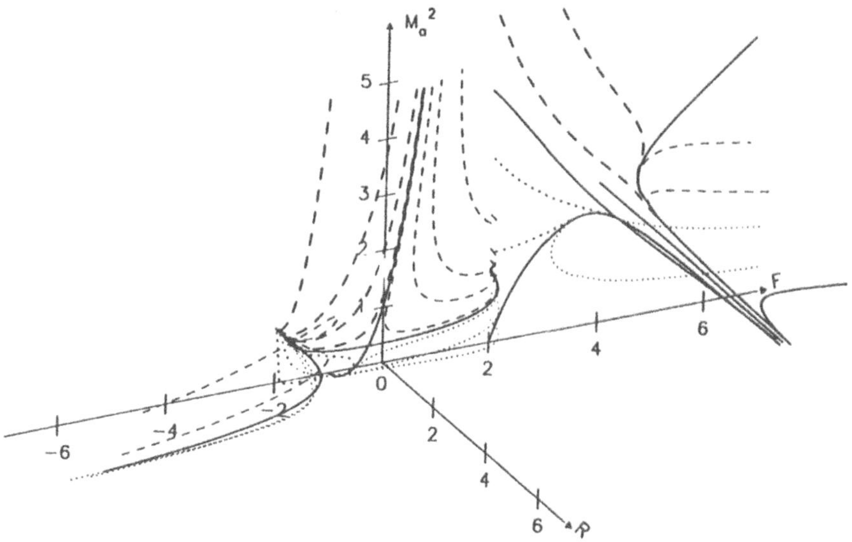

Fig. 2. Topology of the solutions in the three-dimensional space (R, F, M_a^2) for $\delta = \kappa = 0$ and $\nu = 1$.

which we define for convenience to be the logarithmic derivative of the streamlines function $f(R)$, $F \equiv R\mathrm{d}f/f\mathrm{d}R = \mathrm{d}\ln(f)/\mathrm{d}\ln(R)$. Besides its mathematical use for a rather simplified formulation, this quantity F represents physically the deviation of the streamlines from being radial. Thus, the three variables (R, F, M_a^2) should be the natural choice for a study of the solutions and especially their topology. By using these three variables, the r- and θ–components of the momentum balance equation can be written as a system of two equations for $F(R)$ and $M_a(R)$ (Tsinganos and Sauty 1992b).

Considering for simplicity the case wherein the density is spherically symmetric, $\delta = 0$, the following critical points may be found on the (F, M_a^2)–plane (Table 1):

TABLE I

(1)	$(F_1, M_{a1}^2) = (-3+\sqrt{5}, [3\sqrt{5}-5]/2)$	saddle point
(2)	$(F_2, M_{a2}^2) = (-2, 1)$	nodal point
(3)	$(F_3, M_{a3}^2) = (-2, 0)$	saddle point
(4)	$(F_4, M_{a4}^2) = (-1, 0)$	nodal point
(5)	$(F_5, M_{a5}^2) = (+2, 0)$	saddle point
(6)	$(F_6, M_{a6}^2) = (+4, 5/2)$	nodal point

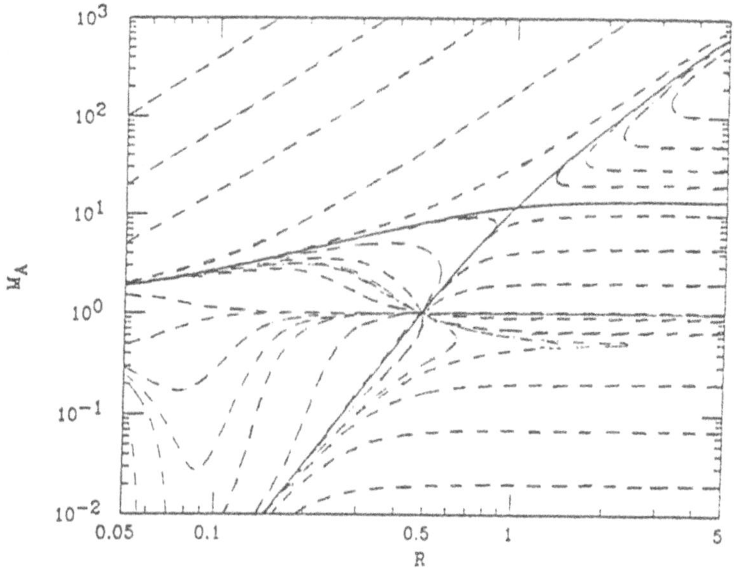

Fig. 3. Topology of the solutions in plane (M_a, R) for a strongly magnetized helicoidal outflow with $M_o = 0.01$, $\delta = 4$, $\nu = 120$ and $\lambda = 0.5$.

From the above figures it is evident that the unique and physically interesting solution has $F(R \longrightarrow \infty) \longrightarrow 0$ with $M_a(R \longrightarrow \infty) \longrightarrow \infty$ and $M_a(R \longrightarrow 0) \longrightarrow 0$. Therefore we obtain a wind-type solution wherein the streamlines become radial at infinity.

4. Asymptotically Focused Outflows in Fast Magnetic Rotators

Let us now turn our attention to the general class of rotating flows ($\lambda \neq 0$), assuming for simplicity that the pressure is spherically symmetric ($\kappa = 0$). Following the philosophy of the previous section, one can draw a topology of the solutions using the variables R, M_a^2, F. As before, a unique physical solution is selected by an X-type critical point. In addition, however, the azimuthal components make the Alfvén point a critical one. An extra regularity condition is emerging now that relates the slope of $M_a(R)$ to the geometry function F at the Alfvén surface ($M_a = 1$). Analogously to what is found for polytropic jets (Heyvaerts and Norman 1989), the governing equations at this point choose a specific geometry in order to avoid a kink in the fieldline structure there.

The topology around the X-type and Alfvén star-type critical points that select a unique physical solution is similar to that illustrated in Figure 3 above for an outflow with only radial and azimuthal components, independent pressure components $Q_o(R)$ and $Q_1(R)$ and an imposed $f(R) = 1$ (Tsinganos and Trussoni 1991).

Once the flow has passed through those two initial critical points and has been accelerated to high transAlfvénic speeds, it may or may not collimate depending

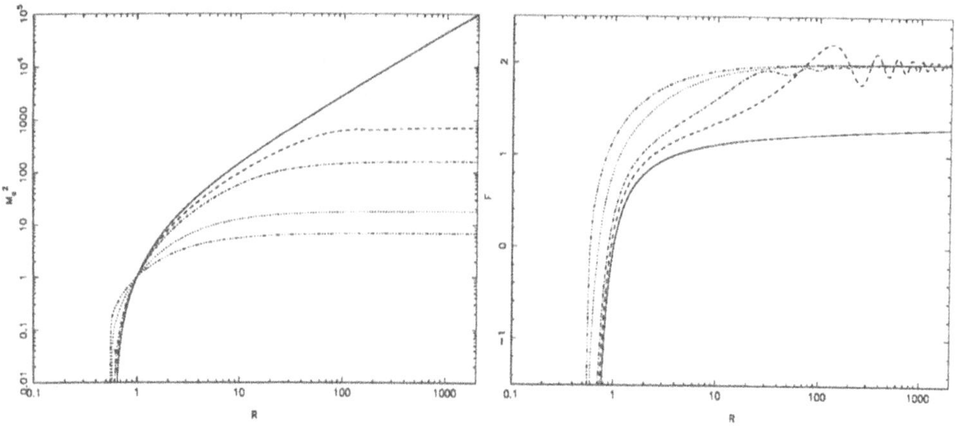

Fig. 4. Critical solutions obtained for $\lambda = 10$ and $k = 0, -10, -20, -50, -70$. Plots are made in the (M_a^2, R) in (a) and (F, R) in (b) planes. An asymptotic value of $F = 2$ corresponds to cylindrical asymptotics.

on whether the outflow is rotationnaly dominated (FMR) or not (SMR). It can be shown that, for this class of solutions, there exists a conserved quantity along each fieldline, k, analogous to the Bernouilli constant for polytropic flows. k is essentially the difference of the total macroscopic energy available on a fieldline of label A, $E(A)$ and the same quantity at the polar streamline $A = 0$,

$$k \equiv \frac{[E(\mathrm{pole}) - E(A)]}{\frac{1}{2}V_o^2}\frac{1+\delta A}{A} . \tag{8}$$

By calculating this quantity in the limit of large R, it appears that only for negative k completely collimated flows can be achieved (i.e., a cylindrical structure). If $k < 0$, the Poynting energy flux dominates the other energy fluxes at the Alfvén point, while if $k > 0$ the Poynting flux is less important than the other energy fluxes at the Alfvén point. In the second case we are in the Slow Magnetic Rotator regime (Cassinelli 1990) and the streamlines become asymptotically conical, or, obtain at most paraboloidal asymptotics. On the other hand, in the first case of negative k (Fast Magnetic Rotators) the wind collimates into a cylindrical jet of finite width and flow speed. By combining the Bernoulli integral k with the transfield equations in the limit of $R \longrightarrow \infty$, analytical expressions may be obtained for the streamline function f^∞, the Alfvén number M_a^∞ and the magnitude of the flow speed V^∞,

$$\frac{f^\infty}{R^2} = -\frac{1 + k/4\lambda^2}{1 + 2\lambda^2/k} , \tag{9}$$

$$M_a^\infty = -\sqrt{2}\left(1 + \frac{2\lambda^2}{k}\right) , \quad V^\infty = -\frac{V_o}{M_o}\left(3 + \frac{4\lambda^2}{k} + \frac{k}{2\lambda^2}\right) . \tag{10}$$

In Figs. 4 we have plotted the critical solutions that satisfy all regularity conditions for a given value of λ ($\lambda = 10$) and various values of the Bernoulli constant k. One can see then that for fixed λ, the faster is the FMR (more negative is k) the more rapidly the streamlines converge to cylindrical asymptotics. The limit wherein the flow converges (very slowly) to cylindrical asymptotics with infinite M_a^∞ is obtained for $k = 0$. As $k < 0$ and becomes more and more negative, the terminal flow speed and Alfvén number, V^∞ and M_a^∞, become lower while they are reaching this value closer to the origin $R = 0$. Finally, at the limiting value $k = -2\lambda^2$, $V^\infty = M_a^\infty = 0$. This last result maybe understood in the sense that with the streamlines focusing very fast into their cylindrical shape, the flow cannot continuously accelerate in such a converging channel cross-sectional area. It is interesting to note that as the flow is focused towards the magnetic and rotational axis for $k \longrightarrow 0$, an oscillatory structure appears naturally in the geometry, making the flow resemble to a sausage with bumps and throats. These oscillations may become quite large when we come close to $k = 0$, making the density to increase locally with R and favouring the development of instabilities in the jets. In addition to the effect of collimation we see that a natural feature of FMR forming jets may be the knotty structure, as observed in jets from YSOs or AGNs.

Acknowledgement: The work reported here was supported in part by NATO grant 870221.

References

Camenzind, M.: 1989, in G. Belvedere, ed(s)., *Accretion Disks and Magnetic Fields in Astrophysics*, Kluwer Academic, 129.

Cassen, P., Shu, F., and Terebey, S.: 1986, in D. Black and M. Matthews, ed(s)., *Protostars and Planets II*, Univ. Arizona Press: Tucson, 448-492.

Cassinelli, J.P.: 1990, in L. A. Wilson and R. Stalio, ed(s)., *Angular Momentum and Mass Loss for Hot Stars*, Kluwer Academic, 135-144.

Hartmann, L., and MacGregor, K.B.: 1982, *Ap. J.* **259**, 180.

Heyvaerts, J., and Norman, C.A.: 1989, *Ap. J.* **347**, 1055.

Hughes, P.A.: 1991, *Beams and Jets in Astrophysics*, Cambridge Astrophysics Series, Cambridge University Press: Oxford.

Kato, M.: 1991, *Ap. J.* **369**, 471.

Königl, A.: 1982, *Ap. J.* **261**, 115.

Low, B.C., and Tsinganos, K.: 1986, *Ap. J.* **302**, 163.

Mundt, R., Ray, T.P., Buehrke, T., and Raga, A.C.: 1990, *Astr. Ap.* **232**, 37.

Orio, M., Trussoni, E., and Ögelman, H.: 1992, *Astr. Ap.* , (in press).

Parker, E.: 1958, *Ap. J.* **128**, 664.

Pudritz, R.E., and Norman, C.A.: 1983, *Ap. J.* **274**, 677.

Ringuelet, A.E., and Iglesias, M.E.: 1991, *Ap. J.* **369**, 463.

Ruggles, C.L.N., and Bath, G.T.: 1979, *Astr. Ap.* **80**, 97.

Sakurai, T.: 1990, *Computer Physics Reports* **12**, 247.

Tsinganos, K., and Low, B.C.: 1989, *Ap. J.* **342**, 1028.

Tsinganos, K., and Sauty, C.: 1992a, *Astr. Ap.* **255**, 405.

Tsinganos, K., and Sauty, C.: 1992b, *Astr. Ap.* **257**, 790.

Tsinganos, K., and Trussoni, E.: 1990, *Astr. Ap.* **231**, 270.

Tsinganos, K., and Trussoni, E.: 1991, *Astr. Ap.* **249**, 156.

Tsinganos, K., Trussoni, E., and Sauty, C.: 1992, in J. Brown and J. Schmelz, ed(s)., *The Sun: A Laboratory for Astrophysics*, Kluwer Academic, (in press).

Wang, Y.M., Sheeley, N.R., Jr., and Nash, A.G.: 1990, *Nature* **347, No 6292**, 726.

Weber, E.J., and Davies, L.: 1967, *Ap. J.* **148**, 217.

Zirker, J.B.: 1977, *Coronal Holes and High Speed Wind Streams*, Colorado Associated University Press: Boulder.

STEADY MHD FLOWS IN UNIFORM GRAVITY

A class of 2-D exact solutions for coronal loops

K. TSINGANOS [1], G. SURLANTZIS [1] and E. P. PRIEST [2]

[1] *Dept. of Physics, Un. of Crete, GR-71409, Heraklion, Crete, GREECE*

[2] *University of St. Andrews, Scotland, UK*

Abstract. We present a class of analytical solar coronal loop models which emerge naturally as the only solutions of the coupled transfield and Bernoulli nonlinear equations for steady MHD flows in uniform gravity. The atmosphere is vertically stratified and horizontally compressible. The topology of the solutions is controlled by a novel saddle critical point, corresponding to a new characteristic speed for MHD wave propagation in an inhomogeneous medium. We find that a special class of loop-like solutions exists only for a mildly stratified atmosphere; for very strong stratification no solutions exist while for moderate stratification only periodic solutions are found. In the case of loop-like solutions, an increase of the magnitude of the flow speed along the loop increases its height, in accordance with solar observations where some loops seem to extend over heights much higher than those predicted by static models. However, for strong field-aligned flows, no equilibrium solutions are found and the loop is possibly disrupted.

Key words: hydromagnetics – sun: magnetic fields – sun:prominences – sun:corona – stars: atmospheres

1. Introduction

High-resolution X-ray and EUV observations have revealed that the solar transition region and corona are not only radially stratified but also horizontally highly inhomogeneous, consisting of magnetized loops of various sizes and properties (Vaiana and Rosner 1978). One basic aspect of the problem posed by the observation of these loops concerns the existence of plasma flows in them. For example, Doppler shifts in transition region lines (Doschek, Feldman and Bohlin 1976) and the fact that most loops extend over heights much higher than those predicted by models without flows (Foucault 1989), indicate that flows of a few km s^{-1} magnitude may be crucial in determining their gross properties. Recently, Peres, Spadaro and Noci (1992) comparing the fitting of the intensities of some EUV transition region emission lines by loops with steady siphon flows relative to those without flows concluded that siphon flow models of compact active-region loops are in better agreement with the observations. However, previous theoretical studies of flows in magnetic loops have considered them either as *rigid* and unperturbed by the flow within the loop (Cargill and Priest 1980, Noci 1981), or have investigated siphon flows in *isolated slender* magnetic flux tubes (Thomas 1988, Degenhardt 1989, Thomas and Montesinos 1990, 1991). The first approximation is certainly valid only for loops whose plasma β is small, while in the second the tube is assumed to be a one-dimensional structure with uniform gas pressure inside and force balance across the magnetic lines. In this article we study a more general family of exact MHD solutions and discuss their relevance to solar coronal loop models.

2. A Class of Exact MHD Solutions with Flows in Uniform Gravity

Consider a compressible plasma structure in dynamical equilibrium in an isothermal atmosphere with a uniform sound speed V_s. Assume that the plasma is compressible

J.F. Linsky and S. Serio (eds.), Physics of Solar and Stellar Coronae, 623–627.

in the horizontal and vertical directions, in a uniform external gravitational field $-g\hat{z}$, and all physical variables depend on the horizontal distance x and the vertical height z in the orthogonal system xyz with z pointing upwards. In particular, search for equilibria where the z-dependence of all physical quantities is exponential, i.e.,

$$\tilde{\rho}(x,z) = \rho(x)e^{-\xi z/H}\,, \quad \tilde{A}(x,z) = A(x)e^{-\xi z/2H}\,, \tag{1}$$

where H is the constant scale height, $H = V_s^2/g$ and $A(x,z)$ the magnetic flux function defining (B_z, B_x). In terms of the total sonic and Alfven Mach numbers, $M(X)$ and $M_a(X)$, respectively, with $X = x/x_o$, force balance across and along the planar streamlines gives then a system of two coupled nonlinear equations,

$$\frac{2}{M_a^2}\frac{\mathrm{d}M_a^2}{\mathrm{d}M^2} = \frac{\xi(M^2/2 + M_a^2) - 1}{[\xi(M^2/2 + M_a^2) - 1] - [M^2 + M_a^2 - M_a^6/\lambda^2 - 1]}\,, \tag{2}$$

$$\pm\frac{2}{M_a^2}\frac{\mathrm{d}M^2}{\mathrm{d}X} = \frac{[\xi(M^2/2 + M_a^2) - 1] - [M^2 + M_a^2 - M_a^6/\lambda^2 - 1]\sqrt{\lambda^2 M^2 - M_a^4}}{(M^2 + M_a^2 - M_a^6/\lambda^2 - 1)} \tag{3}$$

where λ is a dimensionless constant, representing the relative strength of fluid and magnetic effects. For example, small values of λ correspond to magnetically dominated states while large values of λ to dynamically dominated states.

The topology of the solution of Eqs.(2-3) is controlled by a saddle critical point where the numerator and denominator of Eqs.(2-3) vanishes simultaneously. The nature of this critical point is related to the vertical stratification. Also, the flow speed at this critical point seems to be a generalisation of the known *tube-speed*, V_T, encountered in wave propagation in slender flux tubes (Tsinganos *et al* 1992). The critical point is not neccessarily however at the highest point of the tube, as is the case in slender flux tube models with flows (Thomas 1988, Degenhart 1989, Thomas and Montesinos, 1990,1991). Instead, this critical point seems to be similar to an analogous critical point that emerges from analytical solutions of 2-D MHD astrophysical outflows (winds and jets) (Tsinganos and Trussoni, 1991, Tsinganos and Sauty, 1992). In both cases it appears when the two coupled partial differential equations that govern symmetric MHD equilibria – the transfield equation and the equation for force balance along the magnetic and stream lines – are properly solved. The critical point does not appear in this case at the *fast* and *slow* MHD wave speeds, but at a characteristic speed which evidently is controlled by the elasticity of the fieldlines to changes in the flow magnitude.

In Fig. 1 we show a typical topology of the solutions of Eqs. (2-3) in the planes (M_a^2, M^2) and (M, X) for the characteristic value of the parameter $\xi = 1$ such that the magnetic field of the same configuration without flows is force-free. Several groups of curves appear in these planes corresponding to several classes of physical solutions. However, out of this large set of mathematical solutions of Eq. (2-3), only some classes are physically interesting. To see that, we have indicated on the plane (M_a^2, M^2) the shaded area which is bounded by the curve $M^2\lambda^2 - M_a^4 = 0$ that determines the allowed part of the plane (M_a^2, M^2) where $M^2\lambda^2 - M_a^4 \geq 0$ [*c.f.* Eq. (3)]. Note that this curve also determines the locations where the field lines

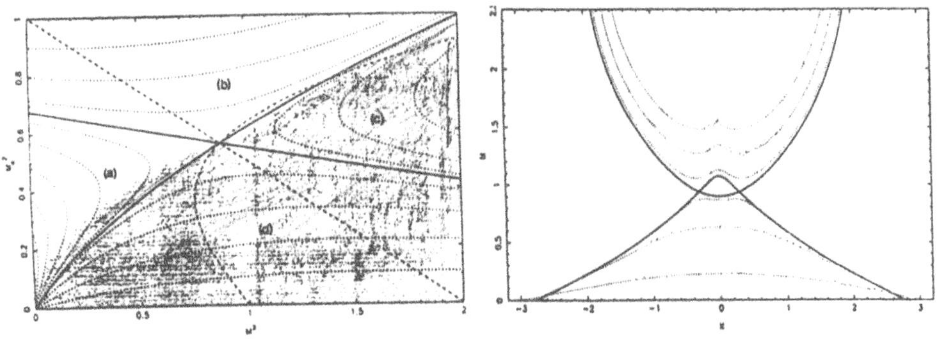

Fig. 1. Topology of the solutions (left) in the $(M_a^2 - M^2)$ plane and (right) the $(M - X)$ plane for $\xi = 1$, $\lambda = 0.32$. The shaded area corresponds to allowed solutions.

are horizontal. Thus, solutions which intersect this curve where $B_z = 0$ may have a valley, or a summit there. Branches with only one such intersection may correspond to loop-like solutions with one summit, or, prominence-like solutions with one valley only, while those with two intersections may correspond to periodic solutions. Two critical branches (solid lines) may be also seen and the rest of the branches which are adjacent to those critical ones may be classified in four groups labeled (a) through (d). Note that along a streamline $\tilde{A}(x,z) = constant$ the density $\tilde{\rho}(x,z)$ is inversely proportional to the square of the Alfven Mach number M_a. Obviously, the density at the summits (valleys) is minimum (maximum) otherwise, the solution would be Raleigh-Taylor unstable in the gravitational field.

With these considerations in mind, we can see that only solutions of group (a) (subcritical solutions) with a decreasing $M_a(X)$ and $M(X)$ and increasing density as we move from the loop summit at $X = 0$ towards its legs (*c.f.* Fig. 2) and (c) (supercritical solutions) with a decreasing $M_a(X)$ and increasing $M(X)$ and density as we move from the loop summit at $X = 0$ towards its legs, together with the two critical solutions are physically acceptable. Of those solutions, however, only branches (a) which correspond to subsonic flows may be relevant to solar applications rather than the supersonic flows of branches (c).

3. Physical Implications

In a previous study Tsinganos and Surlantzis (1992) we searched for 1-D MHD steady states in a uniform solar gravitational field considering variation of the physical quantities only with the horizontal coordinate X. Only periodic solutions were found including valleys and summits. The new element introduced in the present treatment is stratification in the vertical distance Z and therefore dependence of all physical quantities on both Z and X. As a result, a new feature has emerged naturally from this 2-D analysis, namely the existence of loop-like solutions not encountered in the previous unstratified treatments. Nevertheless, in the limit of no stratification ($\xi \longrightarrow 0$) we recover the models of the unstratified state. All these

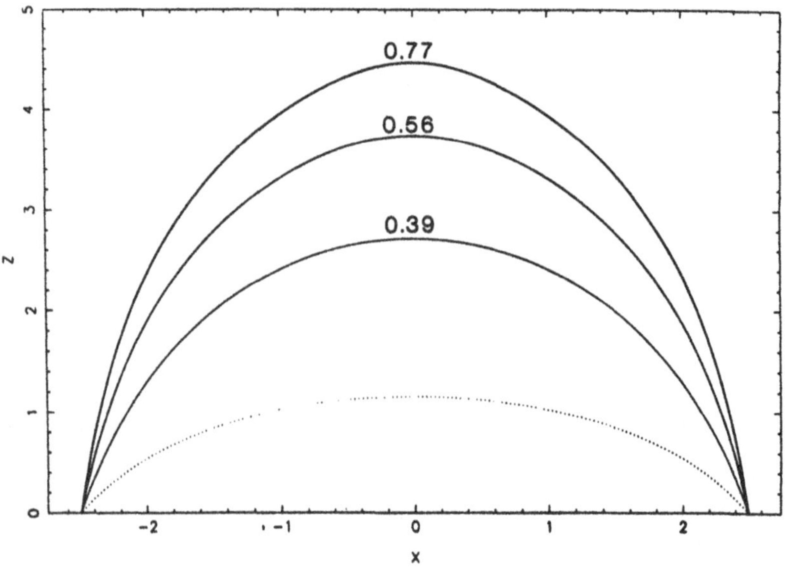

Fig. 2. Change of fieldline shape as value of Alfven number increases at top of loop (progressively thicker solid lines) for $\xi = 1$. Dotted line corresponds to static loop without flows.

loops have similar properties to those corresponding to the characteristic value of $\xi = 1$, a case giving force-free magnetostatic loops in the limit of no flows. In the following, we summarize the main features of our analysis.

(1) The more interesting solutions are those where the flow along the loop is everywhere sub-Alfvénic. The flow speed is also subsonic along most of the loop reaching the sonic transition, and sometimes slightly overcoming it at the top of the loop.

(2) For $\xi = 1$, as the magnitude of the flow at the loop top increases, the loops respond by becoming more curved while the separation of their footpoints decreases and it is smaller compared with the static force-free case. This property may explain the observational fact that most loops extend over heights much larger than those predicted by static models without flows (Foucault 1989).

(3) There is a limit on the magnitude of the flow speed at the loop top beyond which steady solutions do not exist. We conjecture that the loops disrupt when the flow speed exceeds this limit.

(4) For nonsymmetric loops there is the possibility of a shock transition that would connect the $M > 1$ branches with the $M < 1$ branches. The existence and location of this shock along the loop depends of course on the pressure difference between the two symmetric loop footpoints.

(5) The density is a minimum at the top of the loop and increases as we move down in the gravitational field toward the footpoints. Also, the density is a minimum

horizontally at the summit of the arcade.

(6) It is interesting that loop-like solutions exist only for the restricted case of the mildly stratified atmosphere, $1 \leq \xi \leq 2$. This conclusion which is also in agreement with the analysis of MHD wave propagation in a stratified medium, may suggest, for example, that in high gravity stars whose atmosphere is rather highly stratified, loops may not exist.

Acknowledgement: The work reported here was supported in part by NATO grant 870221.

References

Cargill, P.J., Priest, E.R: 1980, *Solar Physics*, **65**, 251
Degenhart, D.: 1989, *Astron. Astrophys.*, **222**, 297
Doschek, G.A., Feldman, U., Bohlin, J.D.: 1976, *Astrophys. J. Letts.* **205**, L177
Foukal, P.: 1989, *Astrophys. J.* **210**, 575
Noci, G.: 1981, *Solar Phys.*, **69**, 63
Parker, E.N.: 1979, *Cosmical Magnetic Fields*, Clarendon Press: Oxford.
Peres, G. Spadaro D. Noci, G.: 1992, *Ap. J. 389*, 777
Thomas, J.H.: 1988, *Astrophys. J.* **333**, 407
Thomas, J.H. Montesinos, B: 1990, *Astrophys. J.* **359**, 550
Thomas, J.H. Montesinos, B: 1991, *Astrophys. J.* **375**, 401
Tsinganos, K.: 1982, *Astrophys. J.* **252**, 775
Tsinganos, K., Trussoni, E.: 1991, *Astron. Astrophys.* **249**, 156.
Tsinganos, K., Sauty, C.: 1992, *Astron. Astrophys.* **257**, 790.
Tsinganos, K., Surlantzis G.: 1992, *Astron. Astrophys.* **259**, 585.
Tsinganos, K., Surlantzis G., Priest E. R.: 1992, *Astron. Astrophys.*, (submitted).
Vaiana, G.S. Rosner, R.: 1978, *ARA& A* **16**, 393.

MHD EQUILIBRIA IN UNIFORM GRAVITY

A general 2-D class of low-β loop solutions

G. SURLANTZIS,[1] P. DÉMOULIN,[2] J. HEYVAERTS[2] and C. SAUTY[2]
[1] *Dep. of Physics, Un. of Crete, GR-71409, Heraklion, Crete, Greece*
[2] *Observatoire de Meudon, F-92195 Meudon Cedex, France*

Abstract. We present a general, two-dimensional, class of low-β steady MHD flow solutions. The method could be used for any type of boundary conditions, not only the one discussed here. In the case of coronal loops, when the distribution of pressure at the footpoints is symmetrical the flow is subsonic. Otherwise it becomes supersonic at the summit and then passes through a shock. In this case the plasma flow is driven by the pressure difference between the two footpoints of the loop and the loop is now asymmetrical. Loops have a significant density contrast against their environment only if their energy flux differs significantly from the background one.

Key words: MHD equilibria – Coronal loops – MHD flow

1. Introduction

It is by now well known that the active region plasma is dynamic, with continual activity in the form of a wide range of flows, with very small Alfvén Mach numbers. At high temperatures the *solar active regions* are seen to consist of *active-region loops* with almost constant temperature, for a wide range of densities. The magnetic field of an active region probably evolves slowly through a series of essentially stationary, usually force-free states. As far as motion normal to the magnetic field is concerned, the plasma is completely dominated by the field, since the plasma β is much less than unity. But, along the field, the plasma is observed to be in continuous motion rather than in static state (Priest 1981).

In this paper we study siphon flows in an isothermal magnetized atmosphere, taking into account the feedback of the flow on the magnetic structure, and looking for as wide a class of boundary conditions as possible. Since however, the solar situation is one of a small β plasma, we take advantage of the fact that the flow is subAlfvénic but not necessarily subsonic, to simplify the search for such solutions.

2. General Formalism

Because in coronal loops we have both the plasma β, and the Alfven-Mach number much less than unity, we assume that in the absence of flow we have a force-free field $\mathbf{B}_0$, and the velocity, pressure and density having little influence on this force-free structure.Thus it is equivalent to say that the force-free structure is some sort of an empty container in which pressure forces, gravity forces and inertial forces are perturbed weakly. We then expand the relevant equations to first order in β. The magnetic field can be written as

$$\mathbf{B} = \mathbf{B}_0 + \mathbf{B}_1 , \tag{2.1}$$

where $\mathbf{B}_1$ is meant to be a small perturbation of order β as compared to $\mathbf{B}_0$. We then linearize the MHD equations with respect to the magnetic field, but not with respect to flow variables, since we want to allow the field-aligned flow to be even

J.F. Linsky and S. Serio (eds.), Physics of Solar and Stellar Coronae, 629–632.

strongly nonlinear. This is again a valid approach for $\beta << 1$ and $\rho V^2 \approx \beta(B_0^2/8\pi)$ or, equivalently $M_A^2 << 1$. An expansion to first order in the small β parameter gives

$$\nabla \cdot \mathbf{B}_1 = 0\,, \tag{2.2a}$$

$$\nabla \cdot (\rho \mathbf{V}) = 0\,, \tag{2.2b}$$

$$\nabla \times (\mathbf{V} \times \mathbf{B}_0) = 0\,, \tag{2.2c}$$

$$\rho \mathbf{V} \cdot \nabla \mathbf{V} = -\nabla P + \frac{(\nabla \times \mathbf{B}_0) \times \mathbf{B}_1}{4\pi} + \frac{(\nabla \times \mathbf{B}_1) \times \mathbf{B}_0}{4\pi} - \rho g \hat{\mathbf{z}} \tag{2.2d}$$

We adopt an energy equation of the form

$$P = P(\rho)\,. \tag{2.2e}$$

Equations (2.2) constitute a closed system for P, ρ, $\mathbf{V}$ and $\mathbf{B_1}$. We stress again that the velocities are not so small as to ignore the left-hand side term in Eq. (2.2d). We still have the plasma β, and the Alvénic Mach number much less than unity.

The coordinate system is yzx wherein coordinate x is ignorable. The general solution of Eqs. (2.2a,b) can be written in terms of a vector potential with a x-component $A_1(y, z)$ and a stream function $\Psi(y, z)$. Following the formalism of Tsinganos (1982), for field aligned flows we find

$$4\pi\rho\mathbf{V} = \Psi_{A_0}\mathbf{B}_0\,, \tag{2.3}$$

where the x-component of the vector potential $A_0(y, z)$ refers to the force-free structure. The subscript A_0 denotes derivation.

From the x-component of the momentum balance equation we get

$$B_x - \Psi_{A_0} V_x = \frac{dB_{0_x}(A_0)}{dA_0} A_1 + b_{1_x}(A_0)\,, \tag{2.4}$$

where $b_{1_x}(A_0)$ is a function of A_0. Note that the force-free contition yields that B_{0_x} is also a function of A_0 (Priest 1981).

Momentum balance equation along stream-lines gives the Bernoulli equation for each stream line

$$\int \frac{dP}{\rho} + U + \frac{V^2}{2} = F(A_0)\,, \tag{2.5}$$

where $F(A_0)$ is another function of A_0.

Now, momentum balance across stream-lines is given by the transfield equation

$$\nabla^2 A_1 + \frac{d^2 {B^2}^2_{0_x}}{dA_0^2} A_1 - \Psi_{A_0} \left[\frac{\partial}{\partial x_1} \frac{\Psi_{A_0}}{4\pi\rho} \frac{\partial A_0}{\partial x_1} + \frac{\partial}{\partial x_2} \frac{\Psi_{A_0}}{4\pi\rho} \frac{\partial A_0}{\partial x_2} \right]$$

$$+ \frac{d(B_{0_x} b_{1_x})}{dA_0} + \frac{1}{8\pi\rho} \frac{d(B_{0_x}^2 \Psi_{A_0}^2)}{dA_0} + 4\pi\rho \frac{dF}{dA_0} = 0\,. \tag{2.6}$$

Equations (2.5) and (2.6) describe both the field-aligned flow and the petrubed magnetic structure respectively. A number of the original equations has been absorbed into the first integrals represented by Ψ (which expresses mass conservation),

B_{0_x} and b_{1_x} (which refers from the force balance equation in the ignorable coordinate), and F which expresses energy conservation. These functions must be defined by consideration of the boundary conditions. First we solve the Bernoulli equation obtaining the density $\rho(y,z)$ and then we solve the linear equation in $A_1(y,z)$.

3. Loop-like Solutions

Let us take the simplest current-free field as the unperturbed configuration

$$A_0 = A_{00} cos\left(\frac{\xi y}{H}\right) e^{-\frac{\xi z}{H}}, \tag{3.1}$$

This is defined in the strip $|y| \leq \pi H/2\xi$, $z \geq 0$ and **H** is the scale height ($H = V_s^2/g$), and ξ is a parameter. We have assumed an isothermal atmosphere, $P = V_s^2 \rho$. Thus, Eq. (2.6) reduces to a Poisson equation with homogeneous Dirichlet boundary conditions (because the flux distribution is not altered at the boundary).

Before we solve eqs.(2.6) it is useful to focus our attention on the Bernoulli equation which can be written for a given field line ($A_0 = const.$) in terms of the Mach number of the flow

$$-lnM + \frac{M^2}{2} + \left(\frac{1}{\xi} - 1\right) ln(\cos\frac{\xi y}{H}) = constant\,. \tag{3.2}$$

This equation is identical with the Bernoulli equation in ordinary pipes if the effective cross-section were to vary with Y as $S(Y) = cos^{1-\frac{1}{\xi}}(\xi y/H)$. The term 1 in the power of $cos(\xi y/H)$ is due to the actual flux tube geometry, while the term $-1/\xi$ is due to the gravity. For $\xi < 1$ the critical point is of the X-type (the effective cross section has a minimum) and for $\xi > 1$ it is of the O-type (the effective cross section has a maximum). If $\xi = 1$ the effective cross section is constant along each field line. Thus, for $\xi > 1$ (loops with lengths smaller than the scale height H) only subsonic solutions are possible, if boundary conditions forbid supersonic flows at the base of the loop, which we assume. But for $\xi < 1$ (loops with lengths greater than the scale height) either subsonic or transonic solutions with shocks are allowed. When $\xi = 1$ obviously the velocity is constant throughout the field lines.

Because of the symmetry of the problem the pressure at both footpoints of the same field line are always equal. Only in the cases of transonic solutions are the pressures unequal (the plasma flows from high pressure to the lower pressure footpoint) (Surlantzis *et al.* 1992).

In order to solve Eq. (2.6) we must prescribe the functional forms of $\Psi_{a_0}(a_0)$ and $F(a_0)$. These functional forms correspond to the boundary conditions at the photospheric level ($z = 0$) of the system of Eqs. (2.3) for the density and the velocity. In the subsonic case, because of the symmetry of the problem, the quantities, density, velocity and magnetic field B_0 at the level $z = 0$, are functions of A_0 alone. Regarding Eq. (2.3) at $z = 0$ we see that, for given Ψ_{A_0}, if we give the boundary distribution of the density (or pressure) for example, the velocity distribution is determined automatically. In the transonic cases with shocks, because we have another free function, which is the post-shock temperature, or equivalently the dis-

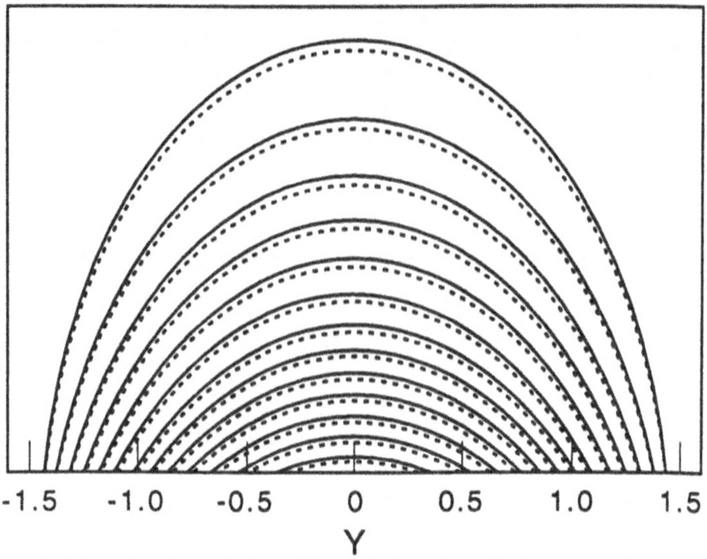

Fig. 1. A typical low β subsonic loop-like solution, in which energy flux is mush greater than its uniform enviroment.

tribution of pressure at $z = 0$, we must also give the velocity distribution. These cases will not be considered here (see Surlantzis *et al.* 1992).

Having in mind that we want to construct solutions for coronal loops the function $\Psi(A_0)$ must be such that flows are not present on all field lines. That means $\Psi_{A_0}(A_0) \neq 0$ if $A_{min} < A_0 < A_{max}$ and $\Psi_{A_0}(A_0) = 0$ otherwise. So, we have flows only in the strip between the magnetic field lines A_{min}, A_{max}. The only requirement is that Ψ_{a_0} and its derivative, must be continuous functions of A_0. In order to have loop-like solutions the flux energy $F(A_0)$ inside the loop must be very different than its enviroment. The final field lines are pushed in because the gradient of pressure dominates on centrifugal forces which point outwards (Fig. 1).

4. Conclusion.

We have constructed two-dimentional low-β isothermal loop solutions and discussed their relevance to coronal loops. We have solved the MHD equations in the low-β limit (assuming that the magnetic structure is approximately force-free) independently of the boundary conditions. Coronal loops are symmetrical when the flow is subsonic and there is no pressure difference between the two footpoints. As soon as a pressure difference is maintained, the flow becomes supersonic at the loop-summit and then decelerates through a shock wave (Surlantzis *et al.* 1992). At this point the loop becomes asymmetrical.

References

Priest, E.: 1981, *Solar Magnetohydrodynamics*, D. Reidel, Dordrecht, Holland.
Tsinganos, K.: 1982, *Astrophys. J.*, **252**, 775
Surlantzis, G., Démoulin P., Heyvaerts J., Sauty C.: 1992, *Astron. Astrophys.*, (submitted)

DISCUSSION ON THEORETICAL DEVELOPMENTS

Discussion following paper by E. Priest

L. Golub: As you said, the X-ray bright points do seem to be associated in large part with locations of opposite polarity magnetic flux which comes together and disappears. However, merely having mixed polarity fields on the surface is not enough. The first problem is that the active region magnetic flux seems to have a $1/e$ decay time of about 10 days, according to Jack Harvey. It is hard to see how this flux can remain until solar minimum. Second, we examined locations on the surface at which magnetic flux from two different active regions with opposite polarities was diffusing into the same area. We found no excess of bright points in such regions. I conclude that some additional factor is missing.

E. Priest: Yes, I agree, we are only just beginning to understand bright point production and the picture of active regions emerging and then spreading to produce all the flux that we see is much too simplistic. Probably, there is continual emergence, cancellation and submergence of flux over the whole solar surface in different ways, and we do not know how the global magnetic flux balance is maintained. It is important to determine what large-scale factors are involved in the production and location of cancelling magnetic features and of bright points. Why do they appear to be uniformly distributed at solar minimum but to have peaks near $\pm 30^\circ$ and $\pm 50^\circ$ at solar maximum? When oppositely directed flux approaches, what determines how much is reconnected and how much heat is released?

A. Maggio: New observational evidence has been presented that nonflaring solar coronal loops have constant cross sections. It has been suggested that this may be due to the magnetic field being twisted somehow along the loop. Can you comment on the possibility that such a magnetic field configuration occurs?

E. Priest: The structure of coronal loops is an important problem, and I feel that a careful observational study is needed to answer the question you pose, including a determination of the photospheric magnetic pattern and a comparison between resulting potential and force-free magnetic fields with observed loop structures. In principle, a potential or force-free field could produce loops with either constant or non-constant cross section depending on the photospheric sources. For example, a point source and a point sink or a dipole source would produce strongly diverging field lines, but a normal photospheric field falling off as r^{-1} from some axis would produce a simple cylindrical arcade with loops having constant cross sections. Thus a constant cross section does not necessarily imply loops with twist. Also, I would not expect the twist to be too great or otherwise the loop would be kink unstable.

B. Somov: At what height does reconnection in bright points occur?

E. Priest: If the field is roughly potential, the reconnection height would be roughly equal to the separation distance between the centers of the magnetic fragments. Initially, this would be at 10–20 Mm, in the lower corona, but as the fragments become closer the reconnection height should decrease.

N. Weiss: Kinematic modeling of photospheric motion by Simon, Title and myself

J.F. Linsky and S. Serio (eds.), Physics of Solar and Stellar Coronae, 633–639.

indicates that a linear magnetic network can only be maintained if flux continually emerges within supergranules and is dissipated in the network. Is this compatible with a model in which X-ray bright points are identified with sites of reconnection?
E. Priest: Yes, certainly. Most of the observed flux is located at the supergranule boundaries and most of the cancelling magnetic features and dark points are indeed located in the network. It would be useful to have more detailed collaborative programs of soft X-ray observations of bright points and high-resolution video magnetograms to build on the excellent results of Karen Harvey and Sara Martin.
Y. Uchida: We find clear evidence in the *Yohkoh* data for magnetic reconnections. There are examples in which the structures are either heated or not heated. This suggests that reconnections can occur either with or without energy release, corresponding to whether the reconnecting region is strongly stressed or not.
C. Chiuderi: I would like to caution people against relying too much on results based on normal mode analysis of linearized equations. We have recently run a numerical simulation (Malara et al., Ap.J., in press) of the time evolution of a 2–D incompressible plasma and have not been able to recognize the normal modes in the results. It seems that the available energy is dissipated before one reaches the time asymptotic normal mode behavior.
G. Field: You have said, "if there is no reconnection, how are we to release magnetic energy?" There may be a way other than reconnection, which may or may not be applicable here. If the magnetic configuration generated by footpoint motion is not a global minimum in the energy, a finite amplitude instability can carry the field from the local minimum to the global minimum with the release of a finite amount of magnetic energy. In ideal MHD, this energy must go into motion of the plasma. When $\beta \ll 1$, as in the corona, the motions are supersonic and shock waves are generated. Shock waves dissipate kinetic energy into heat, so this might be an additional heating mechanism. Do you have any comments on this?
E. Priest: Yes, I agree this seems a good way in principle to convert free magnetic energy into heat by viscous or resistive dissipation via motions. I would like to see specific examples worked out in order to determine how this works in practice.

Discussion following paper by P. Ulmschneider

A. Maggio: For a star of given mass evolving from the main sequence to the giant branch, how does the amount of acoustic heating and the heating scale height change during evolution?
P. Ulmschneider: The acoustic surface flux F_M, computed by Bohn (Ph.D. thesis 1981) as a function of $T_{\rm eff}$ and gravity, is decreased by radiation damping in the photosphere. dF_M/dx is roughly given by the derivative of Eq. (2).
J. Schmitt: How does one find stars without any magnetic flux?
P. Ulmschneider: One looks for basal flux stars which also have very little or no Si IV, C IV and X-ray emission.
J. Linsky: I agree with your conclusion that pure acoustic heating can explain the observed basal emission of chromospheric lines. I would like to know whether the acoustic heating theory also predicts basal emission rates for transition region lines.

Have you made such calculations? Are the basal heating rates at a level that can be detected by HST observations of C IV from very inactive stars?

P. Ulmschneider: Equation (2) shows that the heating is proportional to the gas pressure. The cooling is proportional to the square of the density. Since the density decreases with height, acoustic heating will always overwhelm the cooling and result in an acoustically generated transition layer and even a corona. However, as heat also goes into wind energy, the question is what maximum temperature can one achieve. Except for late-type giants, I believe there is basal level emission in C IV. The height where the pure acoustic transition layer occurs may be large, and thus the number of photons small. We have not yet calculated this.

Discussion following paper by N. Weiss

G. Field: I've never understood the relationship between the theory you described, which is formulated in terms of ensemble-averaged fields, to the fields observed on the Sun, which are organized into flux tubes with fields thousands of times stronger than the mean. Can you explain this to me?

N. Weiss: Mean field electrodynamics can be justified when there is a separation of scales in space or time. In principle, one could average over a strong, intermittent small-scale field to leave a weak, smoothly varying large-scale field. In solar-type stars there is no such separation of scales, so one cannot average over large isolated flux tubes in the convection zone. On the other hand, mean field dynamo theory can provide a helpful qualitative description of more smoothly varying fields in a magnetic layer below the base of the convection zone.

A. Maggio: Do any of the dynamo models predict a threshold rotational velocity, below which the dynamo action ceases abruptly? As a star steadily slows down during its evolution, do you expect that the dynamo also fades slowly, or is there a critical velocity below which the dynamo turns off, even when substantial convective motions still exist?

N. Weiss: Most dynamo models predict a critical value of the dynamo number D (proportional to Ω^2 in the simplest case) below which dynamo action ceases. The strength of the field should typically increase monotonically as D is increased above this value. As a main sequence star spins down, the dynamo number decreases and the field fades (despite the presence of convection). It seems, however, that the rate of spindown itself diminishes, so that the star never quite reaches the critical point. (In post-main sequence evolution D may indeed fall below the critical value.)

P. Byrne: You said that, in solar-type stars, there is a boundary dynamo creating a global field which undergoes cyclic behavior. The fibril dynamo, on the other hand, is chaotic. In the latest-type (late M) main-sequence stars, the stars become fully convective so that only the fibril-type dynamo can operate. Therefore, we only expect periodic cycles in magnetic activity down to early-to-mid M. In fact, cyclic behavior is found only down to early-to-mid K. Why is this?

N. Weiss: I would only expect to find a shell dynamo in stars like the Sun, which are sufficiently slowly rotating for there to be a radial gradient in angular velocity at the base of the convection zone. Perhaps these are the only stars that exhibit cyclic activity. In rapid rotators, or stars with very deep convection zones, things will be different. Perhaps they all have fibril dynamos. At present, we have to be guided by the observations.

E. Antonucci: There is observational evidence for rapid rotation of the solar photospheric magnetic fields and for different rotation rates, in the two hemispheres, as found from an analysis of the Stanford magnetic data performed over one solar cycle. These observations confirm the rapid rotation found for the large-scale structures of the green line (Fe XIV) corona and coronal holes and add the surprising feature of a north-south asymmetry in rotation. How does this observational evidence change the solar dynamo models?

N. Weiss: Large-scale magnetic fields are not found to rotate with the plasma at any level. The convection pattern may rotate as a wave relative to the plasma and nonaxisymmetric field components may drift relative to the convection pattern. Thus there is no reason why magnetic fields should not rotate rapidly, as is observed. Breaking of symmetry about the equatorial plane would lead to different rotation rates in the two hemispheres.

R. Rosner: To my knowledge there is little serious work on how convection "works" in the cores of fully-convective stars such as the dM stars later than $\sim$ dM5; and even less relevant work concerning magnetoconvection. So I suspect that we are at a serious disadvantage as far as even "cartoon models" of dynamos in such stars.

N. Weiss: Clearly such calculations would be valuable and are helped by the large scale-heights and small pressure fluctuations.

B. Foing: How does a close companion in a binary system affect the differential rotation, convection and magnetic dynamo, and introduce preferential longitudes?

N. Weiss: I do not know of any detailed models of dynamo action in close binaries, but I expect that rotation and magnetic fields are stabilized by tidal interactions. There may be strong fields that do not show much variation.

Discussion following paper by G. Field

J. Trümper: Recent results obtained from the ROSAT deep survey (147,000 sec PSPC exposure at the Lockman Hole) with a flux limit of 2×10^{-15} ergs cm^{-2} s^{-1} (10 times deeper than *Einstein*) reveal that $\approx$ 56% of the total flux is resolved into discrete sources (mostly QSO's). Since the log$N$-log$S$ relation is continuing to rise with a (integral) slope of 0.8, the total contribution of discrete sources will be $\approx$ 80% or even higher.

G. Field: Our model of the X-ray background (Rogers & Field, Ap.J. **370**, L57 (1991); **378**, L17 (1991)) requires that virtually all of it be due to AGNs of moderate luminosity such as Seyferts. If by QSOs you mean primarily moderate luminosity objects, your findings seem to be in agreement with our model. However, more luminous AGNs ($L \geq 10^{44}$ ergs s^{-1}) might have very different spectra than we calculate, both because electron-positron pairs become optically thick and because relativistic jets, which are seen in energetic QSOs, contribute a completely different spectrum. The key to which dominates the X-ray background is the luminosity function of AGNs. Our model applies if there are enough moderate-luminosity ob-

jects to overwhelm the less numerous bright ones.
J. Trümper *Einstein*, EXOSAT and ROSAT data show that the X-ray spectra of AGN's (except BL LAC's) steepen in the 0.1 - 1 keV range ("soft excess"), with no indication of the spectral notch predicted by your model. Could you broaden the thermal emission bump or shift it to higher energies in order to fit the data?
G. Field: Our model does not predict soft excess in the 0.1 - 1 keV range. The spectra I presented were calculated for a distance of 24 gravitational radii from the black hole, inside of which half of the luminosity originates: here the temperature reaches 23,800 K. While it is true that the disk reaches a maximum temperature of 39,000 K somewhat further in, that is still far too low a temperature to produce soft X-rays. Within the context of the model, we should investigate whether a transition layer between the cool disk and the corona, heated by energetic ions precipitating from the corona, could account for the excess. Alternatively, one should consider the model advocated by Zdziarski et al. (Ap.J. **363**, L1 (1990)), in which excess soft X-rays arise naturally in a non-relativistic pair plasma.
R. Rosner: To complicate your model a bit more: unless the loop footpoints are at exactly the same disk radius, the loop structure must "slip" across the polar axis, because $\partial\Omega/\partial R \neq 0$. Thus, for a sparse set of loops, it may be that the dominant loop interactions occur during this "polar crossing," as two or more loops cross, and at their loop tops. One other point is that I think there is yet another spectral component, namely Bremsstrahlung from the electrons hitting the disk/loop footpoints (the visibility of that component may depend on the disk-observer orientation).
G. Field: Both points are good ones. On the first point, we assumed – as you guessed – that both footpoints are at the same radius. We find that for loops at $R = 1/2$, the time scale for loops to be destroyed by reconnection with loops at neighboring radii, is shorter for reasonable parameters than the time it would take for the loop footpoints to separate by 180° in azimuth, so that polar crossing would occur. The factor is not a large one, however, and I agree that loop crossings should be investigated. It is true that in our model Bremsstrahlung could occur from those electrons that get all the way down the field lines to the footpoints. Many do not make it, because they lose energy by synchrotron and inverse Compton before reaching the footpoints, but others – particularly those of low energy – do. We have assumed that accelerated ions are thermalized deep in the disk, but as you point out, the ions as well as the electrons may heat a thin region which emits thermal Bremsstrahlung at a temperature far above the thermal peak at ~ 10 eV.

Discussion following paper by B. Montesinos et al.

R. Rosner: Could you explain a bit more what you meant by your conclusion that you would like to understand your scaling laws? Does that mean you actually view the arguments you gave in support of your scaling as inadequate?
B. Montesinos: I did not mean that. The arguments we gave based on the minimum energy loss hypothesis are quite solid and can account for the observed trends shown in our paper. What I really meant is that we should understand the dependences of chromospheric and coronal parameters on the Rossby number, *Ro*.

S. Drake: The values of the coronal magnetic fields that you inferred for your sample ($\sim$ 10 G) are much smaller than the typical values of several hundred gauss that are inferred (by modeling the radio emission) to be present in the corona above solar active regions. What is the reason for this?
B. Montesinos: The magnetic field B_c we compute from the coronal emission measure, the coronal temperature and the stellar gravity is an average for the whole corona, so certainly there must be regions with a stronger-than-average field. What I do not know is whether this field can be as strong as several hundred gauss. Also, I am not sure where the magnetic field inferred from radio emission is placed.
J. Schmitt: How does an extreme case like Altair ($T_c \sim 1-2\times10^6$ K, $v \sin i = 180$ km s^{-1}, spectral type A7 V) fit your scaling relations?
B. Montesinos: Let us assign to Altair typical parameters for an A7 V star: $R = 1.6R_\odot$, $\log g_* = 4.32$ and $B-V = 0.2$. The radius and the $v \sin i$ value imply a rotation period $P_{\rm rot} \simeq 0.45$ days. To compute the Rossby number, $Ro = P_{\rm rot}/\tau$, we could extrapolate the curve given in Figure 5 of Noyes *et al.* (1984), but since we do not know the precise behaviour of τ below $B-V = 0.4$ we can take an upper limit for the turnover time $\log\tau < 0.2$, with τ in days. This implies $Ro > 0.28$. Using the above gravity and expression (17) in the paper one obtains $\log T_c \leq 6.82$, whereas the value you gave to me ranges from 6.0 to 6.3. To match $\log T_c = 6.3$ from our scaling, we require $Ro \simeq 1.66$ which would imply $\log\tau \simeq -0.57$.

Discussion following paper by G. Halberstadt

L. Golub: Could you connect your results with our NIXT observations? In the five minutes of a rocket flight we see several loops brightening, with the heating first appearing along a helical structure. This pattern is transient and within about 100 – 200 sec it evolves into a continuous "standard" loop showing no twist. Is this the type of heating pattern that you expect to see due to resonant absorption?
G. Halberstadt: Although the azimuthal field B_θ does not appear in the expression for the Alfvén continuum in a line-tied loop, the eigenfunctions certainly are determined by B_θ: the ballooning structure is entirely due to it. Hence in a line-tied loop the continuum modes are helical, and we expect a helical heating pattern. At present our model does not exhibit a transition to non-helical modes.
C. Chiuderi: Comparing the two expressions for the combining of Alfvén frequencies, we see that the line-tied case is simply obtained from the unconstrained case by putting $m = 0$. This fact had already been emphasized long ago by Mok and Van Hoven in a paper on line-tied reconnections. Would you comment on this?
G. Halberstadt: The similarity between the line-tied continua and the unconstrained continuum for $m = 0$ is only present in the expressions for the continuum frequencies and not for the eigenfunctions. These two cases are not at all similar. The paper by Mok and Van Hoven does not treat the resonant heating of line-tied plasmas or the effect of line tying on the continuous Alfvén spectrum.

Discussion following paper by R. Hammer

A. Maggio: When you consider chromospheric and transition region lines which form at increasing heights in the atmosphere, the power law correlations with the

X-ray emission have slopes which also increase. Can you comment on this?
R. Hammer: I have restricted myself to the relationship between the transition region and corona, because these two regions are largely coupled together by thermal conduction, so that we can hope to understand their relationship more easily. The chromosphere adds another level of complexity. Here we must also try to understand the location of the transition region along a given magnetic field line, the different geometric constraints, and perhaps different heating mechanisms in the chromosphere and corona. But I agree that even for different transition region lines the power law correlations with the X-ray emission are not exactly constant. I presume that this is due to differences in the importance of the various effects that contribute to the nonlinearity. For example, emission from "cool" low-lying loops might contribute significantly to emission formed at a few times 10^4 K, while it is unimportant for lines formed at several times 10^5 K.

Discussion following paper by R. Dahlburg

B. Somov: In the solar corona, the magnetic field is strong in the sense: $v_s^2 \ll v_a^2$, $v_d^2 \ll v_a^2$. Under these conditions, it seems to be difficult to create MHD turbulence, except in the "neutral" ($B = 0$) plane. How do your results depend on magnetic field components which make this plane "non-neutral"?
R. Dahlburg: We are currently studying this problem. We find that with an additional sheetwise magnetic field the system is still ideally unstable, but that the additional magnetic field reduces the growth rate. There is a critical value of the magnetic field component strength above which the secondary mode is stabilized.

Discussion following paper by R. Tsinganos

R. Rosner: Can one think of the physics underlying the nonexistence of time-independent solutions and conjectured loop disruption for high Alfvén numbers in terms of the failure of the magnetic curvature force to balance the centrifugal force?
K. Tsinganos: Yes. In an isothermal atmosphere any flux tube without flows needs to be anchored at points separated by no more than $\Lambda_o = 2\pi H$, if it is going to be held in equilibrium by magnetic tension against the buoyant forces (Parker 1979). The radius of curvature at the highest point of the loop is $R_o = 2H$ in our model when $\xi = 1$. By allowing flows, the separation becomes $\Lambda < \Lambda_o$ and the radius of curvature becomes $R = 2(1 - M_a^2) < R_o$. Thus as we increase the magnitude of the flow speed along the loop, the magnetic lines respond by becoming more curved (for fixed footpoint separation) and for $M_a = 1$ we get $R = 0$.
S. Serio: You mention that loop-like solutions exist only for mild stratification. Can you explain what you mean by "mild" stratification?
K. Tsinganos: If the parameter ξ is large, the effective scale-height is smaller than V_s^2/g and the density drops very rapidly with the height. Loop-like solutions exist only for $\xi \leq 2$.

AUTHOR INDEX

INDEX OF KEYWORDS AND SUBJECTS

INDEX OF STAR NAMES